Fish & Fisheries Series

Volume 42

Series Editor

Kai Lorenzen, University of Florida, Gainesville, FL, USA

The Fish & Fisheries series aims to publish key books on all aspects of fish and fisheries science, conservation and management: biology and ecology of fishes and invertebrates of fisheries interest, fisheries assessment and management, aquaculture, and the social and economic dimensions of aquatic resource use and conservation. For over thirty years, the Fish & Fisheries series has been at the forefront of providing in-depth syntheses of knowledge in this vibrant and fast-moving field. The series is directed at researchers in the fish and fisheries sciences, senior undergraduate and postgraduate students, and at professionals in fisheries, aquaculture and marine and freshwater conservation.

Prospective authors and/or editors are encouraged to consult the Publishing Editor for more details. Comments or suggestions for future volumes are welcome:

Éva Lőrinczi
Publishing Editor
eva.loerinczi@springer.com

Raymond T. Bauer

Shrimps

Their Diversity, Intriguing Adaptations and Varied Lifestyles

Springer

Raymond T. Bauer
Department of Biology
University of Louisiana
Lafayette, LA, USA

ISSN 2542-890X ISSN 2542-8926 (electronic)
Fish & Fisheries Series
ISBN 978-3-031-20965-9 ISBN 978-3-031-20966-6 (eBook)
https://doi.org/10.1007/978-3-031-20966-6

This Springer imprint is published by the registered company Springer Nature Switzerland AG
The registered company address is: Gewerbestrasse 11, 6330 Cham, Switzerland

to Lydia, of course

Preface

Shrimps are animals known to most people as nutritious and tasty food items. Shrimp fisheries and aquaculture are a vital part of US and world economies. These crustaceans are key ecological and food-web components of marine and freshwater habitats. Shrimps occur in all oceanic depths and latitudes. The 4400 species of decapod shrimps are amazingly diverse in size, shape, coloration, and natural history. Perhaps, it is no wonder that I have spent the last 52 years of my life in the study, research, publication, and teaching of marine science and crustacean biology, with a focus on shrimps. It became obvious to me that a book covering *all* the shrimps was needed and desired by marine biologists, zoologists, and carcinologists (crustacean biologists), as well as by applied scientists and amateur naturalists. Thus, in 2017, I launched into the writing of this book to cover those crustaceans recognized by most persons as "shrimps," i.e., the groups belonging to the crustacean order Decapoda: Dendrobranchiata (penaeoids and sergestoids), Caridea, Procarididea, and Stenopodidea. In this book, I cover important aspects of the organismal biology and natural history of shrimps: taxonomy, ecological diversity, structure and function of shrimp anatomy, antifouling adaptations, coloration and camouflage, reproductive biology, sexual systems, mating systems and behavior, life-history strategies, shrimp symbioses (parasitism, commensalism, and mutualism), and the evolution of shrimps. The book is based on past and present scientific literature (journal publications and books) as well as my own research and experience. I point out remaining questions about shrimps which need investigation and propose hypotheses for testing them. This book is envisioned as a comprehensive reference and source of information, a "go-to" book about shrimps. I have tried to write in a style appealing to a wide audience: advanced undergraduate and postgraduate students, academic scientists, and fisheries and aquaculture professionals, as well as amateur naturalists, aquarists, and interested laypersons.

I would not have been able to do this book without the help and support of numerous persons and key institutions along the way. I had the good fortune of being accepted in 1969 to the Scripps Institution of Oceanography (University of California, San Diego) where I received a doctorate in Marine Biology. William (Bill)

A. Newman mentored my dissertation work. It was at his suggestion that I first investigated and sampled the nearby tidepools in search of possible research ideas. There I encountered fascinating species of shrimps, a group which became my life's work. A class on crustaceans taught by Bill and Robert Hessler cemented my interest in the group. A year's postdoctoral fellowship, sponsored by Raymond B. Manning, at the Smithsonian Institution's National Museum of Natural History (NMNH), was another key influence. There I worked among giants of crustacean biology such as Fenner A. Chace, Jr., Brian Kensley, Austin Williams, Horton H. Hobbs, Jr., Thomas Bowman, and Isabel Pérez Farfante, among others.

My first tenure-track position was at the University of Puerto Rico, Río Piedras, as a marine biologist. The beauty and diversity of this island's tropical reefs, seagrass meadows, cayos, and freshwater streams were the source of many productive studies. This research could not have been done without the help of many enthusiastic and agreeable UPR students. From there, I moved to the University of Louisiana at Lafayette, which became my permanent academic home. The support I have received from ULL throughout the years is gratefully recognized.

In the course of writing this book, I have received considerable help from students, professional colleagues, journal editors, and laypersons who responded to requests for information of various kinds. There are too many to list, but to all of you, my sincere thanks. I have been a member of the Crustacean Society since its inception in late 1979. Its contribution to my work and carcinology has been indispensable. I am indebted to TCS and its *Journal of Crustacean Biology* for its support of this book.

With the saying "a picture is worth a thousand words" in mind, I have included in the book 26 plates, each composed of several color photographs and/or grayscale micrographs. These plates were made possible in large part by generous image contributions from the colleagues listed below. Special credit and thanks are due Arthur Anker, a prominent carcinologist whose photography of shrimps is renowned. Many of the plate photos were contributed by Arthur. Marine biologist Zdeněk Ďuriš, professor of biology at the University of Ostrava, Czech Republic, kindly contributed several valuable photographs from his work in tropical habitats. T.Y. Chan, National Taiwan Ocean University, kindly sent several important shrimp photos taken during his extensive marine research, as did Alexander Vereshchaka, Institute of Oceanology of Russian Academy of Sciences. Other colleagues who provided much appreciated photographic contributions are Kevin Bourdon, Florent Charpin, David Darom, Robert Ditter, William Folsom, Christian Gloor, David Herasimtschuk, David Hillegas, Ivan Hinojosa, Watcharapong Hongjamrassilp, Gregory Jensen, Werner Klotz, Leslie Knoesen, David Knott, Richard Larocque, Benjamin Magana, Jeremy Monroe, Jun Ohtomi, João Pantaleão, Jennifer Rasch, Greg W. Rouse, Bernard Sainte-Marie, Tomomi Saito, Troy Sakihara, David Shale, Al Stokes, María Morales Suazo, Martin Thiel, Cris Harris Toro, Eva Tóth, Sebastian Vazquez, Fernando Vega-Villasante, Mary Wicksten, and Yusuke Yamada. The Louisiana Sea Grant Program's Digital Images Collection was a good source of photographs about shrimp fishery techniques.

The WoRMS (World Registry of Marine Species) website has been invaluable for staying current on the taxonomy and classification of shrimps and other aquatic organisms. Vim De Cock of the WoRMS management team and Sammy De Grave and Charles H.J.M. Fransen (editors, Decapoda) were very helpful with questions and data on species numbers. Many thanks to Rose Gulledge (NMNH) for obtaining copies of hard-to-get journal articles and illustrations. It has been a pleasure working with Springer Nature senior editor Éva Loerinczi and the production teams led by Bibhuti Sharma and Jana Yagnavaragan. Their advice, patience, and help of all kinds have been invaluable—many thanks!!

I will finish the Preface with the most important acknowledgment of all: to my wife Lydia, whose constant encouragement, support, and patience were essential during the production of this book. Lydia, I owe you big time; lots of love, Ray

Lafayette, LA, USA Raymond T. Bauer

Contents

Chapter 1
The Shrimps: An Overview

1.1 What Are Decapod Shrimps?

Shrimps are long-tailed crustaceans, a group of organisms with a jointed external exoskeleton, a trait shared with fellow arthropods such as the familiar insects and arachnids (spiders, mites, scorpions). The crustaceans, a morphologically and ecologically diverse group, are uniquely defined as arthropods with two pairs of antennae. The term "shrimp" is very loosely applied as a common name for a number of long-tailed crustaceans, most of which are not familiar to the layman (see classification of Crustacea, Table 1.1). However, the word "shrimp" is used by most of us for those crustaceans with long antennae, slender legs, and a muscular tail (abdomen) which is good to eat, both nutritious and tasty. These are the shrimps which belong to the crustacean order Decapoda, which includes close relatives such as the crayfishes, lobsters, and crabs. "Prawn" is often used as a synonym of "shrimp" for penaeoidean and caridean shrimps, especially those of large size. However, usage of the terms "shrimp" and "prawn" varies regionally throughout the English-speaking world, being used for small and large shrimps in some areas of the world but the reverse in others, as discussed at length by Lipke Holthuis in his 1980 catalogue of shrimps and prawns of interest to fisheries. This confusing practice has become even more common with increasing globalization and rise of the internet. To avoid this confusion, I will use the more generally utilized and scientifically correct term "shrimp" throughout the book, unless the internationally accepted species common name includes prawn rather than shrimp, e.g., McLaughlin et al. (2005) and the Food and Agriculture Organization of the United Nations (FAO).

A typical shrimp shows a suite of morphological adaptations responsible for their well-developed forward and backward swimming behavior. In 1909, William Calman, a famous zoologist and carcinologist (one who studies crustaceans), proposed that these were primitive features of shrimplike crustaceans of the Class Malacostraca. He coined the term "caridoid facies" for the shrimplike body plan (Fig. 1.1) from which most of the malacostracan taxa have evolved. In the biological

Table 1.1 Position of the decapod shrimps within the classification of the arthropod subphylum Crustacea. The classification is based on Bowman and Abele (1982), updated by Martin and Davis (2001) and Ahyong et al. (2011). Taxonomic groups containing the decapod shrimps are printed in boldface. Common names are enclosed within parentheses

Class Cephalocarida (cephalocarids)
Class Branchiopoda (fairy or brine, tadpole, clam shrimps, water fleas)
Class Remipedia (remipedes)
Class Maxillopoda (e.g., barnacles, copepods)
Class Ostracoda (mussel or seed shrimps)
Class Malacostraca
Subclass Phyllocarida (leptostracans)
Subclass Hoplocarida (stomatopods or mantis shrimps)
Subclass Eumalacostraca
Superorder Syncarida
Superorder Peracarida (e.g., mysidan, lophogastridan shrimps, isopods, amphipods)
Superorder Eucarida
Order Euphausiacea (krill)
Order Decapoda (shrimps, lobsters, hermit crabs, true crabs)
Suborder Dendrobranchiata
Superfamily Penaeoidea
Superfamily Sergestoidea
Suborder Pleocyemata
Infraorder Procarididea
Infraorder Caridea
Infraorder Stenopodidea
Infraorder Gebiidea (mud lobsters, mud shrimps)
Infraorder Axiidea (ghost shrimps, lobster shrimps)
Infraorder Astacidea (crayfishes, clawed lobsters)
Infraorder Achelata (spiny, slipper lobsters)
Infraorder Glypheidea (achelate/semichelate lobster-like decapods)
Infraorder Polychelida (multichelate lobsters)
Infraorder Anomura (e.g., hermit crabs, king crabs)
Infraorder Brachyura (true crabs)

sciences, a "body plan" is used in the sense of a set of functional and structural characteristics of an organism that result from evolutionary processes. Robert Hessler (1983) later expanded on the features and importance of the shrimplike body plan in the evolution of the Malacostraca. Most fully exhibiting the classic caridoid facies are species found in the malacostracan Superorder Eucarida ("true shrimps") composed of the euphausiaceans (krill) and the subjects of this book, the decapod shrimps (Sergestoidea, Penaeoidea, Caridea, Procarididea, Stenopodidea), as well as two orders (Mysida, Lophogastrida) in the Superorder Peracarida. Many penaeoidean and caridean species are the familiar shrimps of culinary and commercial importance.

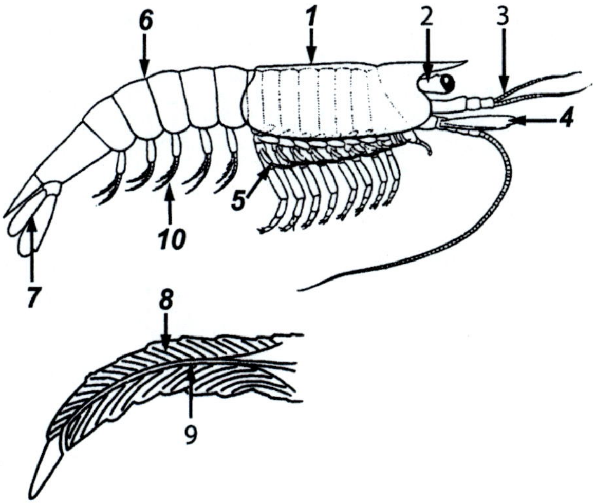

Fig. 1.1 Basic shrimp body plan ("caridoid facies") of the crustacean class Malacostraca, as envisaged by Calman (1909) and expanded by Hessler (1983). Hydrodynamic traits related to forward and backward swimming are in italic boldface. From Bauer (2004) after Calman (1909) and (B) Hessler (1983)

1.2 Highly Adapted Swimmers

The various elements of the caridoid facies are hydrodynamic and natatory (swimming) adaptations that make the quintessential shrimp highly streamlined and proficient in both forward and backward swimming. Forward swimming is powered by the paddlelike abdominal appendages, the pleopods (swimmerets) (Fig. 1.1, #10), and additionally, in some shrimps, by oarlike outer branches of the thoracic legs, the exopods (Fig. 1.1, #5). During forward swimming, shrimps flex their thoracic legs (pereopods) against the body to reduce drag, and the flat scalelike uropods are extended to serve as stabilizing fins during forward motion.

The typical backward tail flipping (retrograde escape response) (Fig. 1.2) of a startled shrimp is another form of swimming to which several caridoid features are related (Fig. 1.1). Propulsion is provided a rapid, often repeated flexion of the abdomen or tail (Fig. 1.1, #6), initiated by giant nerve fibers for super-fast impulse transmission. In caridean shrimps, extra myelination of these and other nerve fibers increases the nerve impulse speed even more (Guerin and Neil 2015). During the tail flip, the extended uropods (Fig. 1.1, #7) form, together with the terminal telson, a broad rowing paddle termed the tail fan, which propels the shrimp swiftly backwards through the water. The abdominal musculature is highly developed and spiraled (Fig. 1.1, #8) to generate a powerful flexion of the tail. A consequence of this massive musculature is that there is little room for internal organs with the exception of the hindgut (Fig. 1.1, #9) and ventral nerve cord. However, in reproductive females, the mature ovaries may extend back into the abdomen from the cephalothorax, especially in penaeoid shrimps. It is this large mass of tail muscle that makes shrimps so highly prized and commercially important as a source of appetizing high-protein food.

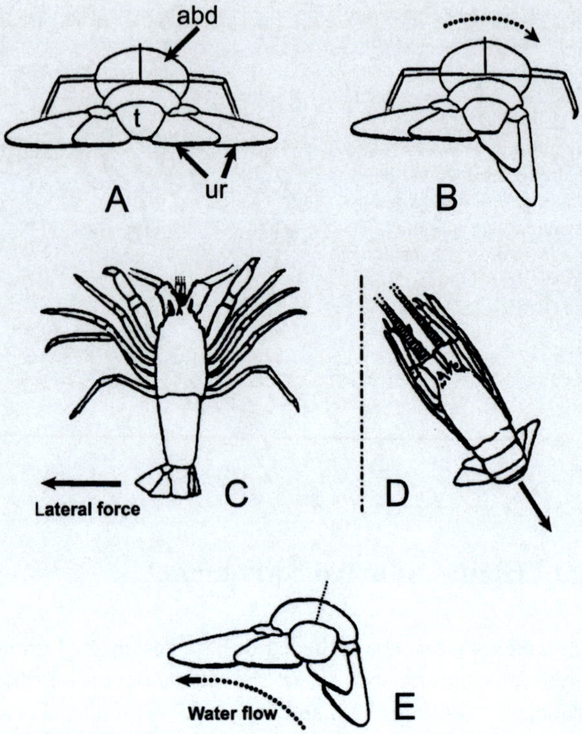

Fig. 1.2 Backward escape response (tail flip) in a lobster (*Jasus edwardsii*) in which the lobster swims backward rapidly to the right. In (**a**), the abdomen is seen from the rear, showing the twin branches (endopods) of the uropods on either side of the telson. In (**a**), the abdomen is horizontally straight, but as the abdominal flexion begins, it rotates towards the right side in this example (**b**) while the uropods of the right side flex from a horizontal to more vertical orientation (**b**, **c**). The asymmetrical position of the uropods forces water forward to the left during flexion (**e**), steering the lobster to the right, going backwards (**d**). abd, abdomen; t, telson; ur, uropods. Modified from Neil and Ansell (1995)

Hydrodynamic elements of the caridoid facies include a cuticular cover, the carapace (Fig. 1.1, #1), which extends back from the cephalon, or head, and covers the individual thoracic somites (segmental body divisions). The dorsal midline of the cephalothorax (front half of the animal) is often produced into a dorsal keel, or carina; typically, this is extended forward into a pointed knife-like blade, the rostrum. The abdomen of many shrimps may also be equipped with carinae. Thus, a typical shrimp is streamlined, somewhat like a hull of a ship, with a solid, semicylindrical form and keel, only turned upside down. The rostra of many shrimps are long, compressed blades which may enhance streamlining and stability, especially during backward swimming. Additionally, the rostrum is often serrated with small spines, which may reduce turbulence and subsequent drag during the backward tail flip. However, a spiny rostrum has also been shown to be a deterrent to predators (Chap. 2, "Structure and Function").

The flat antennal scales (exopods of the second antennae) (Fig. 1.1, #4) act somewhat like the bow of a ship, parting the water ahead of the shrimp in forward swimming (Young 1959). Thanks to the complex musculature associated with them, they are able to variably spread and tilt during both forward and backward swimming, acting as rudders and stabilizing planes. Besides rowing during the tail flip, the uropods of the tail fan are important steering devices during backward swimming, helping to control direction of movement, much like oars in a small boat do

(Fig. 1.2). For example, when the two branches of the right uropods are folded together and those of the left are spread out during a tail flip, the animal will veer off towards the right in a backward direction. Uropod musculature and articulations with the tail permit horizontal, vertical, and rotatory movements serving in fine-scale steering during swimming. The maneuverability of a true shrimp during retrograde escape swimming is truly exceptional. As a graduate student collecting my favorite research subject, the tidepool shrimp *Heptacarpus sitchensis* in southern California tidepools, I was impressed by the shrimp's mastery of escape movements using the tail flip response. When one of these shrimps is approached with a dipnet, it leaps back off through the water. Naturally, the eye of the observer tends to follow this line of movement. However, the escaping shrimp does not always end up in the direction it first appears to go. I soon learned that this shrimp, using a series of tail flips, often made a 180° turn back to near the spot where it began its escape response, which fools the naïve observer. Such a rapid sophisticated movement is the result of the combined action of the shrimp's hydrodynamic features described above, i.e., the streamlined shape, drag-reducing features and the two sets of planing, braking and steering devices, the antennal scales and uropods of the tail fan.

Studies using high-speed video analyses confirm that the decapod shrimps have a more highly refined backward escape behavior when compared to other long-tailed ("macruran") decapods such as crayfishes and lobsters (Daniel and Meyhöfer 1989; Arnott et al. 1998, 1999). In the typical shrimp tail flip, initiated by the shrimp in response to the sight or touch of an approaching predator, a strong contraction of the abdominal muscles flexes the abdomen against the cephalothorax, like a folding jack knife. Simultaneously, the laterally extended antennal scales form a "head fan" which is pressed against the extended uropods + telson ("tail fan"). Both processes squeeze water out forward, which shoots the animal up and backwards at an angle, a form of jet propulsion reminiscent of that utilized by other aquatic animals such as squids. Biomechanical forces cause the shrimp to initially roll on one side or the other, which determines its initial direction of movement. At the end of the tail flip or series of flips, the shrimp rolls back to an upright position. The direction of the escape response is, as one might expect, generally away from the predator (or other stimulus), but with a high degree of unpredictability so that a potential predator cannot easily learn which path the escape trajectory might take in a future attack. Other features of the caridoid character complex are stalked eyes (Fig. 1.1, #2) and a biramous (two-branched) antennule (first antenna) (Fig. 1.1, #3). These features are not directly related to swimming but rather are adaptations of active, mobile animals. These characters, like the carapace, natatory pleopods and thoracic exopods, had already evolved in the crustacean group which gave rise to the malacostracan ancestor of shrimps.

1.3 Reduction of the Caridoid Facies in the Decapoda

Although the adaptive significance of the various hydrodynamic features of the caridoid facies may seem apparent, experimental work testing hypotheses about their exact function in swimming is still somewhat limited. However, natural experiments have been done for us during the evolutionary history of two malacostracan groups. An overriding theme in the evolution of the Order Decapoda and Superorder Peracarida has been the modification of the swimming caridoid facies in ancestral species to a body form better suited for various forms of benthic (bottom) living. Martin Glaessner (1960) pointed out that the long muscular caridoid abdomen is a liability when a primarily crawling crustacean has to drag it through the mud or over an irregular bottom. Thus, the shrimp's tail is reduced in relative size in crawling (reptant) species with the long-tailed "macrurous" morphological level of organization (Fig. 1.3). The hydrodynamic cylindrical shape of the caridoid cephalothorax is less cylindrical and more depressed in the lobsters and crayfishes. However, macrurans such as crayfishes and lobsters are still quite capable of the backward tail flip escape swimming, but it is clumsy in comparison to that of shrimps. In brachyurous ("short-tailed") crabs, rapid walking and running are highly developed, and the abdomen is reduced to a small flap tucked under the cephalothorax (Fig. 1.3). These crab-like decapods have a flattened unstreamlined shape which is more appropriate for these ambulatory (walking, running) species. Although some crabs do swim, like the blue crab *Callinectes sapidus* (Portunidae), their mode of swimming with a pair of secondarily evolved paddlelike legs is not related to the caridoid facies.

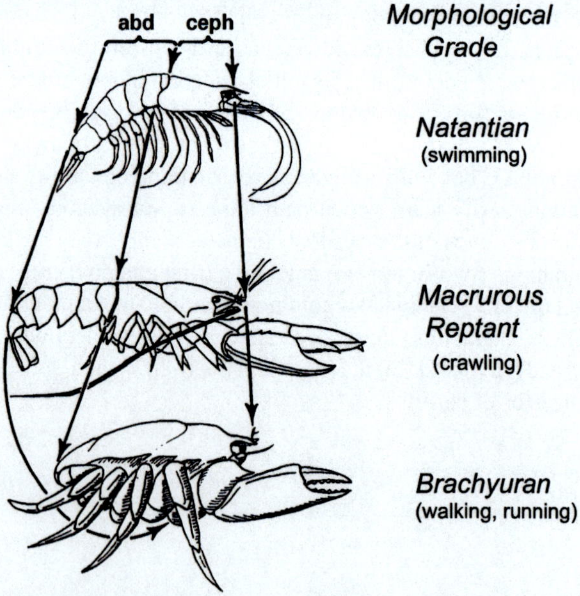

Fig. 1.3 Reduction of the shrimplike ("caridoid") body form in the Decapoda from: (1) a basic swimming (natant) morphology to (2) one better adapted to crawling along the bottom as in the heavy-bodied lobsters and crayfishes (macrurous reptants) to (3) the rapid walking and running seen in crab-like (brachyurous) decapods. Modified from Bauer (2004) after Glaessner (1969)

All of the caridoid accessories associated with swimming, such as dorsal carina, antennal scales, rostrum and uropods, are reduced from their swimming function in reptant macrurous, especially in brachyurous species (Fig. 1.3). The twin branches (rami) of the pleopods, which function as paddles in forward swimming in shrimps, are reduced and relatively useless for swimming in reptant decapods, serving only reproductive functions in true crabs. The exoskeleton of the benthic reptants is generally heavier (thicker and more calcified) than that of truly caridoid shrimps. The walking legs of reptant species are stout and highly muscular, an adaptation for moving their robust, heavy bodies over the bottom. In typical shrimps, in which swimming is a vital mode of locomotion, walking legs are slender and delicate, adapted for easy swimming and perching on objects rather than constantly crawling over and through the bottom. The reduction of swimming morphology from natantian to macrurous reptant to brachyuran morphologies is correlated with a decrease in grooming morphology and behaviors, which keep the general body surfaces clean of fouling by debris and sessile organisms (Chap. 4, "Shrimp Hygiene"). Such fouling interferes with mobility and swimming, much as it does on the hulls of ships. However, as we will see in Chap. 3 ("The Families of Decapod Shrimps"), there is a somewhat analogous reduction of hydrodynamic characters in some shrimp genera and families whose adaptations tie them closely to a benthic, less mobile lifestyle (e.g., the dendrobranchiate Sicyoniidae; the caridean Crangonidae, the stenopodidean Spongicolidae).

A similar evolutionary convergent trend from a pelagic swimming crustacean to a highly benthic one with loss of hydrodynamic features has occurred in the diverse and speciose malacostracan superorder Peracarida. The "mysidaceans, i.e., orders Mysida (Fig. 1.4b) and Lophogastrida, are small coastal and large deep-sea peracaridan shrimps (respectively), clearly with all the usual caridoid features including the backward escape response. They are considered the most primitive members of their Superorder, most similar to the ancestral peracarid. Other descendants of ancestral peracarids have invaded and radiated extensively into the benthic environment with a variable loss of caridoid characters (Fig. 1.4c–e). They have all lost the caridoid escape response, and characters associated it with it are reduced or lost, i.e., carapace, massive complex abdominal musculature, and broad antennal scales and uropods. The convergent reduction of caridoid features from ancestors adapted for swimming to descendants primarily adapted for a crawling, walking, or infaunal benthic lifestyle is good evidence for the hydrodynamic function of caridoid traits in the more primitive Decapoda and Peracarida. The mantis shrimps (stomatopods: Order Hoplocarida) have an incipient caridoid body form (Fig. 1.4f) in which the abdomen is large but with a poorly developed caridoid musculature. Unlike the true shrimps, the stomatopod abdomen has room for internal organs such as the gonads and digestive ceca. However, stomatopods do have large flat antennal scales and uropods. Analysis of their locomotion shows that these features are used for an escape response but not the backward tail flipping of full caridoids (Heitler et al. 2000). When presented with a stimulus, the mantis shrimp rapidly flexes the abdomen but does not flip backward, as in natantians decapods. Rather, it loops vertically, leaving it upside down and facing away from a stimulus. The mantis

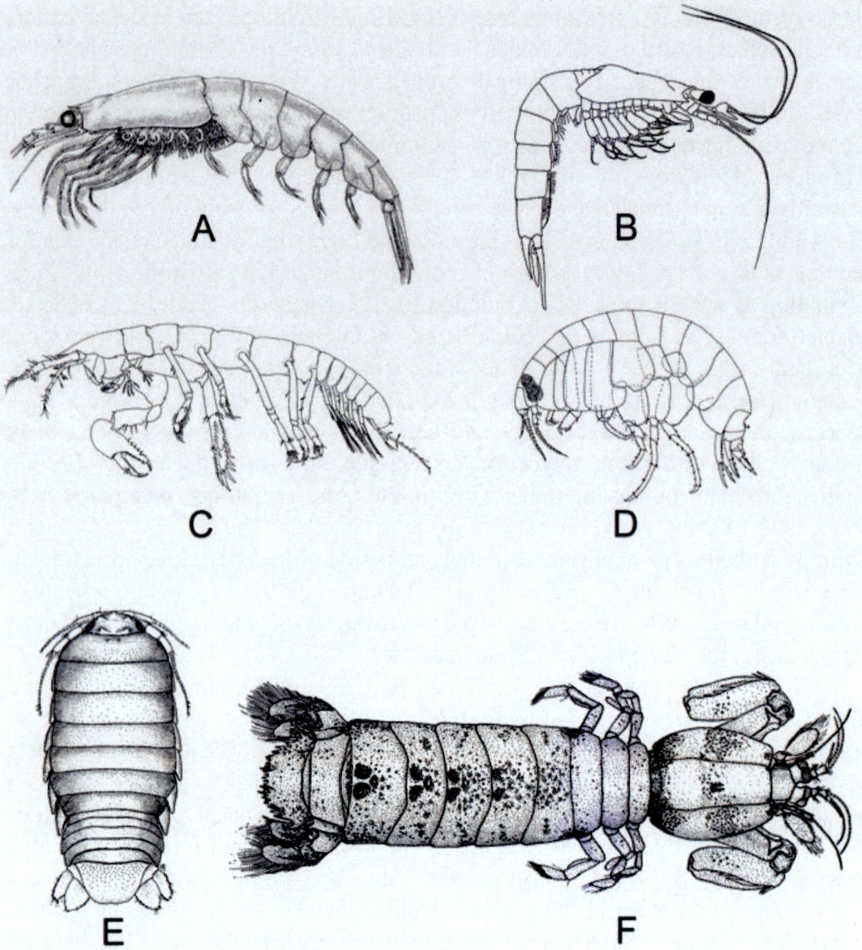

Fig. 1.4 Some non-decapod caridoids. (**a**) *Euphausia superba* (krill), a well-developed caridoid. (**b**) A generalized mysid shrimp. (**c–e**) Peracarids with a modified and reduced caridoid facies: (**c**) the tanaid *Kudinopasternaki siegei,* (**d**) the amphipod *Orchomene limodes*, (**e**) the isopod *Tridentella quinicornis* and (**f**) the stomatopod *Tetrasquilla mccullochae*, a stomatopod with a rudimentary caridoid body plan. From (**a**) FAO species fact sheet; (**b**) Meland et al. (2015) Creative Commons CC BY 4.0; (**c**) Viskup and Heard (1989); (**d**) Meador and Present (1985) and (**e**) Delaney and Brusca (1985), courtesy of Journal of Crustacean Biology and (**f**) Manning and Chace (1990)

shrimp then quickly swims forward while righting itself. In Robert Hessler's view, the ancestral malacostracan was pre-caridoid, giving rise to the hoplocarids, an incipient caridoid, as well as the fully caridoid ancestor.

1.4 Systematic Position of the Decapod Shrimps Within the Crustacea

The decapod shrimps belong to the class Malacostraca (Table 1.1), composed of crustaceans with a "6-8-6 (7)" body plan, that is: a head consisting of a preantennular region bearing the eyes, followed by five somites with appendages (two pairs of antennae, three pairs of mouthparts), a thorax of 8 somites and an abdomen with (usually) six somites all of which bear appendages. Another important feature defining the group is the position of the gonopores (female apertures on the sixth, male apertures on the eighth thoracic somites). With six somites in the abdomen, shrimps belong to the subclass Eumalacostraca. The decapod shrimps of this book (sergestoids, penaeoids, carideans, procarididеans, stenopodideans) and their sister group, the euphausiaceans (krill) (Fig. 1.4a), are part of the superorder Eucarida in which the carapace, a cuticular fold extending backward from the head, extends over and fuses with all thoracic somites. In addition to shrimps, the order Decapoda includes less shrimplike species such as crayfishes, lobsters, hermit crabs, and brachyuran crabs. In all decapods, the first three pairs of thoracic legs (maxillipeds) are functionally associated with the mouth and feeding, leaving the remaining five pairs of legs (Decapoda = "ten feet") for a variety of functions, including locomotion, food gathering, weapons of defense and offense, and grooming. The gills of decapods are enclosed in a branchial chamber, an adaptation with major advantages as well as some disadvantages (Chap. 4).

For many years prior to 1963, the decapod shrimps were classified in a subclass, the Natantia, so named because of their well-developed swimming abilities, with crayfishes, lobsters, and crabs in another subclass, the Reptantia. However, Martin Burkenroad in 1963 proposed a more natural decapod classification, with two suborders, the Dendrobranchiata and the Pleocyemata (Table 1.1; also see Chap. 12, "Evolution"). This classification puts the sergestoidean and penaeoidean shrimps in one suborder, the Dendrobranchiata, with the caridean, procarididean, and stenopodidean shrimps in another suborder, the Pleocyemata, along with all the other decapods, including those with little resemblance to a shrimp, such as hermit crabs and brachyuran crabs. Burkenroad clearly showed that the Natantia is a morphological grade representing a primitive level of decapod body plan. In this view, the Natantia was not a true clade, i.e., a taxon in which all members are more closely related genealogically to each other than to members of other groups. Burkenroad pointed out that the penaeoidean and sergestoidean shrimps have a very different, more primitive mode of reproduction than do other decapods, including the carideans, stenopodideans and (probably, although still unknown) procarididеans. Female dendrobranchiates are free (broadcast) spawners, releasing eggs, fertilized as they leave the gonopores, into the water where all embryonic development takes place before hatching as a nauplius larva (Fig. 1.5a). The nauplius, with only three pairs of appendages, is found in other crustacean classes and is considered the primitive larval stage of the Crustacea.

Fig. 1.5 First larval stages of a dendrobranchiate and a pleocyemate shrimp. (**a**) Nauplius 1 of the penaeid *Metapenaeopsis dalei*. (**b**) Zoea 1 of crangonid caridean *Crangon uritai*. Scale bar is 1.7 mm in (**a**) and 2 mm in (**b**). From (**a**) Choi and Hong (2001); (**b**) Li and Hong (2004), courtesy of the Journal of Crustacean Biology

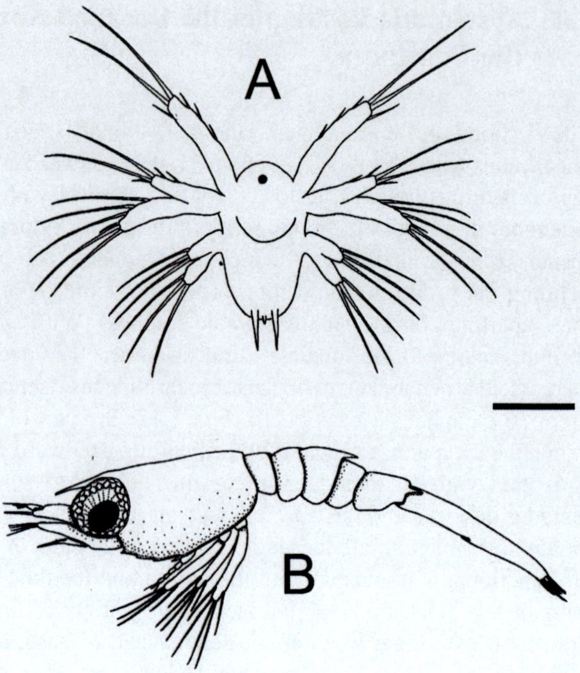

In contrast, females of the Pleocyemata attach fertilized eggs (embryos) to the pleopods (swimmerets) beneath the abdomen to incubate them during their longer embryonic development. The hatching larva, the zoea, is further developed than the nauplius, with several pairs of appendages, and a definite cephalothorax and abdomen (Fig. 1.5b). Incubation is a specialized reproductive character, indicating that all the diverse decapods with incubation are descended from a common ancestor. In Burkenroad's hypothesis, incubation initially evolved in a shrimp group (pleocyemate ancestor) which diverged from the common decapod shrimp ancestor, a broadcast spawner without incubatory care of embryos. This decapod ancestor also gave rise to the dendrobranchiates, which retained the primitive free spawning and the nauplius larva while diverging somewhat otherwise morphologically. Carideans and stenopodideans have retained most of the primitive caridoid body form, while in other pleocyemates all degrees of modification of the caridoid facies, discussed above, have taken place. Modification of the caridoid facies did not occur in the Dendrobranchiata, which received its name from the form of its gills (highly branched or dendritic). Burkenroad named the Pleocyemata (from the Greek "pleo" = "swim," and "cym" = "embryo") for the incubatory habit of its females. As with most new and different hypotheses about taxonomy, there were initially some years of debate and resistance to change, but both morphological and molecular analyses of decapod phylogeny have supported this view throughout the years.

1.5 Biological Success of Shrimp Groups

The biological success organisms can be measured in a number of ways: biological productivity (biomass; numbers of individuals); taxonomic diversity (number of genera and species); morphological diversity (variety of body forms); and ecological diversity (number and kind of habitats; breadth of latitudinal and depth distribution). One assessment of biological productivity is fisheries production, which is estimated by the FAO (Food and Agriculture Organization of the United Nations) for commercially important species. FAO fisheries statistics show that the penaeoids, especially the family Penaeidae, came out on top in 2019 at 1.5 million metric tonnes (t) (FIGIS 2021a). Penaeoid shrimps are benthic shrimps inhabiting shallow waters of the continental platform in subtropical and tropical latitudes. Still very significant but somewhat less productive are the other dendrobranchiate group, the epipelagic Sergestoidea (~488,000 t) and the pleocyemate Caridea (~552,000 t). This ranking in productivity has remained fairly constant since 1950 (Chap. 11,"Fisheries and Aquaculture"). On the other hand, shrimp aquaculture now far surpasses shrimp fisheries in production by weight and monetary value. The Penaeidae make up over 90% of shrimp aquaculture production (~ 6,500,000 t), with a few caridean (*Macrobrachium*) species composing the rest (~500,000 t) (FIGIS 2021b). The stenopodidean decapod shrimps are low in abundance and biomass, and are not fished or reared for food, although there is a small market for them in the ornamental shrimp trade.

Outside of the Decapoda, euphausiid shrimps (krill) are extremely abundant in high latitude and polar seas, serving as an important food source for species at higher trophic levels. The extensive fish, seabird, and marine populations in Antarctic waters depend heavily on the krill populations thriving in the productive waters there. In terms of commercial fisheries production, the krill catch has been below that of the other shrimp groups in recent years (~366,000 t in 2019) and never as great as that of the penaeoid and sergestid shrimps (Chap. 11). Mysidan shrimps are mainly small in size and are important trophic elements in coastal and estuarine waters. However, their total biomass and biological productivity is low compared to other shrimp groups, and they are of relatively little commercial value.

The taxonomic diversity of a group is an indication of its morphological and ecological variability. The malacostracan shrimplike body form is clearly a successful one in aquatic habitats, with a total of at least 700 genera and ~5700 species. The decapod shrimps (penaeoids, sergestoids, procarideans, carideans, stenopodideans are the most successful of these, with nearly 500 genera and over 4400 species (WoRMS 2022). Among decapod shrimps, the Caridea has the highest diversity by far, with ~400 genera and ~3800 species (a substantial increase from the 350 genera and 2800 species reported in my 2004 book on carideans). This reflects the great variety of shape size, and appendage complexity of the group, as shown in Chap. 3. Another factor is the very broad range of habitats in which carideans occur. They are found from tropical to polar waters of the world's oceans, at all depths, in both benthic and pelagic environments, in estuaries, and have done well in freshwater environments as well, with about 25% of caridean species inhabiting rivers and streams. Many species have formed symbiotic relationships with other invertebrates and fishes (Chap. 10, "Symbioses"). A few shrimps are even semiterrestrial.

Although the sergestoid and penaeoid shrimps (Dendrobranchiata) are very successful in terms of biomass and productivity, they are not so variable in their general body morphology as the carideans, reflecting the relatively low number of habitats they occupy. Sergestoid shrimps do not occur in the diverse habitats of the ocean floor with its many ecological niches; these pelagic animals are found up in the water column, both in deeper waters of the open ocean as well as in shallow coastal waters. The pelagic environment is a relatively homogenous environment compared to the benthic, and so there are many fewer ecological niches to radiate into, perhaps accounting for the low sergestoid taxonomic diversity (21 genera, 104 species). Similarly, the penaeoids (currently with 57–62 genera and 428 species) are relatively homogeneous in overall body form. This marine group principally lives in tropical and subtropical (warm temperate) seas, mainly over vast stretches of soft bottom (sand, mud) sea floor. The most diverse family of penaeoids is the Penaeidae, the principal species of fishing and aquaculture, especially those found in estuaries and shallow coastal waters of the continental shelves.

Another taxon of decapod shrimps is the Stenopodidea (13 genera, 95 species), which are restricted to tropical and subtropical seas, free-living in shallow rocky or coral reef habitats, or are in the deep-sea, associated with (symbiotic) with sessile invertebrates, especially sponges. In spite of the fact that they live in ecologically diverse habitats, their taxonomic diversity has remained low, compared to the dendrobranchiate and especially the caridean shrimps. One final decapod shrimp group is the Procarididea with only a single family, the Procarididae, with two genera and seven species. This taxon, closely related to but presently considered separate from the Caridea, is confined to the very specialized and restricted subterranean anchialine habitat (Chap. 3), so the low diversity is understandable.

Although extremely productive in terms of biomass, the euphausiacean shrimps are low in diversity, currently with only 16 genera and 86 species, not surprising in a group in which basic morphology is so uniform. Like the sergestoids, these epi- and mesopelagic shrimps occur in the relatively homogeneous pelagic realm, although they are widely distributed in coastal and open ocean waters of all latitudinal zones. Another group of non-decapod shrimps, the Mysida (Superorder Peracarida), are fairly diverse taxonomically, with ~180 genera and ~1200 species. Like the carideans, species of these shrimps occur throughout the world's oceans and are common in nearshore and estuarine habitats. Some species are purely pelagic, others are benthic, or epibenthic during the day, moving up into the water column at night. Although not occupying the variety of benthic habitats as do many carideans, their assortment of habitats and lifestyles may explain in part their relatively high species diversity.

Besides the current biological success of the shrimplike body plan, the basically pelagic and epibenthic caridoid ancestors have given rise to numerous descendants in which the basic body organization has been modified by an extensive radiation into benthic and even infaunal environments. In the malacostracan superorder

Peracarida (Table 1.1), the benthopelagic mysidaceans are a stem group from which have evolved the body plans of the speciose isopods and amphipods (over 10,000 species each), groups so successful in marine, freshwater, and terrestrial habitats, as well as a number of other benthic peracaridan groups. In the order Decapoda, ~4400 shrimp species have been described. There has been a diversification from a decapod shrimplike ancestor into the benthic environment resulting in over ~13,000 other decapod species such as crayfishes, lobsters, hermit crabs, and "true" (brachyuran) crabs. The caridoid facies and its various modifications' success have truly flourished since evolving hundreds of million years ago.

1.6 Which Group of Shrimps?

When shrimps are first examined, how might they be quickly identified to the major group, or taxon, to which they belong? Separation of non-decapod shrimps from decapod ones is fairly easy. The carapace of mysidan shrimps is not firmly attached to the thorax; if the carapace is gently tilted from the rear, one should be able to easily lift it forward and see the thoracic somites below. In contrast, the carapace of euphausiaceans and decapod shrimps is firmly fused with the thorax. Most mysidaceans (Mysida) have a translucent lens-like statocyst (organ of equilibrium) in the basal part of the inner branch (endopod) of the tail fan uropods, and, in females, a marsupium composed of overlapping plates below the thorax in which embryos are brooded. Neither of these features are found in euphausiaceans nor decapod shrimps. In the euphausiaceans, the gills are exposed, and all the thoracic appendages are similar in structure. In decapod shrimps, the gills are covered by ventrolateral extension of the carapace, the branchiostegite (gill cover). The first three pairs of thoracic legs in decapods are associated with the mouth (maxillipeds), while the remaining five pair, the pereopods, are modified and adapted for a variety of functions such as food gathering, weaponry, or for perching and walking over the bottom.

Once identifying specimens as decapod shrimps, how might one easily distinguish among the different major groups, i.e., the penaeoids, sergestoids, procaridids, carideans, or stenopodideans? If the specimens are in good shape, with the body complete and all appendages intact, one can use the number and general form of chelipeds, which bear chelae (pincers). The sergestoids have minute, almost microscopic, chelae on the second and third pereopods (Fig. 1.6a). The first pereopods may or may not have a chela, depending on the genus. Their fellow dendrobranchiates, the penaeoid shrimps, have three pairs of chelipeds; the chelae are relatively small and similar in form (Fig. 1.6b). The caridean shrimps, however, have only two pairs of chelipeds (pereopods 1 and 2), often quite dissimilar in size and shape (Fig. 1.7a). In a few caridean families, the chelae of the first or second chelipeds may be reduced or absent, but the third pereopod is never chelate in carideans, clearly separating them from the dendrobranchiates and stenopodideans.

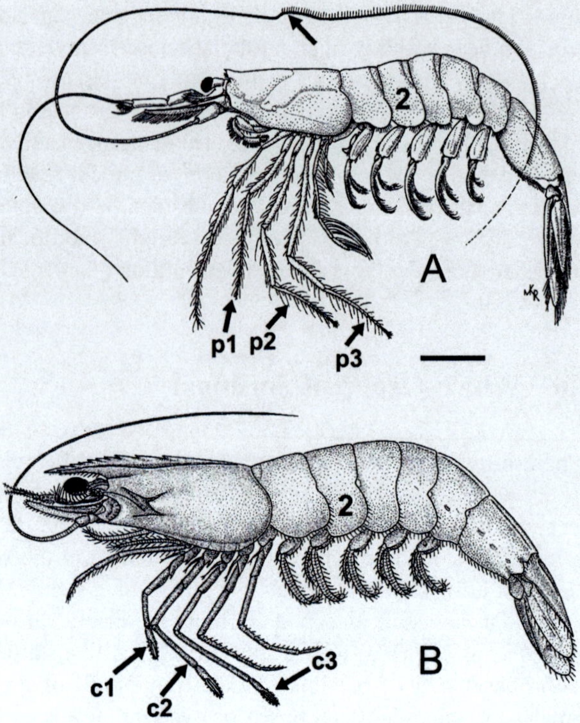

Fig. 1.6 Representatives of dendrobranchiate decapod shrimps: (**a**) *Sergestes atlanticus* (Sergestoidea) and (**b**) *Farfantepenaeus (Penaeus) aztecus* (Penaeoidea). Scale bar represents 3.5 mm in (**a**) and 15 mm in (**b**). p1, p2, p3 are pereopods 1,2,3 in (**a**) (in this species, p1 lacks a chela and p2-3 have minute chelae); c1, c2, c3 are chelipeds 1-3; abdominal pleuron 2 is marked. Arrow in (**a**) points to kink in antennal flagellum. From (**a**) Pérez Farfante and Kensley (1997), courtesy ©Publications Scientifiques du Muséum national d'Histoire naturelle, Paris and (**b**) Pérez Farfante (1988)

Procarididean shrimps are similar morphologically and close phylogenetically (genealogically) to the carideans, and once were included in the same infraorder. However, these rare anchialine shrimps lack chelipeds altogether; all the pereopods are simple (not chelate) (Chap. 3). The stenopodideans have three pairs of chelipeds, but unlike the dendrobranchiates, the third chelipeds are massive compared to the first two slender pairs (Fig. 1.7b).

Collection of shrimps with trawls, nets, and other gear often damages shrimps even before preservation. Most specimens end up in long-term storage immersed in alcohol, usually ethanol, which makes them somewhat fragile. The pereopods (chelipeds and walking legs) are easily snapped off near the base, broken during collection or when handled after preservation. Fortunately, other characteristics can be used to identify a specimen to its major taxon. If specimens are incubating embryos (often termed eggs) below the abdomen, they can only be a caridean or stenopodidean shrimp, since penaeoidean nor sergestoidean females do not incubate their embryos.

The abdomen offers a good clue to identity no matter what the sex of the specimen. In carideans and procaridideans, the second abdominal pleura (side plates) overlap those of the first and third somites (Fig. 1.7a). In most carideans, connected across the dorsum of the shrimp, the second pleura appear very much like a western-style saddle (Fig. 1.7a). In dendrobranchiate shrimps, the first pleura overlaps the

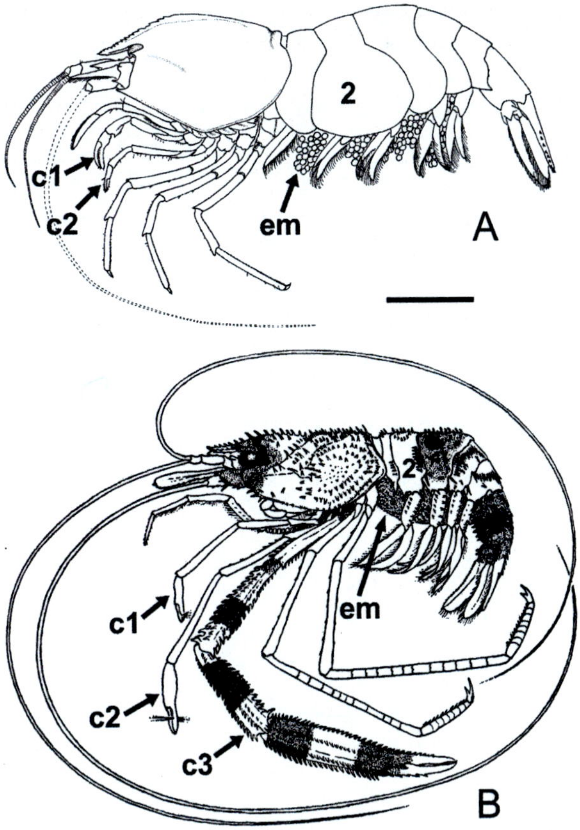

Fig. 1.7 Representatives of pleocyemate decapod shrimps. (**a**) *Alvinocaris alexander* (Caridea) and (**b**) *Stenopus hispidus* (Stenopodidea). Scale bar represents 7 mm in (**a**) and 12 mm in (**b**). c1, c2, c3 are chelipeds 1,2,3; em, embryos; abdominal pleuron 2 is marked. Modified from (**a**) Ahyong (2009) and (**b**) Limbaugh et al. (1961). Courtesy of (**a**) Crustaceana (Brill) and (**b**) Bulletin of Marine Science

second, the second the third, and so on down the abdomen (Fig. 1.6). In stenopodideans, the first abdominal pleura are reduced in size; the second pleura do not overlap the third (Fig. 1.7b).

If the specimen's abdomen is severally damaged or missing (perhaps someone removed it for dinner) and all that remains is the front half (cephalothorax) missing the pereopods, the specimen can still be identified to a major shrimp group. An examination of the gills hidden under the gill cover will reveal its general identity. In carideans and procaridids, the gills are phyllobranchs ("leaf gills") in which a series of platelike lamellae are found on each side of the main gill axis. In a penaeoid or sergestoid, the gills are highly branched dendrobranchs ("tree gills"), from which the name of their suborder is derived. Stenopodideans have trichobranchs ("hairy gills") in which many filaments surround a central stalk. These gill types are more fully described and illustrated in Chap. 2.

In the following chapters, the variation in form and function, biological diversity, reproduction, sexual systems, mating adaptations, life history strategies and evolution will be presented and discussed. In addition to their basic biological importance to world habitats and ecosystems, the many shrimp species make a significant contribution to global food supply, commerce, and culture. An understanding of their adaptations and natural history is a key to pursuing further knowledge about them, as well as conserving them to protect the many and varied services they provide to the world's ecosystem—and to the human race.

References

Ahyong ST (2009) New species and new records of hydrothermal vent shrimps from New Zealand (Caridea: Alvinocarididae, Hippolytidae). Crustaceana 82:775–794

Ahyong ST, Lowry JK, Alonso M, Bamber RN, Boxshall GA, Castro P, Gerken S, Karaman GS, Goy JW, Jones DS, Meland K, Rogers DC, Svavarsson J (2011) Subphylum Crustacea Brünnich, 1772. In: Zhang Z-Q (ed) Animal biodiversity: an outline of higher-level classification and survey of taxonomic richness, vol 3148. Zootaxa, pp 1–237

Arnott SA, Neill DM, Ansell AD (1998) Tail-flip mechanism and size-dependent kinematics of escape swimming in the brown shrimp *Crangon crangon*. J Exp Biol 201:1771–1784

Arnott SA, Neill DM, Ansell AD (1999) Escape trajectories of the brown shrimp *Crangon crangon*, and a theoretical consideration of initial escape angles from predators. J Exp Biol 202:193–209

Bauer RT (2004) Remarkable shrimps: adaptations and natural history of the carideans. University of Oklahoma Press, Norman

Bowman T, Abele LG (1982) Classification of the Crustacea. In: Abele LG (ed) Biology of the Crustacea, vol 1, systematics, the fossil record, and biogeography. Academic Press, New York, pp 1–27

Burkenroad MD (1963) The evolution of the Eucarida (Crustacea: Eumalacostraca) in relation to the fossil record. Tulane Stud Geol 2:3–16

Calman WT (1909) Crustacea. In: Lankester R (ed) A treatise on zoology, Part VII, Appendiculata, third fascicle. Adam and Charles Black, London

Choi JH, Hong SY (2001) Larval development of the kishi velvet shrimp, *Metapenaeopsis dalei* (Rathbun) (Decapoda: Penaeidae), reared in the laboratory. Fish Bull 99:275–291

Daniel TL, Meyhöfer E (1989) Size limits in escape locomotion of caridean shrimp. J Exp Biol 143: 245–265

Delaney PM, Brusca RC (1985) Two new species of *Tridentella* Richardson, 1905 (Isopoda: Flabellifera: Tridentellidae) from California, with a rediagnosis and comments on the family, and a key to the genera of Tridentellidae and Corallanidae. J Crustac Biol 5:728–742

FIGIS (Fisheries Global Information System) (2021a) Global Capture Production 1950-2019. FI Institutional Websites. FAO Fisheries and Aquaculture Department, Rome, Italy. http://www.fao.org/fishery/statistics/global-capture-production/query/en

FIGIS (Fisheries Global Information System) (2021b) Global aquaculture production 1950– 2019. FI Institutional Websites. FAO Fisheries and Aquaculture Department, Rome, Italy. https://www.fao.org/fishery/statistics/global-aquaculture-production/query/en

Glaessner MF (1960) The fossil Decapod Crustacea of New Zealand and the evolution of the Order Decapoda. New Zealand Geological Survey Paleontological Bulletin, No. 31, 63 pp

Glaessner MF (1969) Decapoda. In: Moore RC (ed) A treatise on invertebrate palaeontology, Part R, Arthropoda 4. Geological Society of America, University of Kansas Press, Lawrence, pp 399–533

Guerin AJ, Neil DM (2015) Escape trajectories of the rockpool prawn (*Palaemon elegans*) in response to visual and mechanosensory stimuli. Mar Freshw Behav Physiol 48:145–161

Heitler WJ, Fraser K, Ferrero EA (2000) Escape behaviour in the stomatopod crustacean *Squilla mantis*, and the evolution of the caridoid escape reaction. J Exp Biol 203:183–192

Hessler RR (1983) A defense of the caridoid facies: wherein the early evolution of the Eumalacostraca is discussed. In: Schram FR (ed) Crustacean phylogeny. A.A. Balkema, Rotterdam, pp 145–164

Holthuis LB (1980) FAO species catalogue. Shrimps and prawns of the world. An annotated catalogue of species of interest to fisheries. FAO Fisheries Synopsis 125, vol 1, 261 pp

Li HY, Hong SY (2004) Larval development of *Crangon uritai* (Decapoda: Crangonidae) reared in the laboratory. J Crustac Biol 24:576–591

Limbaugh C, Pederson H, Chace FA Jr (1961) Shrimps that clean fishes. Bull Mar Sci Gulf Caribb 11:237–257

Manning RB, Chace FA Jr (1990) Decapod and stomatopod Crustacea from Ascension Island, South Atlantic Ocean. Smithsonian Contrib Zool 503:1–91

Martin JW, Davis GE (2001) An updated classification of the recent Crustacea. Natural History Museum Los Angeles Contributions to Science No. 39, 124 pp

McLaughlin PA, Camp DK et al (2005) Common and scientific names of aquatic invertebrates from the United States and Canada: Crustaceans (with CD-ROM). American Fisheries Society Special Publication, vol 31, pp 1–533

Meador JP, Present TMC (1985) *Orchomene limodes,* new species, a scavenging amphipod from Scripps Canyon, California; species description and analysis of morphological variation. J Crustac Biol 5:523–538

Meland K, Mees J, Porter M, Wittmann KJ (2015) Taxonomic review of the Orders Mysida and Stygiomysida (Crustacea, Peracarida). PLoS ONE 10:e0124656. https://doi.org/10.1371/journal.pone.0124656

Neil DM, Ansell AD (1995) The orientation of tail-flip escaped swimming in decapod and mysid crustaceans. J Mar Biol Assoc U K 75:55–70

Pérez Farfante I (1988) Illustrated key to penaeoid shrimps of commerce in the Americas. NOAA Technical Reports NMFS 64:1–32

Pérez Farfante I, Kensley BF (1997) Penaeoid and sergestoid shrimps and prawns of the world. Keys and diagnoses for the families and genera. Mémoires du Muséum National d' Histoire Naturelle Paris 175:1–233

Viskup BJ, Heard RW (1989) Tanaidacea (Crustacea: Peracardia) of the Gulf of Mexico. VIII. *Pseudosphyrapus siegi*, N. Sp. (Sphyrapidae) from the continental slope of the northern Gulf of Mexico. Gulf Caribb Res Gulf Res Rep 8:107–115

WoRMS Editorial Board (2022) World Register of Marine Species. Available from https://www.marinespecies.org at VLIZ. Accessed 2022-03-05. https://doi.org/10.14284/170

Young JH (1959) Morphology of the white shrimp *Penaeus setiferus* (Linnaeus 1758). Fish Bull Fish Wildlife Serv 59:1–168

Chapter 2
Structure and Function

2.1 From Stem to Stern

In order to understand the biology and evolution of any organism, its basic morphology needs to be described and understood in terms of its mechanical operation and function. As we have seen in the previous chapter, shrimps are fundamentally swimming animals or have descended from them. Perhaps it is appropriate to organize our exploration of their form and function by using the nautical phrase "from stem to stern," i.e., starting with the anterior or front end and working back to the posterior or tail end of the shrimp. In this chapter, some specific taxa of shrimps will be mentioned; the reader can refer to the next chapter ("The Families of Decapod Shrimps") for more specific information on shrimp families and genera mentioned below and throughout the book.

2.2 Cephalothorax and Carapace

A shrimp's body is divided into two major sections (tagma), the anterior cephalothorax and posterior abdomen (tail) (Fig. 2.1). As the term implies, the cephalothorax is the cephalon (head) fused with the thorax. The carapace refers to the cuticular cover of the head (cephalic shield) which extends backwards over the thoracic somites (segmental divisions of the body). In decapods and euphausiaceans, the carapace covers and fuses to the thoracic somites below it. In shrimps and other decapods, the carapace overhangs and covers the sides of the cephalothorax, forming an enclosed space, the branchial chamber, in which the gills are found.

The carapace may be ornamented with a variety of spines (solid projections of the exoskeleton), grooves (sulci) and ridges or keels (carinae), the latter often bordered by a parallel sulcus (Fig. 2.2). In many shrimps, these features may be variably absent or weakly formed. Carapace grooves may represent the remnants of somite

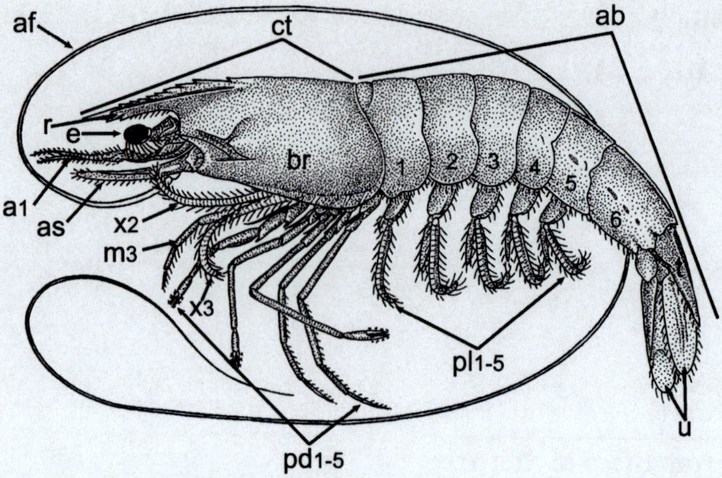

Fig. 2.1 Basic shrimp features illustrated by a penaeoid shrimp, *Litopenaeus setiferus*. a1, antennule (first antenna); ab, abdomen; af, antennal (antenna 2) flagellum; as, scaphocerite (antennal scale); br, branchiostegite (gill cover); ct, cephalothorax; e, eye; m3, maxilliped 3; pd, pereopods; pl, pleopods; r, rostrum; t, telson; u, uropods; x2, x3; exopods of maxillipeds 2 and 3; 1–6, pleura of abdominal somites 1–6. Modified from Pérez Farfante (1988)

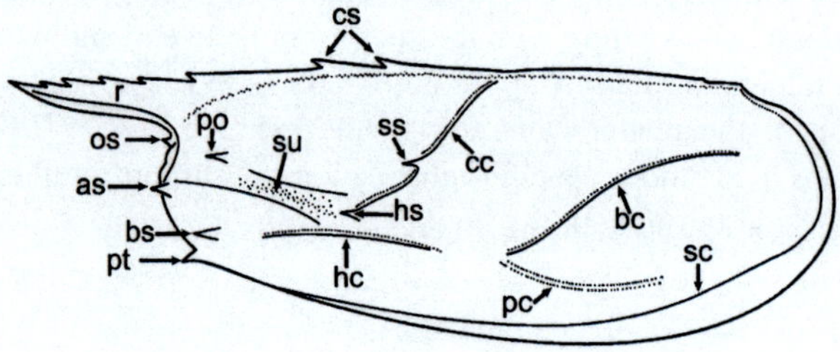

Fig. 2.2 Carapace ornamentation in a penaeoid shrimp: as, antennal spine; bc, branchiocardiac carina; bs, branchiostegal spine; cc, cervical carina; cs, carapace spines; hc, hepatic carina; hs, hepatic spine; os, orbital spine; pc, posthepatic carina; po, postorbital spine; pt, pterygostomial spine; r, rostrum; sc, submarginal carina; ss, suprahepatic spine; su, sulcus. Modified from Pérez Farfante and Kensley (1997), courtesy of ©Publications Scientifiques du Muséum national d'Histoire naturelle, Paris

boundaries, mark internal muscle insertions, correspond to some area of internal anatomy or still unexplained adornment. For example, an obliquely transverse cervical carina and groove mark the border between the anterior gastric and posterior cardiac areas within the cephalothorax, and a branchiocardiac groove marks the dorsal limits of the branchial chamber. Similarly, the orbital spine is found on the eye orbit, and the antennal spine is close to the basal segments of the second antenna, the hepatic spine approximately marks the anterior extent of the hepatic region, and a branchiostegal spine the anterior edge of the branchiostegite. Ridges or keel (carinae) on the carapace may correspond to regions occupied by internal organs, or by their proximity to some feature of the carapace, e.g., the adrostral grooves and carinae extending posteriorly on either side of the rostra in some penaeid shrimps (Pérez Farfante and Kensley 1997). In some shrimps, the carapace ridges, sharp cuticular projections and spines may aid in deterring predators. The presence and degree of carapace sculpturing and spination vary greatly among shrimp taxa and are thus useful taxonomic characters.

The shape of the cephalothorax, as well as the length and stoutness of the anteriorly projecting spiny rostrum, are quite diverse, apparently related to predation pressure and habitat. Rostrum shape and strength are probably linked to the type of escape swimming of shrimps in different kinds of habitats, as outlined by the Russian carcinologist Burukovsky (1972) and studied by Sardà et al. (2005) in a number of deep-sea shrimps. Fully pelagic shrimps are always up in the water column and can swim in any direction. In these shrimps, the keel-like hydrodynamic function of a long highly compressed rostrum, important in the upward and backward escape response, is not as critical as for those shrimps that are just on or near the bottom at least part of the time. Thus, fully pelagic shrimps tend to have a short, sometimes beaklike rostrum (Fig. 2.3a). This is certainly true in sergestid and some pelagic penaeoids and carideans. Interestingly, a short rostrum is characteristic of procarididean species which swim most of the time and are equipped with a functional set of swimming exopods. Although they now are restricted to anchialine subterranean habitats, Chace and Manning (1972), in describing the first procaridid discovered, suggested a pelagic ancestor. In some aristeid species, the more pelagic males have shorter rostra than the larger females, which live nearer the bottom. Thus, sexual dimorphism in rostral length may be related to different locomotory needs and abilities of the two sexes.

Another factor to be considered is that long rostra in both freshwater and marine species studied tend to be associated with proportionately larger abdomens, especially those of larger and/or ovigerous females (Sardà et al. 2005; Anastasiadou and Leonardos 2008; Liasko et al. 2015; Ocasio-Torres et al. 2015a). The correlation of long rostra and larger abdomens, as well as with other features associated with enhanced backward swimming, indicates that both mechanical protection, e.g., long spiny rostra, and hydrodynamic adaptations for enhanced escape swimming are an important defense against predators. However, like all generalizations, there are a number of exceptions to these trends which need to be further explored.

Shrimps that spend part of the day perched on the bottom but are good swimmers often up and off the bottom at others (nektobenthic), more frequently have the long sable-like rostra adaptive in backward and upward escape swimming (Fig. 2.3a). Endobenthic shrimps that burrow directly into the substrate (e.g., sicyoniid penaeoids; crangonid, glyphocrangonid, and processid carideans) or live in constructed burrows (e.g., alpheid carideans) have highly reduced rostra (Fig. 2.3a). This is generally true also for symbiotic shrimps that stay close to their hosts, e.g., spongicolid stenopodideans, numerous alpheid, palaemonid, hippolytid, and other caridean species (see Chap. 10, "Symbioses"). The stenopodideans as a group are highly benthic and none have the long sword-like rostra of nektobenthic carideans and penaeoids (Goy 2010). Penaeoid shrimps that spend time burrowed

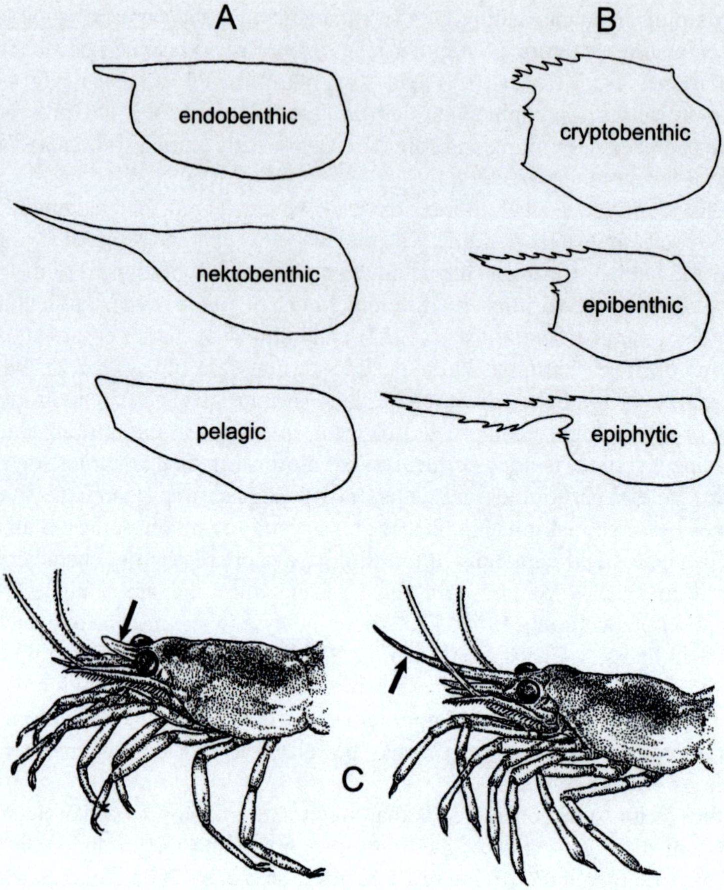

Fig. 2.3 Variation in cephalothorax and rostral shape with general lifestyle. (**a**) Average shape of cephalothoraxes and rostra in a group of 18 penaeoid and caridean deep-sea species. (**b**) Shape of cephalothorax and rostrum in three *Heptacarpus* species (Thoridae) with different lifestyles. (**c**) Intraspecific rostral variation in the tropical mountain-stream dweller *Xiphocaris elongata*; (left) shrimps living in predator-free waters upstream of waterfalls (barriers to fish predators) have short rostra (arrow) while (right) those in downstream exposed to fish predation develop long rostra (arrow). Adapted from (**a**) Sardà et al. (2005), (**b**) Bauer (1984), courtesy of Journal of Crustacean Biology and (**c**) Covich et al. (2009)

within the sediment during part of the day and out active above it at other periods tend to have rostra intermediate in length. Cephalothorax and rostrum shape may be influenced by other habitat characteristics as well. In the caridean genus *Heptacarpus*, more cryptic benthic species have short robust cephalothoraxes and short rostra, while epiphytic species which lie stretched out on substrates such as seagrasses and algal blades and stipes have elongated rostra and bodies; active mobile epibenthic species are intermediate in both characteristics (Fig. 2.3b) (Bauer 1984).

Another hypothesis about a well-developed spiny rostrum is that it may deter a predator from biting or swallowing the shrimp that bears it. Some shrimp species show intraspecific variation in rostrum length. Alan Covich et al. (2009) observed that the freshwater shrimp *Xiphocaris elongata* has short rostra at higher elevations of streams in Puerto Rico, where predatory fishes are excluded by downstream waterfalls. Below these barriers, rostra were noticeably and measurably longer (Fig. 2.3c). Covich's group tested the hypothesis that soluble compounds emanating from the fishes (kairomones) stimulated rostrum growth. They exposed shrimps with amputated rostra to various treatments in which predatory fish cues were and were not present. Those exposed to such cues regrew longer rostra than controls (Ocasio-Torres et al. 2014). Other experiments (Ocasio-Torres et al. 2015b) demonstrated that a dominant predatory fish was less successful in attacking long-rostra than short-rostra shrimps. De Mazancourt et al. (2017) found that freshwater shrimps of the genus *Caridina* on an Indo-Pacific island have longer rostra in downstream populations than those at higher elevation, a phenomenon they labeled the "Pinocchio effect" after the fairy-tale character. Rostra in freshwater cave-dwelling shrimps of the atyid genus *Troglocaris* (Caridea) are longer in caves in which the predatory salamander *Proteus* occurs (Jugovic et al. 2010). Direct observations showed that the salamander is not as successful in attacking and is slower in consuming long-rostra shrimps when approaching from the front of the shrimp, sometimes spitting them out.

2.3 Gills

Shrimps, like all decapod crustaceans, respire through gills (branchiae) which are evaginations (outgrowths) of the thoracic pleura (body wall). Some gills or other branchial structures may develop from the coxae of thoracic appendages (Fig. 2.4). Thus, gill tissues are covered by exoskeleton, albeit a very thin delicate one, which allows exchange of dissolved gases (CO_2 and O_2) between the internal and external environments of the shrimp. In addition, gills are important sites of ion exchange and excretion, especially ammonia, the principal metabolic waste of most aquatic animals. The gills are contained and sheltered within a gill or branchial chamber which is formed by a lateral extension of the carapace (branchiostegite). The inner side of the branchiostegite is covered with a very thin cuticle, and so is also an important respiratory surface.

The terminology of gills may refer to their morphology per se or to their attachment site in the gill chamber. Both gill characteristics are important in the placement of genera and species in taxonomic categories and phylogeny. The basic form of the decapod gill is a central axis (rachis) along which arise filaments (secondary branches). The standard model for the dendrobranchiate gill is that these filaments come off the main axis in opposite pairs, and each filament bears tubular branches which subdivide dichotomously one or more times. This gives the gills the dendritic (tree-like) appearance for which they are named (dendrobranchiate

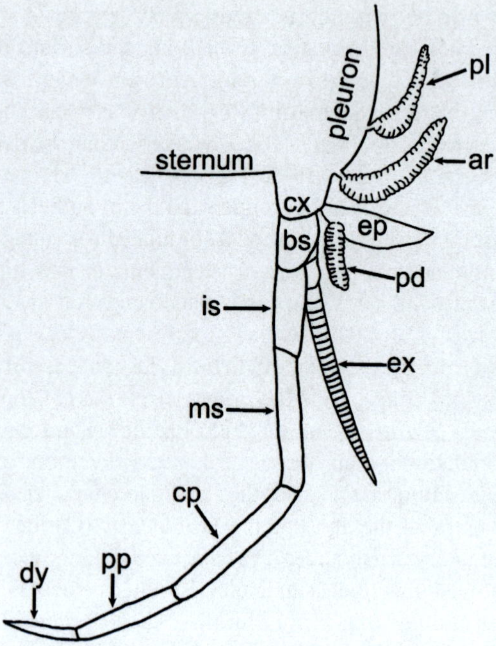

Fig. 2.4 Gill positions. Generalized pereopod and its segments attached to the sternum and body wall (pleuron) of one side, showing origin and position of pleurobranch, arthrobranch and epipod/podobranch gills in the gill (branchial) chamber (gill cover or branchiostegite not figured). ar, arthrobranch; bs, basis; cp, carpus; cx, coxa; dy, dactyl; ep, epipod; ex, exopod; is, ischium; ms, merus; pd, podobranch; pl, pleurobranch; pp, propodus

or dendrobranchs) (Fig. 2.5a, Plate 2.1a, b). This description generally holds true for the penaeoid families Penaeidae, Aristeidae, and Solenoceridae, as shown by the enlightening survey done by Joel Martin et al. (2007). However, they demonstrated that this classic example of dendrobranchiate gills does not hold true in other dendrobranchiate families. In the monogeneric family Sicyoniidae, yes, there is a series of paired gill filaments stemming from the rachis, but the filaments are flattened and platelike, and each gives rise to numerous other lamellar subdivisions (Fig. 2.5b). In the deep-sea family Benthesicymidae, the gills are less dendritic than the classical pattern and with considerable variation among species examined. In sergestid shrimps, gill filaments arise alternately from the rachis, rather than oppositely, and their ultimate subdivisions are oval or platelike, rather than tubular as in the classical pattern (Fig. 2.5c). Members of the other sergestoid family, Luciferidae, characterized by an extremely reduced morphology, have no gills.

Gill structure in procarideans and carideans is much less variable. Their gills are termed "phyllobranchs" (leaf-gills) for the pinnate arrangement of the lamellar filaments on each side of the gill rachis (Fig. 2.5d, Plate 2.1c, d). Similarly, the trichobranchiate gills of stenopodideans are fairly standard as well, with simple filaments arranged in irregular rows around the gill axis (Fig. 2.5e, Plate 2.1e, f) except in one genus, *Globospongicola*, in which the gills are simple finger-like structures without filaments (Goy 2010).

Gills are also categorized by their point of origin from the body wall or appendage. A shrimp's gill is connected to the body wall or appendage by a short narrow stalk. When the gill stalk is attached directly to the body wall (pleuron), it is called a

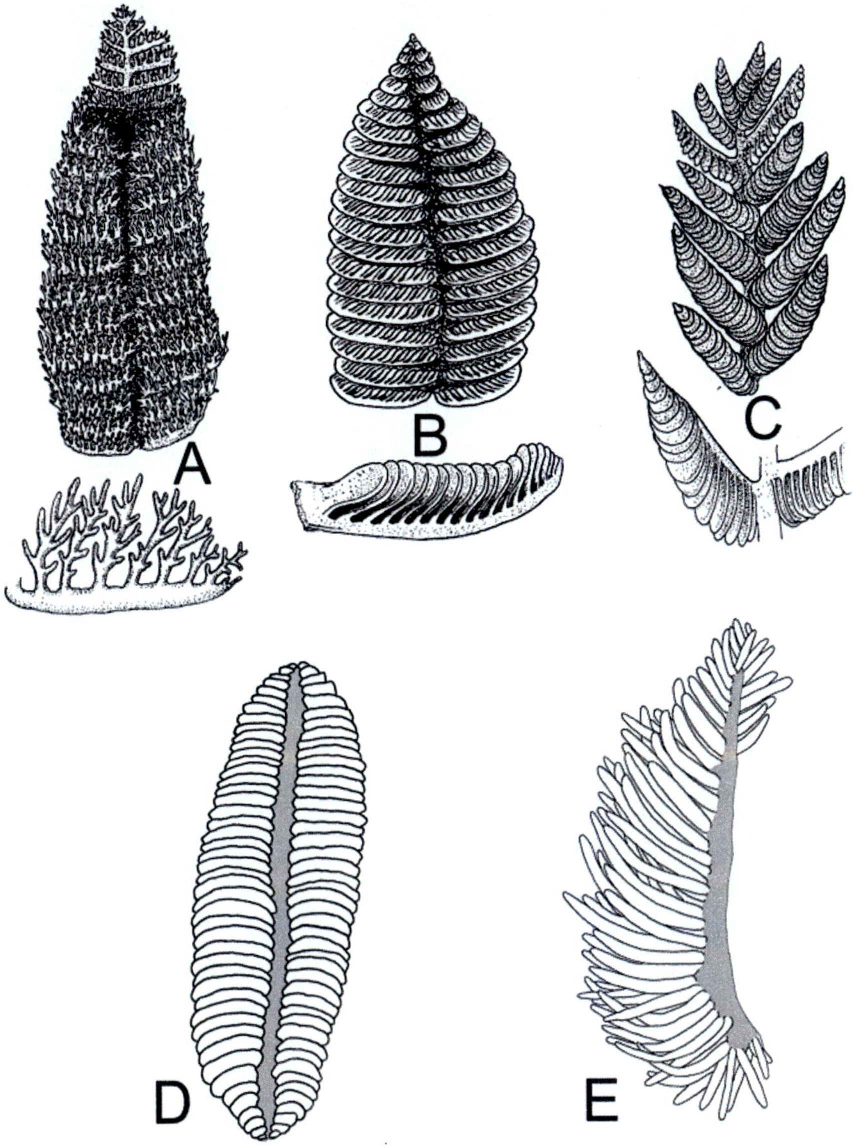

Fig. 2.5 Morphological gill types found in decapod shrimps. (**a**) Classical dendrobranchiate gill of the Penaeidae. (**b**) Sicyoniid variation of dendrobranchiate gill. (**c**) Sergestid variation of dendrobranchiate gill. (**d**) Phyllobranchiate gill (Caridea). (**e**) Trichobranchiate gill of *Stenopus hispidus* (Stenopodidea). In (**a**–**c**), the whole gill is shown above; an enlargement of a gill filament (secondary or side branch) of that gill type is shown below the gill. In (**d**, **e**), the gill axis (rachis) is shaded, gill filaments are unshaded. From (**a**–**c**) Martin et al. (2007) and (**d**, **e**) by R.T. Bauer

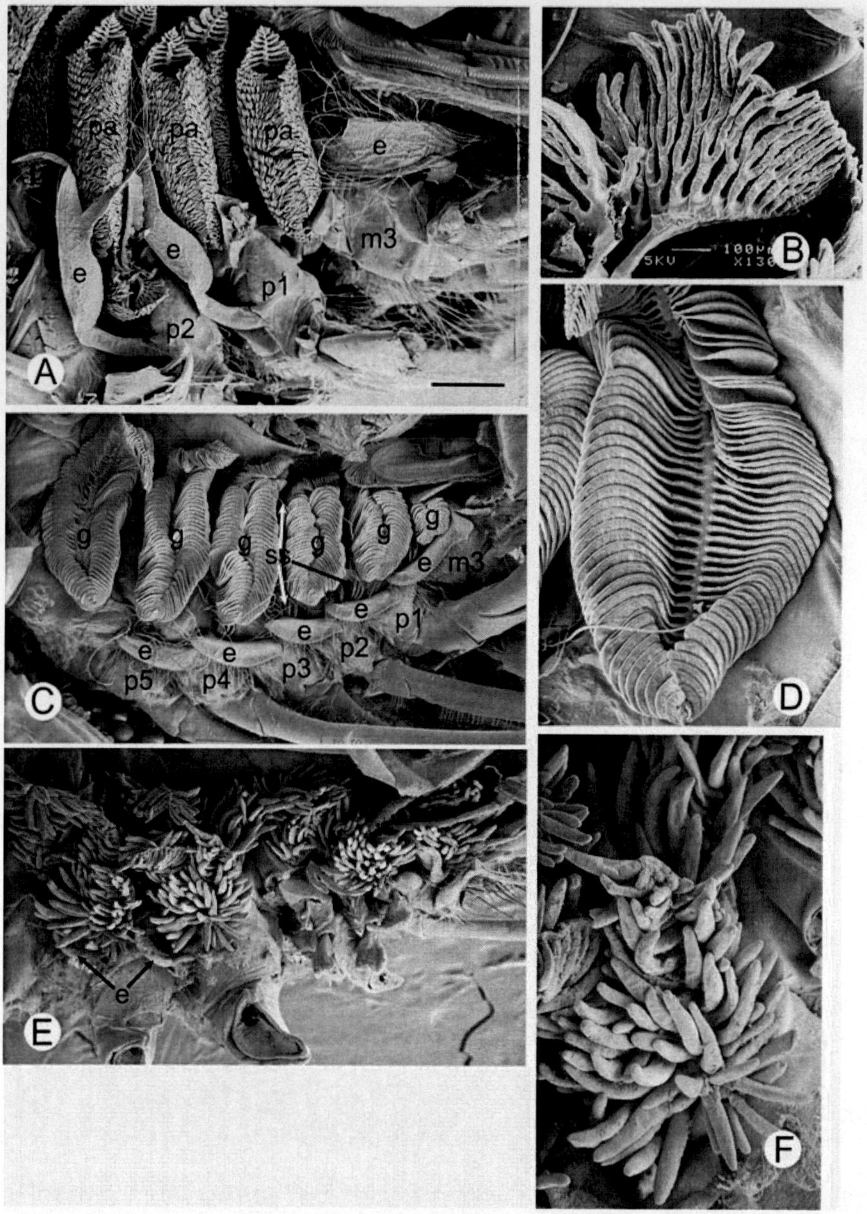

Plate 2.1 Shrimp Gills. (**a**) Partial view of branchial chamber of a dendrobranchiate shrimp, *Rimapenaeus similis* (Penaeoidea, Penaeidae) showing dendrobranch gills; the anterior arthrobranchs have been removed to reveal posterior arthrobranchs, the pleurobranchs behind them and setiferous (gill cleaning) epipods. (**b**) Dichotomous branching of a single gill filament ("secondary branch") of dendrobranch gills shown in (**a**). (**c**) Branchial chamber of the caridean *Lysmata wurdemanni* (Lysmatidae), showing phyllobranch gills and epipods; the spaghetti-like setae above the epipods serve in gill cleaning. (**d**) Phyllobranch gill from (**c**) showing platelike (lamellar) gill filaments stemming from the gill axis (rachis). (**e**) Gill chamber of the stenopodidean *Stenopus hispidus* with trichobranch gills and their numerous digitate filaments; two of the small

pleurobranch; when attached to the articular membrane between the pleuron and coxal (most basal) segment of the appendage, it is an arthrobranch; and when arising directly from the coxa, a podobranch (Fig. 2.4). Podobranchs may be found on several appendages in penaeoids, rarely in sergestoids, only on the second maxillipeds of stenopodideans and some carideans and are absent in the procarididideans. Podobranchs may arise together with epipods from which they are thought to be derived (Calman 1909). The diagram shown in Fig. 2.4 seems to make the identification of pleuro-, arthro-, and podobranchs an easy task. However, in the crowded gill chamber of many shrimps, it is not always simple to distinguish them, as many a novice (and expert!) shrimp biologist can attest (Figs. 2.6 and 2.7). When many gills are present in the branchial chamber, the best way is to resolve gill origin is to observe carefully where the gill stalk is attached to the body.

Another important branchial element that may be carried on the coxa of thoracic appendages is the epipod (sometimes referred to as a "mastigobranch"). It is variable in presence or absence, form and function. In penaeoids, epipods may small or large, variably leaf-or L-shaped, the latter with a proximal horizontal bar to which is attached a foliaceous and membranous vertical extension protruding upward between segmental sets of gills (Figs. 2.6, 2.8a; Plate 2.1a). The thin-walled cuticle of most epipods allows them to serve as additional surfaces for respiration. The epipod of one appendage may separate the gills of its somite from those of the somite posterior. This arrangement creates a series of segmental channels within the gill chamber, making respiratory flow within the gill chamber more efficient. Penaeoid epipods often serve additional duty as gill-cleaning devices (Chap. 4, "Shrimp Hygiene").

The coxae of the third maxillipeds and pereopods of procarididideans and many carideans bear a bar-shaped epipod with a hook which, in association with coxal setae known as setobranchs (Fig. 2.8b, c; Plate 2.1c), is part of a gill-cleaning system. In procarididideans and oplophorid carideans, the epipods are L-shaped with a horizontal bar and vertical extension, as in penaeoids (Fig. 2.8c, d), but with the unique caridean and procarididean hook. In many carideans, hooked epipods have been lost and gill cleaning is accomplished by another mechanism. Epipods of stenopodideans are generally small and can be foliaceous or fingerlike, depending on the species and appendage which bears them (Goy 2010) (Fig. 2.8e). In sergestoids, like much of their morphology, both epipods and gills are highly reduced and limited in number.

The number and arrangement of gills and epipods on the thoracic appendages are important characters in taxonomy and systematics. In his classic work on the

Plate 2.1 (continued) pereopodal epipods are indicated with arrows. (**f**) Higher magnification of gill filaments of trichobranchs shown in (**e**). e, epipods; g, gills; m3, coxa of third maxilliped; p1–5, coxae of pereopods 1–5; pa, posterior arthrobranch gills; ss, setobranch setae. Scale bar represents 850 μm in (**a**), 180 μm in (**b**), 1.1 mm in (**c**), 330 μm in (**d**), 1.5 mm in (**e**) and 360 μm in (**f**). (**a**) from Bauer (1999) courtesy of University of Chicago Press, (**b–d**, **f**) modified from Bauer (2004) and (**e**) R.T. Bauer

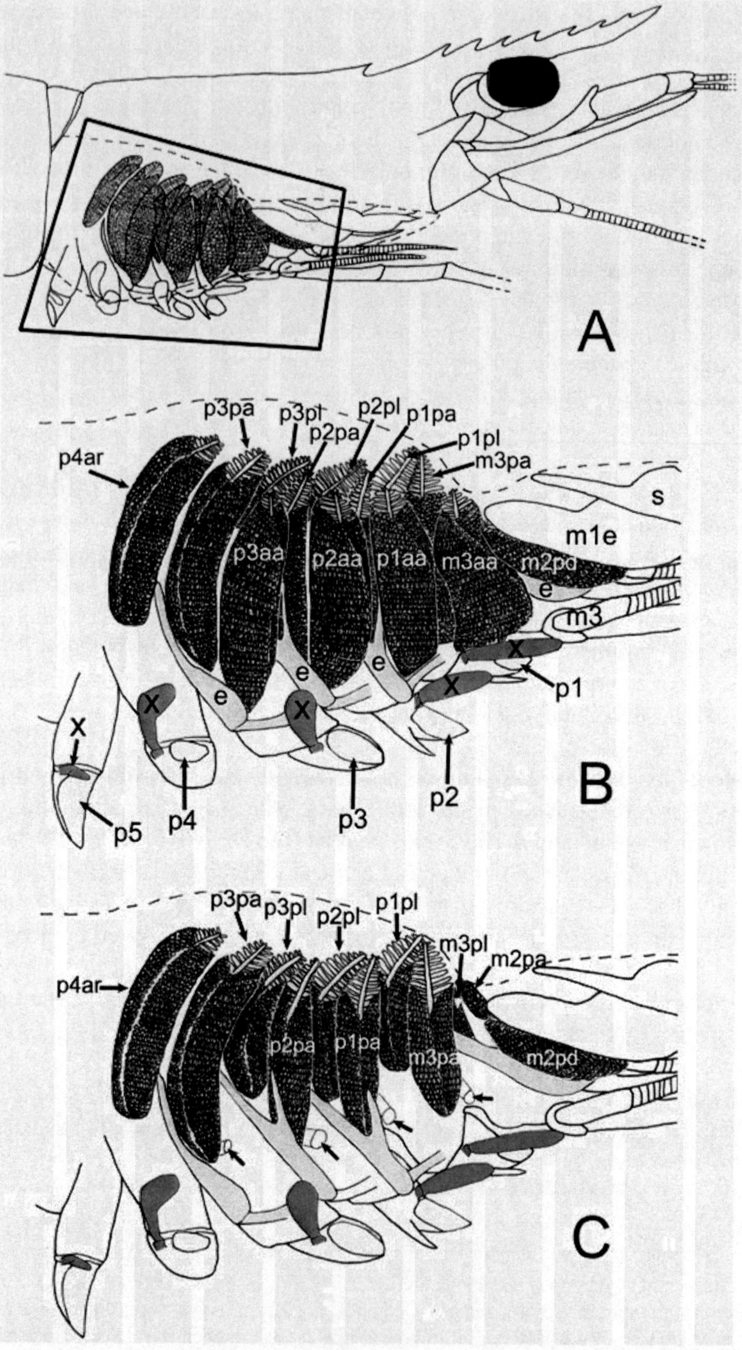

Fig. 2.6 Gills of a penaeoid shrimp, *Rimapenaeus similis* (Penaeidae). (**a**) Anterior end of shrimp with branchiostegite (gill cover) cut away to show gills in the branchial chamber. Area outlined by the box is shown magnified in (**b**) and (**c**). (**b**) Complete set of gills and other branchial structures as they appear from exterior view without the branchiostegite; setae of epipods and exopods not

crayfish, Thomas Huxley (1880) proposed a hypothetical maximum of 32 branchial elements for decapod crustaceans: one pleurobranch, two arthrobranchs, and one podobranch for each of the eight thoracic somites. Huxley did not include epipods in his formulas for various decapod species, as he considered them part of podobranchs or ones which had lost their filaments. Burkenroad (1981) agreed that podobranchs are intimately associated with epipods, but he, like many authors, counted the two elements separately, so the maximum possible number of branchial elements would be 40. Using this criterion, Burkenroad, based on his observations on extant (living) decapods, postulated 34 branchial elements for a hypothetical ancestral decapod. The observed range in the number of branchial elements for the Dendrobranchiata is 7–31. The lowest numbers are found in the sergestoids, in which a reduced morphology is typical—in fact, the two genera of Luciferidae have no gills at all. In the two procarididean genera, the number varies from 11–20, in Caridea from 4–19 and in the Stenopodidea from 13–26.

The positional type and number of branchial elements can be conveniently summarized in branchial formulas for species, genera, and other taxonomic categories. A range of gill formulas for major shrimp groups is given in Table 2.1. A branchial formula with a large number of elements is said to be "rich" or "full" while one with only a few is termed "impoverished." Although not branchial elements, the presence or absence of exopods on thoracic appendages is often reported by many authors as additional serially homologous (segmentally repeated) data for thoracic somites.

The mechanics of water flow through the branchial chamber and gills is powered by the scaphognathite ("gill bailer") (Figs. 2.7a and 2.15), the flat blade-like exopod of the second maxilla. Details of the flow have been described for shallow-water penaeids by Young (1959) and Dall et al. (1990). The anterior and posterior portions of this gill pump rock up and down about a central pivot point, driving water through the branchial chamber. Water is drawn into the anterior end of the chamber as the anterior half of the pump moves down and the posterior up, and then as the movement is reversed, water in the anterior chamber is driven down and posteriorly through the gills, leaving the chamber along the ventral margin of the branchiostegite (Fig. 2.9a). Most shallow-water penaeids burrow into the substrate during the day, completely or with only their eyes and/or antennule tips protruding. Water can enter the branchial chamber via a channel formed dorsally by the opposed antennules and ventrally by the broad antennal scales (scaphocerites), mandibular palps and exites

Fig. 2.6 (continued) shown. (**c**) Gills and other branchial structures with the anterior arthrobranchs of maxilliped 3 and pereopods 1–3 removed to show underlying gills; unlabeled arrows show attachment stalks of anterior arthrobranchs shown in (**a**) that were removed. Anterior arthrobranch of second maxilliped is hidden below (mesial) to its posterior arthrobranch. aa, anterior arthrobranch; ar, single arthrobranch of pereopod 4; e, gill-cleaning epipods; m1e, epipod of maxilliped 1; m2,3, maxilliped 2, maxilliped 3; pa, posterior arthrobranch; pd, podobranch; p1–5, proximal segments of pereopods1–5 (labeled arrows point to basi-ischial breakage planes, distal limb segments not shown); pl, pleurobranch; s, scaphognathite (gill bailer) of maxilla 2; x, pereopodal exopods. From Bauer (1999)

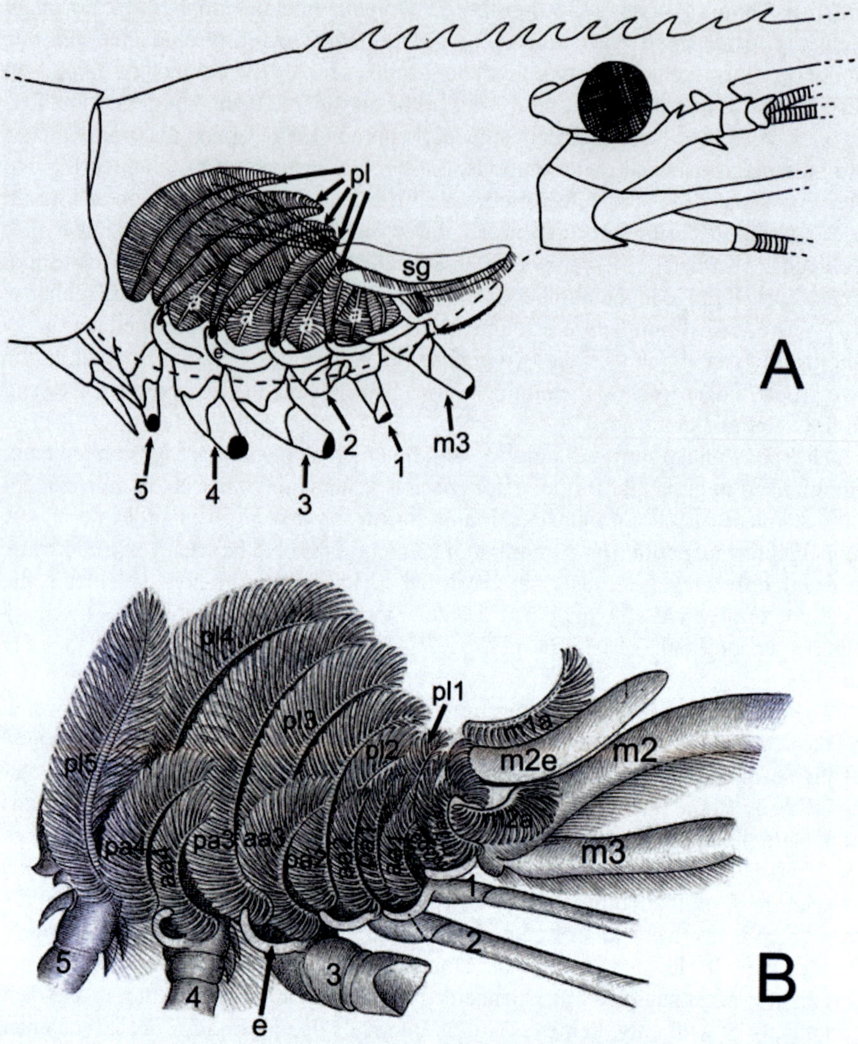

Fig. 2.7 Gills of a caridean and a stenopodidean shrimp. (**a**) Gill chamber of *Pandalus danae* (Caridea) with arthrobranchs and pleurobranchs; dotted line is ventral edge of branchiostegite. Note the long cleaning setae lying over the gills and arising from the posterior border of the scaphognathite. (**b**) Arthrobranchs and pleurobranchs of *Stenopus hispidus* (Stenopodidea). In (**a**), numbered arrows indicate basal segments of pereopods 1–5; a, arthrobranchs of pereopods and third maxillipeds; e, epipod (fingerlike epipod of pereopod 3 is labeled; similar epipods on maxilliped 3 and other pereopods not labeled; epipod of maxilliped 2 flat, leaflike); m3, basal segments of maxilliped 3; pl, pleurobranch gills of pereopods 1–5; sg, scaphognathite. In (**b**), numbered arrows indicate basal segments of pereopods 1–5; aa, anterior arthrobranchs; m1–3, maxillipeds 1–3; pa, posterior arthrobranchs; pl, pleurobranchs. From (**a**) Bauer (1979) and (**b**) Gerstaecker and Ortmann (1909)

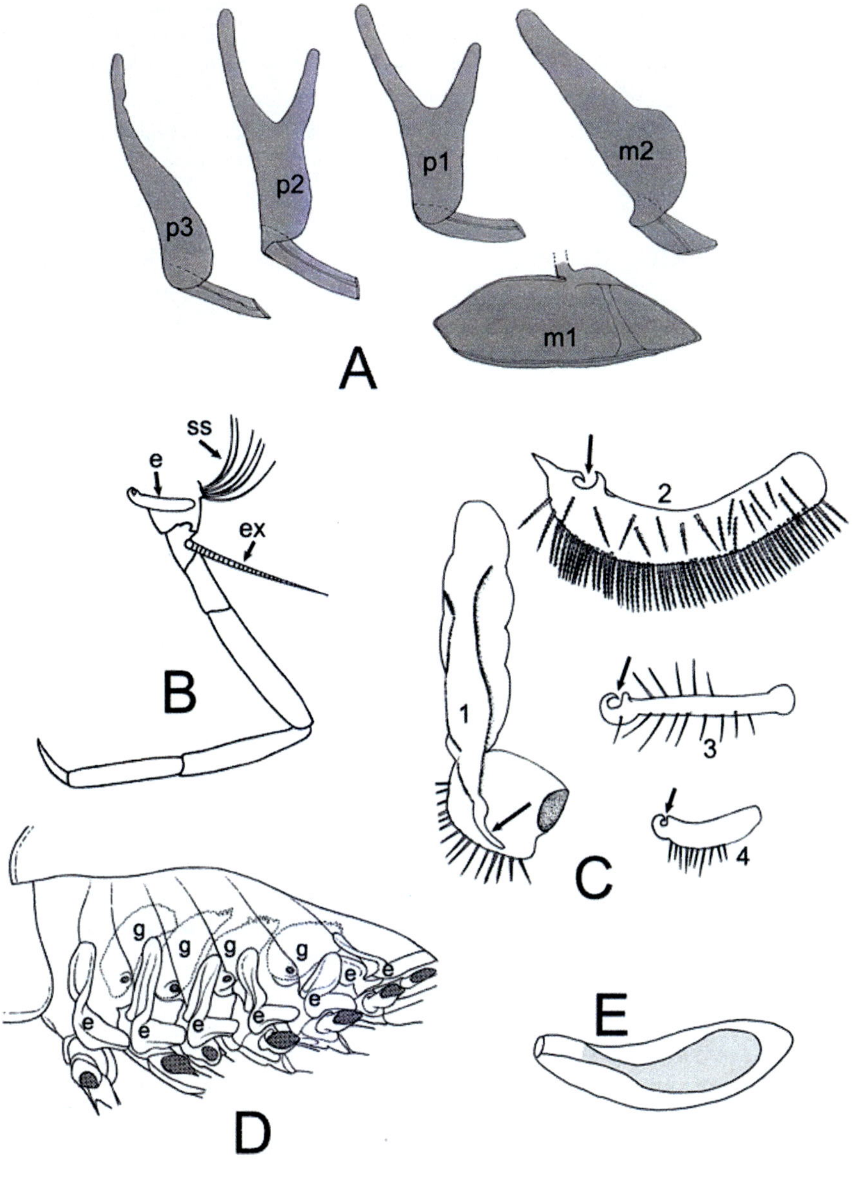

Fig. 2.8 Shrimp epipods. (**a**) Thoracic epipods of the penaeoid *Rimapenaeus similis*. Note the bifurcate shape of pereopodal epipods 1–3; gill-cleaning setae on p1–3 and m2 are not shown (see Chap. 4). (**b**) Ambulatory pereopod of a caridean shrimp showing the position of the hooked epipod and setobranch setae which are part of a gill cleaning mechanism in many carideans (see Chap. 4). (**c**) Hooked epipods found in many caridean and procaridean shrimps:1, *Procaris hawaiana*; 2, *Pandalus danae*; 3. *Betaeus macginitieae*; 4, *Heptacarpus sitchensis*; hooks are indicated by arrows. (**d**) Gill chamber of *Procaris ascensionis* showing epipods with vertical appendix extending up between gills (given in outline); small, hatched areas within gill outlines show sites of gill stalk connections to body wall. (**e**) Small digitate epipod of *Stenopus hispidus* (see also Plate 2.1e). e, epipod; ex, exopod; g, gill; ss, setobranch setae; m1, m2, epipods of maxillipeds 1,2; p1–p3,

of the first maxillipeds. Thus, the shrimps can irrigate the gills while hidden within the sediment. However, even when out and exposed above the bottom at night or during turbid conditions, the same pattern of respiratory flow persists. The anterior to posterior flow pattern is apparently true also for the planktonic sergestoids (Xiao and Greenwood 1993) and solenocerids, as the latter use flattened or subcylindrical antennules (Pérez Farfante and Kensley 1997) that fit together to make a long respiratory siphon when burrowing.

Although powered by the same type of scaphognathite pump, the pattern of water flow over the gills in carideans (and presumably in the closely related procarididеans) is completely the opposite! Water enters the branchial chamber between the pereopodal protopods and exits anteriorly (Fig. 2.9b). Small differences in the morphology of the branchial chamber, basal segments of associated appendages and details of scaphognathite structure may account for this difference in pattern. To my knowledge, there are not yet any reports or published studies on the branchial currents in stenopodideans, a lack which needs to be remedied.

2.4 Eyes

Shrimps have compound eyes composed of a multifaceted cornea atop movable eyestalks or peduncles (Figs. 2.10a and 2.11a, c, e), which may be inclined on each side from 0° to 90° from the midline and held slightly forward when at rest (Fig. 2.10b). In dendrobranchiate shrimps, the eye rests within a cup (optic calathus), which is articulated with the rest of the eyestalk (Fig. 2.10a), permitting additional eye movement. The horizontal extent of the field of vision of each eye is a little over 180° on each side, so that shrimp has a complete 360° view of its surroundings (Fig. 2.10b). The cornea of the eye is composed of many square or hexagonal facets, each of which corresponds to an underlying functional visual unit, the ommatidium. In species which are nocturnal or live in low-light habitats, a tapetum (a layer of reflective pigment) may be present and helpful in detecting light. It also is responsible for the glowing eyes ("eyeshine") that is observed in many shrimps and other nocturnal animals when one shines a light on them in the dark.

There may be a small "accessory eye" on the border between the compound eye and eyestalk, in many caridean species (Figs. 2.10c and 2.11c, e). More technically termed the nebenauge (German for "next to eye"), it is also referred to in the literature as an ocellus or pigment spot. Its corneal facets are hexagonal or irregular in shape rather than square as in some compound eyes, but its role in vison or other sensory function is unclear. A survey of the majority of caridean genera by Johnson et al. (2015) indicated a weak correlation of the accessory eye with lifestyle, more

Fig. 2.8 (continued) epipods of pereopods 1–3. From (**a**) Bauer (1999), (**b**, **c**) from Bauer (1976a), (**d**) from Chace and Manning (1972) and (**e**) by R.T. Bauer

Table 2.1 Branchial formulas of major decapod shrimp taxa. Examples of the minimum, an intermediate, and maximum formula are given for each major group. Overall total of branchial elements is given in boldface. Dendrobranchiata (A–C), Procarididea (D, E), Caridea (F–H) and Stenopodidea (I–K). From (A–C) Pérez Farfante and Kensley (1997), (D) Chace and Manning (1972), (E) Kensley and Williams (1986), (F) Calman (1909), (G) Thompson (1967), (H) Thompson (1966), (I) Alvarez et al. (2006) (with correction on epipod position) and (J, K) Saito and Takeda (2003). r, rudimentary; s, small; v, vestigial

Somite	Mx1	Mx2	Mx3	P1	P2	P3	P4	P5	Totals
Dendrobranchiata									
(A) *Acetes* spp. (Sergestidae)									
Pleurobranchs	–	–	–	–	–	–	–	–	0
Arthrobranchs	–	r (–)	1	1	1	1	1	–	5–6
Podobranchs	–	r (–)	–	–	–	–	–	–	0–1
Epipods	1	1	–	–	–	–	–	–	2
Totals	1	1–3	1	1	1	1	1	0	**7–9**
(B) *Rimapenaeus* spp. (Penaeidae)									
Pleurobranchs	–	–	1	1	1	1	–	–	4
Arthrobranchs	r	2	2	2	2	2	1 + v	–	13
Podobranchs	–	1	–	–	–	–	–	–	1
Epipods	1	1		1	1	1		–	5
Totals	2	4	3	4	4	4	2	0	**23**
(C) *Aristaeomorpha* spp. (Aristeidae)									
Pleurobranchs	–	–	s	1	1	1	1	1	6
Arthrobranchs	s	s + 1	2	2	2	2	2	0	13
Podobranchs	–	1	1	1	1	1	–	–	5
Epipods	1	1	1	1	1	1	1	–	7
Totals	2	4	5	5	5	5	4	1	**31**
Procarididea									
(D) *Procaris ascensionis*									
Pleurobranchs	–	–	–	1	1	1	1	–	4
Arthrobranchs	–	–	–	–	–	–	–	–	0
Podobranchs	–	–	–	–	–	–	–	–	0
Epipods	1	1	1	1	1	1	1	–	7
Totals	1	1	1	2	2	2	2	0	**11**
(E) *Vetericaris chaceorum*									
Pleurobranchs	–	–	1	1	1	1	1	–	5
Arthrobranchs	–	–	1	2	2	2	1	–	8
Podobranchs	–	–	–	–	–	–	–	–	0
Epipods	1	1	1	1	1	1	1	–	7
Totals	1	1	3	4	4	4	3	0	**20**
Caridea									
(F) *Limnocaridina* spp. (Atyidae)									
Pleurobranchs	–	–	–	1	1	1	1	–	4
Arthrobranchs	–	–	–	–	–	–	–	–	0
Podobranchs	–	–	–	–	–	–	–	–	0

(continued)

Table 2.1 (continued)

Somite	Mx1	Mx2	Mx3	P1	P2	P3	P4	P5	Totals
Epipods	–	–	–	–	–	–	–	–	0
Totals	0	0	0	1	1	1	1	0	**4**
(G) *Pontophilus* spp. (Crangonidae)									
Pleurobranchs	–	–	1	1	1	1	1	1	6
Arthrobranchs	–	–	0–1	0	0	0	0	0	0–1
Podobranchs	–	1	–	–	–	–	–	–	1
Epipods	1	1	0–1	0	0	0	0	0	2–3
Totals	1	2	1–3	1	1	1	1	1	**9–11**
(H) *Nematocarcinus ensifer* (Nematocarcinidae)									
Pleurobranchs	–	–	–	1	1	1	1	1	5
Arthrobranchs	–	–	2	1	1	1	1	–	6
Podobranchs	–	1	–	–	–	–	–	–	1
Epipods	1	1	1	1	1	1	1	–	7
Totals	1	2	3	3	3	3	3	1	**19**
Stenopodidea									
(I) *Macromaxillocaris bahamaensis* (Macromaxillocarididae)									
Pleurobranchs	–	–	1	1	1	1	1	1	6
Arthrobranchs	–	–	1	1	1	1	1	–	5
Podobranchs	–	1	–	–	–	–	–	–	1
Epipods	1	–	–	–	–	–	–		1
Totals	1	1	2	2	2	2	2	1	**13**
(J) *Spongicoloides* spp. (Spongicolidae)									
Pleurobranchs	–	–	1	1	1	1	1	1	6
Arthrobranchs	–	1	1	1	1	1	r	–	6
Podobranchs	–	r	–	–	–	–	–	–	1
Epipods	1	1	1	–	–	–	–	–	3
Totals	1	3	3	2	2	2	2	1	**16**
(K) *Stenopus hispidus* (Stenopodidae)									
Pleurobranchs	–	–	1	1	1	1	1	1	6
Arthrobranchs	1	1	2	2	2	2	2	–	12
Podobranchs	–	1	–	–	–	–	–	–	1
Epipods	1	1	1	1	1	1	1	–	7
Totals	2	3	4	4	4	4	4	1	**26**

common in active epibenthic or pelagic than in burrowing or mostly benthic species. The accessory eye seems to be unique to the Caridea, although it is found in only about 40% of genera surveyed. The only mention of such a structure in dendrobranchiates (3 *Penaeus s.l.* species; Ugolini and Borgioli 1993) has not been confirmed, and none has been reported in the classic morphological or taxonomic literature in that group. The world expert on stenopodideans, Joseph Goy, tells me that he has never found accessory eyes in that taxon.

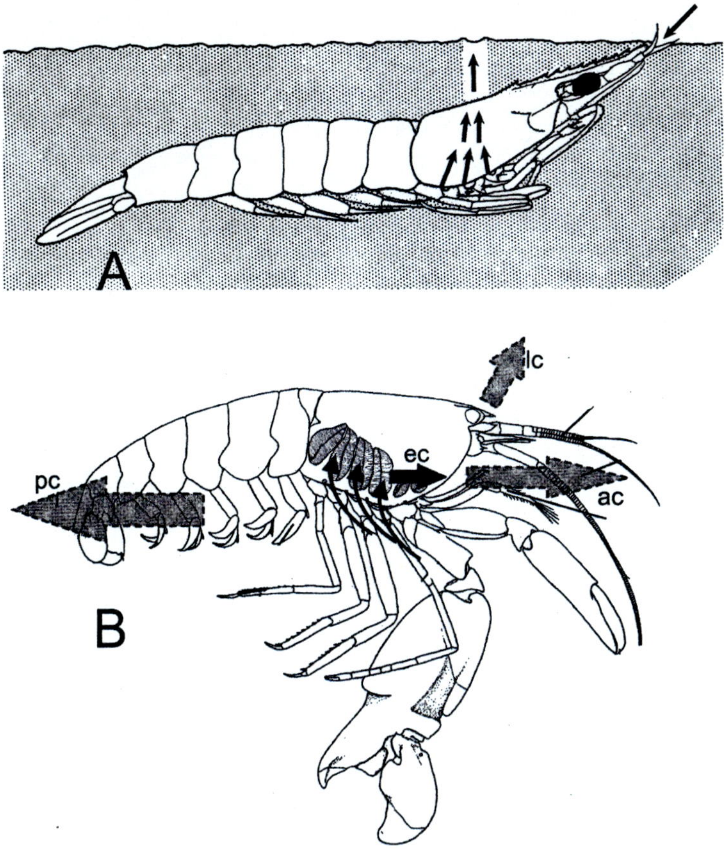

Fig. 2.9 Respiratory flow in and out of the branchial chamber. (**a**) In a burrowed penaeoid (*Metapenaeus* species), water enters the anterior branchial chamber near the exposed antennules (unlabeled single arrow), flows over the gills, and exits at the ventral edge of the branchiostegite (multiple arrows). (**b**) In a caridean (*Alpheus heterochaelis*), flow enters between the leg bases along the ventral edge of the branchiostegite (unmarked black arrows), flows over the gills, and exits the branchial chamber anteriorly. ac, anterior current; ec, exit current; lc, lateral current; pc, posterior current is produced by pleopods. (**a**) modified from Dall et al. (1990) and (**b**) from Herberholz and Schmitz (2001) courtesy of University of Chicago Press

Shrimp eyes tend to be reduced, vestigial or absent in species inhabiting perpetually dark habitats such as caves or other subterranean habitats (Hobbs et al. 1977; Wilkens 1986), as well as in the bathypelagic zone, which is unlighted except for the frequent light flashes from bioluminescent organisms. The eyes of related benthic neritic and slope species, as well as some mesopelagic shrimps, tend to increase with depth to better capture the decreasing light with depth (Herring and Roe 1988). Corneal setae have evolved in some cave shrimps and other troglobitic crustaceans which may be a compensatory chemosensory/tactile adaptation for loss of vision (Mejía-Ortíz and Hartnoll 2006).

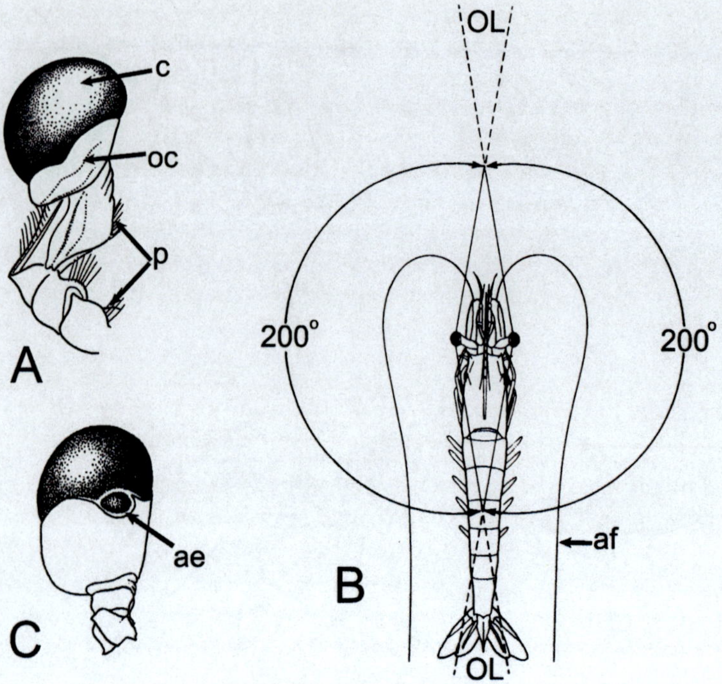

Fig. 2.10 Eyes. (**a**) Penaeoid eye and (**b**) visual field. (**c**) Caridean eye with accessory eye (*Rhynchocinetes albatrossae*). ae, accessory eye ("nebenauge," "pigment spot"); af, antennal flagellum; c. cornea; oc, optic calathus; p, peduncle. (**a**) from Pérez Farfante and Kensley (1997), courtesy of ©Publications Scientifiques du Muséum national d'Histoire naturelle, Paris; (**b**) from Dall et al. (1990); (**c**) from Chace (1997)

2.5 Appendages

Shrimps have 19 pairs of segmental appendages: 5 cephalic, 8 thoracic, and 6 abdominal (Table 2.2). The appendages carry out the shrimps' various activities and interactions with the external environment. Details of shrimp appendage structure and function in the penaeid *Litopenaeus setiferus* were published in a monographic paper by Young (1959) with considerable detail and information rarely found in a single publication. Other important monographs containing information about shrimp morphology are Pérez Farfante and Kensley (1997) on dendrobranchiates; Xiao and Greenwood (1993) on sergestids, specifically *Acetes japonicus*; Bauer (2004) and Wicksten (2010) on carideans; Tavares and Martin (2010) on penaeoids and sergestoids and Goy (2010) on stenopodideans.

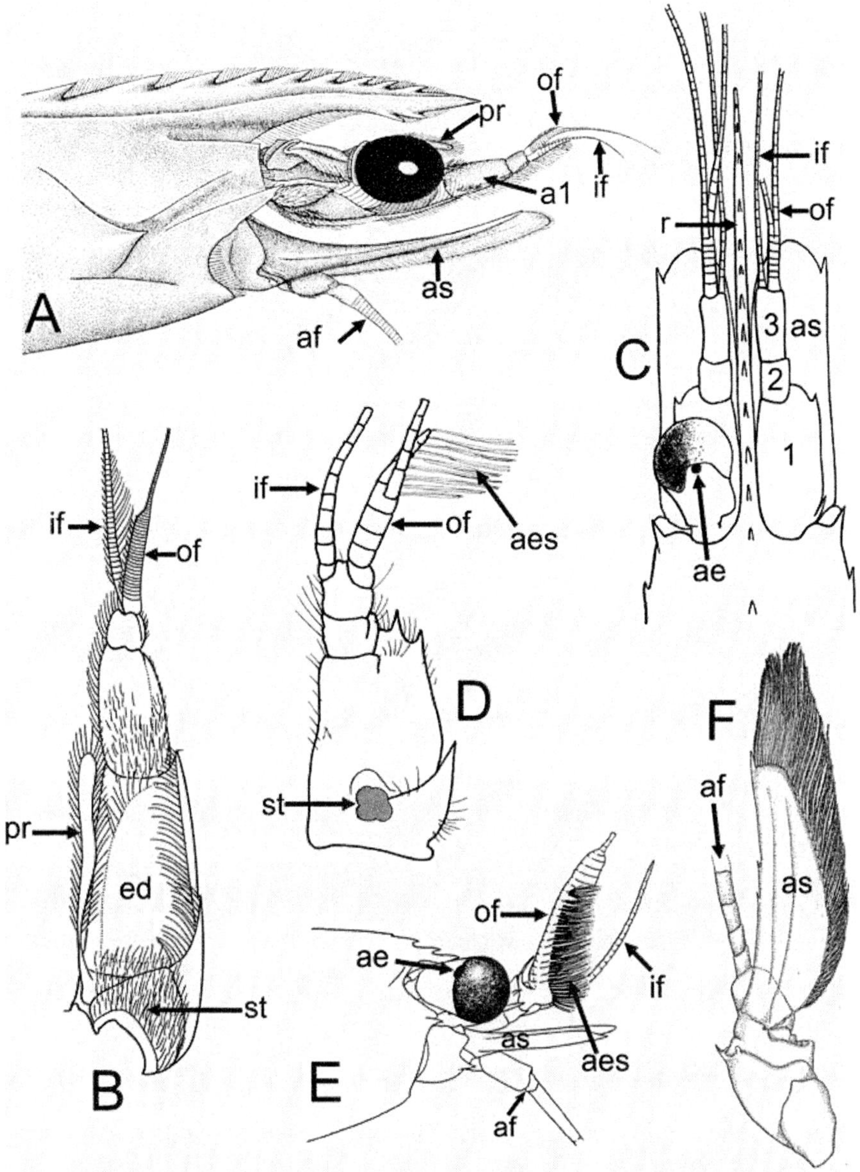

Fig. 2.11 Antennules and antennae. (**a**) Anterior cephalothorax of the penaeid *Farfantepenaeus duorarum*, showing large eye, antennule, antennal scale, and basal part of antennal flagellum. (**b**) Antennule of a penaeoid shrimp (dorsal view, right side). (**c**) Cephalic region of caridean *Leandrites indicus* (Palaemonidae), dorsal view, showing eye of left side with small accessory eye, antennules, and antennal scale. (**d**) Antennule of caridean *Unguicaris pilipes* (Palaemonidae) (dorsal view, right side), with low density of aesthetascs (olfactory setae) on outer flagellum. (**e**) Cephalic region of caridean *Thor amboinensis*, lateral view, showing antennule with high density of aesthetascs on the outer flagellum. (**f**) Basal part of antenna (dorsal view, right side) of the stenopodidean *Macromaxillocaris bahamaensis*, showing peduncle, antennal scale, basal articles of the antennal flagellum. a1, antennule; ae, accessory eye; aes, aesthetascs; af, antennal flagellum; as, antennal

2.6 Antennules

The anterior most pair of head limbs, the first antennae (antennules) are important olfactory and chemotactile appendages (Figs. 2.1 and 2.11) The antennules are composed of a basal peduncle of three segments, the most distal containing two multiflagellate flagella, an outer ("upper," "dorsal," "superior, "lateral") and an inner ("lower," "ventral," "inferior," "mesial") flagellum (Fig. 2.11). The outer flagellum is equipped with delicate, thin-walled setae, the aesthetascs (Fig. 2.11d, e), which are sensitive to a multitude of compounds dissolved in water. Simply put, the aesthetascs house the shrimp's sense of smell (olfaction). When the aesthetascs are rather sparsely spaced (Fig. 2.11d), the antennule is waved slowly to circulate water containing potential chemical stimuli through them. In some carideans, aesthetascs are numerous and arranged in a thick brush on the outer flagellum (Fig. 2.11e). The antennule spins or flutters in a semicircle in rapid back-and-forth bursts forcing water through the aesthetascs. The outer flagellum (and sometimes the inner) may be long and has a variety of tactile (touch) receptors, as well as chemoreceptors sensitive to high concentrations of chemical stimuli (gustatory or taste receptors). In some caridean families, e.g., Lysmatidae, Palaemonidae, and Euryrhynchidae, the outer flagellum may be subdivided into two separate ones (Fig. 2.11c, d). The basal segment of dendrobranchiate shrimps has a cavity (eye depression or cup) whose margins bear long plumose setae (Fig. 2.11b). The eye can be moved into the eye depression for cleaning or when burrowing. In the penaeoid families Penaeidae and Solenoceridae, there is a long and narrow setose lobe, the prosartema (Fig. 2.11b), on the mesial side of the basal segment which serves as an eye brush (Chap. 4). In most shrimps, the basal segment is extended laterally into a usually sharp projection or spine, the stylocerite. The antennular flagella are variable in length, sometimes quite long, but often much shorter in the more benthic species.

The first segment of the antennular peduncle has an important organ of equilibrium or balance, the statocyst (Fig. 2.11b, d) except in stenopodideans (Goy 2010). The statocyst is composed of a hollow chamber containing a concretion, the statolith, composed of very fine sediment particles. The chamber forms during embryonic development from an invagination of the exoskeleton and therefore are lined with cuticle from which many sensory setae project into the interior of the chamber. When the shrimp changes position, the statolith shifts against the sensory setae of some part of the chamber, stimulating the nerve endings inside those setae. This sensory input indicates to the shrimp's central nervous system the shift in motion and spatial

---

Fig. 2.11 (continued) scale; ed, eye depression; if, inner flagellum; of, outer flagellum; pr, prosartema; r, rostrum; st, statocyst; 1,2,3, first (proximal) to third (distal) segments of antennular peduncle. (**a**) from Pérez Farfante (1988); (**b**) modified from Pérez Farfante and Kensley (1997), courtesy of ©Publications Scientifiques du Muséum national d'Histoire naturelle, Paris; (**c**) modified from Holthuis (1950); (**d**) modified from Bruce and Zmarzly (1983); (**e**) from Chace (1972); (**f**) from Alvarez et al. (2006). (**d, f**) courtesy of Journal of Crustacean Biology

Table 2.2 Body somites and their appendages. In shrimps, as in other decapod crustaceans, the head (cephalon) and thorax are fused into a cephalothorax. The two pairs of antennae are sensory, while the mandibles, maxillae, and maxillipeds are primarily mouthparts. The anterior three pairs of pereopods bear chelae (pincers, claws) in stenopodideans, penaeoids and some sergestoids, but only the first two pairs usually have them in carideans, and all pereopods are simple (achelate) in procarididеans. Nonchelate pereopods serve as walking or perching legs. The pleopods and uropods are swimming limbs. Note that some carcinologists recognize the preantennular area as a body somite 1, e.g., Pérez Farfante and Kensley (1997), so that body somite 1 below (antennular somite) becomes somite 2 and so on

Body somite	Region	Regional somite #	Appendages
1	Cephalon	1	Antennules (first antennae)
2	"	2	Antennae (second antennae)
3	"	3	Mandibles
4	"	4	Maxillules (first maxillae)
5	"	5	Maxillae (second maxillae)
6	Thorax	1	1st maxillipeds
7	"	2	2nd maxillipeds
8	"	3	3rd maxillipeds
9	"	4	1st pereopods
10	"	5	2nd pereopods
11	"	6	3rd pereopods
12	"	7	4th pereopods
13	"	8	5th pereopods
14–18	Abdomen	1–5	1st–5th pleopods (swimmerets)
19	"	6	Uropods

orientation, somewhat similar to the workings of the vestibular apparatus of the human inner ear.

Because the lining of the statocyst chamber is cuticular, its lining and contents are shed at each molt along with the rest of the exoskeleton. The shrimp has to supply the now empty chamber with materials for a new statolith after the molt, a process admirably demonstrated by Haywood and Alexander (1982). When the penaeid shrimp *Melicertus plebejus* was maintained through molts in an aquarium with clean water and no sand or sediment, the interior of the statocyst only had a small coagulum of glandular material. Such shrimps swam in a disoriented fashion. Video recordings of shrimps maintained in an aquarium with a natural sandy substrate showed how the shrimp fills the statocyst with new statolith material after a molt. Using its chelipeds, the shrimps digs and pushes sediment forward beneath the antennules (Fig. 2.12a). The shrimp then roils up and flicks some of this sediment onto the dorsal surface of the antennules and into the eye depression (Fig. 2.12b), then tilts its anterior end upwards to move the accumulated sediment in the eye depression towards the statocyst opening. The eyes then flick repeatedly into the eye depression (Fig. 2.12c), forcing sediment particles into the statocyst opening. The particles are then glued together into a statolith by secretions of tegumental glands lining the statocyst chamber. Whether this or other types of behavior occur in

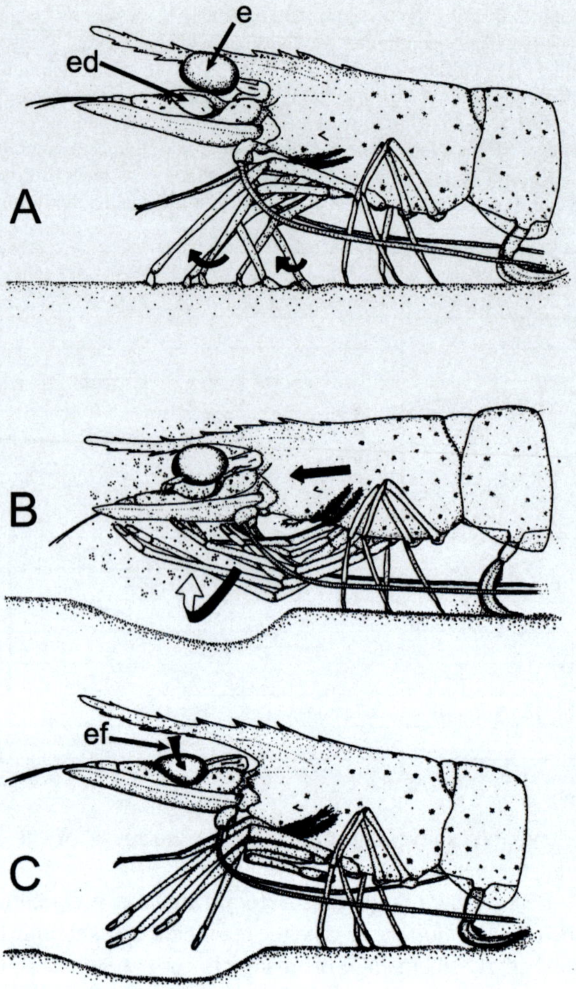

Fig. 2.12 Replacement of statocyst sediment after a molt in the shrimp *Fenneropenaeus merguiensis.* (**a**) Sediment dug out of the substrate by the chelipeds (curved arrows) is pushed forward beneath the shrimp's anterior end and antennules. (**b**) The shrimp tilts downward while the chelipeds toss particles from the accumulated sediment onto the dorsal surface of the eye depression. (**c**) The shrimp tilts back up, moving the sediment in the eye depression towards the opening of the statocyst, then the eyes flick in and out of the eye depression (labeled arrow), forcing sediment particles into the statocyst opening. e, eye; ed, eye depression; ef, eye flicking in and out of the eye depression. From Haywood and Alexander (1982)

statolith replacement of other shrimps is an open question. Likewise is the apparent lack of a statocyst in stenopodideans; is it just a lack of observation or is there a statocyst elsewhere in the body that serves the same function? For example, the statocysts of mysidan (peracaridan) shrimps are located in the uropodal endopods on the end of the animal.

2.7 Second Antennae

The doubly branched (biramous) second antenna (Fig. 2.1), also simply called the antenna, consists of a peduncle of two segments, the coxa and basis, with the latter giving rise to an endopod of three segments plus a long antennal flagellum. Also stemming from the basis is an exopod, the broad and flat antennal scale (scaphocerite) (Fig. 2.11a, c, e, f). The hydrodynamic functions of the scaphocerite in swimming were discussed in the previous chapter. The long antennal flagella (Figs. 2.1 and 2.10b), equipped with an array of tactile and gustatory chemical receptors, sweep out ahead, alongside and behind the shrimp, providing it, together with the antennular flagella, with a sophisticated system for detection of stimuli. Most shrimps are active at night or live in turbid, dark environments where visual cues other than simple detection of light and movement are of little value. The flagella of the second antennae serve the shrimp well in detecting food, predators, conspecifics and location, something like cat whiskers or catfish barbels. Antennal flagella are multiarticulated for flexibility and may be quite long. In pelagic shrimps, the long antennal flagella function somewhat like parachutes, producing drag that retards sinking with subsequent conservation of the swimming energy needed to maintain position. Antennal flagella are usually shorter in more "homebound" carideans, e.g., those living in constructed burrows or those burrowing directly into the sediments during part of the day. However, many penaeoid shrimps are daytime burrowers but are epibenthic or move up into the water column at night and retain very long antennal flagella. The most proximal (coxal) segment of the antenna bears a pore, the opening of the nephridial excretory organ, termed the antennal gland. This gland is an important site of ionic and osmoregulation, and it produces a urine which contains metabolic wastes. The urine may also contain pheromones which are important in intraspecific interactions (Herberholz and Schmitz 2001).

2.8 Mouthparts

Shrimps have an array of appendages around or near to the mouth which work together to process food items. The latter are delivered to the mouthfield primarily by the chelipeds. Just anterior to the mouth and protruding a bit into it is a fleshy, muscular labrum, sort of an upper lip which is not considered a segmental appendage. The first pair of mouthparts are mandibles, which serve as crustacean jaws. Like the first and second antennae and remaining appendages, the mandibles are considered biramous, albeit highly modified from the generalized pattern shown in Fig. 2.4. A mandible consists of a large coxal segment necessary to contain its massive powerful musculature. The mandibles, one on each side of the mouth, bear highly calcified processes that protrude into the mouth. The incisor processes are generally considered cutting tools that chop the food into pieces and help push it towards the broadly ridged molar processes, which pound and mash the food as it is moved into the mouth. However, detailed studies show that such generalizations about mandible function may not always hold (Ashelby et al. 2015). In dendrobranchiate, procarididean, and stenopodidean shrimps, both mandibular processes are generally present, closely spaced or fused into unit (Fig. 2.13a–c, h–j). One notable exception is the penaeid genus *Funchalia* (Fig. 2.13d), in which the molar process is absent, but the wicked-looking incisor is elongated sickle-like blade, a probable adaptation for its predatory lifestyle (Hopkins et al. 1994). A highly developed incisor process of procarideans (Fig. 2.13h) and pasiphaeid carideans indicates a predatory life-style (Bauer 2004). The processid, glyphocrangonid and crangonid carideans have mandibles which lack an incisor and palp completely, but in these groups, the molar has been modified into a piercing or cutting tool (Fig. 2.13g). A large broad molar process (Fig. 2.13e) may indicate a more generalized feeding habitat. In the Caridea, there is considerable variation in mandibular structure among genera and families, with some groups having only one or the other process (Fig. 2.13g), or both but with one process reduced (Fig. 2.13e, f), or both processes subequal in development. This diversity reflects the greater variety of caridean lifestyles and habitats compared to those of the dendrobranchiates and stenopodideans.

In addition to these mandibular processes for chopping and grinding food, many shrimps have masticating structures within the foregut, known collectively as the gastric mill. The foregut of shrimps, as in other crustaceans, is formed from an embryological invagination of the outer cuticle, and as such can be lined numerous chitinous and calcified plates and teeth. The walls of the foregut are supplied with a complex intrinsic and external musculature to power mastication. Patwardhan (1935, 1936) made a survey of the mandibular processes and degree of gastric mill development in decapod shrimps. He found that the molar processes of dendrobranchiate and stenopodidean mandibles were present but not particularly massive, while their gastric mill was well developed compared to that of carideans. In the caridean genera he examined, the gastric mill was absent or reduced. Patwardan hypothesized that the masticating function in carideans lacking the mill was substituted by that of the molar processes of the mandible. However, a more extensive survey of decapod shrimp families showed that a number of caridean families, although not a majority, have well-developed gastric mills (Atyidae, Nematocarcinidae, Rhynchocinetidae, Stylodactylidae; Felgenhauer and Abele 1983). Another factor in mastication, especially those shrimps lacking gastric mills, is that one often finds sand or small gravel in the foregut of shrimps. This ingested material must help in the grinding of food items into the very fine particles needed to enter the hepatopancreas, where final digestion and absorption of food takes place.

In pelagic predators such as the caridean *Pasiphaea* and *Leptochela*, (Pasiphaeidae) and the penaeid *Funchalia* (Renfro and Pearcy 1966; Hopkins et al. 1994), there are no molar processes, only a massive incisor for piercing, cutting, or chopping up prey. However, a well-developed gastric mill is present for mastication. In the caridean *Atya,* the molar process is massive but not involved in grinding (Felgenhauer and Abele 1985) but a gastric mill is present. In *Notostomus* (acanthephyrid caridean), a predator on jellyfish, there is a large slicing incisor,

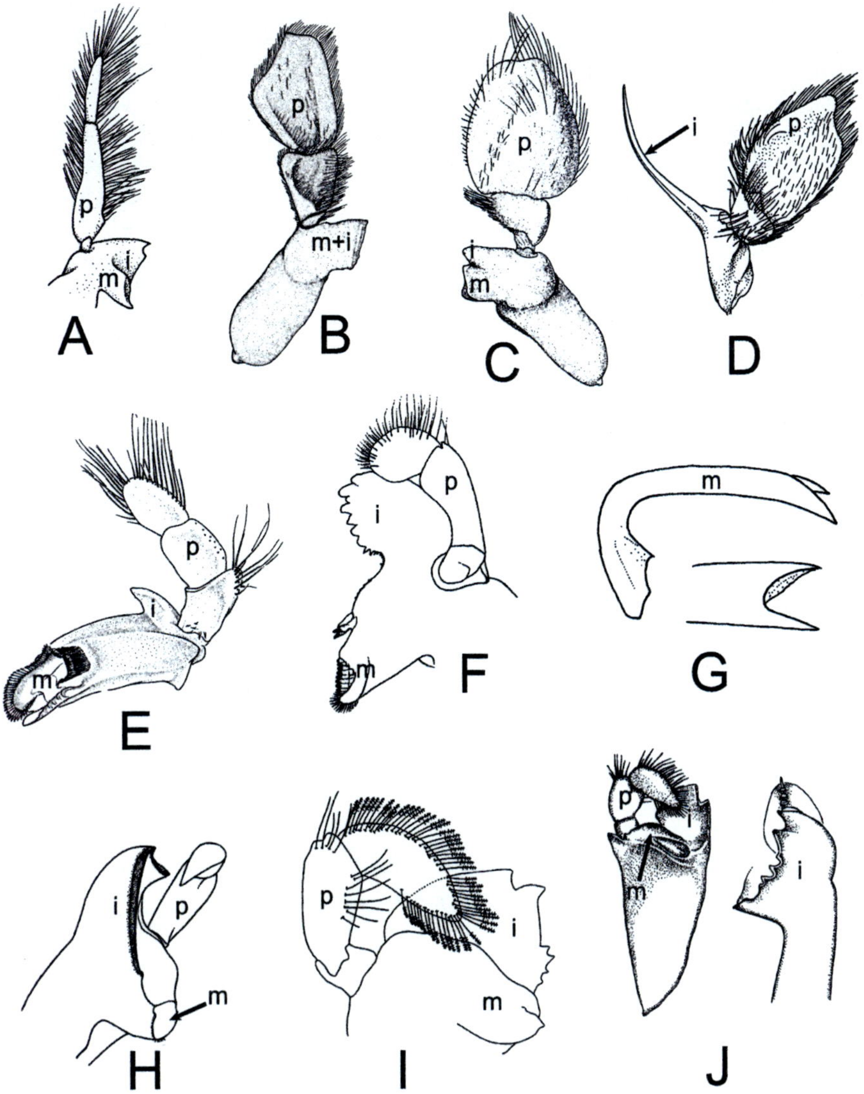

Fig. 2.13 Mandibles. (**a**) *Sergia manningorum* (Sergestidae), with incisor and molar processes. (**b**) *Sicyonia ingentis* (Penaeoidea, Sicyoniidae) with fused molar and incisor processes. (**c**) *Penaeopsis balssi* (Penaeoidea, Penaeidae) with separate molar and incisor. (**d**) *Funchalia woodwardi* (Penaeoidea, Penaeidae) with long scythe-like incisor, molar process absent. (**e**) *Eumanningia pliarthron* (Caridea, Hippolytidae), with massive molar and reduced incisor. (**f**) *Rhynchocinetes albatrossae* (Caridea, Rhynchocinetidae), with massive incisor and reduced molar process. (**g**) *Placopsicrangon formosa* (Caridea, Crangonidae), long cutting/piercing molar process with close-up of tip, incisor process and palp absent. (**h**) *Procaris ascensionis* (Procarididea), with small molar and massive incisor process fused. (**i**) *Odontozona lopheliae* (Stenopodidea, Stenopodidae). (**j**) *Macromaxillocaris bahamaensis* (Stenopodidea, Macromaxillocarididae); left, right mandible, anterior view; right, closeup of incisor process from posterior view. i, incisor process; m, molar process; p, palp. From (**a**) Froglia and Gramitto (2000), (**b**) Pérez Farfante (1985), (**c**) Pérez Farfante (1980), (**d**) Pérez Farfante and Kensley (1997), courtesy of ©Publications Scientifiques du Muséum national d'Histoire naturelle, Paris, (**e**) Crosnier (2000), (**f**) Chace (1997), (**g**) Komai and Chan

only a small molar process, but no gastric mill (Moore et al. 1993); perhaps extensive mastication is not needed for such soft-bodied prey. In the predatory *Procaris ascensionis*, closely related to the carideans, both gastric mill and fused incisor-molar process occur. Thus, there is evidence for and against the Patwardhan hypothesis. For a more definitive test about the correlation of mandible structure and development of gastric mill with diet, further detailed studies on this issue are warranted. One such study is that of Ashelby et al. (2015) which attempted to relate the details of the molar and incisor processes of several palaemonid species to their known or presumed diet. They found in this preliminary study that the structure of the mandible was related more to diet than to ancestry (phylogeny). On the other hand, Felgenhauer and Abele (1983), in an extensive survey of gastric mills in decapod shrimps, suggested, at the family level at least, that the inverse relationship between the degree of development of gastric mill and mandibular structures is more dependent on ancestry than diet.

The mandible has one more part that needs to be mentioned. This is a palp, which may or not be present in the Caridea, varies from 1–3 articles (Fig. 2.13e, f). The other decapod shrimps have a palp of 3 articles; the distal article in penaeoids is large and foliaceous (Fig. 2.13b, c); only in the sergestoid *Lucifer* is it absent. The palp was considered by Calman (1909) to represent the reduced endopod of a basic biramous appendage. The mandibular palp may help push food into the mouth and wipe off the mandible (Garm 2004), but in penaeoid shrimps serves as part of the inhalant siphon of the gill chamber (Dall et al. 1990). Just behind the mandibles, serving as a labium or lower lip to the mouth, are a pair of fleshy, slightly scoop-shaped structures, the paragnaths. Food being passed into the mouth from the posterior mouthparts is entangled by secretions of the paragnaths into a bolus, or food ball, which facilitate its manipulation into the mouth by the muscular labrum (upper lip) (Hunt et al. 1992).

The remaining appendages associated with the mouth (first maxillae to third maxillipeds) are, like the mandibles, basically biramous, but homologies about basic limb segments are often difficult to determine and sometimes controversial but become less so from anterior to posterior mouthparts. Each mouthpart is overlapped by those of appendage just posterior. First behind the mouth and paragnaths are the first maxillae (maxillules). The maxillule is delicate limb consisting of a coxa and basis with setose endites, accompanied by a simple exite termed a palp, which probably represents an endopod (Young 1959) (Fig. 2.14). This is the basic pattern, with variations in the details, for all the decapod shrimps except for some sergestoids such as *Lucifer* and *Acetes*, in which the palp is absent (Pérez Farfante and Kensley 1997).

The second maxilla is both a mouthpart and water pump. Interpretation of the second maxilla's component parts varies greatly (Young 1959). Proximally there are

---

Fig. 2.13 (continued) (2009), (**h**) Chace and Manning (1972), (**i**) Goy and Cardozo (2014), (**j**) Alvarez et al. (2006). (**a, e, g, j**) courtesy Journal of Crustacean Biology, (**i**) courtesy of Zootaxa

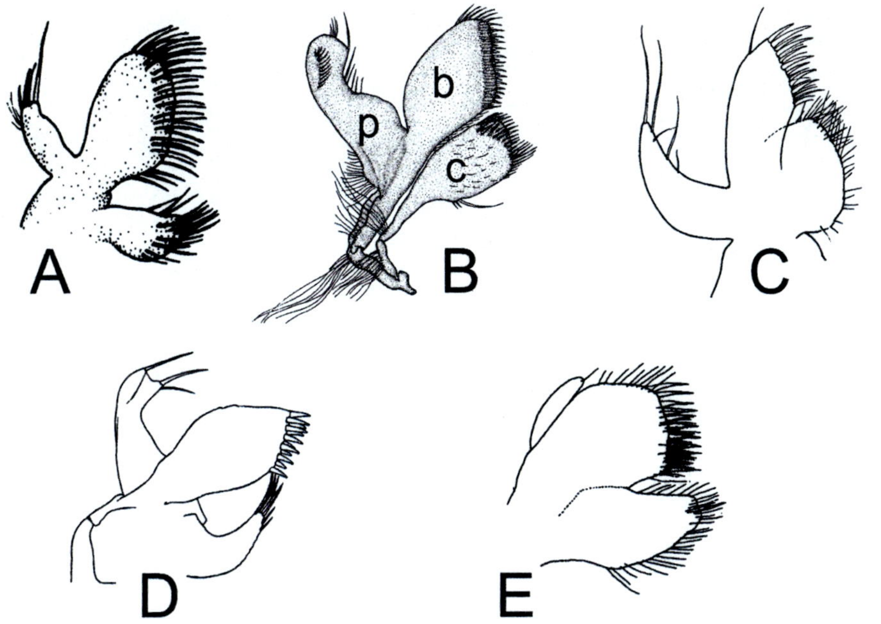

Fig. 2.14 Maxilla 1 (Maxillules). (**a**) *Sergia manningorum* (Sergestidae). (**b**) *Sicyonia ingentis* (Penaeoidea, Sicyoniidae). (**c**) *Vetericaris chaceorum* (Procarididea). (**d**) *Plesionika kensleyi* (Caridea, Pandalidae), (**e**) *Spongicoloides iheyaensis* (Stenopodidea, Spongicolidae). b, basal endite; c, coxal endite; p, palp. From (**a**) Froglia and Gramitto (2000), (**b**) Pérez Farfante (1985), (**c**) Kensley and Williams (1986), (**d**) Chace (1985) and (**e**) Saito et al. (2006). (**a**, **c**, **e**), courtesy of Journal of Crustacean Biology

coxal and basal endites (well-equipped mesially with setae), a palp (endopod), and, on the lateral side of the limb, a scaphognathite. Variously interpreted as an exopod or exopod + epipod), it extends anteriorly and posteriorly as a large, setose, foliaceous blade. The scaphognathite serves, it may be recalled, as the primary pump producing the respiratory current irrigating the gills. The second maxillae of all the shrimp groups are similar (Fig. 2.15), except, as with many sergestoid appendages, it may be reduced (*Acetes*) or absent (*Lucifer*) (Pérez Farfante and Kensley 1997). In some shrimps, the posterior border of the scaphognathite is equipped with long setae (Fig. 2.15e), which are involved in gill cleaning (Chap. 4).

2.9 Maxillipeds

The three pairs of maxillipeds, as their etymology ("jaw feet") implies, are transitional in structure and function from the head appendages (sensory, food handling) and the remaining thoracic appendages, the pereopods, with their variety of locomotory, feeding, grooming, and defense functions. The first maxilliped (Mx1)

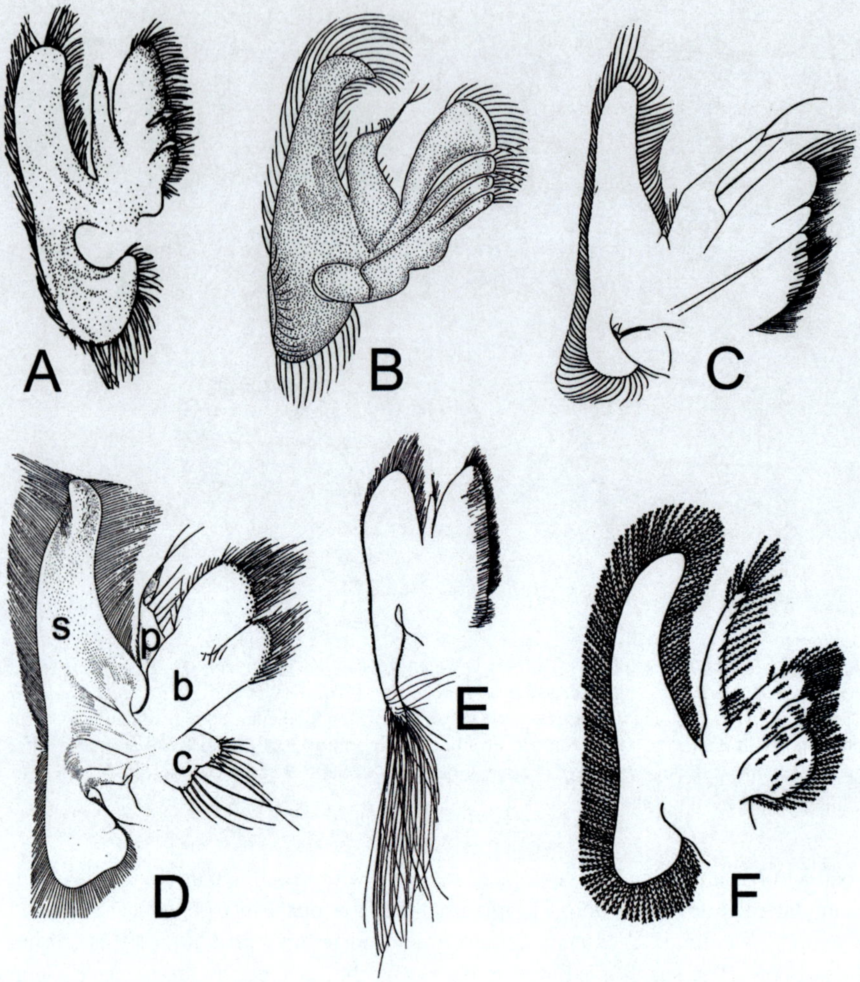

Fig. 2.15 Maxilla 2. (**a**) *Sergia manningorum* (Sergestidae). (**b**) *Pseudaristeus kathleenae* (Penaeoidea, Aristeidae). (**c**) *Procaris ascensionsis* (Procarididea). (**d**) *Eumanningia pliarthron* (Caridea, Hippolytidae). (**e**) *Atya limnetes* (Caridea, Atyidae). (**f**) *Stenopus pyrsonotus* (Stenopodidea, Stenopodidae). Note the long setae extending from the posterior border of the scaphognathite in (**e**) *A. limnetes,* an adaptation for gill cleaning. Components of the appendage are labeled in (**d**). b, basal endites; c, coxal endite; p, palp (endopod); s, scaphognathite. From (**a**) Froglia and Gramitto (2000); (**b**) Pérez Farfante (1987); (**c**) Chace and Manning (1972); (**d**) Crosnier (2000); (**e**) Holthuis (1986) and (**f**) Goy and Devaney (1980). (**a**, **d**, courtesy of Journal of Crustacean Biology)

consists of a protopod (coxa + basis) with well-developed setose endites, a small palp-like endopod (variously segmented), and flagelliform, usually multiarticulate exopod (Fig. 2.16). The maxillipeds often bear epipods which may serve a respiratory or gill-cleaning function. The proximal portion of the exopod of many carideans

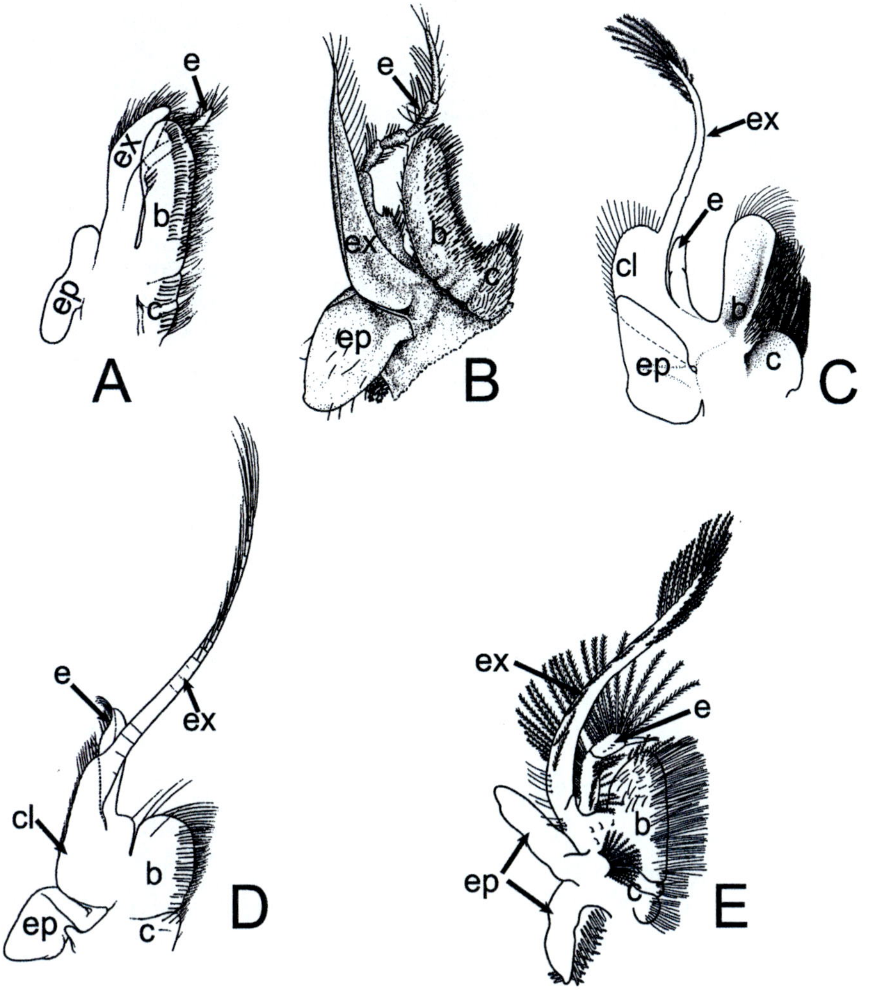

Fig. 2.16 First maxillipeds. (**a**) *Sergia manningorum* (Sergestidae). (**b**) *Penaeopsis balssi* (Penaeoidea, Penaeidae). (**c**) *Palaemon tonkinensis* (Caridea, Palaemonidae). (**d**) *Procaris ascensionis* (Procarididae). (**e**) *Stenopus pyrsonotus* (Stenopodidea, Stenopodidae). b, basal endite; cl, caridean lobe; c, coxal endite; e, endopod; ep, epipod; ex, exopod. From (**a**) Froglia and Gramitto (2000), (**b**) from Pérez Farfante (1980), (**c**) Bruce (1989), (**d**) Chace and Manning (1972) and (**e**) Goy and Devaney (1980). (**a**, **c**), courtesy of Journal of Crustacean Biology

is often expanded into a "caridean lobe" (Fig. 2.16c). In the Procarididea, Chace and Manning (1972) describe an indication of a caridean lobe in the primitive procaridids *Procaris ascensionis* (Fig. 2.16d) and somewhat more developed in *Vetericaris chaceorum* (Kensley and Williams 1986).

The second and third maxillipeds (Mx2, Mx3) are more leglike (Figs. 2.17 and 2.18). In most groups, the Mx2 has a coxa, basis, and endopod of 5 segments

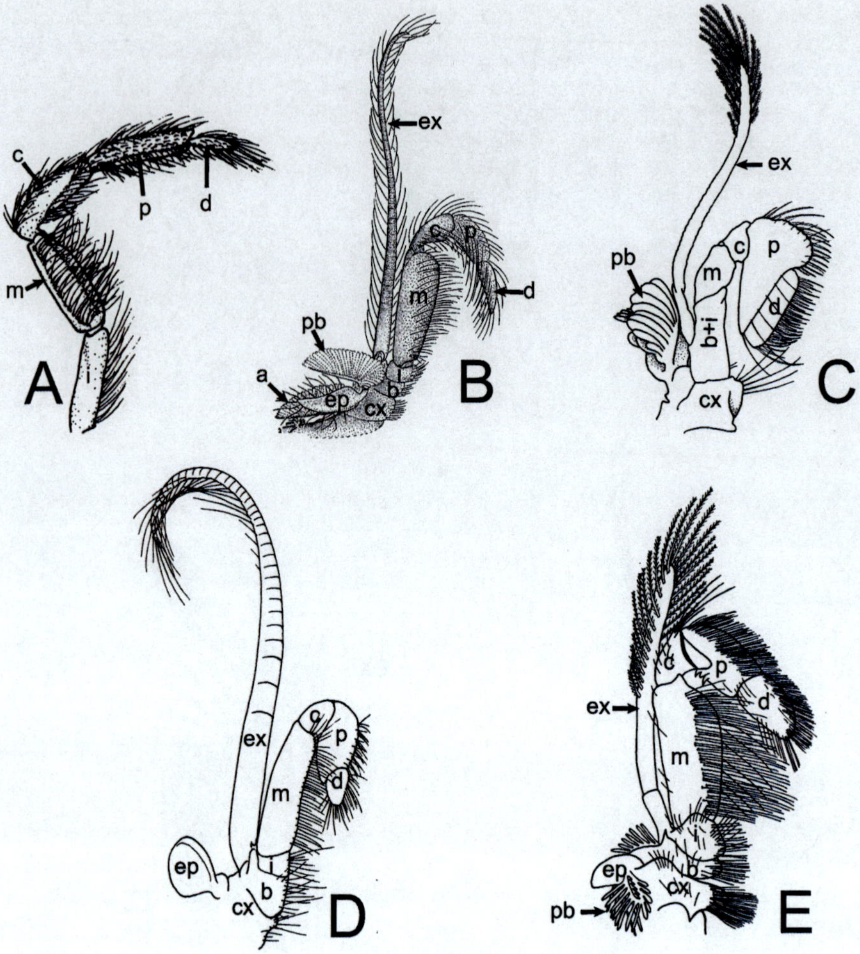

Fig. 2.17 Second maxillipeds. (**a**) *Sergia manningorum* (Sergestidae). (**b**) *Pseudaristeus kathleenae* (Penaeoidea, Aristeidae). (**c**) *Palaemon tonkinensis* (Caridea, Palaemonidae). (**d**) *Vetericaris chaceorum* (Procarididea). (**e**) *Stenopus pyrsonotus* (Stenopodidea, Stenopodidae). a, arthrobranch; b, basal endite; c, carpus; cx, coxal endite; d, dactyl; ep, epipod; ex, exopod; i, ischium m, merus; pb, podobranch; p, propodus. From (**a**) Froglia and Gramitto (2000); (**b**) Pérez Farfante (1987), (**c**) Bruce (1989), (**d**) Kensley and Williams (1986) and (**e**) Goy and Devaney (1980). (**a**, **c**, **d**) courtesy of Journal of Crustacean Biology

(sometimes variously fused) and a well-developed flagelliform exopod except in sergestoids. The Mx2 shows a characteristic bend of the distal endopodal segments against the more proximal ones, giving it a prehensile aspect. The third maxilliped (Mx3) is definitely more leglike—sticklike—, usually with a full complement of segments in penaeoids, sergestoids, stenopodideans, and procaridideans. In carideans there is variable fusion of endopodal segments (Bauer 1977) (Fig. 2.18d). Typically, there is a flagelliform exopod (except in sergestoids), but

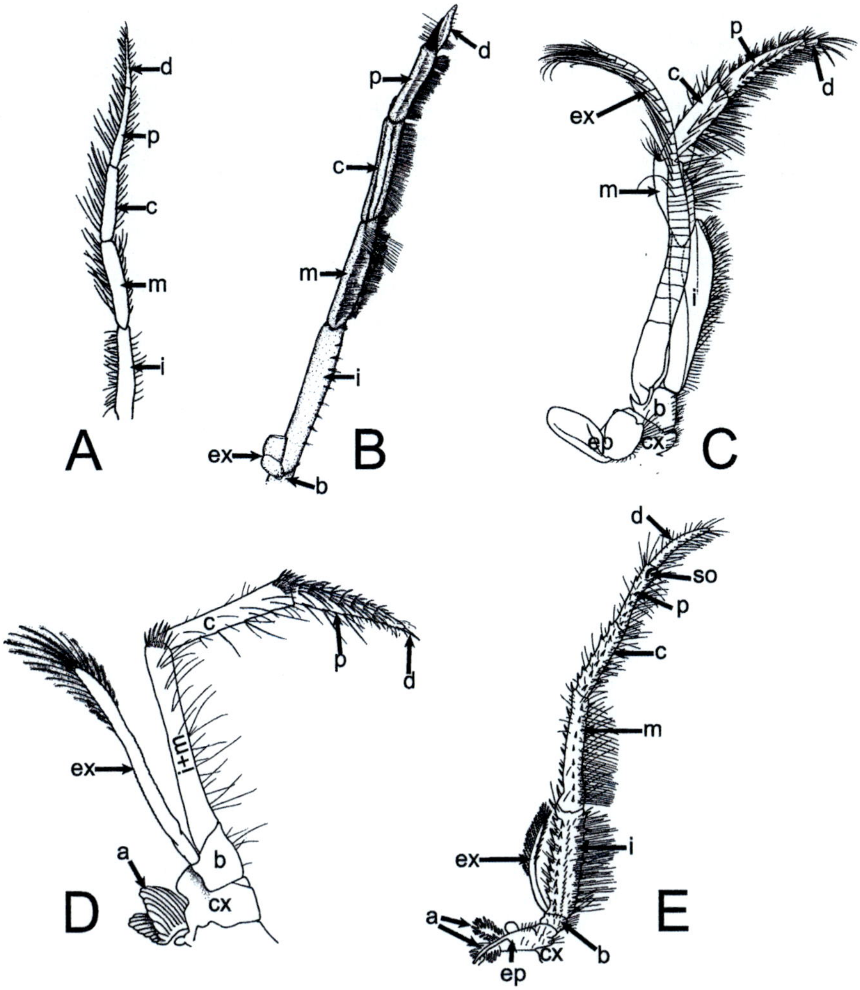

Fig. 2.18 Third maxillipeds. (**a**) *Sergia manningorum* (Sergestidae). (**b**) *Fenneropenaeus merguiensis* (Penaeoidea, Penaeidae); only the proximal two articles of the exopod are shown. (**c**) *Procaris ascensionis* (Procarididea). (**d**) *Palaemon tonkinensis* (Caridea, Palaemonidae). (**e**) *Stenopus pyrsonotus* (Stenopodidea, Stenopodidae). a, arthrobranch; b, basis; c, carpus; cx; coxa; d, dactyl; e, epipod; ex, exopod; i, ischium; m, merus; p, propodus; so, setiferus organ. From (**a**) Froglia and Gramitto (2000), (**b**) Alexander et al. (1980), (**c**) Chace and Manning (1972), (**d**) Bruce (1989) and (**e**) Goy and Devaney (1980). (**a**, **d**), courtesy of Journal of Crustacean Biology

this is highly variable in carideans in which the presence or absence of maxillipedal exopods can be an important taxonomic character. Although the endopods of Mx3 participate in feeding, they are also important grooming appendages (Chap. 4) and may serve as weapons in inter- and intraspecific agonistic (aggressive) behaviors (Chap. 8, "Mating Systems").

How does this plethora of mouthparts and associated appendages interact to pass and process captured or foraged food into the shrimp's mouth? Direct observation of these tightly appressed and overlapping limbs is difficult, but studies aided by high-speed cinematography (e.g., Alexander and Hindley 1985; Garm et al. 2003; Hunt et al. 1992) have been revealing. When feeding on the bottom, the Mx3 may pin down a larger food item so that the chelipeds can pick off pieces of food and insert them into the inner mouthparts. The shrimp may press its mouthfield directly on a large piece of food and grip it with the proximal segments of the Mx3 and mandibles. It then lifts off its body and mouthfield away from bottom, tearing off with Mx3 a chunk from the food mass which has been pinned to the substrate. Smaller food items placed by chelipeds, Mx3, or Mx2 into the mouthfield are sequentially ratcheted up towards the mouth by the interdigitating proximal endites of Mx2, Mx 1 and maxillae which finally push the food between the mandibles and then into the mouth. Rejected food particles and debris are cleaned off endite setae and swept away from the body by the reverse action of these inner mouthparts. The setal ratchet of these mouthparts may also filter out particulate matter in the water, which is then passed up between the paragnaths, whose glandular secretions form a food ball pushed into the mouth by the labrum without the action of the mandibles.

All three maxillipeds may have flagelliform exopods; these structures flutter and beat intermittently during different activities. The primitive function of these type of exopods was participation in forward swimming, but they now have other functions in decapod shrimps. They generate currents that may deflect or steer water away from the front end of the body (Fig. 2.9b). Food debris generated during feeding may foul the shrimp. Water that has passed through the branchial chamber over the gills has a high concentration of respiratory and excretory wastes. Thus, the exopods help deflect fouled water away from the olfactory antennules and from the gills (Borradaile 1917; Fryer 1960; Bauer 2004). Behavioral studies (Herberholz and Schmitz 2001) have also shown that currents leaving the branchial chamber bear important pheromonal information that the maxillipedal exopods may accelerate and direct towards other individuals during intraspecific interaction and communication.

2.10 Pereopods

The five pairs of posterior thoracic limbs are the pereopods (also termed pereiopods, peraeopods). These appendages show the typical structure of a biramous crustacean appendage (Fig. 2.4), with a basal protopod of coxa and basis, the latter bearing an endopod of 5 segments (ischium, merus, carpus, propodus, and dactylus, often shortened to dactyl). The basis may also give rise to a branch, the exopod, which is multiarticulate and flagellar when well developed and used to aid forward swimming, as in the procaridid, some oplophoroid and pasiphaeid carideans. However, exopods are often reduced or absent in decapod shrimps. The coxal segment may or may not bear an epipod, as discussed earlier. Pereopods 1 and 2 (carideans) or 1–3 (dendrobranchiates) usually terminate in a pincer or claw, which is properly termed a "chela." Pereopods which have a chela are chelipeds. The posterior three pereopods (P3–5) are the walking legs in carideans; the last two pairs (P4–5) serve this function in penaeoids and stenopodideans. In the primitive group Procarididea, none of the pereopods are chelate. *Procaris ascensionis*, which swims much of the time, rests on the bottom or perches on hard substrates with all five pairs of pereopods (Bruce Felgenhauer, pers. comm.).

2.10.1 Chelae: Shrimp Pincers

Chelae have evolved from modifications of the distal two limb segments, the propodus and dactyl. A chela consists of a movable (dactylar) finger which can open and close against a fixed finger, derived from a distal extension of the propodal segment (Fig. 2.19a). The fingers of the chelae of many species terminate in stout setae which interconnect to form a chela lock that assures proper closure of the chela fingers.

Chelae serve as the toolbox of shrimps; they may serve in defensive or aggressive behavior, feeding, grooming, sensory exploration of their surroundings and in intraspecific interactions. The chelipeds increase in length from P1–P3 in penaeoids; chelae are generally small and not particularly robust (Fig. 2.19b). The opposing fingers may have cutting edges, teeth, sensory setae, peg-like setae, pads, as well as other modifications for gripping objects (Hindley and Alexander 1978). Observations on subtle differences in claw structure among penaeoid genera, together with data on dietary preferences, may reveal distinctions between predatory, specialized or generalist feeding behaviors in penaeoids. P1–3 chelae of sergestoidean genera are nearly microscopic or absent (Chap. 1, Fig. 1.6a; Chap. 3), again following the evolutionary trend of reduced morphology in that group. In the stenopodideans, the first two pairs of chelipeds are slender with small chelae, used as sensory, feeding, and grooming structures. However, in most members of this group, the third pair of chelipeds are massive weapons (Chap. 1: Fig. 1.7b; Chap. 3). The procaridideans lack chelae completely, with P1–2 essentially the same as P3–5, i.e., simple pereopods terminating in a small dactyl (Fig. 2.19c).

It is in the Caridea, with chelae only on the first two pereopods, that we find the greatest variation in chela structure. In the Crangonidae, the first pereopods have subchelae, in which the propodal segment is expanded vertically but not extended distally except for a sharp stabbing spine below (Fig. 2.19d). The long, curved dactyl folds against the distal edge of the propodus, somewhat like the blade of a jackknife against the handle. The first pereopod of the Glyphocrangonidae is often described as a subchela but by others as simply prehensile. The dactyl does double back against the propodus as in a subchela, but the distal end of the latter is not expanded vertically as in a subchela (Chap. 3). Thus, the glyphocrangonid pereopod 1 is somewhat between a simple nonchelate and a subchelate limb. Analogous prehensile pereopods can occur in species from other caridean families for gripping slender

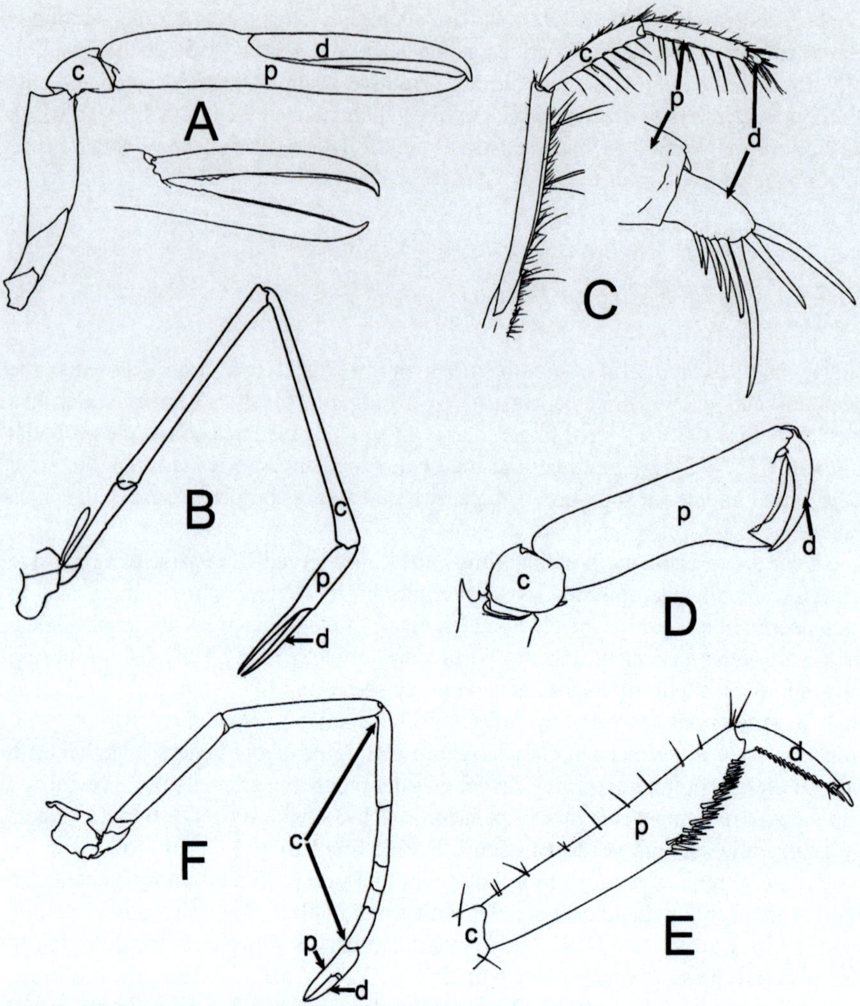

Fig. 2.19 Chelae, subchelae and prehensile pereopods. (**a**) Basic chela structure (closed above, open below) illustrated with a minor (cutting) chela of *Alpheus malabaricus* (Caridea, Alpheidae). The dactyl forms the movable finger while an extension of the propodus is the fixed (immovable) finger of the chela. (**b**) Penaeoid chelate pereopod. (**c**) Nonchelate pereopod 1 of *Procaris ascensionis* (Procarididea); below, magnified terminal end of limb. (**d**) Pereopod 1 subchela of *Pontophilus gracilis* (Caridea, Crangonidae). (**e**) Third pereopod of *Thor dobkini* (Caridea, Thoridae) with a prehensile structure. (**f**) Second cheliped of *Alpheus suluensis* (Caridea, Alpheidae) with a multiarticulate carpus. c, carpus; d, dactyl; p, propodus. From (**a**, **f**) Chace (1988), (**b**) Pérez Farfante and Kensley (1997), courtesy ©Publications Scientifiques du Muséum national d'Histoire naturelle, Paris, (**c**) Chace and Manning (1972), (**d**) Chace (1984), (**e**) Bauer and VanHoy (1996) courtesy Bulletin of Marine Science and (**f**) Chace (1988)

blades of vegetation or branches of sessile animals, or in males for grasping females during copulation (Fig. 2.19e).

In some caridean families, the carpus of the second pair of chelipeds is multiarticulate, with the number of carpal articles varying from a few to many (Fig. 2.19f). This multiarticulation bestows flexibility in movement, especially helpful in grooming and cleaning of the body (Chap. 4). Chelipeds of a pair may be dissimilar in size and function, as in pandalids and processids (Chap. 3). For example, in pandalid carideans, the left cheliped is more slender and longer, with more carpal articles, than the right one, while the reverse is true in some processid species (Chap. 3). In the pandalids, at least, the left cheliped is more often used in grooming, while the right is more often employed in feeding (Bauer 1975). In most processids, the right first pereopod is chelate, but the left is not, a simple pereopod terminating in a stout dactyl (Chap. 3). The chelae of a pair may be polymorphic, i.e., may differ greatly not only in size (major and minor chelae) but also in morphology and function. For example, in alpheids and some pontoniine palaemonids, one chela is large and swollen, adapted for snapping, an aggressive behavior (see below), while the other may be adapted for cutting and grasping. In pandalid carideans, the first pereopods lack a functional chela, as the dactylar segment is either absent or almost microscopic, but obviously the remnant of a once-developed chela (Chap. 3). Another twist in the saga of chelae is found in the family Atyidae in which the fingers of both chelae bear lush groups of specialized setae used in filtering currents or scraping microbial growth from hard surfaces (Chap. 3).

Although chelae serve many roles in shrimp biology, almost all chelae can be used to some extent as weapons. The chelae of penaeoids are not particularly impressive in size and armature (Chap. 1: Fig. 1.6b; Chap. 3). However, they certainly can be utilized in subduing small prey items. In captivity, most penaeoids will eagerly attack and try to feed on molting or otherwise impaired shrimps, including those of its own species. But it is in the carideans (e.g., Fig. 2.20a, b; Fig. 3.31b; Chap. 3) and stenopodideans (Chap. 1, Fig. 1.7b; Chap. 3) that we find the most impressive cheliped weapons. The stenopodideans live in male-female pairs, and their massive third chelipeds have been shown to be important in defense of pair partner and territory, at least in *Stenopus hispidus* (Johnson 1969). Sexual dimorphism in cheliped weaponry is common in caridean species, with males of some species having much larger and stouter cheliped weapons than females (Chap. 8).

One of the most striking adaptations of a chela is the snapping claw of some alpheid and palaemonid carideans. When one goes out to a rocky intertidal zone at low tide, frequent snaps, crackles and pops may be heard. Likewise, when diving over a rocky or coral reef, similar snaps and popping sounds may be perceived if one's breath is held to avoid exhalation bubbles and noise. Hydrophones lowered into shallow sublittoral waters, especially at low latitudes, reveal noise similar to the sound of burning dry twigs or fat. This undersea noise was loud enough to interfere with equipment being developed by the USA navy during World War II for detecting and analyzing submarine sound. The Navy turned to Dr. Martin Johnson, noted marine biologist at the Scripps Institution of Oceanography, who realized at once the

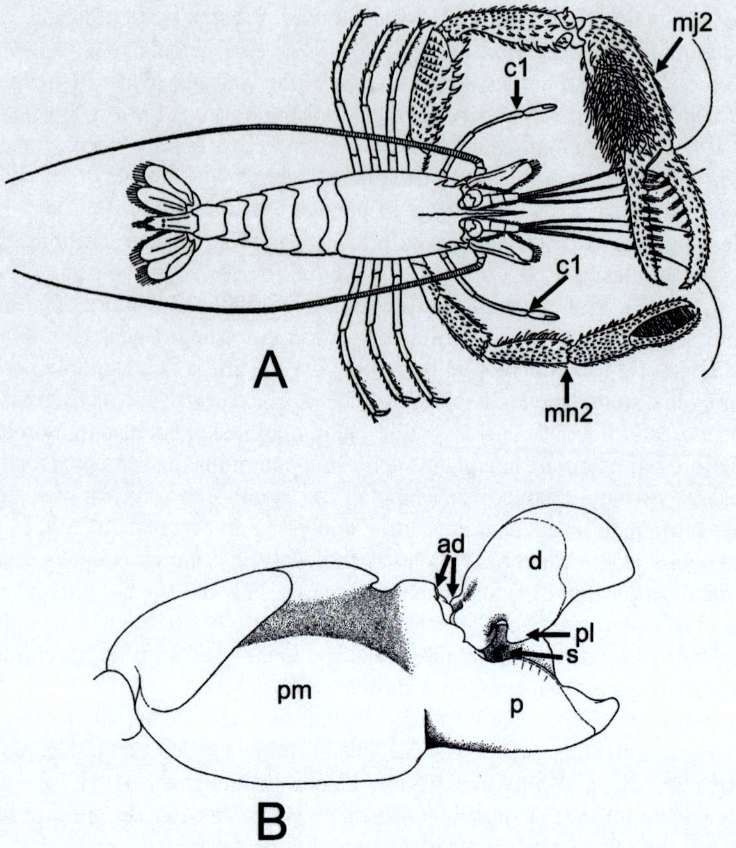

Fig. 2.20 Cheliped weaponry. (**a**) *Macrobrachium faustinum* dominant male, showing major and minor second chelipeds. (**b**) Snapping (major) first chela of *Alpheus suluensis* (Caridea, Alpheidae) in a semi-open position. ad, adhesive disks; c1, cheliped 1; d, dactylar finger; mj2, major cheliped 2; mn2, minor cheliped 2; p, propodal (fixed) finger; pl, plunger; pm, propodal palm; s, socket. From (**a**) Holthuis (1952) and (**b**) Chace (1988)

source. He showed that the noise was generated by the large benthic populations of alpheid snapping shrimps (*Alpheus* and other genera) (Johnson et al. 1947).

The functional morphology, sound production and true function of snapping claws has been best studied in the genus *Alpheus* (Anker et al. 2006). Snapping claws and behavior are also found in *Synalpheus*, in some pontoniine (Palaemonidae) genera and to various degrees in other alpheid genera (Kaji et al. 2018). In the snapping alpheid genera, one of the first pereopods has a minor ("cutting") chela and a major ("snapping") claw (Figs. 2.19a and 2.20b, respectively). Early laboratory observations showed that *Alpheus* snapping shrimps could stun small fishes, which are then eaten (MacGinitie 1937). However, many subsequent studies have shown that snapping behavior is mainly used in intraspecific interactions such as defense of a burrow or mate. Johnson and colleagues determined that the movable (dactylar) finger of a snapping claw has a projection (plunger) that when the chela is closed, fits snugly into a depression (socket) on the fixed (propodal) finger (Fig. 2.20b). Snapping shrimps are sometimes referred to as "pistol shrimps" because of the analogous structure of the snapping claw to a pistol. The dactylar finger (hammer) is first cocked and then when ready, is "fired" by the powerful contraction of a massive closer muscle within the propodus. The forcible blow of the plunger into the socket forces a high-velocity jet of water (the "bullet") forward from the chela which can stun or otherwise damage a prey or opponent.

This snapping behavior can be rather easily observed in the laboratory. If a male-female pair of *Alpheus* snapping shrimps is maintained in an ordinary saltwater aquarium, they will establish and defend a burrow. When another snapping shrimp is introduced into the aquarium, it may attempt to take over the burrow or drive out the same-sex member of the established pair. One member of the pair will advance on the stranger and cock its chela, as will the intruder. They may parry and thrust forward and backward several times before one either backs down or the encounter may escalate until one or both discharge their weapons with much snapping and crackling.

It was long thought that the forceful impact of the dactylar plunger into the propodal socket produced the snapping sound (Knowlton and Moulton 1963). However, Versluis et al. (2000), in an elegant analysis of snapping using hydrophones and high-speed video, found that a cavitation bubble is formed as the high-speed water jet shoots forward after a snap. The pressure within the bubble is much lower than ambient pressure, so that the bubble implodes violently with a loud pop. Lohse et al. (2001) demonstrated that a high-magnitude sound pressure wave is generated by bubble collapse; it is this concussive force which stuns or otherwise damages a prey or opponent. The bubble collapse is also accompanied by a brief flash of light, but its biological significance, if any, is unknown.

How does the snapping claw generate the considerable forces necessary for producing the high-velocity water jet and subsequent pressure and sound waves? The snapping claw is stout and swollen due to the large dactylar closer muscles contained within the main body (palm) of the propodus. Different mechanisms cock the dactylar finger before firing, producing resistance to dactylar closure which obliges the closer muscles to contract forcefully and energetically close. (Ritzmann 1974; Kaji et al. 2018). Ritzmann found that when the dactylar finger of *Alpheus californiensis* is abducted away from the fixed finger, a smooth disk (adhesive plaques) at the base of the dactyl opposes a similar disc on the main body of the propodus. These adhesive discs stick tightly together by the cohesive forces formed by the thin film of water trapped between them. Thus, the closer muscle must contract powerfully enough, so these cohesive forces are overcome. The dactylar plunger then smacks into the propodal socket with great force, generating the water jet and subsequent cavitation bubble, its collapse, and the sound pressure wave. In another species, *A. heterochaelis*, the discs are quite small and incapable of generating any significant cohesive forces between them. In this species, Ritzmann described another mechanism. When the dactyl is closed, the insertion of the dactylar closer muscle is forward of the pivot point around which the dactyl swings during opening and closing. When cocked open, this insertion is moved above the pivot point. When the main body of the closer muscle contracts, the dactyl continues to be held open by this latching arrangement, but high tension (stored energy) becomes generated as the closer muscle tries to close the dactyl. At a certain level of tension, a small distal slip of the closer muscle contracts, pulling the apodeme insertion back below the dactylar pivot point. This allows the dactyl to be slammed into the propodal plunger with the substantial force generated by the prolonged contraction of the closer muscle.

More recently, the complex mechanics and evolution of chela cocking, closure and snapping in carideans have been analyzed and updated using technology such as microcomputed tomography, confocal imaging, high-speed video, and kinematic experiments with selected 3D-printed models (Kaji et al. 2018). The reader is referred to this elegant and informative publication for more detailed information on this subject.

Snapping has evolved in some genera of another caridean group, the palaemonid pontoniines (Bruce 1976; Anker et al. 2006; Almeida et al. 2014; Kaji et al. 2018). Unlike the alpheids, the major chelipeds are the second pereopods, not the first. In *Periclimenaeus, Paraclimenaeus* and *Typton* species, the second chelae are unequal, and the movable finger of the major (larger) second cheliped has the molar-shaped tooth, with socket on the fixed finger, as in alpheids. However, in *Coralliocaris graminea*, in which both major chelae can snap, it is just the opposite, i.e., the plunger is on the fixed finger with the socket on the dactylar finger. The major chelae of other pontoniine genera may produce sound without the plunger-socket system using other morphological modifications of their chelae. In the Alpheidae, adhesive discs for claw cocking, as well as dactylar tooth and socket systems, have evolved independently in different phylogenetic clades. These represent preadaptations that led to the evolution of snapping in some alpheid genera (Anker et al. 2006; Kaji et al. 2018).

2.10.2 Walking Legs

The posterior pereopods are the appendages which primarily support the body weight of shrimps and function in walking about over the bottom. They are simple, nonchelate appendages (e.g., similar to Fig. 2.4) which terminate in a curved or straight nail-like dactyl which serves as a foot or grip when stepping about on the surface. Walking legs are employed in other functions besides ambulation, such as perching on hard substrates or plants, burrowing into soft bottom habitats, gripping prey and grooming the body, depending on the taxa involved. In penaeoids and stenopodideans, pereopods 4 (P4) and (P5) are the walking legs which support the body at rest, with P4s placed forward and P5s rearwards to form a tetrapod of support. In the carideans, there is a sexapod of support with P3s placed anteriorly and P5s posteriorly with P4s in an intermediate position. All procaridean pereopods are simple (e.g., Fig. 2.19c); *Procaris* shrimps swim much of the time but when on the bottom or perching on some structure, they use all five pairs of pereopods (Bruce Felgenhauer, pers. obs.).

The basic structure of walking legs is similar, i.e., a nonchelate appendage which may bear an exopod but in most extant (living today) shrimps, those exopods are either reduced and nonfunctional in swimming or absent altogether. The dactyl is usually simple or single but may be bifid (biunguiculate), i.e., double or divided into two, or otherwise variously specialized, e.g., a "pseudochela" in a caridean family named after this trait (Chap. 3). In some carideans and the family Stenopodidae, the propodal, carpal and more proximal limb segments may be multiarticulate, giving the limb extra flexibility. Deep-sea or other shrimps living on muddy bottoms often have delicate elongated walking legs to prevent sinking into the soft substrate by distributing the body weight over a greater surface area (e.g., Nematocarcinidae, Chap. 3). Alternately, highly benthic species with stout calcified bodies tend to have short powerful pereopods for burrowing (e.g., sicyoniid penaeoids, crangonid carideans) or crawling over and up surfaces. Some carideans, such as some crayfish-looking *Macrobrachium* spp. and atyid carideans, live in the rushing waters of waterfalls in tropical mountain streams. They are capable of clinging to rocks and even climbing up cascades against the flow of water using their powerful P3–P5. Caridean species that cling to narrow elongate structures, such as seagrass blades, sea urchin spines, branches of algae or sessile invertebrates such as gorgonians, often have short stout walking legs. There are many other interesting variations in the morphology and function of both chelae and walking legs; these will be covered in more detail on a family by family basis in the next chapter.

2.11 The Abdomen and Its Appendages

The abdomen (pleon) consists of 6 somites and terminates in the telson, a nonsegmental structure upon which the anus opens. As noted earlier, the abdomen contains the well-developed musculature that powers the backward escape response in shrimps. Like the cephalothorax, some or all abdominal somites and the telson may have a dorsal carina or keel, sometimes extending posteriorly into spine-like projections, which act as hydrodynamic keels during forward swimming. The abdominal somites act together as a coordinated unit during this forceful ventral flexion. Side to side lateral movement (wriggling) is prevented by locking adjacent somites using cuticular connections, the pleonic hinges (Burkenroad 1963). These structures have been best described in the penaeid *Litopenaeus setiferus* by Young (1959) with observations on a number of shrimp groups by Felgenhauer and Abele (1983). The pleuron of one somite may have a prominent condyle or protuberance on its anterior border which locks into a socket on the posterior margin of the pleuron just anterior to it. In dendrobranchiates, such hinges occur between all adjacent abdominal somites except that the hinge between the 3rd and 4th somites is covered, somewhat ill-defined, allowing some lateral flexibility. In procaridideans and carideans, the hinge is completely lacking between the 3rd and 4th somites, allowing the abdomen to bend noticeably upward in many species, giving rise to the moniker of "humpbacked shrimps" to carideans. In the stenopodideans, there are locking joints only between abdominal somites 4–5 and 5–6. Although the function of pleonic hinges is known, the significance of the differences in the number and position of them is open to hypothesis proposal and testing. Does the lack of a pleonic hinge between abdominal somites 3–4 confer on carideans (and the closely related but more primitive procaridideans) some benefit that may contribute to the considerable ecological diversity of the Caridea?

The abdomen has swimming paddles, the pleopods, on somites 1–5. A pleopod consists of a proximal coxa and a usually longer distal basis, together termed the pleopodal peduncle or protopod. At the tip of the basis are usually two flattened paddle-like branches, the mesial endopod and lateral exopod (Fig. 2.21a–c). These structures are edged with long plumose (feather-like) setae which increase the effective surface area of the pleopods in forward swimming. In dendrobranchiates and procaridideans the exopods are much larger than the endopods. In the penaeoids, the exopods are splayed out laterally during the power stroke, but bend and move forward vertically during the return stroke (Young 1959). This lateral displacement is, in part, an adaptation for burrowing in soft substrates, as described in the next chapter. The pleopods of carideans which burrow in mud and sands, e.g., the crangonids, also may have a similar orientation. However, in most carideans and stenopodideans, the pleopods are oriented vertically beneath the abdomen.

Caridean shrimps and some procaridideans have a unique adaptation, the appendix interna (AI), for linking the pleopods together for a coordinated beating of a pleopod pair. In most carideans, AI are located on the inner edge of endopods from pleopods 2–5 (Pl2–5) (Fig. 2.21c–e, Plate 2.2a, b) and on the first pleopods of males in many species (Fig. 6.6, Chap. 6). In the procarididean *Vetericaris chaceorum*, appendices internae are found only on PL3–5 (Fig. 2.21a) but not in *Procaris* spp. The AI is usually composed of a stalk which, in carideans at least, is adorned at its tip by a number of hooked or comma-shaped setae termed cincinnuli (Latin for "little curls") (Fig. 2.21e, Plate 2.2c).When the tips of the appendices internae of a pleopod pair make contact, they stick together, Velcro-like. Many plant and animal species have evolved comparable structures as fasteners. Appendices internae are completely lacking in stenopodideans and found only on Pl2 of males in the penaeoid families Aristeidae, Benthesicymidae, and Solenoceridae (Fig. 2.21f, g). However, dendrobranchiate AI, when present, does not have cincinnuli or link the second pleopods. Burkenroad (1963) proposed that appendices internae were a characteristic of the ancestral decapod. If so, they may have been selected against and lost in various taxa. For example, the AI have been lost in some crangonid carideans such as *Crangon* species (Christoffersen 1988). In that genus and many penaeoids, the exopods are splayed out laterally and can be used in burrowing. As the endopods of dendrobranchiate pleopods are much smaller than the exopods, appendices internae, if they were present, could not couple together. However, the loss of AI in the epibenthic stenopodideans still needs to be explained.

There is sexual dimorphism in the pleopods of carideans and dendrobranchiates, but little to none in stenopodideans and procaridideans. The Pl2 endopod of caridean

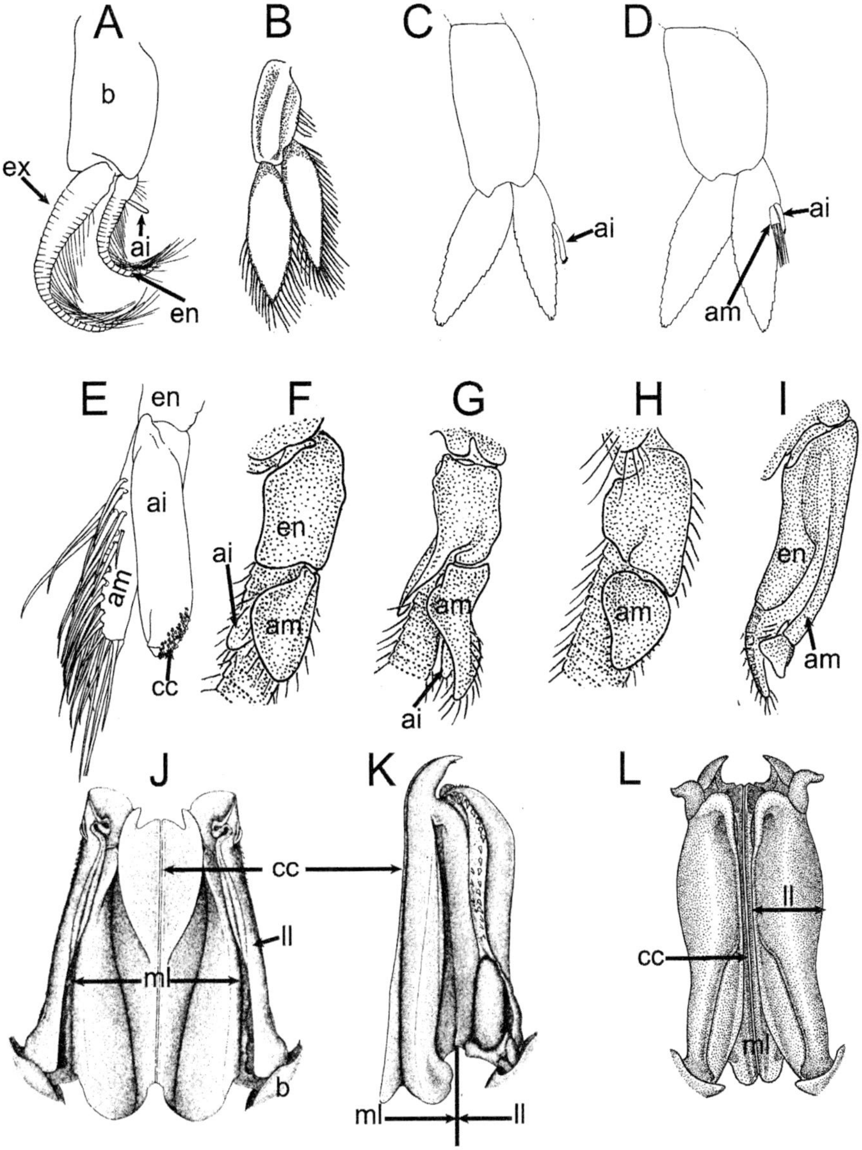

Fig. 2.21 Pleopods. (**a**) third pleopod, *Vetericaris chaceorum* (Procarididea). (**b**) second pleopod, *Macromaxillocaris bahamaensis* (Stenopodidea, Macromaxillocarididae). (**c**) third pleopod and (**d**) second pleopod of *Heptacarpus sitchensis* (Caridea, Thoridae). (**e**) Appendix masculina and appendix interna, *Plesionika kensleyi* (Caridea, Pandalidae). (**f**, **g**, **h**, **i**) Appendices masculinae and internae of penaeoid families Aristeidae, Solenoceridae, Penaeidae, and Sicyoniidae, respectively. (**j**) "Semi-open" petasma of *Litopenaeus setiferus* (Penaeoidea, Penaeidae), ventral (interior) view, with lateral lobes spread apart to show interior (ventral) view of the medial lobes. (**k**) "Semi-closed" petasma of *Farfantepenaeus duorarum* (Penaeoidea, Penaeidae), lateral view, with lateral lobes in natural position (not spread apart). (**l**) "Closed" petasma of *Sicyonia brevirostris* (Penaeoidea, Sicyoniidae), ventral view. ai, appendix interna; am, appendix masculina; b, basipod; cc, cincinnuli; en, endopod; ex, exopod; ll, lateral lobe; ml, median lobe. From (**a**) Kensley and

males usually carries, side by side with the AI, an additional spinous structure, the appendix masculina (AM) (Fig. 2.21d, e). This AM is usually a simple stalk bearing long spines, but also may be large and complex in structure. There is considerable variation in the AM among carideans, as will be shown in Chap. 6 (Reproduction). There is an AM on Pl2 of dendrobranchiates but is quite different from that of carideans (Fig. 2.21f–i, Plate 2.2e). An AM in lacking in males of procaridideans and stenopodideans. In dendrobranchiates, the Pl1 endopod is absent in females but in males is represented by a very complex and variable structure, the petasma (Fig. 2.21j–l, Pl. 2.2f). This structure is a union of the endopods of Pl1 joined along their mesial edges by cincinnuli (Plate 2.2d). The bewildering variation in these sexual specializations in carideans and especially dendrobranchiates will be treated in greater detail in Chap. 6.

It is appropriate that we exit this examination of shrimp structure and function with the uropods and telson, located at the very end of the shrimp (Fig. 2.22). The uropods are the appendages of the 6th abdominal segment, but instead of being positioned beneath the shrimplike the pleopods, they are oriented horizontally alongside the telson. The protopod of a uropod is short and gives rise to a large-scale or paddle-like endopod and exopod, both fringed with plumose setae. In procaridideans and many carideans, the blade of the exopod has a suture, the diaeresis (Fig. 2.22b, c), which may be superficial or an articulation permitting some bending. The end piece of the body, the telson, varies in shape and ornamentation. It may be long and pointed posteriorly, short, rectangular, or rounded (Fig. 2.2a–d). It may bear sharp projections ("fixed spines") (Fig. 2.22a), spine-like setae ("movable spines") dorsally and can be bordered posteriorly by setae of a variety of shapes and sizes. The details of these telsonic ornaments, as all the structures described in this chapter, are often in important in the taxonomy and identification of shrimp species. Together, the uropods and telson make up the large paddle that pushes water forward and the shrimp backward during the backward escape response.

I end this chapter with the hope it raises many questions about the adaptive function of shrimp morphology and stimulates testing of hypotheses by present and future readers.

Fig. 2.21 (continued) Williams (1986), (**b**) Alvarez et al. (2006), (**c**, **d**) Bauer (1976b), (**e**) Chace (1985), (**f–i**) Pérez Farfante (1988), (**j–l**) Pérez Farfante (1969). (**a**, **b**, courtesy of Journal of Crustacean Biology)

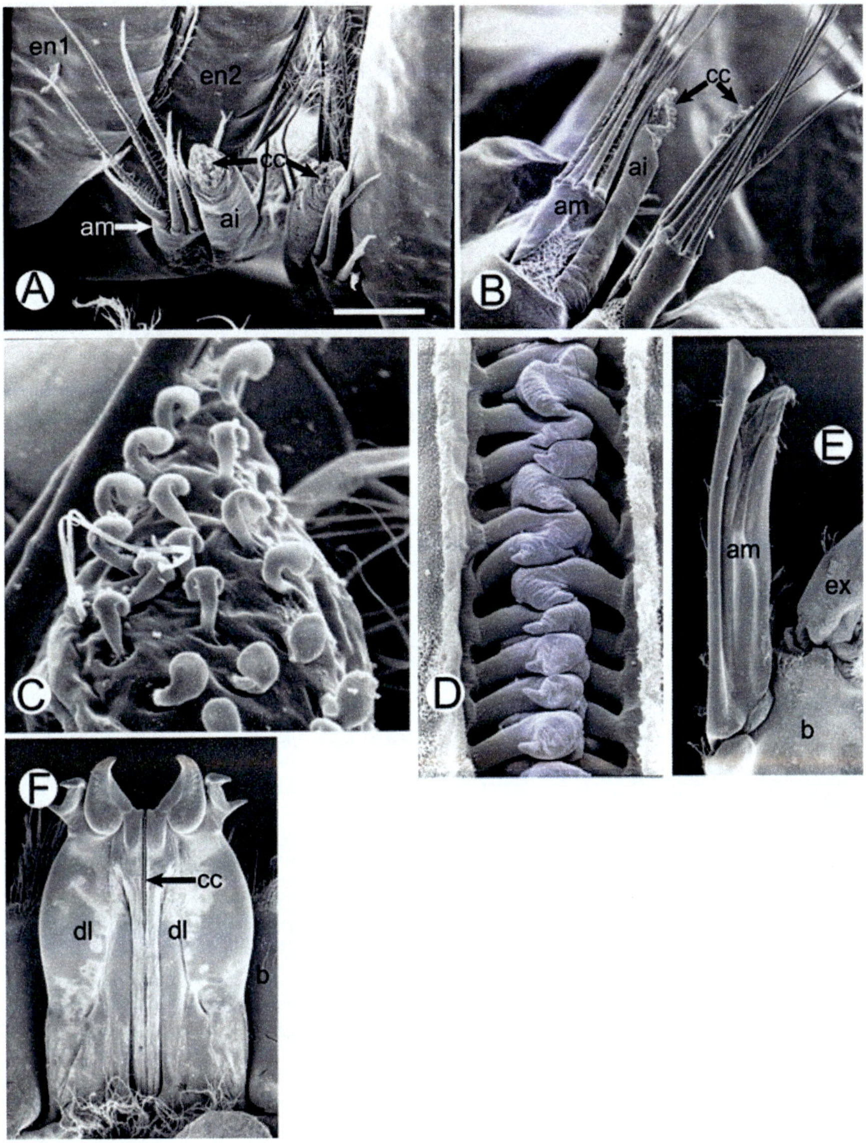

Plate 2.2 Reproductive appendices and structures. (**a**) Appendices internae and masculinae on second pleopod (anterior view) of *Heptacarpus sitchensis* (Caridea, Thoridae). (**b**) Posterior view of appendices in (**a**) at higher magnification. (**c**) Cincinnuli at high magnification on appendix of masculinae show in (**a**, **b**). (**d**) Cincinnuli connecting petasmal endopods of *Sicyonia dorsalis* (Penaeoidea, Sicyoniidae). (**e**) Appendix masculina of *S. dorsalis*. (**f**) Petasma of *S. dorsalis*, dorsal view. ai, appendix interna; am, appendix masculina; b, basipod; cc, cincinnuli; dl, dorsal (median) lobe of petasma; en1, en2, endopods of pleopod 1–2; ex, exopod. Scale bar in (**a**) represents 60 μm in (**a**, **b**), 13 μm in (**c**), 23 μm in (**d**), 750 μm in E, and 1.0 mm in F. From (**a**–**c**) Bauer (1976b); (**d**–**f**) Bauer (1996)

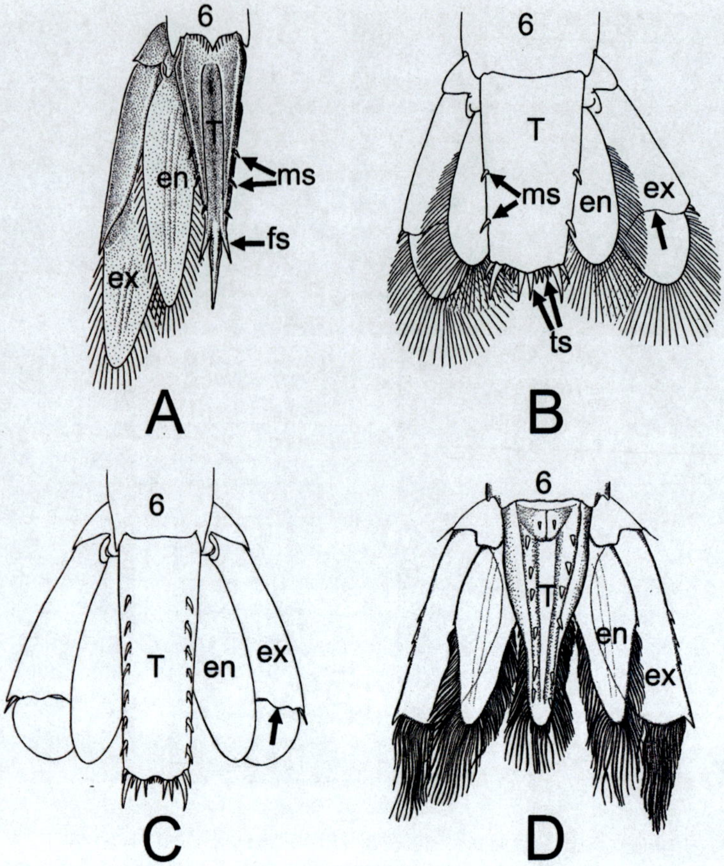

Fig. 2.22 Uropods and telson. (**a**) *Penaeopsis balsii* (Penaeoidea, Penaeidae). (**b**) *Vetericaris chaceorum* (Procarididea). (**c**) *Alvinocaris lusca* (Caridea, Alvinocarididae). (**d**) *Macromaxillocaris bahamaensis* (Stenopodidea, Macromaxillocarididae). diaeresis, unlabeled arrows in B and C; en, endopod; ex, exopod; fs, fixed spine; ms, movable spine (seta); T, telson; ts, terminal spines (setae); 6, sixth abdominal segment. From (**a**) Pérez Farfante (1980), (**b**) Kensley and Williams (1986), (**c**) Williams and Chace (1982) and (**d**) Alvarez et al. (2006). (**b–d**, courtesy of Journal of Crustacean Biology)

References

Alexander CG, Hindley JPR (1985) The mechanism of food ingestion by the banana prawn, *Penaeus merguiensis*. Mar Behav Physiol 12:33–46

Alexander CG, Hindley JPR, Jones SG (1980) Structure and function of the third maxillipeds of the banana prawn, *Penaeus merguiensis*. Mar Biol 58:245–249

Almeida AO, Anker A, Mantelatto FL (2014) A new snapping species of the shrimp genus *Typton* Costa, 1844 (Decapoda: Palaemonidae) from the coast of São Paulo, southeastern Brazil. Zootaxa 3835:110–120

Alvarez F, Iliffe TM, Villalobos JL (2006) Macromaxillocarididae, a new family of stenopodidean shrimp from an anchialine cave in the Bahamas, with the description of *Macromaxillocaris bahamaensis*, n. gen., n. sp. J Crustac Biol 26:366–378

Anastasiadou C, Leonardos ID (2008) Morphological variation among populations of *Atyaephyra desmarestii* (Millet, 1831) (Decapoda: Caridea: Atyidae) from freshwater habitats of northwestern Greece. J Crustac Biol 28:240–247

Anker A, Ahyong ST, Noel PY, Palmer R (2006) Morphological phylogeny of alpheid shrimps: parallel preadaptation and the origin of a key morphological innovation, the snapping claw. Evolution 60:2507–2528

Ashelby CW, De Grave S, Johnson ML (2015) Preliminary observations on the mandibles of palaemonoid shrimp (Crustacea: Decapoda: Caridea: Palaemonoidea). PeerJ 3:e846. https://doi.org/10.7717/peerj.846

Bauer RT (1975) Grooming behaviour and morphology of the caridean shrimp *Pandalus danae* Stimpson (Decapoda: Natantia: Pandalidae). Zool J Linn Soc 56:45–71

Bauer RT (1976a) Antifouling adaptations of caridean shrimp: grooming morphology and behavior. Doctoral dissertation, Scripps Institution of Oceanography, University of California, San Diego, 269 pp

Bauer RT (1976b) Mating behaviour and spermatophore transfer in the shrimp *Heptacarpus pictus* (Stimpson) (Decapoda: Caridea: Hippolytidae). J Nat Hist 10:415–440

Bauer RT (1977) Antifouling adaptations of marine shrimp (Decapoda: Caridea): functional morphology and adaptive significance of antennular preening by the third maxillipeds. Mar Biol 40:260–276

Bauer RT (1979) Antifouling adaptations of marine shrimp (Decapoda: Caridea): gill cleaning mechanisms and grooming of brooded embryos. Zool J Linn Soc 65:281–303

Bauer RT (1984) Morphological trends in the genus *Heptacarpus* (Decapoda, Caridea) and their phylogenetic significance. J Crustac Biol 4:201–225

Bauer RT (1996) Role of the petasma and appendices masculinae during copulation and insemination in the penaeoid shrimp, *Sicyonia dorsalis* (Crustacea: Decapoda: Dendrobranchiata). Invertebr Reprod Dev 29:173–184

Bauer RT (1999) Gill-cleaning mechanisms of a dendrobranchiate shrimp, *Rimapenaeus similis* (Decapoda: Penaeidae): description and experimental testing of function. J Morphol 242:125–139

Bauer RT (2004) Remarkable shrimps: adaptations and natural history of the carideans. University of Oklahoma Press, Norman

Bauer RT, VanHoy R (1996) Variation in sexual systems (protandry, gonochorism) and reproductive biology among three species of the shrimp genus *Thor* (Decapoda: Caridea). Bull Mar Sci 59:53–73

Borradaile LA (1917) On the structure and function of the mouthparts of the palaemonid prawns. Proc Zool Soc Lond 1917:37–71

Bruce AJ (1976) Shrimps and prawns of coral reefs, with special reference to commensalism. In: Jones O, Endean R (eds) Biology and geology of coral reefs, vol III, biology 2. Academic Press, New York, pp 38–94

Bruce AJ (1989) Reestablishment of the genus *Coutierella* Sollaud, 1914 (Decapoda: Palaemonidae), with a redescription of *C. tonkinensis* from the Mai Po marshes, Hong Kong. J Crustac Biol 9:176–187

Bruce AJ, Zmarzly DL (1983) *Periclimenes pilipes*, new species, a crinoid associate from Enewetak Atoll, Marshall Islands (Crustacea: Decapoda: Pontoniinae). J Crustac Biol 3:644–654

Burkenroad MD (1963) The evolution of the Eucarida (Crustacea: Eumalacostraca) in relation to the fossil record. Tulane Stud Geol 2:3–16

Burkenroad MD (1981) The higher taxonomy and evolution of Decapoda (Crustacea). Trans San Diego Soc Nat Hist 19:251–268

Burukovsky RN (1972) On the function of the rostrum in shrimps. Fishery investigations in Atlantic science. Proceedings of Atlantic Investigation Institution of Fishery Industry 42:176–179. (English translation of Russian original)

Calman WT (1909) Crustacea. In: Lankester R (ed) A treatise on zoology, Part VII, Appendiculata, third fascicle. Adam and Charles Black, London, 346 pp

Chace FA Jr (1972) The shrimps of the Smithsonian-Bredin Expeditions with a summary of the West Indian shallow-water species (Crustacea: Decapoda: Natantia). Smithson. Contrib. Zool., No. 98, 179 pp

Chace FA Jr (1984) The caridean shrimps (Crustacea: Decapoda) of the Albatross Philippine Expedition, 1907-1910, Part 2: Families Glyphocrangonidae and Crangonidae. Smithsonian Contributions to Zoology No. 397, 63 pp

Chace FA Jr (1985) The caridean shrimps (Crustacea: Decapoda) of the Albatross Philippine Expedition, 1907-1910, Part 3: Families Thalassocarididae and Pandalidae. Smithson. Contrib. Zool., No. 411, 143 pp

Chace FA Jr (1988) The caridean shrimps (Crustacea: Decapoda) of the Albatross Philippine Expedition, 1907-1910, Part 5: Family Alpheidae. Smithson. Contrib. Zool., No. 466, 99 pp

Chace FA Jr (1997) The caridean shrimps (Crustacea: Decapoda) of the Albatross Philippine Expedition, 1907-1910, Part 7: Families Atyidae, Eugonatonotidae, Rhynchocinetidae, Bathypalaemonellidae, Processidae, and Hippolytidae. Smithson. Contrib. Zool., No. 587, 106 pp

Chace FA Jr, Manning RB (1972) Two new caridean shrimps, one representing a new family, from marine pools on Ascension Island (Crustacea: Decapoda: Natantia). Smithson. Contrib. Zool., No. 131, 18 pp

Christoffersen ML (1988) Genealogy and phylogenetic classification of the world Crangonidae (Crustacea, Caridea), with a new species and new records for the southwestern Atlantic. Revista Nordest Biol 6:43–59

Covich AP, Crowl TA, Hein CL, Townsend MJ, McDowall WH (2009) Predator–prey interactions in river networks: comparing shrimp spatial refugia in two drainage basins. Freshw Biol 54: 450–465

Crosnier A (2000) A new genus and species of hippolytid shrimp (Decapoda, Caridea) from Wallis Island, South-West Pacific. J Crustac Biol 20(Sp. No. 2):109–115

Dall W, Hill BC, Rothlisberg PC, Sharples DJ (1990) The biology of the Penaeidae. Adv Mar Biol 27:1–489

de Mazancourt V, Marquet F, Keith P (2017) The "Pinocchio-shrimp effect": first evidence of variation in rostrum length with the environment in *Caridina* H. Milne-Edwards, 1837 (Decapoda: Caridea: Atyidae). J Crustac Biol 37:249–257

Felgenhauer BE, Abele LG (1983) Phylogenetic relationships among shrimp-like decapods. In: Schram FR (ed) Crustacean phylogeny. A.A. Balkema, Rotterdam, pp 291–311

Felgenhauer BE, Abele LG (1985) Feeding structures of two atyid shrimps, with comments on caridean phylogeny. J Crustac Biol 5:397–419

Froglia C, Gramitto ME (2000) A new pelagic shrimp of the genus *Sergia* (Decapoda, Sergestidae) from the Atlantic Ocean. J Crustac Biol 20:71–77

Fryer G (1960) The feeding mechanism of some atyid prawns of the genus *Caridina.* Trans R Soc Edinb 64:217–244

Garm A (2004) Mechanical functions of setae from the mouth apparatus of seven species of decapod crustaceans. J Morphol 260:85–100

Garm A, Hallberg E, Høeg JT (2003) Role of maxilla 2 and its setae during feeding in the shrimp *Palaemon adspersus*(Crustacea: Decapoda). Biol Bull 204:126–137

Gerstaecker A, Ortmann AE (1909) Die Klassen und Ordnungen der Arthropoden: wissenschaftlich dargestellt in Word und Bild, 13 (2), Crustacea. (second half of pls.)

Goy JW (2010) Infraorder Stenopodidea Claus, 1872. In: Schram FR, von Vaupel Klein JC (eds) Treatise on zoology - anatomy, taxonomy, biology -the Crustacea volume 9 Part A. Koninklijke Brill NV, Leiden, pp 215–265

Goy JW, Cardozo IA (2014) Redescription of *Odontozona edwardsi* (Bouvier, 1908) (Decapoda: Stenopodidea: Stenopodidae) and description of a new species of *Odontozona* commensal on the deep-water coral, *Lophelia pertusa*. Zootaxa 3774:522–566

Goy JW, Devaney DM (1980) *Stenopus pyrsonotus*, a new species of stenopodidean shrimp from the Indo-West Pacific region (Crustacea: Decapoda). Proc Biol Soc Wash 93:781–796

Haywood M, Alexander CG (1982) Statolith implantation in the banana prawn *Penaeus merguiensis*. Mar Biol 72:113–118

Herberholz J, Schmitz B (2001) Signaling via water currents in behavioral interactions of snapping shrimp (*Alpheus heterochaelis*). Biol Bull 201:6–16

Herring PJ, Roe HSJ (1988) The photoecology of pelagic oceanic decapods. Symp Zool Soc Lond 59:263–290

Hindley JPR, Alexander CG (1978) Structure and function of the chelate pereiopods of the banana prawn, *Penaeus merguiensis*. Mar Biol 48:153–160

Hobbs HH Jr, Hobbs HH III, Daniel MA (1977) A review of the troglobitic decapod crustaceans of the Americas. Smithsonian Contributions to Zoology, 244, 183 pp

Holthuis LB (1950) The Decapoda of the Siboga Expedition. Part X. The Palaemonidae collected by the Siboga and Snellius Expeditions with remarks on other species. I. Subfamily Palaemoninae. Siboga Expéditie 39(A9):1–268

Holthuis LB (1952) A general revision of the Palaemonidae (Crustacea Decapoda Natantia) of the Americas. Part II. The subfamily Palaemoninae. Allan Hancock Found Publ Occas Pap 12:1–396

Holthuis LB (1986) Fresh-water shrimps of the family Atyidae (Crustacea: Decapoda) from western Colombia. J Crustac Biol 6:438–445

Hopkins TL, Flock ME, Gartner JV Jr, Torres JJ (1994) Structure and trophic ecology of a low latitude midwater decapod and mysid assemblage. Mar Ecol Prog Ser 109:143–156

Hunt MJ, Winsor H, Alexander CG (1992) Feeding by the penaeid prawns: the role of the anterior mouthparts. J Exp Mar Biol Ecol 160:33–46

Huxley TH (1880) The Crayfish. C. Kegan Paul and Co., London

Johnson VR Jr (1969) Behavior associated with pair formation in the banded shrimp *Stenopus hispidus* (Olivier). Pac Sci 23:40–50

Johnson MW, Everest FA, Young RW (1947) The role of snapping shrimp (*Crangon* and *Synalpheus*) in the production of underwater noise in the sea. Biol Bull 93:122–138

Johnson ML, Dobson N, De Grave S (2015) External morphology of eyes and nebenaugen of caridean decapods–ecological and systematic considerations. PeerJ 3:e1176. https://doi.org/10.7717/peerj.1176

Jugovic J, Prevorcnik S, Aljancic G, Sketa B (2010) The atyid shrimp (Crustacea: Decapoda: Atyidae) rostrum: phylogeny versus adaptation, taxonomy versus trophic ecology. J Nat Hist 44:2509–2533

Kaji T, Anker A, Wirkner CS, Palmer AR (2018) Parallel saltational evolution of ultrafast movements in snapping shrimp claws. Curr Biol 28:106–113

Kensley B, Williams D (1986) New shrimps (families Procarididae and Atyidae) from a submerged lava tube on Hawaii. J Crustac Biol 6:417–437

Knowlton RE, Moulton JM (1963) Sound production in the snapping shrimps *Alpheus* (*Crangon*) and *Synalpheus*. Biol Bull 125:311–331

Komai T, Chan TY (2009) New genus and species of Crangonidae (Decapoda: Caridea) with a large plate-like eye from the abyssal zone off Taiwan, Northwestern Pacific. J Crustac Biol 29:254–265

Liasko R, Anastasiadou C, Ntakis A, Leonardos ID (2015) How a sharp rostral dimorphism affects the life history, population structure and adaptability of a small shrimp: the case study of *Hippolyte sapphica*. Mar Ecol 36:400–407

Lohse D, Schmitz B, Versluis M (2001) Snapping shrimp make flashing bubbles. Nature 413:477–478

MacGinitie GE (1937) Notes on the natural history of several marine Crustacea. Am Midl Nat 18: 1031–1037

Martin JWE, Liub M, Striley D (2007) Morphological observations on the gills of dendrobranchiate shrimps. Zoologischer Anzeiger 246:115–125

Mejía-Ortíz LM, Hartnoll RG (2006) A new use for useless eyes in cave crustaceans. Crustaceana 79:593–600

Moore PG, Rainbow PS, Larson RJ (1993) The mesopelagic Shrimp *Notostomus robustus* Smith (Decapoda: Oplophoridae) observed in situ feeding on the medusan *Atolla wyvillei* Haeckel in the Northwest Atlantic, with notes on gut contents and mouthpart morphology. J Crustac Biol 13:690–696

Ocasio-Torres ME, Crowl TA, Sabat AM (2014) Long rostrum in an amphidromous shrimp induced by chemical signals from a predatory fish. Freshw Sci 33:451–458

Ocasio-Torres ME, Crowl TA, Sabat AM (2015a) Allometric differences between two phenotypes of the amphidromous shrimp *Xiphocaris elongata*. J Crustac Biol 35:747–752

Ocasio-Torres ME, Giray T, Crowl TA, Sabat AM (2015b) Antipredator defence mechanism in the amphidromous shrimp *Xiphocaris elongata* (Decapoda: Xiphocarididae): rostrum length. J Nat Hist 49:25–26

Patwardhan SS (1935) On the structure and mechanism of the gastric mill in Decapoda. 5. The structure of the gastric mill in natantous Macrura-Caridea. Proc Indian Acad Sci (B) 1:693–704

Patwardhan SS (1936) On the structure and mechanism of the gastric mill in Decapoda. VI. The structure of the gastric mill in the Natantous Macrura - Penaeidea and Stenopidea; conclusion. Proc Indian Acad Sci (B) 2:155–174

Pérez Farfante I (1969) Western Atlantic shrimps of the genus *Penaeus*. Fish Bull Fish Wildl Serv 67:461–591

Pérez Farfante I (1980) Revision of the penaeid shrimp genus *Penaeopsis* (Crustacea: Decapoda). Fish Bull 77:721–763

Pérez Farfante I (1985) The rock shrimp genus *Sicyonia* (Crustacea: Decapoda: Penaeoidea) in the eastern Pacific. Fish Bull US 83:1–79

Pérez Farfante I (1987) Revision of the gamba prawn genus *Pseudaristeus*, with description of two new species (Crustacea: Decapoda: Penaeoidea). Fish Bull 87:311–338

Pérez Farfante I (1988) Illustrated key to penaeoid shrimps of commerce in the Americas. NOAA Technical Reports NMFS 64:1–32

Pérez Farfante I, Kensley BF (1997) Penaeoid and sergestoid shrimps and prawns of the world. Keys and diagnoses for the families and genera. Mémoires du Muséum National d' Histoire Naturelle Paris 175:1–233

Renfro WC, Pearcy WG (1966) Food and feeding apparatus of two pelagic shrimps. J Fish Res Board Canada 23:1971–1975

Ritzmann RE (1974) Mechanisms for the snapping behavior of two alpheid shrimp, *Alpheus californiensis* and *Alpheus heterochelis*. J Comp Physiol 95:217–236

Saito TP, Takeda M (2003) Phylogeny of the family Spongicolidae (Crustacea: Stenopodidea): evolutionary trend from shallow-water free-living to deep-water sponge-associated habitat. J Mar Biol Assoc U K 83:119–131

Saito T, Shinji T, Yamamoto T (2006) *Spongicoloides Iheyaensis*, a new species of deep-sea sponge-associated shrimp from the Iheya Ridge, Ryukyu Islands, southern Japan (Decapoda: Stenopodidea: Spongicolidae). J Crustac Biol 26:224–233

Sardà F, Company JB, Costa C (2005) Morphological approach for relating decapod crustacean cephalothorax shape with distribution in the water column. Mar Biol 147:611–618

Tavares C, Martin JW (2010) Suborder Dendrobranchiata Bate, 1888. In: Schram FR, von Vaupel Klein JC (eds) Treatise on zoology - anatomy, taxonomy, biology -the Crustacea volume 9 Part A. Koninklijke Brill NV, Leiden, pp 99–164

Thompson JR (1966) The Caridean Superfamily Bresilioidea (Decapoda Natantia). A revision and a discussion of its validity and affinities. Crustaceana 11:129–140

Thompson JR (1967) Comments on phylogeny of section Caridea (Decapoda Natantia) and the phylogenetic importance of the Oplophoroidea. Proceedings Symposium on Crustacea, Ernakulam, Marine Biological Association India 1:314–326

Ugolini A, Borgioli E (1993) On the "accessory eye" of shrimps (Decapoda, Natantia). Crustaceana 63:112–115

Versluis M, Schmitz B, von d. Heydt A, Lohse D (2000) How snapping shrimp snap: through cavitating bubbles. Science 289:2114–2117

Wicksten MK (2010) Infraorder Caridea Dana, 1852. In: Schram FR, von Vaupel Klein JC (eds) Treatise on zoology - anatomy, taxonomy, biology -the Crustacea. Volume 9 Part A. Koninklijke Brill NV, Leiden, pp 165–206

Wilkens H (1986) The tempo of regressive evolution: studies of the eye reduction in stygobiont fishes and decapod crustaceans of the Gulf coast and West Atlantic region. Stygologia 2:130–143

Williams AB, Chace FA Jr (1982) A new caridean shrimp of the family Bresiliidae from thermal vents of the Galapagos Rift. J Crustac Biol 2:136–147

Xiao Y, Greenwood JG (1993) The biology of *Acetes* (Crustacea, Sergestidae). Oceanogr Mar Biol: Annu Rev 31:259–444

Young JH (1959) Morphology of the white shrimp *Penaeus setiferus* (Linnaeus 1758). Fish Bull Fish Wildl Serv 59:1–168

Chapter 3
The Families of Decapod Shrimps

3.1 Taxonomic Considerations

Before delving into the major subject areas of shrimp biology in the following chapters, we should have some overall familiarity with shrimp taxa. The family sketches below will provide some general information about taxonomy, diversity, morphology, distribution, ecology, and behavior of the many decapod shrimp families. This information is not static; the number of new publications on shrimps is continually growing. For example, the number of described species in all major shrimp groups keeps increasing with little sign of reaching a plateau (Fransen and De Grave 2009). Within the Dendrobranchiata, the Penaeoidea have 57–62 genera (opinions vary) and 428 species; the Sergestoidea 21 genera and 104 species; the Procarididea only 2 genera and 7 species; the Caridea a whopping 407 genera and 3777 species; and the Stenopodidea 13 genera and 95 species. Where do these figures come from? Fenner Chace, Jr., a renowned carcinologist whose work is widely cited in this book, initially made counts of decapod species using the bibliographic publication "Zoological Record" as well as a variety of other sources (Chace 1951). In the predecessor to this book (Bauer 2004), I continued for the Caridea where Fenner stopped counting (1990) and added new species through 2001. In 2011, De Grave and Fransen (2011) published a checklist of the names and numbers of shrimp species. The numbers of shrimp (and other) taxa have since been updated continually on the World Registry of Marine Species (WoRMS) website (which also includes freshwater shrimps). The database managers have kindly supplied me with the most up-to-date counts at the writing of this manuscript. Sammy De Grave and Charles Fransen are expert shrimp biologists who are the current editors of the shrimp data on that website.

The reader should be aware that taxonomy (the naming of organisms) and systematics (the classification of species considering their genealogy = phylogeny) are constantly changing as new information and techniques of study become available. New genera and species are constantly being described. The names and composition of taxonomic groups vary with new information and new interpretation. For example, in Chace (1951), the number of caridean families, genera, and species was given as 21,171, and 1590, respectively. In Bauer (2004), these numbers had risen to 28, 35, and 2818, but the current count as of this writing is now 38 families, 407 genera, and 3777 species. During the writing of this book, I have had to rearrange and modify this chapter several times in a 5–6 year period because of new taxonomic information and opinions. Although the WoRMS website gives excellent taxonomical information with numerical data on diversity, there are always disagreements among biologists about the composition and nomenclature of taxa. This is certainly true in the world of shrimps. New ideas and proposals are often questioned and may or may not survive the test of time. I will mention disputes of particular interest in the family profiles below.

Taxonomic classifications reflect or imply the phylogenetic relationships among its component taxa to varying degrees. The consensus classification of Martin and Davis (2001) for the Decapoda (including the shrimp groups) and that of Chace (1992) for the Caridea are good examples. In the classification of shrimp groups in Table 3.1, the order of the major groups implies the more primitive (closer to the ancestral taxa) nature of the dendrobranchiate shrimps to the other groups. The Stenopodidea is likely more advanced, i.e., further removed genealogically from the stem decapod shrimp group). Within these major groups, I have arranged the superfamilies and families alphabetically, in the order I discuss them below. There are some exceptions, e.g., the more primitive Procarididea is placed after the Caridea; reverse alphabetical order of stenopodidean families, which was done for ease of comparison. The phylogenetic relationships among major shrimp taxa and families will be discussed at length in Chap. 12 (Shrimp Evolution).

3.2 Suborder Dendrobranchiata

It is appropriate to begin our survey of shrimp families with the dendrobranchiates, which are considered closest to the decapod ancestor, especially in their primitive mode of reproduction. Females spawn fertilized eggs directly into the water, which soon hatch as naupliar larvae at the beginning of an extended development in the plankton. Although the dendrobranchiates do not come near the carideans in terms of biological and ecological diversity, they are considerably more important economically in fisheries and aquaculture. They are primarily a tropical and subtropical marine group, although some coastal species may range into brackish or even fresh water during some point in their life cycle. The morphological diversity of male and female genitalia in this group is both astounding and bewildering, as the role of these structures in spermatophore transfer and insemination is still largely unresolved (Chap. 6). In the superfamily Penaeoidea, most species are epibenthic or benthic over soft sediments (sand, mud) in continental shelf (neritic) seas (Penaeidae, Sicyoniidae). Other penaeoid families are nektobenthic in the colder deep waters of the continental slopes (bathyal zone, ~200–3000 m depth) and beyond, spending

Table 3.1 Superfamilies and families of the crustacean order Decapoda

Suborder Dendrobranchiata
Superfamily Penaeoidea
Family Aristeidae
Family Benthesicymidae
Family Penaeidae
Family Sicyoniidae
Family Solenoceridae
Superfamily Sergestoidea
Family Luciferidae
Family Sergestidae
Suborder Pleocyemata
Infraorder Caridea
Superfamily Alpheoidea
Family Alpheidae
Family Barbouriidae
Family Bythocarididae
Family Hippolytidae
Family Lysmatidae
Family Merguiidae
Family Ogyrididae
Family Thoridae
Superfamily Atyoidea
Family Atyidae
Superfamily Bresilioidea
Family Agostocarididae
Family Alvinocarididae
Family Anchialocarididae
Family Bresiliidae
Family Disciadidae
Family Pseudochelidae
Superfamily Campylonotoidea
Family Bathypalaemonellidae
Family Campylonotidae
Superfamily Crangonoidea
Family Crangonidae
Family Glyphocrangonidae
Superfamily Nematocarcinoidea
Family Eugonatonotidae
Family Lipkiidae
Family Nematocarcinidae
Family Rhynchocinetidae
Family Xiphocarididae
Superfamily Oplophoroidea

(continued)

Table 3.1 (continued)

Family Acanthephyridae
Family Oplophoridae
Superfamily Palaemonoidea
Family Desmocarididae
Family Euryrhynchidae
Family Palaemonidae
Family Typhlocarididae
Superfamily Pandaloidea
Family Chlorotocellidae
Family Pandalidae
Superfamily Pasiphaeoidea
Family Pasiphaeidae
Superfamily Physetocaridoidea
Family Physetocarididae
Superfamily Processoidea
Family Processidae
Superfamily Psalidopodoidea
Family Psalidopodidae
Superfamily Stylodactyloidea
Family Stylodactylidae
Taxon inquirendum (uncertain placement)
Amphionides
Infraorder Procarididea
Family Procarididae
Infraorder Stenopodidea
Family Macromaxillocarididae
Family Spongicolidae
Family Stenopodidae

time on the bottom but capable of ranging well up into the water column, (Solenoceridae, Aristeidae, some Penaeidae and Benthesicymidae). Several species of benthesicymids are truly meso- or bathypelagic. On the other hand, members of the superfamily Sergestoidea are fully pelagic or planktonic and are characterized by a reduced dendrobranchiate morphology, an adaptation to their pelagic environments.

3.3 Superfamily Penaeoidea

3.3.1 Aristeidae

Shrimps of this family (9 genera, 29 species), along with the families Benthesicymidae and Solenoceridae, were aptly termed "oceanic" penaeoids by the brilliant (and eccentric) carcinologist Martin Burkenroad (1936), who contributed so much to the knowledge of penaeoid shrimps. Aristeids are nektobenthic shrimps that occur in the deep waters of the continental slopes (bathyal zone, 200–3000 m) and great expanses of abyssal (> ~3000 to 6000 m depth) sea floor, although some species may range up into the deeper parts of the continental shelf. The bottoms over which they live are usually composed of mud, sand, or some mix of the two. As in many mobile deep-sea organisms, aristeid species have broad, sometimes cosmopolitan geographic distributions. They are usually some shade of pink, orange or red, either uniformly or variably distributed over the body (Crosnier 1978; Dall 2001), a common characteristic of deep-water shrimps (Plate 3.1a) (see Chap. 5). *Aristeus* species and *Aristaeomorpha foliacea* bear photophores (light-producing organs) distributed over various parts of their ventral surface and pereopods (e.g., Crosnier 1978; Chan et al. 2017). The light produced by these organs helps to conceal their body silhouettes when viewed from below (counterillumination; countershading; see Chap. 5). Like most penaeoids, aristeids are good-sized shrimps, ranging as adults from about 10 up to a very large 33 cm total length in female *Aristaeopsis edwardsiana* (Holthuis 1980). Not surprisingly, species such as *Aristeus antennatus* and *A. antillensis, Aristaeomorpha foliacea,* and *Aristaeopsis* (formerly *Plesiopenae*us) *edwardsiana* are important components of deep-water shrimp fisheries (FIGIS 2021; Bauer 2020).

Aristeid shrimps show a fairly typical penaeoid general morphology (Fig. 3.1). Most have long saber-like rostra with variable numbers of teeth, and carapaces adorned with several carinae and grooves. Their distinguishing character from the benthesicymids, which they most resemble and were once grouped, is a short upper (lateral or outer) antennular flagella, which is quite long in the benthesicymids. As with all penaeoids, it is the genitalia that show the most complex morphology. Aristeid males have an "open" petasma in which the flexible lateral lobes are extended laterally and not folded to enclose a space ventrally (posteriorly), as in closed petasmata (Pérez and Kensley 1997). Although it has always been assumed that the petasma functions in spermatophore transfer in some way, the mechanics of the diverse petasmata of penaeoids is still in the hypothesis stage, as will be discussed in Chap. 6. The female genitalia of penaeoids are termed the "thelycum," which can be defined as modifications of the posterior two (or sometimes three) thoracic sternites.

Thelyca serve either for (a) temporary attachment or prolonged storage of externally applied spermatophores ("open thelycum") or (b) as an area of genital contact with the male, which leads to the apertures of a median or twin seminal receptacles (spermathecae) for enclosed sperm storage (Bauer 1994; Pérez and Kensley 1997).

Plate 3.1 Dendrobranchiates. (**a**) *Aristeus virilis* (Aristeidae). (**b**) *Gennadas talismani* (Benthesicymidae). (**c**) *Penaeus monodon* (Penaeidae). (**d**) *Sicyonia laevigata* (Sicyoniidae) burrowing into substrate. (**e**) *Sicyonia dorsalis*. (**f**) *Solenocera* sp., Phillipines (Solenoceridae) (**g**) *Belzebub* sp. (Luciferidae). Photo acknowledgments: Courtesy of (**a**) T.-Y. Chan, (**b**) A.L. Vereshchaka, (**c**) David Knott, South Carolina Department Natural Resources, (**d–e**) R.T. Bauer, (**f**) Florent Charpin, (**g**) Arthur Anker

In aristeids females, the thelyca are open and to which twin spermatophores have been commonly observed attached in *Aristeus* species and *Aristaeomorpha foliacea* (Orsi Relini and Tunesi 1987; Tunesi 1987; Kapiris and Thessalou-Legaki 2009).

The general ecology and life history of *Aristeus* species and *Aristaeomorpha foliacea* is better known than other aristeids due to their importance in Mediterranean

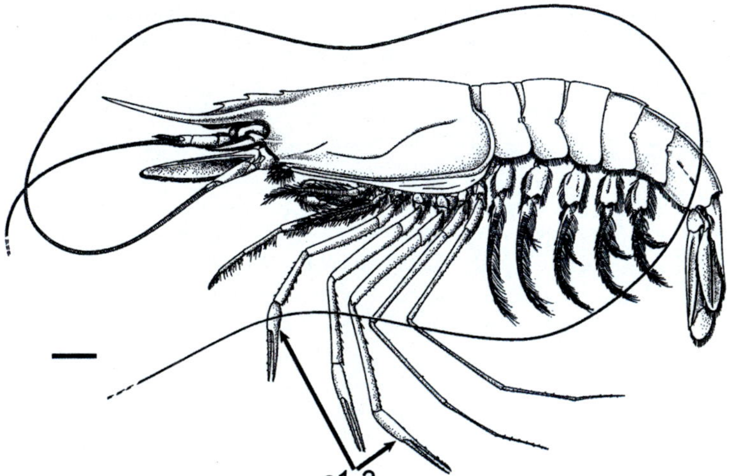

Fig. 3.1 Aristeidae. *Aristeus antennatus*. Scale bar represents 10 mm. c1–3, chelipeds 1–3. From Pérez and Kensley (1997), ©Publications Scientifiques du Muséum national d'Histoire naturelle, Paris

fisheries. As in most penaeoid species, males are smaller than females. In mating systems in which males do not guard females until mating, large male size is not an advantage (see Chap. 8). However, large female size is beneficial as larger females produce and spawn a greater number of offspring. Although they are deeper-dwelling species, reproduction is seasonal in the late spring to early fall when many females have mature (prespawning) ovaries and are inseminated, with twin spermatophores attached to the open thelycum (Demestre 1995; D'Onghia et al. 1998; Kapiris and Thessalou-Legaki 2009). These large shrimps have well-developed chelae for a penaeoid, and so it is not surprising that studies on diet show them to be deep-sea predators. They prey on benthic invertebrates such as polychaetes, small molluscs, and other crustaceans. However, their diet may also include a diverse assortment of species from the mesopelagic zone (~200–1000 m), such as myctophid fishes, shrimps, hyperiid amphipods, and pteropods (Chartosia et al. 2005; Cartes et al. 2008). The latter study suggested a strong link between primary and subsequent secondary productivity of surface waters and the summer reproductive season of the aristeid species far below. These nektobenthic shrimps are strong swimmers that move up the continental slope at night and swim up into the water column when the diel distributions of predator and mesopelagic prey most closely coincide. Life span of *Aristaeomorpha foliacea* has been estimated by population studies at 7–8 years for females and 5–6 years in males (D'Onghia et al. 1998). In *Aristeus antennatus*, life span is similarly long, calculated by various workers working at different sites at 4–9 years (Orsi Relini and Relini 1998). These are very long life spans when compared with some coastal penaeids which reach comparable sizes but do so within a 1–1.5 year life span. A slower growth rate and longer life span is a common life history pattern of shrimps living at depth or at higher latitudes (Chap. 9).

3.3.2 Benthesicymidae

The benthesicymids (12 genera, 41 species) are a generally deep-water group which contains both nektobenthic (e.g., *Benthesicymus, Bentheonectes, Altelatipes*) and truly pelagic (*Gennadas, Amalopenaeus, Notogennema* genera (Vereshchaka et al. 2019, 2021). In the sampling of deep-sea fauna with benthic trawls, it is sometimes difficult to know if the organisms are taken on the bottom or up in the water column as the net is being retrieved by the ship. In an interesting confirmation of a benthopelagic versus a truly pelagic lifestyle, Burkenroad (1936) examined the contents of the antennular statocysts (balance organs) of various benthesicymid species. As might be recalled, benthic shrimps need to refill their statocysts after a molt with some fine sediment which is mixed with statocyst gland secretions to form the statolith within. Burkenroad verified a purely pelagic lifestyle in *Gennadas* and *Bentheogennema* species by observing that their statoliths were formed only by secretions of exoskeletal glands inside the statocyst. In contrast, statocysts of the benthopelagic *Benthesicymus* had bottom sediments as well, confirming their contact with the sea floor during part of the day. After Burkenroad's observations, midwater trawls have confirmed the pelagic lifestyle of various benthesicymid species. Rostra of benthesicymids are short and somewhat expanded vertically, a characteristic of many pelagic species (Sardà et al. 2005) (Fig. 3.2a, b).

Benthesicymids, like many pelagic and deep-sea shrimps, have a thin, soft integument with relatively little calcareous content. Two selective pressures are at work. Food is relatively scarce in deep-sea habitats compared to neritic ones, and a thick, heavily calcareous exoskeleton requires energy to make. Additionally, a shrimp with a heavy exoskeleton has to work harder to maintain its position in the water column without sinking. The lightweight cuticles of pelagic shrimps help them to hover or move easily up and down in the water column with less swimming effort. The very long antennal flagella of benthesicymids and other pelagic shrimps are additional flotation devices, greatly increasing the surface area of the shrimp with little increase in weight or volume (Omori 1974). Shrimps with such features are able to hang nearly motionless in the water column with little expenditure of energy. In *Gennadas* species, the antennal flagellum has a bend or kink, dividing the flagellum into a naked proximal part and setose distal section (Fig. 3.2b, Plate 3.1b) (Foxton 1969). The antennal flagella are held out at a 45°–90° angle from the body with the setose distal portion streaming out parallel and at some distance behind the body of the shrimp. Foxton pointed out that this is a common feature of other epi- and mesopelagic shrimps, particularly the sergestids (Fig. 3.6b) (Cowles 1994). Upper oceanic waters are usually less dense (warmer) than deeper (colder) waters, perhaps explaining why this additional buoyancy adaptation has not been selected for in deeper-dwelling pelagic or benthopelagic species. Foxton also suggested that setose

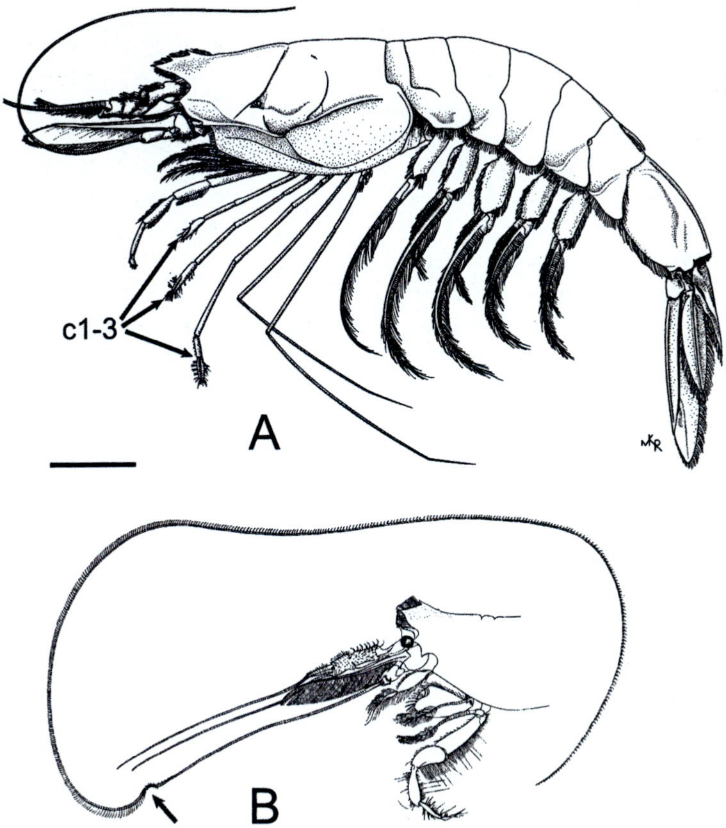

Fig. 3.2 Benthesicymidae. (**a**) *Benthesicymus bartletti.* (**b**) Anterior end of *Gennadas valens*, showing a very long antennal flagellum with kink or bend (arrow) dividing the flagellum into a setose distal and naked proximal portion. Scale bar represents 10 mm in (**a**) and 7 mm in (**b**). c1–3, chelipeds 1–3. From (**a**) Pérez and Kensley (1997), ©Publications Scientifiques du Muséum national d'Histoire naturelle, Paris and (B) Foxton (1969), courtesy of Crustaceana (Brill)

antennal flagella may serve a sensory function as vibration receptors, alerting the shrimp to potential predators or prey.

Benthesicymids are not particularly large shrimps compared to many other penaeoids, with the pelagic genera only in the 2–4 cm total length range, while the benthopelagic genera are larger, ~4–11 cm in total length. Due to their smaller size and relatively low abundance, they are of little interest to the fishing industry, although *Benthesicymus tanneri* has been mentioned as a possibility because of its large size for this family (Papiol and Hendrickx 2016). With their pelagic and deep benthopelagic habitats, benthesicymids have broad, sometimes cosmopolitan, geographic distributions. According to Dall (2001), most benthesicymid, aristeid, and solenocerid species occur between 40° N and 40° S latitudes, which is generally true of all dendrobranchiates. Dall hypothesized that the near-surface waters of these latitudes present more favorable environmental conditions for larval development, which takes place in the upper water column, than in higher latitude seas. Benthesicymids are also members of the "red shrimp, black fish" deep-sea community, with coloration varying from some shade of reddish or orange to deep red, increasing in intensity with depth. Pelagic species show diel (daily) vertical migrations, moving up more shallowly at night, and down deeper during the day. Diel vertical migration is a common feature of pelagic shrimps in which the shrimp hides from visually hunting predators by day in the darkness of the depths but migrates up into shallower waters by night where food supplies are more abundant. This movement is perhaps a compromise between different selective pressures. One is for the shrimps to follow light intensities optimal for their body (camouflage) coloration; another is the pursuit of prey that also migrate daily, up near the surface at night, deeper during the day.

As with any penaeoid, the genitalia of benthesicymids are of interest both in terms of systematics and reproductive morphology. Open female thelyca and open male petasmata are found in *Benthonectes*, *Altelatipes*, and one group of *Benthesicymus* species. Males of another group of *Benthesicymus* species and all *Bentheogennema* and *Gennadas* species have open petasmata. Atypically, females have closed thelyca guarding the openings to paired seminal receptacles. These sperm storage structures are invaginations between the 6th and 7th thoracic sternites, unlike those of species of Penaeidae and Sicyoniidae, in which seminal receptacles open between the 7th and 8th thoracic somites. Burkenroad (1936) suggested that penaeoid species without seminal receptacles are more primitive and gave rise to species which have them. However, the difference in the location of the seminal receptacles in the Benthesicymidae and Penaeidae/Sicyoniidae show that these sperm storage organs evolved independently in the two groups, a case of convergent evolution.

3.3.3 *Penaeidae*

This family might be considered the "crown jewel" of the superfamily Penaeoidea in a number of ways. The Penaeidae, with 27–32 genera and 229 species, is the most diverse of the superfamily. Its member species dominate the shrimp fauna of the extensive soft-bottom (sand/mud) communities of tropical and subtropical continental shelves of the world's oceans. Penaeid general morphology is rather conservative and typically dendrobranchiate (Fig. 3.3a, Plate 3.1c). Most species have a rather dull grey, olive-green, or brown overall coloration, a useful camouflage over soft-bottom habitats. However, a few, such as *Penaeus monodon*, the "giant tiger prawn," is a bit more colorful (Plate 3.1c). Although penaeids are a benthic, neritic group, two closely related genera, *Funchalia* and *Pelagopenaeus*, are mesopelagic (Omori 1974; Burkenroad 1936). These two genera share features associated with some other pelagic penaeoids such as the benthesicymid *Gennadas*, e.g., very long antennal flagella with a naked proximal and a setose distal portion (Figs. 3.2b, 3.3b).

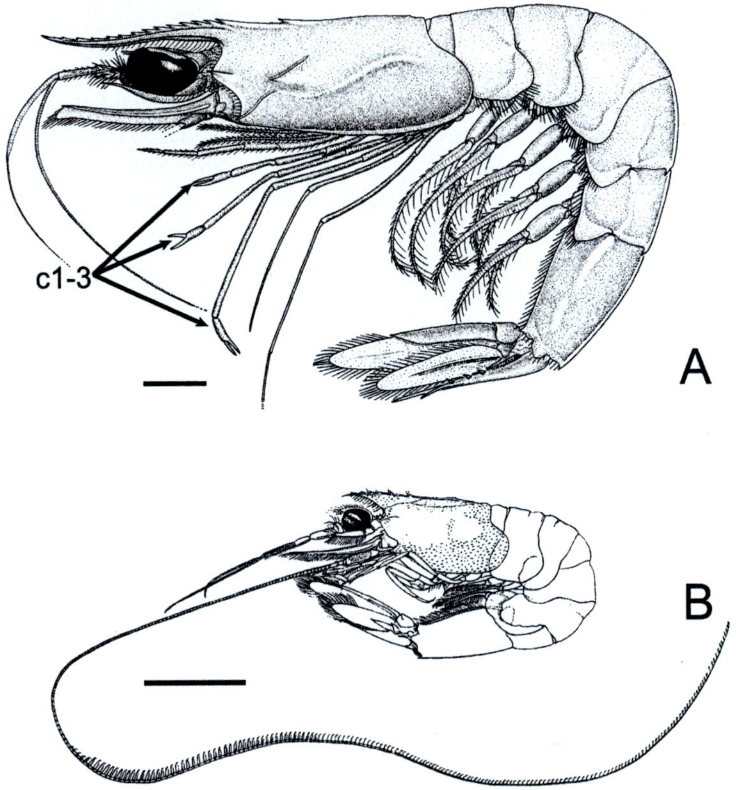

Fig. 3.3 Penaeidae. (**a**) *Parapenaeopsis jerryi*. (**b**) *Funchalia villosa*. Scale bar represents 5 mm in (**a**) and 10 mm in (**b**). c1–3, chelipeds 1–3. From (**a**) Pérez Farfante (1980) and (**b**) Foxton (1969), courtesy of Crustaceana (Brill)

Most penaeids are aggregated to a certain extent, with some species (e.g., *Fenneropenaeus merguiensis*) often occurring in great numbers at high density ("schools"). The high fecundity and fast growth of some species give rise to abundant populations of large shrimps, which account for their dominance in the world's shrimp fisheries and shrimp aquaculture (Chap. 11). Although primarily a marine group, juveniles of several of the commercially important species take advantage of the high productivity of brackish-water estuarine environments to feed and grow before moving out to sea to reproduce. A major volume of Advances in Marine Biology (Dall et al. 1990) is devoted to the biology of the family, and the recent review of Tavares and Martin (2010) on the Dendrobranchiata has more recent and valuable information on the family as well. The reproduction, life histories, fisheries, aquaculture, and evolution of this important family will be particularly emphasized in the following chapters of this volume.

As pointed out earlier, taxonomy and systematics of most groups are works in progress, with frequent changes as new information becomes available or new interpretations of existing data are made. Some taxonomists are "lumpers," i.e., are conservative in grouping apparently related species in the same taxon. Others are "splitters," dividing up one taxon into two or more, e.g., dividing the species of one genus into two or more based on what they see as separate evolutionary lineages within the original genus. This often generates controversy, which certainly has been the case with the genus *Penaeus* (*s.l.* = "*sensu lato*" = "in the broad sense"). Pérez and Kensley (1997) raised already recognized subgenera to generic status, so that the former *Penaeus* (*s.l.*) became 6 genera: *Farfantepenaeus, Fenneropenaeus, Litopenaeus, Marsupenaeus, Melicertus,* and *Penaeus* ss. (= "*sensu stricto*", restricted to some of the former *Penaeus s.l.* species). The change was done on the basis of morphological characteristics, especially genitalia and reproductive biology. A firestorm of taxonomic controversy ensued with some against the division of *Penaeus (s.l.)* into several genera (Dall 2007; Flegel 2007) but some supporting the change (McLaughlin et al. 2008). A number of studies based on molecular characters have yielded varying results and interpretations, depending on the number of species sampled, gene sequences, and techniques used. There is a lack of concordance of phylogenetic relationships between traditional morphological characters and molecular characters. Molecular phylogenies have concluded that subdivision of *Penaeus* (*s.l.*) is not yet warranted (Ma et al. 2011; Katneni et al. 2021) and this is currently accepted by the WoRMS taxonomic database. However, the issue is not completely settled. The molecular studies cited above point out that there are generic-level clades in *Penaeus* (*s.l.*) which might be recognized as genera, but that their recognition should be delayed until morphological and molecular characters are concordant. The six genera proposed by Pérez Farfante and Kensley have been (and still are) widely recognized for more than two decades in the scientific literature. The use of those six generic names conveys more biological information about these taxa, so that I will continue to use those generic names in this book for convenience and clarity.

Most penaeid shrimps are burrowing species, and the major environmental factor stimulating burrowing behavior is light intensity. Most shallow-water species are markedly nocturnal, active at night and burying themselves in soft bottoms during the day. This has been well demonstrated by collections made in scientific studies on catch success by commercial fishermen (e.g., Penn 1984; Bauer 1985a; Carvalho-Batista et al. 2012). However, there is variation in this behavior among species, depending on water clarity and other factors (Chap. 11).

Penaeids bury themselves just under or a few centimeters below the sediment surface. The shallow burrows are temporary, as these shrimps are highly mobile and remake a new burrow in another location each day. Burrowing behavior has been described for some penaeid species (Fuss 1964; Dall et al. 1990). Penaeids frequently begin by standing on the bottom and beating the pleopods vigorously, which scours out a furrow in the sediment into which the shrimp begins to settle. The shrimp then uses its pereopods to dig into the bottom, moving sediment outwards and upwards, which pushes the shrimp forward and downward. The antennal scales and uropods flick sand over the body. The shrimp thus buries itself in the bottom, sometimes entirely or only with the tip of the respiratory inhalant siphon or the tops of the eyes exposed. When the shrimp is buried completely, there may be small openings in the sediment marking the location of the inhalant and exhalant flows to and from the gill chamber. If the sediment is sufficiently coarse to allow rapid flow through it, these openings may not form. One adaptation related to burrowing is a well-developed prosartema or eye brush (Chap. 2), which in conjunction with the ocular cup, cleans the eyes free of sediment when not burrowed (Chap. 4).

While overall penaeid morphology is rather conservative, the genitalia of penaeid species are quite diverse. Females of some species have an open thelycum to which the males attach external spermatophores (e.g., *Litopenaeus* species, Pérez Farfante 1975). In other species, thelyca are closed, with two sternal plates covering a median seminal receptacle (e.g., *Penaeus monodon,* Pérez and Kensley 1997) or the apertures of twin seminal receptacles invaginated into the cephalothorax (e.g., *Rimapenaeus* species, Bauer and Lin 1993) (Chap. 6). In open-thelycum species, spawning occurs just after mating without a female molt (e.g., *Litopenaeus* species, Misamore and Browdy 1996). In closed-thelycum penaeoid species, mating occurs after a female molt, but multiple ovarian maturations and spawnings take place later during the female intermolt (Anderson et al. 1985; Dall et al. 1990; Scelso 1991).

3.3.4 Sicyoniidae

These penaeoids are characterized by their stout, robust bodies with a heavily sclerotized and calcified exoskeleton, short rostra and, compared to other penaeoids, relatively short flagella on both pairs of antennules (Fig. 3.4, Plate 3.1d, e). All 52 species are grouped in the same and only genus in the family (*Sicyonia*), as there is relatively little morphological diversity among species. They are the most benthic of the penaeoids, staying on or close to the bottom. One unique feature is that they

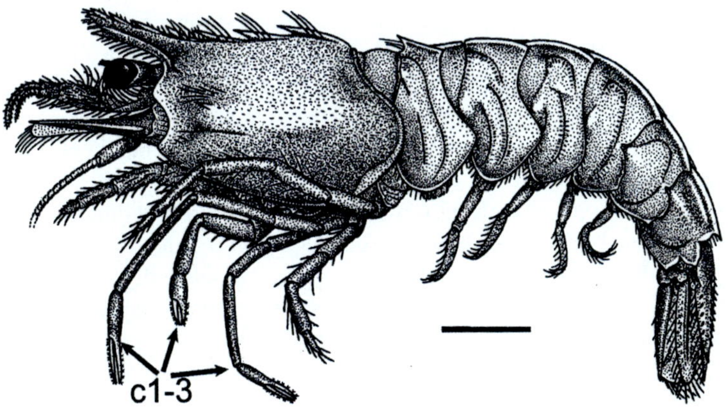

Fig. 3.4 Sicyoniidae. *Sicyonia disparri*. Scale bar represents 5 mm. c1–3, chelipeds 1–3. From Pérez Farfante (1985)

lack endopods on all the pleopods except for the first and second of males, which are modified into sexual structures as in other penaeoids. The lack of these endopods and resultant poorer swimming ability is related to their benthic way of life. Most species occur in shallower depths of the continental shelf, with some species in very shallow-water habitats such as seagrass meadows (e.g., Bauer 1985a) and others ranging down into upper slope depths (Pérez Farfante 1985).

Sicyoniids occur on various substrates, e.g., mud sand, rubble, and seagrass meadows in tropical and subtropical waters. One basic requirement is a substrate in which the shrimps can bury themselves. During the day, they can be found just under the sediment surface, but they emerge and are active at night. Their marked nocturnal activity has been noted and documented by day-night collections of various workers (e.g., Bauer 1985a). When working on seagrass shrimps in Puerto Rico, I used to demonstrate this nocturnal activity to students in the laboratory. Aquaria with sand bottoms with imbedded *Thalassia* seagrass plants were set up and stocked with *Sicyonia parri* and *S. laevigata*, the two most abundant penaeoids in the meadows studied. During the day, the shrimps buried themselves; with camera equipment set up, the room lights would be extinguished. We soon photographed some shrimps walking over the bottom, but most had crawled up and were clinging to the seagrass blades by their walking legs. Their chelipeds were extended and waved about, as the shrimp waited for some prey or other food item to drift by. After the room lights were turned on again, the shrimps would descend within a short time to the bottom and bury themselves (Bauer 1985a). Like other burrowing penaeoids (Penaeidae and Solenoceridae), sicyoniids have a prosartema (eye brush), but it is puzzling that it is rudimentary and not well developed.

The ground color of these shrimps is rather drab, with some shades of brown, grey, or olive green. However, many species superimpose bold stripes of black and white (Plate 3.1e) or other colors on the abdomen, with bright ocelli (eyespots) on the branchiostegal region of the carapace (Pérez Farfante 1985).

The thelycum of *Sicyonia* females is considered a closed one, but it is not very elaborate as in many penaeoids (Chap. 6). The apertures to the twin seminal receptacles are located there. Studies on the mating system and insemination of *Sicyonia dorsalis* (Bauer 1992a, 1996a, b) show that males have a short courtship similar to that of penaeids, but it all occurs on the bottom and does not involve swimming. More than one male can inseminate a female. Some research indicates that the petasma may be a male device for attaching to the genitalia of the female during copulation, rather than a sperm injection system (Bauer 1996b).

As in many shrimps, reproduction is continuous, occurring throughout the year, in tropical habitats but becomes more seasonal at higher latitudes, a result of factors such as water temperature, photoperiod and primary productivity which are correlated with latitude (Bauer 1992b; Chap. 9). Most sicyoniids are relatively small (<10 cm total length), and so only minor local fisheries for them occur for the few larger species, principally *S. brevirostris* (Gulf of Mexico, southeastern USA) and *S. ingentis* (southern California) (FIGIS 2021). In addition to the smaller size of sicyoniid species, the hard exoskeleton makes it a bit difficult to peel the tail. However, it is worth the effort as they are quite tasty (personal observation).

Sicyoniids display cataleptic behavior when threatened or handled. The cephalothorax and abdomen bend upwards toward each other, so that the shrimp forms a rigid "U." This makes this hard spiny shrimp a difficult mouthful for some predators. Wally Clark, Jr. and his many student colleagues took advantage of this cataleptic behavior in their numerous studies on sicyoniid sperm structure and sperm-egg interactions (e.g., Clark and Griffin 1988; Pillai et al. 1988). They could place a female ready to spawn on top of a beaker of seawater, and it would assume the cataleptic posture and not flip off, all the while spawning eggs into the beaker below.

3.3.5 Solenoceridae

Solenocerids (Fig. 3.5a, Plate 3.1f) are a benthic, deep-water family (10 genera, 81 species) generally inhabiting deep shelf and upper slope depths (Pérez Farfante 1977; Holthuis 1980). *Solenocera* and *Pleoticus* are burrowers, burying themselves in the sand/mud substrate during the day and emerging at night (Heegaard 1967; Pérez Farfante 1977). In *Solenocera*, the inner (ventral) pair of antennular flagella are subcylindrical and fit together to form a trough that, when covered by the flattened outer (dorsal) pair of flagella, make up the respiratory tube or siphon (Pérez and Kensley 1997). As most of the other genera of solenocerids have similar antennular flagella, they are likely burrowers as well. Heegaard (1967) was able to make observations on burrowing and other behaviors in *Solenocera membranacea*, a widely distributed species in the Atlantic Ocean at depths of 50–400 m (Fig. 3.5b). It is difficult to bring up and maintain alive deeper-dwelling species like solenocerids, aristeids, and benthesicymids. They are damaged by the weight of material in the net in the long haul to the surface and by the change in temperature and perhaps pressure. Heegaard collected them at about 100 m during a time of the year when temperatures between the bottom and surface temperature were similar. He was thus able to keep them in laboratory aquaria to observe them. Their method of burrowing is somewhat different than that of penaeids described above. They begin by laying on the sediment horizontally and use the anterior chelipeds and third maxillipeds to shovel sediment away from themselves to form a shallow furrow. The anterior end is then raised back at right angles to the abdomen. These shrimps are able to do so by a unique adaptation: the tergites (dorsal part of the segmental exoskeleton) of the first and second abdominal segments are shaped so that the first can slide back over the second, allowing the right angle orientation of the abdomen to the cephalothorax. The shrimp then forcefully repeatedly plunges the anterior end of the body forward into the sediment several times while still digging with the second pereopods, forming a horizontal tunnel into which the rest of the body sinks At the end of all this activity, the shrimp is about 1–2 cm under the muddy sediment with the antennular respiratory siphon vertically upright above the sediment surface (Fig. 3.5b-1). Unlike the aristeids and benthesicymids, members of this family may have a well-developed prosartema, the eye-cleaning adaptation also found in the Penaeidae.

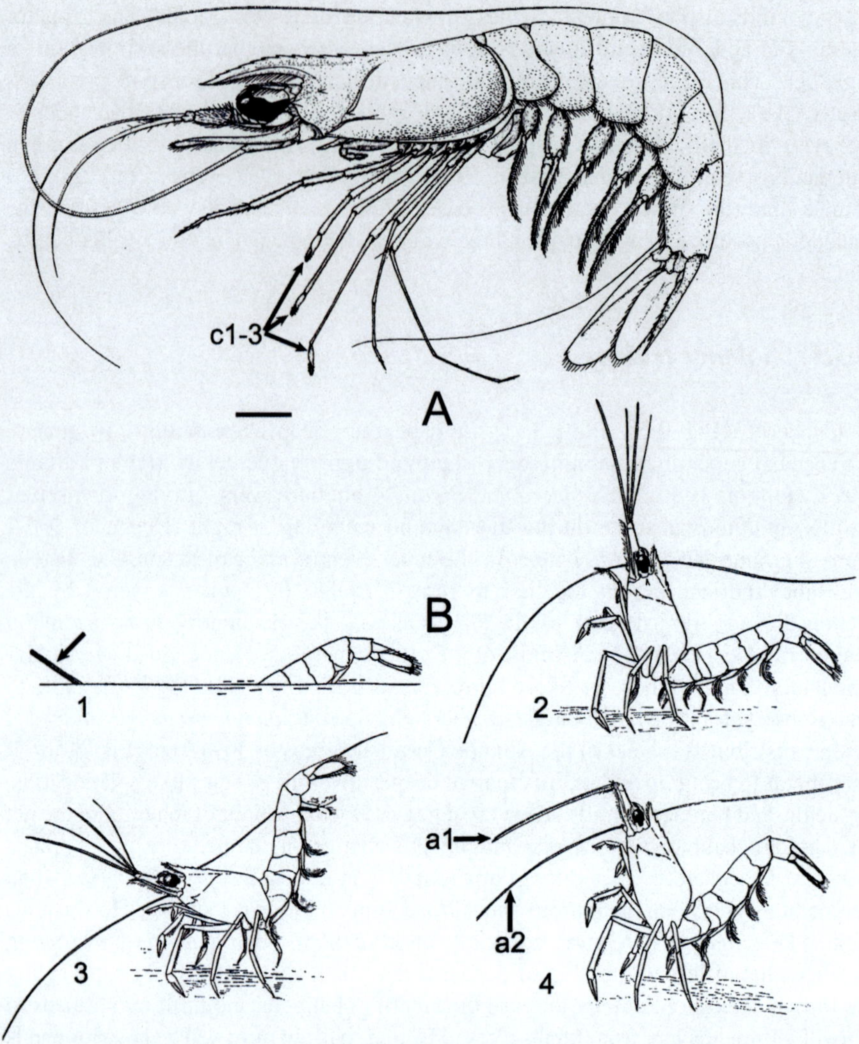

Fig. 3.5 Solenoceridae. (**a**) *Pleoticus robustus*. (**b**) *Solenocera membranacea* behaviors: (1) burrowing behavior, cepthalothorax buried with abdomen soon to follow, antennular respiratory siphon (arrow) protruding above the sediment; (2) food search; (3) "helicopter" food search, with shrimp rotating fifth pleopods, abdomen nearly at right angles to cephalothorax; (4) cataleptic posture, body stiff and unresponsive to stimuli. Scale bar in (**a**) represents 10 mm. a1, antennular flagellum; a2, antennal flagellum; c1–3, chelipeds 1–3. From (**a**) Pérez Farfante (1977) and (**b**) Heegaard (1967) courtesy of Crustaceana (Brill)

As adults, solenocerids range in size from 5 to 20 cm in total length. Some of the larger-sized species, such *Solenocera agassizii*, *Haliporoides diomedeae, H. triarthrus,* and especially *Pleoticus muelleri,* are important to commercial fisheries. Solenocerids typically show a coloration ranging from orange to deep red, with the latter color more highly developed in deeper-living species. Species of the genera *Hadropenaeus, Hymenopenaeus, Mesopenaeus,* and *Solenocera* may have photophores (Noël and Chassard-Bouchaud 1994; Pérez Farfante 1977), probable camouflage adaptations. Diet-content studies (e.g., Cartes 1995; Dineshbabu and Manissery 2009) indicate that they are active hunters of polychaetes and other invertebrates, detecting prey with sweeps of their long antennular and antennal flagella (Fig. 3.5b-2). Heegaard observed that, when hunting, *S. membranacea* may also assume an upright tail posture, with the last pair of pleopods beating in rotary fashion (Fig. 3.5b-3). The shrimp just barely touches the bottom, skimming over it helicopter-fashion while sweeping over it with the very long antennal flagella, which can be up to three times the length of body.

This species shows marked nocturnal activity patterns, burying at sunrise and emerging at sunset. Their nocturnal activity was first shown in laboratory observations by Heegaard (1967) and then confirmed by trawling studies by Aguzzi et al. (2006). Populations living at shelf depths were abundant in night trawls but absent from day-time trawls. At slope depths, the pattern was not as sharp, with some activity throughout the day but increasing considerably from late afternoon until early morning. This result is perhaps not surprising as light attenuates more rapidly in greater depths as sunset approaches.

An interesting defensive behavior in *Solenocera membranacea* described by Heegaard is catalepsis, in which the shrimp assumes a stiff rigid stance (Fig. 3.5b-4)) when alarmed; if touched in this position, the shrimp will fall on its side motionless. Such cataleptic behavior has evolved independently in a number of benthic shrimps, in the Sicyoniidae and a number of caridean species.

Some aspects of the reproductive biology of solenocerids can be deduced by observations on external and internal sexual biology. Solenocerid females have open thelyca, and males have complex open petasmata (Chap. 6). Intricately structured spermatophores are often observed attached to the open theylca of females in collected specimens (Burkenroad 1936; Pérez Farfante 1977). Observations on mating behavior have not yet been made, which is not surprising given the difficulty of maintaining deep-water shrimps alive. However, similar to most penaeids, males are smaller than females, indicating a pure-search (promiscuous) mating system (Chap. 8). As in tropical and subtropical but deep-water species, their reproductive and spawning patterns ranging from extended seasonal (e.g., June-December, Ohtomi and Irieda 1997) to continuous with seasonal peaks (e.g., Anderson and Lindner 1971; Castilho et al. 2008). Growth is slower and life spans longer (2–3 years) than most shallow-water penaeids and reaching to at least 5 years in *Pleoticus (Hymenopenaeus) robustus* (Anderson and Lindner 1971).

3.4 Superfamily Sergestoidea

This is a pelagic group of relatively small-sized, laterally compressed dendrobranchiates with a reduced morphology compared to that of penaeoids. The number of gills varies from none to 7–9 compared to the much higher numbers found in penaeoids (e.g., Table 2.1a–c, Chap. 2). The 4th and 5th pereopods are usually reduced in size, number of articles, or absent. The chelae of pereopods 1–3 are small, almost microscopic, or incomplete. These characteristics are favorable for a planktonic/pelagic existence, e.g., a small, somewhat flimsy body and a reduced morphology compared to penaeoids.

This group is very important in the pelagic habitat in both neritic and oceanic waters, often reaching large abundances. They are an important link from lower to higher trophic levels of pelagic food webs—predators of smaller pelagic organism and prey of fishes, squids, seabirds, and great whales. A few species, especially *Acetes japonicus*, are a major component of the commercial shrimp-fishing industry (Chap. 11).

3.4.1 Luciferidae

The species in this group (2 genera, 7 species) are extreme, bizarre examples of a reduced (or perhaps rudimentary) paedomorphic sergestoid morphology (Burkenroad 1983). They are small, often minute, only several mm in size; extremely compressed laterally with an elongate anterior cephalothorax; a thin, transparent cuticle; pereopods 1–2 lacking chelae, the third with a small, imperfect chela, and the 4th and 5th pereopods completely lacking. Gills are absent, and females incubate embryos briefly on the third pereopods. This minimal morphology is an extreme set of adaptations for planktonic existence (Pérez and Kensley 1997). With these features, and the tubular elongate eyestalks with outsized eyes of the type genus, *Lucifer* (Fig. 3.6a), the generic name and family name do appear appropriate. However, Thompson's (1829) etymology of the type genus, *Lucifer* ("light-bearer"), is not only based on the mythological Devil but also on its still unexplained luminescence, as Vereshchaka et al. (2016) have pointed out. But given the usual association of "*Lucifer*" with the Devil, these authors, when proposing a second genus for *Lucifer* (*s.l.*) species with very short eyestalks, deemed it appropriate to use a synonym, *Belzebub* (Plate 3.1g).

Luciferids are all epipelagic and planktonic, with only very weak swimming power. Both *Lucifer* species are oceanic, but *Belzebub* species are neritic in coastal waters and bays. They may go near the bottom during the day and return near the surface at night (Woodmansee 1966). Lee et al. (1992) showed that *Belzebub faxoni* has a very short life span (30–40 days), and females can produce 3–5 broods during this time. Luciferids are carnivorous, feeding on small planktonic organisms captured with their third pair of pereopods (Vereshchaka et al. 2016). During spawning,

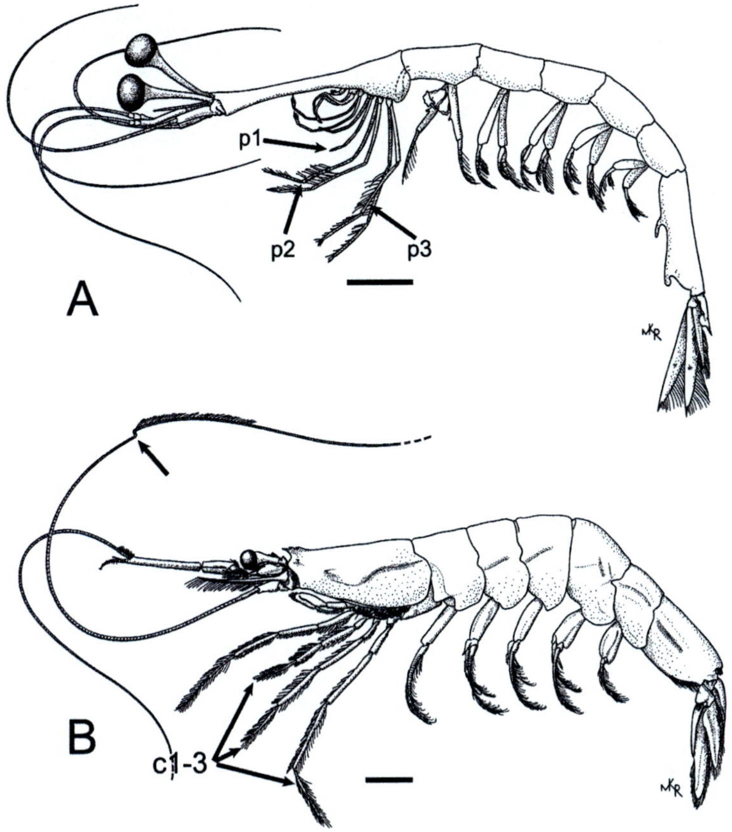

Fig. 3.6 Sergestoidea. (**a**) *Lucifer typus* (Luciferidae). (**b**) *Acetes indicus* (Sergestidae). Note kink (arrow) in antennal flagellum, dividing it into proximally naked basal and distal setose parts. Scale bars represent 1.0 mm in (**a**) and 2 mm in (**b**). c1–3, minutely chelate pereopods 1–3; p1–p3, nonchelate pereopods 1–3. From Pérez and Kensley (1997), ©Publications Scientifiques du Muséum national d'Histoire naturelle, Paris

fertilized eggs are attached to the third pair of pereopods for development prior to hatching as nauplii. Luciferids are the only dendrobranchiates to brood their embryos but note that they do not do so under the abdomen attached to the pleopods, as do pleocyemate shrimps.

Burkenroad (1934) made important observations on the genital morphology of luciferids, noting that females have twin seminal receptacles invaginated, as in all sergestoids, into the posterior part of 6th thoracic sternite (Chap. 6) just mesial to the coxae of the third pereopods. Although the male testes empty via an ejaculatory duct and gonopore on each side of the body, the large gourd-shaped spermatophore is only developed on one side of the body at a time. Burkenroad proposed that a male can only pass one spermatophore at a mating, so that the female would have to mate with at least two different males to fill both her seminal receptacles. This multiple mating per brood is supported by observations in Lee et al. (1992), who reported that females had multiple copulations between broods. The males of luciferids (as in other sergestoids) have a very complicated petasma (Chap. 6). As in the penaeoids, the actual manner in which it is utilized during mating and insemination is still largely unknown.

3.4.2 *Sergestidae*

This is the major group of sergestoids, with 19 genera and 97 species. Although a somewhat stripped-down version of a dendrobranchiate compared to penaeoids, they are very definitely shrimp-like in general form (Fig. 3.6b, Plate 3.2a, b). Like many pelagic crustaceans, their integument is thin or membranous. Pereopods 4–5 are reduced to some extent, and absent in *Acetes*. Unlike luciferids, sergestids do have gills, although very few compared to penaeoids (Table 2.1a). Sergestids are relatively small in size (~3–6 cm total length) but much more robust than luciferids. The taxonomy of the group has fluctuated, with the number of genera generally increasing as previous genera are subdivided or, in some cases, synonymized with others. Pérez and Kensley (1997) recognized 6 genera, which were then increased to 11 by Judkins and Kensley (2008). Alexander Vereshchaka and colleagues have made important contributions about the group in recent years based on a number of morphological and phylogenetic studies (Vereshchaka 2017 and others cited therein). Currently, WoRMS (2022) recognizes 19 genera. Most of the newer genera are derived from the two major genera treated by Pérez and Kensley (1997), *Sergia* and *Sergestes*. For simplicity in this chapter, I will use *Sergia* (*s.l.*) and *Sergestes* (*s. l.*) to refer to sergestid species listed by Pérez Farfante and Kensley. When referring to a particular species, I will include the generic name proposed by Vereshchaka (2017) in parentheses if the original generic name is no longer used by those authors.

Sergestids are basically a marine pelagic group found in tropical, subtropical and temperate latitudes, although *Acetes paraguayensis* is a freshwater species, occurring in rivers of tropical South America. Two genera (*Acetes*, *Sicyonella*) are neritic while other genera are oceanic in distribution (Vereshchaka 2017). Most *Acetes* species inhabit near-shore coastal waters, estuaries and river mouths in large swarms. *A. japonicus* is a commercially important species (Chap. 11) although relatively small in size compared to fished penaeoids (Xiao and Greenwood 1993; FIGIS 2021). Their vast schools in shallow water make them easy to fish in large quantities with simple nets. They are primarily used to make the very popular shrimp paste utilized in Asian cooking. *Sergestes* (*s.l.*) species are oceanic and mesopelagic, as are most *Sergia* (*s.l.*) species. Some of the latter are benthopelagic, found near the sea floor of seamounts and continental slopes (Vereshchaka 2017).

Sergestids are important trophic links in epipelagic and mesopelagic habitats, often occurring in large schools, serving as food for fishes, squids, and great whales. On the other hand, sergestids are predators on smaller zooplankters such as copepods, euphausiids, other smaller crustaceans and invertebrate larvae, as well as

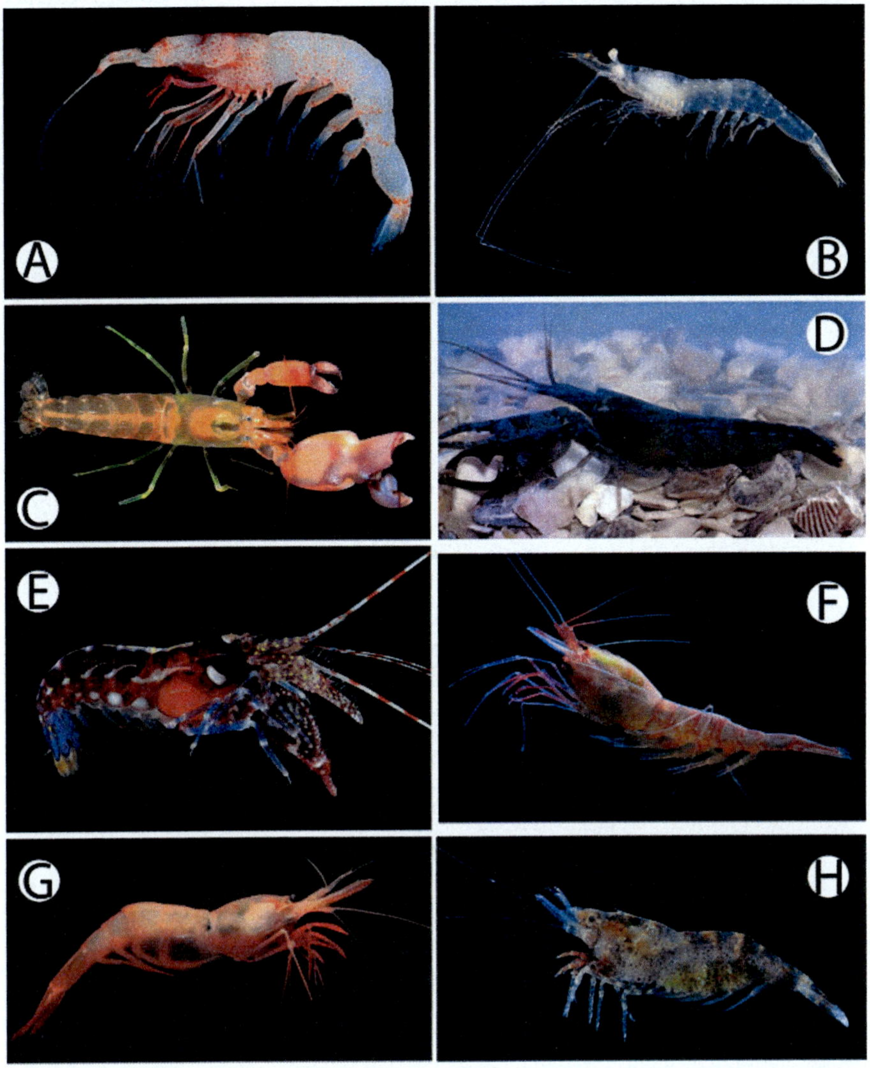

Plate 3.2 Sergestidae to Hippolytidae. (**a**) *Sergia (Challengerosergia) talismani* (Sergestidae). (**b**) *Sicyonella maldivensis (Sergestidae).* (**c**) *Alpheus amarillo* (Alpheidae). (**d**) *Betaeus longidactylus* (Alpheidae) (**e**) unidentified alpheid, Panama. (**f**) *Barbouria cubensis* (Barbouriidae) (**g**) *Bythocaris payeri* (Bythocarididae). (**h**) *Latreutes anoplonyx* (Hippolytidae). Photo acknowledgments: Courtesy of (**a**, **b**) T.-Y. Chan, (**c**, **e**, **h**) Arthur Anker, (**d**) R.T. Bauer, (**f**) Robert Ditter, (**g**) A.L. Vereshchaka

feeding on decomposing matter in the water column. Vereshchaka et al. (2014) note that the morphology of some sergestids appears adapted for an "offensive" (predatory) lifestyle, with elongated third maxillipeds and other adaptations for sneaking up on and grasping prey. As might be expected for predators, these species have a relatively low abundance relative to other sergestids. These latter, important prey for mesopelagic predators, have a "protective" (defensive) morphology, with a more streamlined body and larger stronger uropod paddles for a strong backward escape response.

Mesopelagic sergestids undergo diel vertical migrations from deeper waters where they dwell during the day and then migrate closer to the surface at night, protected from visually hunting predators by the cover of darkness. They make the upward migration to feed on the more abundant food supplies there. They are protected during the day in the depths of the mesopelagic zones by their coloration and countershading provided by light-producing photophores or organs of Pesta (Chap. 5). Shrimps in the upper mesopelagic are often termed the "half-red" shrimps because of their pale reddish translucent coloration (Plate 3.2a), while those of the lower mesopelagic and bathypelagic (~600–4000 m) zones are much more intensely red, part of the "all-red" shrimp community. At night, bathypelagic sergestids may make only limited migrations within the bathypelagic and into the lower mesopelagic, and thus they do not need countershading and so lack light-producing organs (e.g., Vereshchaka et al. 2014). Epipelagic sergestids, living in the lighted, clear-water photic zone, find camouflage in transparency (Plate 3.2b), as do most epipelagic invertebrates and larvae.

Life span varies from 3 to 10 months in the neritic *Acetes japonicus* in the summer and winter generations, respectively (Xiao and Greenwood 1993). In the larger mesopelagic *Sergestes similis*, living in the colder waters of the North Pacific, the life span is 1–2 years (Omori 1979). Sergestids have a similar reproductive system as do luciferids, i.e., females with twin seminal receptacles invaginated into the posterior 6th thoracic sternite (Burkenroad 1934). Male spermatophores are similar to those of luciferids; however, unlike them, males can apparently deposit two spermatophores during a single mating, as indicated by observations made by Burkenroad and Xiao and Greenwood (1993).

3.5 Infraorder Caridea

As noted previously, the carideans are by far the most diverse in terms of numbers of species, ecological niches, habitats occupied, and overall general morphology than the dendrobranchiates. The latter are confined mainly to soft bottom or pelagic habitats in subtropical and tropical latitudes, while the caridean species can be found at nearly all latitudes in marine waters and up into temperate latitudes in many rivers, streams, and lakes. They occur in most pelagic zones and benthic substrates in both neritic and oceanic marine waters. This ecological and morphological diversity is mirrored by their taxonomic diversity, which is several times that of the Dendrobranchiata and Stenopodidea. As noted below and in Chap. 12, the composition of superfamilies and families has been the subject of much controversy and new methods of phylogenetic analysis (e.g., Christofferson 1990; Martin and Davis 2001; Li et al. 2011).

3.6 Superfamily Alpheoidea

This superfamily is one of the five "polycarpidean" superfamilies (Alpheoidea, Pandaloidea, Processoidea, Physetocaridoidea, Crangonoidea) in which the carpus of the second pereopod, a cheliped in most carideans, is subdivided into a few to many articles. This feature gives the limb extra flexibility for grooming (next chapter) and/or probing over the substratum. The first pair of chelae in alpheoids is somewhat much larger and more robust than the second. Throughout the years, the makeup of the Alpheoidea (Table 3.1) has varied a lot, particularly within the composition of the family Hippolytidae (*s.l.*) which is now subdivided into several families (Christofferson 1990; Martin and Davis 2001; De Grave et al. 2014; WoRMS 2022).

3.6.1 Alpheidae

This family is the second most speciose of the carideans, with currently 49 genera and 763 species, 324 of these are in the namesake of the family, the genus *Alpheus*. In alpheids, the typical caridoid body form is modified for benthic life in burrows and shelters, or as symbionts with other marine invertebrates. Their body and legs are often stout and heavy, antennal flagella are relatively short, the rostrum is reduced or absent, and, in many genera, the eyes are partially or completely covered by the anterior edge of the cephalothorax (Fig. 3.7, Plate 3.2c–e). Alpheid first chelae are large, often more so in males. In two genera, *Alpheus and Synalpheus*, one of the first pair is modified into a massive snapping claw while the other remains a generalized cutting or gripping chela (Fig. 3.7a, Plate 3.2c, e). The form and mechanics of these formidable weapons, used primarily in defensive and agonistic behavior with other snapping shrimps, were discussed in the previous chapter. In some genera, the claw is inverted with the dactyl below (e.g., *Betaeus* species, Fig. 3.7b, Plate 3.2d). As in other alpheoid families, the second pair of chelipeds are slender with a subdivided carpus. Although this type of limb is used for gill cleaning in some other families, most alpheids have well-developed epipod-setobranch complexes for this purpose (Chap. 4).

Alpheids are completely benthic and mostly inhabit shallow-water coastal marine habitats. However, some *Potamalpheops* species live in fresh- to brackish water in mangrove swamps (Powell 1979a; Bruce 1991; Yeo and Ng 1997; Anker 2003; De Grave et al. 2008). There are a few truly freshwater species, such as *Alpheus cyanoteles* and *Potamalpheops amnicus* from Southeast Asia (Yeo and Ng 1996). Many species are live symbiotically with marine invertebrates such as sponges, anemones, corals, sea urchins, comatulids (feather stars), echiuroids, and other shrimps (e.g., Bruce 1976; Boltana and Thiel 2001). Hooded" or "visored" shrimps (*Betaeus* species), the only alpheid genus with most of its species living in cool to cold-water latitudes, associates with various invertebrates such as sea urchins, abalones, and even other crustaceans (Hart 1964; Anker and Baeza 2012). Some

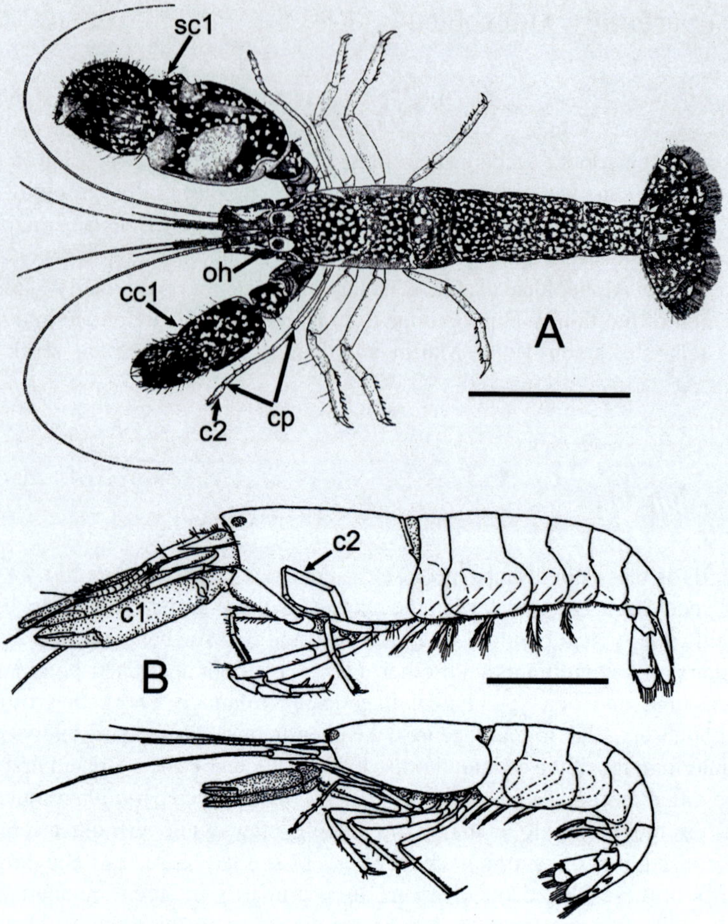

Fig. 3.7 Alpheidae. (**a**) *Alpheus nuttingi*, showing complex color pattern and snapping chela. (**b**) *Betaeus levifrons*. c1, first cheliped; c2, second cheliped; cc1, cutting claw (minor cheliped 1); cp, carpus of pereopod 2; oh, orbital hood covering eye; sc1, snapping claw (major cheliped 1). Scale bar represents 10 mm in (**a**) and mm in (**b**). From (**a**) Anker et al. (2007) and Marin (2010), both courtesy of Zootaxa

alpheid species form associations with gobiid fishes in which the two species live in the same burrow which the shrimp digs and maintains. However, the shrimp is alerted to danger by its goby partner (e.g., Karplus 1987; Lyons 2013). These symbiotic relationships will be discussed in greater depth in Chap. 10.

Alpheid species often occur in male-female pairs, which does not guarantee sexual fidelity, (e.g., Knowlton 1980; Mathews 2007). Others do not occur in pairs, but dominant males may temporarily guard and defend a receptive female in order to be her only mating partner (e.g., *Arete dorsalis* (Nakashima 1987)). In pair-living species, males and females are similar in body size, as are their cheliped weapons. In species in which males only guard females when the female is approaching receptivity, males are often larger in body size and relative size of the major chelipeds. In *Synalpheus*, some species occur in pairs when symbiotic with invertebrates such as feather stars. However, as Emmett Duffy (1996) discovered and later studied extensively with colleagues (e.g., Duffy and Macdonald 2010), numerous *Synalpheus* species live in sponges as eusocial colonies, with a single reproducing "queen." She produces all the offspring ("helpers") which defend the sponge and its shrimp colony from intruders (Chap. 10). Although most alpheids studied have separate sexes, some *Arete* species are protandric, i.e., change sex from male to female (Suzuki 1970; Nakashima 1987; Gherardi and Calloni 1993). There is growing evidence that the pair-living *Salmoneus* species are simultaneous hermaphrodites (Anker 2011; Oliveira et al. 2015) (Chap. 7).

The snapping claw of *Alpheus* and *Synalpheus* is a key adaptation which is largely responsible for the great ecological success and diversity of these genera (Anker et al. 2006). Together, these two genera account for 64 % of the species in the family. Various adaptations associated with snapping claws can be found in more rudimentary conditions in other alpheid genera. For example, the eyes of most alpheids are partially or completely shielded by the anterior margin of carapace. Anker et al. (2006) hypothesized that this adaptation, elaborated in some genera as orbital hoods (Fig. 3.7a), shields the eyes from the forces generated by a snap of the claw. Thus, the eye protection variously developed in other genera was a preadaptation in an ancestor of the *Alpheus* + *Synalpheus* clade which allowed, in part, the evolution of the snapping claw. However, in my opinion, orbital hoods are an adaptation to protect the eyes from mechanical injuries as most alpheids occur within restricted spaces, e.g., in burrows, under stones (Fig. 3.7b). Other such possible preadaptations for snapping which occur in other alpheid genera included adhesive discs (for enhanced claw cocking) and tooth-cavity systems on opposing claw fingers. However, in these other genera, such features did not lead to the evolution of a snapping claw.

3.6.2 Hippolytidae (s.l.) Families

These 6 alpheoid families, with a total of 37 genera and 382 species, have been at one time or another variously grouped under the family Hippolytidae (*s.l.*) (Chace 1992; Christofferson 1990; Martin and Davis 2001; De Grave et al. 2014). Following the latter authors' study of molecular phylogeny and morphological characters, WoRMS (2022) now lists all these families as separate (Barbouriidae, Bythocarididae, Hippolytidae, Lysmatidae, Merguiidae, Ogyrididae, Thoridae). Again, this might be viewed as a result of the perpetual differences of opinion between the taxonomic "splitters" and "lumpers." However, as long as decisions are made on sound evolutionary rationales and phylogenetic analyses, such changes are inevitable as new information or interpretations become available. De Grave et al. (2014) provides a good review of the different opinions of hippolytid (*s.l.*)

systematics, and point out, as did Chace (1992), that with further description of characters in more genera, more changes may be forthcoming in the future. The detailed morphological characters which separate them may seem esoteric and meaningful only to professional taxonomists, so below I will emphasize other aspects of their biology of greater interest to the non-specialist.

A number of generalizations about the hippolytid (*s.l.*) families can be made before giving details on individual families. They are a completely marine group, with most species found in a variety of benthic habitats on the continental shelf, although some species are found in the deep sea. Hippolytid (*s.l.*) species occur from tropical to high-latitude seas. Generally, hippolytid (*s.l.*) shrimps have a typical epibenthic cardioid morphology for perching on or in a variety of substrates and habitats, e.g., sand, mud or rocky bottom, seagrass beds, and coral reefs. They do not burrow directly in the sediment nor live in excavated burrows as do many alpheids. There is certainly a lot of variation in overall shape, varying from short and robust to elongated and slender, with most intermediate between these extremes. All have the slender alpheoid all-purpose second cheliped with subdivided flexible carpus for grooming and other functions, and the first chelipeds are more robust than the second, although generally not so much so as in the Alpheidae. Males are generally smaller than females and their first chelipeds are similar in relative shape and size in most species, but certainly, there are genera in which the reverse is true. In step with this generalization in sexual dimorphism, the mating system of most species studied is pure-search polygamy (promiscuity) (Chap. 8). However, temporary mate guarding and social monogamy are found in some species. Although many, perhaps most, species have separate sexes, a number of them are protandric sex changers or simultaneous hermaphrodites (Bauer 2000; Chiba 2007).

3.6.3 Barbouriidae

This is a group of 3 genera and 10 species that inhabit marine caves and anchialine habitats. The latter are defined as caves and land-locked bodies of water with subterranean connections to the sea, and thus influenced by the rise and fall of the tides (Holthuis 1973). The three genera, *Barbouria, Calliasmata,* and *Parhippolyte* have short reduced rostra, eyes varying from normal sized to much reduced in size and pigmentation (e.g., Fig. 3.8, *Calliasmata rimoli*; Chace 1975; *Barbouria cubensis*, Plate 3.2f). The carpus of the second pereopod is highly subdivided with ~20–30 articles (Fig. 3.8), and multiarticulation may extend to the merus and ischium as well. In some barbouriids, the long walking legs may have subdivided propodi and carpi which may give them considerable flexibility in perching onto subterranean walls and substrates. Coloration of these shrimps, like many subterranean species, tends to vary or mix transparency with light to deep reddish coloration (e.g., Holthuis 1973; Mejía et al. 2008). Although the biology of most species is unknown, the sexual system of at least one species (*Parhippolyte misticia*) has been shown to be a protandric simultaneous hermaphrodite (PSH), with individuals first

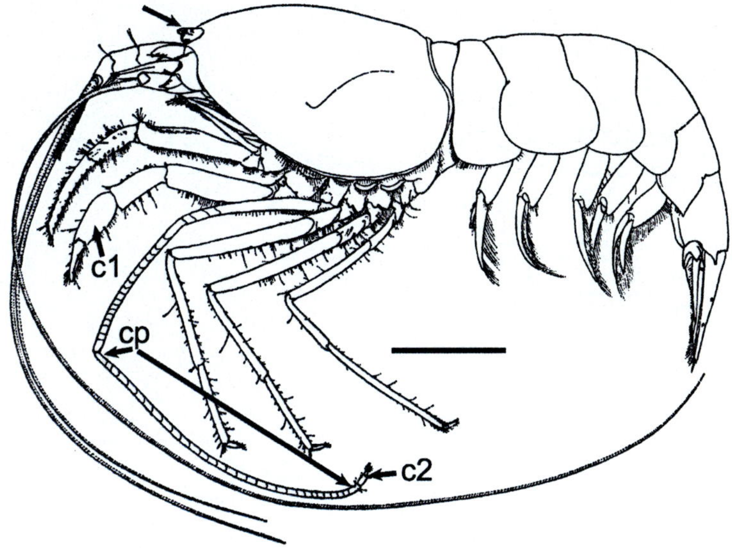

Fig. 3.8 Barbouriidae. *Calliasmata rimoli*. c1, robust first cheliped; c2, slender second cheliped; cp, multiarticulate carpus of the second pereopod. Note that the more proximal merus and ischial segments are multiarticulate as well. Scale bar represents 5 mm. From Chace (1975)

maturing sexually as males (Onaga et al. 2012). When older and larger, individuals of PSH species mature and reproduce as a female while still retaining male sexual function (Bauer 2000). Another genus of the family, *Calliasmata*, may also have a PSH sexual system (Baeza 2013; Chap. 7).

3.6.4 Bythocarididae

Four hippolytid (*s.l.*) genera, *Bythocaris, Bythocarides, Bathyhippolyte,* and *Cryptocheles*, with a total of 20 species, currently compose this family of deep-sea (bathyal) shrimps. The principal genus is *Bythocaris*, with 17 species, and much of the knowledge of this genus is due to the Russian scientist V. Sokolov (2000), who also described the single species of *Bythocarides*. Common features of these genera include a short, unarmed rostrum, with a second cheliped of 8–12 articles (Wicksten 2010). Their coloration is mostly red (Plate 3.2g), as is common with shrimps living at bathyal depths, although those from Arctic waters are somewhat pale (Alexander Vereshchaka, personal observation). One tendency in the group is reduction of the eyes. Some species of *Bythocaris* have eyes of normal size with pigmented corneas, while others have small eyes without pigmentation or even eyestalks or cornea (Sokolov 2000) (Fig. 3.9). The eyes of *Bathyhippolyte* and *Cryptocheles* are also highly reduced (Hayashi and Miyake 1970). Shrimps with such eyes must be nearly or completely blind, not much of a handicap in the darkness of bathyal depths, and

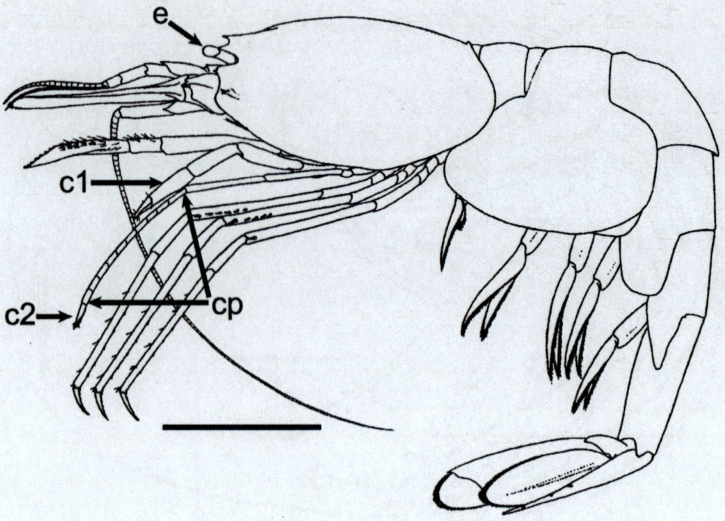

Fig. 3.9 Bythocarididae. *Bythocaris irene* (note reduced eye without corneal pigmentation). e, eye; c1, robust first chela; c2, small second chela; cp, multiarticulate carpus of the second pereopod. Scale bar represents 10 mm. From Sokolov (2000)

with the benefit of saving the metabolic energy needed to develop and maintain eyes, as in many cave animals. (Rétaux and Casane 2013). Female bythocarids spawn and incubate very large eggs (Hayashi and Miyake 1970; De Grave et al. 2014), indicating that they have greatly abbreviated or direct development. As in many hippolytid (*s.l.*) species, males are smaller than females. The sexual system of several *Bythocaris* species was investigated by Sokolov and Spiridonov (2006) by examination of the male and female sexual characters. They found no indications of protandric sex change, so that we know that at least these species have separate sexes. Much remains to be discovered about the biology of this these shrimps.

3.6.5 Hippolytidae (s.s.)

This family (16 genera, 106 species) has in recent years been sometimes called a "wastebasket" of hippolytid (*s.l.*) genera because of its morphological variability (Plates 3.2h, 3.3a–c) and probable polyphyletic composition (Martin and Davis 2001; De Grave et al. 2014). The chelipeds are of the typical hippolytid (*s.l.*) morphology, with the second pereopod carpus subdivided into from just 2–3 to usually several but up to 90 articles (De Grave et al. 2014), with more robust chelae on the first than on the second pereopods. They live within or perched on both shallow-water and deep neritic epibenthic assemblages of plants (seagrasses, algae) or sessile invertebrates such as sponges, hexacorals, octocorals, and bryozoans. Most range from tropical to temperate seas but two genera, *Chorismus* and *Nauticaris*,

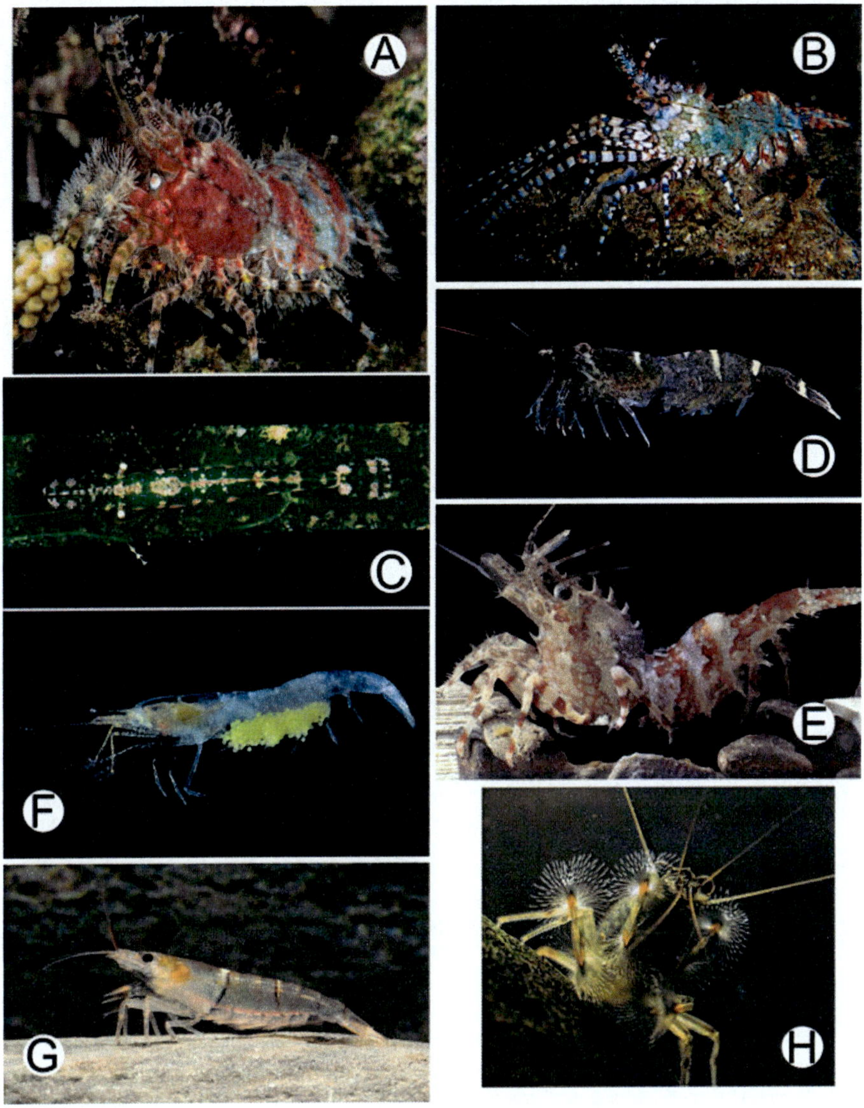

Plate 3.3 Hippolytidae to Atyidae. (**a**) *Saron marmoratus* female (Hippolytidae). (**b**) *S. marmoratus* male. (**c**) *Hippolyte inermis* on *Posidonia oceanica* seagrass blade. (**d**) *Merguia rhizophorae* (Merguiidae). (**e**) *Lebbeus groenlandicus* (Thoridae). (**f**) *Ogyrides orientalis* (Ogyrididae). (**g**) *Australatya obscura* (Atyidae). (**h**) *Atya lanipes* (Atyidae), frontal view of filter feeding with open cheliped setal fans. Photo acknowledgments: Courtesy of (**a**) Kevin Bourdon, Creative Commons CC BY 2.0; (**b**) Sébastien Vasquez, Wikipedia Commons; (**c**) Zdenek Ďuris; (**d, f**), Arthur Anker; (**e**) R.T. Bauer; (**g**) Werner Klotz; (**h**) Jeremy Monroe, Freshwaters Illustrated

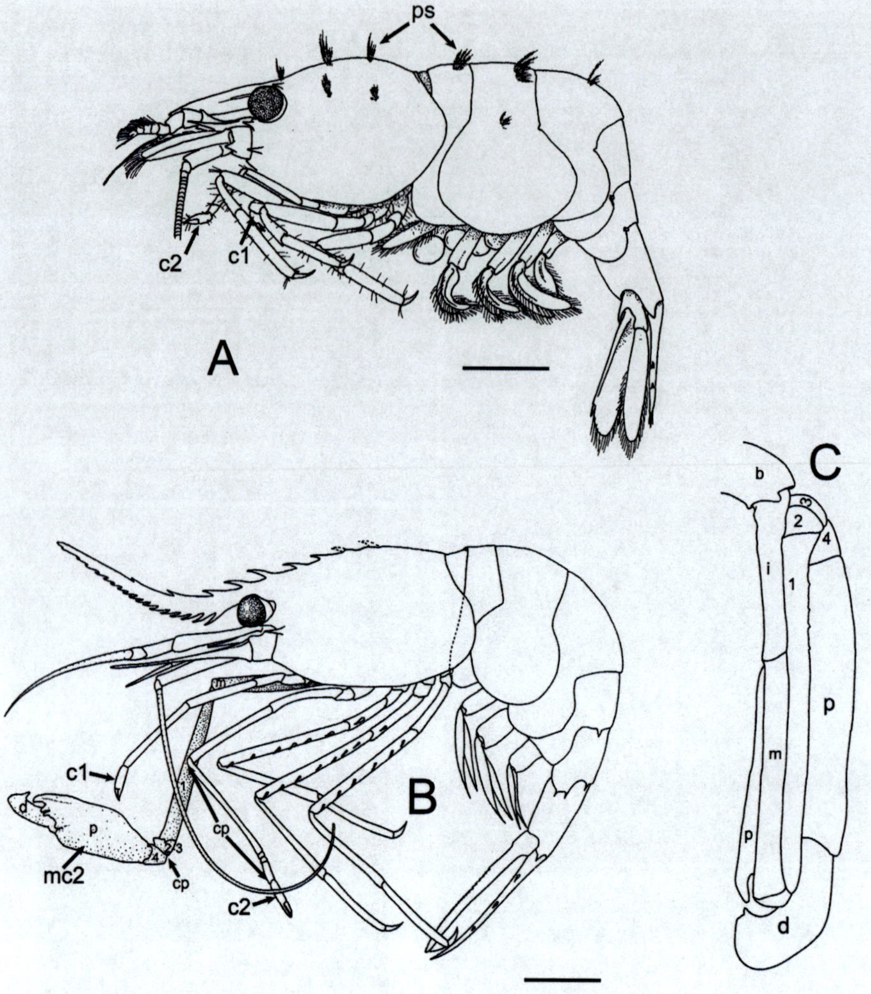

Fig. 3.10 Hippolytidae. (**a**) *Hippolyte commensalis,* carpus of pereopod 2 with 3 articles. (**b**) *Leontocaris yarramundi*, carpus of pereopod 2 with 4 articles. (**c**) Major cheliped 2 of *L. alexander* in folded position, with merus and ischium lying parallel to the long first proximal article of the carpus, which is locked into a lateral groove on the propodus. b, basis; c1, first chela; c2, second chela in (**a**), minor second chela in (**b**); cp, carpal segments of pereopod 2; d, dactyl; i, ischium; m, merus; p, propodus; ps, tufts of pilose setae; 1, 2, 3, 4, proximal to distal carpal articles of major cheliped 2 in (**b**, **c**). Scale bar represents 1.0 mm in (**a**) and 4 mm in (**b**). From (**a**) Marin et al. (2011), courtesy of Zootaxa, (**b**) Taylor and Poore (1998), and (**c**) Poore (2009), courtesy of Crustaceana (Brill)

occur in sub-Antarctic and Antarctic waters with epibenthic communities dominated by sessile organisms such as sponges, red algae, and bryozoans, (Mascetti et al. 1997; Pakhomov et al. 2000; Haley et al. 2017). Some *Hippolyte* species have elongated bodies (Plate 3.3c) for lying along blades of seagrasses or algae (d'Udekem d'Acoz 1996), as does *Tozeuma*, which also inhabits gorgonians and other branching invertebrates (Voss 1956). Both of these genera are noted for their ability to match the coloration of the seagrass, alga, or other substrates upon which they perch (Chap. 5). Species of *Hippolyte* (Fig. 3.10a) and *Latreutes* (Plate 3.2h) are ubiquitous and often abundant around the world in seagrass meadows (e.g., *Zostera, Thalassia, Posidonia*) and on a diverse assortment of algal species (Bauer 1985b; d'Udekem d'Acoz 1996; Jensen 2014). Like a number of epiphytic shrimps, some *Latreutes* and *Hippolyte* species (e.g, Fig. 3.10a) sport tufts of pilose (feathery) setae which resemble filamentous epiphytic algae like those growing on seagrass blades and other substrates. Along with the shrimp's coloration, the setal tufts augment the shrimp's camouflage. *Hippolyte coerulescens* is an important component of the floating *Sargassum* community, as is *Latreutes fucorum* (Hacker and Madin 1991), but the latter is equally at home in *Thalassia* seagrass meadows (Bauer 1985b).

An unusual genus currently in the Hippolytidae (*s.s.*) is *Leontocaris*, whose species inhabit deep-sea coralline communities of seamounts and continental slopes (Poore 2009; Cardoso and Fransen 2012). The second chelipeds are asymmetrical, with one second cheliped of the usual slender hippolytid pattern, while that of the other side is quite large and heavy. The shrimp prevents this major cheliped from dragging along the bottom by folding the proximal segments into a groove on the robust propodus of the chela (Fig. 3.10b, c). The major cheliped segments can unlock and unfold, allowing the large chela to thrust forward when needed. The propodus of the major chela has a small elongated or rounded area of membranous cuticle surrounded by similarly soft villi or papillae. Poore (2009) proposed that this is part of a hydraulic system for unlocking the zigzag-arranged cheliped segments during the extension of the limb. He also hypothesized that the enlarged chela, with sinister-looking dactylar finger and teeth, which is found in both males and females, is some sort of predatory weapon. The first pair of chelipeds, usually more robust than the second in hippolytids, are similar in size to the slender minor second cheliped. All of the slender chelipeds bear tufts of setae typical of grooming chelipeds.

Most species of this family follow the usual hippolytid (*s.l.*) pattern of males smaller than females, with similar relative size in cheliped weaponry. The usual mating system associated with this type of sexual dimorphism is pure-search polygamy, in which males and females have little interaction before or after mating. However, males of at least some species of *Saron* have hypertrophied, massive first chelipeds and elongated, lance-like third maxillipeds, similar to some species of rhynchocinetids (Plate 3.3a, b) (see below). Martin Thiel and colleagues have shown that the latter have a mating system in which males with similarly enlarged weapons fight for the privilege of mating with females (see Chap. 8). Most hippolytid (*s.s.*) species that have been studied have separate sexes, but male to female (protandrous) sex change has been demonstrated in *Chorismus antarcticus* (Mascetti et al. 1997), *Nauticaris marionis* (Pakhomov et al. 2000), and *Hippolyte inermis* (Zupo et al. 2008) (Chap. 7).

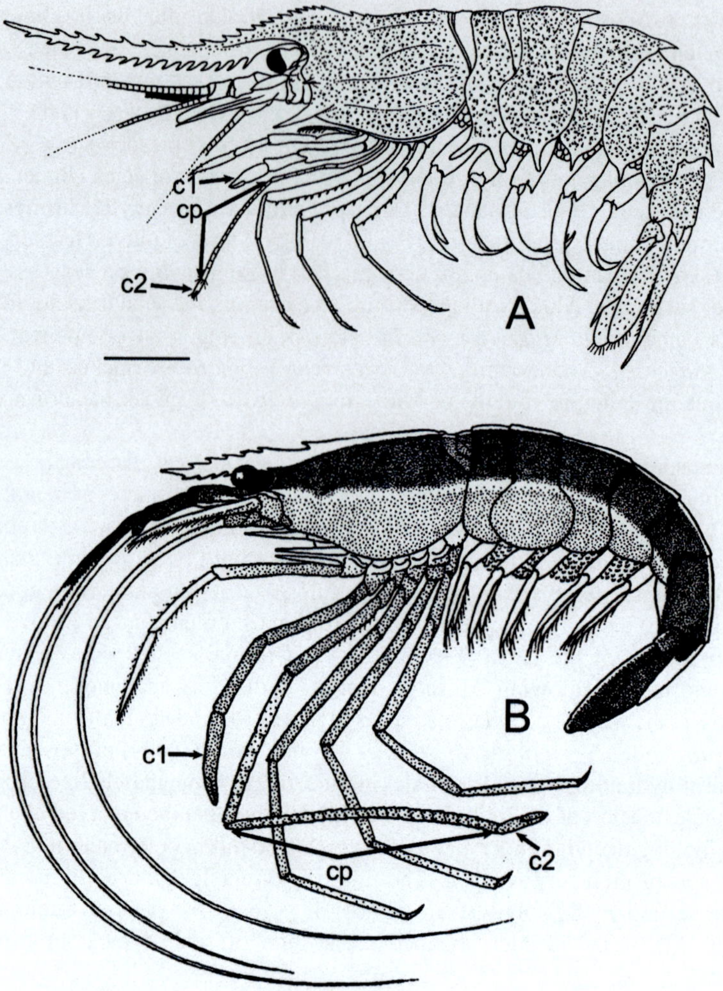

Fig. 3.11 Lysmatidae. (**a**) *Mimocaris heterocarpoides*. (**b**) *Lysmata grabhami*. Its striking color pattern is indicated by shading: no shading = bright white; dark shading = red; stipple = orange-yellow. c1, c2, chela 1, 2; cp, multiarticulate carpus of second pereopod. Scale bar in (**a**) represents 3 mm and 6 mm in (**b**). From (**a**) Xuan (2000) courtesy of Crustaceana (Brill), and (**b**) Limbaugh et al. (1961) courtesy Bulletin of Marine Science

3.6.6 Lysmatidae

This is a rather easy group to characterize morphologically, as the 5 genera and 55 species are similar in overall typical hippolytid (*s.l.*) form (Fig. 3.11; Plate 7.1e, f). They have a short to long spiny rostrum, and small first chelipeds that are more robust than the slender second ones, in which the carpus is subdivided into usually numerous articles, ranging from 12 to 50. Most species inhabit fully marine sublittoral habitats with some structure, such as rocky or coral reefs, epibenthic algal or sessile invertebrate communities, and seagrass meadows. Some exceptions are the monospecific *Mimocaris heterocarpoides,* residing in shallow estuaries with sand/mud bottoms in Southeast Asia (Xuan 2000), and the also the monospecific *Ligur ensirostris,* from bathyal depths in the Mediterranean and Atlantic Ocean (Chace 1997). *Exhippolysmata oplophoroides* from southeastern Brazil occurs over sandy/mud sublittoral habitats but is most abundant where there is some shelter, e.g., plant or other debris of terrestrial origin (Fransozo et al. 2005).

The majority of the 48 species of this family are in the genus *Lysmata,* which inhabit the typical lysmatid sublittoral habitats given above. This genus has garnered considerable attention in the last few decades with the discovery and exploration of its unique system of protandric simultaneous hermaphroditism (PSH) (Chap. 7). This was first experimentally confirmed in *L. amboinensis* (Fiedler 1998) and *L. wurdemanni* (Bauer and Holt 1998). Subsequently, all other *Lysmata* species investigated have shown PSH, such as *L. californica* (Bauer and Newman 2004) and several others studied by J. Antonio Baeza (2009 and papers cited therein). Some species of *Lysmata* are very attractively colored (Fig. 3.11b, Plate 7.1f) and show interesting behaviors in captivity, and therefore are popular with marine aquarists as one of the "ornamental shrimps" (Calado 2008). These factors undoubtedly focused attention on this group, stimulating numerous studies on its taxonomy and phylogeny (e.g., Rhyne and Lin 2006; Baeza 2013). Two species of *Exhippolysmata*, *E. ensiferus* and *E. oplophoroides,* also are definitely PSH species (Kagwade 1982; Laubenheimer and Rhyne 2008; Braga et al. 2009). The sole species of *Lysmatella, L. prima,* appears to be a simultaneous hermaphrodite but apparently skips the male phase, i.e., goes directly from juvenile to the adult hermaphrodite (Murray et al. 2012; A. Rhyne, personal observation,).

The sociobiology and natural history of *Lysmata* and *Exhippolysmata* have also been of considerable interest. In spite of having the same unusual sexual system, the social organization and ecology of these species varies widely. Bauer (2000) pointed out that some species live in hermaphroditic pairs, while others live in aggregations or "crowds," a term expanded by Baeza (2013) to include "groups" (several individuals) and "swarms" (very large aggregations). The pair-living species are brightly colored (Plate 7.1f), diurnal, and gain their living by cleaning parasites, dead tissue, and mucus of fishes (Limbaugh et al. 1961). The species living in aggregations are nocturnal, not so brilliantly colored (Plate 7.1e), and are not specialized fish cleaners (Bauer 2000; Baeza 2009). A question explored in Chap. 7 is how such a specialized sexual system could have evolved and been maintained in species with such different sociobiology and natural history.

3.6.7 Merguiidae

Although other shrimps are completely aquatic animals, the members of this monogeneric family have evolved a semiterrestrial existence in some of the tropical mangrove forests of the Atlantic and Indo-Pacific (Bruce 1993). The two species,

Merguia oligodon and *M. rhizophorae*, can be found crawling and leaping around the roots and limbs of mangroves trees at night. The latter species may also inhabit the burrows of other crustaceans, e.g., the mud shrimp *Upogebia* (Alves et al. 2021). Their unusual natural history and habitats were first brought to light by Lawrence Abele (1970). The shrimp is typically hippolytid in general morphology (Plate 3.3d). It is not noticeably adapted for terrestrial activity except for somewhat robust pereopods which support their bodies when walking. The stout third maxillipeds are used for crawling by anchoring their tips on a mangrove root and then contracting them, thus dragging the body forward. When startled or pursued, these shrimps can jump repeatedly up or in a forward direction ten times or more their 1–2 cm body length. This shrimp holds the abdomen arched above the substrate, leading Abele (1970) to propose that it uses the telson for purchase and the abdominal musculature to power the leaps.

The shrimp and its gills stay moist by being active at night, when humidity is highest, and during the day it hides in moist cracks, crevices, and crab or mud shrimp burrows. Although semiterrestrial, they are strong swimmers and can remain submerged indefinitely, demonstrating that they are still closely tied to the marine habitat (Gillikin et al. 2001). Baeza (2010) showed that these shrimps are protandric sex changers, maturing first as males and then as females.

3.6.8 Thoridae

This completely marine group, with 8 genera and 189 species, is particularly close to my heart, as much of my earlier work, including dissertation and postdoctoral research, was done on genera of this family. These are typical epibenthic shrimps with short to long rostra, first chelipeds more robust than the second, and the usual slender hippolytid (*s.l.*) second chelipeds with carpus subdivided into 6 or more usually 7 articles. Thorids often display beautiful camouflage coloration to mimic or blend into their background (Bauer 1981, 1982; Butler 1980; Jensen 2014) (see Chap. 5, Plate 5.3a–c). The bulk of species belongs to four related genera (*Eualus*, *Heptacarpus*, *Lebbeus*, and *Spirontocaris*) which were classified together for many years as the genus *Spirontocaris* (Holthuis 1947) (Plates 3.3e, 5.3a–d). The spirontocarids are highly diverse in temperate and higher latitudes on seas of continental shelves of the North Atlantic and especially the North Pacific. They occur in a number of shelf habitats such as sand/mud to rocky bottoms, seagrass, and algal epibenthic communities. The nicely illustrated books by Butler (1980) and more recently by Jensen (2014) are excellent sources on the taxonomy and biology on these and other West Pacific coast shrimps.

The only diverse genus of the family restricted to the tropics and other warm-water areas is *Thor* named, perhaps facetiously in these often tiny shrimps, for the Norse god of thunder and war *Thor*, suggested by the shield-like plate at the distal end of the antennular peduncle. Species of *Thor* are abundant in seagrass meadows and algal-covered substrates, and often perch on and associate with sessile

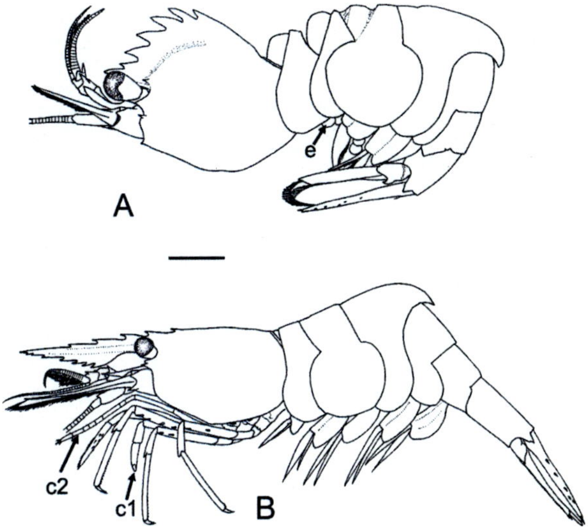

Fig. 3.12 Thoridae. (**a**) *Lebbeus compressus*, a "spirontocarid" with a short stocky body shape. Cephalothoracic appendages not figured. (**b**) *Heptacarpus camtschaticus*, another "spirontocarid" thorid with a somewhat elongate body shape. c1, c2, chela 1, 2, e, embryo. Scale bar represents 2 mm in (**a**) and 3.5 mm in (**b**). From (**a**) Komai et al. (2021), courtesy of Journal of Crustacean Biology and (**b**) Komai and Ivanov (2008) courtesy of Zootaxa

invertebrates or rubble. One definite *Thor* symbiont is the cosmopolitan *T. amboinensis*, found in association with anemones and corals around the world (Chace 1997; Baeza and Piantoni 2010).

An important morphological trend in the spirontocarids was first indicated by Butler (1980) and then studied in depth by Bauer (1984). Within each of the genera *Heptacarpus*, *Eualus*, and *Spirontocaris*, there is a morphocline from species with short, stout bodies to elongate species with other species intermediate between these two extremes (Fig. 2.3b in Chap. 2). The stocky species (e.g., Fig. 3.12a) have the highest number of thoracic epipods, and the more elongate species fewer or none at all (e.g., Fig. 3.12b). I used epipod number as an evolutionary marker to test the hypothesis that the short robust species are primitive (ancestral) and the elongate species more advanced (derived). The different morphotypes tend to be associated with different lifestyles: the short robust species have somewhat sluggish and cryptic behavior, clinging tightly and hiding within or under rocks, algae, and sessile invertebrates. Elongate species, such as *Heptacarpus stylus*, may cling or lay flat, like some *Hippolyte* grass shrimps, along blades of seagrasses and algae, or stand vertically head down (Bauer 1984; Jensen 2014). Species with intermediate morphotypes are typically epibenthic, active shrimps perching and moving around on a variety of substrates. Interestingly, some short-bodied *Lebbeus* species show cataleptic behavior; indeed, *L. catalepsis* and *L. lagunae* usually remain in the cataleptic posture (Jensen 2014).

The spirontocarids are typically small male—larger female species, and no studies have yet revealed hermaphroditic tendencies. On the other hand, the sexual systems of the tropical/subtropical genus *Thor* are quite variable, as some species have separate sexes (Bauer and VanHoy 1996) while *T. amboinensis* is a straightforward male-to-female (protandric) sex changer (Baeza and Piantoni 2010). *T. manningi* is partially protandric, i.e., about half of the population is composed of primary males that never change sex, while the other half is protandric (Bauer 1986).

3.6.9 *Ogyrididae*

The Ogyrididae consists of a single distinctive genus, *Ogyrides*, currently with 13 species. Ogyridid species are all found in the shallow waters of estuaries, near marine beaches, and even dug into the exposed low intertidal. Studies on these somewhat bizarre-looking shrimps are lacking, and largely confined to taxonomic descriptions and a few natural history observations (Hay and Shore 1918; Pearse et al. 1942; Williams 1984; Ayón-Parente and Salgado-Barragán 2013). They are small, fragile shrimps, generally so colorless that the internal organs are clearly visible through the transparent exoskeleton, which may be decorated with scattered small yellow and red spots and blotches (Plate 3.3f). They burrow shallowly into the sediment by day and emerge at night to join the plankton. What they do there, i.e., feeding habits and other behaviors, is practically unknown.

Morphologically, they are characterized by a very short rostrum and elongated eyestalks, which bear small eyes (Fig. 3.13). The first two pairs of chelae are small and similar, and not especially notable except that the carpus of the second cheliped is divided into 4–5 articles. The 3rd–5th pereopods are slender, delicate, and adorned with long plumose setae, which is probably retards sinking when they are up in the plankton at night. Pearse et al. (1942) were able to observe the burrowing behavior in the laboratory. The third maxillipeds are much stronger and more robust than any of the pereopods and are used to dig into the sediment, so that the shrimp plunges into

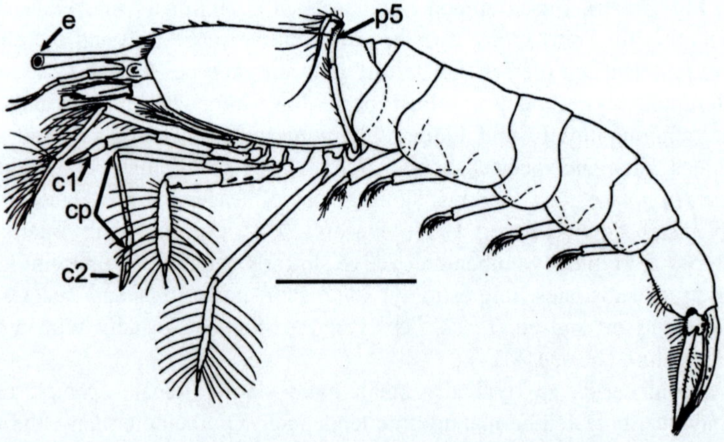

Fig. 3.13 Ogyrididae. *Ogyrides orientalis*. Scale bar represents 2 mm. e, small eye borne at end of elongate eyestalk; c1, c2, cheliped 1, 2; cp, carpus of second cheliped; p5, fifth pereopod. Modified from Bauer (2004) from Bruce (1986c)

the bottom headfirst, with the other pereopods moving sediment forward and above, while the fifth pereopods are held up dorsally and move sediment posteriorly above the body. Burrowing is aided by the beating of the pleopods. This genus is unusual for a caridean in that females have a thelycum-like structure (Hay and Shore 1918) which may serve for temporary spermatophore storage between mating and spawning, as has been shown in processid carideans (Rasch and Bauer 2016). Clearly, further studies on the natural history and behavior of the ogyridids are called for.

3.7 Superfamily Atyoidea: Atyidae

This is a freshwater family of shrimps, whose 46 genera and 527 species have invaded and radiated into rivers, streams, lakes, and other freshwater bodies of water in tropical and subtropical latitudes around the world. The most important feature of this group is their microphagous feeding mechanism, made possible by their unique chelae, which are not pincer-like weapons as in most shrimps (Fig. 3.14a–c, Plate 3.3g–h). The blunt dactylar and propodal fingers of both pairs of chelae are similar in length and bear long brushes of complex setae which scrape detritus and periphyton (a film of bacteria, microalgae, and detritus) from stream rocks and other substrates (Fryer 1977; Bailey-Brock and Brock 1993; Covich and McDowell 1996). Alternately, the shrimp may direct the chelae forward into a current, open the chelae wide, causing their setal brushes to snap open into conical fans (Plate 3.3h) which filter detrital particles, microplankton, and other organic matter from the water. The end of the carpus is scoop-shaped to accommodate the proximal end of the chela, allowing it considerable flexibility to open widely and to buttress the chela during scraping or passive filtration. During a feeding bout, the chelae periodically bend back towards the mouthfield whose various appendages remove the catch and pass it into the mouth. This passive filtration mechanism was first documented and its functional morphology analyzed in an elegant monographic paper by Geoffrey Fryer (1977). This feeding mode has allowed atyids to flourish in niches that had been previously dominated by other invertebrates such as snails and insects, as both Fryer (1977) and Walker (1992) have noted. The freshwater stream systems of tropical rainforests and habitats are rich in organic matter from leaf fall, twigs, and fruit, which sustains a productive detritus-based web of which atyid shrimps are key components (Chace and Hobbs 1969; Covich and McDowell 1996; Covich 2006). Given their basal position in stream food webs and their lack of defensive weapons, it is not surprising that atyid shrimps are important prey items for fish and predatory shrimps (Covich 2009). Other morphological features include the presence of variably developed pereopodal exopods, a primitive characteristic, and a setal comb on the dactyl of the last walking leg (Fig. 3.14c, d) used in grooming the body (Bauer 1989) (Chap. 4).

Notable genera include *Caridina*, whose 332 species comprise 63% of the family's diversity. This genus occurs in streams and rivers on islands or continents

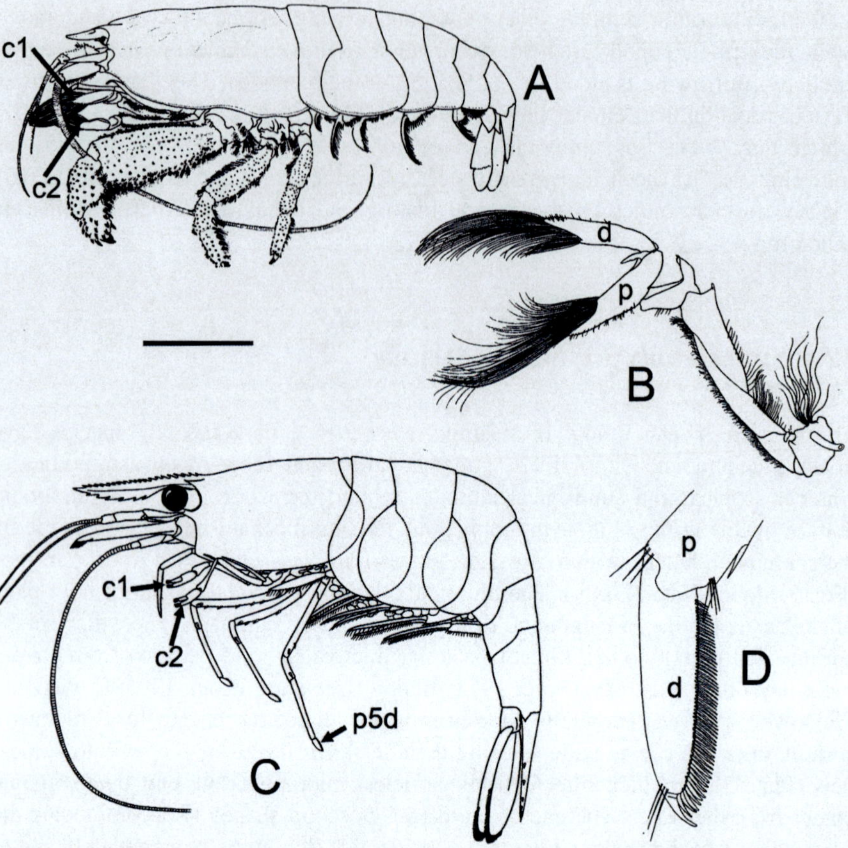

Fig. 3.14 Atyidae. (**a**) *Atya gabonensis*. (**b**) Cheliped of *Atya innocuous*. (**c**) *Atyaephyra thyamisensis*. (**d**) Dactylar comb of *Caridina grandirostris*. Scale bar represents 5 mm in (**a**) and 18 mm in (**c**). c1, c2, chelipeds 1, 2; d, dactyl; p, propodus; p5d, dactyl of pereopod 5. From (**a**, **b**) Hobbs and Hart (1982), (**c**) Christodoulou and Anastasiadou (2017) and (**d**) Cai et al. (2006), courtesy of the Journal of Crustacean Biology

bordering the Indo-West Pacific region (Chace 1997). However, von Rintelen et al. (2012) demonstrated that this massive group of species consists of a number of monophyletic clades that may well be elevated to separate genera in the future. Adaptive radiation has given rise to "species flocks" of colorful *Caridina* habitat specialists in ancient lakes in Africa and Indonesia (von Rintelen and Cai 2007). On the other side of the world, the genus *Atya,* although with only 13 species, is an important component of stream systems on the mountainous islands of the Antilles, as well as coastal slopes of the tropical Atlantic and Eastern Pacific (Hobbs and Hart 1982). The two genera show the range of variation in overall atyid body morphology. *Atya* species have a robust, crayfish-like morphology, with highly sclerotized exoskeletons, short spiny robust pereopods for gripping rocks and hard substrates (Fig. 3.14a). These features come in handy for maintaining position in mountainous streams with fast water flow and waterfalls. Some species, e.g., *A. scabra* in the Caribbean, may cling to rocks and crevices in such waterfalls. Atyids in calmer waters, e.g., *Atyaephyra* and *Caridina* species, tend to have a more slender body morphology typical of an epibenthic shrimp (Fig. 3.14c).

Although the atyids are an ancient freshwater group, their ancestors were marine, a conclusion reached from their morphological similarity to other primitive carideans (Ortmann 1894; Hedgpeth 1968; Bauer 2011), as well as by their basal position in most morphological and molecular phylogenies of (Christofferson 1990; Bracken et al. 2009; Li et al. 2011). Availability of unoccupied freshwater niches and escape from marine predators may have been key factors in the invasion of freshwater habitat by atyid ancestors (McDowall 2007; Bauer 2013). Although atyids are freshwater inhabitants, many species have maintained their connections to the sea via marine larvae. Such species have an amphidromous lifestyle, i.e., a type of diadromy in which juveniles grow and mature, and adults mate and spawn, in rivers and streams but produce larvae which have extended larval development in the sea. Females release larvae upstream to drift downstream into estuaries or the open sea to complete development (Chap. 9). Postlarvae must find the mouths of rivers or streams and then migrate upstream, sometimes very long distances, to upstream freshwater habitats. Atyids which occur in lakes or other land-locked bodies of water usually have evolved abbreviated or direct development, in which females spawn large embryos which can develop and hatch into advanced nonfeeding larvae or directly to small juveniles within the adult habitat. The advantages and disadvantages of these two extremes of life-history strategies will be discussed in greater depth in Chap. 9.

The vast majority of atyids are freshwater inhabitants in subtropical and tropical areas, but there are some exceptions. In North America, vestigial populations of *Syncaris pacifica* are found in some northern California streams (Hedgpeth 1968; Martin and Wicksten 2004), as are two cave-dwellers, *Palaemonias ganteri* and *P. alabamae* from the Mammoth Cave system in Kentucky (Lisowski 1983). The original distribution of the genus *Atyaephyra* is the rivers and streams of the Middle East, North Africa, and southern Europe. Formerly considered a monotypic genus (*A. desmarestii*), it is now considered to consist of several closely related species. *Atyaephyra desmarestii* (*s.s.*) has now spread into northern and central Europe via canals constructed to connect the main European rivers (Christodoulou et al. 2012).

A number of anchialine atyids (*Halocaridina, Typhlatya*, and *Antecaridina* species) comprise important components of cave ecosystems, primarily as detrital grazers which are important prey items of predatory cave palaemonids, hippolytids, and procaridids (Abele and Felgenhauer 1985; Chávez-Solís et al. 2018). Unlike other adult atyids, these shrimps occur in the nearly fresh to brackish to euhaline salinities that may occur in anchialine pools and their subterranean connections to the sea. As in many subterranean shrimps, the color of these atyids varies from pale to bright red (Holthuis 1973; Bailey-Brock and Brock 1993; Weese and Santos 2009).

3.8 Superfamily Bresilioidea

Opinions about the composition and validity of this superfamily have been variable and controversial over the years (e.g., Thompson 1967; Christofferson 1990; Holthuis 1993; Martin and Davis 2001). The latter authors expressed the common opinion that its component member families are perhaps not very closely related and therefore the superfamily is artificial. However, grouping the following families into this superfamily will be followed here as a convenience subject to future change. Probably the most constant characters applicable to bresilioid families are that the first chelipeds are more robust and usually shorter than the second, and the carpus of the latter is entire, without subdivision into articles.

3.8.1 Family Agostocarididae

The four species of this monotypic genus have a very limited distribution in anchialine caves on the Western Atlantic islands of Bahamas, Turks and Caicos, and Yucatan (Hart and Manning 1986; Kensley 1988; Alvarez et al. 2004). These small shrimps with short triangular rostra are characterized by their unique cheliped morphology (Fig. 3.15a, b). The carpus of the first cheliped articulates about midway along the ventral edge of the propodus instead of the usual proximal arrangement, a unique feature in the Caridea (Fig. 3.15b). The fingers of the chela have the appearance of short, curved shears with tiny cutting setae on their edges, and the dactyl bears proximally a tuft of long setae. Kensley (1988) observed that the gut contents of two *Agostocaris bozanici* specimens were variously filled with filaments of *Oscillatoria*, a cyanobacteria; fecal pellets (presumably of other cave organisms) and detrital material. The second chela is strange also, described by Hart and Manning as vestigial, weakly chelate with a flabby digitiform dactyl (Fig. 3.15a). As in many cave shrimps, the eyes are greatly reduced. In *Agostocaris*, the eye is represented by a conical stalk without differentiation into separate eyestalk and cornea, with little or no pigmentation (Fig. 3.15a). The little information to date on *Agostocaris* is made possible by the daring of cave divers who collect them for study. There is much to learn about the biology and evolution of these enigmatic shrimps.

3.8.2 Family Alvinocarididae

At the beginning of my career, one of the most remarkable discoveries made in marine biology was that of deep-sea hydrothermal vent communities located along the mid-oceanic ridges and other sea-floor spreading sites. First reported by Corliss and Ballard (1977) and Ballard and Grassle (1979) in National Geographic

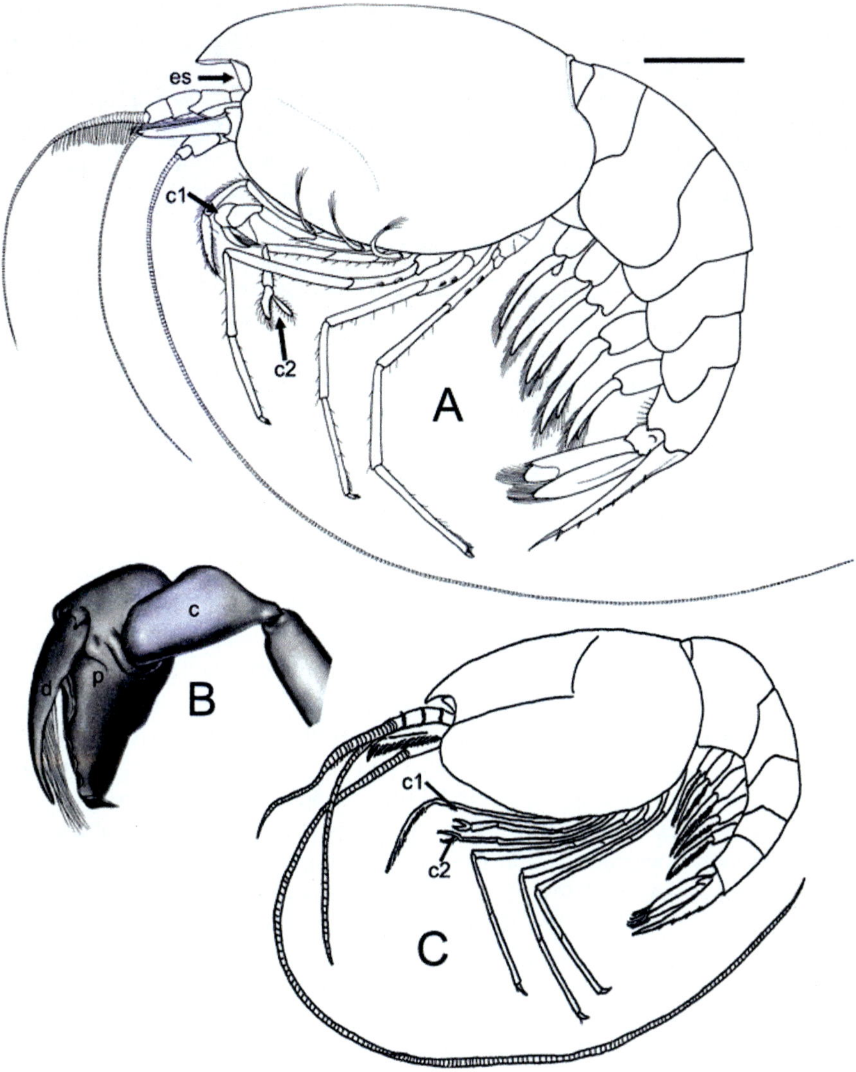

Fig. 3.15 Agostocarididae and Anchialocarididae. (**a**) *Agostocaris bozanici*. (**b**) First chela of *A. williamsi*. (**c**) *Anchialocaris paulini*. Scale bar in (**a**) represents 2 mm in (**a**) and 3 mm in (**c**). c1, c2, chelipeds 1, 2; c, carpus; d, dactyl; es, eyestalk (no cornea, pigmentation); p, propodus. From (**a**) Kensley (1988), (**b**) Hart and Manning (1986), and (**c**) Mejía-Ortíz et al. (2017) courtesy of Crustaceana (Brill)

magazine, observations made from the deep-sea submersible *Alvin* revealed a completely new community of bizarre giant annelidan tube worms (Siboglinidae), bivalve molluscs, crabs, and shrimp species unknown to science. Hydrothermal vents are volcanically active sites in which magma and superheated water spew out of the ocean floor. This water is rich with minerals and sulfides which precipitate out as they meet the very cold deep ocean water. Such areas are perfect habitats for chemoautotrophic sulfide-oxidizing bacteria, which are the primary producers in this ecosystem, much like the photosynthetic plants in the lighted areas of the ocean. These bacteria support an abundant, unique macroinvertebrate and fish fauna.

Among this fauna, there are often very abundant populations or swarms of shrimps which belong to a bresilioid family which was unknown until sampled from hydrothermal vents. First placed in the family Bresiliidae (*s.l.*), their uniqueness and differences from other bresilioids were formally recognized as a separate family Alvinocarididae by Christoffersen (1986). He named the family in honor of the submersible *Alvin* and its role in their discovery. The very first member of the family, *Alvinocaris lusca*, was described by those carcinological greats, Williams and Chace (1982). Description of *A. lusca* was followed by that of the even stranger genus *Rimicaris* (Williams and Rona 1986), which now includes other species previously classified as *Chorocaris* (Vereshchaka et al. 2015). Currently, the family has 9 genera and 38 species, all from hydrothermal vent and "cold seep" hydrocarbon deep-sea communities in which the primary producers are chemoautotrophic bacteria. Vereshchaka et al. (2015) published an extensive and useful taxonomic and phylogenetic analysis of the family.

Alvinocaridids have the basic bresilioid cheliped features (first cheliped more robust but shorter than the second, carpus of both entire) (Fig. 3.16, Plate 3.4a). Unlike other bresilioids, their pereopods completely lack exopods. Morphological trends in the family include reduction or loss of: pigmented faceted eyes; a short rostrum; hook-like epipods (important in gill cleaning of many carideans); and cincinnuli (Velcro-like coupling hooks) of the appendices internae, necessary to link a pleopod pair together for coordinated paddling. In some species (especially *Rimicaris*), there is an expansion of the maxilla 2 scaphognathite (gill bailer) and a large foliaceous maxilliped 1 exopod which, along with anterior thoracic appendages, are densely covered to various degrees by plumose setae on which symbiotic chemoautotrophic bacteria grow. The biology of alvinocarids might be best viewed as a continuum of species with, at one extreme, a typical epibenthic shrimp morphology (e.g., *Alvinocaris* (Chap. 1, Fig. 1.7a) and a scavenging/predatory lifestyle. At the other extreme are shrimps adapted to obtain their nutrition primarily from epibiotic chemoautrophic bacteria.

Rimicaris exoculata is the best example of the extremes in alvinocaridid morphology and lifestyle. It has completely lost the rostrum, and the eyes are absent, represented by a broad ocular plate (a vestige of the fused eyestalks) and no sign of a pigmented faceted cornea (Williams and Rona 1986) (Fig. 3.16a, b). The anterior part of the branchiostegite (gill cover) is expanded forward on each side to fuse with the head. The ventral edges almost meet along the midline to form a chamber in which the mouthparts and anterior pereopods are enclosed. The greatly expanded scaphognathite and maxilliped 1 exopod take up considerable space in the inflated branchiostegal chamber, and bear abundant plumose setae, as does the underside of the chamber and anterior thoracopods. These setae are covered with symbiotic chemoautrophic (sulfur-oxidizing) bacteria. The bacteria are harvested by the chelae's recurved spatulate fingers which bear comb-like tiny teeth (Fig. 3.16c). Studies

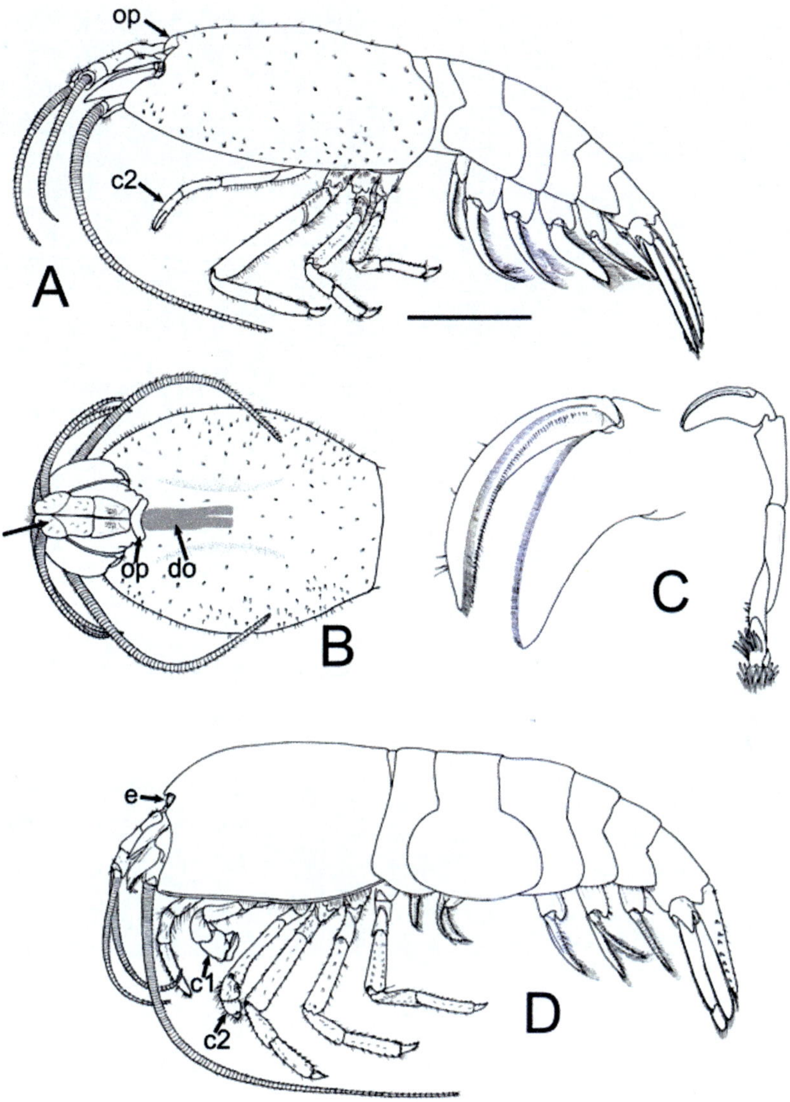

Fig. 3.16 Alvinocarididae (see also Fig. 1.7a, Chap. 1). (**a**) *Rimicaris exoculata* (first cheliped hidden within expanded carapace chamber); eyes are absent. (**b**) Dorsal view of *R. exoculata* cephalothorax, showing expanded antennular peduncles, ocular plate, and approximate position of dorsal organ beneath the carapace. (**c**) First cheliped (right) and higher magnification of first chela (left) of *R. exoculata*. (**d**) *R. chacei*, with general morphology intermediate between *R. exoculata* and other more typically epibenthic alvinocaridids. Scale bar in (**a**) represents 10 mm in (**a, d**). a1p, antennular peduncle; c1, c2. chela 1, 2; do, dorsal organ; e, vestigial eye; op, optical plate. From Williams and Rona (1986), courtesy of the Journal of Crustacean Biology

Plate 3.4 Alvinocarididae to Glyphocrangonidae. (**a**) *Alvinocaris costaricensis* (Alvinocarididae). (**b**) *Bresilia rufioculus* (Bresiliidae). (**c**) *Campylonotus vagans* (Campylonotidae). (**d**) *Crangon nigricauda* (Crangonidae). (**e**) *C. stylirostris,* with coloration matching that of the sand substrate. (F) *Paracrangon echinata* (Crangonidae). (**g**) *Glyphocrangon* sp. (Glyphocrangonidae). Photo acknowledgments: Courtesy of (**a**) G.W. Rouse, (**b**) Yusuke Yamada, (**c**) Gustavo Lovrich, (**d–e**) Gregory Jensen, (**f**) R.T. Bauer and (**g**) Arthur Anker

by various workers using stomach content, stable carbon isotope and other chemical analyses of tissues show that these shrimps obtain their nutrition from chemoautotrophic bacteria. These bacteria may be those growing on their bodies and also those covering sulfide mineral particles ingested by the shrimp from the substrate (Segonzac et al. 1993; Gebruk et al. 1993; Van Dover et al. 1988; Rieley et al. 1999). *Rimicaris exoculata* is found in dense swarms adjacent to the black chimney smokers (vents) from which superheated (~350–400 °C) water gushes out. The shrimps often show burn marks from approaching the vents too closely (Segonzac et al. 1993). They apparently take the risk of such injuries to ensure that their symbiotic bacteria are in contact with the sulfide-rich water they need to grow.

Other alvinocarids are little or not dependent on chemoautrophic bacteria. *Alvinocaris markensis* was found much less abundantly than *R. exoculata* and relatively far from the vents. Based on gut content analysis, *A. markensis* was shown to be necrophagous (carrion-eating scavenger) and/or predatory by Segonzac et al. (1993). The latter authors and Gebruk et al. (1993) found that *R. chacei* (Fig. 3.16d) is intermediate in both diet and morphology between *A. markensis* and *R. exoculata*, with its nutrition coming from both scavenging/predation as well as from epibiotic bacteria. De Busserolles et al. (2009), using carbon and nitrogen isotopic analysis, confirmed the latter in *R. chacei* by showing that it and another alvinocaridid, *Mirocaris fortunata,* include scavenging and predation in their feeding habits.

How does *R. exoculata* maintain its preferred position very near the dangerous superheated waters emitted by the hydrothermal vents? Zbinden et al. (2017) investigated chemosensory (antennular) morphology in four genera of alvinocarids and found no evidence of specialized chemosensory adaptations. Ultrastructural examination of alvinocaridid eyes has confirmed their vestigial nature and blindness (Lakin et al. 1997; Wharton et al. 1997), perhaps not surprising given their dark deep-sea environment. However, Van Dover et al. (1989) described a novel organ, commonly termed the "dorsal organ," just below the exoskeleton of the cephalic area in *R. exoculata*, behind the ocular plate (Fig. 3.16b). They found that the large bilobed dorsal organ is sensory, as evidenced by the large nerve tracts leading to the brain, and by the presence of rhodopsin, a visual pigment found in the eyes of crustaceans and other animals. They concluded that the dorsal organ is derived from the ancestral optical system. There is no light in the deep-sea environment except for bioluminescent light of other deep-sea animals. However, the wavelengths of bioluminescent light (blue-green) are too short to be detected by the *R. exoculata* pigments of the dorsal organ. Van Dover and colleagues proposed that the dorsal organ might be detecting the infrared wavelengths coming from the superheated vent water. Pelli and Chamberlain (1989) confirmed that such black-body radiation would not be detected by submersible operators but could be by the dorsal organ's visual pigments. This sensory capacity would be very useful to *R. exoculata,* which must stay very near (but not too near) the vents to provide their chemoautotrophic symbionts with the sulfides needed to grow and reproduce. Although best developed in *R. exoculata*, a smaller version of the dorsal organ is present in other *Rimicaris* and in *Mirocaris* species (Vereshchaka et al. 2015).

Basic life history information about reproduction and life span on alvinocaridids is available despite the difficulty of sampling the hydrothermal vent and cold seep communities of the deep-sea. With an increase in submersible, ROV (remotely operated vehicles) and hydrobaric technology, it is hoped that many aspects of this fascinating family will be revealed by present and future deep-sea biologists.

3.8.3 Family Anchialocarididae

This newly proposed family consists of just one species, *Anchialocaris paulini*, discovered and described by Mejía-Ortíz et al. (2017) (Fig. 3.15c). It was collected from a cenote cave on Cozumel Island, Gulf of Mexico. This is an anchialine environment with water of ocean salinity and a somewhat high sulfur content. The shrimp is very similar to *Agostocaris* species in color (white, transparent exoskeleton), highly reduced eye with no pigment (i.e., blind) and overall general morphology. The only major difference is in the form of the first chela, in which the carpus articulates with the proximal end of the propodus, as in most shrimps, and not with the middle of the propodus, as in *Agostocaris*, which is the latter's unique feature. The first chela fingers are slender, unlike those of *Agostocaris* (Fig. 3.15b, c). However, the second chela is very much like that of *Agostocaris*. For the present, at least, it is accepted by WoRMS as a separate family with uncertain superfamily designation. However, I will keep it here in the Bresilioidea due to its similarity to *Agostocaris.*

3.8.4 Family Bresiliidae (s.s.)

The Bresiliidae, as now narrowly defined (Calado et al. 2004; De Grave and Fransen 2011; Hendrickx 2014) consists of 10 species of *Bresilia* and one species of *Encantada,* the latter only known from a partial specimen (Wicksten 1989). These shrimps are small (6–29 mm total length; Hendrickx 2014) and characterized by a marked caridean hump in the abdomen (Fig. 3.17a; Plate 3.4b). They have the typical bresilioid cheliped morphology of two pairs of chelae, the first more robust than the second, and the second with undivided carpus. The structure of the first

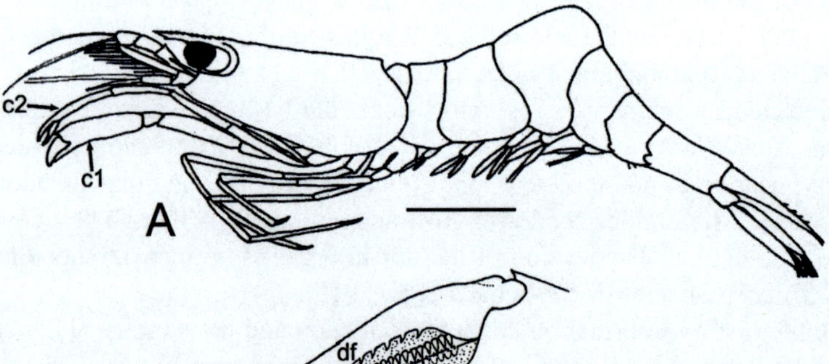

Fig. 3.17 Bresiliidae. (**a**) *Bresilia saldanhai*. (**b**) fingers of chela 1, *B. cinctus*. Scale bar in (**a**) represents 1 mm. c1, c2, chela 1, 2; df, dactylar finger; pf, propodal finger. From (**a**) Calado et al. (2004) and (**b**) Komai and Kohtsuka (2017), courtesy of Zootaxa

cheliped is reminiscent of the alvinocaridids, with the fingers somewhat deflexed and inner margins with minute cutting teeth, appearing somewhat like a pair of shears, indicating a grazing feeding mode (Fig. 3.17b). As they are difficult to sample and rare, little is known about their biology. Calado et al. (2004), based on the non-overlap in size in a limited sample of small males and large females, suggested protandry as the possible sexual system of *B. saldanhai.*

Bresilia is a genus with an intriguing ecological distribution with two extremes: one group of species occurs in shallow marine caves and the other group in deep-water continental platform and slope habitats (Komai and Kohtsuka 2017). They show characteristics of shrimps from habitats with a constant absence of light, low temperature, and food availability, i.e., subterranean and deep-sea environments (Calado et al. 2004). Thus, like many cave shrimps, some species have eyes with reduced corneas without dark pigmentation and a coloration in which the body has variable areas of red or orange, translucence, or white streaks. Their ecological distribution (shallow cave, or deep-water) calls to mind the possible origin of other anchialine cave shrimps, such as the barbouriids and procaridids, a hypothesis presented in Hart et al. (1985). These workers proposed that many of the marine cave/anchialine shrimps are relicts of deep-sea ancestors which were dispersed away from the original deep-water habitat by sea-floor spreading. Thus, the anchialine species might be relict populations, with the subterranean habitats serving as refugia with basic habitat characteristics similar to those of deep-sea bresilioid ancestors.

3.8.5 Disciadidae

There are but 4 genera and 12 species in the family, with most species (8) in the genus *Discias* (Fig. 3.18a). The latter are tropical and subtropical in shallow-water habitats, with most found in association with sponges, corals, and worm tubes. *Tridiscias transkei* was found somewhat deeper in subtropical waters off the south-eastern coast of Africa (Kensley 1983). On the other hand, the two *Kirnasia* species were taken in slope depths north and east of the Azores islands (Burukovsky 1988). *Lucaya bigelowi* was collected off the Bahamas in very deep water (nearly 5000 m) (Chace 1939). *Discias* and *Tridiscias* species are small in size (usually less than 1 cm in total length), while *Kirnasia* species and *L. bigelowi* are about double that size. Color is not usually reported, probably because these shrimps are found by going through field samples that have already been preserved. However, in a major work on Disciadidae, Kensley (1983) described color in *D. serratirostris* as variable (bright purple in one specimen, dark green cephalothorax and yellowish abdomen in another).

The disciadidid's unique first chelae set them apart from other bresilioids. The dactylar finger of the chela is shaped like a disk with a sharp edge that fits into a slot in the propodal finger (Fig. 3.18b). Most species of *Discias* are found within sponges. Kensley (1983) proposed that the first chela is used to scrape off the energy-rich sponge mucus, usually loaded with microbial organisms such as diatoms

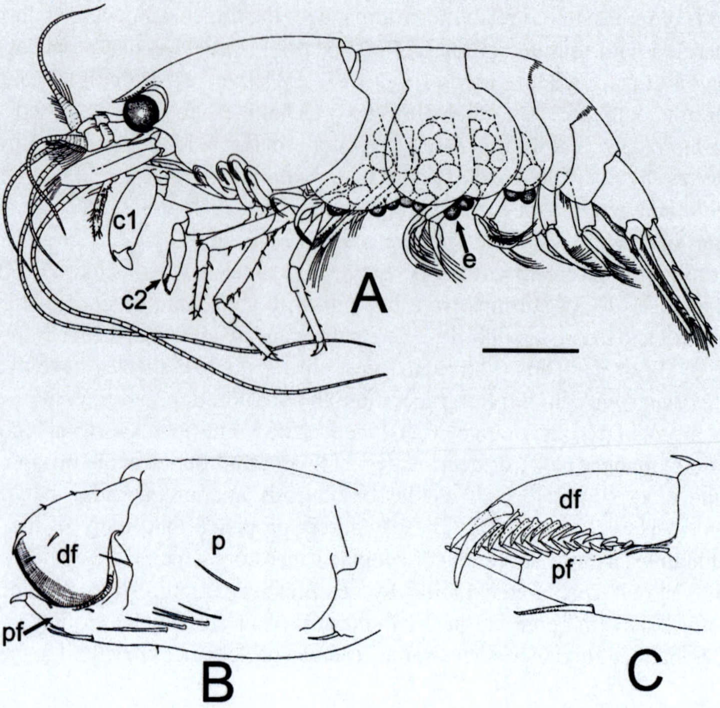

Fig. 3.18 Disciadidae. (**a**) *Discias vernbergi*. (**b**) Chela 1. (**c**) Chela 2. Scale bar in (**a**) represents 2 mm. c1, c2, chela 1, 2; df, dactylar finger; e, embryo; p, propodus; pf, propodal finger. (**a–d**) From Boothe and Heard (1987)

and bacteria, which abounds in sponge canals. The second chela is a typical chela overall, but it should be noted that the chela fingers have long stout spines along their inner margins (Fig. 3.18c). Although the above is about the extent of information on *Discias*, there is even less known about the other species in the family, a deficiency that needs to be remedied.

3.8.6 *Pseudochelidae*

When Chace and Brown (1978) described the first species (*Pseudocheles enigma*) of this monogeneric family, they recognized that it had features in common both with the Disciadidae and Bresiliidae. However, it was clearly distinguished from them and all other carideans by the possession of apparently functional chelae on the third to fifth pereopods in addition to those on the first two pereopods (Fig. 3.19a). These structures were termed "pseudochelae" because their composition is different from most caridean chelae in that the "finger" of the pseudochela is not an extension of the propodus proper, but instead is a long curved propodal seta that opposes the dactyl,

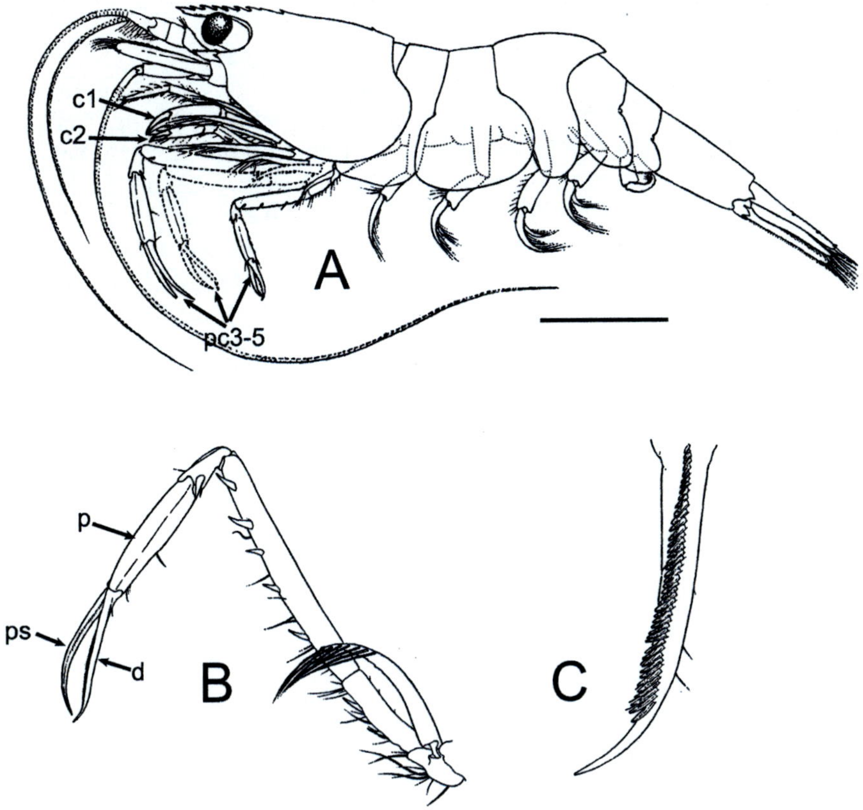

Fig. 3.19 Pseudochelidae. (**a**) *Pseudocheles enigma*. (**b**) *P. enigma* pereopod 4. (**c**) dactyl of *P. enigma* pereopod 4. Scale bar in (**a**) represents 2 mm. c1, c2, chela 1, 2; d, dactyl; pc3–5, pseudochelae; pereopods 3–5; ps, propodal seta. From Chace and Brown (1978)

the latter equipped with a series of spines (Fig. 3.19b, c). Chace and Brown proposed that the pseudochelae were functional, possibly used to grip and cling to marine plants and sessile invertebrates (e.g., soft corals). Subsequently, three other *Pseudocheles* species have been described, and De Grave and Moosa (2004) separated *Pseudocheles* from other bresilioids and placed them in their own family.

Three of the four species have been found in shallow coral reef habits and all are small in size (~1 cm or less in total length). Chace and Brown (1978) noted that specimens of *P. enigma* were colorless (transparent). Anker (2012) has described another species from deeper waters (70 m), 160 m offshore of the Rio Doce estuary in Brazil. *Pseudocheles* species are rarely collected, perhaps often overlooked because of their small size and transparency. Observations on living specimens are very much needed to determine what the function of the pseudochelae actually is in the natural history of these shrimps.

3.9 Superfamily Campylonotoidea

Composed of two deep-water families, the Campylonotidae and Bathypalemonellidae (Martin and Davis 2001), this superfamily has been traditionally linked to the Palaemonoidea (Chace 1992; Holthuis 1993). Like the latter, the second chelipeds are larger than the first, and neither pair has a subdivided carpus. The first chelipeds are equal in size in *Campylonotus* but in bathypalaemonellids, one first cheliped is massive in size and unusual in structure, while the other is much smaller and typical in form. *Campylonotus* species have pereopodal epipods while the bathypalaemonids lack them. Both families are deep-sea inhabitants of continental slope depths, and like most deep-sea shrimps, have a primarily reddish coloration.

3.9.1 Bathypalaemonellidae

Members of his pantropical family (Chace 1997), with 2 genera and 11 species, are medium-sized shrimps up to approximately 80 mm in total length, including the prominent spiny rostrum (Bruce 1966; Cleva 2001, 2004) (Fig. 3.20a). They are rare in collections as are observations from their deep-sea habitats, so there is little information on their natural history. However, Wicksten and Heathman (2015) reported valuable ROV video observations on *Bathypalaemonella serratipalma* confirming the few previous observations that this species is associated with deep-sea octocoral colonies.

Bathypalaemonellids have one massive major second cheliped on either the right or left side while the other is slender and much smaller, similar to the first pair of chelipeds, which are equal in size (Cleva 2001) (Fig. 3.20a). The major second cheliped is reminiscent of the equally bulky second cheliped of the unrelated hippolytid *Leontocaris*. Like the latter, there is a mechanism for folding and locking the chela with the rest of the cheliped to carry it close to the body. However, the system is very different in its details, and given that these two genera are not closely related, is a good example of convergent evolution. When the bathypalaemonellid major second cheliped is folded, a large patch of long recurved setae on the propodal palm of its chela (Fig. 3.20b, c) interlocks with another group of setae on the proximal part (ischium and/or basis) of the appendage (Bruce 1966, 1986a; Cleva 2001). The flexed major cheliped is tucked away beneath the cephalothorax in a channel on the same side between the enlarged proximal segments of the thoracic limbs and the thoracic sternites. Bruce concluded that this was an adaptation, as in the hippolytid *Leontocaris*, for maintaining the hydrodynamic shape during swimming in spite of the large bulky cheliped. As in *Leontocaris*, the major second chela of bathypalaemonellids, present in both males and females, might suggest that these shrimps are predators. These workers observed only 1 bathypalaemonellid per host colony; perhaps the major second cheliped is a defensive weapon against

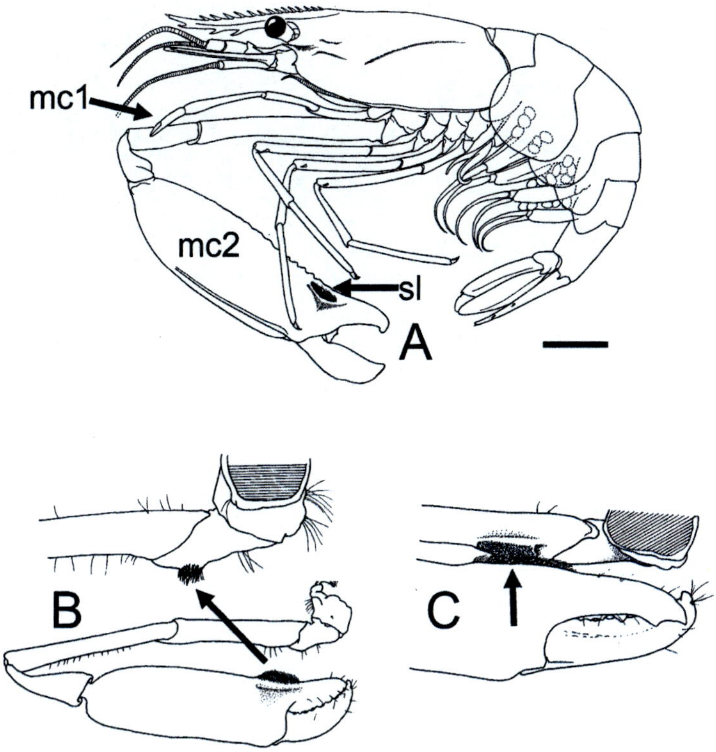

Fig. 3.20 Bathypalaemonellidae. (**a**) *Bathypalaemonella hayashi*. (**b**) Major second cheliped of *Bathypalaemonetes pilosipes*. Basal (upper) and distal (lower) views of the cheliped. Unmarked arrow indicates the setal patches which connect in (**c**) to lock the folded cheliped in place. Scale bar in (**a**) represents 5 mm. mc1, minor chela 1; mc2, major chela 2; sl, setae which form part of mechanism for locking the limb in a flexed position below the cephalothorax. From (**a**) Cleva (2001) and (**b-c**) Bruce (1986a)

intraspecific or other intruders. Wicksten and Heathman (2015) hypothesized that the short curved dactylus of pereopod 3 (first walking leg) in *Bathypalaemonella serratipalma* is an adaptation for clinging onto its host, as in some other associates of octocorals.

3.9.2 Family Campylonotidae

This monogeneric family with only five species is restricted to the subantarctic regions on the southern hemisphere, and primarily occurs at deep continental shelf and slope depths over sand and mud bottoms (Thatje 2003). The coloration of campylonotids is some shade or variation of red (e.g., Torti and Boschi 1973;

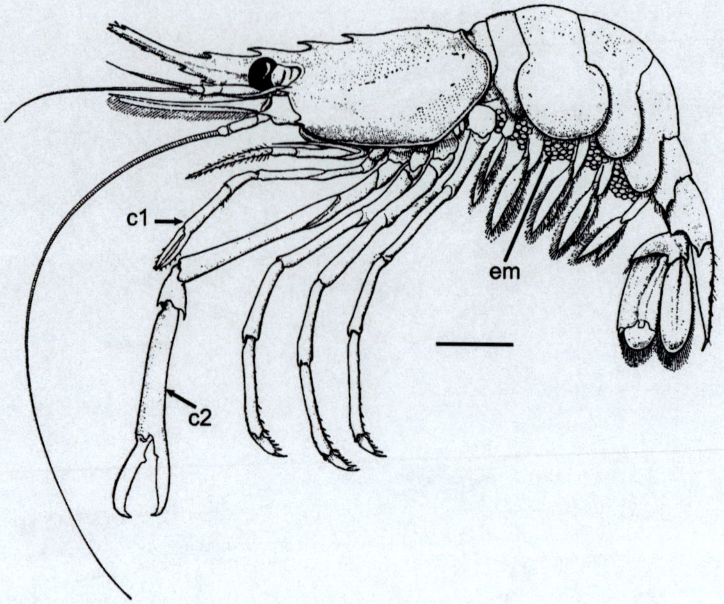

Fig. 3.21 Campylonotidae. *Campylonotus vagans* female. Scale bar represents 10 mm. c1, c2, chela 1, 2; em, embryos. From Bauer (2004) modified from Torti and Boschi (1973)

Arana and Ahumada 2006). However, *C. vagans*, living in shallower more lighted waters, has bright violet and yellow stripes as well (Meyer et al. 2009) (Plate 3.4c).

The second chelipeds of campylonotids are very well developed in both sexes, more robust and longer than the first pair (Fig. 3.21). These second chelae are impressive-looking weapons, indicative of a predatory lifestyle. Although little is known about the natural history of the family (Thatje 2003), it was well established by Yaldwyn (1960) and Torti and Boschi (1973) that these shrimps are protandrous sex changers, first male and then female. The campylonotids are the only caridean family in which the first pair of male pleopods carry gonopods very much like that of the petasma of penaeoid shrimps, a relatively unrelated group. This is another example of convergent evolution among the decapod shrimps. Campylonotid females produce large embryos which undergo an abbreviated embryonic development with only two planktonic larval stages (Thatje et al. 2004). The abbreviated development is a common but not universal characteristic of high-latitude species (Chap. 9).

3.10 Superfamily Crangonoidea

These two marine benthic families Crangonidae and Glyphocrangonidae have been considered closely related to each other chiefly because of the structure of the first pereopods (e.g., Chace 1992; Martin and Davis 2001). These limbs are definitely subchelate in the Crangonidae while those of the Glyphocrangonidae are considered as subchelate or "imperfectly" so (Holthuis 1971; Chace 1984). However, other authors have suggested the first pereopods are simply prehensile (Bauer 2004; Wood and De Grave 2015) so that this important phylogenetic character may not be homologous in the two families after all. Shrimps in this superfamily tend to be somewhat dorsoventrally depressed with often spiny and hard, calcified exoskeletons. The superfamily contains over 300 species and is an important component of benthic ecosystems.

3.10.1 Crangonidae

The crangonids, with 23 genera and 234 species, are highly benthic and nocturnal shrimps that primarily live in soft-bottom communities in all depths of the ocean. Their bodies are generally somewhat stout and dorsoventrally depressed with a short or no rostrum (Fig. 3.22a, b; Plate 3.4d), a departure from the typical cardioid body form, a result of their benthic lifestyle. Many genera are heavily armored by a thick, highly sclerotized, and calcified exoskeleton equipped with a variety of spines and tubercles (Fig. 3.22b). Some species, such as *Metacrangon munita* and *Paracrangon echinata*, enhance this defensive armature with cataleptic behavior (Jensen 2014) (Plate 3.4f). However, less heavily armored members of the family, such as *Crangon* species, are good swimmers and can get up into the water column at night to hunt for prey (Jensen 2014). This family is a cold-water group, with its highest diversity and abundance in temperate and boreal habitats at continental shelf depths of the northern hemisphere, particularly the North Pacific (Butler 1980; Campos et al. 2012). As a cold-water group, they are found worldwide in the deep sea, including equatorial latitudes.

The first pereopods bear subchelae in which the long sharply pointed dactyl can fold back against the curved expanded distal end of the propodus, which bears a sharp spine on the flexor margin (Fig. 3.22a, c). The ultrastructure of the propodal distal margin (PDM) has been studied in detail with scanning electron microscopy (SEM) by Wood and De Grave (2015) in a number of crangonid genera. SEM showed that the PDM fine structure varies greatly among genera and that these differences may be related to sensory function, or more likely serve as "friction-enhancing" devices that aid in firmly gripping prey. Thus, a crangonid attacks a prey by using the subchela as a folding knife, trapping the prey with the dactyl against the PDM, with the PDM spine puncturing and helping to hold the prey. An excellent description of crangonid predation has been made by Jensen (2011) on *Paracrangon*

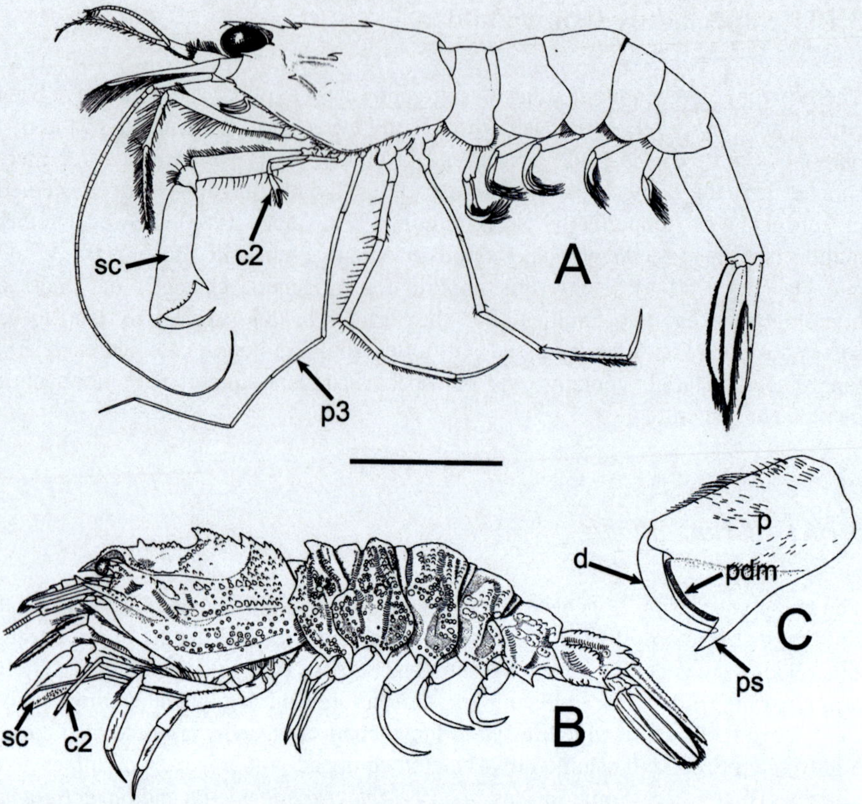

Fig. 3.22 Crangonidae. (**a**) *Parapontophilus gracilis*. (**b**) *Sclerocrangon salebrosa*. (**c**) *S. salebrosa* subchela. Scale bar represents 5 mm in (**a**) and 11 mm in (**b**). c2, second cheliped; d, dactyl; p, propodus; pdm, propodal distal margin; ps, propodal spine; p3, third pereopod; sc, subchela. From (**a**) Dardeau and Heard (1983) and (**b**, **c**) Kim and Komai (2002) courtesy of Crustaceana (Brill)

echinata which primarily feeds on small crustaceans, especially amphipods and shrimps. The prey is captured using the subchela as described above; Jensen noted that the PDM spine is used to repeatedly stab the victim as it is held in the subchela. The prey is ingested whole. Crangonids in general feed on small crustaceans, worms, small bivalves, and fish (Jensen 2011; Campos et al. 2012).

There is considerable variation in the second pereopod (cheliped) among crangonid genera. It is small and slender, may have all its segments and with a small chela; or reduced in size and number of segments and without a chela, or lacking altogether (*Paracrangon*) (Bauer 1979; Holthuis 1993). When chelate, it can be used as a gill-cleaning appendage (*Crangon*; Bauer 1979). If the second cheliped is reduced or absent, the gills are cleaned with long gill-bailer (scaphognathite) setae, as in *Paracrangon* and *Rhynocrangon* (Bauer 1979; Jensen 2014). The third pereopods of some crangonid genera are long and slender (Fig. 3.22a) and may be used as sensory appendages for detecting prey, as in *P. echinata* (Jensen 2011).

Most crangonids bury themselves just under the sediment surface when inactive during the day. They burrow rapidly and efficiently, as described by Lloyd and Yonge (1947) and Jensen (2014). The shrimp makes a shallow trough in the sediment by beating its pleopods rapidly and paddling the pereopods over the surface. Then the shrimp repeatedly expands and contacts the branchiostegites, forcing water from the gill chambers into the sediment. These actions liquefy the sediment, so that the shrimp can cover itself in seconds, with the long antennal flagella sweeping sand over any dorsal surface left exposed. The shrimp always keeps the olfactory flagella of the antennules above the sediment surface to maintain sensory contact with the environment above. In the genus *Argis*, the dactyls of the posterior two pereopods are flattened, an adaptation for burrowing.

Crangonids are well known as masters of camouflage, using color patterns and rapid color change to match the substrate when they are out and exposed (Plate 3.4e) (Chap. 5). Although most species have separate sexes, some species of *Crangon* and *Argis* are protandric hermaphrodites (Chap. 7). The crangonid whose biology is best known is *Crangon crangon*, an ecologically and commercially important species in European waters (e.g., Boddeke et al. 1991; Hufnagl et al. 2014) (Chap. 11). Some good general references on crangonid biology include Chace (1984), Williams (1984), Haefner (1979), Campos et al. (2012) and Jensen (2014).

3.10.2 *Glyphocrangonidae*

The morphological diversity of this speciose family is low, reflected in the taxonomy of the group, with only a single genus, *Glyphocrangon*, but which has numerous (93) described species. Glyphocrangonids are found in the deep-sea soft bottoms from the continental slope to abyssal depths, and they are common in dredges and trawls from these depths. In spite of this, relatively little is known about them aside from their well-described morphology and taxonomy. Glyphocrangonids are large (up to at least 11 cm total length), robust, and heavily armored shrimps (Holthuis 1971) (Fig. 3.23a, b, Plate 3.4g). The exoskeleton is firm and heavily calcified, and covered with ridges (carinae), spines, and tubercles, as in the crangonid *Sclerocrangon* (Fig. 3.22b), although the glyphocrangonid cuticular armor is even more extreme. The fairly long rostrum is dorsoventrally depressed and somewhat broad with a concave, hollowed-out shape. It bears a few to several pairs of lateral teeth. The carapace has several pronounced longitudinal ridges to which labels have been given in order to aid comparison among species (Holthuis 1971; Chace 1984). The shrimp bristles with sharp spines. Projecting anteriorly are the large and robust antennal and branchiostegal spines plus 1–2 dorsoventrally flattened spines behind the branchiostegal ones. The abdominal pleura are rectangular, with the long sides vertical, with little or no overlap. Unlike most carideans, the second abdominal pleura do not overlap those of the first and third abdominal segments. The dorsal

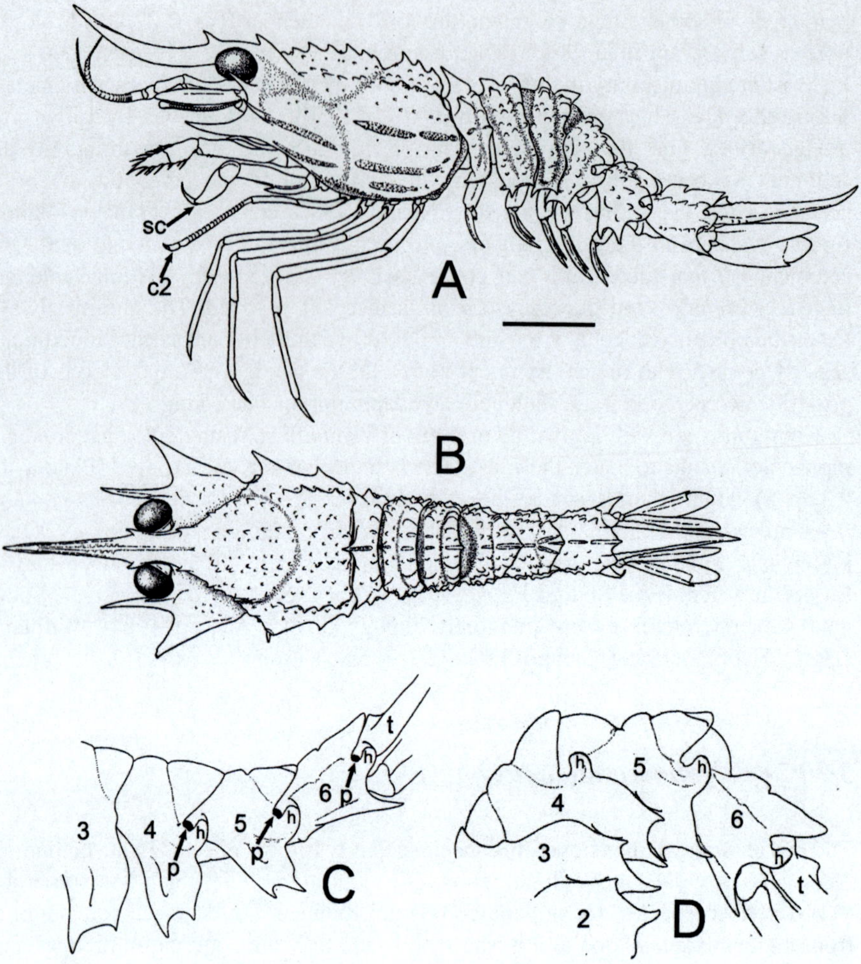

Fig. 3.23 Glyphocrangonidae. (**a**) *Glyphocrangon regalis*, lateral view. (**b**) *G. regalis*, dorsal view. (**c**) *G. longirostris,* locked abdomen, hooks linked to pegs (black fill). (**d**) *G. longirostris*, abdominal segments unlocked and flexed, pegs not visible. Scale bar represents 5 mm in (**a**). c2, second chela; h, abdominal hook; p, abdominal peg (tips in black fill), sc, subchela of pereopod 1; t, telson; numbers in (**c**, **d**) correspond with abdominal somites. (**a**, **b**) from Chace (1984); (**c**, **d**) modified from Rice (1981) courtesy of Crustaceana (Brill)

carinae of the abdominal segments and the ventral edges of the abdominal pleura terminate in strong spines; the telson is triangular and ends in a sharp point.

In addition to these prickly defenses, the glyphocrangonid are able to lock the posterior abdominal somites and the telson into a tightly rigid unit that bends upward towards the cephalothorax so that the body forms a "U" shape (Rice 1981). Together with all the spines and the hard, rigid exoskeleton, this cataleptic posture must present a very difficult mouthful for fish predators. Rice (1981) showed that posterior processes (hooks) of the 4th–6th abdominal segments can attach and lock around small kidney-shaped pegs (Fig. 3.23c) on the side of each of the 5th–6th segments and telson, forming a single functional unit. At least in preserved specimens, an attempt to flex the abdomen results in it breaking into pieces. The hooks can be disengaged from the pegs, presumably by contraction of specific muscles within the abdomen, and the pegs are hidden from view when the abdomen is unlocked and flexed (Fig. 3.23c, d).

The first pereopod has been described as subchelate or "imperfectly subchelate" (Holthuis 1971; Chace 1984). The direction of dactylar flexure in glyphocrangonids is the opposite of that in the crangonid subchela. The dactyl is recurved or inversely flexed back up against the propodus. There is little or no extension of the propodus into a finger (Wood and De Grave 2015) nor is it expanded vertically as in crangonids, so the pereopod might equally be described as prehensile, similar to the third pereopod of primary males in *Thor manningi* (Bauer 1986) and some other carideans. The structure and relatively small size of the subchelae indicate that glyphocrangonid are predators on small invertebrates. The second pereopods have a carpus subdivided into many articles (~15–30), and are slightly asymmetrical, with the slightly longer and more slender right pereopod 2 containing a few more carpal articles that the left (Holthuis 1971). The eyes of glyphocrangonids are relatively large, especially for deep-sea shrimps. The color (pigmentation) of the eyes is variable and somewhat of a mystery. Holthuis (1971), among others, found that the eyes of some species are black and pigmented, while those of others are white or some light opaque color, both in preserved and freshly dredged specimens. Furthermore, the proportion of individuals with nonpigmented but structurally large eyes is more common in juvenile specimens. What this means in terms of visual ability and physiology remains to be determined.

Holthuis (1971) made excellent notes on the color of freshly dredged specimens of several Atlantic species. He described how various parts of the body were tinted with various shades of pink, orange, red, or white, or without specific color. Preserved specimens have provided some information about glyphocrangonid reproductive biology, including the important finding that they produce large embryos that undergo abbreviated development (Chap. 9). Since the publication of Bauer (2004), there has been relatively little new information on glyphocrangonids except for the description of many new species, most of them by Tomoyuki Komai and coworkers (e.g., Komai 2004, 2005). Previous major references to the group include Holthuis (1971) and Chace (1984). An increased effort and continued technological advances on deep-sea instrumentation will be necessary to improve our knowledge of their biology and natural history.

3.11 Superfamily Nematocarcinoidea

The families of this group have bounced around from one superfamily to another, and some genera from family to family, throughout the years (Holthuis 1993; Bauer 2004). The most enduring grouping into one superfamily is that of Chace (1986), the Nematocarcinoidea, in which he placed the Nematocarcinidae, Xiphocarididae, Eugonatonotidae, and Rhynchocinetidae. This superfamily is currently recognized by WoRMS (2022), as is the elevation and inclusion of a former nematocarcinid genus (*Lipkius*) to form another family, the Lipkiidae. Molecular phylogenetic studies (Baeza et al. 2014; d'Udekem d'Acoz and De Grave 2018) have indicated that the Nematocarcinoidea may not be a monophyletic assemblage, but firmer interpretations about the true phylogenetic relationships among its constituent families will require further study (Chap. 12). In nematocarcinoid families, the first chelipeds are more robust although shorter than the second pereopods, the latter with carpus entire. All members of this superfamily have a more or less lengthy and spiny rostrum. Other characters defining the group are specialized traits related to mouthparts and other appendages (e.g., Yaldwyn 1960; Chace 1986, 1997; Burukovsky 2012; d'Udekem d'Acoz and De Grave 2018). Many nematocarcinoids are deep-sea shrimps, but even the shallower members of the group (Rhynchocinetidae) have some sort of pink, orange, or reddish coloration. One exception is the freshwater family Xiphocarididae. Unlike other members of this primarily marine superfamily, its two species live in streams and rivers as adults, but their planktonic larval development takes place in estuaries or the sea (amphidromous life cycle; Chap. 9).

3.11.1 Eugonatonotidae

This group has only two species at present, *Eugonatonotus crassus* and *E. chacei*; both species are found (rarely) in the Indo-Pacific and Western Atlantic oceans (Chace 1997). Like most nematocarcinoid species, these are medium to large-sized shrimps (~5–10 cm total length, excluding the rostrum). They are stout shrimps with a hard exoskeleton, a long deeply curved rostrum with large sharp spines; the rostrum continues to nearly the end of the carapace as a carina equipped with a series of small spines (Fig. 3.24a, Plate 3.5a). The carapace has pronounced longitudinal ridges (carinae). The pereopods bear both exopods and hooked epipods (mastigobranchs). Their beautiful rose-to-lilac color with pale bluish-to-white splotches has been described by Chan and Yu (1988, 1991) who helpfully included excellent photographs of them (Plate 3.5a). They are found on the deeper continental shelf down into shallow bathyal depths (~50–600 m; Chace 1997), and perhaps it is not surprising they do not display the more intense reds of deeper-dwelling shrimps. Their chelipeds are short but robust, especially the first, and the chela fingers of both ends in long, highly sclerotized stout setae ("spines") that form a basket at the end of the chela (Fig. 3.24b). This armature is very much like that of the rhynchocinetids

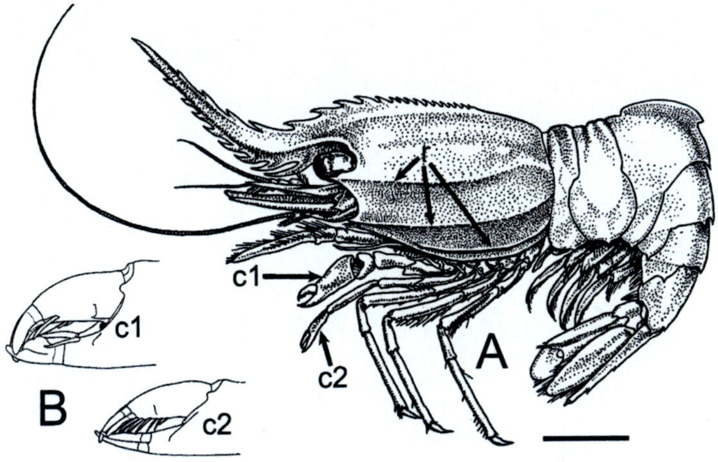

Fig. 3.24 Eugonatonotidae. (**a**) *Eugonatonotus crassus*. (**b**) *E. chacei*, tips of chela 1 (upper) and chela 2 (lower), showing the stout sclerotized setae (spines) which form a gripping or scraping basket. Scale bar represents 10 mm in (**a**). c1, c2, chela 1, 2; r. carapace ridges (carinae). From (**a**) Boone (1927), courtesy of Bulletin of the Bingham Oceanographic Collection and (**b**) Chace (1997)

(see below), with which they have been considered closely related by some workers (e.g., Yaldwyn 1960; Chace 1992). This may be a clue to their lifestyle, as the rhynchocinetids are grazers on sessile epifauna. *Eugonatonotus* species are rarely collected and very little is known about their biology as yet.

The family Galatheacarididae was proposed by Vereshchaka (1997) for *Galatheacaris abyssalis* based on a single specimen. Several more individuals were found in the gut contents of a deep-sea fish by Chow et al. (2000). The primitive morphological characters of these specimens suggested that they might be related to the Procarididae. However, in some careful detective work by De Grave et al. (2010), a comparison of mitochondrial DNA sequences of *Galatheacaris* specimens with those of *Eugonatontous chacei* showed them to be identical. The morphological characters of *Galatheacaris* led De Grave and colleagues to conclude that these specimens are megalopal (decapodid) larval stages of *E. chacei.* Thus, *G. abyssalis* and the Galatheacarididae are no longer considered valid taxa. This is a great example of how molecular evidence, now widely employed, is so important in helping to make taxonomic and phylogenetic interpretations. However, the final proof would be rearing of *E. chacei* larvae to see if the last stage is the now defunct *G. abyssalis*. In this deep-sea shrimp, this will be no easy task.

3.11.2 Lipkiidae

A very small family, the Lipkiidae has only two species, *Lipkius holthuisi* and *Fresnerhynchus crozeti* (Fig. 3.25). The namesake genus was placed in the

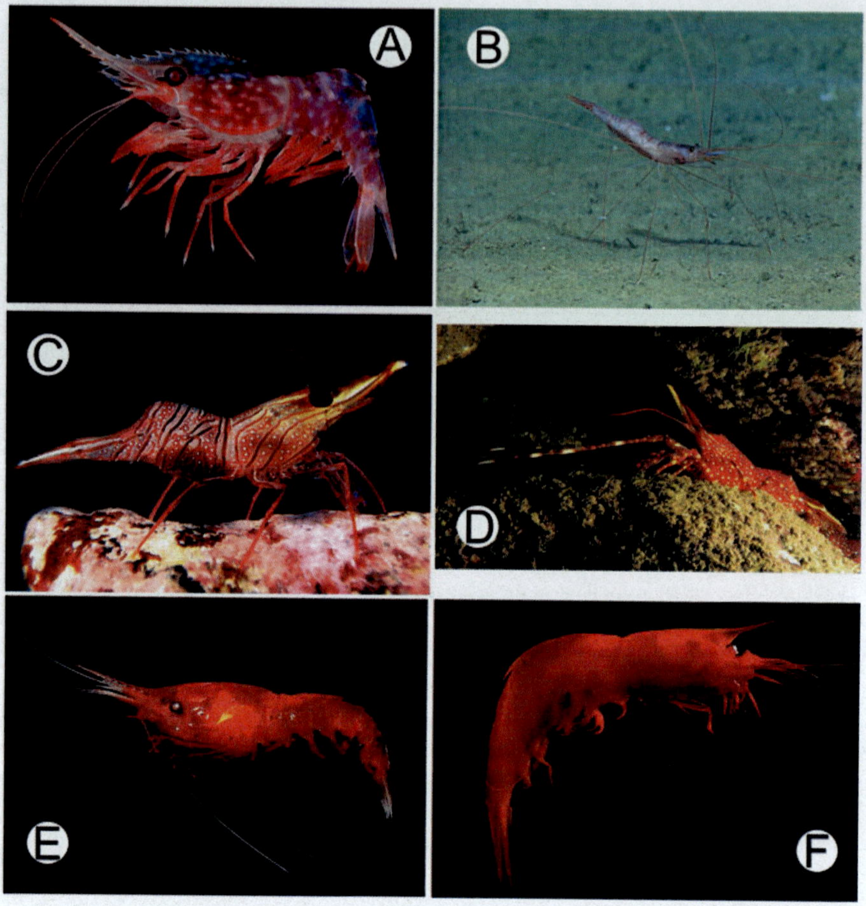

Plate 3.5 Eugonatonotidae to Oplophoroidea. (**a**) *Eugonatonotus crassus* (Eugonatonotidae). (**b**) *Nematocarcinus ensifer* (Nematocarcinidae). (**c**) *Rhynchocinetes uritai* female (Rhynchocinetidae). (**d**) *R. typus* dominant male. (**e**) *Oplophorus spinosus* (Oplophoridae). (**f**) *Acanthephyra pelagica* (Acanthephyridae). Photo acknowledgements: Courtesy of (**a**) T.Y. Chan, (**b**) Mary Wicksten and the U.S. National Oceanographic and Atmospheric Administration Office of Exploration and Research, (**c**) Martin Thiel (**d**) Ivan Hinojosa, (**e**, **f**) A.L. Vereshchaka

Rhynchocinetidae by Yaldwyn (1960), who first described it, along with *Eugonatonotus*, which now is in its own family. *L. holthuisi* was later placed in the Nematocarcinidae by Burukovsky (2012), but in its own subfamily because of its considerably shorter and stouter pereopods 2–5 in which the carpus is not extremely elongated as in other nematocarcinoids. Baeza et al. (2014) tentatively elevated it to familial status based on molecular data. This shows how morphologically interrelated these genera are, and how confusing and unsettled are the systematics of the nematocarcinoids. In the Lipkiidae, pereopodal exopods are rudimentary (*Lipkius*) or absent (*Fresnerhychus*), but both species have well-developed strap-shaped

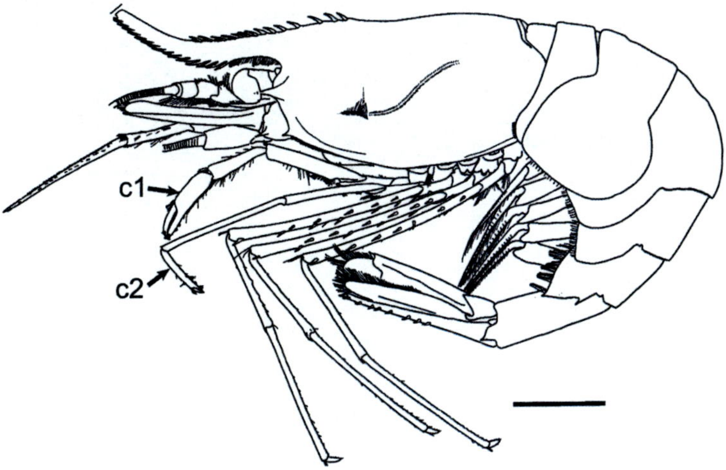

Fig. 3.25 Lipkiidae. *Fresnerhynchus crozeti*. Scale bar represents 10 mm. c1, c2, chela. From d'Udekem d'Acoz and De Grave (2018), courtesy of Zootaxa

epipods on pereopods 1–4 (d'Udekem d'Acoz and De Grave 2018). Both lipkiid species are from the southern hemisphere and from bathyal depths. Both are robust, medium- to large-sized shrimps (~5–10 cm total length, excluding the rostrum). *L. holthuisi* is found on mud bottoms while *F. crozeti* was collected from a hard bottom slope of a seamount.

3.11.3 Nematocarcinidae

This family currently contains 5 genera and 48 species, most of them (43) in the genus *Nematocarcinus*, which is abundant and ecologically important in deep-sea communities in bathyal (slope) depths but also ranges into abyssal depths in all oceans except the Arctic (Chace 1986; Burukovsky 2012; d'Udekem d'Acoz and De Grave 2018). These are relatively medium to large-sized shrimps ~5–10 cm total length, and not unexpectedly for deep-sea shrimps, have some shade of reddish color. They have long antennular and antennal flagella, and the rostrum is spiny above and variable in length. The most striking characteristic about them is their very long, stilt-like legs (pereopods 2–5), with which they perch, spider-like, above the bottom (Fig. 3.26a, Plate 3.5b). This distributes their body weight over a large surface area to prevent them from sinking into the very soft muds or oozes on which they live. In addition, the distal ends of the propodi of their legs, especially pereopods 3–5, are equipped with long stiff setae that appear to function as snowshoe-like "mudshoes" to walk or stand above the soft mud bottom (Fig. 3.26b). These pereopods have elongated ischial and meral segments but especially long carpal segments responsible for their stilt-like structure. Both chelae

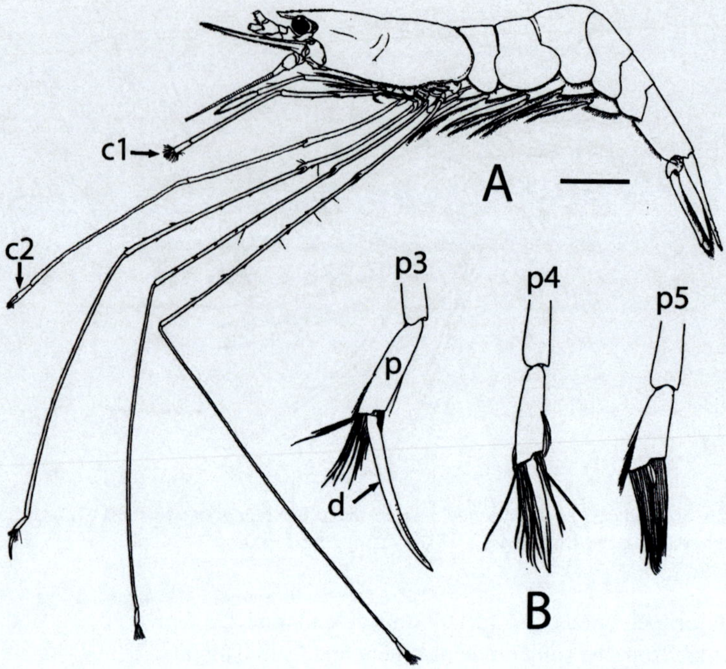

Fig. 3.26 Nematocarcinidae. (**a**) *Nematocarcinus undulatipes*; antennular flagella are broken off in this specimen, and the antennal flagellum partially so. See also Plate 3.5b (*N. ensifer*). (**b**) Propodus and dactylus of pereopods 3–5 (p3–5), showing long setal brushes at end of propodus. Scale bar represents 10 mm in (**a**). c1, chela 1; c2, chela 2; d, dactyl; p, propodus. Modified from Chace (1986)

are small but the second is more so than the first. The pereopods have exopods and the first four have strap-shaped epipods as well, although the latter lack the terminal hook of the epipod-setobranch cleaning complex (Chap. 4).

The eminent carcinologist and deep-sea biologist Rudolf Burukovsky has published extensively on this family, culminating in his 2012 monograph of the group, which includes much useful ecological information. These shrimps are detritivorous and necrophagous, feeding on small detrital particles as well as the occasional carrion windfalls. Members of the family mainly live on continental slopes and in shallow abyssal depths in which much of the detritus is terrestrial in origin. The terrigenous detritus is especially abundant in tropical and subtropical latitudes where organic production is high, and this may account for the pattern of nematocarcinid diversity, which is highest in the tropics and subtropics. Burukovsky also points out that they are also found near hydrothermal vents which are "oases" of organic productivity in the deep sea. Nematocarcinids serve as prey for deep-sea predators, including aristeid shrimps which are found in the same deep-sea communities. In addition to these observations, Burukovsky has described a "distoventral" setal organ on the ventrolateral side of the sixth abdominal somite. No function has yet been proposed for this structure, although the sensory perception of some sort seems likely.

3.11.4 Rhynchocinetidae

Members of this family (2 genera, 27 species) are found in rocky and coral reef habitats most commonly at shallow depths, although some species occur deeper on the continental shelf. Their geographic distribution is in tropical/subtropical waters of the Indo-West Pacific and Western Atlantic temperate waters, as well as temperate zone latitudes of the southeastern Pacific (Peru-Chile; New Zealand, Australia) (Chace 1997; Okuno 1997a, b). There has been an explosion of research on various aspects of the biology of this family since the year 2000, thanks to the efforts of Martin Thiel, his students, and other colleagues. The many papers generated by this group have greatly increased our knowledge of Rhynchocinetidae, as will be seen below and in the chapters on sexual biology and mating systems.

A unique characteristic of these shrimps is their hinged rostrum which is completely movable in *Rhynchocinetes* (Fig. 3.27b). In *Cinetorhynchus* species, however, the rostral articulation is partially fused to the carapace and so the rostrum cannot move. The function of rostral mobility in *Rhychocinetes* is unknown, and its loss in *Cinetorhychus* presents an interesting research opportunity. The first pair of chelae of rhynchocinetids are robust, and the chela fingers end in stout corneous spine-like setae that form a basket or cage similar to that of the Eugonatonotidae (Fig. 3.27a, d). *Rhynchocinetes typus* (and probably other rhynchocinetids) are important grazers (predators) of epifauna of rocky and coral reefs (Dumont et al. 2009). The highly sclerotized setal basket of the chelae is an adaptation for tearing and scraping sessile organisms from hard-bottom substrates. Rhynchocinetids have a somewhat hardened exoskeleton and hooked epipods on the first four pereopods. Pereopodal exopods are absent in this group. Most rhynchocinetids are highly nocturnal, hiding in cracks and crevices of rocky and coral reefs during the day, leaving at night to graze and carry out other activities. As in many shrimps, nocturnal activity is an escape from visual predators, especially fishes, during the day (Ory et al. 2014). Rhynchocinetids are all reddish in general coloration mixed with white stripes and blotches (Plate 3.5c, d). Some are very strikingly and brightly colored and that trait, plus their high degree of activity, make some species attractive to marine aquarists. Common names for rhynchocinetids include "hinge-beak" and "camel" shrimps because of their movable rostra and the pronounced caridean abdominal hump, respectively.

Two strikingly different types of sexual dimorphism and associated mating systems are found among the Rhynchocinetidae. In some species of both genera, as in many caridean species, males are small and females grow larger, a common adaptation for increasing female fecundity (e.g., *Rhynchocinetes uritai*; Bauer and Thiel 2011; Osawa et al. 2015). In other rhynchocinetids, females and younger reproductive males are similar in size and morphology. However, if they live long

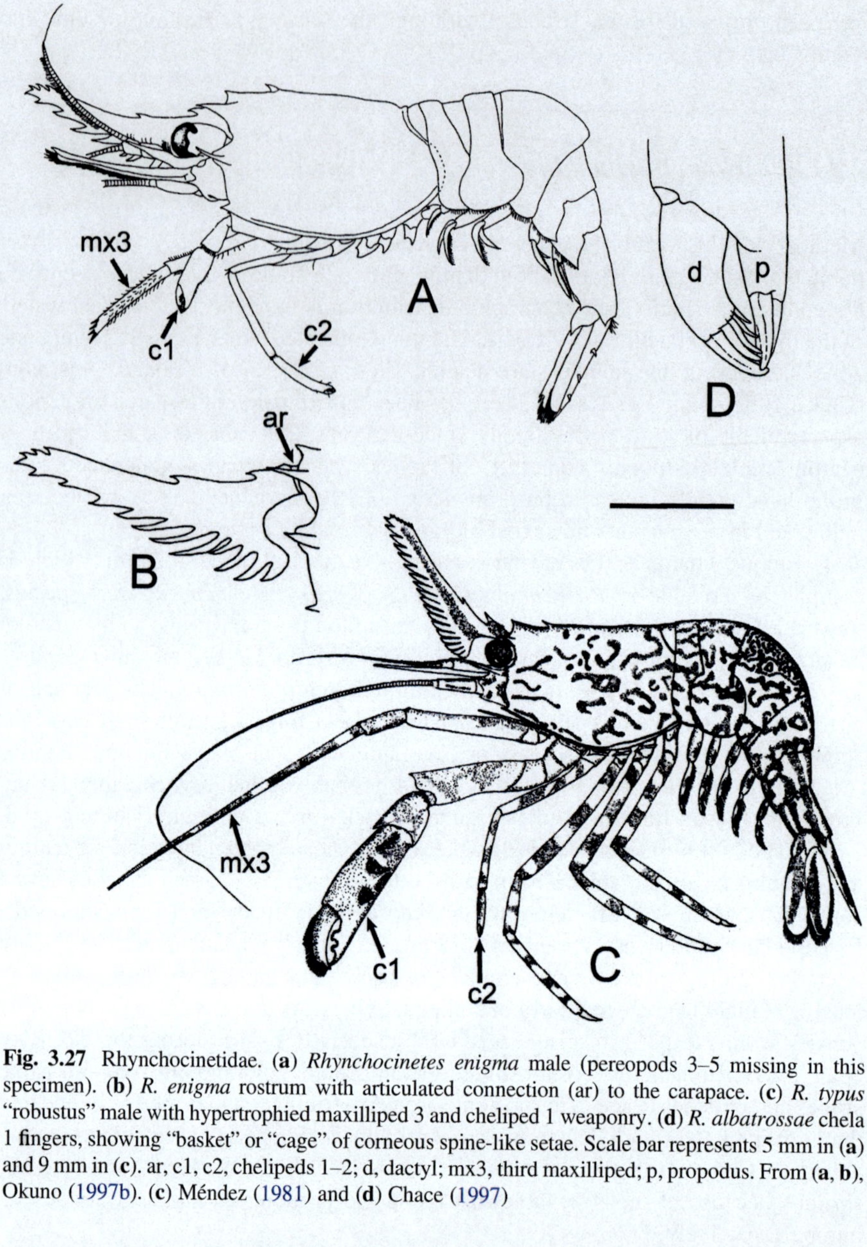

Fig. 3.27 Rhynchocinetidae. (**a**) *Rhynchocinetes enigma* male (pereopods 3–5 missing in this specimen). (**b**) *R. enigma* rostrum with articulated connection (ar) to the carapace. (**c**) *R. typus* "robustus" male with hypertrophied maxilliped 3 and cheliped 1 weaponry. (**d**) *R. albatrossae* chela 1 fingers, showing "basket" or "cage" of corneous spine-like setae. Scale bar represents 5 mm in (**a**) and 9 mm in (**c**). ar, c1, c2, chelipeds 1–2; d, dactyl; mx3, third maxilliped; p, propodus. From (**a**, **b**), Okuno (1997b). (**c**) Méndez (1981) and (**d**) Chace (1997)

enough, some males mature into large "robustus" individuals with hypertrophied weaponry (Correa et al. 2000; Thiel et al. 2010; Prakash et al. 2016). In *Rhynchocinetes typus* robustus males, the first chelipeds are relatively much larger than in smaller males, have specialized setal tufts, and the third maxillipeds become extremely elongated, tipped with stout sclerotized spines (Fig. 3.27c, Plate 3.5d). In

Cinetorhynchus species with robustus males, the third maxillipeds are not hypertrophied but the first chelipeds are extremely elongated and their chelae change in shape with size to a wrench-like structure (Bauer et al. 2014; Chap. 8). Observations by Correa et al. (2003) show that *R. typus* robustus males fight with one another and subordinate males over sexually receptive females; serious injuries are inflicted in these fights. In *Cinetorhynchus* species (Bauer et al. 2014), fighting has not been observed directly, but puncture marks and scars are common on the hypertrophied chelipeds of robustus males. Species with robustus males indicate a female guarding/defense mating strategy. In contrast, the mating strategy of rhynchocinetid species with small males and larger females is likely a "pure-search" or promiscuous mating system in which males do not defend females, but rather search constantly for receptive females to be the first male to inseminate them. This is a common sexual strategy of many caridean species with small males and larger females (Wickler and Seibt 1981; Correa and Thiel 2003a; Bauer 2004). One such rhynchocinetid species, *R. uritai*, is also a protandric hermaphrodite (Bauer and Thiel 2011; Osawa et al. 2015), raising the possibility that other similar rhynchocinetids are protandric. So far, species with robustus males do not show hermaphroditism, as deduced from population structure (Correa and Thiel 2003b; Bauer et al. 2014). More details on sexual and mating systems of rhynchocinetids are discussed in Chaps. 7 and 8.

3.11.5 Xiphocarididae

This is a freshwater nematocarcinoid with a single genus and two species. The better-known *Xiphocaris elongata* is widely distributed in streams of mountainous islands of the West Indies (Chace and Hobbs 1969). The other species, *X. gomezi*, was described from Cuba (Juarrero de Varona 1993). The family was classified for many years as a primitive atyid (e.g., Chace and Hobbs 1969; Fryer 1977), mainly because of the dactylar grooming comb of the last walking legs (pereopod 5), a morphological character shared with atyids (e.g., Fig. 3.14d) in addition to its freshwater amphidromous lifestyle. Other shared characters with atyids are relatively few, and most importantly, *Xiphocaris* does not have the setal brushes on the chelae for sweep- and/or filtering feeding as do atyids. Chace (1992) tentatively placed *Xiphocaris* in the Nematocarcinoidea based on morphological similarity to genera such as *Lipkius* and *Nematocarcinus*, and that is the classification followed at present (e.g., Martin and Davis 2001; WoRMS 2022).

Xiphocaris elongata is an abundant and active shrimp in island streams, and it has been much studied. The best descriptions of its morphology and general behavior are found in Chace and Hobbs (1969) and Fryer (1977). Its chelae have a rather generalized nematocarcinoid form. It is a translucent, delicately built streamlined caridean with rostrum of variable length, well-developed exopods and hooked epipods, and long slender legs (Fig. 3.28) with which it stands high and deftly "tip-toes" above the substrate. It is a swift agile swimmer (Plate 3.6d), capable of both hovering and swimming backward, leaping out of the water when startled and,

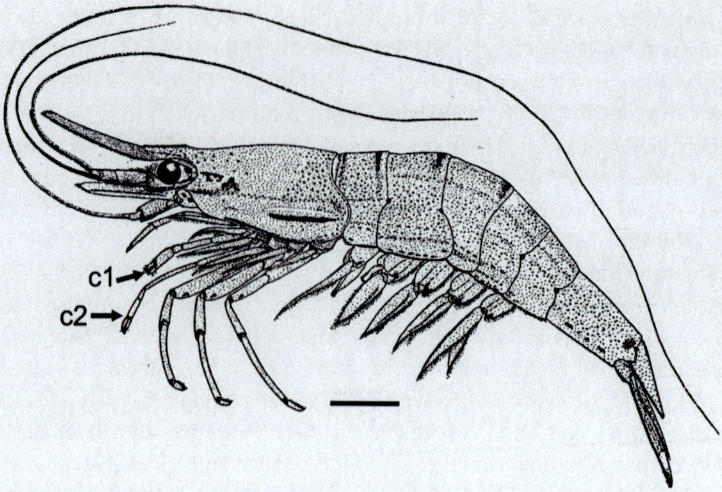

Fig. 3.28 Xiphocarididae. *Xiphocaris elongata.* Scale bar represents 4.4 mm. c1, c2, chela 1, 2. From Chace and Hobbs (1969)

in spite of its spindly legs, crawling back into the water if it jumps out of the stream. Although a strong swimmer, it prefers quiet pools or seeks shelter in the lee of rocks or vegetation in swifter-flowing streams. Its chief mode of nutrition is shredding and ingesting small particles of leaves which fall into the stream (Covich et al. 1999; Crowl et al. 2006). It is a bit of a generalist which can feed on detritus, prey on stream insects, and make good use of carrion or other windfalls, e.g., a bit of picnic lunch tossed into the water will quickly attract them and stimulate a *Xiphocaris* feeding frenzy.

Knowledge about the ecology and life history of *Xiphocaris elongata*, like that of its fellow West Indian stream atyids and *Macrobrachium* species, has been greatly advanced in recent years by freshwater ecologist Alan Covich and his students. It was long suspected that *X. elongata* larvae were released upstream to travel to the sea, where planktonic development takes place, and then migrate as postlarvae/juveniles back up into freshwater streams (e.g., Chace and Hobbs 1969). Kikkert et al. (2009) have confirmed this lifestyle in a detailed study that provided important behavioral information and environmental cues that are involved. *X. elongata* exports, with its high reproductive output in the form of larvae, a large quantity of organic matter downstream to the sea (Cross et al. 2008). Most larvae do not survive their marine development to return as postlarvae. This species' leaf-shredding activities during feeding are important in the breakdown of terrestrial plant leaves, which are the major organic input into West Indian streams. Their leaf shredding aids the decomposition of leaves into smaller detrital particles that become available to filter-feeding shrimps and other stream invertebrates (Crowl et al. 2006; Covich 2009). Long-term mark-recapture experiments with *X. elongata* have revealed that these shrimps grow rapidly in their first 2–3 years, but subsequently, growth slows down. *Xiphocaris elongata* can live remarkably long for a shrimp, from 5 to at least

Plate 3.6 Palaemonidae. (**a**) Unidentified palaemonid from Heron Island, Australia, showing a typical palaemonid body form. (**b**) Palaemonine *Creaseria morleyi* from freshwater cave waters, Yucatan, Mexico. (**c**) Dominant male of the palaemonine *Cryphiops caementarius*. (**d**) Dominant male of the palaemonine *Macrobrachium carcinus* emerging from shelter in a stream, Puerto Rico. The smaller shrimps swimming above are a favorite prey, *Xiphocaris elongata* (Xiphocarididae). (**e**) Pontoniine *Periclimenes yucatanicus*. (**f**) Pontoniine *Periclimenaeus pachydentatus*, one of the pontoniine snapping shrimps (see Chap. 2). (**g**) *Coralliocaris graminea*, a pontoniine snapping shrimp which snaps with both major claws (see Chap. 2). Photo acknowledgments: Courtesy of (**a**, **c**, **e**–**g**) Arthur Anker, (**b**) Benjamin Magana, (**d**) David Herasimtschuk, Freshwaters Illustrated

11 years, depending on local stream conditions and predatory pressures (Cross et al. 2008; Pérez Reyes et al. 2015).

3.12 Superfamily Oplophoroidea

The oplophoroids are a group of pelagic marine shrimps which reside in the deep sea, ranging from ~300 to 5300 m (Chace 1986; Chan et al. 2010). They occur in all oceans of the world (Lunina et al. 2019). The mesopelagic species are vertical migraters, moving up at night nearer to surface waters where food is more abundant than in deeper waters, and then back into protective low-lighted waters by day. Most bathypelagic species migrate vertically little or not at all (Omori 1974). Several species are benthopelagic, residing on or near the sea floor during the day but moving up into the water column at night. The mesopelagic species have a translucent reddish coloration from chromatosomes ("chromatophores": see Chap. 5) with red pigment, while the bathypelagic species are a deep red from pigments in the integument as well as from red chromatosomes (Plate 3.5e, f) (Chap. 5). When alarmed, oplophoroids can emit a blueish cloud of bioluminescent (455–460 nm) secretion from the mouth which originates from hepatopancreas (Herring and Roe 1988). This defensive response, which also occurs in some deep-dwelling shrimps such as the pandalid shrimp *Heterocarpus* (Plate 5.3e), serves to startle and distract predators or other intruders, much like the inky cloud emitted by disturbed cephalopods. The primarily mesopelagic species may have cuticular photophores which, as in the mesopelagic sergestids, produce a bioluminescent light that matches the light reaching their depths from the surface. This is a countershading camouflage which masks the silhouette of the shrimp from visual predators below (Chap. 5).

Adaptations for swimming and maneuvering in the water column are particularly well developed in the oplophoroids, as might be expected in pelagic shrimps (Fig. 3.29a, b). Well-developed pereopodal exopods, which function as rotary propellors to aid forward swimming and agility, are characteristic of these shrimps. Long spiny rostra, longitudinal carapace keels, abdominal carinae ending in sharp spines are common hydrodynamic features. The sixth abdominal segment may be elongated (several *Systellaspis* species) to enable a more forceful backward escape response (Lunina et al. 2019). The latter authors reasonably hypothesize that the variably developed spiny projections (rostrum, abdomen, telson) also function in defense from predators. All oplophoroids have the pereopodal epipods involved in caridean gill cleaning (Chap. 4) but, unusual for a caridean, the epipods bear a small vertical appendix extending somewhat into the branchial chamber, a reduced version of those found as in penaeoid and procaridid shrimps. Oplophoroids are medium to large shrimps from 3 cm up to 18 cm in total length, excluding rostra (Omori 1974; Chace 1986). There is considerable variation in eye size, but the mesopelagic species usually have large eyes while the bathypelagic species have smaller ones. Embryo ("egg") size varies among species with depth as well (e.g., Omori 1974; Chace 1986; Chan et al. 2010). Mesopelagic species usually produce many small embryos while

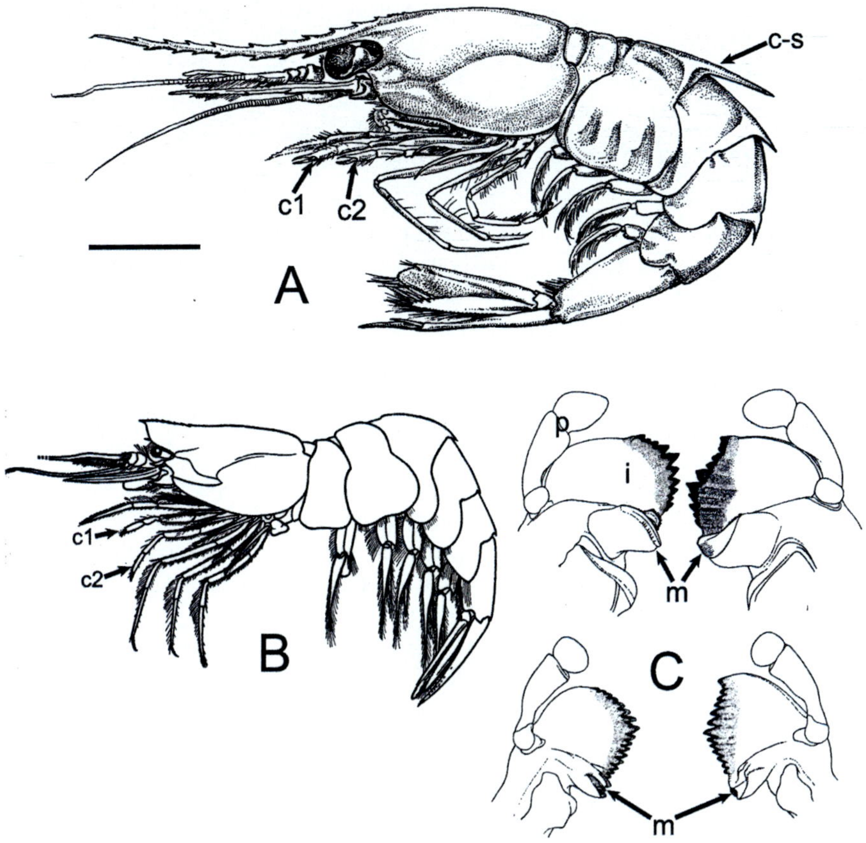

Fig. 3.29 Oplophoroidea. (**a**) *Janicella spincauda* (Oplophoridae). (**b**) *Acanthephyra curtirostris* (Acanthephyridae). (**c**) Dorsal view of mandibles of (above) *Acanthephyra stylorostratis* (large molar processes typical of Acanthephyridae) and (below) *Oplophorus typus* with reduced molar processes, divided on that of left mandible. Scale bar represents 4.5 mm in (**a**) and 12.0 mm in (**b**). c1, c2, chelipeds 1, 2; c-s, dorsal median carina and spine; i, incisor process; m, molar process; p, mandibular palp. From (**a**) Boone 1927 and (**b**, **c**) Chace (1986)

bathypelagic species often incubate fewer but larger embryos. Oplophorids also have relatively long life spans (3–8 years; Omori 1974). The adaptive significance of variation in embryo size and life span is discussed in Chap. 9.

The systematics of this group, as in many other shrimp taxa, has changed back and forth through time. Chace (1986) recognized a single family, the Oplophoridae. More recently, on the basis of both morphological (mainly the mandible) and molecular characters, Chan et al. (2010) recognized two families, the Acanthephyridae and Oplophoridae (*s.s.*). On the other hand, Wong et al. (2015), although showing two major clades within the group, concluded that the oplophoroids are monophyletic, consisting of but one family, the Oplophoridae (*s.

l.). Both Lunina et al. (2019) and currently WoRMS (2022) agree with the Chan et al. (2010) two-family arrangement, which will be followed here.

3.12.1 Oplophoridae and Acanthephyridae

Given that these two oplophoroids are so closely related and have many characteristics in common, it might be best to discuss them in a compare-and-contrast style (Chan et al. 2010). Both the Oplophoridae (*s.s.*; 3 genera, 16 species) (Fig. 3.29a) and the Acanthephyridae (7 genera, 55 species) (Fig. 3.29b) have strong mandibles with highly developed incisor processes, indicative of their carnivorous lifestyle (e.g., Moore et al. 1993; Hopkins et al. 1994; Burukovsky and Falkenhaug 2015; Lunina et al. 2019). However, in the Acanthephyridae, the molar process is well developed while that of the Oplophoridae is reduced and/or divided by a split or channel (Fig. 3.29c) (Chace 1986). The most speciose oplophorid genera are *Oplophorus* and *Systellaspis*, and like all members of the family, they are mesopelagic (50–1000 m depth), bear cuticular photophores, migrate toward the surface at night and have relatively large eyes. Females produce and incubate numerous small embryos. Acanthephyrid species are mainly bathypelagic, although a few are deep mesopelagic, or benthopelagic, residing near or on the bottom during the day, as also do some *Systellaspsis* (Oplophoridae *s.s.*) (Chace 1986). Like oplophorid species, acanthephyrids can emit a defensive bioluminescent cloud but do not bear cuticular photophores, as these countershading structures would serve no purpose in the dark waters where most acathephyrids reside. Oplophorids, living at shallower depths where some surface light penetrates, have large eyes, while acanthephyrids, living in bathyal depths where the only source of light is bioluminescence, tend to have smaller eyes. Acanthephyrids have thin to soft, membranous integuments, which serve in buoyancy (Chace 1940), conserving swimming energy needed to maintain position, an important adaptation in the food-poor bathypelagic zone. Deeper-living acanthephyrids migrate vertically little or usually not at all (e.g., Omori 1974; Hopkins et al. 1994), while the mesopelagic oplophorids do move up into near-surface waters at night.

This is another shrimp group about which, because of the difficulty of making observations and collection of live material in meso- and bathypelagic seas, little is known about much of its general biology and behavior. As with other pelagic and deep-sea taxa, advances in ROV and submersible technology, along with increased funding and research effort on these and other shrimps from the vast and understudied areas of the world's oceans will help remedy this situation.

3.13 Superfamily Palaemonoidea

This superfamily is quite a large and important one, as it contains about 30% (1112) of all caridean species. Its systematics can be confusing, as numerous proposals and changes about the number of families and the genera within them have taken place over the years. The superfamily has contained as many as eight separate families (Martin and Davis 2001) and, as of this writing, has been whittled down to four (WoRMS 2022) as a result of close examination of morphological characters combined with molecular data (e.g., Kou et al. 2013; De Grave et al. 2015a; Chow et al. 2020). Without doubt, further study will produce more change with the goal, as always, of finding a stable and phylogenetically meaningful arrangement.

As its ecological and species diversity shows, the Palaemonoidea is a highly successful group of shrimps. They have radiated into most tropical (especially) and temperate latitude aquatic habitats in shallow continental shelf waters of the world oceans, with relatively few deep-sea species. The group is also well represented in freshwater rivers, streams, lakes, freshwater, and anchialine subterranean habitats. The habitus of its member species ranges from that of a typical epibenthic caridoid to one highly modified for living in symbiotic associations with other invertebrates (Chap. 10). Palaemonoids have small first chelipeds modified for food search, habitat exploration and grooming, while the second pair (carpus entire), are much more robust and modified as cheliped weapons for agonistic behavior and predation (Figs. 3.30, 3.31, and 3.32). These weapons can range from relatively modest structures with forcep-like fingers (Fig. 3.31a, c) to highly hypertrophied massive ones (Fig. 3.31b), and even to snapping claws as discussed in Chap. 2. The second chelipeds are asymmetrical in many genera (e.g., Fig. 3.31b, d) with a major and a minor cheliped differing in morphology and function. The technical details which unite the group (Chace 1992) include a first antenna in which the outer (upper) ramus bears an accessory branch and a mandible in which the incisor process, if present, is distinctly separated from the molar process. The pereopods never have arthrobranch gills nor exopods, the latter trait in keeping with their overall benthic lifestyle, and most do not have gill-cleaning hooked epipods on the pereopods. All types of sexual dimorphism and sexual systems occur in this very large group. Some species have small males and larger females with pure search (promiscuous) mating behavior, some species similar in size live in pairs (social monogamy), and other species in which some sexually dominant males with hypertrophied cheliped weaponry are much larger than females. Intriguingly, although this is a very speciose taxon, no hermaphroditic (protandric) species have been yet found (Chap. 7).

3.13.1 Desmocarididae

This palaemonoid family consists of but one genus and two species (Fig. 3.30a) and are found only in Central and West African swamp forests and streams (*Desmocaris*

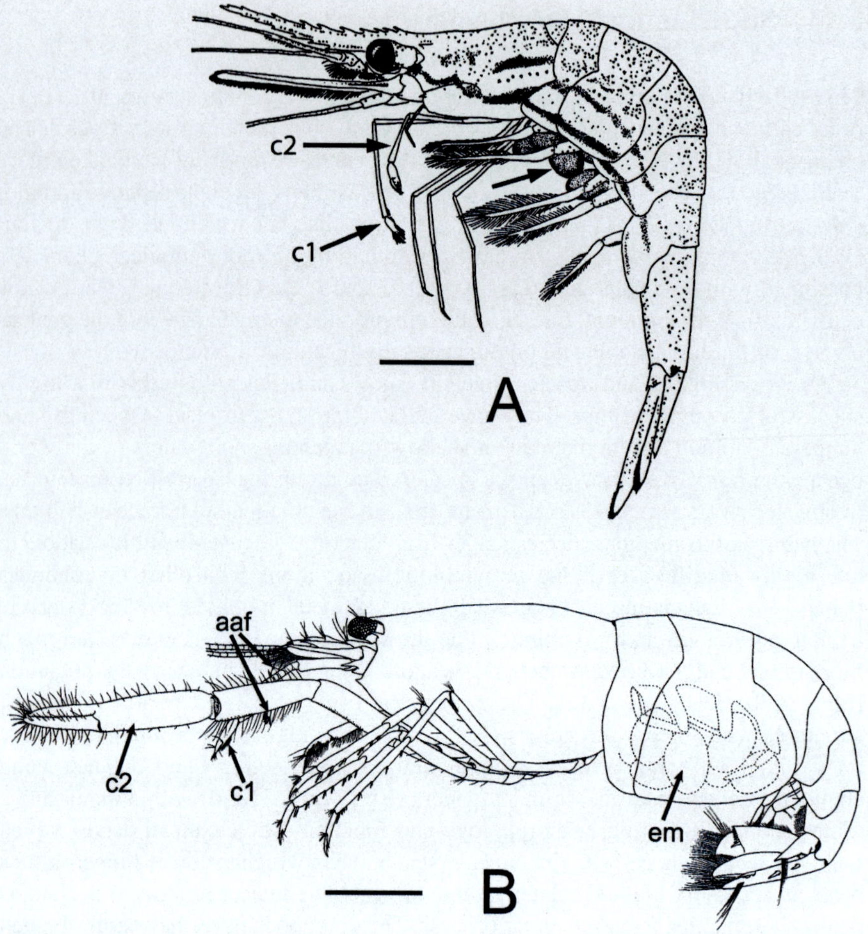

Fig. 3.30 Two palaemonoid families: Desmocarididae and Euryrhynchidae. (**a**) *Desmocaris bislineata*. (**b**) *Euryrhynchus amazoniensis* female incubating large-sized embryos. Scale bar represents 5 mm in (**a**) and 2.0 mm in (**b**). aaf, accessory antennular flagellum which flexes ventrally to test ("sniff") to surrounding water below; em, embryo (unmarked arrow in **a**); c1, 2, cheliped 1, 2. (**a**) from Powell (1977); (**b**) illustration courtesy of Paulo Pachelle

trispinosa) as well as tidal freshwater grass mats (*D. bislineata*). What is known about the biology of this genus is due to the intrepid British zoologist C.B. Powell (1977, 1979b), who spent most of his career living in Africa and studying its fauna (Williamson 2002). He supported recognition of the genus into its own family (Desmocarididae) owning to such characters as the absence of appendices internae on the pleopods, spoon-shaped second chelae adapted for collecting fine detritus from aquatic vegetation and other surfaces, and other specialized morphological traits (Powell 1977). The pleopods are adapted for an unusual shrimp swimming behavior: hovering. The pleopodal endopods are reduced in size while exopods are

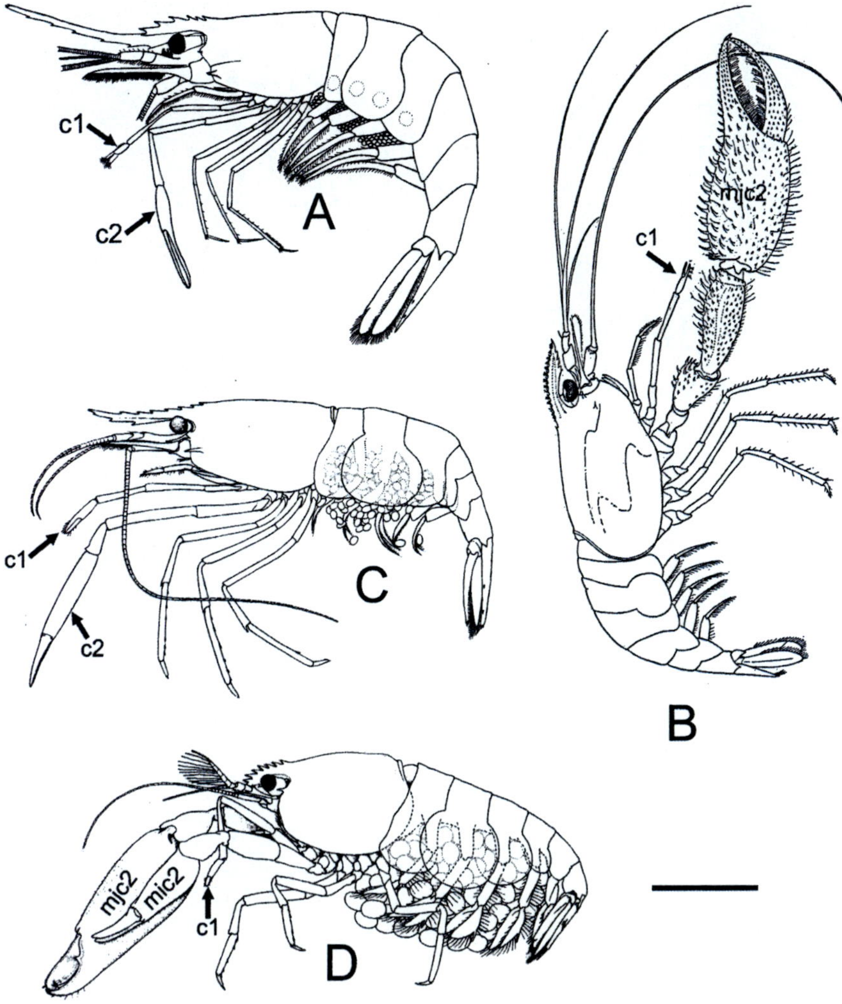

Fig. 3.31 Palaemonidae. (**a**, **b**) "palaemonines", (**c**, **d**), "pontoniines". (**a**) *Palaemon kwangtung*. (**b**) *Macrobrachium olfersii* (large dominant male). (**c**) *Echinopericlimenes dentidactylus*; (**d**) *Epipontonia anceps*. Scale bar represents 14 m in (**a**), 28 mm in (**b**), 6.5 mm in (**c**), and 3 mm in (**d**). c1, 2, chela 1, 2; mic2, minor chela 2; mjc2, major chela 2. From (**a**) Guo et al. (2005) courtesy of Crustaceana (Brill), (**b**) Holthuis (1952a) courtesy USC Digital Library, Allan Hancock Foundation Collection, (**c**) Chace and Bruce (1993) and (**d**) Bruce (1983b)

elongated, held anterolaterally and beat in the horizontal plane, which allows the shrimp to hover stationary in the water column. *D. trispinosa* lives in isolated, stagnant pools of swamp forests, where the oxygen content is so low that shrimps of other taxa cannot survive. *D. trispinosa* often hover just below the water surface, where oxygen content is highest due to the proximity of aerial oxygen. The hovering habit explains the loss of appendices internae which normally link pleopod pairs

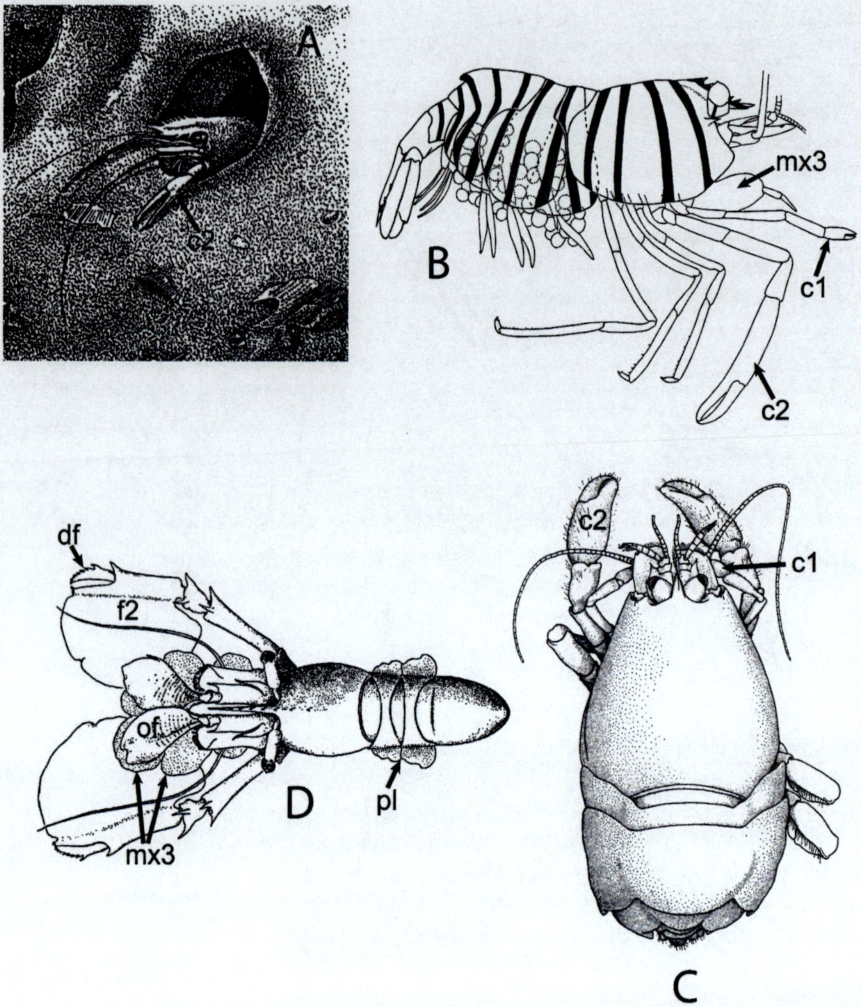

Fig. 3.32 Other palaemonoids. (**a**) *Anchistioides willeyi* in sponge. (**b**) *Gnathophyllum taylori*. (**c**) *Pycnocaris chagoae*. (**d**) *Hymenocera picta*. Scale bar represents 13 mm in (**a**), 6 mm in (**b**), 4 mm in (**c**), 3 mm in (**d**) and 12 mm in (**e**). c1, c2, chela 1. 2; df, dactylar (movable) finger of chela 1; f2, propodal (fixed) finger of chela 2; mx3; expanded basal segment(s) of maxilliped 3; of, outer flagellum of antennule; pl, pleopods. From (**a**) Holthuis (1952b), (**b**) Ahyong (2003), (**c**) Bruce (1972) courtesy of Crustaceana (Brill) and (**d**) Barnard (1950)

together in a vertical position. By changes in position of the pleopods, forward swimming is also possible. Unlike most shrimps, they will not attack weaker organisms but readily will consume carrion in addition to their detritivorous (microphagous) food habits. Individuals of *Desmocaris* are largely transparent with fine markings composed of small chromatosomes. Similar to the euryrhynchinids discussed below, *Desmocaris* females produce very large yolk-filled embryos (Fig. 3.30a, unmarked arrow) so that all development occurs within the incubated embryos, which hatch as small juveniles. The oxygen- and plankton-poor habitats in which *Desmocaris* lives would be difficult for extended planktonic larval development, so direct development has evolved to avoid these problems (Chap. 9).

3.13.2 Euryrhynchidae

The euryrhynchids are a small freshwater palaemonoid group (4 genera, 11 species) found in streams, rivers, inundated tropical forests, as well as epigean habitats (e.g., deep wells) in the Amazon Basin, West Africa, and India (Powell 1976; De Grave 2007; Magalhães et al. 2016; Pachelle and Tavares 2018; De Grave et al. 2018a, b). The seven species of *Euryrhynchus* are found in northern South America (Amazon Basin), the two species of *Euryrhynchina* and one of *Euryrhynchoide*s occur in West Africa, while *Eurindicus bhugarbha* was discovered in a homestead well in southwestern India. The distribution of these closely related genera on these three continents indicates that the family originated in the ancient supercontinent of Gondwana. This continent broke up into several parts, leading to the separation of South America, Africa, and India about 65 million years ago (Dietz and Holden 1972). The euryrhynchinid body is stout, robust, somewhat crayfish-like, and large second chelipeds belie their small size from 2 to 4 mm carapace length in the tiny *Euryrhynchina* species up to 10 mm in *Euryrhynchoides holthuisi* (Powell 1976; Pachelle and Tavares 2018) (Fig. 3.30b). There are excellent photographs of various *Euryrhynchus* species in Pachelle and Tavares (2018).

As palaemonoids, the second pair of chelipeds are considerably larger and stouter than the first (Fig. 3.30b). Euryrhynchids have variously been classified as a subfamily of Palaemonidae, or together with the subterranean brackish water *Typhlocaris* species (see below) in a family Typhlocarididae on the basis of similarity of mouthparts (e.g., acute, well-developed caridean lobe of the first maxilliped; Chace 1992; De Grave 2007). However, euryrhynchids are distinguished from other palaemonoids, including *Typhlocaris*, by the complete separation of the inner accessory branch from the outer branch of the lateral antennular flagellum (Fig. 3.30b). Powell (1976) observed that on each side of the shrimp, this accessory branch, which bears the olfactory hairs, flexes ventrally from time to time into the anterolaterally directed currents generated by the maxillipedal exopods. It appears that the shrimp is "sniffing" (chemically testing) its surroundings.

Other euryrhynchinid traits include a short rostrum, a smooth carapace, and somewhat reduced eyes. The latter may be a result of their sedentary and cryptic behavior in low-light habitats. Some *Euryrhynchus* and *Eurindicus* species have been collected in ground water habitats (wells) (De Grave et al. 2018a, b) but other euryrhynchinids are more often found in surface waters hiding under leaf litter in shaded forest swamps (Kensley and Walker 1982). The euryrhynchinid exoskeleton lacks color, but these shrimps have a blue or greenish tinge, perhaps due to the

Tyndall effect on underlying tissues (Chap. 5). Males of *Euryrhynchina* and *Eurindicus* have a typical appendix masculina on the second pleopod, while those of the other genera have highly modified and unusual gonopods for a caridean (Powell 1976; Pachelle and Tavares 2018) (Chap. 6).

Important observations and reviews of euryrhynchid ecology and life history are those of Powell (1976) for African species, Kensley and Walker (1982) and especially Magalhães et al. (2016) for *Euryrhynchus* (South America). Euryrhynchinids are typically found in tropical forests which are seasonally flooded by surrounding rivers and streams. It is in these swampy shaded pools or lakes that euryrhynchids reproduce and become abundant. As the dry season arrives and these pools dry up, the shrimps retreat into the streams, rivers, and canals which flowed into the forest to create temporary pools and lakes. Euryrhynchids have highly abbreviated larval development so that embryos hatch nearly as miniature adults (Magalhães 1988). The flooded areas and swamp pools in which euryrhynchinids reproduce are acidic and poor in nutrients. These are poor environments for feeding larvae as there is no or little planktonic food in these waters. The euryrhynchids produce a small number of very large yolky eggs (Fig. 3.30b), as do desmocaridids which live in similar habitats. The large supply of yolk sustains the development to hatching of a benthic juvenile which can feed and survive in the same habitat as the adult. In these nutrient-poor environments, fungi that decompose leaf and plant litter are the basis of the food chain. Adults and probably juveniles have a similar diet—some mix of fungi, leaf detritus, and small invertebrates, especially insect larvae (Kensley and Walker 1982).

3.13.3 Palaemonidae (s.l.)

This is the major family of the Palaemonoidea, with 156 genera and 1095 species. Most of these species are composed of the Palaemonidae (*s.s.*) (Chace 1992; De Grave and Fransen 2011) which traditionally has been divided for many years into two subfamilies, the Palaemoninae and Pontoniinae. According to De Grave et al. (2015a), the characters supporting this division, such as telson ornamentation and presence/absence of a mandibular palp, are more variable than once believed. Together with molecular data, these researchers concluded that the two subfamilies do not represent true phylogenetic clades and therefore they no longer recognize them. However, as these two groups of species do share many morphological and ecological traits, for ease of discussion, I will refer to them in the following paragraphs as informal assemblages (palaemonines, pontoniines), and I will do the same with other allied palaemonoids (e.g., gnathophyllids, anchistioids, hymenocerids) not currently recognized as valid families by De Grave et al. (2015a) and Chow et al. (2020).

The "palaemonines," which make up a little over a third of the Palaemonidae (*s.l.*) species, have radiated from shallow-water marine habitats into estuarine and freshwater environments; relatively few are found much deeper than the continental shelf.

Plate 3.7 Other palaemonoids to Pandalidae. (**a**) Prespawning female of unidentified pontoniine female from Panama with large mature ovaries extending from cephalothorax far into abdomen. (**b**) Recently spawned female of same species as (**a**) bearing embryos beneath the abdomen and with empty ovaries (transparent, not visible) in cephalothorax. (**c**) *Gnathophyllum* sp. (Panama). (**d**) gnathophyllid *Pycnocaris chagoae*. (**e**) *Typhlocaris ayyaloni* (Typhlocarididae). (**f**) *Pandalus danae* (Pandalidae). Photo acknowledgments: Courtesy of (**a–d**) Arthur Anker, (**e**) David Darom (**f**) Gregory Jensen

Morphologically, they have the basic epibenthic shrimp body form with moderate to long spiny rostra and laterally compressed bodies (e.g., Fig. 3.31a, b, Plate 3.6a–d). They are generally active, aggressive shrimps always ready to attack and prey on weaker organisms, e.g., the subterranean *Creaseria morleyi* (Plate 3.6b), a predator on the smaller microphagous atyid cave shrimp *Typhlatya* sp. (Chávez-Solís et al.

2018). In addition to their predatory tendencies, palaemonines are feeding generalists attracted to carrion and/or are detritovores, feeding on bits of decaying plant material to digest the decomposing microbes growing there. Ecologically important marine genera include *Leander* and *Brachycarpus*, small to medium-sized shrimps found in tropical and subtropical seagrass meadows, algae, and coral reefs. *Leander tenuicornis* is equally at home in seagrass meadows worldwide (Bauer 1985b; Schaffmeister et al. 2006) as amid masses of floating *Sargassum* seaweed community, of which it is an important member (Jobe and Brooks 2009; RTB personal observation). All stages of invasion into freshwater habitats are seen in the major genus, *Palaemon* (89 species) which occurs in marine, estuarine, and fully freshwater habitats. The genus includes a group of species formerly recognized as the genus *Palaemonetes* (Plate 6.1a, b), which has been synonymized with *Palaemon* (De Grave and Ashelby 2013). *Palemonetes* (*s.s.*) species occur primarily in brackish (e.g., salt marshes) and freshwater habitats around the world. Often quite abundant and accessible to investigators because of their ease of sampling in shallow waters, several species have been the focus of many ecological, physiological and behavioral studies throughout the years, some of which will be cited throughout this book.

The most successful, in terms of species diversity, of the "palaemonines" is the genus *Macrobrachium*, with a massive 264 species of freshwater shrimps. As might be expected from such a large genus, there is considerable variation in size and natural history. In many species of this genus, the male may grow much larger than the female, with relatively heavy weaponry (Fig. 3.31b), as in the closely related genus, *Cryphiops* from the west coast of South America (Plate 3.6c). An extreme example of this sexual dimorphism is the large older males of *Macrobrachium carcinus*, (Plate 3.6d) which can reach the size of a medium-sized American lobster (*Homarus americanus*) (Hedgpeth 1949; RTB personal observation of specimens in the U.S. National Museum of Natural History). As might be expected, sexual interactions are dominated by such males (e.g., Kuris et al. 1987) (Chap. 7). However, there are also numerous *Macrobrachium* species in which reproductive males are smaller on average than females, as in *M. ohione* (Bowles et al. 2000).

Some *Macrobrachium* carry out their entire life cycle in freshwater bodies of water. Most of such species put their reproductive effort into a small number of large yolky eggs (oocytes) that, after fertilization and incubation, hatch as advanced nonfeeding larvae or small juveniles. This reproductive strategy avoids the dangers (but also benefits) of extended larval development in the plankton (see Chap. 9). Other species which live in rivers and streams produce large numbers of small embryos which hatch as first stage larvae that are swept downstream into estuaries or the sea to complete development, as in many atyid shrimps. The resultant postlarvae must then find and migrate up freshwater streams to the adult habitat, a lifestyle known as amphidromy (Bauer 2013) (Chap. 9).

The pontoniines, a group of palaemonid species traditionally but not currently classified as a palaemonid subfamily (Pontoniinae) (De Grave et al. 2015a), make up about two-thirds of Palaemonidae (*s.l.*) recognized today (WoRMS 2022) (Plates 3.6e–g, 3.7a, b). They are overwhelmingly tropical and subtropical shallow-water species which are quite diverse in the extensive coral reefs and associated habitats of the Indo-Pacific Ocean. Many live intimately associated with other invertebrates such as sponges, corals, anemones, bivalve molluscs, and echinoderms (Chap. 10). Free-living pontoniines are morphologically and behaviorally typical epibenthic shrimps with spiny rostra and laterally compressed bodies (Fig. 3.31c), while many others have a symbiotic relationship with invertebrate hosts which has modified the typical shrimp body form in various ways, ranging from squat and stout (Fig. 3.31d) to highly laterally compressed bodies (Bruce 1976). Male and females of many symbiotic species live in socially (and perhaps sexually) monogamous pairs. The large genus *Periclimenes* (115 species) (e.g., Plate 3.6e), as well as the related genera *Cuapetes* and *Ancylomenes* species formerly included in that genus, has both free-living and symbiotic species. *Cuapetes americanus*, for example, is very abundant and ecologically important in seagrass and other shallow-water habitats of the tropical Western Atlantic (e.g., Bauer 1985b).

The *Periclimenes s.l.* species complex now includes 15 recognized genera. A phylogenetic analysis by Kou et al. (2013) indicates that this group is an evolutionary link between an ancestral "palaemonine" and both the many shallow-water clade of symbiotic species as well as another deep-sea clade. Although most palaemonines and pontoniines live in shallow coastal seas, there are at least 80 deep-sea species (100–740 m) which display the light reddish coloration which might be expected from shrimps of these depths (Li et al. 2008). As a group, the deep-sea species are somewhat stout and subcylindrical, indicating a less natant lifestyle. Kou et al. (2013) hypothesized that lower predation pressures in deeper waters has led to this modification from the typical natant form of a primitive palaemonine ancestor. I will leave the palaemonines and pontoniines by saying that this brief treatment does not do them justice; a large volume on their biology could (and should) be written!

Other allied palaemonoids formerly with family status, i.e., the Anchistioididae, Gnathophyllidae, and Hymenoceridae, will be discussed as informal assemblages within the Palaemonidae (*s.l.*). The first such group is the genus *Anchistioides* which currently contains 3 species. These shrimps were placed in their own family or subfamily by various authors (e.g., Bruce 1986b; Chace and Bruce 1993), mainly on the basis of their unusual larval characters as well as other specialized morphological traits. *Anchistioides* species are found in shallow-water habitats in the Red Sea, Indo-West Pacific, and West Indies, either living free or within sponges (Fig. 3.32a). They have typical palaemonoid morphology with large second chelipeds. They lack much coloration, i.e., are transparent, and are 30–40 mm in total length. Little is known about them as they are rarely collected. However, they may be locally abundant. Wheeler and Brown (1936) described swarming behavior up near the surface at night, which was somewhat correlated with the lunar cycle (from last to first quarters of the moon). These authors further observed, based on gut contents, that specimens examined had been feeding on polychaete worms (Williams 1984). An intriguing observation on members of this genus is that females have an appendix interna on the first pleopod, a feature normally found only in males of many carideans. This unique feature has been used, along with their unusual larval morphology, to support separate familial status (Bruce 1986b). Another possibility that needs to be investigated, should sufficient collections be obtained, is that this genus is protandric, and that the pleopod 1 appendix interna of the females is a vestige of a male-phase structure retained after sex change as in some other protandric carideans (see Chap. 7).

Five genera (*Gnathophyllum, Gnathophylloides, Gnathophylleptum, Levicaris, Pycnocaris*) traditionally placed in the family Gnathophyllidae are mainly associates of other invertebrates, especially sea urchins and other echinoids. The typical caridean body form of these small (~1–3 cm total length) shrimps has been modified by these associations, and typical descriptions of body shape include "stout," "robust," "plump," "fusiform," "stocky," and the like (Fig. 3.32b, c, Plate 3.7c, d). This feature and others (reduced rostrum) can be explained by their relatively sedentary status in which streamlined swimming is not a necessity. There has been much discussion in the past of their relationship with other palaemonids, especially the pontoniines (Bruce 1986b). The characters used in support of a familial status are the absence of or vestigial structure of the mandibular incisor process and the expansion of the antepenultimate segment of the third maxilliped into a buccal operculum (function unstudied) (Chace and Bruce 1993). Bruce (1986b) argued that there is some variability in these characters among gnathophyllid genera; importantly, he found that the first larva of *Gnathophyllum* was typically pontoniine, with the conclusion that separate family status was not warranted, a conclusion supported more recently by molecular phylogeny (De Grave et al. 2015a).

Some of the 10 species of *Gnathophyllum* are known echinoid associates, and others have been only collected free-living. An outstanding feature is their spectacular coloration, with a variable rainbow of brightly colored spots and stripes emphasized by dark contrasting outlines (3.32B). The pattern of thin white, yellow, and black lines on the plump body of *Gnathophyllum americanum* has earned them the common name of "bumblebee shrimps" (Calado 2008). Although drawings showing pattern of coloration are valuable (e.g., Manning 1963; Chace and Fuller 1971; Ahyong 2003), only color photos can do gnathophyllids justice, e.g., Plate 3.7c. The biology of the tiny pantropical urchin associate *Gnathophylloides mineri* has been more studied than most gnathophyllids. In the West Indies, it lives among and along the spines of both short- and long-spined urchins *Tripneustes ventricosus* and *Lytechinus variegatus* (Maciá and Robinson 2009). Patton et al. (1985) showed experimentally that these shrimps scrape and feed on epithelial tissue from the urchin host's spines with rasping toothed setae on the second maxillipeds; they also observed what appeared to be a filter-feeding behavior by the mouthparts. Another gnathophyllid, *Levicaris mamillatus*, is even better adapted for feeding on epithelium covering the spines of its host *Heterocentrotus mammillata*. In this shrimp, the second maxillipeds are greatly elongated with stout rasping setae on the dactylus. Both Bruce (1973) and Fujino and Takeda (1977) observed that flexion of the second maxilliped would cause the dactylar setae to scrape over the urchin spine. They suspected that this appendage is used to scrape and ingest urchin epithelial tissue and/or debris on the spines.

The last species of the former gnathophyllid genera to be discussed is *Pycnocaris chagoae*, which is a symbiont of holothurians (Bruce 1972; Prakash et al. 2011). It is a very highly modified shrimp, more crab-like than shrimp-like in overall appearance (Fig. 3.32c, Plate 3.7d). The stout cephalothorax is wide, depressed, and flattened ventrally; the posterior part of the abdomen is tucked under the rest of the body. Like many symbiotic pontoniines, it has short antennae and robust pereopods with hooked tips to cling onto the host. It has the major features used to characterize the gnathophyllids, and so is obviously closely related to them. Only juveniles (individuals that can't be sexed) and ovigerous females have been collected; their color is similar to that of their holothurian hosts: "pink" (Bruce 1972); "creamy white or pale yellow" (Prakash et al. 2011).

Hymenocera picta and the two species of *Phyllognathia* have been grouped in the family Hymenoceridae, the status of which has been controversial. Both genera have bizarre morphologies. *H. picta* has wide expanded, plate-like second chelipeds, as well as similarly expanded outer antennular flagella, foliaceous maxilliped 3 segments, and expanded abdominal pleura (Fig. 3.32d) (see Plate 3.3d in Bauer 2004). Its bright coloration, with a generally creamy-white background onto which there are large splotches and spots of brownish-purple or rose outlined with blue or yellow margins (Calado 2008), has earned it common names such as the "harlequin" or "clown" shrimp. *Phyllognathia* is ornamented with camouflaging bushy setae which may be mimicking epiphytic algae or microbial growth, has club-like extensions of the eyes and expanded and distally pointed abdominal pleura (see T-Y Chan photograph, Fig. 3.34 in De Grave and Fransen 2011). In the hymenocerid species, the antepenultimate segment of the third maxilliped is much more widely expanded than that of gnathophyllids and, unlike them, is clearly articulated with the limb segment just proximal to it. This difference from the gnathophyllids was considered sufficient to place *Hymenocera* and *Phyllognathia* together in their own family. However, this designation has not been supported by De Grave et al. (2015a) who considered both morphological and molecular characters.

These two genera occur in the coral reefs of the Indian and Pacific oceans (Chace and Bruce 1993). The biology of *Hymenocera picta* is much better known than that of the rarer *Phyllognathia* because of its greater availability, interesting biology, and coloration, all of which make it attractive to the ornamental shrimp trade and aquaculture (Calado 2008). German zoologists Uta Seibt and Wolfgang Wickler did pioneering work on several aspects of their natural history and sexual biology (Wickler and Seibt 1970; Seibt and Wickler 1979). This shrimp occurs in socially monogamous male-female pairs (Chap. 8). These shrimps are predators which attack small starfishes, especially those with circular rays or arms in cross-section, especially *Linckia* species (Wickler 1973; Pratoomyot et al. 2018). Their attack and feeding behavior have been described by Wickler (1973), Seibt and Wickler (1979), Prakash and Kumar (2013), and Pratoomyot et al. (2018). The shrimp detects its prey by olfaction (Seibt 1973), approaches and assesses its suitability, and if chosen, begins to feed on the epidermis and tube feet around one of the rays (arms) of the starfish. After removing the marginal tube feet with the small but sharp first chelae, the shrimp uses its plate-like second chelae as levers to push up the ray and eventually turn over the starfish on its aboral (upper) side. The starfish may turn back over, and if so, the shrimp will repeat the attack until the starfish tires, after which the shrimp feeds on the tube feet on the oral side, and then onto the internal organs of the ray. Several starfish species simply autotomize the attacked arm, and rapidly move away from the shrimps, surviving to live another day. The injured starfish is capable of regenerating the lost arm. Some smaller individuals may be killed and consumed completely. There is some disagreement about whether the male and female shrimp partners cooperate in the attack and feeding on the starfish (Seibt and Wickler 1979; Prakash and Kumar 2013).

Outbreaks (relatively sudden increases in density) of the coral-eating starfish *Acanthaster planci* periodically occur in Pacific coral reefs (e.g., Dana et al. 1973; Kayal et al. 2012). It was once suggested that *H. picta* might be used as a biological control agent of *Acanthaster* (Seibt and Wickler 1979). Alas, although *H. picta* can now be bred and reared in captivity, *A. planci* is not a preferred prey item (Pratoomyot et al. 2018). The shrimp itself is endangered by overcollecting for the ornamental shrimp trade, although cultivation by aquaculturists may alleviate this problem. Once last intriguing fact about these shrimps is that they are diurnal, openly displaying their bright coloration to potential fish predators on the reef. However, Bruce (1975) noted anecdotally that predatory reef fishes which readily prey on other small shrimps do not attack *H. picta*. Thus, its bright, openly displayed coloration may be aposematic, a warning to fish predators of a distasteful or poisonous shrimp. This is certainly an interesting hypothesis that needs to be tested.

3.13.4 Typhlocarididae

The family is currently restricted to four species of *Typhlocaris*, although in the past the family has also included the euryrhynchines, as noted above. They are found in isolated caves with brackish water high in hydrogen sulfide content near the Mediterranean Sea (Tsurnamal 2008; Guy-Haim et al. 2018). This enigmatic group of subterranean groundwater shrimps (Plate 3.7e) has a very restricted and disjunct distribution around the eastern Mediterranean Basin. *T. galilea* and *T. ayyaloni* are found only in Israel, *T. salentina* from the bootheel of Italy, and *T. lethaea* from Libya near Benghazi.

Transgressions and regressions of the Mediterranean forced their marine ancestors into and to adapt to the rigorous environmental conditions of their refugial habitats (Guy-Haim et al. 2018). The knowledge we have on the biology of these rare and subterranean species is mainly due to the contributions of Moshe Tsurnamal at the Hebrew University of Jerusalem (Tsurnamal 1978, 2008). These shrimps, from 2 to 5 cm total length, have an overall palaemonoid habitus, but they are distinguished from others in that group by a vestigial accessory branch of only few articles arising from the 25th–38th articles of the outer antennular flagellum and distinct longitudinal sutures or lines on the carapace, a primitive feature seen in some penaeoids and reptant crustaceans ("linea thalassinica") (Calman 1909; Tsurnamal 2008). Typhlocarids have a striking white to pale yellow color. The exoskeleton is transparent, and the white color comes from the underlying tissues. Some individuals are yellowish, probably due to a light growth of sulfide bacteria which the shrimp has not been able to remove by grooming. These shrimps are totally blind, with eyes and eyestalks only represented by a pair of ocular scales without any pigmentation. The second chelipeds, which are quite large with long sharp chela fingers (Plate 3.7e), belie their probable feeding habits, which include grazing with their small first chelipeds on sulfide-oxidizing bacterial mats in their cave habitat, as well as preying on small cave invertebrates. The shrimps may occur in large numbers in upper, warmer parts of their caves where mats of sulfide bacteria are most abundant, while females must incubate their embryos in slightly cooler, lower parts of the cave. Development is direct, with females producing relatively few (~30–40) large eggs which hatch as small juveniles rather than planktonic larvae.

3.14 Superfamily Pandaloidea

This superfamily consists of two families, the Pandalidae and Chlorotocellidae, which share the unique character of a first pereopod which terminates in a microscopic chela, or nonchelate with a tiny dactyl, or nonchelate with dactyl absent. These are two of the polycarpidean families in which the carpus of the second cheliped (pereopod) is usually subdivided into a few to many articles. The one exception is *Thalassocaris* in which the carpus is entire. This latter genus and *Chlorotocoide*s were formerly separated from the other pandalids as the Thalassocarididae (e.g., Chace 1985; Bauer 2004). However, based on both new molecular and morphological evidence, these two "thalassocarid" genera have been incorporated into the Pandalidae, while four other pandalid genera were placed in a new family, the Chlorotocellidae (Liao et al. 2019; Komai et al. 2019).

3.14.1 Pandalidae

Pandalids are medium to large-sized shrimps with a classically natant caridean body form (Fig. 3.33, Plate 3.7f). Most have laterally compressed bodies, long spiny rostra, and large conspicuous eyes. The family consists of benthic and benthonektonic species, with at least one truly pelagic species, *Stylopandalus richardi*. This family of 19 genera contains a large number of species (194), 94 of them in the deep-water genus *Plesionika* (WoRMS 2022). Pandalids are mostly a cold-water family distributed in the shallow to deep continental shelf in cool temperate and higher latitudes, and in the chilly waters of slope depths around the world. The related genera *Pandalus* and *Pandalopsis* (Fig. 3.33a) occur in northern hemisphere waters and most species of these two genera occur in shelf waters around the North Pacific (Komai 1999; Bergström 2000; Jensen 2014). The chelate second pereopods of species in several pandalid genera (e.g., *Pandalus*, *Heterocarpus*, several *Plesionika* species) are asymmetrical, with the cheliped of one side much longer with more carpal articles than the other. This asymmetry indicates a division of labor between these two chelipeds. Bauer (1975) studied the grooming behavior of *Pandalus danae* and found that the longer, more slender left cheliped was used more frequently in cleaning the body than the shorter, stouter right cheliped, which was employed more in feeding. The slender, functionally nonchelate first pereopods appear to be exploratory and sensory, probing here and there into the bottom, and, like most carideans, are used to clean the antennal flagellum (Chap. 4).

The biology of the genus *Pandalus* has been extensively studied because of its importance in fisheries and its abundance in continental shelf benthic communities. The North Atlantic *Pandalus borealis* is the most productive caridean species in shrimp fisheries (Shumway et al. 1985; Bauer 2020), while several *Pandalus* and

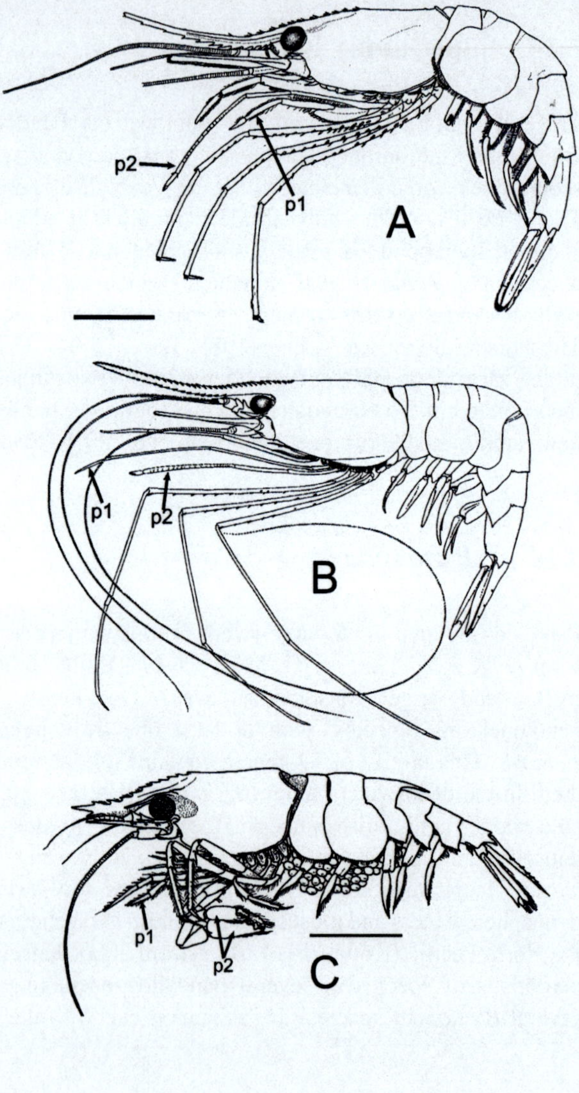

Fig. 3.33 Pandalidae. (**a**) *Pandalopsis longipes*. (**b**) *Plesionika grandis*. (**c**) *Thalassocaris crinita*; Scale bar represents 20 mm in (**a**), 9 mm in (**b**) and 3 mm in (**c**). p1, pereopod 1 (nonchelate); p2, pereopod 2 (chelate); From (**a**) Komai (1994), courtesy of the Journal of Crustacean Biology, (**b**) Chace (1985) and (**c**) Marin and Chan (2011) courtesy of Crustaceana (Brill)

Pandalopsis species are the major fished caridean species in the North Pacific (Butler 1980; Bauer 2020) (Chap. 11). These species are primarily schooling species on continental shelf bottoms. This characteristic and their large size make them a natural target of fisheries. Bo Bergström (2000) published an extensive and quite useful review on all aspects of the biology of *Pandalus,* and Komai (1999) revised the taxonomy of the genus. Valuable species accounts with illustrations and photographs of North Pacific pandalids can be found in Butler (1980) and Jensen (2014), respectively. The sexual system of *Pandalus* and *Pandalopsis* has been of special interest, as all species are protandrous hermaphrodites, first maturing sexually as males and then changing to females. But the details of sex change vary greatly among different species. Some are purely protandric, but in others, only part of the population goes through sex change. In the latter, some individuals may begin to mature as males but change to female before they have mated as males, while others develop directly as females (Berkeley 1930; Butler 1964; Bergström 2000; Chiba 2007) (Chap. 7). However, other pandalids are dioecious and do not change sex.

Plesionika (Fig. 3.33b) and *Heterocarpus* are deep shelf- to slope-depth species, some of which are targets of local deep-water fisheries (Holthuis 1980; Bauer 2020). Members of these genera are benthonektonic species that occur on the bottom during the day but rise up into the water column at night, feeding in both habitats. *Plesionika* is widely distributed in subtropical to temperate latitudes but is most diverse in the Indo-West Pacific (Ahamed et al. 2017). Both Kitsos et al. (2008) and Burukovsky (2017) found that two different *Plesionika* species feed on micronekton and macroplankton (e.g., euphausiids, pelagic shrimps, other invertebrates) when up in the water column, but are detritivorous, predatory and necrophagous (carrion eaters) when on the bottom. Rainer (1992) also verified that two species of *Heterocarpus* feed on midwater prey as well as on benthic prey. They must be necrophagous as well, as they are attracted to baited traps (Bauer 2020). Like these deeper species, the shallower-occurring *Pandalus* species are similar in food habits, feeding as opportunistic scavengers and predators on small crustaceans and other invertebrates on the bottom by day, but often rising into the water column to feed on pelagic prey at night (Bergström 2000; Jensen 2014).

The three species of *Thalassocaris* (Fig. 3.33c) and *Chlorotocoides spinicauda* were formerly considered as a separate pandaloid family because the endopods of the first male pleopods are unusually large and complex (Chace 1985: "petasma-like"). Unlike most pandalids, these are small shrimps, in the 4–10 mm carapace length range. All four species are bioluminescent, capable of lighting up at night with pairs of internalized hepatic photophores shining through the translucent exoskeleton. (Menon and Williamson 1971; Herring and Barnes 1976). *Thalassocaris* species are also capable of emitting luminous clouds of bioluminescence similar to the pandalid *Heterocarpus* and some oplophorids, apparently produced by the hepatic tubules. These organs are reminiscent of the sergestid "Organs of Pesta" but the internal structure is quite different (Herring 1981) (Chap. 5). The two genera were thought to be completely pelagic in both coastal (*C. spinicauda*) and deeper oceanic waters (*Thalassocaris*) (Menon and Williamson 1971; Chace 1985). However, Marin and Chan (2011) showed that *T. crinita*, at least, is a shallow reef species that burrows in the bottom or hides in reef cracks and crevices by day. At night, it swims about near the surface at night, probably preying on small planktonic invertebrates with their relatively large chelipeds. However, the other *Thalassocaris* species are important components of the mesopelagic deep scattering layer (DSL) (Karuppasamy et al. (2006), which shows up as a false bottom on the sonar of ships. The DSL is composed of a variety of mesopelagic animals that dwell in ~300–500 m depths by day and which rise near the surface at night.

3.14.2 Chlorotocellidae

Unlike many pandalid genera, which mostly inhabit colder waters, the seven species of these four Indo-Pacific genera occur in the shallow warm waters of the Indo-West Pacific. They are small in size and live as symbionts of cnidarian hosts; the carpus of the second pereopod only has 3 segments, unlike any pandalid. *Chlorotocella gracilis* occurs on jellyfish and gorgonians, while *Miropandalus hardingi* and the three species of *Anachlorocurtis* are symbionts of antipatharian corals (Bruce 1983a; Horká et al. 2014). The latter two genera have a cryptic coloration resembling that of their host, and the carapace has three dorsal projections or lobes (Fig. 3.34) which appear to mimic the shape of the tiny, retracted zooids along the branches of the host. In *Anachlorocurtis* species, the chela fingers are somewhat spatulate and denticulate at their tips (Horká et al. 2014), suggesting a scraping or grazing function. Embryos carried by females are quite large indicating abbreviated or direct development.

3.15 Superfamily Pasiphaeoidea: Pasiphaeidae

This superfamily has but one family, the Pasiphaeidae, currently with 7 genera and 102 species. These are pelagic shrimps principally in mesopelagic and bathypelagic waters of the open ocean. The speciose genus *Leptochela* is an exception in that it occurs more shallowly in continental shelf depths, up in the water column by day and

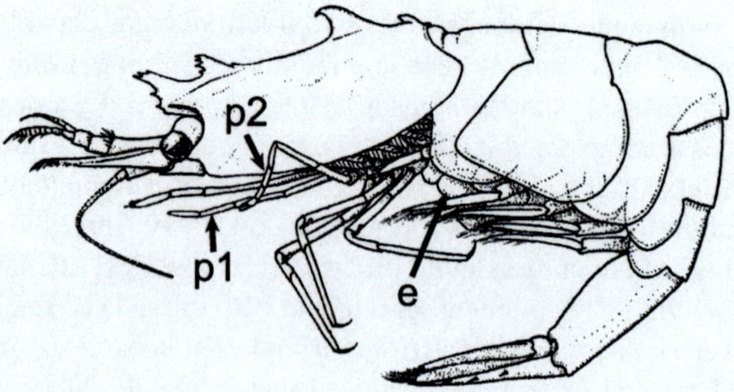

Fig. 3.34 Chlorotocellidae. *Anachlorocurtis occidentalis*. Scale bar represents 1.5 mm. e, large embryos; p1, pereopod 1 (nonchelate); p2, pereopod 2 (chelate). From Horká et al. (2014) Creative Commons BY 4.0

near or on the bottom at night (Chace 1976). Although most other pasiphaeids are pelagic both day and night, some of the deeper pasiphaeids, especially the large, heavy-bodied *Glyphus* species, may live close to the sea bottom (Crosnier and Forest 1973). With the exception of *Leptochela* species, with a carapace length of only 3–10 mm (Chace 1976), pasiphaeids are medium to large-sized shrimps, with the large *Glypus marsupialis* reaching up to 6 cm carapace length (16 cm total length) (Holthuis 1980). The phylogeny of the family has been studied using molecular data by Liao et al. (2017).

Pasiphaeids are characterized by a very short rostrum, laterally compressed bodies, and a soft, membranous exoskeleton. As we have seen, the latter feature occurs in some other pelagic shrimps to increase buoyancy and conserve energy spent in the calcification of the cuticle. The pereopods bear well-developed, multiarticulate, and functional exopods, which assist in swimming and maneuvering (Fig. 3.35a, b). These shrimps have two large pairs of chelipeds with long slender fingers equipped with dense rows of sharp teeth (pectinate), an adaptation to their predatory feeding habits (Cartes 1993; Hopkins et al. 1994) (Fig. 3.35a, b). Another such adaptation is their mandibles in which a grinding molar process is absent, but the incisor process is large with heavy teeth, an apparent tool for chopping up prey, which consists of other shrimps, small fishes, and other pelagic animals. The carpus of the second pereopod is entire. A common name for some of the shallower-occurring species is "glass shrimps," given for their overall transparency and lack of coloration except for a scattering of red chromatosomes here and there on the body. Meso- and bathypelagic species which are at depth during the day are half-red

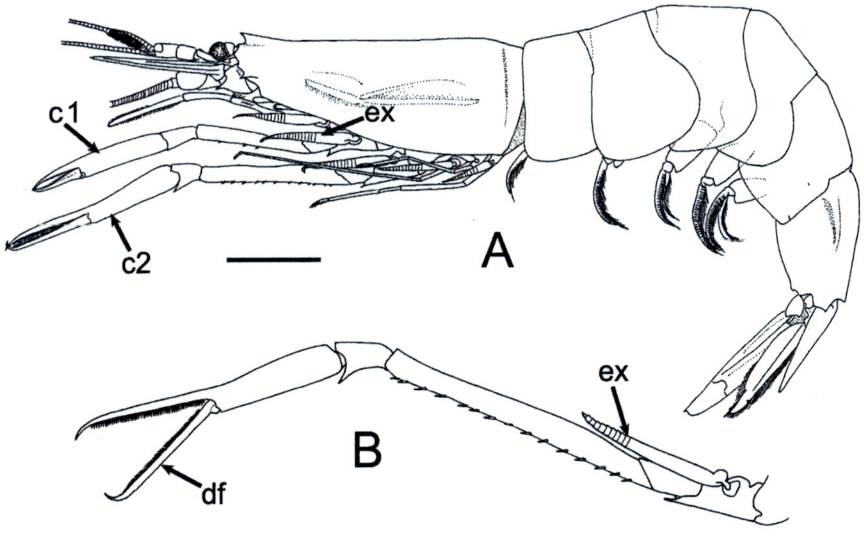

Fig. 3.35 Pasiphaeidae. (**a**) *Pasiphaea aequus*. (**b**) Second pereopod (cheliped) of *P. orientalis*, with large chela and densely toothed fingers; note large exopod. Scale bar represents 10 mm in (**a**). c1, c2, chelipeds 1, 2; df, dactylar finger; ex, exopod. From Komai et al. (2012), courtesy of the Journal of Crustacean Biology

to fully red, with the extent of red coloration roughly increasing from cephalothorax to abdomen with depth (Fig. 21 in Komai et al. 2012). Although pasiphaeids are mainly meso- and bathypelagic, photophores are not known in pasiphaeids, although *Glyphus* species can emit bioluminescent secretions from the mouth (Chan et al. 2008). Pasiphaeids typically make vertical migrations toward the surface at night to feed on the plentiful food resources there (e.g., Omori 1974). They are one of the shrimps such as *Thalassocaris*, Sergestidae and Oplophoridae which compose, in part, the deep-scattering layer (DSL) (Karuppasamy et al. 2006). Like all deep-sea shrimps, relatively little is known about their general biology.

3.16 Superfamily Physetocaridoidea: Physetocarididae

The lack of information about this family is due to its scarcity in the world's oceans, with only 35 specimens reported by Wasmer (1985). There were no substantial updates cited in Bauer (2004) or since. The family's only genus and species, *Physetocaris microphthalma*, was described from 2 specimens by Chace (1940) with further useful observations by Foxton and Herring (1970). This species occurs in the 1000–2000 m range, as shown by sampling with open-closing nets. It is thus a deep-mesopelagic to shallow-bathypelagic species, collected mainly in Northern Hemisphere waters (Wasmer 1985). This small-sized shrimp (~6–9 mm carapace length) has a highly inflated carapace and rostrum with dorsal keel (Fig. 3.36a), the exoskeleton is membranous, adaptations for buoyancy and energy conservation. This hypothesis is supported by Foxton and Herring's observations on a live specimen, which could remain motionless in place for long periods with only a few occasional and feeble beats of its pleopods.

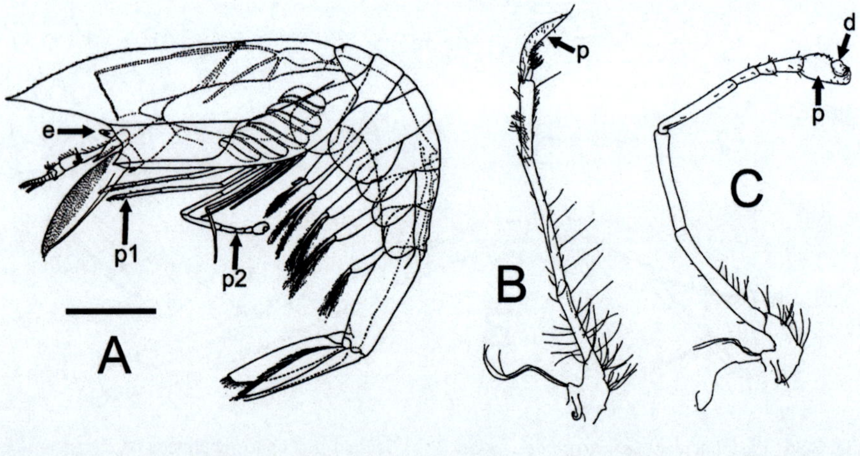

Fig. 3.36 Physetocarididae. (**a**) *Physetocaris microphthalma*. (**b**) Pereopod 1. (**c**) Chelate pereopod 2. Scale bar represents 3 mm in (**a**). e, eye; d, dactyl; p1,2, pereopods 1,2; p, propodus. Modified from Foxton and Herring (1970), courtesy of Crustaceana (Brill)

The pereopods of physetocarids are unusual: the first is nonchelate with only 6 segments, the most distal, which I interpret as a propodus, is sickle-shaped (Fig. 3.36b). The second pereopod is unique as well, with a chela composed of a flattened, broad propodus and a disc-shaped dactyl (Fig. 3.36c). The pereopod 2 carpus is multiarticulate, with 4 articles, making the physetocarids one of the polycarpidean families. As in many other carideans, the chela is equipped with a setal lock formed by interdigitating stout setae on the tips of the fingers. The mandible only has a small molar process, and the other mouthparts are reduced as well. The form of the second pereopod chelae and the reduction of mouthparts indicates a specialized diet, which remains to be described. The eyes have pigmented corneas but are quite small, unlike those of their early-stage larvae.

Foxton and Herring maintained a female alive long enough for it to hatch small, incubated embryos into zoeae which later molted into the second larval stage. Both stages had the typical morphology of carideans with extended planktonic development. Chace, as well as Foxton and Herring, described the coloration of the adults, i.e., mainly transparent with streaks and bands of color shades of orange to red to scarlet of variable intensity, appropriate for a deep pelagic shrimp. Continued exploration of the deep-sea fauna will hopefully yield more specimens so that the biology of this group can be better known.

3.17 Superfamily Processoidea: Processidae

If one samples seagrass meadows (e.g., *Thalassia, Posidonia, Cyomodocea, Zostera*) at night in any tropical to temperate region of the world, the collector will almost surely bring up some small, mostly transparent, large-eyed shrimps of the family Processidae. Members of this family are also found in sand/mud bottoms of continental shelves and into shallow slope depths as well. As in any group, there are exceptions: *Maryprocessa pippinae* is a deep pelagic species, taken by open-closing midwater trawls from about 200 m down to ~800 m (Hendrickx 2012). It is also collected by benthic sledges, so it may spend some time on or near the bottom. There are 69 species to date in 6 genera, most of them (51) in the genus *Processa* and 10 in the genus *Nikoides*. Important taxonomic works on the family include Nouvel and Holthuis (1957), Manning and Chace (1971), Hayashi (1975), and Chace (1997).

Processids are easily recognized by their large eyes and short, usually notched rostra (Fig. 3.37a, Plate 3.8a, b). Their bodies are mainly transparent or opaque but often speckled with red chromatophores. Different researchers have variously described the coloration of some species as pale shades of white, pink, green, or blue (e.g., Smaldon 1979; Manning and Chace 1971; Abele 1972; personal observation). Before 1971, the unique feature used to characterize processids was their asymmetrical first pereopods: on one side, usually the left, a simple stout pereopod terminating in a strong dactyl (achelate), while that of the right side ends in a robust chela (Fig. 3.37b, c). However, in 1971, Manning and Chace described the first species of *Ambidexter*, which upset the easy definition of a processid in that both first

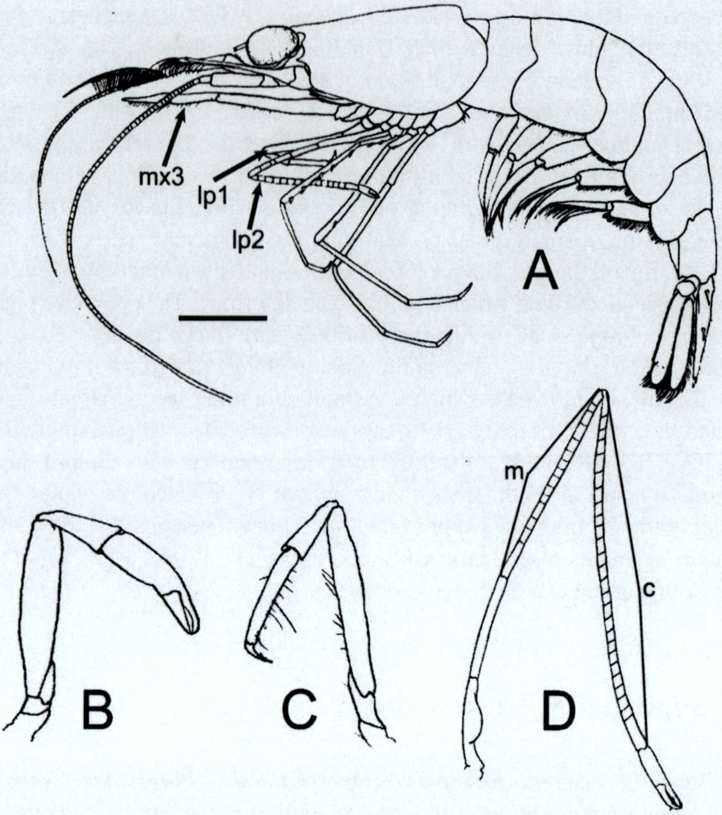

Fig. 3.37 Processidae. (**a**) *Processa famelica*. (**b**) *P. famelica*, chelate right pereopod 1, ventral view. (**c**) *P. famelica*, achelate left pereopod 1, ventral view. (**d**) *P. famelica* right pereopod 2, ventral view; note greater number of articles in carpus and merus than in left pereopod 2 shown in (**a**). Scale bar represents 3 mm in (**a**). c, carpus; lp1, lp2, left pereopods 1,2; m, merus; mx3, third maxilliped. Modified from Manning and Hart (1991)

pereopods in this genus are chelate. So now a combination of processid characters is used to define the group. The second pereopods, equipped with small chelae, are asymmetrical in most species, with that of one side (the right) longer than the left with a greater number of articles subdividing the carpus (Fig. 3.37a, d). It is interesting to note that this asymmetry is the opposite of that found in some pandalid species. The asymmetry may be related to differences in function, as in *Pandalus danae* (Bauer 1975), in which the longer more multiarticulate second pereopod is used more often than the shorter more robust one in grooming and other duties such as exploration and feeding. As in pandalids, the ischium and sometimes the merus of the processid pereopod 2 may have multiple articles, giving the limb even greater flexibility. Both the stout pereopod-like third maxillipeds and achelate pereopod 1 appear to be used to explore and dig into the sediments in search of small infaunal prey (personal observation). Processid females are unusual for carideans in having a

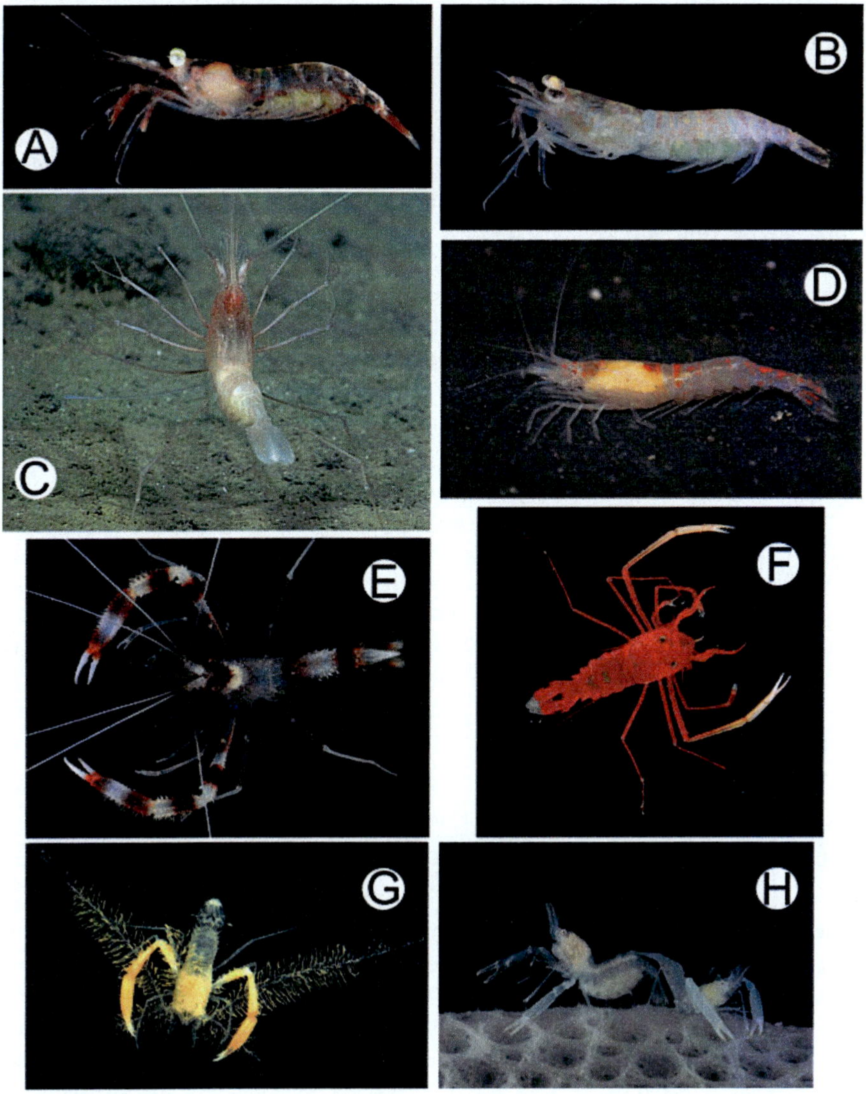

Plate 3.8 Processidae to Stenopodidea. (**a**) *Ambidexter symmetricus* (Processidae). (**b**) cf. *Processa* sp. (Panama). (**c**) *Bathystylodactylus bathyalis* (Stylodactylidae). (**d**) *Procaris hawaiana* (Procarididae). (**e**) *Stenopus hispidus* (Stenopodidae). (**f**) *Odontozona* cf. *ensifera* (Stenopodidae). (**g**) *Microprosthema plumicorne* (Spongicolidae). (**h**) Female and male *Spongiocaris tuerkayi* (Spongicolidae) on hexactinellid (glass) sponge. Photo acknowledgments: Courtesy of (**a**) Rasch and Bauer (2016), (**b**, **e–g**) Arthur Anker, (**c**) Mary Wicksten and the U.S. National Oceanographic and Atmospheric Administration Office of Exploration and Research, (**d**) Troy Sakihara, Hawaii Department of Land and Natural Resources, (h) David Shale (photographer) from Komai et al. (2016) courtesy of Zootaxa

thelycum-like pouch on the ventral part of the cephalothorax which serves as a temporary storage of spermatophores between mating and spawning (Rasch and Bauer 2016).

These are burrowing shrimps, quickly burying themselves just under the sand in the laboratory when suddenly exposed to light. Collections taken in habitats with processids have shown a pattern of scarcity during the day but significantly higher abundances at dusk and night (e.g., Bauer 1985b; De Grave et al. 2006). As with other day-burying species, this is an adaptation that protects the shrimps from visually hunting predators. However, ovigerous females may emerge periodically to protect the embryos from extended exposure to anoxic conditions during burial (Aguzzi et al. 2008). When not buried, the female can oxygenate the embryos by fanning its pleopods, a characteristic caridean incubation behavior that circulates water around and among the embryos.

Males are smaller than females, sometimes with little overlap between large male and small female size classes, which is an indication of protandric hermaphroditism, a hypothesis proposed by Noël (1973, 1976) for *Processa edulis*. However, careful investigation of *P. nouveli* (Gros 1979), *P. bermudensis*, and *P. riveroi* (Bauer and Conner 2011) showed no protandry in these species (Chap. 7). Tropical processids show the typical continuous reproduction of shallow-water tropical shrimps, but subtropical- to temperate-water species usually have an extended seasonal breeding period (Noël 1973, 1976; Bauer and Conner 2011; Rasch and Bauer 2015). Similar to most shallow-water marine species, processid females incubate small embryos that hatch into typical first-stage caridean larvae with extended larval development in the plankton (Williamson 1980; Ortega et al. 2005) (Chap. 9).

3.18 Superfamily Psalidopodoidea: Psalidopodidae

Chace and Holthuis (1978), who, between the two of them, worked on just about any kind of shrimp imaginable, described these medium-sized (2–4 cm carapace length), rare deep-sea shrimps as bizarre, which is certainly justifiable. Many long sharp spines cover the body and rostrum, giving them the appearance of a hedgehog or pin cushion in the shape of a shrimp (Fig. 3.38a). Their general color is pink, and the unpigmented reduced eyes with obscure corneal facets are white (Chace and Holthuis 1978; Toriyama and Horikawa 1993). Members of this family, with only one genus and 3 species, have first and second pereopods which are among the most specialized of decapod shrimps. The first pereopod is chelate but uniquely so, as not only the dactylar but also the propodal finger is movable so that fingers can cross (Fig. 3.38a, b); both fingers are equipped with sharp broad sclerotized teeth. Thus, the sawtoothed fingers give the chela the appearance of a pair of pinking shears, giving rise to the scientific name of the family, which translates into "scissor-footed." Chace and Holthuis carefully described the massive complex musculature and tendon arrangements which allows the fingers to cross upon contraction of the adductor muscles. The dactylar finger has the usual condylar articulation with the

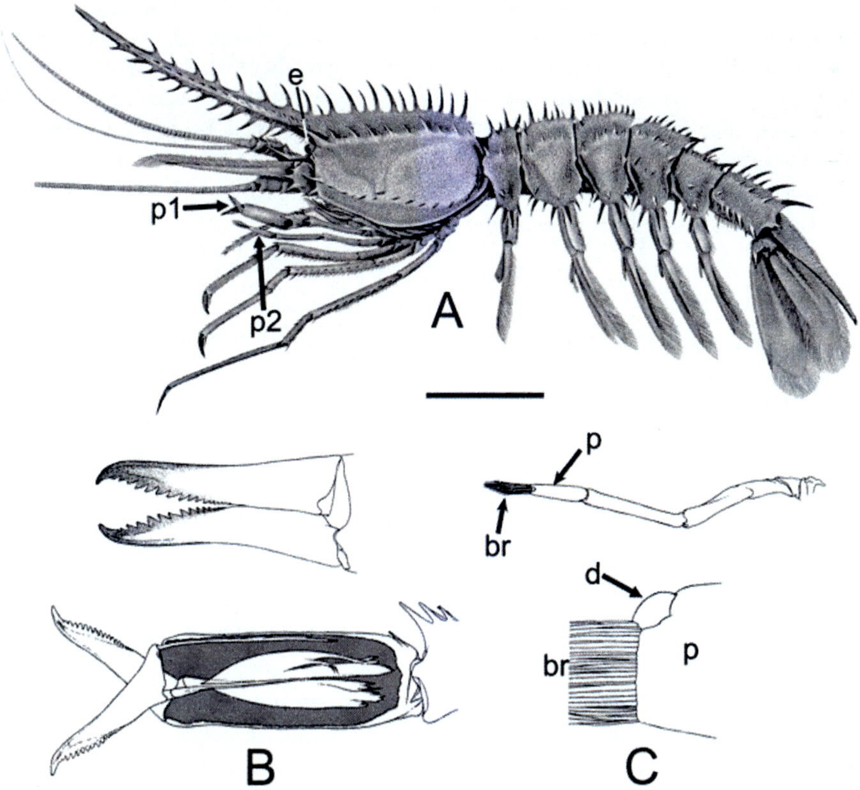

Fig. 3.38 Psalidopodidae. (**a**) *Psalidopus barbouri*. (**b**) Pereopod 1 chela. Above, chela with fingers uncrossed (at rest); below, chela with fingers crossed (adducted), with adductor muscles and tendons shown inside the propodus. (**c**) Above, pereopod 1 with setal brush; below, close-up of vestigial dactyl and distal end of propodus, with proximal part of setal brush protruding from them. Scale bar represents 12 mm in (**a**). br, setal brush; d, dactyl; e, eye; p, propodus; p1,2, pereopods 1,2. Modified from Chace and Holthuis (1978)

propodus, but the propodal finger articulates with the propodal palm only by a cuticular membrane. The second pereopod is described as a degenerate chela and looks like a cleaning whisk (Fig. 3.38c). The dactyl, which can no longer be termed a finger, is highly reduced to a small semicircular cap on the distal end of the propodus, and both it and the propodus are squared off distally from which projects a thick brush of wiry setae (Fig 3.38a, c).

These benthic shrimps are found on soft mud bottoms on insular and continental slopes (*P. barbouri)* but extending into abyssal depths of nearly 3000 m as well (*P. tosaenis*) (Chace and Holthuis 1978; Toriyama and Horikawa 1993, respectively). Embryonic development is highly abbreviated. The very large late-stage embryos have nearly all their appendages well developed before hatching (Chace and Holthuis 1978; Martin 1986), indicating that they emerge as nearly complete juveniles. The usually conservative taxonomists Chace and Holthuis could not resist the temptation to hypothesize about the unusual pereopods 1–2. They speculated that the function of the chelate first pereopod was either to grasp objects, presumably upon closure, or to probe into prey burrows, dislodging them with a sawing action of the chela fingers. Possible functions posited for the whisk-like second pereopods were as brushes for cleaning the body of the fine mud of their habitat, or perhaps to sweep away sea floor sediments under which prey might be hiding. As these authors pointed out, observations on live animals are needed to test these or other hypotheses about the unusual limbs of *Psalidopus*.

3.19 Superfamily Stylodactyloidea: Stylodactylidae

The deeper parts of the ocean abound with strange and enigmatic taxa, including some families of shrimps, as we have seen, and the Stylodactylidae is yet another example. Its 5 genera and 33 species are mainly distributed in deep continental shelf and bathyal (slope) waters of subtropical and tropical seas (Chace 1983). However, they can be taken in more shallow depths and one genus, *Bathystylodactylus*, is an abyssal group, occurring down to at least 5000 m (Wicksten and Martin 2004; Wicksten et al. 2017). Unlike other members of the family, members of this abyssal genus have highly reduced, unpigmented eyes. As is typical of deep-sea species, stylodactylid coloration is reddish (orange, rose, red) with some parts of the body pale or deeply white, as shown in photographs and descriptions in Cleva (1990) and other workers. They are small- to large-sized benthic shrimps (~2–15 cm total length; e.g., Hayashi and Miyake 1968; Cleva 1990) with very long, almost straight to curved dentate rostra, equipped with numerous small spine-like setae (Fig. 3.39a). A unique taxonomic feature is that the last two segments of the second maxilliped are side by side, both arising from the distal end of the preceding segment (Fig. 3.39b).

Another, more interesting unique feature of the group is their chelipeds. These lengthy limbs are subequal in size, the carpus is entire, and each consists of two elongate, stick-like fingers arising from a very short propodal palm (Fig. 3.39c, d). These fingers are equipped, as are the third maxillipeds and first two pereopods, with long fringing setae, giving the shrimp a shaggy appearance. In Bauer (2004), I said that the function of these hairy appendages was completely open to speculation. No longer—thanks to Wicksten et al. (2017), who published a photo (Plate 3.8c) of *Bathystylodactylus bathyalis* with these appendages outstretched into a weak bottom current, which has spread out their elongate fringing setae into a setal net. It is likely that when a passing small organism touches the net, it is detected, grasped, and passed to the mouthparts for consumption. This behavior is similar in function, but not in morphology, to the setal fans used by some atyid shrimps to filter currents of food items (see Atyidae, above). The observations by Wicksten and colleagues show how important ROV and deep-sea photographic techniques are and will be in working out the biology of rare, deep-dwelling species. But this is just a start, albeit an important one, in understanding the biology of stylodactylids.

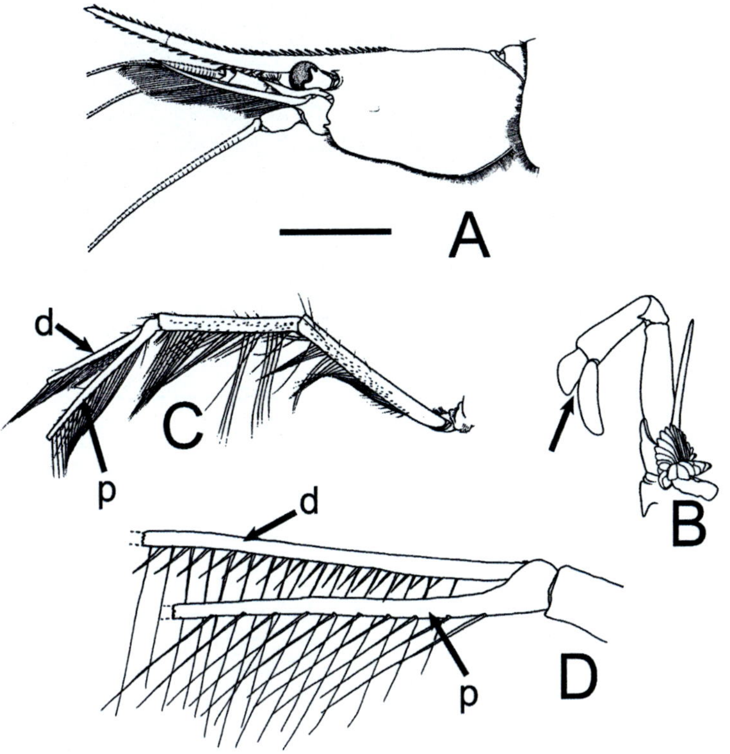

Fig. 3.39 Stylodactylidae. (**a**) *Stylodactylus licinus.* (**b**) *S. licinus* second maxilliped; unmarked arrow indicates two terminal segments side by side, both arising from preceding segment. (**c**). *S. licinus* first (chelate) pereopod 1. (**d**) chela 1 of *Bathystylodactylus echinus*, showing fingers with fringing setae (tips of fingers broken off during collection). Scale bar represents 2 mm in (**a**). d, dactylar finger; p, propodal finger. (**a–c**) Modified from Chace (1983) and (**d**) modified from Wicksten and Martin (2004)

3.20 *Amphionides reynaudii*: Taxon Inquirendum or Pandalidae?

Amphionides reynaudii is a relatively rare pelagic crustacean whose unusual morphology was judged different enough to warrant its status as a separate order (Amphionidacea), equal in rank with the Decapoda and Euphausiacea among the shrimp-like crustaceans (e.g., Williamson 1973; Martin and Davis 2001; Fransen 2010; Martin and Kutschera 2014) (Fig. 3.40). However, De Grave et al. (2015b), after reviewing morphological and new but limited molecular evidence, suggested that this species is a larval form of as yet unknown parentage, but likely some species of Pandalidae. They suggested that it be classified as "*incertae sedis*," a Latin term meaning "of uncertain placement." A more recent molecular phylogenetic analysis

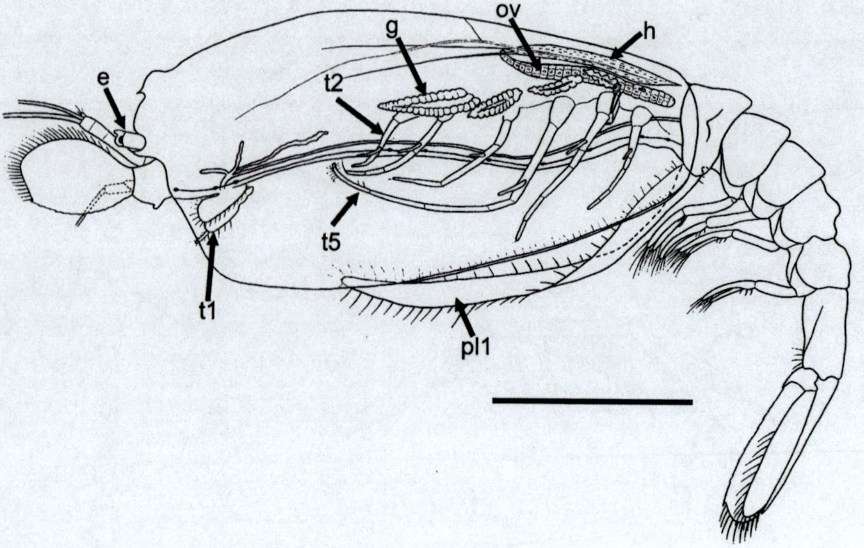

Fig. 3.40 *Amphionides reynaudii*: taxon inquirendum or Pandalidae? Specimen identified by Williamson (1973) as an adult female. Scale bar represents 5 mm. e, eye; g, pleurobranch gill; h, heart; t1, thoracic appendage 1 (maxilliped); t2,5, thoracic appendages 2, 5; ov, ovary; pl1, expanded endopod of pleopod 1, possibly forming the floor of a carapacial brood chamber. Modified from Williamson (1973) courtesy of Crustaceana (Brill)

by Wang et al. (2021) supported the pandalid connection. Up until recently, they were classified in WoRMS (2022) as a "taxon inquirendum," i.e., an incompletely defined genus whose validity is uncertain or disputed which needs additional information and study. However, WoRMS (2022) now puts them in the Pandalidae, apparently based only on molecular evidence. Given the lack of any connection between these purported larval forms and a breeding adult, this conclusion needs verification from other evidence.

Knowledge about *A. reynaudii* was nicely gathered and reviewed by Williamson (1973) and Fransen (2010). The specimens collected to date are pelagic shrimps taken from tropical and subtropical waters (36°N to 36°S) worldwide. The zoeal larvae (perhaps as many as 20 stages; Kutschera et al. 2012) have been taken at less than 100 m while specimens considered to be metamorphosed adults occur much deeper, from 700 to 2000 m. Determination of sex in later stages is based on the position of gonopores which, as in all shrimps and other malacostracans, are located on the 6th thoracic segment in females and on the 8th in males. However, none of the specimens identified as males show an appendix masculina, as in male carideans, or a petasma, as in male dendrobranchs. The majority of adults collected have been females, which Williamson (1973) hypothesized might be due to better swimming and net avoidance abilities of males. Their body shape appears to be more hydrodynamically efficient (less inflated carapace), or perhaps males have a shorter life span and are thus less abundant than females.

The following description is based on what most workers have considered adults; the gradual (anamorphic) larval stages are well described in Kutschera et al. (2012) and Martin and Kutschera (2014). Adults are up to 3 cm in total length, with a very thin, membranous exoskeleton, bulbous inflated carapace, and reduced mouthparts (Fig. 3.40). The mandibles and first maxillae are vestigial, indicating that the adults may not feed. Only the first thoracic limb is considered a maxilliped (instead of first three, a key diagnostic feature of decapods), and the other thoracic limbs (7 in the male, 6 in the female) consist of an endopod with less than the usual 7 segments and have very small, functionless exopods. The eyes of adults are reduced but are quite well developed in the various larval stages. The gills of *A. reynaudii* are by position pleurobranchs (Fig. 3.40) but their taxonomically important morphological type (phyllo-, tricho- or dendrobranch) has not been determined.

The endopods of the female first pleopods are greatly expanded (Fig. 3.40) and may form the floor of a carapacial brood chamber—definitely not a decapod shrimp feature. However, eggs in this chamber have never been observed. The fifth thoracic limb of females appears to be a cleaning appendage, perhaps to clean and jostle developing embryos if indeed they are incubated by the female (Williamson 1973). If the alternative hypothesis that the "adults" of *A. reynaudii* are indeed advanced larvae of a pandalid caridean (De Grave et al. 2015b) is accepted, then the ovaries and testes described previously in *A. reynuadii* specimens may be in an incipient stage of development. The alternative hypothesis proposed by De Grave and coworkers is that *A. reynuadii* is a paedomorphic species in which a larval form has precocious development of the gonads and sexual maturity but retains somatic features of a larval form. Which hypotheses will stand up to later scrutiny, as well as the question of gill type, is of utmost importance in determining the systematic position of this enigmatic species.

3.21 Infraorder Procarididea: Procarididae

The discovery and description of a procaridid shrimp, *Procaris ascensionis* (Fig. 3.41a, Plate 3.8d), by Chace and Manning (1972) excited both considerable interest and controversy in the carcinological community. These shrimps, from anchialine pools on Ascension Island in the mid-Atlantic, have a mix of primitive and advanced characters, with some traits found in penaeoid and stenopodidean shrimps, and others allying them more closely to the carideans. As will be recalled, the anchialine habitat refers to subterranean passages and caves that are connected to the sea on one end and may open some distance from the shore inland. The underground world in which anchialine animals live is mixohaline, with creatures living there variously encountering fresh, brackish, and fully marine water. In the subterranean part of the system, it is totally dark, with waters low in oxygen and often with high sulfur content. Yet the procaridids, like other anchialine animals, have adapted and often thrived in such habitats. They are among the few decapod shrimps with large, fully functional swimming exopods (Fig. 3.41a, b).

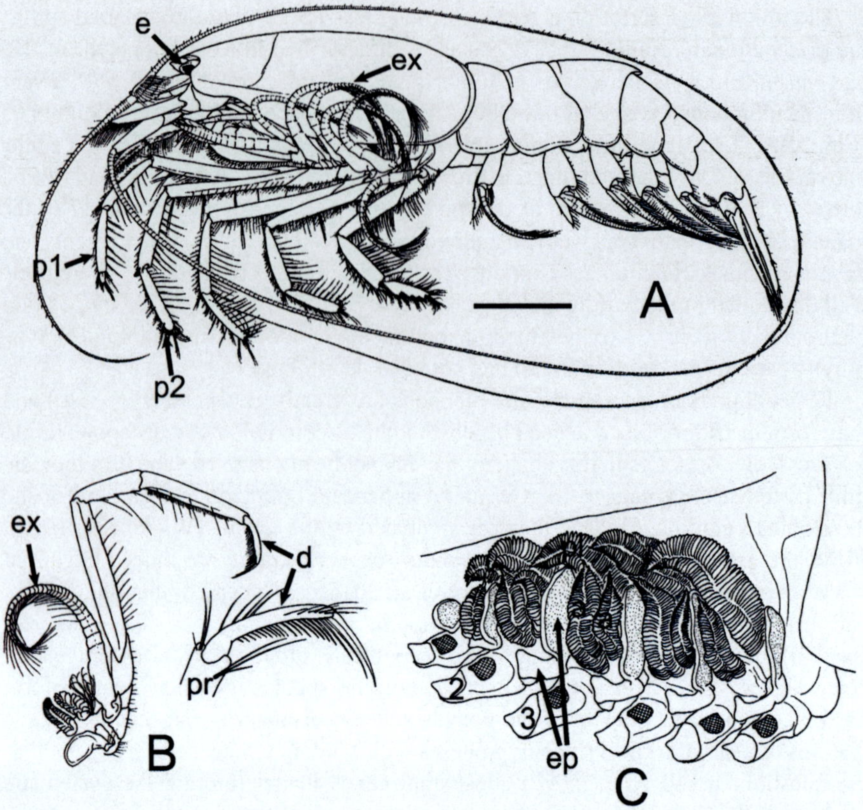

Fig. 3.41 Procarididea. (**a**) *Procaris ascensionis*. (**b**) Achelate pereopod 1 (left) with higher magnification of dactyl at the end of the propodus. (**c**) View of gill chamber from left side with gill cover removed to show gills and epipods. The epipod of pereopod 2 is labeled to show the horizontal bar (left arrow) and vertical appendix (right arrow) separating the gills of pereopod 2 from those of pereopod 3. Scale bar represents 3 mm in (**a**). a, arthrobranch; d, dactyl; e, eye; ep, epipod; ex, exopod; p1,2, pereopods 1, 2; pl, pleurobranch; pr, propodus; 2,3, basal segments of pereopods 2 and 3. From (**a**) Chace and Manning (1972) and (**b**, **c**) Kensley and Williams (1986), courtesy of Journal of Crustacean Biology

What is the mixture of characteristics from each of the three major shrimp taxa alluded to above? John Maciolek, a fisheries biologist from Hawaii, kindly sent me a specimen of *Procaris hawaiana,* the second procaridid described (Holthuis 1973) when I was in graduate school. On first glance. I thought, hmm, this is a caridean, as it plainly has the caridean saddle-like second abdominal pleuron overlapping the first and third pleura. It also had the epipod-setobranch gill cleaning system that only occurs in the Caridea (Chap. 4). However, as Chace and Manning emphasized, procaridids have a number of penaeoidean and stenopodideans traits, e.g., the third maxilliped has a full 7 segments, not the 4–5 found in carideans. The pereopodal epipods, although with hooked ends connected to setobranchs, have massive vertical extensions that separated the gills of different pereopodal segments (Fig. 3.41c), like those seen in penaeoid shrimps (Fig. 2.6, Chap. 2). Only the primitive oplophorid carideans have similar epipodal appendices, although much smaller than in the procaridids and penaeoids. Felgenhauer et al. (1988) showed in *P. ascensionis* that the mature ovaries of females may extend from the cephalothorax into the entire length of the abdomen, seen nicely also in the photo by Troy Sakihara in the procaridid *Vetericaris chaceorum* (Plate 6.1c). This is a mainly penaeoid shrimp trait; in caridean females, mature ovaries usually extend only as far as the 2nd–3rd abdominal segment (Chap. 6).

The gills of procaridids are phyllobranchiate, a trait found only in caridean decapod shrimps, and not dendrobranchiate, as in the penaeoids and sergestoids, nor trichobranchiate, as in the stenopodideans. However, the first and second pereopods are not chelate as in normal carideans; instead, they are just like the other three pairs of pereopods, terminating in small (*Procaris*) to well-developed dactyls (*V. chaceorum*) (Fig. 3.41b). Procaridids are the only decapod shrimps without chelae, a very primitive decapod crustacean feature.

Since the discovery of the first procaridid, 5 other species of *Procaris* and 1 species of another, even more primitive, procaridid genus, *Vetericaris chaceorum* (Kensley and Williams 1986), have been described from anchialine habitats in subtropical and tropical areas in the Atlantic and Pacific. Chace and Manning (1972) classified them as a superfamily within the Caridea—but with reservations. Felgenhauer and Abele (1983) made a careful comparison of morphological characters between dendrobranchiates, carideans, stenopodideans, and procaridideans. They concluded that the latter were a separate group from the others and recognized this by using the term Procarididea, the "-idea" ending implying infraorder status. Thus, within the order Decapoda, these shrimps are classified as in the suborder Pleocyemata, Infraorder Procarididea, equal in rank with the Caridea and Stenopodidea; the dendrobranchiates are a separate decapod suborder. These conclusions were supported by Bracken et al. (2009) and Wang et al. (2021) with molecular characters.

Procaridid shrimps have a number of traits shared with deep-sea pelagic shrimps. The rostrum in short, the exoskeleton is thin, sometimes fragile, and all the thoracic limbs are equipped with large, multiarticulate exopods. These are swimming shrimps, using both the exopods and pleopods to swim agilely and constantly in the water column, rarely resting on the bottom or perching on cave walls (e.g., Abele and Felgenhauer 1985; Kensley and Williams 1986). The eyes of procaridids are highly reduced, with a small patch or spot of pigment at the end of the eyestalk, and no sign of corneal facets. Thus, they are blind or nearly so, as are many shrimps and other animals living in the constant dark of caves. Their slightly reddish translucent exoskeleton is a trait found in deep-sea shrimps and some cave shrimps. The combination of traits from deep-sea pelagic shrimps suggested to Hart et al. (1985) that procaridids represent relict populations of deep-sea shrimps forced into cave refugia by elimination of their original habitat by sea-floor spreading, as hypothesized for bresiliid shrimps.

Procaridids are medium-sized (*Procaris* species, 2–3 cm total length; *V. chaceorum*, 4–5 cm). But they are ferocious predators on smaller animals. Provenzano (1978) and Abele and Felgenhauer (1985) describe, and the latter authors nicely illustrate, the capture of other smaller cave crustaceans by these shrimps which, without chelae, gather their prey item into a basket formed by the outstretched pereopods. The shrimps may search along the bottom or in the water column with their pereopods, grasping a prey item within the pereopodal basket. The massive, protruding mandibles with well-developed incisor and molar processes are used to bite and consume the prey. The anchialine caves that procaridids inhabit are the home of smaller atyid shrimps, e.g., *Typhlatya* and *Halocaridina* species. Atyid shrimps, as will be recalled, are lower down in the food web, grazing on microalgae, microbial growth, and detritus in the caves. As might be expected, they are much more abundant than their procaridid predators, about ten times as much (Kensley and Williams 1986).

Procaridids are considered pleocyemates at present, as the only observations on reproduction (Felgenhauer et al. 1988) indicate that embryos are attached to the pleopods and incubated. The newly spawned embryo size is about 0.8-0.9 mm in diameter, suggesting to the prior authors that development is zoeal (Chap. 9). But other than those observations, nothing more is known about reproduction. No female procaridids have been collected with embryos attached to their pleopods. Does mating and spawning occur deeper within cave habitats? Most collections have been made near surface pools down to about 30 meters or so. Is it possible that species of this group, close to carideans but not one of them, attaches its eggs to walls or bottoms in cave environment, or is something else unexpected occurring? Another anomaly is that there is no obvious sexual dimorphism other than the malacostracan trait of gonopores at the base of the third pereopods in females and the fifth pereopods in males (Felgenhauer et al. 1988). All caridean males, except those of *Synalpheus* species (Dardeau 1984; Toth and Bauer 2007) and the oplophorid *Janicella* (Chace 1986), have an appendix masculina on the second pleopod. Procaridid males do not (Chace and Manning 1972; Kensley and Williams 1986); however, their sperm has the typical thumb-tack shape of a majority of carideans (Chap. 6). Small submarine drones (ROVs) may be the answer to making observations on this question in addition to the rather dangerous observations by cave divers.

3.22 Infraorder Stenopodidea

Although this group of shrimps is low in taxonomic diversity, with only 13 genera and 95 species, they are not completely unknown to the public. Just about any major marine aquarium or tropical aquarium shop will have on display living members of this group, particularly species of *Stenopus hispidus*, commonly known as "red banded coral or banded boxing" shrimps. Their combination of bright colors, long white antennular and antennal flagella, and active interesting behaviors make *S. hispidus* and other related stenopodids an important part of the marine ornamental shrimp trade (Calado 2008). In Japan, a dried and prepared specimen of the beautiful "Venus flower basket" sponge (*Euplectella* species), with a permanently entrapped male-female pair of spongicolid stenopodideans, is often given as a wedding gift as a symbol of eternal monogamous love and devotion (Saito et al. 2002).

Stenopodideans are pleocyemate decapod shrimps, i.e., they incubate spawned and fertilized embryos below the abdomen. However, they are considered closer genealogically to the reptant decapods (e.g., lobsters, crayfishes, anomuran, and brachyuran crabs) than to dendrobranchiate and caridean shrimps (e.g., Goy 2010; Chen et al. 2016; Chap. 12). They are highly benthic, completely marine shrimps in which the first three pairs of pereopods are chelate, with the third massively developed weapons (Fig. 3.42a, b; Chap. 1, Fig. 1.7b). All stenopodideans occur in male-female pairs, and thus are socially and probably sexually monogamous, the latter unusual in decapod shrimps (Chap. 8). Their first abdominal segment is reduced in size, and the abdominal pleura usually do not overlap; in some species, the first may overlap the second. Only the last three abdominal somites are interlocked by cuticular hinges. The second to fifth pleopods are turned inwards about 90°, the adaptive value of which is still unexplained. None of pleopods have appendices internae or masculinae. Importantly, the gills are trichobranchiate (Chap. 2) in structure, not dendrobranchiate as in the sergestoids and penaeoids nor phyllobranchiate as in the carideans and procarideans.

An excellent detailed review of the group has been done by Joseph Goy (2010) who is the world's expert on the group. There are currently three recognized families: the shallow tropical water Stenopodidae; the mostly deep-sea Spongicolidae, which are symbiotic mainly with hexactinellid "glass" sponges, and the Macromaxillocarididae, anchialine cave dwellers with but a single species. A study of molecular phylogeny of the group was done by Chen et al. (2016) which suggests that the current classification (Goy 2010) may not hold as further work is done, but that remains to be seen.

For the Stenopodidea, I change course a bit and discuss the families in reverse alphabetical order, as the Stenopodidae is the best-known family, and morphologically a more typical decapod shrimp. The characteristics and biology of the other two families appear derived compared to that of the Stenopodidae.

3.22.1 *Stenopodidae*

Members of this family, with 4 genera and 39 species, are mostly inhabitants of shallow tropical marine waters. Morphologically, they are characterized by laterally compressed, moderately to highly spinose bodies (Fig. 3.42a, Plate 3.8e, f); elongate third chelipeds; moderately long rostra; and well-developed grooming brushes ("setiferous organs") on the third maxillipeds and first pereopods (Goy 2010). The propodus and carpus of the last two pereopods are multiarticulate, probably an adaptation for flexibility, useful in perching and clinging to irregular surfaces, or grooming.

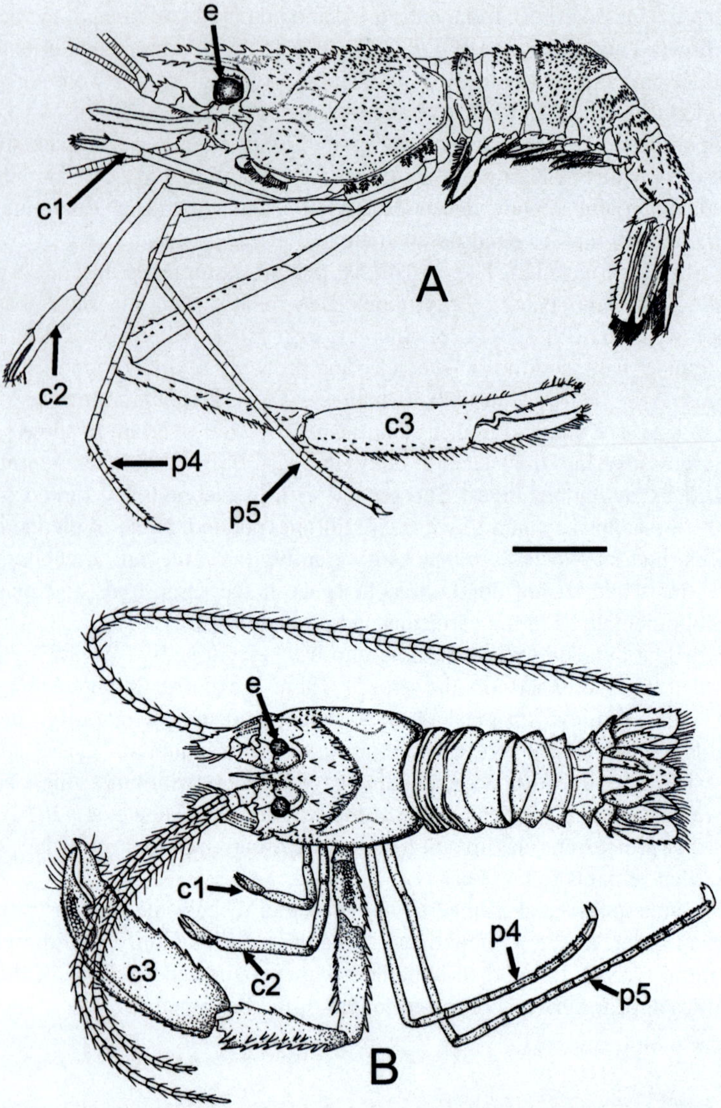

Fig. 3.42 Stenopodidea 1. (**a**) *Stenopus goyi* (Stenopodidae). (**b**) *Microprosthema granatense* (Spongicolidae). For clarity, the antennular flagella of the right side, the antennal flagellum of the left side, and the pereopods of the right side were not illustrated in the original publication. Scale bar represents 5 mm in (**a**) and 3.3 mm in (**b**). c1,2,3, chelipeds 1,2,3; e, eye; p4,5, pereopods 4,5. Note pigmented eyes and multiarticulated pereopods 4–5 in both species. From (**a**) Saito et al. (2009), courtesy Raffles Bulletin of Zoology, Lee Kong Chian Natural History Museum) and (**b**) Criales (1997), courtesy of the Journal of Crustacean Biology

The most well-known and studied genus is *Stenopus* (11 species) (Plate 3.8e), which inhabits coral and rocky reefs as well as other tropical marine shallow-water habitats with some type of hard substrate (e.g., isolated coral heads or boulders in seagrass meadows). The brightly hued *Stenopus hispidus* and some other members of its genus are favorites of the ornamental shrimp trade (Calado 2008; Chap. 11) and some are known cleaners of fishes (Limbaugh et al. 1961). The cleaning symbiosis (Chap. 10) is a mutualistic association (Vaughan et al. 2016) in which the shrimp hops on the body of the fish and obtains food (parasites, dead tissue, excess mucus) that it picks off the fish. The fish being cleaned of these harmful materials does not prey on the shrimp, at least not when it is being cleaned. *Stenopus* species live in socially and probably sexually monogamous male-female pairs which may form when they are still juveniles. These pairs are highly territorial in *S. hispidus* and *S. scutellatus*, staying and guarding a location ("cleaning station") which is only a meter at most and usually much less in extent (Limbaugh et al. 1961; Chockley and St. Mary 2003). Their high degree of aggressiveness toward other conspecifics is due to their defense of territory and mate (Johnson 1969, 1977). Their pugnacious behavior and large cheliped weaponry have earned them the common name of "boxing" shrimps in the ornamental shrimp trade (Calado 2008). Important studies on the reproductive and population ecology of *S. hispidus* have been done by Chockley and St. Mary (2003) and Chockley et al. (2008). Gregati et al. (2010, 2014) have detailed the reproductive morphology and mating behavior, to be discussed at greater length in Chap. 6.

The genus *Odontozona* (Plate 3.8f), with 22 species, is the most diverse genus of the Stenopodidae but its biology is much less known. Most species are commensals with invertebrates hosts (Goy 2010). Although mainly a shallow-water genus, some species occur in bathyal depths (Goy 2010), with *O. foresti* collected at 1200 m over a muddy bottom (Hendrickx 2002).

3.22.2 Spongicolidae

This family, with 7 genera and 54 species, has a basic division in its morphology and biology. On the one hand, the genus *Microprosthema* is much like the stenopodids in basic ecology and reproduction. It also occurs free-living in shallow tropical waters of the seas and reproduces like most shrimps from such habitats. Females produce numerous small eggs which hatch as zoeal larvae with subsequent extended development in the plankton, as do the stenopodids. *Microprosthema* morphology (Fig. 3.42b, Plate 3.8g) is intermediate or transitional between that of the stenopodid and symbiotic spongicolids (Saito and Takeda 2003). Its cephalothorax is only slightly depressed, but the abdomen is dorsoventrally compressed as in other spongicolids (Criales 1997), its body is somewhat spiny, and the rostrum of medium length and typical natant form. The other spongicolids live as symbionts mainly with glass sponges on the outer continental shelf and down into bathyal depths and beyond in tropical and warm-temperate seas (Saito and Takeda 2003; Saito and

Komai 2008). The bodies of these sponge-dwelling spongicolids often differ morphologically from stenopodids in that they are dorsoventrally rather than laterally compressed. Their major (third) chela tend to be broader and/or more robust than in the stenopodids, their bodies have only moderate to little spination, their rostra are short and/or relatively unarmed, and their eyes tend to be reduced with little pigmentation (Fig. 3.43a). Unlike the more colorful stenopodids, the symbiotic spongicolids have little color (translucent, whitish, slightly yellowish; Goy 2010) (Plate 3.8h). In contrast, *Microprosthema* species have hues of white, tan, yellow, or red (e.g., Criales 1997) (Plate 3.8g). Unlike *Microprosthema* species, female of symbiotic spongicolids produce few large eggs which develop and hatch as juveniles without any larval stages (direct development) (Saito and Konishi 1999), a reproductive mode common in deep-sea shrimps (Chap. 9). The sponge-dwelling spongicolids have interesting adaptations to their unusual deep-sea habitat which are given in some detail in Chap. 10.

3.22.3 *Macromaxillocarididae*

This family, with its long but appropriate name, has the key stenopodidean traits but is quite unusual in several ways, as pointed out by Fernando Alvarez et al. (2006) who described the first known species, *Macromaxillocaris bahamaensis*. It is the only stygobitic (cave-dwelling) species known in the Stenopodidea, discovered in an anchialine cave system in the Bahamas. It apparently resides hidden within cracks and crevices in the ceiling and walls of the subterranean areas of the cave. The shrimp shows characteristics shared with many other cave shrimps, e.g., reduced eyes without pigment or corneal facets (blind), a transparent exoskeleton lacking pigmentation, and long slender pereopods (Fig. 3.43b). The species name is derived from its key morphological trait, a greatly enlarged and elongated pair of third maxillipeds. These limbs are equipped with long articulated setae which the authors believe is a sieving apparatus for collecting food. Very few specimens have been collected, so gut content analyses have not been done to determine what kind of food they actually consume. The shrimp's body is only slightly compressed laterally compared to the Stenopodidae. Another trait that is unique to this species among the Stenopodidea is its lack of grooming brushes on the third maxilliped and first pereopods, which are well developed particularly in the Stenopodidae and are also found in the Spongicolidae (Bauer 1989; Goy 2010). The branchial formula is the most highly reduced in this infraorder. A lower number of gills and other branchial elements is also found in the spongicolids (below), shrimps which live inside sponges as adults. Their movement (and need for oxygen) is probably minimal, and so unnecessary gills have been lost. Alvarez et al. hypothesize that *M. bahamaensis* may limit activity, movement, and overall metabolic activity, and thus have less necessity for a fuller branchial formula. However, this hypothesis and the shrimp's biology in general still need to be explored with further study. This may not be an easy task, given this shrimp's rarity and the difficulty of collection and

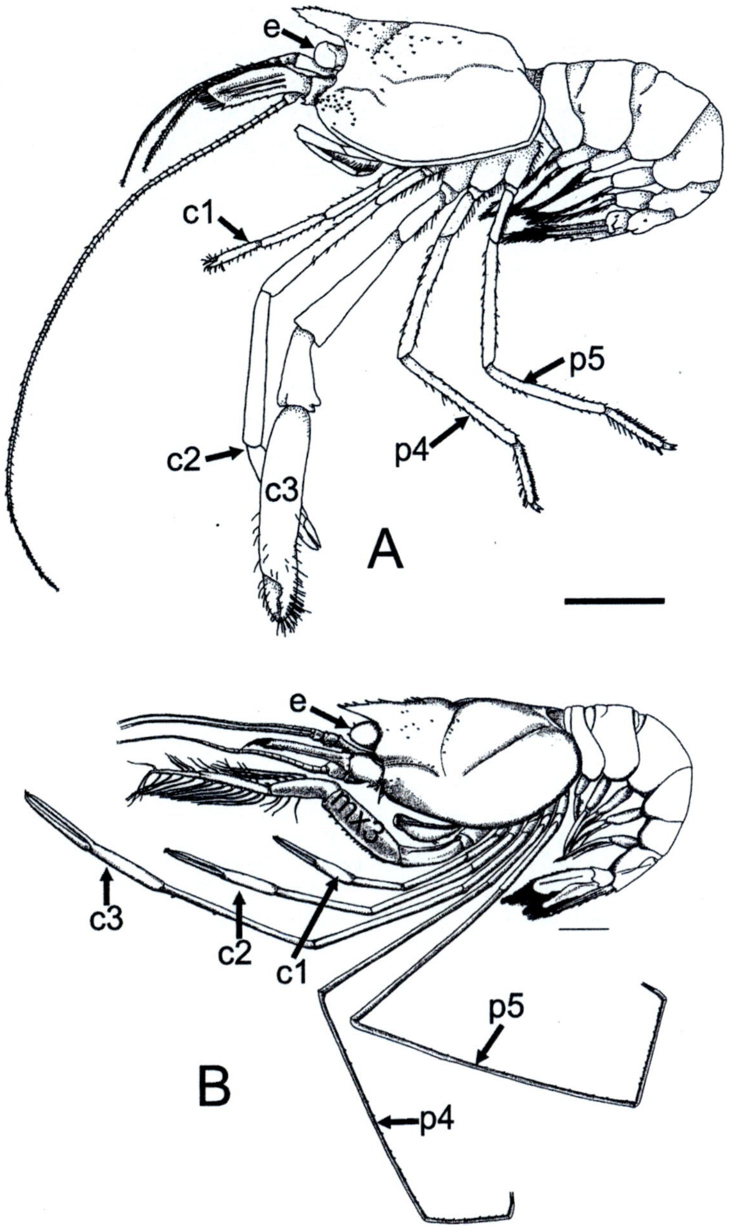

Fig. 3.43 Stenopodidea 2. (**a**) *Spongicoloides iheyaensis* (Spongicolidae). (**b**) *Macromaxillocaris bahamaense* (Macromaxillocarididae). Scale bar represents 8 mm in (**a**) and 2 mm in (**b**). c1,2,3, chelipeds 1,2,3; e, eye; mx3, third maxillipeds; p4,5, pereopods 4,5. Note unpigmented eyes and lack of multiarticulation in pereopods 4 and 5 in both species. From (**a**) Saito et al. (2006) courtesy Raffles Bulletin of Zoology (Lee Kong Chian Natural History Museum) and (**b**) Alvarez et al. (2006) courtesy of the Journal of Crustacean Biology

observation in the anchialine cave habitat. Recently, a reddish shrimp was collected from shallow sandstone reefs in northeastern Brazil. This species, *Chicosciencea pernambucensis*, appears more closely allied to *Macromaxillocaris* morphologically and genetically than to other stenopodidean shrimps (Bochini et al. 2020). However, it is a free-living rather than a cave species like *Macromaxillocaris*. Hopefully, more information will be forthcoming to shed light on its lifestyle and phylogenetic relationships.

3.23 Last Word

In Chap. 1, shrimps were introduced; in Chap. 2, their morphology and the adaptive value of their various body parts were described, and with this present chapter, synopses of all the shrimps families have been given. Equipped with this knowledge, let us now go on to explore the rather remarkable adaptations and natural history of the decapod shrimps.

References

Abele LG (1970) Semi-terrestrial shrimp (*Merguia rhizophorae*). Nature 226:661–662

Abele LG (1972) A review of the genus *Ambidexter* (Crustacea: Decapoda: Processidae) in Panama. Bull Mar Sci 22:365–380

Abele LG, Felgenhauer BE (1985) Observations on the ecology and feeding behavior of the anchialine shrimp *Procaris ascensionis*. J Crustac Biol 5:15–24

Aguzzi JB, Company JB, Abelló P, García JA (2006) Daily activity patterns of the mud shrimp *Solenocera membranacea* (Decapoda: Penaeoidea: Solenoceridae) in the western Mediterranean: a comparison by depth and season. Bull Mar Sci 79:353–364

Aguzzi J, Company JB, García JA (2008) The circadian behavioural regulation of the shrimp, *Processa canaliculata* Leach, 1815 (Decapoda, Processidae) in relation to depth, ontogeny, and the reproductive cycle. Crustaceana 81:1301–1316

Ahamed F, Cardoso IA, Ahmed ZF, Hossain Y, Ohtomi J (2017) An overview of the genus *Plesionika* Bate, 1888 (Decapoda, Caridea, Pandalidae) in Asian waters. Zootaxa 4221:575–593

Ahyong ST (2003) *Gnathophyllum taylori*, a new species of caridean shrimp from south-eastern Australia (Crustacea: Decapoda: Gnathophyllidae). Mem Mus Vic 60:237–242

Alvarez F, Villalobos JL, Iliffe TM (2004) A new species of *Agostocaris* (Caridea: Agostocarididae) from Acklins Island, Bahamas. Proc Biol Soc Wash 117:382–390

Alvarez F, Iliffe TM, Villalobos JL (2006) Macromaxillocarididae, a new family of stenopodidean shrimp from an anchialine cave in the Bahamas, with the description of *Macromaxillocaris bahamaensis*, n. gen., n. sp. J Crustac Biol 26:366–378

Alves D, Barros-Alves S, Almeida A (2021) Sex change and reproductive investment of the protandric shrimp *Merguia rhizophorae* (Rathbun, 1900) (Decapoda, Merguiidae). J Nat Hist 55:2673–2690

Anderson WW, Lindner MJ (1971) Contributions to the biology of the royal red shrimp, *Hymenopenaeus robustus*, Smith. Fish Bull 69:313–336

Anderson SL, Clark WH, Chang ES (1985) Multiple spawning and molt synchrony in a free spawning shrimp (*Sicyonia ingentis*: Penaeoidea). Biol Bull 168:377–394

Anker A (2003) Alpheid shrimps from the mangroves and mudflats of Singapore. Part 1. Genera *Salmoneus, Athanas* and *Potamalpheops*, with the description of two new species (Crustacea: Decapoda: Caridea). Raffles Bull Zool 51:283–314

Anker A (2011) Three new species of the alpheid shrimp genus *Salmoneus* Holthuis, 1955 (Crustacea, Decapoda) from the tropical western Pacific. Zootaxa 2839:67–84

Anker A (2012) The shrimp family Pseudochelidae De Grave & Moosa, 2004 (Crustacea, Decapoda, Caridea) in Brazil, with the description of a new species. Zootaxa 3544:79–87

Anker A, Baeza JA (2012) Morphological and molecular phylogeny of hooded shrimps, genera *Betaeus* and *Betaeopsis* (Decapoda, Caridea, Alpheidae): testing the Center of Origin biogeographic model and evolution of life habits. Mol Phylogenet Evol 64:401–415

Anker A, Ahyong ST, Noël PY, Palmer AR (2006) Morphological phylogeny of alpheid shrimps: parallel preadaptation and the origin of a key morphological innovation, the snapping claw. Evolution 60:2507–2528

Anker A, Hurt C, Knowlton N (2007) Revision of the *Alpheus nuttingi* (Schmitt) species complex (Crustacea: Decapoda: Alpheidae), with description of a new species from the tropical eastern Pacific. Zootaxa 1577:41–60

Arana P, Ahumada M (2006) Camarón navaja (*Campylonotus semistriatus*), crustáceo de aguas profundas frente a la costa central de Chile (Crustacea, Decapoda, Campylonotidae). Investigaciones Marinas Valparaíso 34:3–14

Ayón-Parente M, Salgado-Barragán J (2013) A new species of the caridean shrimp genus *Ogyrides* Stebbing, 1914, (Decapoda: Ogyrididae) from the eastern tropical Pacific. Zootaxa 3683:589–594

Baeza JA (2009) Protandric simultaneous hermaphroditism is a conserved trait in *Lysmata* (Caridea: Lysmatidae): implications for the evolution of hermaphroditism in the genus. Smithson Contrib Mar Sci 38:95–110

Baeza JA (2010) Observations on the sexual system and the natural history of the semi-terrestrial shrimp *Merguia rhizophorae* (Rathbun, 1900). Invertebr Biol 129:266–276

Baeza JA (2013) Multi-locus molecular phylogeny of brokenback shrimps (genus *Lysmata* and allies): a test of the 'Tomlinson-Ghiselin' hypothesis explaining the evolution of simultaneous hermaphroditism. Mol Phylogenet Evol 69:46–62

Baeza JA, Piantoni C (2010) Sexual system, sex ratio and group living in the shrimp *Thor amboinensis* (De Man): relevance to resource-monopolization and sex-allocation theories. Biol Bull 219:151–165

Baeza JA, Bauer RT, Okuno J, Thiel M (2014) Molecular phylogeny of hinge-beak shrimps (Decapoda: Caridea: *Rhynchocinetes* and *Cinetorhynchus*) and allies: a formal test of familiar and generic monophyly using a multilocus phylogeny. Zool J Linnean Soc 172:426–450

Bailey-Brock JH, Brock RE (1993) Feeding, reproduction, and sense organs of the Hawaiian anchialine shrimp *Halocaridina rubra* (Atyidae). Pac Sci 47:338–355

Ballard RD, Grassle JF (1979) Return to oases of the deep. Natl Geogr Mag 156:689–703

Barnard KH (1950) Descriptive catalogue of South African decapod Crustacea (crabs and shrimps). Ann South Afr Mus 38:1–837

Bauer RT (1975) Grooming behaviour and morphology of the caridean shrimp *Pandalus danae* Stimpson (Decapoda: Natantia: Pandalidae). Zool J Linnean Soc 56:45–71

Bauer RT (1979) Antifouling adaptations of marine shrimp (Decapoda: Caridea): gill cleaning mechanisms and grooming of brooded embryos. Zool J Linnean Soc 65:281–303

Bauer RT (1981) Color patterns of the shrimps *Heptacarpus pictus* and *H. paludicola*. Mar Biol 46: 141–152

Bauer RT (1982) Polymorphism of colour pattern in the caridean shrimps *Heptacarpus pictus* and *H. paludicola*. Mar Behav Physiol 8:249–265

Bauer RT (1984) Morphological trends in the genus *Heptacarpus* (Decapoda, Caridea) and their phylogenetic significance. J Crustac Biol 4:201–225

Bauer RT (1985a) Penaeoid shrimp fauna from tropical seagrass meadows: species composition, diurnal and seasonal variation in abundance. Proc Biol Soc Wash 98:177–190

Bauer RT (1985b) Diel and seasonal variation in species composition and abundance of the caridean shrimps (Crustacea, Decapoda) from seagrass meadows in northern Puerto Rico. Bull Mar Sci 36:150–162

Bauer RT (1986) Sex change and life history pattern in the shrimp *Thor manningi* (Decopoda: Caridea): a novel case of partial protandric hermaphroditism. Biol Bull 170:11–31

Bauer RT (1989) Functional morphology, adaptive value, and phylogenetic significance of grooming in Decapod Crustacea. In: Felgenhauer B, Watling L (eds) Functional morphology of grooming and feeding appendages. Balkema Press, Boca Raton, FL, pp 49–73

Bauer RT (1992a) Repetitive copulation and variable success of insemination in the marine shrimp *Sicyonia dorsalis* (Decapoda, Penaeoidea). J Crustac Biol 12:153–160

Bauer RT (1992b) Testing generalizations about latitudinal variation in reproductive and recruitment patterns with caridean and sicyoniid shrimps. J Invertebr Reprod Dev 22:193–202

Bauer RT (1994) Usage of the terms thelycum and spermatheca in the reproductive morphology of the Decapoda, with special reference to the Penaeoidea. J Crustac Biol 14:715–721

Bauer RT (1996a) Role of the petasma and appendices masculinae during copulation and insemination in the penaeoid shrimp, *Sicyonia dorsalis* (Crustacea: Decapoda: Dendrobranchiata). Invertebr Reprod Dev 29:173–184

Bauer RT (1996b) A test of hypotheses on male mating systems and female molting in decapod shrimp, using *Sicyonia dorsalis* (Decapoda: Penaeoidea). J Crustac Res 16:429–437

Bauer RT (2000) Simultaneous hermaphroditism in caridean shrimps: a unique and puzzling sexual system in the Decapoda. J Crustac Biol 20(2):116–128

Bauer RT (2004) Remarkable shrimps: adaptations and natural history of the Carideans. University of Oklahoma Press, Norman

Bauer RT (2011) Amphidromy and migrations of freshwater shrimps. I. Costs, benefits, evolutionary origins, and an unusual case of amphidromy. In: Asakura A (ed) New frontiers in crustacean biology. Proceedings of the TCS Summer Meeting, Tokyo, 20–24 September 2009. Brill, Leiden, The Netherlands, pp 145–156

Bauer RT (2013) Amphidromy in shrimps: a life history pattern between rivers the sea. Lat Am J Aquat Res 41:633–650

Bauer RT (2020) Shrimp fisheries. In: Thiel M, Lovrich G (eds) The natural history of the crustacea: Volume 8. Fisheries and aquaculture. Oxford University Press, Oxford, pp 91–116

Bauer RT, Conner SL (2011) Gonochoric sexual system in the caridean shrimps *Processa riveroi* and *P. bermudensis* (Decapoda: Processidae) inhabiting a tropical seagrass meadow. J Mar Biol Assoc U K 92:521–529

Bauer RT, Holt GJ (1998) Simultaneous hermaphroditism in the marine shrimp *Lysmata wurdemanni* (Caridea: Hippolytidae): an undescribed sexual system in the decapod Crustacea. Mar Biol 132:223–235

Bauer RT, Lin JM (1993) Spermatophores and plug substance of the marine shrimp *Trachypenaeus similis* (Crustacea: Decapoda: Penaeidae): formation in the male reproductive tract and disposition in the inseminated female. Biol Bull 185:174–185

Bauer RT, Newman WA (2004) Protandric simultaneous hermaphroditism in the marine shrimp *Lysmata californica* (Caridea: Hippolytidae). J Crustac Biol 24:131–139

Bauer RT, Thiel M (2011) First description of a pure-search mating system and protandry in the shrimp *Rhynchocinetes uritai* (Decapoda: Caridea). J Crustac Biol 31:286–295

Bauer RT, VanHoy R (1996) Variation in sexual systems (protandry, gonochorism) and reproductive biology among three species of the shrimp genus *Thor* (Decopoda: Caridea). Bull Mar Sci 59:53–73

Bauer RT, Okuno J, Thiel M (2014) Inferences on mating and sexual systems of two Pacific *Cinetorhynchus* shrimps (Decapoda, Rhynchocinetidae) based on sexual dimorphism in body size and cheliped weaponry. Zookeys 457:187–209

Bergström B (2000) The biology of *Pandalus*. Adv Mar Biol 38:1–245

Berkeley AA (1930) The post-embryonic development of the common pandalids of British Columbia. In: Contributions to Canadian Biology New Series, vol 6. University of Toronto Press, Toronto, pp 79–163

Bochini GL, Cunha AM, Terossi M, Almeida AO (2020) A new genus and species from Brazil of the resurrected family Macromaxillocarididae Alvarez, Iliffe & Villalobos, 2006 and a world-wide list of Stenopodidea (Decapoda). J Crustac Biol 40:704–714

Boddeke R, Bosschieter JR, Goudswaard PC (1991) Sex change, mating, and sperm transfer in *Crangon crangon* (L.). In: Bauer RT, Martin JW (eds) Crustacean sexual biology. Columbia University Press, New York, pp 164–182

Boltana S, Thiel M (2001) Associations between two species of snapping shrimp, *Alpheus inca* and *Alpheopsis chilensis* (Decapoda: Caridea: Alpheidae). J Mar Biol Assoc U K 81:633–638

Boone L (1927) Crustacea from tropical east American seas. Scientific results of the first oceanographic expedition of the "Pawnee," 1925. In: Bulletin Bingham oceanographic collection, vol 1. Yale University, New Haven, CT, pp 1–147

Boothe BB, Heard RW (1987) *Discias vernbergi,* new species, a caridean shrimp (Crustacea: Decapoda: Bresiliidae) from the northwestern Atlantic. Proc Biol Soc Wash 100:506–514

Bowles DE, Aziz K, Knight CL (2000) *Macrobrachium* (Decapoda: Caridea: Palaemonidae) in the contiguous United States a review of the species and assessment of threats to their survival. J Crustac Biol 20:158–171

Bracken HD, Toon A, Felder DL, Martin JW, Finley M, Rasmussen J, Palero F, Crandall KA (2009) The Decapod tree of life: compiling the data and moving toward a consensus of Decapod evolution. Arthropod Syst Phylogeny 67:99–116

Braga AA, Lopez Greco LS, Santos DC, Fransozo A (2009) Morphological evidence for protandric simultaneous hermaphroditism in the caridean *Exhippolysmata oplophoroides*. J Crustac Biol 29:34–41

Bruce AJ (1966) *Bathypalaemonella humilis* sp. nov., a new species of shrimp from the South China Sea (Decapoda, Campylonotidae). Crustaceana 11:277–287

Bruce AJ (1972) *Pycnocaris chagoae* gen. nov., sp. nov., a new shrimp from the Chagos Archipelago (Decapoda Natantia, Gnathophyllidae). Crustaceana 23:50–64

Bruce AJ (1973) *Gnathophylloides robustus* sp. nov., a new commensal gnathophyllid shrimp from western Australia, with the designation of a new genus *Levicaris* (Decapoda, Caridea). Crustaceana 24:17–32

Bruce AJ (1975) Coral reef shrimps and their colour patterns. Endeavour 34:23–27

Bruce AJ (1976) Shrimps and prawns of coral reefs, with special reference to commensalism. In: Jones O, Endean R (eds) Biology and geology of coral reefs, Biology 2, vol III. New York, Academic Press, pp 38–94

Bruce AJ (1983a) *Miropandalus hardingi*, new genus, new species, a bizarre commensal pandalid shrimp from the Marshall Islands. J Crustac Biol 3:482–490

Bruce AJ (1983b) *Epipontonia anceps* n. sp., a sponge-associated pontoniine shrimp from Heron Island, Queensland. (Crustacea: Decapoda: Palaemonidae). Rec Aust Mus 35:19–28

Bruce AJ (1986a) Two new species of *Bathypalaemonella* Balss (Crustacea, Decapoda, Campylonotidae) from the Australian Northwest Shelf. Zool Scr 15:251–264

Bruce AJ (1986b) Observations on the family Gnathophyllidae Dana, 1852 (Crustacea: Decapoda). J Crustac Biol 6:463–470

Bruce AJ (1986c) Redescriptions of five Hong Kong carideans first described by William Stimpson, 1860. In: Morton B (ed) Proceedings of the second international marine biological workshop: the marine flora and fauna of Hong Kong and Southern China. Hong Kong University Press, Hong Kong, pp 569–610

Bruce AJ (1991) The 'African' shrimp genus *Potamalpheops* in Australia, with the description of *P. hanleyi*, new species (Decapoda: Alpheidae). J Crustac Biol 11:629–638

Bruce AJ (1993) The occurrence of the semi-terrestrial shrimps *Merguia oligodon* (De Man 1888) and *M. rhizophorae* (Rathbun 1900) (Crustacea Decapoda Hippolytidae) in Africa. Trop Zool 6: 179–187

Burkenroad MD (1934) The Penaeidea of Louisiana with a discussion of their world relationships. Bull Am Mus Nat Hist 68:61–143

Burkenroad MD (1936) The Aristaeinae, Solenocerinae and pelagic Penaeinae of the Bingham Oceanographic collection. Bull Bingham Oceanogr Collect 5:1–151

Burkenroad MD (1983) Natural classification of Dendrobranchiata, with a key to recent genera. In: Schram FR (ed) Crustacean phylogeny. Crustacean issues, vol 1. A.A. Balkema, Rotterdam, pp 279–290

Burukovsky RN (1988) New representatives of the family Bresiliidae (Crustacea, Decapoda) from the north-east Atlantic [in Russian]. Zoologicheskii Zhurnal 67:456–460

Burukovsky RN (2012) Deep sea shrimps of the family Nematocarcinidae. Prospekt Nauki Publishing House, Saint Petersburg, Russia, pp 1–288

Burukovsky RN, Falkenhaug T (2015) Feeding of the pelagic shrimp *Acanthephyra pelagica* (Risso, 1816) (Crustacea: Decapoda: Oplophoridae) in the northern Mid-Atlantic Ridge area in 1984 and 2004. [In Russian with English summary]. Arthropoda Selecta 24:303–316

Burukovsky RN (2017) Distribution, reproduction, and feeding of West African shrimp *Plesionika carinata* Holthuis, 1951 (Decapoda, Pandalidae). Oceanology 57:668–674

Butler TH (1964) Growth, reproduction, and distribution of pandalid shrimps in British Columbia. J Fish Res Board Can 21:1403–1452

Butler TH (1980) Shrimps of the Pacific coast of Canada. Can Bull Fish Aquat Sci 202:1–80

Cai Y, Ng PKL, Shokita S, Satake K (2006) On the species of Japanese atyid shrimps (Decapoda: Caridea) described by William Stimpson (1860). J Crustac Biol 26:392–419

Calado R (2008) Marine ornamental shrimp: biology, aquaculture and conservation. Wiley and Blackwell, Oxford

Calado R, Chevaldonne P, dos Santos A (2004) A new species of the deep-sea genus *Bresilia* (Crustacea: Decapoda: Bresiliidae) discovered from a shallow-water cave in Madeira. J Mar Biol Assoc UK 84:191–199

Calman WT (1909) On a blind prawn from the Sea of Galilee (*Typhlocaris galilea* g. et sp. n.). Transactions of the Linnean Society London, Zoology 11:93–97

Campos JC, Moreira C, Freitas F, van der Veer HW (2012) Short review of the eco-geography of *Crangon*. J Crustac Biol 32:159–169

Cardoso IA, Fransen CHJM (2012) A new species of the deepwater shrimp genus *Leontocaris* (Hippolytidae: Caridea) from the South Mid-Atlantic Ridge. J Mar Biol Assoc U K 92:1083–1088

Cartes JE (1993) Feeding habits of pasiphaeid shrimps close to the bottom on the western Mediterranean slope. Mar Biol 117:459–468

Cartes JE (1995) Diets of, and trophic resources exploited by, bathyal penaeoidean shrimps from the Western Mediterranean. Mar Freshw Res 46:889–896

Cartes JE, Papiol V, Guijarro B (2008) The feeding and diet of the deep-sea shrimp *Aristeus antennatus* off the Balearic Islands (Western Mediterranean): influence of environmental factors and relationship with the biological cycle. Prog Oceanogr 79:37–54

Carvalho-Batista A, Castilho AL, Fransozo A, Costa RC (2012) Diel comparison of the catch and size of the shrimp *Artemesia longinaris* (Dendrobranchiata, Penaeidae) in the Ubatuba region, northern coast of the state of São Paulo. Crustaceana 85:1179–1191

Castilho AL, Costa RC, Fransozo A, Negreiros-Fransozo ML (2008) Reproduction and recruitment of the South American red shrimp, *Pleoticus muelleri* (Crustacea: Solenoceridae), from the southeastern coast of Brazil. Mar Biol Res 4:361–368

Chace FA (1939) Reports on the scientific results of the first Atlantis Expedition to the West Indies, under the joint auspices of the University of Havana and Harvard University preliminary descriptions of one new genus and seventeen new species of decapod and stomatopod Crustacea. Memorias de la Sociedad Cubana de Historia Natural 13:31–54

Chace FA (1940) Plankton of the Bermuda Oceanographic Expedition. IX. The bathypelagic caridean Crustacea. Zool N Y 25:117–209

Chace FA (1951) The number of species of decapod and stomatopod Crustacea. J Wash Acad Sci 41:370–372

Chace FA (1975) Cave shrimps (Decapoda: Caridea) from the Dominican Republic. Proc Biol Soc Wash 88:29–44

Chace FA (1976) Shrimps of the pasiphaeid genus *Leptochela* with descriptions of three new species (Crustacea: Decapoda: Caridea). Smithsonian Contributions to Zoology No. 222, p 51

Chace FA (1983) The caridean shrimps (Crustacea: Decapoda) of the Albatross Philippine Expedition, 1907-1910, Part 1: Family Stylodactylidae. Smithsonian Contributions to Zoology No. 381, p 21

Chace FA (1984) The caridean shrimps (Crustacea: Decapoda) of the Albatross Philippine Expedition, 1907-1910, Part 2: Families Glyphocrangonidae and Crangonidae. Smithsonian Contributions to Zoology No. 397, p 63

Chace FA (1985) The caridean shrimps (Crustacea: Decapoda) of the Albatross Philippine Expedition, 1907-1910, Part 3: Families Thalassocarididae and Pandalidae. Smithsonian Contributions to Zoology, No. 411, p 143

Chace FA (1986) The caridean shrimps (Crustacea: Decapoda) of the Albatross Philippine Expedition, 1907-1910, Part 4: Families Oplophoridae and Nematocarcinidae. Smithsonian Contributions to Zoology, No. 432, p 82

Chace FA (1992) On the classification of the Caridea (Decapoda). Crustaceana 63:70–80

Chace FA (1997) The caridean shrimps (Crustacea: Decapoda) of the *Albatross* Philippine Expedition, 1907–1910, Part 7: Families Atyidae, Eugonatonotidae, Rhynchocinetidae, Bathypalaemonellidae, Processidae, and Hippolytidae. Smithsonian Contributions to Zoology 587:1–106

Chace FA, Brown DE (1978) A new polychelate shrimp from the Great Barrier Reef of Australia and its bearing on the family Bresiliidae (Crustacea: Decapoda: Caridea). Proc Biol Soc Wash 91:756–766

Chace FA, Bruce AJ (1993) The Caridean Shrimps (Crustacea: Decapoda) of the Albatross Philippine Expedition, 1907-1910, Part 6: Superfamily Palaemonoidea. Smithsonian Contributions to Zoology, No. 543, p 152

Chace FA, Fuller SLH (1971) A new shrimp of the genus *Gnathophyllum* (Decapoda, Caridea) from Puerto Rico. Proc Biol Soc Wash 83:493–504

Chace FA, Hobbs HH (1969) The freshwater and terrestrial decapod crustaceans of the West Indies with special reference to Dominica. U S Natl Mus Bull 292:1–258

Chace FA, Holthuis LB (1978) *Psalidopus*: the scissor-foot shrimps (Crustacea: Decapoda: Caridea). Smithsonian Contributions to Zoology No. 277, p 22

Chace FA, Manning RB (1972) Two new caridean shrimps, one representing a new family, from marine pools on Ascension Island (Crustacea: Decapoda: Natantia). Smithsonian Contributions to Zoology No. 131, p 18

Chan T-Y, Yu H-P (1988) An uncommon deep-sea shrimp *Eugonatonotus crassus* (A. Milne Edwards, 1881) (Crustacea: Decapoda: Eugonatonotidae) from Taiwan. Bull Inst Zool Acad Sin 27:259–263

Chan T-Y, Yu H-P (1991) *Eugonatonotus chacei* sp. nov., second species of the genus (Crustacea, Decapoda, Eugonatonotidae), Bulletin du Muséum national d'Histoire naturelle (4) section A, Zoologie. Biologie et Écologie animales 13:143–152

Chan BKK, Lin I-C, Shin T-W, Chan TY (2008) Bioluminescent emissions of the deep-water pandalid shrimp, *Heterocarpus sibogae* De Man, 1917 (Decapoda, Caridea, Pandalidae) under laboratory conditions. Crustaceana 81:341–350

Chan T-Y, Lei HC, Li CP, Chu KH (2010) Phylogenetic analysis using rDNA reveals polyphyly of Oplophoridae (Decapoda: Caridea). Invertebr Syst 24:172–181

Chan TY, Kumar AB, Yang CH (2017) Photophore counts in the deep-sea commercial shrimp *Aristeus alcocki* Ramadan, 1938 (Crustacea: Decapoda: Aristeidae), with a revised key to the Indo-West Pacific species of the genus. Zootaxa 4329:392–400

Chartosia N, Tzomos T, Kitso M-S, Karani I, Tselepides A, Koukouras A (2005) Diet comparison of the bathyal shrimps, *Aristeus antennatus* (Risso, 1816) and *Aristaeomorpha foliacea* (Risso, 1827) (Decapoda, Aristeidae) in the eastern Mediterranean. Crustaceana 78:273–284

Chávez-Solís EM, Mejía-Ortíz LM, Simões N (2018) Predatory behavior of the cave shrimp *Creaseria morleyi* (Creaser, 1936) (Caridea: Palaemonidae), the blind hunter of the Yucatán cenotes, Mexico. J Crustac Biol 38:1–7

Chen C-L, Goy JW, Bracken-Grissom HD, Felder DL, Tsang LM, Chan T-Y (2016) Phylogeny of Stenopodidea (Crustacea: Decapoda) shrimps inferred from nuclear and mitochondrial genes reveals non-monophyly of the families Spongicolidae and Stenopididae and most of their composite genera. Invertebr Syst 30:479–490

Chiba S (2007) A review of ecological and evolutionary studies on hermaphroditic decapod crustaceans. Plankton Benthos Res 2:107–119

Chockley BR, St. Mary CM (2003) Effect of body size on growth, survivorship and reproduction in the banded coral shrimp *Stenopus hispidus*. J Crustac Biol 23:836–848

Chockley BR, St. Mary CM, Osenberg CW (2008) Population sinks in the Upper Florida Keys: the importance of demographic variation in population dynamics of the marine shrimp *Stenopus hispidus*. Mar Ecol Prog Ser 360:135–145

Chow S, Okazaki M, Takeda M, Kubota T (2000) A rare abyssal shrimp, *Galatheocaris abyssalis*, found in the stomach of a lancetfish. Crustaceana 73:243–246

Chow LH, De Grave S, Tsang LM (2020) The family Anchistioididae Borradaile, 1915 (Decapoda: Caridea) is a synonym of Palaemonidae Rafinesque, 1815 based on molecular and morphological evidence. J Crustac Biol 40:277–287

Christodoulou M, Anastasiadou C (2017) Sexual dimorphism in the shrimp genus *Atyaephyra* (Caridea: Atyidae): the case study of *Atyaephyra thyamisensis*. J Crustac Biol 37:588–601

Christodoulou M, Antoniou A, Magoulas A, Koukouras A (2012) Revision of the freshwater genus *Atyaephyra* (Crustacea, Decapoda, Atyidae) based on morphological and molecular data. Zookeys 229:53–110

Christoffersen ML (1986) Phylogenetic relationships between Oplophoridae, Atyidae, Pasiphaeidae, Alvinocarididae Fam. N., Bresiliidae, Psalidopodidae and Disciadidae (Crustacea Caridea Atyoidea). Bol Zool 10:273–281

Christofferson ML (1990) A new superfamily classification of the Caridea (Crustacea: Pleocyemata) based on phylogenetic pattern. Zeitschrift für Zoologische Systematik und Evolutionsforschung 28:94–106

Clark WH, Griffin FJ (1988) The morphology and physiology of the acrosome reaction in the sperm of the decapod, *Sicyonia ingentis*. Dev Growth Differ 30:451–462

Cleva R (1990) Crustacea Decapoda: Les genres et les espèces indo-ouest pacifiques de Stylodactylidae. In: Crosnier A (ed) Résultats des Campagnes Musorstom, vol 6. Mémoires du Muséum national d'Histoire naturelle, Paris, pp 71–136

Cleva R (2001) Les Bathypalaemonellidae de Saint Laurent, 1985 (Crustacea, Decapoda, Caridea) avec description d'une espèce nouvelle et définition d'un genre nouveau. Zoosystema 23:757–782

Cleva R (2004) Stylodactylidae and Bathypalaemonellidae from Taiwan (Crustacea: Decapoda: Caridea). Raffles Bull Zool 52:497–511

Corliss JB, Ballard RD (1977) Oases of life in the cold abyss. Natl Geogr Mag 152:441–453

Correa C, Thiel M (2003a) Mating systems in caridean shrimp (Decapoda: Caridea) and their evolutionary consequences for sexual dimorphism and reproductive biology. Rev Chil Hist Nat 76:187–203

Correa C, Thiel M (2003b) Population structure and operational sex ratio in the rock shrimp *Rhynchocinetes typus* (Decapoda: Caridea). J Crustac Biol 23:849–861

Correa C, Baeza JA, Dupré E, Hinojosa IA, Thiel M (2000) Mating behavior and fertilization success of three ontogenetic stages of male rock shrimp *Rhynchocinetes typus* (Decapoda: Caridea). J Crustac Biol 20:628–641

Correa C, Baeza JA, Hinojosa IA, Thiel M (2003) Male dominance hierarchy and mating tactics in the rock shrimp *Rhynchocinetes typus* (Decapoda: Caridea). J Crustac Biol 23:33–45

Covich AP (2006) Dispersal-limited biodiversity of tropical insular streams. Pol J Ecol 54:523–547

Covich AP (2009) Freshwater ecology. In: Gillespie RG, Clague DA (eds) Encyclopedia of islands. University of California Press, Berkeley, pp 343–347

Covich AP, McDowell WH (1996) The stream community. In: Reagan DP, Waide RB (eds) The food web of a tropical rain forest. University of Chicago Press, Chicago, pp 433–459

Covich AP, Palmer MA, Crowl TA (1999) Role of benthic invertebrate species in freshwater ecosystems. Bioscience 49:119–127

Cowles DL (1994) Swimming dynamics of the mesopelagic vertically migrating penaeid shrimp *Sergestes similis*: modes and speed of swimming. J Crustac Biol 14:247–257

Criales MM (1997) *Microprosthema granatense* new species, from the southern Caribbean, with a key to shrimps of the genus *Microprosthema* from the western Atlantic and a new record of *Odontozona liberate* (Decapoda: Stenopodidea). J Crustac Biol 17:538–545

Crosnier A (1978) Crustacés Décapodes pénéides Aristeidae (Benthesicyminae, Aristeinae, Solenocerinae). Faune de Madagascar 46:1–197

Crosnier A, Forest J (1973) Les crevettes profondes de l'Atlantique oriental tropical. Faune Tropicale (Cahiers de l'Office de la Recherche Scientifique et Technique Outre Mer), No. 19, p 409

Cross WF, Covich AP, Crowl TA, Benstead JP, Ramirez A (2008) Secondary production, longevity, and consumption rates of freshwater shrimps in two tropical streams with contrasting geomorphology and food web structure. Freshw Biol 53:2504–2519

Crowl TA, Welsh V, Heartsill-Scalley T, Covich AP (2006) Effects of different types of conditioning on rates of leaf-litter shredding by *Xiphocaris elongata*, a neotropical freshwater shrimp. J N Am Benthol Soc 25:198–208

D'Onghia G, Maiorano P, Matarrese A, Tursi A (1998) Distribution, biology, and population dynamics of *Aristaeomorpha foliacea* (Risso, 1827) (Decapoda, Natantia, Aristeidae) in the northwestern Ionian Sea (Mediterranean Sea). Crustaceana 71:518–544

d'Udekem d'Acoz C (1996) The genus Hippolyte Leach, 1814 (Crustacea: Decapoda: Caridea: Hippolytidae) in the East Atlantic Ocean and the Mediterranean Sea, with a checklist of all species in the genus. Zoologische Verhandelingen, Leiden 303:1–133. https://repository.naturalis.nl/pub/317714

d'Udekem d'Acoz C, Degrave S (2018) A new genus and species of large-bodied caridean shrimp from the Crozet Islands Southern Ocean (Crustacea Decapoda Lipkiidae) with a checklist of Antarctic and sub-Antarctic shrimps. Zootaxa 4392(2):201. https://doi.org/10.11646/zootaxa.4392.2.1

Dall W (2001) Australian species of Aristeidae and Benthesicymidae (Penaeoidea: Decapoda). Mem Queensl Mus 46:409–441

Dall W (2007) Recent molecular research on *Penaeus sensu* lato. J Crustac Biol 27:380–382

Dall W, Hill BC, Rothlisberg PC, Sharples DJ (1990) The biology of the Penaeidae. Adv Mar Biol 27:1–489

Dana TF, Newman WA, Fager EW (1973) *Acanthaster* aggregations: interpreted as primarily responses to natural phenomena. Pac Sci 26:355–372

Dardeau MR (1984) *Synalpheus* shrimps (Crustacea: Decapoda: Alpheidae). I. The Gambarelloides group, with a description of a new species. Mem Hourglass Cruises 7:1–125

Dardeau MR, Heard RW (1983) Crangonid shrimps (Crustacea: Caridea), with a description of a new species of *Pontocaris*. Mem Hourglass Cruises 6:39

De Busserolles FJ, Gauthier O, Gélinas Y, Fabri MC, Sarradin PM, Desbruyères D (2009) Are spatial variations in the diets of hydrothermal fauna linked to local environmental conditions? Deep Sea Res II 56:1649–1664

De Grave S (2007) A new species of *Euryrhynchus* Miers, with a discussion of the systematic position of the Euryrhynchidae Holthuis (Crustacea, Decapoda). Zool Anz 246:193–203

De Grave S, Ashelby CW (2013) A re-appraisal of the systematic status of selected genera in Palaemoninae (Crustacea: Decapoda: Palaemonidae). Zootaxa 3734:331–344

De Grave S, Fransen CHJM (2011) Carideorum catalogus: the recent species of the dendrobranchiate, stenopodidean, procarididean and caridean shrimps, vol 85. Zoologische Mededellingen, Leiden, pp 195–589

De Grave S, Moosa MK (2004) A new species of the enigmatic shrimp genus *Pseudocheles* (Decapoda: Bresiliidae) from Sulawesi (Indonesia), with the designation of a new family Pseudochelidae. Crust Res 33:1–9

De Grave S, Livingston D, Speight MR (2006) Diel variation in sea grass dwelling shrimp: when to sample at night? J Mar Biol Assoc 86:1421–1422

De Grave S, Cai Y, Anker A (2008) Global diversity of shrimps (Crustacea: Decapoda: Caridea) in freshwater. Hydrobiologia 595:287–293

De Grave S, Chu KH, Chan T-Y (2010) On the systematic position of *Galatheacaris abyssalis* (Decapoda: Galatheacaridoidea). J Crustac Biol 30:521–527

De Grave S, Li CP, Tsang LM, Chu KH, Chan T-Y (2014) Unweaving hippolytoid systematics (Crustacea, Decapoda, Hippolytidae): resurrection of several families. Zool Scr 43:496–507

De Grave S, Fransen CHJM, Page TJ (2015a) Let's be pals again: major systematic changes in Palaemonidae (Crustacea: Decapoda). PeerJ 3:e1167. https://doi.org/10.7717/peerj.1167

De Grave S, Chan T-Y, Chu KH, Yang CH, Landeira JM (2015b) Phylogenetics reveals the crustacean order Amphionidacea to be larval shrimps (Decapoda: Caridea). Sci Rep 5:17464. https://doi.org/10.1038/srep17464

De Grave S, Arjun CP, Raghavan R (2018a) The discovery of Euryrhynchidae (Crustacea: Decapoda) in India, with the description of a new genus and species. Zootaxa 4462:367–378

De Grave S, Piscart C, Kayo RPT, Anker A (2018b) A new groundwater-dwelling species of *Euryrhynchina* from Cameroon (Malacostraca, Decapoda, Euryrhynchidae). Zootaxa 4254: 120–126

Demestre M (1995) Moult activity-related spawning success in the Mediterranean deep-water shrimp *Aristeus antennatus* (Decapoda: Dendrobranchiata). Mar Ecol Prog Ser 127:57–64

Dietz RS, Holden JC (1972) The breakup of Pangaea. In: Wilson JT (ed) Scientific American, Continents Adrift. W.H. Freeman and Company, San Francisco, pp 102–113

Dineshbabu AP, Manissery JK (2009) Food and feeding of the ridgeback shrimp, *Solenocera choprai* Nataraj, along Karnataka coast. Indian Journal of Fisheries 56:21–26

Duffy JE (1996) Eusociality in a coral-reef shrimp. Nature 381:512–514

Duffy JE, Macdonald KS (2010) Kin structure, ecology and the evolution of social organization in shrimp: a comparative analysis. Proc R Soc Lond B 277:1–13

Dumont CP, Urriagoa JD, Abarca A, Gaymer CF, Thiel M (2009) The native rock shrimp *Rhynchocinetes typus* as a biological control of fouling in suspended scallop cultures. Aquaculture 292:74–79

Felgenhauer BE, Abele LG (1983) Phylogenetic relationships among shrimp-like decapods. In: Schram FR (ed) Crustacean phylogeny. A.A. Balkema, Rotterdam, pp 291–311

Felgenhauer BE, Abele LG, Kim W (1988) Reproductive morphology of the anchialine shrimp *Procaris ascensionis*. J Crustac Biol 8:333–349

Fiedler GC (1998) Functional, simultaneous hermaphroditism in female-phase *Lysmata amboinensis* (Decapoda: Caridea: Hippolytidae). Pac Sci 52:161–169

FIGIS (Fisheries Global Information System) (2021) Global Capture Production 1950-2019. FI Institutional Websites. FAO Fisheries and Aquaculture Department, Rome, Italy. http://www.fao.org/fishery/statistics/global-capture-production/query/en

Flegel TW (2007) The right to refuse revision in the genus *Penaeus*. Aquaculture 264:2–8

Foxton P (1969) The morphology of the antennal flagellum of certain of the Penaeidea (Decapoda, Natantia). Crustaceana 16:33–43

Foxton P, Herring PJ (1970) Recent records of *Physetocaris microphthalma* Chace with notes on the male and description of the early larvae (Decapoda, Caridea). Crustaceana 18:93–104

Fransen CHJM (2010) Order Amphionidacea Williamson. 1973. In: Schram FR, von Vaupel Klein JC (eds) Treatise on zoology-anatomy, taxonomy, biology. The Crustacea, vol 9A. Brill, Leiden, pp 83–95

Fransen CHJM, De Grave S (2009) Evolution and radiation of the shrimp-like Decapoda. In: Martin JW, Crandall KA, Felder DF (eds) Decapod phylogenetics. Crustacean issues 18. CRC Press, Boca Raton, FL, pp 245–260

Fransozo V, Costa RC, Bertini G, Cobos VJ (2005) Population biology of the spine shrimp *Exhippolysmata oplophoroides* in a subtropical region, São Paulo Brazil. Revista Brazileira de Zoologia 22:1078–1084

Fryer G (1977) Studies of the functional morphology and ecology of the atyid prawns of Dominica. Philos Trans R Soc Lond B Biol Sci 277:57–129

Fujino T, Takeda M (1977) *Levicaris mammillata* (Edmondsn), a gnathophyllid shrimp associated with slate-pencil sea urchin, *Heterocentrotus mammillatus* (Linnaeus), from the Ogasawara and Ryukyu Islands. Bull Natl Mus Sci Ser A (Zool) 3:131–140

Fuss CM (1964) Observations on burrowing behavior of the pink shrimp, *Penaeus duorarum* Burkenroad. Bull Mar Sci Gulf Caribb 14:62–73

Gebruk AV, Pimenov NV, Savichev AS (1993) Feeding specialization of bresiliid shrimps in the TAG site hydrothermal community. Mar Ecol Prog Ser 98:247–253

Gherardi F, Calloni C (1993) Protandrous hermaphroditism in the tropical shrimp *Athanas indicus* (Decapoda: Caridea), a symbiont of sea urchins. J Crustac Biol 13:675–689

Gillikin DP, De Grave S, Tack JF (2001) The occurrence of the semi-terrestrial shrimp *Merguia oligodon* (De Man, 1888) in *Neosarmatium smithi* H. Milne Edwards, 1853 burrows in Kenyan mangroves. Crustaceana 74:505–508

Goy JW (2010) Infraorder Stenopodidea Claus, 1872. In: Schram FR, von Vaupel Klein JC, Forest J, Charmantier-Daures M (eds) Treatise on zoology – Anatomy, taxonomy, biology. The Crustacea, Decapoda. Vol 9. Part A: Eucarida: Euphausiacea, Amphionidacea, and Decapoda (partim). Koninklijke Brill NV, Leiden, pp 215–265

Gregati RA, Fransozo VL, Lopez-Greco LS, Negreiros-Fransozo ML (2010) Reproductive cycle and ovarian development of the marine ornamental shrimp *Stenopus hispidus* in captivity. Aquaculture 206:185–190

Gregati RA, Fransozo V, López-Greco LS, Negreiros-Fransozo ML, Bauer RT (2014) Functional morphology of the reproductive system and sperm transfer in *Stenopus hispidus* (Crustacea: Decapoda: Stenopodidea) and their relation to mating system. Invertebr Biol 133:381–393

Gros P (1979) Biologie de *Processa nouveli* Holthuisi, Al-Adhub et Williamson (Crustacea, Decapoda) dans le secteur nord du Golfe de Gascogne. Annals de l'Institut Océanographique Paris 55:49–70

Guo ZL, Wang XQ, Zhang JP (2005) On the genus *Exopalaemon* (Decapoda, Caridea, Palaemonidae) in Guangdong Province, southern China. Crustaceana 78:839–850

Guy-Haim T, Simon-Blecher N, Frumkin A, Naaman I, Achituv Y (2018) Multiple transgressions and slow evolution shape the phylogeographic pattern of the blind cave-dwelling shrimp *Typhlocaris*. PeerJ 6:e26616v1. https://doi.org/10.7287/peerj.preprints.26616v1

Hacker SD, Madin LP (1991) Why habitat architecture and color are important to shrimps living in pelagic *Sargassum*: use of camouflage and plant-part mimicry. Mar Ecol Prog Ser 70:143–155

Haefner PA (1979) Comparative review of the biology of North Atlantic shrimps (*Crangon*), with emphasis on *C. septemspinosa*. Bull Biol Soc Wash 3:1–40

Haley C, von der Meden C, Atkinson L, Reed C (2017) Habitat associations and distribution of the hyperbenthic shrimp, *Nauticaris marionis*, around the sub-Antarctic Prince Edward Islands. Deep Sea Res I 127:41–48

Hart JFL (1964) Shrimps of the genus *Betaeus* on the Pacific coast of North America with description of three new species. Proc U S Natl Mus 115:431–466

Hart CW, Manning RB (1986) Two new shrimps (Procarididae and Agostocarididae, new family) from marine caves of the western North Atlantic. J Crustac Biol 6:408–416

Hart CW, Manning RB, Iliffe TM (1985) The fauna of Atlantic marine caves: evidence of dispersal by sea floor spreading while maintaining ties to deep waters. Proc Biol Soc Wash 98:288–292
Hay WP, Shore CA (1918) The decapod crustaceans of Beaufort, NC, and the surrounding region. Bull U S Bur Fish 35:369–476
Hayashi K-I (1975) The Indo-West Pacific Processidae (Crustacea, Decapoda, Caridea). J Shimonoseki Univ Fish 24:47–145
Hayashi K-I, Miyake S (1968) Notes on the family Stylodactylidae with the description of a new genus *Neostylodactylus*. J Fac Agric Kyushu Univ 14(4):583–611
Hayashi K-I, Miyake S (1970) *Bathyhippolyte yaldwyni* n. gen., n. sp., a deepsea hippolytid (Decapoda, Natantia) from New Zealand. R Soc N Z Biol Sci 12:41–47
Hedgpeth JW (1949) The North American species of *Macrobrachium*. Tex J Sci 1:28–38
Hedgpeth JW (1968) The atyid shrimps of the genus *Syncaris* in California. Internationale Revue der gesamten Hydrobiologie und Hydrographie 53:511–524
Heegaard P (1967) On the behaviour, sex-ratio and growth of *Solenocera membranacea* (Risso) (Decapoda, Penaeidae). Crustaceana 13:227–237
Hendrickx ME (2002) A new deep water species of *Odontozona* Holthuis (Decapoda, Stenopodidae) from the southern Gulf Of California, Mexico. Crustaceana 75:405–412
Hendrickx ME (2012) Distribution and abundance of the pelagic processid, *Processa pippinae* Wicksten & Méndez, 1985 (Decapoda, Caridea, Processidae), collected during the Talud XIV cruise in the Gulf Of California, Mexico, and description of a new genus. Crustaceana 85:463–472
Hendrickx ME (2014) First record of the caridean shrimp genus *Bresilia* Calman, 1896 (Crustacea: Decapoda: Bresiliidae) from the East Pacific and description of a new species. Zootaxa 3878: 185–195
Herring PJ (1981) The comparative morphology of hepatic photophores in decapod Crustacea. J Mar Biol Assoc U K 61:723–737
Herring PJ, Barnes AT (1976) Light stimulated bioluminescence of *Thalassocaris crinita* (Crustacea: Caridea). Crustaceana 31:107–109
Herring PJ, Roe HSJ (1988) The photoecology of pelagic oceanic decapods. Symp Zool Soc Lond 59:263–290
Hobbs HH, Hart CW (1982) The shrimp genus *Atya* (Decapoda: Atyidae). Smithsonian Contributions to Zoology, No. 364, p 143
Holthuis LB (1947) The Decapoda of the Siboga Expedition, part IX. The Hippolytidae and Rhynchocinetidae collected by the Siboga and Snellius Expeditions with remarks on other species. Siboga Expeditie 39a8:100
Holthuis LB (1952a) A general revision of the Palaemonidae (Crustacea Decapoda Natantia) of the Americas. II. The subfamily Palaemoninae. Allan Hancock Foundation Publications, Occasional Papers 12:1–396
Holthuis LB (1952b) The Decapoda of the Siboga Expedition. Part XI The Palaemonidae collected by the Siboga and Snellius Expeditions with remarks on other species II Subfamily Pontoniinae. Siboga Expéditie 39(A10):1–253
Holthuis LB (1971) The Atlantic shrimps of the deep-sea genus *Glyphocrangon*. A. Milne Edwards, 1881. Bull Mar Sci 21:267–373
Holthuis LB (1973) Caridean shrimps found in land-locked pools at four Indo-West Pacific localities (Sinai Peninsula, Funafuti Atoll, Maui and Hawaii islands), with the description of one new genus and four new species. Zoologische Verhandelingen 128:1–48
Holthuis LB (1980) FAO species catalogue. Vol. 1. Shrimps and prawns of the world. An annotated catalogue of species of interest to fisheries. FAO Fisheries Synopsis 125, vol 1, p 261
Holthuis LB (1993) The recent genera of the Caridean and Stenopodidean Shrimps (Crustacea, Decapoda): with an Appendix on the Order Amphionidacea. Nationaal Natuurhistorisch Museum, Leiden
Hopkins TL, Flock ME, Gartner JV, Torres JJ (1994) Structure and trophic ecology of a low latitude midwater decapod and mysid assemblage. Mar Prog Ecol Ser 109:143–156

Horká I, De Grave S, Ďuriš Z (2014) A new species of shrimp of the genus *Anachlorocurtis* Hayashi, 1975 from the Red Sea, with range extension of *A. commensalis* Hayashi, 1975 (Crustacea, Decapoda, Pandalidae). ZooKeys 407:9–28

Hufnagl M, Temming A, Pohlmann T (2014) The missing link: tidal-influenced activity a likely candidate to close the migration triangle in brown shrimp *Crangon crangon* (Crustacea, Decapoda). Fish Oceanogr 23:242–257

Jensen GC (2011) Feeding behavior of the horned shrimp, *Paracrangon echinata* (Caridea: Crangonidae). J Crustac Biol 31:246–248

Jensen GC (2014) Crabs and Shrimps of the Pacific Coast. A guide to shallow-water Decapods from Southeastern Alaska to the Mexican Border. Mola Marine, Bremerton, Washington

Jobe CF, Brooks WR (2009) Habitat selection and host location by symbiotic shrimps associated with *Sargassum* communities: the role of chemical and visual cues. Symbiosis 49:77–85

Johnson VR (1969) Behavior associated with pair formation in banded shrimp *Stenopus hispidus* (Olivier). Pac Sci 23:40–50

Johnson VR (1977) Individual recognition in the banded shrimp *Stenopus hispidus* (Olivier). Anim Behav 25:418–428

Juarrero de Varona A (1993) Nueva especie del género *Xiphocaris* (Crustacea: Atyidae) de Cuba. Poeyana 440:1–12

Judkins D, Kensley B (2008) New genera in the family Sergestidae (Crustacea: Decapoda: Penaeidea). Proc Biol Soc Wash 121:72–84

Kagwade PV (1982) The hermaphrodite prawn *Hippolysmata ensirostris* Kemp. Indian J Fish 28: 189–194

Kapiris K, Thessalou-Legaki M (2009) Comparative reproduction aspects of the deep-water shrimps *Aristaeomorpha foliacea* and *Aristeus antennatus* (Decapoda, Aristeidae) in the Greek Ionian Sea (Eastern Mediterranean). Int J Zool 2009:979512. https://doi.org/10.1155/2009/979512

Karplus I (1987) The association between gobiid fishes and burrowing alpheid shrimps. Oceanogr Mar Biol Annu Rev 25:507–562

Karuppasamy PK, Menon NG, Nair KKC, Achuthankutty CT (2006) Distribution and abundance of pelagic shrimps from the deep scattering layer of the eastern Arabian Sea. J Shellfish Res 25: 1013–1019

Katneni VK, Mudagandur SL, Jangam AK, Paran BC, Selvaraj A, Krishnan K, Kaikkolante N, Prabhudas SK, Gopalapillai G, Koyadan VK (2021) Phylogenetic relations and mitogenome-wide similarity metrics reveal monophyly of *Penaeus sensu lato*. Ecol Evol. https://doi.org/10.1002/ece3.7148

Kayal M, Vercelloni J, Lison de Loma T, Bosserelle P, Chancerelle Y, Geoffroy S, Stievenart C, Michonneau F, Penin L, Planes S, Adjeroud M (2012) Predator crown-of-thorns starfish (*Acanthaster planci*) outbreak, mass mortality of corals, and cascading effects on reef fish and benthic communities. PLoS One 7(10):e47363. https://doi.org/10.1371/journal.pone.0047363

Kensley B (1983) New records of bresiliid shrimp from Australia, South Africa, Caribbean, and Gulf of Mexico (Decapoda: Natantia: Caridea). Smithson Contrib Zool 394:1–31

Kensley B (1988) New species and records of cave shrimps from the Yucatan peninsula (Decapoda: Agostocarididae and Hippolytidae). J Crustac Biol 8:688–699

Kensley B, Walker U (1982) Palaemonid shrimps from the Amazon Basin, Brazil (Crustacea: Decapoda: Natantia). Smithson Contrib Zool 362:28

Kensley B, Williams D (1986) New shrimps (families Procarididae and Atyidae) from a submerged lava tube on Hawaii. J Crustac Biol 6:417–437

Kikkert DA, Crowl TA, Covich AP (2009) Upstream migration of amphidromous shrimp in the Luquillo Experimental Forest, Puerto Rico: temporal patterns and environmental cues. J N Am Benthol Soc 28:233–246

Kim JN, Komai T (2002) *Sclerocrangon gasuyebi* Yokoya synonymized with *Sclerocrangon salebrosa* (Owen) (Decapoda, Caridea, Crangonidae). Crustaceana 74:1335–1347

Kitsos MS, Tzomos T, Anagnostopoulou L, Koukouras A (2008) Diet composition of the pandalid shrimp, *Plesionika narval* (Fabricius, 1787) (Decapoda, Pandalidae) in the Aegean sea. Crustaceana 81:23–33

Knowlton N (1980) Sexual selection and dimorphism in two demes of a symbiotic, pair-bonding snapping shrimp. Evolution 34:161–173

Komai T (1994) Deep-Sea shrimps of the genus *Pandalopsis* (Decapoda: Caridea: Pandalidae) from the Pacific coast of eastern Hokkaido, Japan, with the description of two new species. J Crustac Biol 14:538–555

Komai T (1999) A revision of the genus *Pandalus* (Crustacea: Decapoda: Caridea: Pandalidae). J Nat Hist 33:1265–1372

Komai T (2004) Deep-sea shrimps of the genus *Glyphocrangon* A. Milne-Edwards (Crustacea, Decapoda, Caridea, Glyphocrangonidae) from off southeastern coast of Brazil collected during the Revizee Program. Arquivos do Museu Nacional (Rio de Janeiro) 62:31–44

Komai T (2005) Current status of taxonomy of the family Glyphocrangonidae (Crustacea: Decapoda: Caridea). Taxa 19:8–19

Komai T, Ivanov BG (2008) Identities of three taxa of the hippolytid shrimp genus *Heptacarpus* (Crustacea: Decapoda: Caridea), with description of a new species from East Asian waters. Zootaxa 1684:1–34

Komai T, Kohtsuka H (2017) A new deep-sea species of the caridean genus *Bresilia* Calman, 1896 (Crustacea: Decapoda: Bresiliidae) from Sagami Bay, central Japan. Zootaxa 4299:405–414

Komai T, Lin CW, Chan TY (2012) Bathypelagic Shrimp of the genus *Pasiphaea* (Decapoda: Caridea: Pasiphaeidae) from waters around Taiwan, with descriptions of four new species. J Crustac Biol 32:295–325

Komai T, de Grave S, Saito T (2016) Two new species of the stenopodidean shrimp genus *Spongiocaris* Bruce & Baba, 1973 (Crustacea: Decapoda: Spongicolidae) from the Indo-West Pacific. Zootaxa 4111:421–447

Komai T, Chan T-Y, De Grave S (2019) Establishment of a new shrimp family Chlorotocellidae for four genera previously assigned to Pandalidae (Decapoda, Caridea, Pandaloidea). Zoosyst Evol 95:391–402

Komai T, Chan T-Y, Chang SC (2021) Redescription of the colorful deep-water shrimp *Lebbeus compressus* Holthuis, 1947 and description of a closely allied new species from Taiwan (Decapoda: Caridea: Thoridae). J Crustac Biol 41:1–13

Kou Q, Li X, Chan T-Y, Chu KH, Gan Z (2013) Molecular phylogeny of the superfamily Palaemonoidea (Crustacea:Decapoda: Caridea) based on mitochondrial and nuclear DNA reveals discrepancies with the current classification. Invertebr Syst 27:502–514

Kuris AM, Ra'anan Z, Sagi A, Cohen D (1987) Morphotypic differentiaton of male Malaysian giant prawns, *Macrobrachium rosenbergii*. J Crustac Biol 7:219–237

Kutschera V, Maas A, Waloszek D, C, Haug and J, T. Haug. (2012) Re-study of larval stages of *Amphionides reynaudii* (Malacostraca: Eucarida) with modern imaging techniques. J Crustac Biol 32:916–930

Lakin RC, Jinks RN, Battelle BA, Herzog ED, Kass L, Renninger GH, Chamberlain SC (1997) Retinal anatomy of *Chorocaris chacei*, a deep-sea hydrothermal vent shrimp from the Mid-Atlantic Ridge. J Comp Neurol 385:503–514

Laubenheimer H, Rhyne AL (2008) Experimental confirmation of protandric simultaneous hermaphroditism in a caridean shrimp outside of the genus *Lysmata*. J Mar Biol Assoc U K 88:301–305

Lee WY, Omori M, Peck RW (1992) Growth, reproduction and feeding behavior of the planktonic shrimp, *Lucifer faxoni* Borradaile, off the Texas coast. J Plankton Res 14:61–69

Li X, Mitsuhashi M, Chan T-Y (2008) Deep-sea pontoniines (Decapoda: Palaemonidae) from the Philippine "Panglao 2005" expedition, with descriptions of four new species. J Crustac Biol 28: 385–411

Li CP, De Grave S, Chan T-Y, Lei HC, Chu KH (2011) Molecular systematics of caridean shrimps based on five nuclear genes: implications for superfamily classification. Zool Anz 250:270–279

Liao YS, De Grave TW, Hoa BHY, Ip LM, Tsang TYC, Chua KH (2017) Molecular phylogeny of Pasiphaeidae (Crustacea, Decapoda, Caridea) reveals systematic incongruence of the current classification. Mol Phylogenet Evol 115:171–180

Liao Y, Ma KY, De Grave S, Komai T, Chan T-Y, Chu KH (2019) Systematic analysis of the caridean superfamily Pandaloidea (Crustacea: Decapoda) based on molecular and morphological evidence. Mol Phylogenet Evol 134:200–210

Limbaugh C, Pederson H, Chace FA (1961) Shrimps that clean fishes. Bull Mar Sci 11:237–257

Lisowski EA (1983) Distribution, habitat, and behavior of the Kentucky cave shrimp *Palaemonias ganteri* Hay. J Crustac Biol 3:88–92

Lloyd AJ, Yonge CM (1947) The biology of *Crangon vulgaris* L. in the Bristol Channel and Severn Estuary. J Mar Biol Soc U K 26:626–661

Lunina AA, Kulagin DN, Vereshchaka AL (2019) Oplophoridae (Decapoda: Crustacea): phylogeny, taxonomy, and evolution studied by a combination of morphological and molecular methods. Zool J Linn Soc 186:213–232

Lyons PJ (2013) The benefit of obligate versus facultative strategies in a shrimp–goby mutualism. Behav Ecol Sociobiol. https://doi.org/10.1007/s00265-013-1497-6

Ma KY, Chan T-Y, Chu KH (2011) Refuting the six-genus classification of *Penaeus s.l.* (Dendrobranchiata, Penaeidae): a combined analysis of mitochondrial and nuclear genes. Zool Scr 40:498–508

Maciá S, Robinson MP (2009) Why be cryptic? Choice of host urchin is not based on camouflage in the caridean shrimp *Gnathophylloides mineri*. Acta Ethol 12:105–113

Magalhães C (1988) The larval development of palaemonid shrimps from the Amazon region reared in the laboratory. II. Extremely abbreviated larval development in *Euryrhynchus* Miers, 1877 (Decapoda, Euryrhynchinae). Crustaceana 55:39–52

Magalhães C, Campos MR, Collins PA, Mantelatto FL (2016) Diversity, distribution and conservation of freshwater crabs and shrimps in South America. In: Kawai T, Cumberlidge N (eds) A global overview of the conservation of freshwater decapod crustaceans. Springer, Cham, pp 303–322. https://doi.org/10.1007/978-3-319-42527-6_10

Manning RB (1963) The east American species of *Gnathophyllum* (Decapoda, Caridea), with the description of a new species. Crustaceana 5:47–63

Manning RB, Chace FA (1971) Shrimps of the family Processidae from the Northwestern Atlantic Ocean (Crustacea: Decapoda: Caridea). Smithson Contrib Zool 89:41

Manning RB, Hart CW (1991) A new species of *Processa* from Bermuda (Crustacea: Decapoda: Caridea). Proc Biol Soc Wash 104:317–321

Marin I (2010) Redescription of the alpheid shrimp *Betaeus levifrons* Vinogradov, 1950 (Crustacea, Decapoda, Alpheidae) from Peter the Great Bay, Russian coast of the Sea of Japan. Zootaxa 2613:51–60

Marin I, Chan T-Y (2011) New records of the caridean shrimp family Thalassocarididae Bate, 1888 (Decapoda, Caridea) from Asia. Crustaceana 84:243–249

Marin I, Okuno J, Chan T-Y (2011) On the "*Hippolyte commensalis* Kemp, 1925" species complex (Decapoda, Caridea, Hippolytidae), with the designation of a new genus and description of two new species from the Indo-West Pacific. Zootaxa 2768:32–54

Martin JW (1986) A late embryo of the deep water shrimp *Psalidopus barbouri* Chace, 1939 (Decapoda, Caridea). Crustaceana 51:299–302

Martin JW, Davis GE (2001) An updated classification of the recent Crustacea. Nat Hist Mus Los Ang Contrib Sci 39:124

Martin JW, Kutschera V (2014) Amphionidacea. In: Martin JW, Olesen J, Høeg JT (eds) Atlas of crustacean larvae. Johns Hopkins University Press, Baltimore, pp 226–229

Martin JW, Wicksten MK (2004) Review and redescription of the freshwater atyid shrimp genus *Syncaris* Holmes, 1900, in California. J Crustac Biol 24:447–462

Mascetti P, Fernandez R, Albornoz L, Oyarzún S, Gorny M, Wehrtmann I (1997) Gonopore development and sex change in the Antarctic shrimp *Chorismus antarcticus* (Caridea: Hippolytidae). Polar Biol 17:384–388

Mathews LM (2007) Evidence for high rates of in-pair paternity in the socially monogamous snapping shrimp *Alpheus angulosus*. Aquat Biol 1:55–62

McDowall RM (2007) On amphidromy, a distinct form of diadromy in aquatic organisms. Fish Fish 8:1–13

McLaughlin PA, Lemaitre R, Ferrari FD, Felder DL, Bauer RT (2008) A reply to T.W. Flegel. Aquaculture 275:370–373

Mejía LM, Zarza E, López M (2008) *Barbouria yanezi* sp. nov., a new species of cave shrimp (Decapoda, Barbouriidae) from Cozumel island Mexico. Crustaceana 81:663–672

Mejía-Ortíz LM, Yañez G, López-Mejía M (2017) Anchialocarididae, a new family of anchialine decapods and a new species of the genus *Agostocaris* from Cozumel Island, Mexico. Crustaceana 90:381–398

Méndez MG (1981) Claves para la identificación y distribución de los langostinos y camarones (Crustacea, Decapoda) del mar y ríos de la costa del Perú. Boletín Instituto del Mar del Perú 5:1–170

Menon PG, Williamson DI (1971) Decapod Crustacea from the International Indian Ocean Expedition; the species of *Thalassocaris* and their larvae. J Zool (Lond) 165:27–51

Meyer R, Lochner R, Melzer R (2009) Decapoda – crabs, shrimps & lobsters. In: Häussermann V, Försterra G (eds) Marine Benthic Fauna of Chilean Patagonia. Nature in Focus, Santiago, Chile, pp 623–670

Misamore MJ, Browdy CL (1996) Mating behavior in the white *shrimps Penaeus setiferus* and *P. vannamei*: a generalized model for mating in *Penaeus*. J Crustac Biol 16:61–70

Moore PG, Rainbow PS, Larson RJ (1993) The mesopelagic *Notostomus robustus* Smith (Decapoda: Oplophoridae) observed in situ feeding on the medusan *Atolla wyvillei* Haeckel in the northwest Atlantic, with notes on gut contents and mouthpart morphology. J Crustac Biol 13:690–696

Murray KB, Demko AM, Ward BK, Delillo CA, Balazadeh K, Bourque BD, Rhyne AL (2012) The absence of male features and functionality in the monotypic shrimp genus *Lysmatella*, a simultaneous hermaphrodite. Meeting Abstract, Society for Integrative and Comparative Biology

Nakashima Y (1987) Reproductive strategies in a partially protandric shrimp, *Athanas kominatoensis* (Decapoda: Alpheidae): sex change as the best of a bad situation for subordinates. J Ethol 5:145–159

Noël P (1973) Cycle biologique et inversion sexuelle du crustacés Décapode Natantia *Processa edulis*. Cah Biol Mar 14:217–227

Noël P (1976) L'évolution des caractères sexuels chez *Processa edulis* (Risso) (Décapode, Natantia). Vie et Milieu 26:65–104

Noël P, Chassard-Bouchaud C (1994) Chromatophores et pigmentation. In: Forest J (ed) Traité de Zoologie. Tome VII. Crustacés. Fascicule I. Masson, Paris, pp 91–105

Nouvel H, Holthuis LB (1957) Les Processidae (Crustacea Decapoda Natantia) des eaux européennes, vol 32. Zoologische Verhandelingen, Leiden, pp 1–53

Ohtomi J, Irieda S (1997) Growth of the deep-water mud shrimp *Solenocera melanthos* De Man, 1907 (Decapoda, Penaeoidea, Solenoceridae) in Kagoshima Bay, Southern Japan. Crustaceana 70:45–58

Okuno J (1997a) Crustacea Decapoda: review on the genus *Cinetorhynchus* Holthuis, 1995 from the Indo-West Pacific (Caridea: Rhynchocinetidae). In: Richer de Forges B (ed) Les Fond Meubles Des Lagons De Nouvelle-Calédonie (Sédimentologie, Benthos). Études et Thèses, vol 3. ORSTOM, Paris, pp 31–58

Okuno J (1997b) A new shrimp of the genus *Rhynchocinetes* from the great Australian Bight (Crustacea: Decapoda: Rhynchocinetidae). Records South Austr Mus 30:13–18

Oliveira MV, Costa-Souza AC, Guimarães F, Almeida AO, Baeza JA (2015) Observations on the life history of a rare shrimp, *Salmoneus carvachoi* (Crustacea: Caridea: Alpheidae), a possible simultaneous hermaphrodite. Mar Biodivers Rec. https://doi.org/10.1017/S1755267215001219

Omori M (1974) The biology of pelagic shrimps in the ocean. Adv Mar Biol 12:233–324

Omori M (1979) Growth, feeding, and mortality of larval and early postlarval stages of the oceanic shrimp *Sergestes similis* Hansen. Limnol Oceanogr 24:273–288

Onaga T, Fiedler GC, Baeza JA (2012) Protandric simultaneous hermaphroditism in *Parahippolyte misticia* (Crustacea: Decapoda: Hippolytidae): implications for the evolution of mixed sexual systems in marine shrimps. J Crustac Biol 32:383–394

Orsi Relini L, Relini G (1998) Seventeen instars of adult life in female *Aristeus antennatus* (Crustaeea: Decapoda: Aristeidae). A new interpretation of life span and growth. J Nat Hist 32:1719–1734

Orsi Relini L, Tunesi L (1987) The structure of the spermatophore in *Aristeus antennatus* (Risso, 1816). Investig Pesq 51(1):461–470

Ortega A, Queiroga H, González-Gordillo JI (2005) Planktonic stages of *Processa macrodactyla* (Decapoda: Caridea: Processidae) reared in the laboratory. J Mar Biol Assoc U K 85:1449–1460

Ortmann AE (1894) A study of systematic and geographical distribution of the decapod family Atyidae Kingsley. Proc Acad Natl Sci Phila 46:397–416

Ory NC, Dudgeon D, Duprey N, Thiel M (2014) Effects of predation on diel activity and habitat use of the coral-reef shrimp *Cinetorhynchus hendersoni* (Rhynchocinetidae). Coral Reefs 33:639–650

Osawa Y, Aoki MN, Thiel M, Bauer RT (2015) Analysis of life-history traits in a sex-changing marine shrimp (Decapoda: Caridea: Rhynchocinetidae). Biol Bull 228:125–136

Pachelle PPG, Tavares M (2018) The freshwater shrimp family Euryrhynchidae (Crustacea: Decapoda: Caridea) revisited, with a taxonomic revision of the genus *Euryrhynchus* Miers, 1878. Zootaxa 4380:1–110

Pakhomov EA, Kuun PJ, McQuaid CD (2000) Biology of the bottom-dwelling shrimp *Nauticaris marionis* Bate, 1888 at the sub-Antarctic Prince Edward Islands. Polar Biol 23:522–530

Papiol V, Hendrickx ME (2016) Ecology of *Benthesicymus tanneri* Faxon, 1893 (Dendrobranchiata: Benthesicymidae) from the Mexican Pacific slope. J Crustac Biol 36:50–60

Patton WK, Patton RJ, Barnes A (1985) On the biology of *Gnathophylloides mineri*, a shrimp inhabiting the sea urchin *Tripneustes ventricosus*. J Crustac Biol 5:616–626

Pearse AS, Humm HJ, Wharton GW (1942) Ecology of sand beaches at Beaufort. N C Ecological Monographs 12:135–190

Pelli DG, Chamberlain SC (1989) The visibility of 350 °C black-body radiation by the shrimp *Rimicaris exoculata* and man. Nature 337:460–461

Penn JW (1984) The behavior and catchability of some commercially exploited penaeids and their relationship to stock and recruitment. In: Gulland JA, Rothschild BJ (eds) Penaeid shrimps: their biology and management. Fishing News Books, Farnham, Surrey, pp 173–186

Pérez Farfante I (1975) Spermatophores and thelyca of the American white shrimps, genus *Penaeus*, subgenus *Litopenaeus*. Fish Bull 73:463–486

Pérez Farfante I (1977) American solenocerid shrimps of the genera *Hymenopenaeus*, *Haliporoides*, *Pleoticus*, *Hadropenaeus* new genus, and *Mesopenaeus* new genus. Fish Bull 75:261–346

Pérez Farfante I (1980) Revision of the penaeid shrimp genus P*enaeopsis* (Crustacea: Decapoda). Fish Bull 77:721–773

Pérez Farfante I (1985) The rock shrimp genus *Sicyonia* (Crustacea: Decapoda: Penaeoidea) in the Eastern Atlantic. Fish Bull 83:1–79

Pérez FI, Kensley B (1997) Penaeoid and sergestoid shrimps and prawns of the world. Keys and diagnoses for the families and genera. Mémoires du Muséum National d'Histoire Naturelle 175: 1–233

Pérez Reyes O, Crowl TA, Covich AP (2015) Effects of food supplies and water temperature on growth rates of two species of freshwater tropical shrimps. Freshw Biol. https://doi.org/10.1111/fwb.12584

Pillai MC, Griffin FJ, Clark WB Jr (1988) Induced spawning of the decapod crustacean *Sicyonia ingentis*. Biol Bull 174:181–185

Poore GCB (2009) *Leontocaris alexander*, a new deepwater hippolytid shrimp from Tasmanian seamounts and a phylogeny of the genus (Decapoda, Caridea). Crustaceana 82:951–967

Powell CB (1976) Two new freshwater shrimps from West Africa: the first euryrhynchinids (Decapoda Palaemonidae) reported from the Old World. Revue de Zoologie Africaine 90: 883–902

Powell CB (1977) A revision of the African freshwater shrimp genus *Desmocaris* Sollaud, with ecological notes and description of a new species (Crustacea, Decapoda, Palaemonidae). Revue Zoologique Africaine 91:649–674

Powell CB (1979a) Three alpheid shrimp species of a new genus from West African fresh and brackish waters: taxonomy and ecological zonation (Crustacea Decapoda Natantia). Revue Zoologique Africaine 93:116–150

Powell CB (1979b) Suppression of larval development in the African freshwater shrimp *Desmocaris trispinosa* (Aurivillius) (Decapoda Palaemonidae). Crustaceana 5:185–194

Prakash S, Kumar TTA (2013) Feeding behavior of harlequin shrimp *Hymenocera picta* Dana, 1852 (Hymenoceridae) on sea star *Linckia laevigata* (Ophidiasteridae). J Threat Taxa 5:4819–4821

Prakash S, Babu I, Gopi M, Thangappanpillai T, Kumar A, Balasubramanian T (2011) Discovery of the shrimp *Pycnocaris chagoae* Bruce, 1972 (Decapoda: Caridea: Gnathophyllidae) in the Lakshadweep Archipelago, India. Zootaxa 2998:66–68

Prakash S, Ajithkumar TT, Bauer R, Thiel M, Subramoniam T (2016) Reproductive morphology and mating behavior in the hingebeak shrimp *Rhynchocinetes durbanensis* Gordon, 1936 (Decapoda: Caridea: Rhynchocinetidae) in India. J Mar Biol Assoc U K 96:1331–1340

Pratoomyot J, Choosria S, Muthuwan V, Luangoon N, Charoendee W, Phuangsanthia W, Shinn AP (2018) Sand star, *Astropecten indicus* Döderlein, 1888, as an alternative live diet for captive harlequin shrimp, *Hymenocera picta* Dana, 1852. Aquaculture 484:51–36

Provenzano AJ (1978) Feeding behavior of the primitive shrimp, *Procaris* (Decapoda, Procarididae). Crustaceana 35(2):170–176

Rainer SF (1992) Diet of prawns from the continental slope of north-western Australia. Bull Mar Sci 50:258–274

Rasch JA, Bauer RT (2015) Reproductive pattern and sexual system of the nocturnal seagrass shrimp *Ambidexter symmetricus* (Decapoda: Caridea: Processidae) in a Florida bay. Mar Freshw Res 67:1141–1152

Rasch JA, Bauer RT (2016) The functional morphology and role of the thelycum in insemination, and its relation to the mating system in the seagrass shrimp *Ambidexter symmetricus* (Decapoda: Processidae). Invertebr Biol 135:163–173

Rétaux S, Casane D (2013) Evolution of eye development in the darkness of caves: adaptation, drift, or both? EvoDevo 4:26. https://doi.org/10.1186/2041-9139-4-26

Rhyne AL, Lin J (2006) A western Atlantic peppermint shrimp complex: redescription of *Lysmata wurdemanni*, description of four new species, and remarks on *Lysmata rathbunae* (Crustacea: Decapoda: Hippolytidae). Bull Mar Sci 79:165–204

Rice AL (1981) The abdominal locking mechanism in the deep-sea shrimp genus *Glyphocrangon* (Decapoda, Glyphocrangonidae). Crustaceana 40:316–319

Rieley G, Van Dover CL, Hedrick DB, Eglinton G (1999) Trophic ecology of *Rimicaris exoculata*: a combined lipid abundance/stable isotope approach. Mar Biol 133:495–499

Saito T, Komai T (2008) A review of species of the genera *Spongicola* de Haan, 1844 and *Paraspongicola* de Saint Laurent & Cleva, 1981 (Crustacea, Decapoda, Stenopodidea, Spongicolidae). Zoosystema 30:87–147

Saito T, Konishi K (1999) Direct development in the sponge-associated deep-sea shrimp *Spongicola japonica* (Decapoda: Spongicolidae). J Crustac Biol 19:46–52

Saito T, Takeda M (2003) Phylogeny of the Spongicolidae (Crustacea: Stenopodidae): evolutionary trend from shallow-water free-living to deep-water sponge-associated habitat. J Mar Biol Assoc U K 83:119–131

Saito T, Uchida I, Takeda M (2002) Skeletal growth of the deep-sea hexactinellid sponge *Euplectella oweni*, and host selection by the symbiotic shrimp *Spongicola japonica* (Crustacea: Decapoda: Spongicolidae). J Zool 258:521–529

Saito T, Tsuchida S, Yamamoto T (2006) *Spongicoloides iheyaensis*, a new species of dee-sea sponge-associated shrimp from the Iheya Ridge, Ryukyu Islands, southern Japan (Decapoda: Stenopodidea: Spongicolidae). J Crustac Biol 29:224–233

Saito T, Okuno J, Chan T-Y (2009) A new species of *Stenopus* (Crustacea: Decapoda: Stenopodidae) from the Indo-West Pacific, with a redefinition of the genus. Raffles Bull Zool 20:109–120

Sardà F, Company JB, Costa C (2005) Morphological approach for relating decapod crustacean cephalothorax shape with distribution in the water column. Mar Biol 147:611–618

Scelso MA (1991) Cópula en cautiverio y desove del camarón *Artemesia longinarus* Bate, 1888 (Decapoda, Penaeidae). Informes Técnicos de Scientia Marina 164:16

Schaffmeister BE, Hiddink JG, Wim JW (2006) Habitat use of shrimps in the intertidal and shallow subtidal seagrass beds of the tropical Banc d'Arguin, Mauritania. J Sea Res 55:230–243

Segonzac M, de Saint Laurent M, Casanova B (1993) L'énigme du compartement ophique des crevettes Alvinocarididae des sites hydrothermaux de la dorsale médio-Atlantique. Cah Biol Mar 34:535–571

Seibt U (1973) Sense of smell and pair-bond in *Hymenocera picta*. Micronesica 9:231–236

Seibt U, Wickler W (1979) The biological significance of the pair-bond in the shrimp *Hymenocera picta*. Zeitschrift fur Tierpyschologie 50:166–179

Shumway SE, Perkins HC, Schick DF, Stickney AP (1985) Synopsis of biological data on the pink shrimp, *Pandalus borealis* Kröyer, 1838. NOAA Technical Report NMFS 30. FAO Fisheries Synopsis No 144:1–57

Smaldon G (1979) British coastal shrimps and prawns. Academic Press, London

Sokolov V (2000) Deep-sea shrimps of the genus *Bythocaris* G.O. Sars in the collections of Russian museums, with the description of a new species (Crustacea: Decapoda: Hippolytidae). Zool Meded 74:403–468

Sokolov VI, Spiridonov VA (2006) Development of external sexual characters of deep sea *Bythocaris* shrimps (Crustacea: Decapoda: Hippolytidae). J Mar Biol Assoc UK 86:349–359

Suzuki H (1970) Taxonomic review of four alpheid shrimps belonging to the genus *Athanas* with reference to their sexual phenomena. Sci Rep Yokohama Natl Univ Sect 2(17):1–37

Tavares C, Martin JW (2010) Suborder Dendrobranchiata Bate, 1888. In: Schram FR, von Vaupel Klein JC (eds) Treatise on zoology - Anatomy, taxonomy, biology - The Crustacea Volume 9 Part A. Koninklijke Brill NV, Leiden, pp 99–164

Taylor J, Poore GCB (1998) A review of the genus *Leontocaris* (Crustacea: Caridea: Hippolytidae) with descriptions of three species from southeastern Australian seamounts and slope. Mem Mus Vic 57:57–69

Thatje S (2003) *Campylonotus arntzianus*, a new species of the Campylonotidae (Crustacea: Decapoda: Caridea) from the Scotia Sea (Antarctica). Polar Biol 26:242–248

Thatje S, Lovrich GA, Anger K (2004) Egg production, hatching rates, and abbreviated larval development of *Campylonotus vagans* Bate, 1888 (Crustacea: Decapoda: Caridea) in subantarctic waters. J Exp Mar Biol Ecol 301:15–27

Thiel M, Chak STC, Dumont CP (2010) Male morphotypes and mating behavior of the dancing shrimp *Rhynchocinetes brucei* (Decapoda: Caridea). J Crustac Biol 30:580–588

Thompson JV (1829) On the luminosity of the ocean, with descriptions of some remarkable species of luminous animals (*Pyrosoma pigmaea* and *Sappharina indicator*) and particularly of the four new genera, *Noctiluca*, *Cynthia*, *Lucifer*, and *Podopsis*, of the Shizopodae. In: Hennessy J (ed) Zoological researches, and illustrations; or, natural history of nondescript or imperfectly known animals, in a series of memoirs, illustrated by numerous figures. Memoir 3. French Church Street Press, Cork, pp 37–66. Plates 5–8

Thompson JR (1967) Comments on phylogeny of section Caridea (Decapoda Natantia) and the phylogenetic importance of the Oplophoroidea. In: Proceedings Symposium on Crustacea,

Ernakulam, Marine Biological Association India, vol 1. Marine Biological Association of India, Cochin, pp 314–326

Toriyama M, Horikawa H (1993) A New Caridean Shrimp, *Psalidopus tosaensis*, from Tosa Bay, Japan (Decapoda: Caridea, Psalidopodidae). Bull Nansei Natl Fish Res Inst 26:1–8

Torti MR, Boschi EE (1973) Nuevos aportos al conociemento de los crustáceos Decápodos Caridea del género *Campylonotus* Bate 1988. Physis Sección A 32:65–84

Toth E, Bauer RT (2007) Gonopore sexing technique allows determination of sex ratios and helper composition in eusocial shrimps. Mar Biol 151:1875–1886

Tsurnamal M (1978) The biology and ecology of the blind prawn, *Typhlocaris galilea* Calman (Decapoda, Caridea). Crustaceana 34:195–213

Tsurnamal M (2008) A new species of the stygobiotic blind prawn *Typhlocaris* Calman, 1909 (Decapoda, Palaemonidae, Typhlocaridinae) from Israel. Crustaceana 81:487–501

Tunesi L (1987) La formation du spermatophore chez *Aristaeomorpha foliacea* (Risso, 1827). Investig Pesq 51(1):471–476

Van Dover CL, Fry B, Grassle JF, Humphries S, Rona P (1988) Feeding biology of the shrimp *Rimicaris exoculata* at hydrothermal vents on the Mid-Atlantic Ridge. Mar Biol 98:209–216

Van Dover CL, Szuts EZ, Chamberlain SC, Cann JR (1989) A novel eye in "eyeless" shrimp from hydrothermal vents of the Mid-Atlantic Ridge. Nature 337:458–460

Vaughan DB, Grutter AS, Costello MJ, Hutson KS (2016) Cleaner fishes and shrimp diversity and a re-evaluation of cleaning symbioses. Fish Fish 2016:1–19. https://doi.org/10.1111/faf.12198

Vereshchaka AL (1997) New family and superfamily for a deep-sea caridean shrimp from the Galathea collections. J Crustac Biol 17:361–373

Vereshchaka AL (2017) The shrimp superfamily Sergestoidea: a global phylogeny with definition of new families and an assessment of the pathways into principal biotopes. R Soc Open Sci 4: 170221. https://doi.org/10.1098/rsos.170221

Vereshchaka AL, Olesen J, Lunina AA (2014) Global diversity and phylogeny of pelagic shrimps of the former genera *Sergestes* and *Sergia* (Crustacea, Dendrobranchiata, Sergestidae), with definition of eight new genera. PLoS One 9:e112057. https://doi.org/10.1371/journal.pone.0112057

Vereshchaka AL, Kulagin DN, Lunina AA (2015) Phylogeny and new classification of hydrothermal vent and seep shrimps of the family Alvinocarididae (Decapoda). PLoS One 10(7): e0129975. https://doi.org/10.1371/journal.pone.0129975

Vereshchaka AL, Olesen J, Lunina AA (2016) A phylogeny-based revision of the family Luciferidae De Haan, 1849 (Crustacea: Decapoda). Zool J Linn Soc 178:15–32

Vereshchaka AL, Corbari L, Kulagin DN, Lunina AA, Olesen J (2019) A phylogeny-based revision of the shrimp genera (Crustacea: Decapoda: Benthesicymidae). Zool J Linn Soc 189:207–227

Vereshchaka AL, Kulagin DN, Lunina AA (2021) Across the benthic and pelagic realms: a species-level phylogeny of Benthesicymidae (Crustacea: Decapoda). Invertebr Syst 35:776–796

von Rintelen K, Cai Y (2007) Radiation of endemic species flocks in ancient lakes: systematic revision of the freshwater shrimp *Caridina* H. Milne Edwards, 1837 (Crustacea: Decapoda: Atyidae) from the ancient lakes of Sulawesi, Indonesia, with the description of eight new species. Raffles Bull Zool 57:343–452

von Rintelen K, Page TJ, Cai Y, Roe K, Stelbrink B (2012) Drawn to the dark side: a molecular phylogeny of freshwater shrimps (Crustacea: Decapoda: Caridea: Atyidae) reveals frequent cave invasions and challenges current taxonomic hypotheses. Mol Phylogenet Evol 63:82–96

Voss GL (1956) Protective coloration and habitat of the shrimp *Tozeuma carolinensis* Kingsley (Caridea: Hippolytidae). Bull Mar Sci 6:359–363

Walker I (1992) Life history traits of shrimps (Decapoda: Palaemonldae) of Amazonian inland waters and their phylogenetic interpretation. Stud Neotropical Fauna Environ 27:131–143

Wang Y, Ma KY, Tsang LM, Wakabayashi K, Chan T-Y, De Grave S, Chu KH (2021) Confirming the systematic position of two enigmatic shrimps, *Amphionides* and Procarididae (Crustacea: Decapoda). Zool Scr 50:812–823

Wasmer RA (1985) Records for *Physetocaris microphthalma* Chace (Decapoda, Caridea, Physetocarididae). Crustaceana 49:315–321

Weese DA, Santos SR (2009) Genetic identification of source populations for an aquarium-traded invertebrate. Anim Conserv 12:13–19

Wharton DN, Jinks RN, Herzog ED, Battelle BA, Kass L, Renninger GH, Chamberlain SC (1997) Morphology of the eye of the hydrothermal vent shrimp, *Alvinocaris markensis*. J Mar Biol Assoc U K 77:1097–1108

Wheeler JFG, Brown FA (1936) The periodic swarming of *Anchistioides antiguensis* (Schmitt) (Crustacea Decapoda) at Bermuda. J Linn Soc Lond Zool 39:413–428

Wickler W (1973) Biology of *Hymenocera picta* Dana. Micronesica 9:225–230

Wickler W, Seibt U (1970) Das Verhalten von *Hymenocera picta* Dana, einer Seesterne fressenden garnele (Decapod, Natantia, Gnathophyllidae). Z Tierpsychol 27:352–368

Wickler W, Seibt U (1981) Monogamy in Crustacea and Man. Z Tierpsychol 57:215–234

Wicksten MK (1989) *Encantada spinoculata*, a new genus and species of shrimp from the Galapagos Islands (Caridea: Bresiliidae). J Crustac Biol 9:667–671

Wicksten MK (2010) Infraorder Caridea. In: Schram FR, von Vaupel Klein JC (eds) Treatise on zoology–anatomy, taxonomy, biology. The crustacea. Volume 9, Part A. Eucarida: Euphausiacea, Amphionidacea and Decapoda (partim). Koninklijke Brill NV, Leiden, pp 165–206

Wicksten MK, Heathman T (2015) New host records for *Bathypalaemonella serratipalma* (Caridea: Bathypalaemonellidae) in the northern Gulf of Mexico. Mar Biodivers Rec 8:1–4

Wicksten MK, Martin JW (2004) A new species of caridean shrimp of the family Stylodactylidae from the eastern Pacific Ocean. Proc Biol Soc Wash 117:377–384

Wicksten M, De Grave S, France S, Kelley C (2017) Presumed filter-feeding in a deep-sea benthic shrimp (Decapoda, Caridea, Stylodactylidae), with records of the deepest occurrence of carideans. ZooKeys 646:17–23

Williams AB (1984) Shrimps, lobsters, and crabs of the Atlantic Coast of the Eastern United States, Maine to Florida. Smithsonian Institution Press, Washington, DC

Williams AB, Chace FA (1982) A new caridean shrimp of the family Bresiliidae from thermal vents of the Galapagos Rift. J Crustac Biol 2:136–147

Williams AB, Rona PA (1986) Two new caridean shrimps (Bresiliidae) from a hydrothermal vent on the mid-atlantic ridge. J Crustac Biol 6:446–462

Williamson DI (1973) *Amphionides reynaudii* (H. Milne Edwards), representative of a proposed new order of Eucaridan Malacostraca. Crustaceana 25:35–50

Williamson DI (1980) Larval development in a species of *Ambidexter* Manning & Chace (Decapoda, Caridea, Processidae). Crustaceana 39:235–246

Williamson K (2002) Charles Bruce Powell 1943-1998. Crustaceana 75:1275–1278

Wong JM, Pérez-Moreno JL, Chan T-Y, Frank TM, Bracken-Grissom HD (2015) Phylogenetic and transcriptomic analyses reveal the evolution of bioluminescence and light detection in marine deep-sea shrimps of the family Oplophoridae (Crustacea: Decapoda). Mol Phylogenet Evol 83: 278–292

Wood LE, De Grave S (2015) Functional morphology of the first pereiopod in crangonoid shrimps (Crustacea, Decapoda, Caridea, Crangonoidea). Zoomorphology 134:469–486

Woodmansee RA (1966) Daily vertical migration of *Lucifer*. Planktonic numbers in relation to solar and tidal cycles. Ecology 47:847–850

WoRMS (World Register of Marine Species) (2022). https://www.marinespecies.org. Accessed 08 November 2021

Xiao Y, Greenwood JG (1993) The biology of *Acetes* (Crustacea, Sergestidae). Oceanogr Mar Biol Annu Rev 31:259–444

Xuan NV (2000) Redescription of *Mimocaris heterocarpoides* Nobili, 1903 (Decapoda, Caridea, Hippolytidae) with remarks on its occurrence in Vietnam. Crustaceana 73:857–868

Yaldwyn JC (1960) Crustacea Decapoda Natantia from the Chatham Rise: a deep water bottom fauna from New Zealand. New Zealand Department of Scientific And. N Z Dept Sci Ind Res 139:13–53

Yeo DCJ, Ng PKL (1996) A new species of freshwater snapping shrimp, *Alpheus cyanoteles* (Decapoda: Caridea: Alpheidae) from peninsular Malaysia and a redescription of *Alpheus paludicola*. Raffles Bull Zool 44:37–63

Yeo DCJ, Ng PKL (1997) The alpheid shrimp *Potamalpheops* Powell, 1979, (Crustacea: Decapoda: Caridea: Alpheidae) from Southeast Asia, with description of three new species. J Nat Hist 31:163–190

Zbinden M, Berthod C, Montagné N, Machon J, Léger N, Chertemps T, Rabet N, Shillito B, Ravaux J (2017) Comparative study of chemosensory organs of shrimp from hydrothermal vent and coastal environments. Chem Senses 42:319–331

Zupo V, Messina P, Carcaterra A, Aflalo ED, Sagi A (2008) Experimental evidence of a sex reversal process in the shrimp *Hippolyte inermis*. Invertebr Reprod Dev 2:93–100

Chapter 4
Shrimp Hygiene: Antifouling Adaptations

4.1 Fouling and Antifouling Countermeasures

We are all familiar with behaviors by which commonly observed terrestrial animals clean themselves of dirt and parasites (autogrooming). Flies, ants, and other insects frequently preen antennae and other body parts; birds use their beaks to clean and arrange their plumage; mammals use feet, tongue, and teeth to maintain and clean plumage and skin. Aquatic animals similarly need to keep themselves clean to function, survive, and reproduce. The waters in rivers, streams, lakes, and oceanic habitats have suspended fine particles of sediment washed in from the terrestrial environment. Waves and currents stir up similar materials from the sea bottom. Detritus, small fragments of decomposed plants, invertebrates, and fishes, as well as their cast-off mucus, tissues, and fecal matter, is generated in situ in the sea, in addition to that carried in by rivers and streams. The water surrounding aquatic organisms also bears the spores and larval stages of bacteria, fungi, parasites, sessile plants and animals, all of which need a surface upon which to settle and grow. Such surfaces in marine and freshwater habitats are often in short supply.

Fouling by sessile organisms such as algae, sponges, hydroids, anemones, bryozoans, ascidians, tubiculous worms, bivalve molluscs, and barnacles is particularly intense in the marine environment. Inanimate objects and surfaces, both natural (rocky bottoms) and man-made (e.g., bottles, docks, pier pilings, the bottoms of ships) can quickly become covered with a community of fouling organisms. Studies have shown that clean hard surfaces immersed in the sea first become covered with a film of organic matter. This organic matter attracts the settlement of bacteria, fungi, and diatoms termed "biofilm," followed in time by sessile macroscopic organisms (Bauer 1975; Wahl 1989).

Both freshwater and marine animals use a variety of mechanisms to clean themselves. Many marine animals such as polychaete worms, corals, and anemones secrete mucus which entraps fouling material that epidermal cilia cast off the body (e.g., Hyman 1940). These and other soft-bodied invertebrates can periodically slough off fouled epidermal tissues. Many fishes have developed a mutualistic relationship with "cleaner" fishes and shrimps which pick off and feed on the excess mucus, dead tissues, and parasites of their client fishes without themselves being consumed (Chap. 10, Symbioses). Starfishes and sea urchins fend off settling stages of fouling organisms with numerous tiny jaw-like pedicellariae that cover their exterior (Hyman 1955). Similar structures, the avicularia (modified zooids), serve the same function in bryozoan colonies. Bryozoans also may possess other modified zooids, the vibracula, which sweep over the colony and brush off fouling particles and the settling stages of sessile organisms (Ryland 1970; Brusca et al. 2016). Other sessile marine invertebrates such as gorgonians (e.g., sea fans and whips), ascidians ("sea squirts"), and sponges secrete or contain noxious chemical substances that discourage fouling by other organisms (Mayzel et al. 2014).

Crustaceans are covered by a non-living chitinous calcified exoskeleton which is an attractive surface for the development of a fouling community. There is no epidermal tissue that can secrete mucus to entrap and shed fouling materials. There are no equivalents to the pedicellariae, avicularia, and vibracula of echinoderms and bryozoans that can attack and eliminate the larvae of settling organisms. Currently, there is no convincing evidence that antifouling compounds are secreted onto the surface of the cuticle to ward off fouling organisms (e.g., Becker and Wahl 1996). Thus, the crustacean cuticle is a very suitable hard or stiff surface for particulate matter to settle on and for fouling organisms to attach and grow. It is true that to grow, crustaceans must molt the entire exoskeleton, casting off the old one with replacement by a new clean one secreted and formed below it. However, the time between molts increases with age, and many crustaceans, including some shrimps, undergo a terminal molt with an exoskeleton that will be with them until death. In shrimps living in temperate or high latitude environments, molting may decrease or stop altogether during the colder months. Females incubating embryos cannot molt until their young hatch out. Furthermore, deleterious fouling can develop quite rapidly between molts and cause serious problems for the crustacean. Thus, it is not surprising that many crustaceans, especially shrimps and other decapods, have evolved various antifouling (grooming, cleaning) structures and behaviors. Besides grooming behavior per se, crustaceans may have other behavioral strategies to prevent or solve fouling. A summary of the environmental fouling pressures acting on aquatic crustaceans and the various solutions to prevent or remove fouling is given in Fig. 4.1.

I became interested in the subject of antifouling adaptations as a graduate student (Scripps Institution of Oceanography: SIO) looking for a suitable dissertation project. My interest in shrimps was stimulated by a common and colorful caridean species from the southern California intertidal pools, *Heptacarpus sitchensis* (at that time, *H. pictus*). These day-and night-active shrimps were fascinating to observe. I kept them in an aquarium on my desk to provide some occasional distraction and entertainment while studying or otherwise occupied. I was fortunate not only that this shrimp was abundant and easily collected in the intertidal pools just down the beach from SIO, but also that the shrimp displayed such fascinating and readily observable behavior. I quickly noticed how much time and energy it spent in

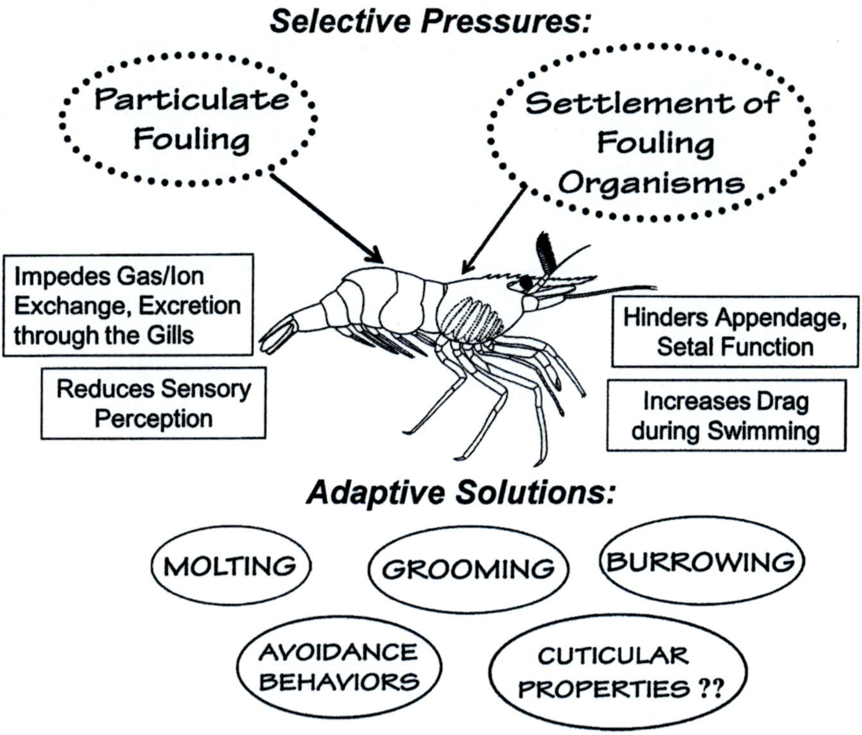

Fig. 4.1 Sources and deleterious effects of fouling on shrimps, and the adaptive solutions for preventing or removing it. Modified from Bauer (2013)

preening and apparently cleaning its body with one pair of chelae and various setal brushes and combs on other appendages. These structures and their setae were frequently illustrated and described in species descriptions and other taxonomic publications. However, the behavior and adaptive value of these setal complexes were rarely mentioned.

Martin Burkenroad, an eminent (and eccentric) carcinologist cited several times in this book, came to do some research in the SIO invertebrate zoology museum where I had my research and study space. He was quite amiable and interesting, and we frequently discussed crustaceans. One day he asked me what I was thinking of doing for my dissertation. I told him what I had in mind, something worthy of study but a bit ordinary, and he did not appear very impressed. He told me that yes, I would be able to get a doctorate with that project but that it would not garner any special attention—it just be another similar study among many. He kindly advised me to choose a different project, something new and different, that would attract attention. I was a bit crushed at first but after he moved on, I thought about how interesting were the grooming behaviors and structures that I had noticed in *H. sitchensis*. I found very little literature on cleaning (autogrooming) morphology and its adaptive value. At about the same time, SIO obtained a then new technology, scanning electron microscopy (SEM), with which I could reveal the intricate and beautiful ultrastructure of grooming structures. I took Martin's advice and made antifouling adaptations of shrimps and other decapods my dissertation project, a choice never regretted. My dissertation publications initiated a new line of investigation about shrimps and other decapod crustaceans by fellow carcinologists (reviewed in Bauer 1981, 1989, 2013 and other publications cited therein). My aquarium observations on *H. sitchensis* also stimulated studies on shrimp coloration and camouflage (Chap. 5, Coloration) as well as on mating behavior and sexual systems. I feel fortunate indeed that this species was available to excite and to jump start so much of my research career as well as that of other colleagues!

4.2 Grooming of Chemosensory Appendages

Shrimps, like other decapod crustaceans, depend on information from their surrounding environment about possible food sources, predators, noxious substances, conspecifics, and potential mating partners. The chemical senses of olfaction (smell) and gustation (taste) are perhaps most important to shrimps for obtaining this data. In olfaction, soluble (dissolved) substances quite low in concentration can be detected at some distance from their source by shrimps (Bauer 2011; Breithaupt and Thiel 2011). In shrimps and other decapod crustaceans, the primary olfactory receptors are specialized thin-walled setae, the aesthetascs, which are always located basally on the outer (lateral, upper) flagellum of the antennule (first antenna) (Fig. 2.11 in Chap. 2). Chemical stimuli at very low concentrations can pass through the thin cuticle of the aesthetasc to contact sensory neurons contained within. Gustatory or taste receptors, on the other hand, detect compounds that are relatively insoluble and high in concentration. Taste receptors must be very near or make physical contact with the chemical stimulus to be perceived. Both the antennular and antennal flagella bear gustatory receptors. These flagella also have tactile (touch) receptors which respond to sensations such as pressure, movement, and texture when the shrimp makes physical contact with an object in the environment. As a result, all shrimps and other decapod crustaceans groom (scrape and brush) the antennules and antennae to keep their receptors clean and free to detect stimuli.

4.2.1 Antennular Grooming

Frequent grooming of the antennules (A1) is done with the third maxillipeds (Mx3). It is a behavior found in all decapod crustaceans, including shrimps (Bauer 1975, 1977, 1981, 1989, 2013; Felgenhauer and Schram 1979; Alexander et al. 1980; Van Maurik and Wortham 2011, 2014). In both penaeoids and carideans, the more distal segments of the Mx3 have rows of serrate or otherwise complex setae adapted for rasping (Fig. 4.2, Plate 4.1 A-E). The exact placement of these cleaning setae depends on the length and flexibility of the Mx3, but they are always situated so

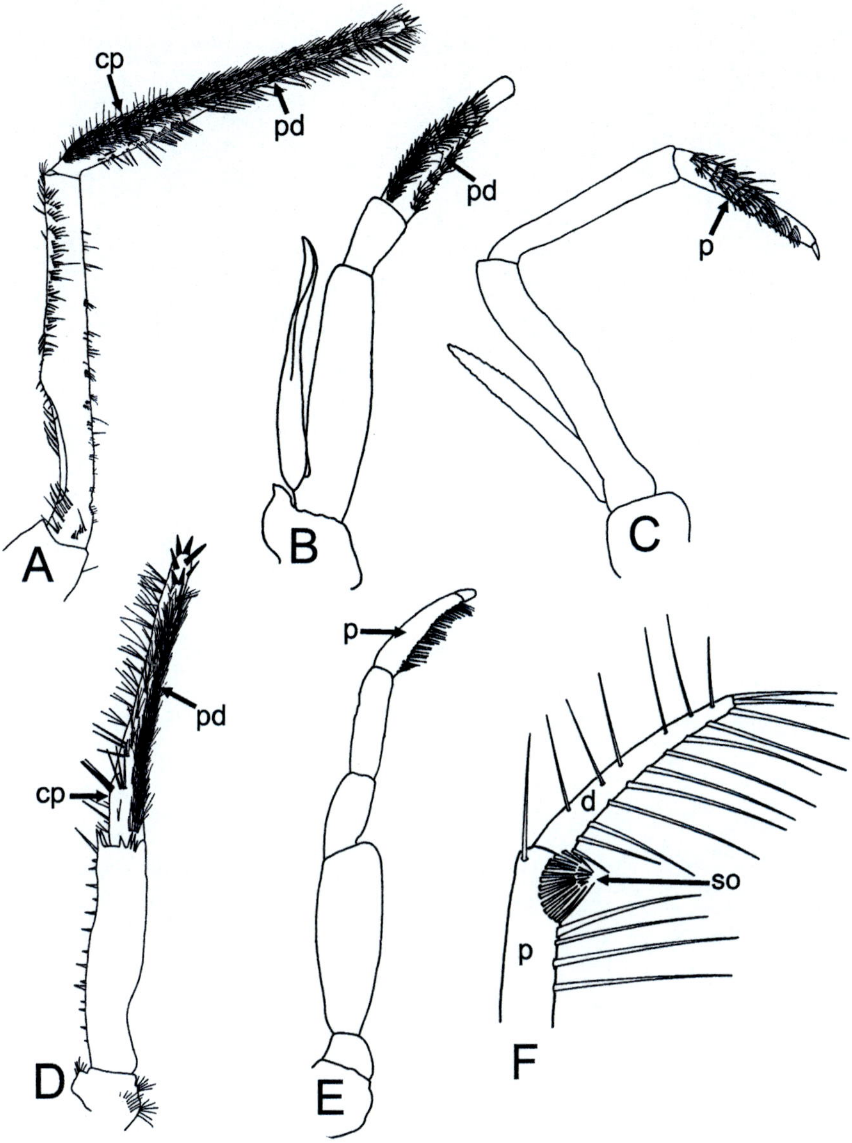

Fig. 4.2 Distribution of grooming setae on third maxillipeds. (**a**) *Pandalus danae* (Pandalidae). (**b**) *Alpheus* sp. (Alpheidae). (**c**) *Palaemon ritteri* (Palaemonidae). (**d**) *Heptacarpus sitchensis* (Thoridae). (**e**) *Procaris hawaiana* (Procarididae), and (**f**) *Stenopus hispidus* (Stenopodidae). See Chap. 2, Fig. 2.18B, for distribution of grooming setae on the propodus and carpus of *Fenneropenaeus merguiensis* (Penaeidae). *cp,* carpus; *d,* dactyl; *p,* propodus; *pd,* fused propodus-dactyl; *so,* setiferous organ. From (**a**) Bauer (1975), (**b-e**) Bauer (1976) and (**f**) Bauer (1989)

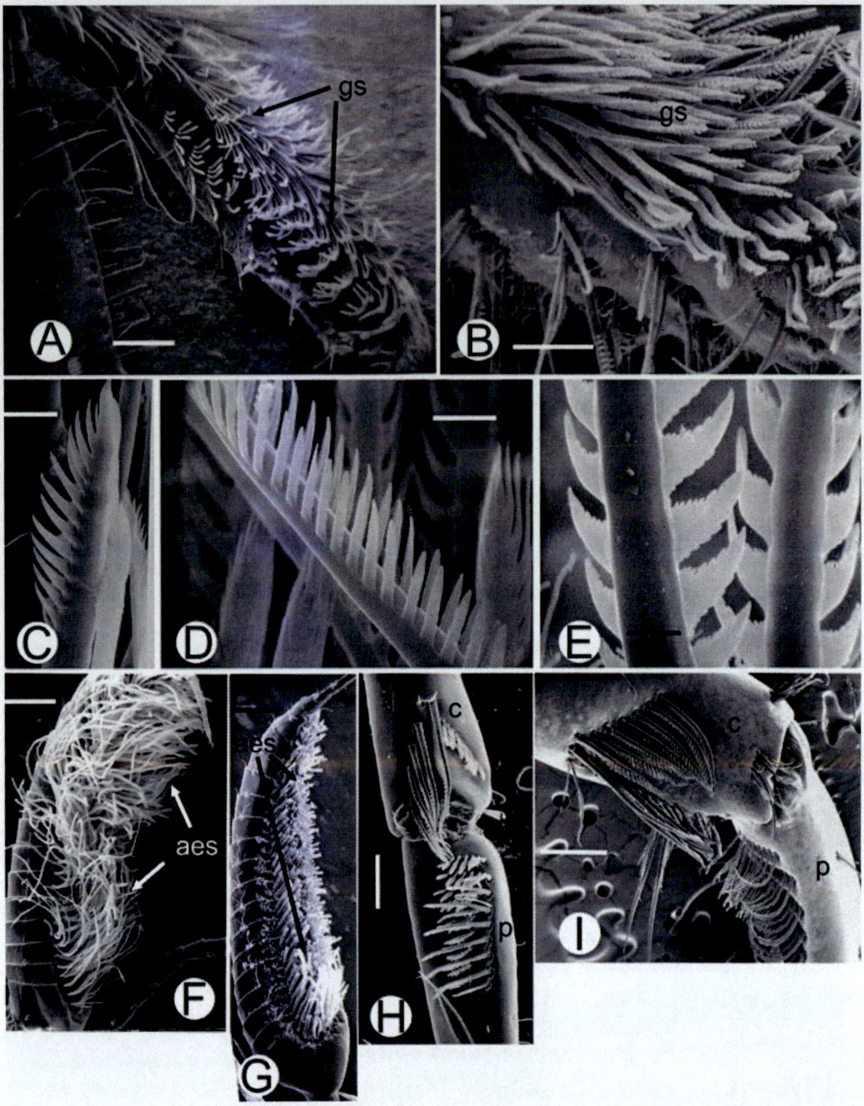

Plate 4.1 Antennular and antennal grooming structures. Rows of serrate grooming setae on the propodus-dactylus of the third maxillipeds of (A) *Betaeus macginitieae* (Alpheidae) and (B) *Palaemon ritteri* (Palaemonidae). (C-E) Ultrastructure of serrate grooming setae on third maxilliped of *Heptacarpus sitchensis* (Thoridae). (F) Control and (G) experimental antennules of *H. sitchensis* in antennular ablation experiment (Bauer 1976, 1977). Note breakage and loss of aesthetascs in (G) experimental shrimps without antennular grooming limbs. (H) *P. ritteri* and (I) *Pandalus danae* (Pandalidae) P1-CP (propodal-carpal) antennal grooming brushes. aes, aesthetascs; c, carpus; p, propodus; gs, grooming setae. Scale bars represent 400 μm in (A), 300 μm in (B), 10 μm in (C-D), 5 μm in (E), 200 μm in (F-G), 750 μm in (H) and 400 μm in (I). From (A-H) Bauer (1976) and (I) Bauer (1975)

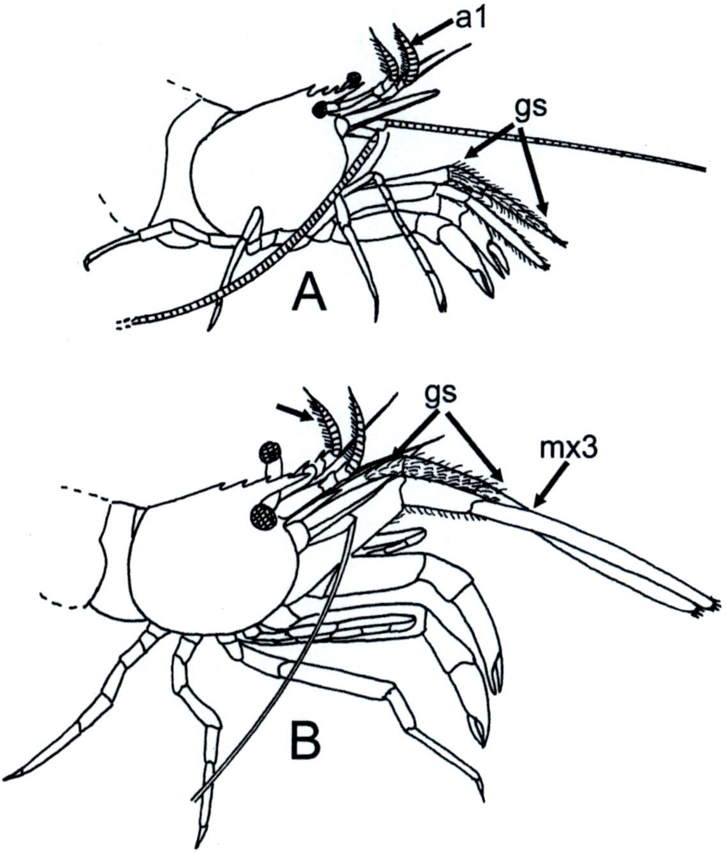

Fig. 4.3 Distribution of grooming setae on (**a**) the third maxillipeds of a female of *Heptacarpus palpator* (Thoridae) and (**b**) the hypertrophied third maxillipeds of a large male. Note that antennular grooming setae are on the terminal segments of the third maxilliped of the female while in the large male the setae have moved proximally to the carpus during ontogeny. *a1*, antennule; *gs*, grooming setae; *mx3*, third maxilliped. From Bauer (1976)

that they can reach and grasp the depressed A1 during its cleaning. In the penaeid *Fenneropenaeus merguiensis*, for example, the propodus and carpus have comb-like row of setae used in grooming (Chap. 2, Fig. 2.18b) (Alexander et al. 1980). In some caridean males, e.g., some species of *Heptacarpus*, *Saron*, and *Rhynchocinetes*, the relative size of third maxilliped increases in length, an adaptation related to mate acquisition (Chap. 8, Mating Systems). In juveniles and smaller males of *H. taylori* and *H. palpator*, the distribution of grooming setae on Mx3 is similar to that of the female, i.e., on the carpus and fused propodus-dactylus (Fig. 4.3a). As the male grows and matures to more dominant status, the third maxillipeds become enlarged (hypertrophied). The position of the grooming setae moves back to more proximal segments so that these setae stay in the path of the A1 during grooming (Fig. 4.3b) (Bauer 1977).

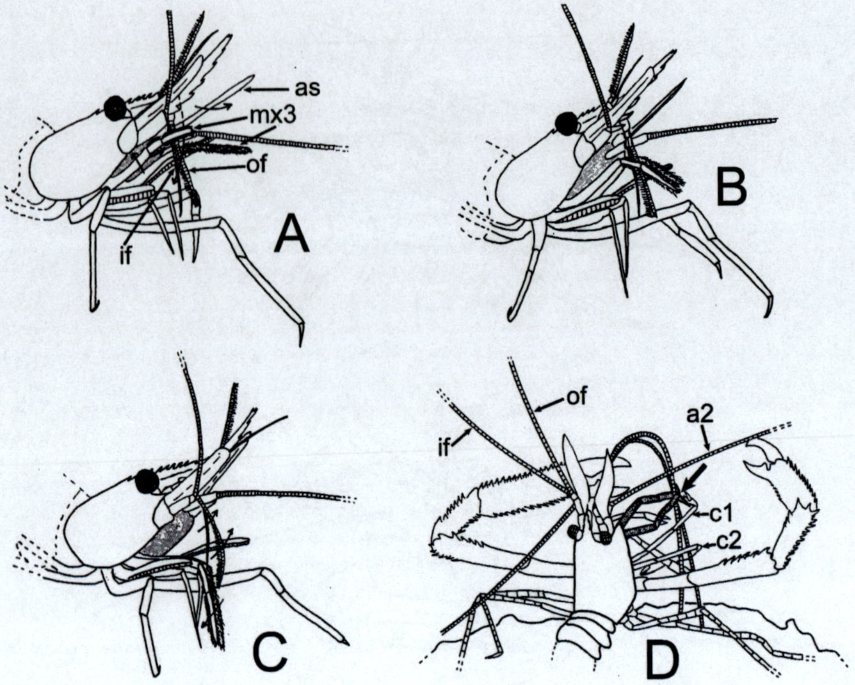

Fig. 4.4 Antennular grooming in (**a-c**) *Pandalus danae* (Caridea) and (**d**) *Stenopus hispidus* (Stenopodidae). In (**a**), the inner and outer antennular flagella are lowered between the outstretched third maxillipeds (Mx3) which (**b**) grasp it. (**c**) The flagella are then returned to their upright position as the Mx3 are lowered, rasping and cleaning the flagella with their medial grooming setae. In (**d**), the flagella are grasped and fit into the setiferous organs on the propodi of the Mx3 (unmarked arrow), and the flagella are run through and scraped by the denticulate setae of the setiferous organ. The first and second chelipeds are seen depressing the very long flagella so that they can be grasped by the third maxilliped (stippled). *as*, antennal scale; *a2*, antennal flagellum; *c1*, *c2*, chelipeds 1, 2; *if*, inner flagellum of the antennule; *of*, outer flagellum of the antennule. From (**a-c**) Bauer (1975) and (**d**) Bauer (1989)

The ultrastructure of Mx3 grooming setae is adapted for rasping and scraping the cuticle of the A1 and also the pereopods. Its setae are set at an angle on a limb segment so that they are directed distally (Fig. 4.2; Plate 4.1 A-B). Simple serrate setae have a double row of tooth denticles on one side which curve toward each other and point toward the tip of the seta (Plate 4.1 C-E). Another common setal type in Mx3 grooming brushes has the tooth denticles on one side of the seta, while the other side is equipped with multidenticulate scales with sharp tips also oriented toward the setal tip (Plate 4.3 J); these are also found on other grooming structures (Bauer 1975, 1977).

During antennular grooming of decapod shrimps, the A1 twists so that the aesthetascs of the outer flagellum face forward; the antennal scale is rotated laterally so that the A1 can be lowered between the outstretched pair of Mx3 which grasp the A1 proximally (Fig. 4.4a-b). Then these actions are reversed: the A1 then flexes back upward toward its normal position as the Mx3 slide down its length; scraping adhering material from its flagella and aesthetascs; and the antennal scale moves medially back to its normal position (Fig. 4.4c). A single act or a bout of A1 or other Mx3 grooming is usually followed by Mx3 autogrooming, in which the Mx3 rub back and forth together to clean themselves of material groomed off the A1 (Fig. 4.5a-c) (Bauer 1975, 1977; Alexander et al. 1980). The distally slanted orientation of the grooming setae ensures that debris moves toward the tips of the setae and then down the length of limb. Debris on one Mx3 moves back and forth between the Mx3 but always distally to drop off the tip of the limbs (Fig. 4.5c). A bout of A1 grooming may include several repeated acts of grooming of either or both A1.

In *Stenopus hispidus*, the A1 flagella are groomed by a specific cup-shaped group of setae on the propodus of Mx3 (Fig. 4.2f). This cleaning brush, termed the "setiferous organ" in the stenopodidean literature, is composed of serrate setae in which the shaft of the seta bears a double row of tooth- or trident-like denticles at the setal tip (Bauer 1989; Goy 2010). This brush is present in most stenopodideans, although it may be reduced in some species and is absent in the Macromaxillocarididae (Alvarez et al. 2006; Goy 2010). As in all shrimps, the A1 are cleaned one at a time. Both Mx3 reach up toward to the base of the A1 flagella, grasp and lower it with an outer flagellum fitting neatly into the concavity of the grooming organ of one Mx3 and the inner flagellum into that of the other. The flagella are then pulled through the grooming setae as the Mx3 moves down the length of flagella (Fig. 4.4d). Thus, both the olfactory aesthetascs and flagella in general are scraped clean. The second and third chelipeds may drape around the antennular flagella to help depress them during cleaning (Fig. 4.4d). The tips of the antennular flagella are sometimes grasped and vigorously rubbed by the second maxillipeds.

4.2.2 Adaptive Value of Antennular Grooming

What exactly does frequent antennular grooming actually accomplish? All shrimps either flick or rapidly rotate their antennules to test their surroundings for compounds of possible interest—the crustacean equivalent of mammalian sniffing (Koehl 2011). Both behavioral and morphological evidence suggest that an important function of routine grooming is to remove fouling material. Shrimps and other decapods increase their rate of A1 flicking and grooming when detecting food or other chemical compounds of interest. Grooming may also clean macromolecular substances off the aesthetascs to improve chemoreception at such critical times (Pearson and Olla 1977).

However, the function of a particular morphology is easily proposed, but such hypotheses should be tested with experiments when possible. Fortunately, this was somewhat easy to do with A1 grooming in the shrimp *Heptacarpus sitchensis* (Bauer 1977). This species has a thick tuft of aesthetascs on the A1 outer flagellum which it

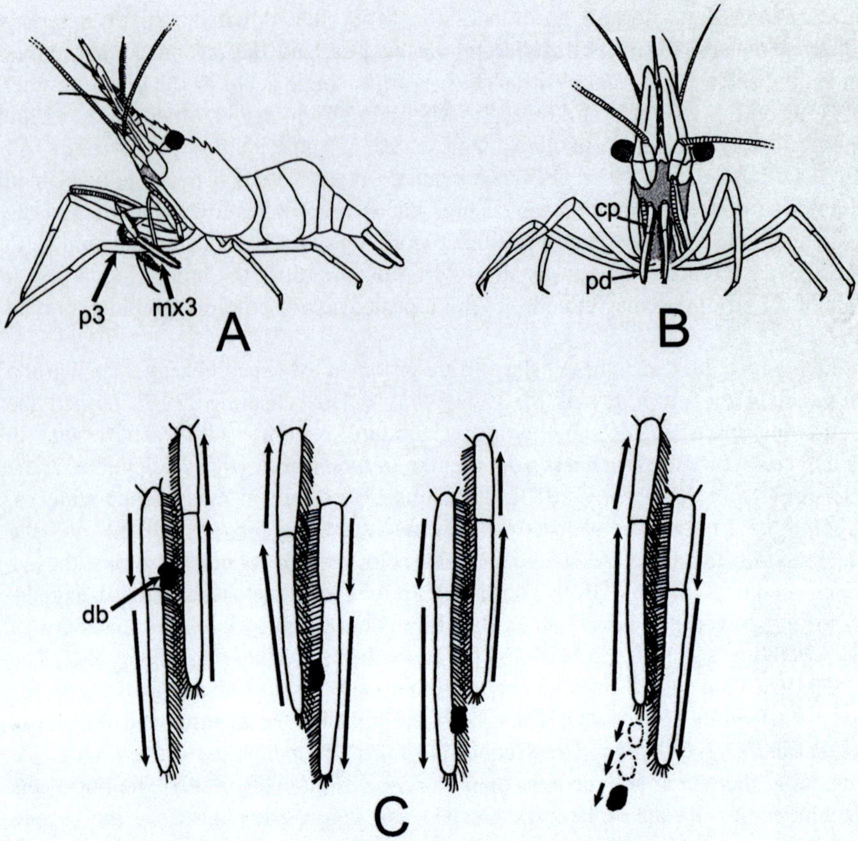

Fig. 4.5 Grooming by the third maxillipeds (Mx3) in *Pandalus danae* (Pandalidae). (**a**) Grooming of a walking leg (pereopod 3). (**b**) Position of third maxillipeds prior to grooming. (**c**) Autogrooming, in which the third maxillipeds rub together, with slanted setal rows moving a piece of debris cleaned from another appendage from one side to the other distally; the fouling particles finally drop off the end of the appendage. *cp,* carpus of mx3; *db,* debris (fouling); *mx3,* third maxilliped; *pd,* propodus of mx3 *p3,* third pereopod. From Bauer (1975)

spins rapidly in bursts through the water. This forces water into and through the tuft, circulating water with soluble chemical information to the sensory hairs. But this water also contains the spores and settling stages of microbial and other organisms as well as particulate matter that can settle and dirty the aesthetascs. Fouling must be removed so that ambient compounds can pass through the aesthetasc cuticle to contact the sensory neurons within. I exposed groups of these shrimps to fouling from natural, unfiltered seawater constantly pumped in from the end of the SIO pier for periods ranging from 2–4 weeks. The shrimps were in buckets perforated to let water constantly flow through, and they were provided with a substrate of natural rocks and shell. The Mx3 (grooming limbs) of the experimental group were ablated, while the first walking legs of the control group were removed. Thus, the experimental group could not groom the A1 while the control group could.

The results were very dramatic. Within a few to several days, the aesthetasc tufts of the experimental group darkened due to fouling. After a few more days, the aesthetascs became shorter and more sparse, and finally disappeared in the experimentals. Antennules of the control group did not darken and remained intact (Plate 4.1 F). Amazingly, the aesthetascs of experimentals which had not molted were broken off at their bases (Plate 4.1 G). At the end of the experiment, the antennules were examined with light microscopy to quantify the degree of aesthetasc damage (Fig. 4.6) (Bauer 1976, 1977). In those experimentals which had not molted, or which had molted early in the experiment, the aesthetascs were fouled by a variety of microbial organisms such as diatoms, microalgae, and sessile protists. A filamentous bacterium, *Leucothrix,* a ubiquitous epibiotic fouling organism in the sea, was especially abundant. Sediment and detritus accumulated around the bases of the sensory hairs. In the experimentals, breakage of aesthetascs increased with days of exposure to fouling (Fig. 4.6a). In the controls, the aesthetascs were almost unaffected with little or no damage. The experiment was replicated three times from winter to early summer. As the ocean waters warmed, favoring the growth of microfoulers, the fouling and subsequent damage became more intense (Fig. 4.6b).

I also did the same type of experiment with another caridean, *Pandalus danae* (Bauer 1975). In this species, the aesthetascs are not in a thick tuft but instead rather sparsely spaced on the antennular flagellum. *P. danae* constantly flicks its A1 back and forth to circulate water through the aesthetascs but the movement is not so rapid or forceful as in *H. sitchensis.* I similarly had arranged control and experimental groups and exposed them to natural fouling in cages hung off the pier of the Friday Harbor, Washington (USA) marine laboratory. The experiments showed similar results in that there was darkening and fouling of aesthetascs in the experimentals but not in the controls. However, the aesthetascs of experimentals did not break off as in the *H. sitchensis* experiment. In both the *H. sitchensis* and *P. danae* experiments, settling plates (glass slides or asbestos boards) placed alongside the shrimp cages experienced similar fouling as that found on ungroomed antennules.

These experiments show that A1 grooming successfully removes and controls fouling. Ungroomed antennules suffer significant fouling and probably diminished aesthetasc chemoreception—the *H. sitchensis* that lost their aesthetascs must have completely lost their sense of smell! A loss or impairment of olfactory ability is indeed a major problem for shrimps as chemoreception is probably their most important means of obtaining information about their environment. It is true that those individuals that molted sometime during the experiment had a lower degree of fouling than those that did not, and those that molted near the end of the experiment completely regenerated their aesthestascs. But compromised olfaction of any degree must reduce the chances of survival and successful reproduction. One might ask, why did the aesthetascs of *H. sitchensis* break off when heavily fouled and those of *P. danae* did not? The heavy fouling must have produced an increased hydrodynamic drag on the rapidly spun *H. sitchensis* aesthetascs that caused them to break. Such drag was not so great on the *P. danae* aesthetascs, which are flicked (waved),

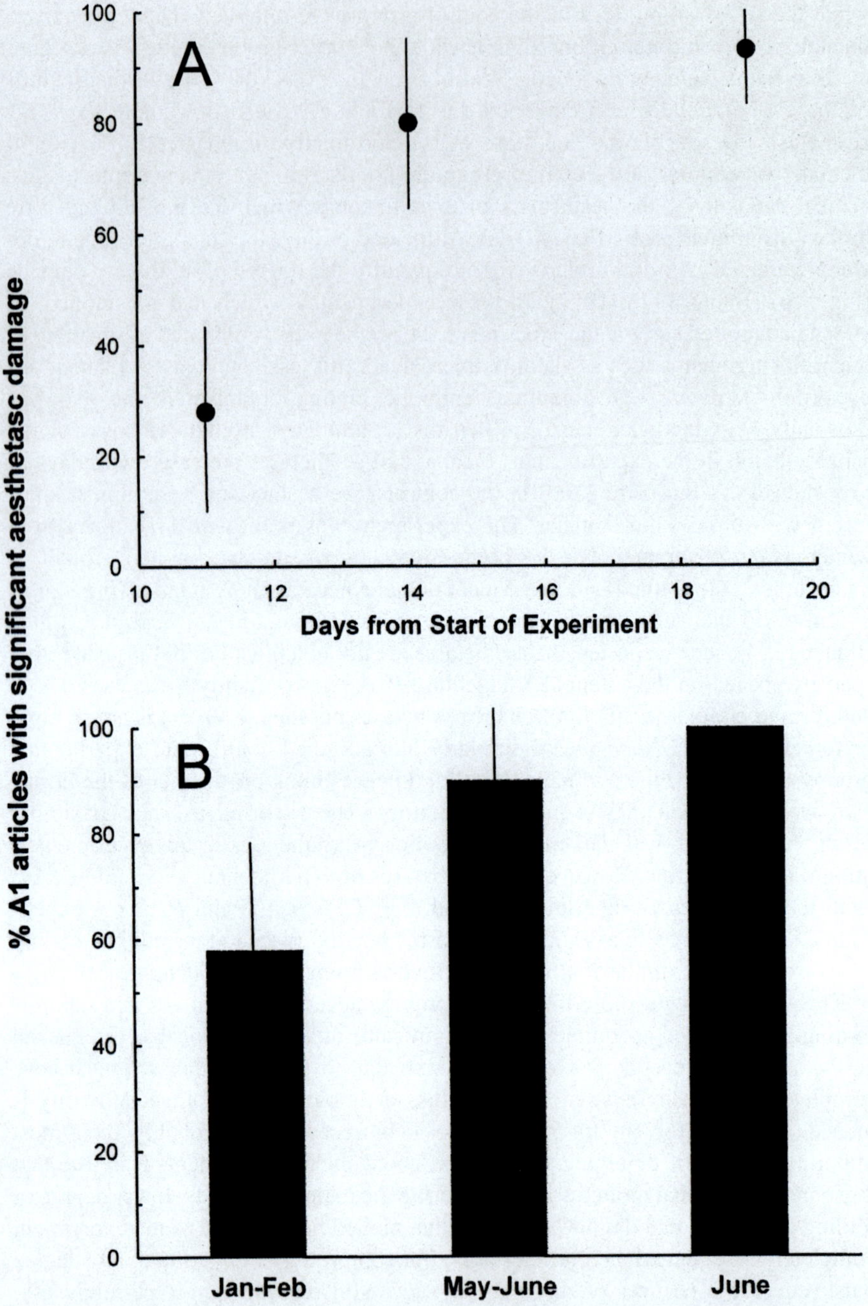

Fig. 4.6 Measurement of antennular fouling as the percentage of antennular articles with significant aesthetasc breakage. (**a**) Time course of fouling in an experiment. (**b**) Intensity of antennular fouling in experiments in three different periods of the year from mid-winter to early summer. See Bauer (1977) for methods and details. From data in Bauer (1976)

not furiously rotated. It should also be pointed out that fouling pressures vary with environmental conditions. Although experiments such as these show that fouling on ungroomed antennules does occur, further experiments need to be conducted to see if and how much olfactory impairment occurs because of fouling.

4.2.3 Cleaning of the Antennal Flagellum: The Pereopod 1 Carpal-Propodal Brush

All decapod shrimps have moderately to very long second antennae (A2), usually simply termed the "antenna" to distinguish it from the first antenna (antennule). These flagella are important chemotactile organs carrying gustatory (taste), tactile (touch), and pressure (vibration) receptors with helps the shrimp find food, recognize conspecifics and predators. The A2 sensory receptors are particularly useful in turbid waters, at night, or in the unlighted regions of the sea. As with A1 sensilla, antennal receptors must remain clean to function well. A specific structure, the carpal-propodal brush of the first pereopod (P1-CP brush), has evolved in decapod shrimps with the principal task of cleaning the A2 flagella (Bauer 1981, 1989). The P1-CP brushes are present in the vast majority of species in all shrimp groups (Bauer 1978, 1981, 1989; De Grave and Goulding 2011). The appendage, limb segments, setal groups and behaviors of P1-CP A2 flagellar cleaning are comparable to those used for antenna grooming in insects (Jander 1966; Wilson 1971; Hlavac 1975; Hackmann et al. 2015).

The P1-CP brushes of penaeoid shrimps (Fig. 4.7a) are typical. At the distal end of the chelate pereopod 1 carpus is a "V"- shaped or similar group of serrate setae which arches over the carpal-propodal joint when the propodus is flexed. Rows of similar setae are situated on the proximal end of the propodus. The same general arrangement of setal groups is found in sergestoid shrimps (e.g., Judkins 1978: Fig. 2m), procarididean, carideans, and stenopodideans (Fig. 4.7b-g; Plate 4.1 H-I). A2 flagellar cleaning by the P1-CP brushes (Fig. 4.8) was first described in carideans by Bauer (1975, 1978) and Felgenhauer and Schram (1979). In grooming of the A2 flagellum, the peduncle of an A2 rotates so that the flagellum is brought toward the front of the shrimp. The first pereopod 1 (P1), usually a cheliped, can now drape the carpal-propodal joint around it, with the setal groups on each side trapping the flagellum between them (Fig. 4.8a). As the P1 is depressed (Fig. 4.8b), the A2 flagellum is pulled back upward, forming a characteristic loop (Fig. 4.8c), so that it is scraped and cleaned by the setae of the P1-CP brush. While the A2 flagellum is depressed by P1, it may be also grasped by the third maxillipeds (Fig. 4.8a), which rub back and forth on it, also cleaning the flagellum as it is drawn through the P1-CP brushes. A2 cleaning is similar in the stenopodideans except that the second and sometimes third chelae help to depress the very long flagellum so that the P1-CP brush can be brought into play (Fig. 4.8d).

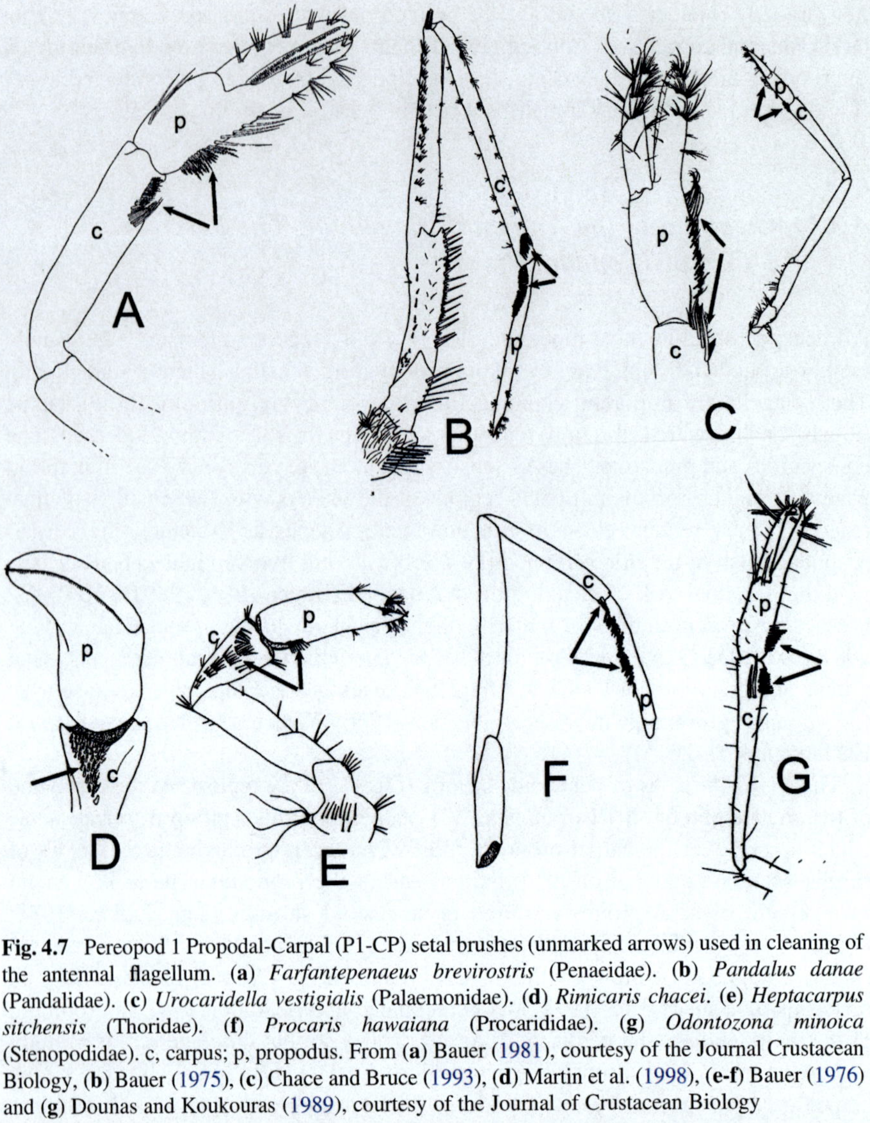

Fig. 4.7 Pereopod 1 Propodal-Carpal (P1-CP) setal brushes (unmarked arrows) used in cleaning of the antennal flagellum. (**a**) *Farfantepenaeus brevirostris* (Penaeidae). (**b**) *Pandalus danae* (Pandalidae). (**c**) *Urocaridella vestigialis* (Palaemonidae). (**d**) *Rimicaris chacei*. (**e**) *Heptacarpus sitchensis* (Thoridae). (**f**) *Procaris hawaiana* (Procarididae). (**g**) *Odontozona minoica* (Stenopodidae). c, carpus; p, propodus. From (**a**) Bauer (1981), courtesy of the Journal Crustacean Biology, (**b**) Bauer (1975), (**c**) Chace and Bruce (1993), (**d**) Martin et al. (1998), (**e-f**) Bauer (1976) and (**g**) Dounas and Koukouras (1989), courtesy of the Journal of Crustacean Biology

P1-CP brushes are present in most decapod shrimps but are by no means universal. De Grave and Goulding (2011) found that in species from 9 of 38 caridean families, a P1-CP brush was completely absent, while in three others only a P1 carpal brush was observed (e.g., Fig. 4.7d). However, P1-CP brushes are fairly conservative in presence and structure within a family. But there are exceptions, such as in the Pasiphaeidae and Alvinocarididae, in which some genera have the P1-CP brushes, but others do not (De Grave and Goulding 2011; Martin et al. 1998). On the list of 9 families without the P1-CP brushes was the Euryrhynchidae, but the structure since

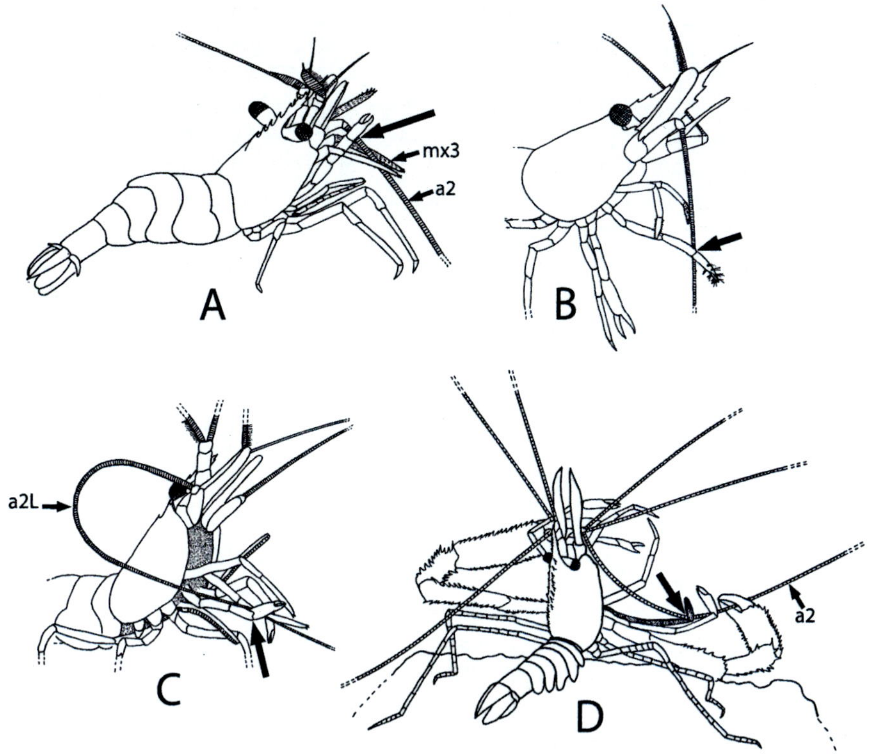

Fig. 4.8 Cleaning of the antennal (A2) flagellum by the first pereopodal P1-CP brushes. (**a**) *Heptacarpus sitchensis* (Thoridae). (**b**) *Palaemon ritteri* (Palaemonidae). (**c**) *Lysmata californica* (Lysmatidae). (**d**) *Stenopus hispidus* (Stenopodidae). In (**a**), the shrimp reaches up to the base of the A2 to flex the P1 carpal-propodal joint around it (unmarked arrow), and then slides the P1 down the A2 (**b-c**), with the A2 forming a characteristic loop. In *Stenopus hispidus* (**d**), the second and third chelipeds may assist by pulling the A2 down so that the first cheliped (stippled) may grasp and clean it with the P1-CP brushes. *a2* antennal flagellum, *a2L* loop formed by the A2 flagellum during its cleaning, *mx3* third maxillipeds. From (**a-c**) Bauer (1976) and (**d**) Bauer (1989)

has been found since in the newly described genus *Eurindicus* (De Grave et al. 2018). The antennal cleaning brushes are present in the Procarididea (Bauer 1978; De Grave and Goulding 2011) and all penaeoid families (Bauer 1981, 1989). However, they are a sexually dimorphic character in the sicyoniid penaeoids: males have the brushes, but females do not (Burkenroad 1934; Bauer 1989). Among the stenopodideans, the P1-CP brushes are present in all the Stenopodidae, some genera of the Spongicolidae, and are absent in the Macromaxillocarididae (Goy 2010). The P1-CP setal ultrastructure, although always with adaptations for scraping and rasping the A2 flagellum, may vary as well (De Grave and Goulding 2011).

How are the A2 flagella cleaned in families which completely lack a P1-CP brush? In those atyids and alpheids observed alive, the relatively short antennal flagella are simply scrubbed with the third maxillipeds (Bauer 1978; Felgenhauer and Abele 1983) as in reptantian decapods such as crayfishes, lobsters, and crabs (Bauer 1981). In reptantians, as in many alpheid shrimp species, the first pereopods are enlarged cheliped weapons or otherwise robust appendages too large and clumsy to make the quick precise movements necessary for P1-CP cleaning. Additionally, the A2 flagella of many reptantians are short and can be easily cleaned by Mx3. The widespread distribution of the P1-CP brushes in decapod shrimps indicates that it is a primitive (plesiomorphic) trait that must have been present in the decapod shrimp ancestor. It apparently evolved as a quick and efficient mechanism to clean A2 flagella, which are long and delicate in shrimps compared to those of other decapods. The structure and behavior was apparently lost in some shrimp taxa and in the ancestral reptantians when Mx3 grooming of the A2 flagella was sufficient to do the job—inefficient or useless structures are usually eliminated by natural selection.

4.3 Gill-Cleaning Mechanisms

A unique characteristic of decapod crustaceans—and a very significant factor behind their diversity and biological success—is the enclosure of the gills within a branchial chamber. The branchiostegite, an extension of the carapace, covers the gills and protects them from mechanical injury. The pumping action of the second maxilla's exopod, the scaphognathite (gill bailer), forces a relatively large volume of water rapidly into and through the limited space of the gill chamber. This high velocity flow facilitates the vital gill functions of gas and ion exchange as well as excretion. Carbon dioxide, a major waste product of metabolism, passes out of the body through the thin cuticle of gills and other branchial surfaces, and then is whisked away by the respiratory current. Oxygen from water flowing by the gills enters the blood (hemolymph). Decapods have specific nephridial excretory organs, the antennal glands, which open at the base of the antennal (A2) peduncle (Brusca et al. 2016). Although these organs are important, as much as or more excretion of nitrogenous wastes (as ammonia) and ion exchange occurs through the gills. Thus, the gill surfaces must remain clean to carry out these essential functions.

The benefits of a decapod branchial chamber also come with some serious costs. The mass of gill filaments within the narrow space of the gill chamber acts as a sediment trap, inadvertently trapping sediment and other particles within the respiratory stream. The constant flow of oxygenated water makes the gills and gill chamber a great place for microfouling organisms to flourish. Consequently, a variety of mechanisms have evolved in shrimps and other decapod crustaceans for the upkeep and cleaning of the gills and branchial chamber.

4.3.1 *Reversal and Filtering of the Respiratory Current*

Shrimps and other decapods periodically flush out the gill chamber by reversing or altering the beat of the scaphognathite (Vuillemin 1967; Wilkins and Mcmahon 1972; Bauer 1979). During a respiratory current reversal, the carapace may rapidly adduct (press) against the gill chamber. Somewhat analogous to a cough, these actions help to dislodge particles stuck on the gills and back into the respiratory stream, which carries them out of the branchial chamber. Some shrimps, like many other decapods, have numerous setae fringing the entrances into the branchial chamber, e.g., along the ventral edge of the branchiostegite and epipods in some carideans (Bauer 1984), which may help filter the incoming current. In addition, such setae may also be sensory, signaling the scaphognathite to stop pumping water when a high concentration of fouling particles or noxious substances are detected (Wilkins and Mcmahon 1972). Reversals may help to blow off material that accumulate on these filtering/sensory setae. However, reversals and filtering are usually not sufficient to keep the gills clean of particle and microbial fouling, and a number of other more efficient means of gill cleaning have evolved in shrimps and other decapods.

In Table 4.1, I have an overview of these gill-cleaning mechanisms. This table is by no means complete; it is based on my own studies (Bauer 1979, 1981, 1989) and information from the literature to date. Some of the variation in these mechanisms is indicated in the table, but many more studies on the functional morphology and behavior of shrimps on more taxa are needed to make such a survey complete. In a number of taxa, the characteristics vary little, e.g., the sergestoids, penaeoids, stenopodideans, procarideans, and palaemonoid families. In the carideans, there is some to considerable variation described or to be determined. The data in Table 4.1 will hopefully stimulate future work in this area. Behavioral observations, in particular, are needed in many taxa.

4.3.2 *Setiferous Epipods*

In the penaeoid shrimps, as in some other decapods such as crayfishes and lobsters, the gill chamber is divided up into channels by long, blade-like, and often forked epipods that arise from the coxae of the pereopods and Mx2 or 3 (Young 1959; Dall et al. 1990; Bauer 1989, 1998, 1999; Batang and Suzuki 2003) (Fig. 4.9a; see also Chap. 2: Figs. 2.6b-c, 2.8a) (Plate 4.2 A-B). These epipods bear many multidenticulate scale setae that intertwine and are inserted among the gill filaments. As will be seen, such setae are typical of any kind of structure that cleans the gills in shrimps and other decapods (Plate 4.2 B-C). When the limbs that bear these epipods move during locomotion or other activities, the attached epipods jostle, scrape, and brush the gill filaments. I did experiments with groups of the penaeid *Rimapenaeus similis* to determine if these epipodal setae did indeed keep the gills clean. I carefully

Table 4.1 Overview of major gill-cleaning mechanisms in decapod shrimps

Taxon	1: Setiferous epipods	2: Epipod-Setobranchs	3: Multidenticulate scaphognathite setae	4: Cheliped gill brushing	Source
DENDROBRANCHIATA:					
Penaeoidea					
Aristeidae	P	A	A	A	1–4 M (a,b)
Benthesicymidae	P or A	A	A	A or P1–3?	1–4 M (a,b)
Penaeidae	P	A	A	A	1–4 M (a-c),1B (c)
Sicyoniidae	P	A	A	A	1–4 M (a,d)
Solenoceridae	P	A	A	A	1–4 M (a, p.o.)
Sergestoidea					
Luciferidae	A	A	A	A	1–4 M (lack gills (a)
Sergestidae	A	A	A	A	1–4 M (a,b)
CARIDEA:					
Alpheoidea					
Alpheidae	A	P or A	A	A or P2?	1–4 M, 4B (e, f-g)
Barbouriidae	A	P	A	A	1–4 M (h,i)
Bythocarididae	A	A	A	P2?	1–4 M (j,k)
Hippolytidae	A	P or A	A	A or P2??	1–4 M, 4B (e,l,m)
Lysmatidae	A	P or A	A	A or P2?	1–4 M, 4B (e,l)
Merguiidae	A	A	A	P2?	1–4, M (l)
Ogyrididae	A	A	A	P2?	1–4, M (p.o.)
Thoridae	A	P or A	A	A?, P2	1–4 M, 4B (e,n)
Atyoidea					
Atyidae	A	P or A	P	A or P1–2?	1–4 M (l, o,p)
Bresilioidea					
Agostocarididae	A	A	P?	P2?	1–4 M (q,r)

Alvinocarididae	A	P or A	P?	P2?	1–4 M (s,t)
Anchialocarididae	A	A	P?	P2?	1–4 M (u)
Bresiliidae	A	A	P?	P1 or P2?	1–4 M (v,w)
Disciadidae	A	A	A	P2?	1–4 M (x)
Pseudochelidae	A	A	A	P1?, P2?	1–4 M (x,y)
Campylonotoidea					
Campylonotidae	A	P	P?	A	1–4 M (z,a1)
Bathypalaemonellidae	A	A	A	P1?	1–4 M (1,b1)
Crangonoidea					
Crangonidae	A	A	P or A	P2 or A	1–4 M, 4B (e, c1, d1)
Glyphocrangonidae	A	A	A	P2?	1-4 M (d1, e1)
Nematocarcinoidea					
Eugonatonotidae	A	P	A	A	1–4 M (l)
Lipkiidae	A	P	P?	A	1–4 M (f1)
Nematocarcinidae	A	A	P	P1?	1–4 M (e,g1)
Rhynchocinetidae	A	P	P	A	1–4 M (e, l, h1, p. o.)
Xiphocarididae	A	P	P	A	1–4 M, (p.o.)
Oplophoroidea					
Acanthephyridae	A	P	A	P5	1–4 M (a13, e,g1,p. o.)
Ophlophoridae	A	P	A	A	1–4 M (e,g1,p.o.)
Palaemonoidea					
Desmocarididae	A	A	A	P1?	1–4 (a14)
Euryrhynchidae	A	A	A	P1?	1–4 M (l, i1)
Palaemonidae	A	A	A	P1	1–4 M, 4B (e,j1,k1)
Typhlocarididae	A	A	A	P1?	1–4 M (j1,m1)

(continued)

Table 4.1 (continued)

Taxon	1: Setiferous epipods	2: Epipod-Setobranchs	3: Multidenticulate scaphognathite setae	4: Cheliped gill brushing	Source
Pandaloidea					
Pandalidae	A	P or A	P? Or A	A or P2?	1–4 M, 4B (e,n1, o1)
Thalassocarididae	A	P	A	A	1–4 M (n1)
Pasiphaeoidea					
Pasiphaeidae	A	A	A	P4?	1–4, M (e,p1,q1, p. o.)
Physetocaridoidea					
Physetocarididae	A	P	A	A	1–4 M (r1)
Processoidea					
Processidae	A	A	P?	P2?	1–4 M (l,s1)
Psalidopodoidea					
Psalipodidae	A	A	A?	P2?	1–4 M (t1)
Stylodactyloidea					
Stylodactylidae	A	A	P?	A	1–4 M (e,u1,v1)
INCERTAE SEDIS					
Amphionididae	A	A	A	P5?	1–4 M (w1)
PROCARIDIDEA					
Procarididae	A	P	A	A	1–4 M (e,x1,y1)
STENOPODIDEA					
Macromaxillocarididae	A	A	A?	P1–2?	1–4 M (z1)

Spongicolidae	A	A	P?	P1–2	1-4 M,4B (b,a11, a12)
Stenopodidae	A	A	A	P1–2	1-4 M, 4B (b,a12)

P (present) or A (absent) shows that at least one species of the family has been found with (P) or without (A) without that mechanism. P1,2,3 refers to pereopods (chelipeds) 1–3, respectively. Type and source of observation: (B) behavioral; (M) inferred from morphological data in the literature and previously uncited R.T. Bauer personal observations (p.o). A question mark shows that the presence or absence of a particular gill-cleaning mode is uncertain or not yet reported. Lower case letters under "Source" column refers to literature citations given below. See the text for more details

References: (a) Pérez and Kensley (1997); (b) Bauer (1981); (c) Bauer (1999); (d) Bauer (1989); (e) Bauer (1979); (f) Chace (1988); (g) Read et al. (1991); (h) Manning and Hart (1984); (i) Chace (1975); (j) Abele and Martin (1989); (k) Sokolov (2000); (l) Chace (1997); (m) d'Udekem d'Acoz (1996); (n) Bauer (1984); (o) Suzuki and McLay (1998); (p) Felgenhauer and Abele (1983); (q) Alvarez et al. (2004); (r) Hart and Manning (1986); (s) Williams and Chace (1982); (t) Vereshchaka et al. (2015); (u) Mejía-Ortíz et al. (2017); (v) Komai and Kohtsuka (2017); (w) Wicksten (1989); (x) Kensley (1983); (y) Chace and Brown (1978); (z) Torti and Boschi (1973); (a1) Yaldwyn (1960); (b1) Cleva (2004); (c1) Butler (1980); (d1) Chace (1984); (e1) Holthuis (1971); (f1) d'Udekem d'Acoz and De Grave (2018); (g1) Chace (1986); (h1) Okuno (1996, 1997); (i1) De Grave et al. (2018); (j1) Chace and Bruce (1993); (k1) Wickler and Seibt (1970); (l1) Van Maurik and Wortham (2014); (m1) Calman (1909); (n1) Chace (1985); (o1) Komai (1999); (p1) Chace (1976); (q1) Thompson (1967); (r1) Foxton and Herring (1970); (s1) Manning and Chace (1971); (t1) Chace and Holthuis (1978); (u1) Cleva (1990); (v1) Hanamura and Masatsune Takeda (1996); (w1) Williamson (1973); (x1) Chace and Manning (1972); (y1) Kensley and Williams (1986); (z1) Alvarez et al. (2006); (a11) Saito and Komai (2008); (a12) Goy (2010); (a13) Vereshchaka et al. (2021); (a14) Powell (1977); p.o., previously uncited RTB personal observation

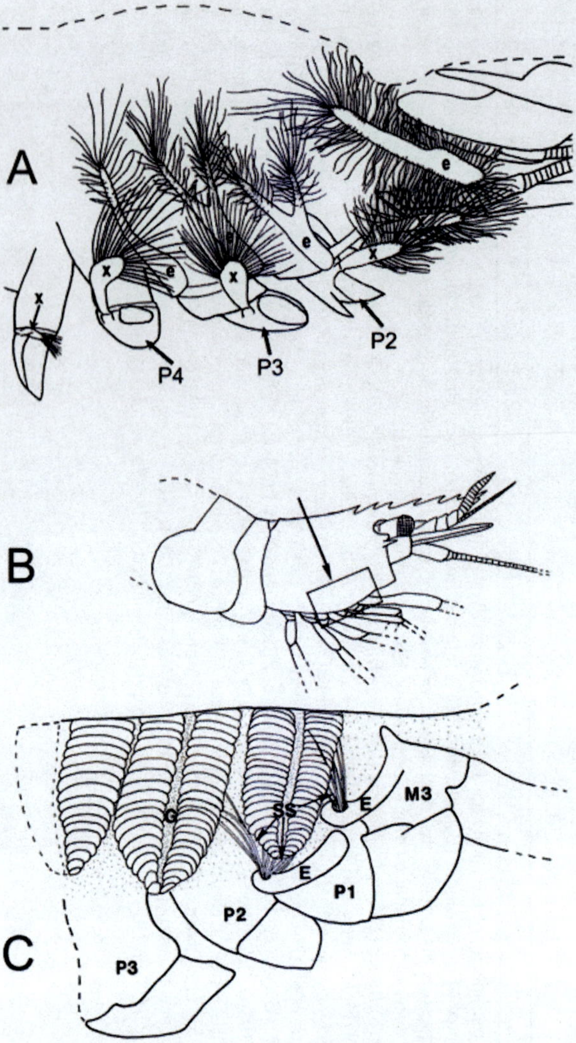

Fig. 4.9 Penaeoid setiferous epipods and caridean epipod-setobranch complexes. (**a**) Setiferous epipods and exopods in the gill chamber of the *Rimapenaeus similis* (Penaeidae) (gills removed; see Chap. 2, Fig. 2.6 for disposition of gills with respect to epipods and exopods). (**b-c**) Epipod-setobranch complexes on the anterior pereopods of *Heptacarpus paludicola* (Thoridae). (**b**) Upper, box indicated by arrow shows portion of the uncovered branchial chamber shown in (**c**). The epipod hook of the third maxilliped is fastened around the setobranch setae of pereopod 1, as is the epipod hook of pereopod 1 with the setobranch setae of pereopod 2. *e,* penaeid setiferous epipod; *E,* caridean epipod with hook; *G,* gill; *P2-P4,* basal segments of pereopods 1–4 in *R. similis*; *M3, P1-P3,* coxae of maxilliped 3, pereopods 1–3 in *H. paludicol; SS*, setobranch setae; *x,* setiferous exopod. From (**a**) Bauer (1999) and (**b-c**) Bauer (1984)

removed the epipods from the branchial chamber of one side (experimental) but not the other (control). Shrimps were maintained for two weeks on a recirculating water table. The water was visually clean but the normal waste, food, and other particulate matter in such a system provided sufficient fouling pressure for the experiment. When shrimps' gills were examined at the end of the experiment, the gills in the control gill chambers were clean but those on the experimental side were fouled with sediment particles (Plate 4.2 D-E). This fouling can be measured quantitatively by mounting a gill on a microscope slide in a standard way, viewing it with transmitted light on a microscope equipped with a light meter, recording the relative transparency of the gill. Fouled gills showed lower transmission of light (Fig. 4.10a). The

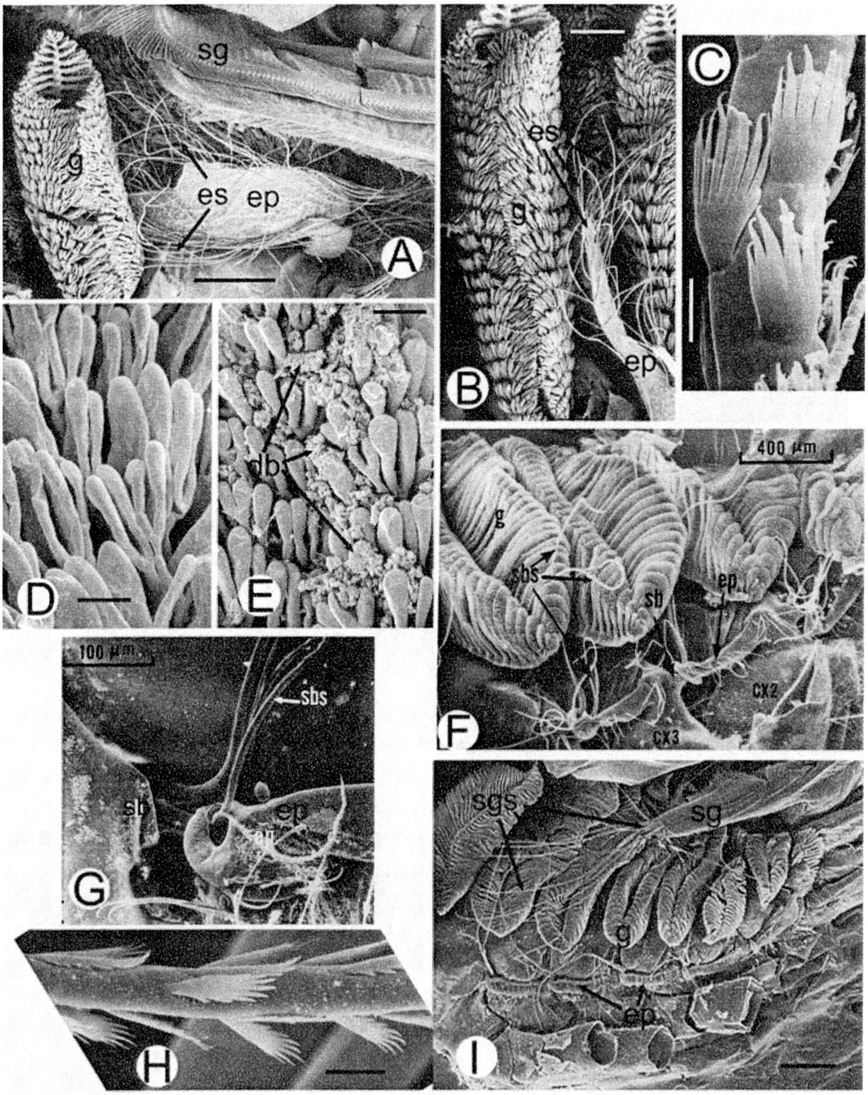

Plate 4.2 Gill-cleaning structures. (A) *Rimapenaeus similis* (Penaeidae) setiferous epipod of maxilliped 2 and (B) fork of pereopod 1 epipod with setiferous setae extending to gills; other gills lying above the setiferous branches of the epipods were removed to reveal the cleaning setae. (C) Ultrastructure of a multidenticulate scale seta of a setiferous epipod (*Sicyonia parri*, Sicyoniidae); note denticulate scale setules. (D) Highly magnified gill filaments from control (clean) and (E) experimental gill chambers (fouled with particulate matter) in epipod ablation experiment with *Rimapenaeus similis*. (F) Gill chamber of *Betaeus macginitieae* (Alpheidae) showing hooked epipods and setobranch setae, some of which are up among the gill lamellae. (G) close-up of epipod hook, setobranch, and its setae. In (F) and (G), the epipod hooks are somewhat disengaged from the bases of the setobranch setae, and the latter from the gills, due to the SEM preparation. (H) Ultrastructure of setobranch seta of *B. macginitieae*, showing multidenticulate scale setules projecting from setal shaft. (I) Gill chamber of *Caridina typus* (Atyidae) showing long multidenticulate scaphognathite setae extending over lateral side of gills.

penaeoid epipods and their multidenticulate setae are indeed effective in keeping the gills clean.

The informative branchial formulas of Pérez and Kensley (1997) show that setiferous epipods are widely distributed in all penaeoid families. Although a systematic survey of their gill-cleaning function has not been done, it is quite probable that they operate similarly to those of *R. similis*. However, Bauer (1981) reported that epipods were reduced and not setiferous in a penaeoid (benthesicymid) *Gennadas* species. Likewise, although most species of Penaeidae have setiferous epipods, some species have a reduced number, raising the question of how they keep gills clean. On the other hand, sergestoid shrimps lack setiferous (or any) epipods on the Mx3 and the pereopods. Perhaps this is not too surprising; as with many sergestoid traits, the gills are highly reduced or rudimentary in the group. It is possible that the anterior pereopods, as in some carideans and stenopodideans (Table 4.1), or some unknown mechanism clean what few gills sergestoids have. It is more likely that simple flushing of the gill chamber by scaphognathite reversals and periodic molting, which replaces the gill exoskeleton with a clean one, is sufficient to take care of the reduced gill-cleaning needs in this group.

4.3.3 Epipod-Setobranch Complexes

In many decapods, there are papillae on the coxae (basal segments) of Mx3 and the pereopods which bear long multidenticulate scale setae that reach up into and among the gills (Bauer 1979, 1981, 1989, 1998, 2013; Batang and Suzuki 2003) (Plates 2.2 C, 4.2 F-H). Similar to the setiferous epipods of penaeoids, when the limbs that bear setobranch setae move about during their normal locomotory or other behaviors, these setae jostle among and rasp the gill filaments. This dislodges and rasps off much of the fouling material that might accumulate or attached to the gill filaments. In decapod shrimps, setobranchs and their cleaning setae occur only in procaridids and several caridean families (Table 4.1). In these taxa, the setobranchs are always associated with unique hooked epipods (Plate 4.2 F-G; Chap. 2: Fig. 2.8b-d). The epipod hook of one appendage clasps around the setobranch setae of the appendage just posterior to it, e.g., the epipod on pereopod 1 hooks around the setobranch setae of pereopod 2 (Fig. 4.9b-c; Plate 4.2 F-G). Therefore, in a complete epipod-setobranch complex, there are hooked epipods on Mx3-P4 and setobranchs with setobranch setae on P1–5 (Bauer 1979). I have observed individuals of *Pandalus danae* swaying the body back and forth while perched on stationary walking legs.

Plate 4.2 (continued) cx2,3, coxae of pereopods 2,3; db, particulate debris; ep, epipod; es, epipod setae; g, gill; sb, setobranch; sb(s), setobranch seta(e); sg, scaphognathite; sgs, scaphognathite setae. Scale bars represents 600 μm in (A), 250 μm in (B), 3 μm in (C), 50 μm in (D-E), 400 μm in (F), 100 μm in (G), 5 μm in (H), 200 μm in (I). From (A-B) Bauer (1999), (C) Bauer (1989), (D-H) Bauer (1976) and (I) Suzuki and McLay (1998), courtesy of the Journal of Crustacean Biology

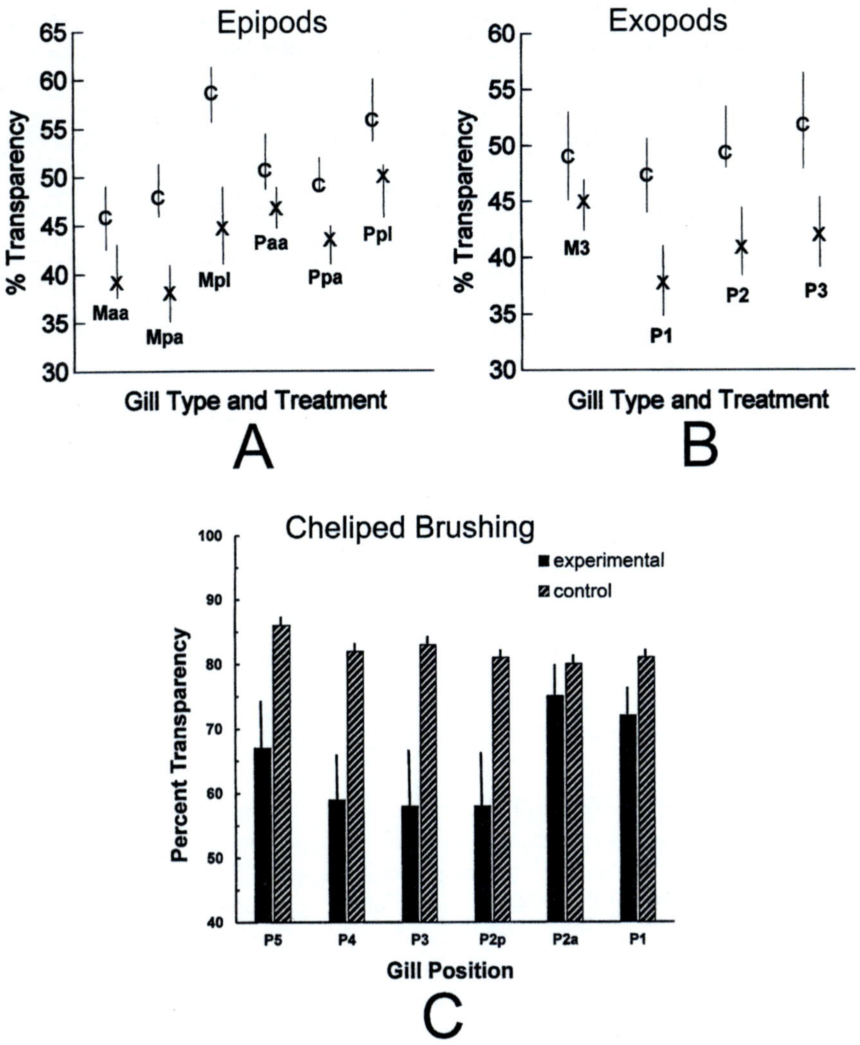

Fig. 4.10 Experiments on adaptive value of gill cleaning mechanisms. Decrease in gill transparency shows gill fouling in shrimps in which the gill-cleaning mechanism had been removed (experimentals) compared to shrimps with complete gill-cleaning structures (controls). (**a**) Setiferous epipod ablation experiment in *Rimapenaeus similis* (Penaeidae). (**b**) Setiferous exopod removal experiment in *R. similis*. (**c**) Gill-brushing cheliped removal experiments in *Heptacarpus sitchensis* (Thoridae). *a* anterior; *c,* mean gill transparency in control shrimps; *x,* mean transparency in experimental shrimps; *M3,* third maxillipeds; *Maa, Mpa, and Mpl,* anterior arthrobranchs, posterior arthrobranchs, and pleurobranchs, respectively, of third maxillipeds; *p,* posterior; *P1-P5*, pereopods 1–5; *Paa Ppa, and Ppl*, anterior arthrobranchs, posterior arthrobranchs, and pleurobranchs, respectively, of third pereopods. From (**a-b**) Bauer (1999) and (**c**) data in Bauer (1976)

This action brings the setobranch setae attached to its limb coxae to move in and among the gills. Such "limb rocking" is also a common behavior in the crayfish *Procambarus clarkii* which uses a profusion of setobranchs for gill cleaning (Bauer 1998). Further investigation may show that limb rocking is a common behavior associated with epipod-setobranch complexes in those shrimps that have them.

Setobranchs occur in a number of decapod groups (Batang and Suzuki 2003; Bauer 1989, 2013) but are not linked with hooked epipods as in carideans and procarididеans. So why are the epipods hooked in the latter groups and linked with setobranch setae? In the gill chamber of crayfishes, there is a large tuft of tangled setobranch setae reaching into and wrapped around and among the gills (Huxley 1880; Bauer 1998). In shrimps with epipod-setobranchs, there are many fewer setobranch setae. When the coxa of a limb in these shrimps moves backward, its setobranch setae are drawn down away from the gills, but when the limb goes back forward, they are guided back into and among the gills by the epipod hook of the limb anterior. The reduction of these setae in carideans seems to be correlated with the more precise manner in which they are maintained in proper position with the gills they serve. In crayfishes, experiments showed that setobranch setae are very effective in keeping the gills clean from sediment but not microbial fouling (Bauer 1998). The gill-cleaning efficacy of the epipod-setobranch complexes still needs to be tested in procaridids and those carideans which possess them.

4.3.4 *Scaphognathite Gill Cleaning*

When I was doing a survey of gill cleaning in shrimps during my doctoral work, I noticed that one of the crangonids, *Paracrangon echinata*, lacked other mechanisms that I had described for other taxa (epipod-setobranchs, gill brushing). On closer examination, I found that the posterior border of the scaphognathite had very long multidenticulate scale setae that extended over the lateral surface of the gills. As the scaphognathite rocked up and down during its usual function of pumping water through the branchial chamber, these setae brushed up and down over the gills, presumably to clean them. This mechanism was explored in much greater depth in a comparative study in atyid carideans by Suzuki and McLay (1998). In these species, the scaphognathite is elongated posteriorly and equipped with long multidenticulate scale setae which extended over the lateral surface or most or all the gills (Fig. 4.11a; Plate 4.2 I). Most Atyidae have a well-developed epipod-setobranch complex as well, but Suzuki and McLay found that there was a negative correlation between the number of these long scaphognathite setae and the number of setobranch setae. Thus, when one mechanism was well developed, the other was reduced. Together, they must do the work of gill cleaning in most atyids. Suzuki and McLay reported that the scaphognathites of several caridean families have similar long multidenticulate scale setae which may clean or assist in cleaning the gills and I have added a few to the list (Table 4.1) (e.g., Fig. 4.11b-c). This mode of gill cleaning appears to be absent in sergestoids and penaeoids. This is also true of most stenopodideans (Goy 2010),

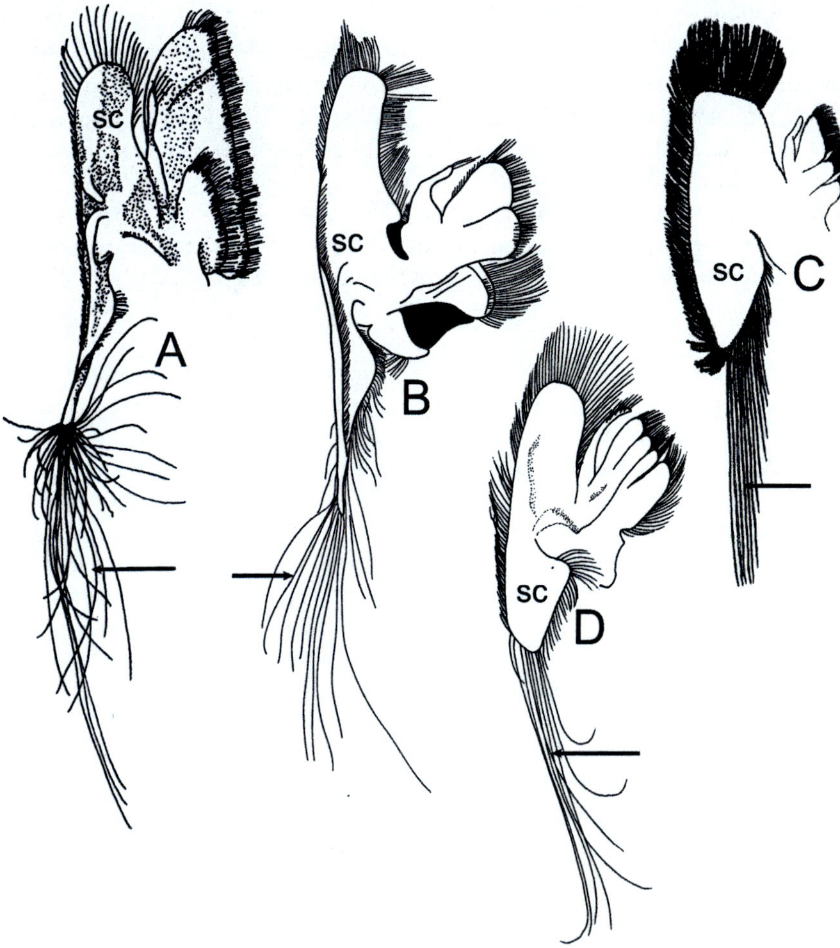

Fig. 4.11 Scaphognathite setae that clean (**a**) or possibly clean gills (**b–d**). The second maxillae and their long posterior scaphognathite setae (unmarked arrows) are shown in (**a**) *Atyopsis spinipes* (Atyidae), (**b**) *Fresnerhynchus crozeti* (Lipkiidae), (**c**) *Agostocaris acklinsensis* (Agostocarididae) and (**d**) *Spongicola venustus* (Spongicolidae). *sc,* scaphognathite. From (**a**) Suzuki and McLay (1998), courtesy of Journal of Crustacean Biology; (**b**) d'Udekem d'Acoz and De Grave (2018), courtesy Zootaxa; (**c**) Alvarez et al. (2004) and (**D**) Saito and Komai (2008)

although *Spongicola* species do have long posterior scaphognathite setae (Saito and Komai 2008) (Fig. 4.11d). The caridean *Pandalus danae* has long posterior scaphognathite setae (Chap. 2: Fig. 2.7a). In this species, the setae do not sweep over the gills during gill bailer movement, but instead anchor the posterior end of that structure within the gill chamber (Bauer 1979). However, these setae may still assist cleaning of gills that they do move over. Nonetheless, most pandalids have a full epipod-setobranch complex.

4.3.5 Gill Cleaning by Chelipeds

Stenopodidean and many caridean shrimps have dispensed with all of the gill-cleaning mechanisms described so far, except for scaphognathite reversals of respiratory flow. Instead, they clean the gills with usually one pair (carideans) of specialized chelipeds (Fig. 4.12) or two (stenopodideans, chelipeds 1,2) (Table 4.1) (Bauer 1979, 1989). Chelipeds which brush the gills or other parts of the body are slender, delicate, and flexible (Fig. 4.12). In the "polycarpidean" caridean families, flexibility for grooming is imbued on the second cheliped by the subdivision of the carpus (Fig. 4.12b–e). Muscles and their apodemes (tendons) in each separate carpal article allows articulation and bending of each with its neighbors (Berkeley 1928; Read et al. 1991). On the other hand, palaemonoids and perhaps some other caridean taxa use the first pair of chelipeds for gill and other grooming. The first pair of chelipeds in these taxa always has an undivided carpus (Fig. 4.12a), but their basal segments have considerable mobility (Bauer 1979). Grooming chelae are usually equipped with thick tufts or brushes of the multidenticulate setae found in other types of gill cleaning structures (Fig. 4.12a–d; Plate 4.3 A-B). However, in the only stenopodidean examined for this trait, *Stenopus hispidus*, they are inexplicably smooth. In glyphocrangonids, the second (chelate) pereopod has a multiarticulated carpus, but no setal tufts. Grooming (or any other behavior) has not been observed in this deep-sea group, so perhaps the multiarticulate cheliped is used primarily in other behaviors, e.g. general body grooming, feeding.

Chelae used for grooming gills or in general body grooming (see below) usually have some sort of chela lock or clasp (Bauer 1975, 1979; Felgenhauer and Schram 1979) that allows the pincer fingers to close together very precisely and without slipping off one another. In a chela lock, one or two stout setae on one finger fit between similar setae on the other, bringing the tips and cutting edges of the chela fingers tightly together (Fig. 4.13; Plate 4.3 C-D). The opposing edges of cleaning chela fingers may bear a series of ridges and pegs that closely contact and overlap when the chela is closed (penaeids: Hindley and Alexander 1978 and Dall et al. 1990; stenopodideans: Goy 2010). These adaptations allow the shrimp to grasp, pick, or remove very small objects from body or other surfaces.

In gill cleaning, the shrimp reaches back and inserts one or both grooming chelipeds inside a branchial chamber and vigorously brushes the gills with setal tufts on the chelae (Fig. 4.14). This may be interspersed with nipping and tugging at some small object on the cuticle of a gill. I have observed with carideans such as *Heptacarpus sitchensis*, *Palaemon ritteri*, and the stenopodidean *S. hispidus* that shrimps may spend considerable time and energy in this behavior. In a few caridean families with no other apparent gill-cleaning device, other pereopods might participate in grooming the gills, e.g. the short pereopod 4 with a comb-like dactyl of the Pasiphaeidae (e.g., *Leptochela*; Chace 1976) and a chelate cleaning structure on pereopod 5 of *Sclerodora* and other acanthephyrids (Vereshchaka et al. 2021).

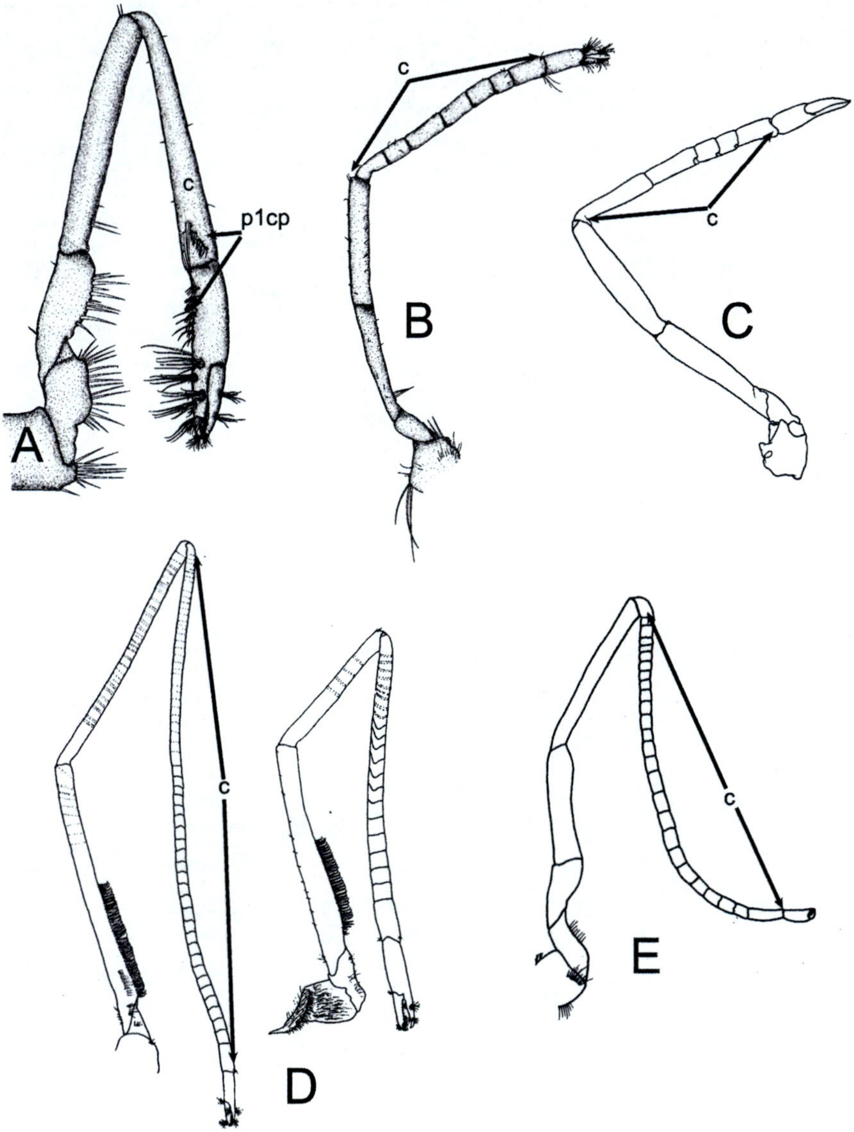

Fig. 4.12 Grooming chelipeds of caridean shrimps. (**a**) *Palaemon ritteri* (Palaemonidae) cheliped 1. (**b-e**) second chelipeds of (**b**) *Heptacarpus sitchensis* (Thoridae). (**c**) *Alpheus hyphalus* (Alpheidae), (**d**) *Pandalus danae* (Pandalidae), and (**e**) *Glyphocrangon* species. Complete horizontal lines in the second chelipeds of *P. danae* indicate carpal articles (joints); light dotted lines on carpus and more proximal articles are annulations of the cuticle, not complete articles. Note tufts of cleaning setae in (**a-b**, **d**) (setae were not illustrated for (**c**) by Chace (1988). c, carpus; p1cp; pereopod 1 propodal-carpal brush). From (**a-b**, **e**) Bauer 1976, (**c**) Chace (1988), (**d**) Bauer (1975)

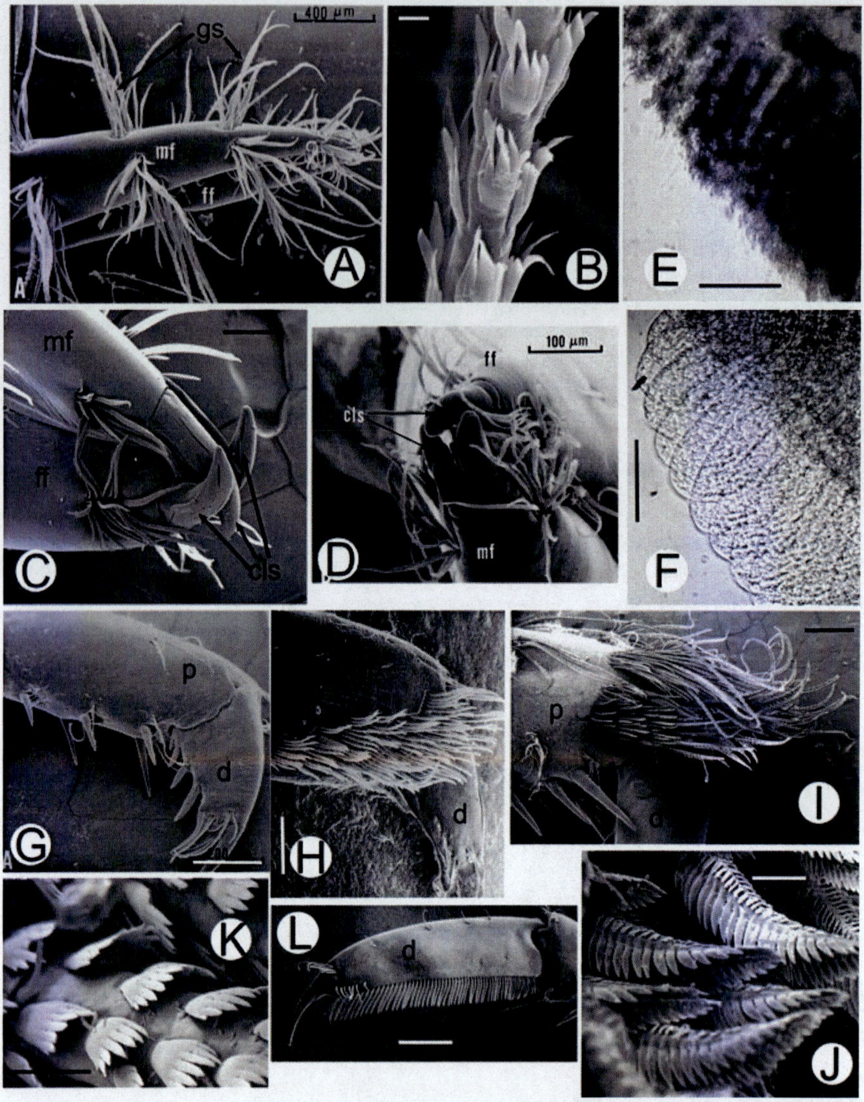

Plate 4.3 Chela and pereopod 5 cleaning structures. (A) Tip of first chela of *Palaemon ritteri* (Palaemonidae), with groups of multidenticulate grooming setae projecting from the chela fingers. (B) Portion of multidenticulate scale seta (grooming) seta of *P. ritteri*. (C) Chela 2 locks of *Pandalus danae* (Pandalidae) and (D) *Heptacarpus sitchensis* (Thoridae). (E) Gill lamellae of experimental (fouled) and (F) control (clean) gills from *H. sitchensis* in the grooming cheliped ablation experiment (Bauer 1976, 1977). (G) and (H), distal end of pereopods 4 and 5, respectively, of *H. sitchensis* showing absence of grooming brush on pereopod 4 and its presence on pereopod 5. (I) Multiple grooming brushes on the distal end of pereopod 5 of *P. danae*. (J) Serrate setae with multidenticulate scale setules from a pereopod 5 grooming brush of *P. danae*. (K) multidenticulate scale setules from a portion of a pereopod 5 grooming seta in *P. danae*. (L) Pereopod 5 dactyl with setal grooming comb in *Xiphocaris elongata* (Xiphocarididae). cls, chela lock setae; d, dactyl; ff, fixed (propodal) chela finger; gs, grooming setae; mf, movable (dactylar) chela finger; p, propodus. Scale bar represents 400 μm in (A), 25 μm in (B), 100 μm in (C-D), 400 μm in (E-F), 200 μm in (G-I), 20 μm in (J), 8 μm in (K), and 1 mm in (L). From (A-H) Bauer (1976), (J-K) Bauer (1975) and (L) Bauer (1989)

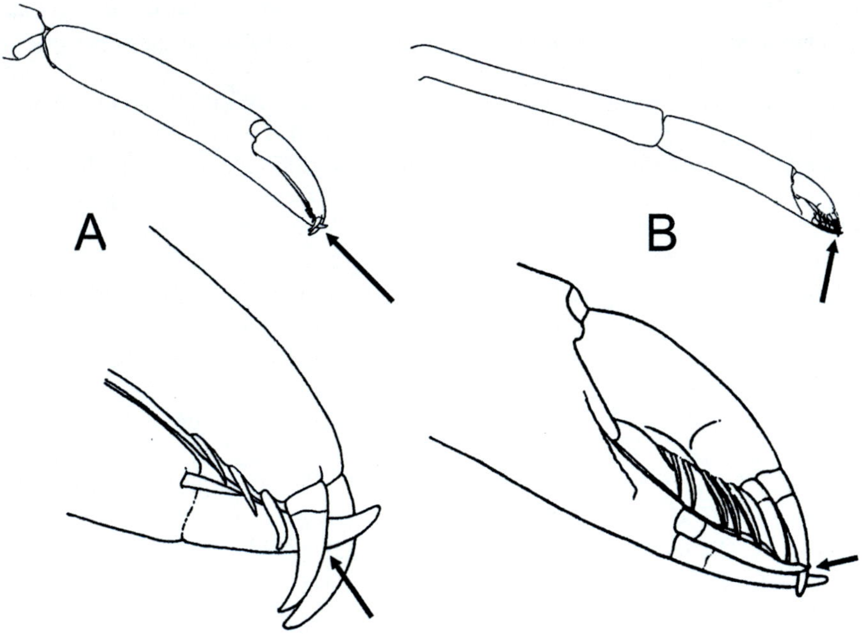

Fig. 4.13 Caridean chela locks: (upper) shows terminal end of the chelipeds and (lower) tips of the chela fingers with interlocking stout terminal setae. (**a**) Cheliped 1 of *Kemphyra corallina* (Acanthephyridae). (**b**) Cheliped 2 of *Eugonatonotus chacei* (Eugonatonotidae). Arrows indicate intermeshing of chela lock setae. From (**a**) Chace (1986) and (**b**) Chace (1997)

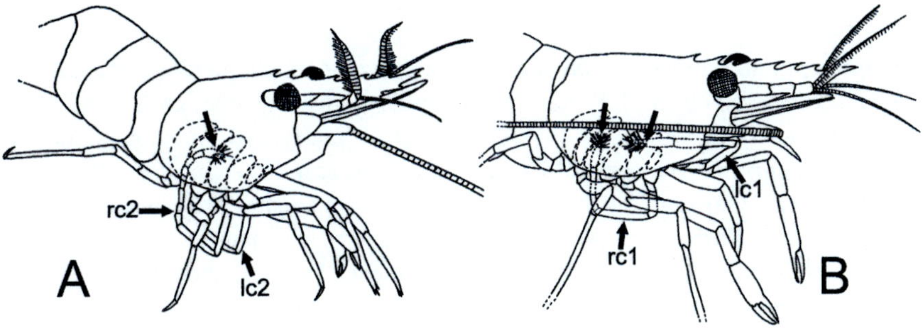

Fig. 4.14 Cheliped gill-cleaning in carideans. (**a**) *Heptacarpus sitchensis* (Thoridae) inserting the multiarticulate second chelipeds into branchial chamber of each side, brushing and cleaning the gills. (**b**) *Palaemon ritteri* (Palaemonidae) inserting both first chelipeds into the right branchial chamber to clean the gills there. *lc1,* left cheliped 1; *lc2,* left cheliped 2; *rc1,* right cheliped 1; *rc2,* right cheliped 2. Unmarked arrows indicate the position of the grooming chelae within the branchial chamber; dotted oval outlines represent the gills. From Bauer (1976)

Gill cleaning with chelipeds is very efficient at maintaining the gills in a pristine condition. This statement is supported by experiments that are rather easily done with ablation (amputation) experiments similar to those done on antennular cleaning.

In experimental groups of *H. sitchensis* and *P. danae*, I removed the second (grooming) chelipeds, while in the control group the first walking legs, which are not involved with grooming, were removed (Bauer 1975, 1979). Such treatment may seem a bit harsh, but there was no noticeable effect on survival or overall activity. The pereopods of shrimps, like those of other decapods, have a natural amputation or breakage plane associated with the basi-ischial joint (McVean 1982). It allows the limb to autotomize and the wound to close with little or no blood loss when a predator or aggressive conspecific grabs the limb. After limb ablation in the Bauer (1979) experiment, the shrimps were placed in cages in the natural environment for two weeks. In *H. sitchensis,* the results were again dramatic; in the experimental treatments, ungroomed gills soon visibly darkened with a heavy load of sediment while those of controls remained clean (Plate 4.3 E-F). When gill fouling was quantified as in the *R. similis* experiment described above, the control gills, groomed by the cleaning chelipeds, were significantly less opaque than those darkened by fouling (Fig. 4.10c). When shrimps with fouled gills were placed in oxygen-poor water, they became sluggish and immobile, while control shrimps seemed unaffected. In the *P. danae* experiment, however, the gills of both groups remained visibly clean (Bauer 1979). This species, like most pandalids, has a complete set of epipod-setobranchs. Cheliped gill grooming was never observed (Bauer 1975) and has not been reported in *P. danae*. In pandalids, the epipod-setobranch complex, perhaps with the help of scaphognathite setae, appears quite effective at controlling gill fouling as in the Atyidae discussed above.

4.3.6 Passive Versus Active Gill Cleaning

Setiferous epipods, epipod-setobranchs, and multiscaled scaphognathite setae can be called "passive" gill cleaning. Cleaning of the gills by these devices goes on more or less automatically in the course of the normal respiratory and locomotory movements of the shrimp. The shrimp does not halt food search, feeding or interaction with conspecifics to employ these gill-cleaning mechanisms; they will continue during these other behaviors. Another minor passive mechanism not yet mentioned is the brushing of the lateral side of the gills by the back and forth movement of small pereopodal exopods, equipped with long multidenticulate setae, in some penaeoid shrimps, e.g. *Rimapenaeus similis* and *Litopenaeus setiferus* (Bauer 1999) (Fig. 4.9a). These setiferous exopods brush the outer (lateral) side of the gills and are effective in removing sediment from the gills, as amputation experiments show (Fig. 4.10b). Pereopodal exopods are present in most penaeoids except for sicyoniids (Pérez and Kensley 1997). The involvement of setiferous exopods in procaridids or carideans, if any, has not yet been examined. Pereopodal exopods are completely lacking in stenopodideans (Goy 2010).

In the crayfish *Procambarus clarkii*, the lateral sides of the gills are cleaned during body movements by multidenticulate scale setae on the inside of the gill cover (branchiostegite) (Bauer 1998). I found a very limited amount of such inner branchiostegal setae in *R. similis* adjacent to a posterior gill which did not seem to be reached by setiferous epipods or exopods in that species (Bauer 1998). Perhaps this mechanism is more widespread, but it has not yet been reported in other shrimps.

Active and passive gill-cleaning are, in general, mutually exclusive in the Decapoda. It is often the case that more than one passive mechanism works in concert with another one in those shrimps that do not clean the gills with chelipeds, e.g., the scaphognathite setae and epipod-setobranch setae of atyid and pandalid shrimps, the various mechanisms described above in *R. similis*, the crayfish *P. clarkii* and a number of thalassanidean decapods (Batang and Suzuki 2003). However, in most shrimps and other decapods that actively clean the gills with chelipeds (e.g., anomurans, some thalassanideans), there is no passive mechanism used, only cheliped grooming (Bauer 1989; Pohle 1989; Batang and Suzuki 2003). The distribution of gill cleaning devices indicates that ancestral decapods probably cleaned the gills with passive setal devices. The more efficient cheliped gill cleaning evolved independently and replaced passive cleaning in a number of shrimp families.

The process of substitution of passive with active gill cleaning is further demonstrated in the hippolytoid families, in which all stages of replacement are found. *Lysmata californica* (Lysmatidae) has a full set of epipod-setobranchs like most members of its family (Lysmatidae) However, it infrequently reaches into the branchial chamber and briefly cleans its gills with its grooming chelipeds (Bauer 1979). In the lysmatid *Lysmatella prima*, pereopodal epipods are absent; has cheliped gill grooming completely replaced them? In *Eualus and Heptacarpus* (Thoridae), some species have a full set of epipod-setobranchs but in others these structures are being lost (from posterior to anterior) or are completely absent. In the latter case, gills are cleaned only by the chelipeds. In species with reduced epipod-setobranchs, an epipod may lack a hook or be reduced to a nub, and there are no setobranch setae on the following limb. These epipods are not functional and clearly in the process of being lost (Bauer 1984). These observations support the view that the more efficient gill brushing is replacing epipod-setobranchs in this family.

4.4 General Body Grooming

4.4.1 Morphology and Behavior

Up to this point, I have covered rather specialized or specific cleaning activities and behaviors: cleaning of antennules, antennae, and gills. But many shrimps spend extended periods in bouts of cleaning a variety of other body surfaces: the eyes, broad expanses of carapace and abdomen, pereopods, pleopods, and articular areas between appendage segments and body somites. The major cleaning appendages for these body parts are grooming chelipeds and/or posterior (usually the fifth) pereopods. Grooming by these limbs maintains a clean exoskeleton free of fouling organisms and particulate matter that can interfere with the shrimp's activities. Fouling within articulations between limb segments and body somites can interfere with locomotion and movement. Cleaning of these areas of the body, as well as that of the eyes, will be grouped under the general heading of "general body grooming" (GBG). In those shrimps that clean the gills, the same pair or pairs of chelipeds that do the gill brushing are also employed in GBG.

Let us start with the penaeoids. This is one group in which there are relatively few reports of grooming in the literature. However, this is also a group in which specialized structures for cleaning the eyes are found. The basal segment of the antennule has a depression in which the eye can be moved into (Fig. 2.11b). This "eye cup" is densely lined with setae (Dall et al. 1990; Pérez and Kensley 1997). The eye is normally held out at an angle from the antennular peduncle but is occasionally flicked into the eye cup whose setae help brush off debris. Additionally, in two penaeoid families, the Penaeidae and Solenoceridae, an eye brush, the prosartema, is well developed (Fig. 2.11b). The prosartema is a long setose lobe or process that can flick over the eye when the latter is inside the eye cup, giving additional cleaning. In another burrowing family, Sicyoniidae, the prosartema is reduced to a setose knob, while in the more pelagic families Aristeidae and Benthesicymidae, it is reduced or vestigial. Penaeids and probably most other decapod shrimps often twitch their eyes, which helps to shake off particles which have settled on the corneal surface (Dall et al. 1990; pers. obs,)

There are relatively few reports of GBG by the chelipeds or posterior appendages in the penaeoids. Certainly, the first three chelipeds do engage in GBG in *Fenneropenaeus merguiensis* (Alexander et al. 1980). The first three chelipeds clean the abdomen as the shrimp bends it forward beneath the body so that it can be reached by the chelipeds. I observed GBG in the benthesicymid *Gennadas* (Bauer 1981). Alexander and colleagues also reported that the pereopods themselves were cleaned by the third maxillipeds and the P1-CP brushes used in antennal grooming. It may be, though, that the burrowing habit of many penaeoids may make much GBG unnecessary. Bouts of grooming in sergestoids have been reported (Cowles 1994) but no details were given. The small, almost microscopic chelipeds appear to be the likely candidates as GBG limbs in this group. However, observations on this aspect of sergestoid biology on living individuals, not an easy task in sergestoids, need to be made.

In the Caridea, however, GBG occurs frequently and extensively in some species, e.g., in some palaemonoids (Bauer 1978; Felgenhauer and Schram 1979; Van Maurik and Wortham 2011, 2014), pandalids (Bauer 1975), alpheids (Bauer 1979; Read et al. 1991; Palomar et al. 2005), and hippolytids (Bauer 1979, 1984). It is very probable that all the polycarpidean families, i.e., those with a multiarticulate carpus on cheliped 2 (Fig. 4.12b-e), engage in GBG with that limb. The flexibility of this limb is an obvious advantage in those families in which GBG has been reported. The cheliped can flex and twist in many ways (even do a coil!) in order to reach some out-of-the-way location. However, the first chelipeds are efficient grooming devices in those families using it for GBG (e.g., Palaemonidae, Fig. 4.12a). Similar to cheliped gill cleaning, the preening chelipeds reach to some body part (including the eyes) and rapidly rub, pick, and nip at the cuticular surface with its chelae (Fig. 4.15a-c). Grooming chelipeds can reach to almost all parts of the body, but

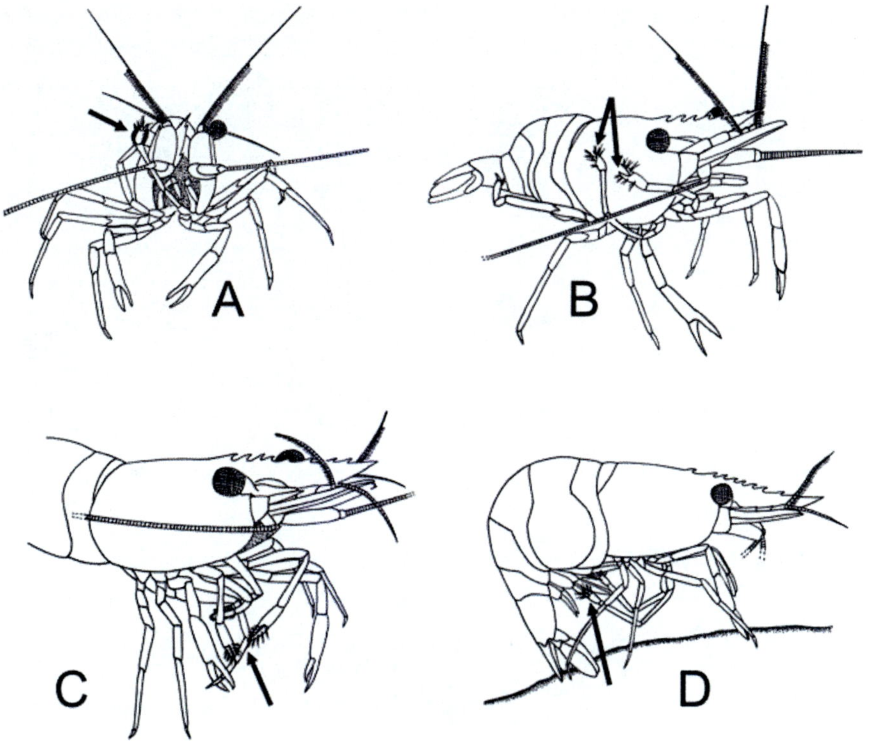

Fig. 4.15 General body grooming by chelipeds (arrows). (**a**) *Heptacarpus sitchensis* (Thoridae) brushing an eye with both second chelipeds. (**b**) *Palaemon ritteri* (Palaemonidae) grooming the carapace, (**c**) a pereopod, and (**d**) pleopods with the first chelipeds. Note that in (**c**) the pereopod being groomed is also being cleaned more basally by the third maxillipeds. From Bauer (1976)

primarily groom the cephalothorax and beneath the abdomen. To preen the abdomen, the shrimps bends the abdomen forward to expose the ventral surface of the abdomen and pleopods to give the grooming chelipeds access (Fig. 4.15d). After pleopods are cleaned, they beat rapidly, presumably to shake off dislodged material and return displaced setae to their usual positions. In some families, neither pair of chelipeds appears to be specialized for cleaning, yet one or both chelipeds may be used in GBG, e.g., the atyid *Caridina nilotica* (Fryer 1960) and xiphocaridid *X. elongata* (Fryer 1977). In some processids and pandalids, the multiarticulate second chelipeds are unequal in length, stoutness, and number of carpal articles (Manning and Chace 1971; Chace 1985) (e.g., Fig. 4.12d). In *Pandalus danae*, I found that the longer and slenderer member of the pair groomed more frequently than of the shorter and stouter one (Bauer 1975). This difference in function between asymmetrical left and right second chelipeds may be common but needs to be determined in other species. In the Stenopodidea, the slender first and second chelipeds, with their tufts of setae, engage in both gill cleaning and GBG (Bauer 1981, 1989; Goy 2010).

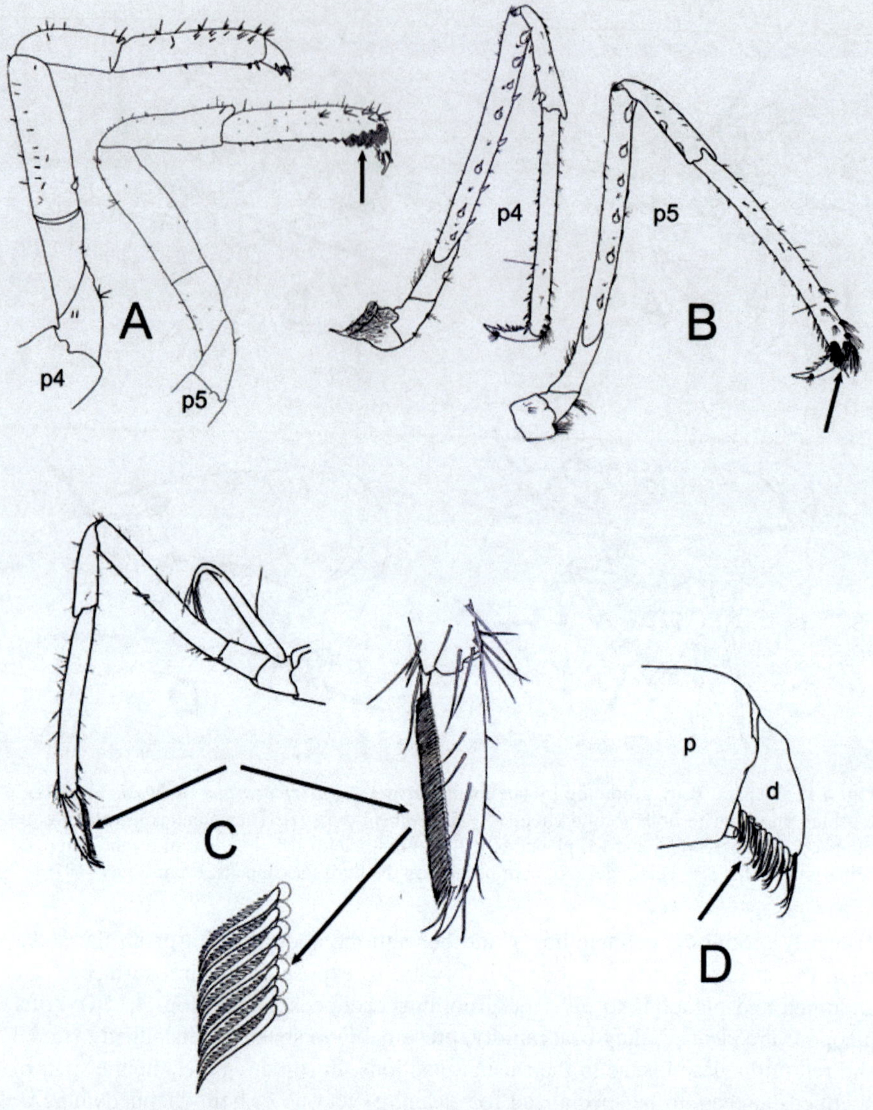

Fig. 4.16 Pereopod 5 grooming structures. Pereopods 4 and 5 are shown for (**a**) *Betaeus macginitieae* (Alpheidae) and (**b**) *Pandalus danae* (Pandalidae); note grooming setae and brushes (arrows) on pereopod 5 and their absence on pereopod 4. (**c**) Pereopod 5 of *Typhlatya rogersi* (Atyidae) (left) with dactylar comb (arrow) and (right) enlarged and setae magnified (bottom left). (**d**) Dactylar comb (arrow) of pereopod 5 of *Kemphyra corallina* (Acanthephyridae). *d*, dactyl; *p*, propodus; *p4, p5*, pereopods 4,5. From (**a**) Bauer (1976, (**b**) Bauer 1975, (**c**) Chace and Manning (1972) and (**d**) Chace (1986)

In a number of caridean families, the fifth pereopods have cleaning brushes or combs (Fig. 4.16). The brushes of serrate and/or multidenticulate scale setae are located at the end of the propodal segment of P5 (e.g., Alpheidae, Pandalidae,

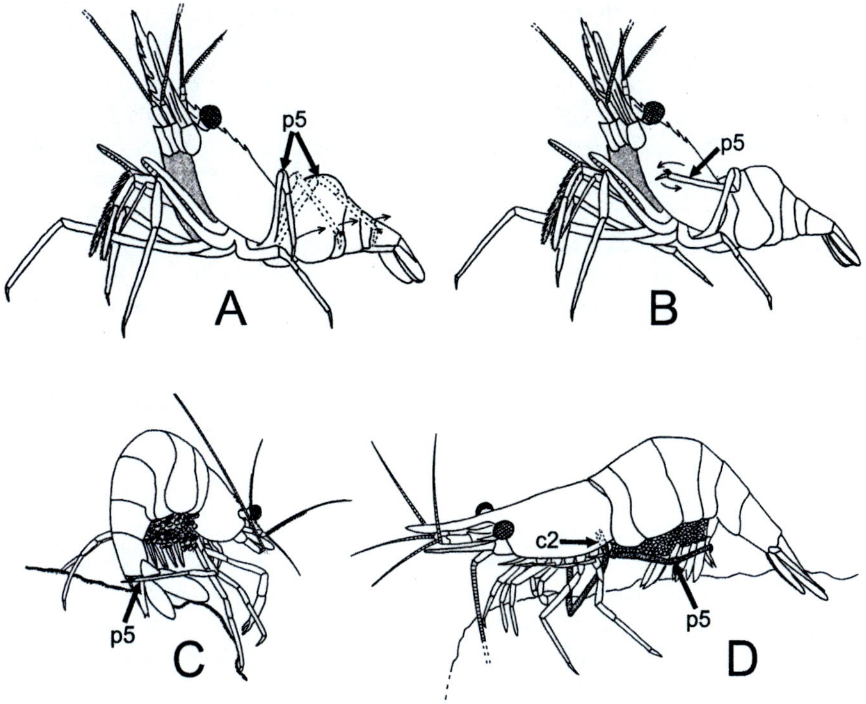

Fig. 4.17 General body grooming with the fifth pereopods (P5). *Pandalus danae* (Pandalidae) grooming the abdominal pleura (**a**) and carapace (**b**) with P5 brush. (**c**) *Palaemon ritteri* (Palaemonidae) cleaning the dorsal side of the telson with P5 brush. (**d**) *Xiphocaris elongata* (Xiphocarididae) cleaning the pleopods with P5 dactylar comb. Note also that the second cheliped is cleaning just inside the branchial chamber. *c2,* second cheliped; *p5* fifth pereopod. From (**a–b**) Bauer (1975), (**c**) Bauer (1976) and Bauer (1989)

Palaemonidae) (Fig. 4.16a-b; Plate 4.3 G-K). The P5 is capable of cleaning the entire abdomen (Fig. 4.17a, c-d), but also can twist around to groom the posterior part of the cephalothorax (Fig. 4.17b). In a few families, e.g., Atyidae, Xiphocarididae, Acanthephyridae, the dactyl has a series of stout serrate setae arranged like the teeth of a comb (Fig. 4.16c-d; Plate 4.3 L) (Bauer 1979; Hobbs and Hart 1982; Chace 1986). In the atyids and xiphocaridids, grooming by the P5 comb has been observed (Fryer 1977; Bauer 1989) (Fig. 4.17d). In the Pasiphaeidae, the dactyl of the relatively short fourth (and sometimes fifth) pereopod has a series of sharp spine-like setae which might serve as a grooming structure. Pereopod 5 cleaning brushes have not been reported in sergestoids, penaeoids, or stenopodideans. In the latter, GBG by the chelipeds may be supplemented by the P1-CP brushes (Goy 2010). This is a possibility in carideans as well; however, it was not reported nor apparently observed in the study of GBG by Bauer (1979).

4.4.2 Adaptive Value of General Body Grooming

Ablation experiments with *Pandalus danae* and *Heptacarpus sitchensis* (Bauer1975, 1978) showed the selective pressures and adaptive value of GBG. In the *H. sitchensis* GBG experiments, I removed only the grooming chelipeds of experimentals. With *P. danae*, I had three treatments with removal of (a) only grooming chelipeds, (b) only fifth pereopods, and (c) both sets of grooming appendages. For controls, the first walking legs (third pereopods) were removed in both sets of experiments. The shrimps were exposed in flow-through cages hung in the sea for two weeks in the smaller and faster molting *H. sitchensis* but for one month in the larger *P. danae*. Settling plates for sampling ambient fouling on surfaces were exposed along with the cages.

In both species, fouling of the general body surfaces occurred when GBG was not allowed. With *H. sitchensis*, fouling of experimentals was similar to that found on ungroomed gills and antennules: long strands of filamentous bacteria, diatoms, and sessile protists, especially vorticellid ciliates. Sediment did not accumulate on smooth surfaces of the body but did so in areas of complex topography. There was a noticeable accumulation of material in cuticular grooves and crevices underneath the thorax, around the bases of the legs, edges of the gill covers, and the pleopods. Fouling on controls was negligible (Fig. 4.18a). In the *P. danae* experiment, which lasted an entire month, epizoic fouling was more extensive. In addition to stalked protozoans and filamentous bacteria, there was the beginning of more abundant fouling by sessile invertebrates such as hydroids. The cephalothorax, primarily groomed by the chelipeds, had significantly greater fouling than controls. A similar difference between controls and experimental individuals occurred on their abdomens, mainly groomed by the fifth pereopods, and those without either chelipeds or pereopod 5 grooming had the greatest fouling of all (Bauer 1975; Fig. 4.8 in Bauer 2004).

Clearly, GBG is effective at keeping most of the body clean. When it does not occur, fouling can be much more serious than that shown by these relatively short-term experiments. Periods between molts may be prolonged, especially in temperate and high latitudes. Females incubating embryos cannot molt or they will lose their broods, and incubation times can be prolonged at higher latitudes. In the long-living *Sclerocrangon boreas* (Crangonidae), some females may not molt for as long as two years; fouling by barnacles, algae, and other sessile organisms can be quite severe (Plate 5.1A) (Sainte-Marie et al. 2006). The heavy load experienced by the female shown in Plate 5.1A must have been quite an impediment to further survival and reproduction. Grooming may be depressed by environmental stress. Giri and Wicksten (2001) showed serious barnacle and other epizoic fouling on *Lysmata wurdemanni* after an extensive period of lower than normal salinity, which may have caused physiological stress and reduced grooming.

However, even a small amount of macroscopic fouling or extensive microbial fouling on a swimming organism such as a shrimp might increase hydrodynamic drag sufficiently to impede forward or backward swimming. As any boat or ship

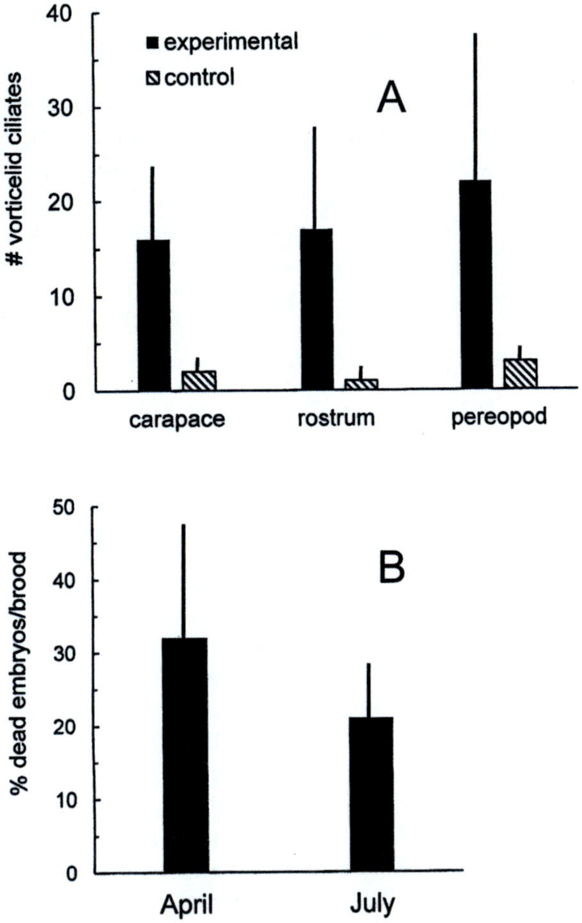

Fig. 4.18 General body and embryo grooming experiments in *Heptacarpus sitchensis* (Thoridae) with (controls) and without (experimental) grooming chelipeds. (**a**) Comparison of vorticellid ciliate fouling on carapace, rostrum, and a pereopod between experimentals and controls. (**b**) Embryo death in broods carried by females without grooming chelipeds. No embryos appeared dead in controls. Mean (bars) and 95% confidence limit lines are shown. From data in Bauer (1976)

owner knows, fouling by settling organisms such as barnacles, bivalves, algae, and other macroscopic foulers can cause a significant drag and concomitant increase in fuel (energy) costs. Although the size and amount of fouling on shrimps without GBG may seem small compared to that experienced by ships, the hydrodynamic costs may be proportionately as severe (e.g., Lane et al. 2018).

The importance of GBG in shrimps is seen in the low degree of grooming morphology and behavior in non-natant decapods such as lobsters and crabs. Earlier authors noted anecdotally that severe macroscopic fouling is relatively rare in penaeoid shrimps, lobsters, and crabs (Dawson 1957; Eldred 1962). Schöne (1961) noted the high frequency of GBG grooming in *Palaemon* (*s.l.*) species while rare in crabs. I found that the crayfish *Procambarus clarkii* groomed the body with its minor chelipeds, but the behavior was only occasional and likely ineffective when fouling intensities were high (Bauer 2002). The macruran decapods, in which swimming is clumsy or lacking, typically lack specialized GBG grooming appendages and behavior. As pointed out in Chap. 1 (Fig. 1.3), the hydrodynamic shape and swimming ability decline from high in the natant decapods to very low in the true crabs. The tendency is not completely linear as the variably shaped anomurans, although usually highly benthic and non-natant, often have a chelate grooming appendage (pereopod 5) with which they clean the body and gills (Bauer 1981, 1989; Pohle 1989). Although observations of frequent GBG are rare in penaeoids, the burrowing habits of many might be an alternate behavior which controls fouling (along with its antipredator function). It would be of considerable interest to measure the actual hydrodynamic costs experienced by naturally or experimentally fouled shrimps.

4.5 Embryo Care

With exception of the luciferids, females of penaeoid and sergestoid shrimps spawn fertilized eggs into the water and let them fend for themselves during their embryonic development. On the other hand, females of caridean and stenopodidean shrimps, like other pleocyemate decapods, incubate fertilized eggs attached to their pleopods beneath the abdomen. The female cares for the embryos until they hatch. Water needs to be circulated through the mass of developing embryos to supply them with oxygen and removed carbon dioxide and other metabolic wastes. To accomplish this task, the female periodically raises her abdomen and briskly beats the pleopods (“fanning behavior”) (Phillips 1971; Bauer 1975, 1979; Baeza et al. 2019). However, this behavior is not sufficient care by itself. Fanning can be simulated in the laboratory by bubbling and circulating water via aquarium aerators among early-stage embryos removed from a female. The embryos seem to do well initially, but eventually will become covered with microfoulers such as sessile protozoa and filamentous and other bacteria. Fungi may attack them as well. Few embryos suffering such fouling make it through to hatching (Phillips 1971; Bauer 1979; Fisher 1983a, 1983b). The embryos may suffer simple suffocation by the sediment and heavy bacterial coating over the embryo’s outer membrane. Chitinolytic bacteria may break down the surface membrane, which is catastrophic for the embryo. Dead embryos soon attract saprophagous protozoa and nematode worms, i.e., carrion feeders. Presterilization of culture water with ultraviolet light, and treatment with antibacterial compounds such as low concentrations of copper sulfate, malachite green, and methylene blue can greatly reduce embryo death (e.g., Phillips 1971; Fisher 1976; Ituarte et al. 2005).

Besides fanning behavior, females of many caridean species brush and clean their embryos with the grooming chelipeds (Bauer 1979; Goy 2010; Baeza et al. 2019). In families which primarily use the last walking legs in grooming also use them to clean the embryos (Atyidae, Xiphocarididae), as in the atyids *Caridina nilotica* and *C. africana* (Fryer 1960). Females may groom embryos for long periods of time, brushing up and down vigorously within the embryo mass, snapping the chelae open and shut to bring the tufts of cleaning setae into play, nipping and picking at

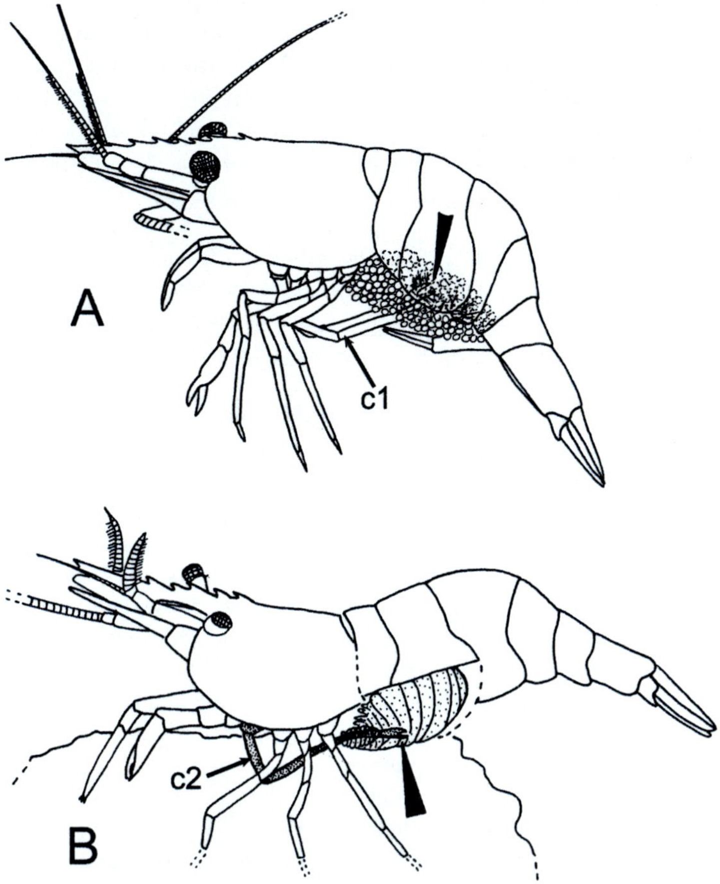

Fig. 4.19 Brooding behavior (embryo grooming). (**a**) *Palaemon ritteri* (Palaemonidae) cleaning and brushing (arrow) brooded embryos with its grooming chelipeds. (**b**) *Heptacarpus paludicola* (Thoridae) grooming (arrow) an abdominal isopod parasite, *Hemiarthrus abdominalis* (Epicaridea, Bopyridae) which may be mimicking a brood of embryos. *c1, c2*, chelipeds 1,2. From (**a**) Bauer (1976) and (**b**) Bauer (1989)

the surface of embryos (Fig. 4.19a). What does all this grooming accomplish? To test the hypothesis that grooming cleans the embryos and reduces mortality, I performed the simple experiment of removing the grooming chelipeds (experimentals) or first walking legs (controls) from *H. sitchensis* females carrying early-stage embryos (Bauer 1979). Females in cages were immersed in water flowing in from the natural environment until embryos were nearing hatching, when females were sacrificed (mercifully, by chilling). I then examined the embryo masses for fouling and embryo death. I did replicates of the experiment in winter, spring, and summer to examine the effects of temperature on grooming and embryo mortality.

Examination of embryo masses of experimentals showed noticeable accumulations of sediment, while controls had little or none. Individual embryos of experimentals were heavily fouled with sessile protozoans, filamentous and nonfilamentous bacteria. Dead embryos could be recognized by a change of color and obvious decomposition. Saprophagous nematodes and ciliates could be seen crawling in an around the dead embryos. In controls, sediment and microbial fouling was light, when it occurred. Dead or decaying embryos were never seen in control broods. In the experimental broods, mortality was negligible in the cool water temperatures experienced by shrimps in the winter but was significantly higher in the warmer waters of the spring and summer replicates (Fig. 4.18b). Accelerated microbial fouling and increased metabolic demands of embryos at higher water temperatures were the probable cause of increased embryo mortalities in the spring and summer experiments.

Females of anomuran crabs and some thalassanideans (e.g., ghost shrimps) groom their embryos with specialized cleaning limbs, the chelate fifth pereopods, which also clean their gills, (e.g., Pohle 1989; Batang and Suzuki 2003). A fellow graduate student and lab mate, Larry Ritchie, worked on grooming and parasitism in one such anomuran, a porcellanid crab (Ritchie and Hoeg 1981). When the grooming limbs of females were removed in experiments unrelated to embryo survival, he found that their embryo masses suffered considerable fouling and mortality. Similar results were reported for porcellanid crab and lithodid crabs (Anomura) by Förster and Baeza (2001) and (Pohle 1989), respectively. Other decapods such as brachyuran ("true") crabs do not have specialized grooming chelipeds or other GBG/gill grooming appendages. Heavy mortality of embryos has been reported in crabs due to epibiotic fouling (e.g., Fisher 1976; Fisher and Wickham 1976). The embryos of decapods such as lobsters and brachyuran crabs are often attacked by tiny predatory nemertean worms (*Carcinonemertes*) which can cause severe mortality in broods. These predators have never been reported in carideans and stenopodideans, and rarely in anomurans (Wickham and Kuris 1985), which shows the efficacy and adaptive advantage of embryo grooming.

4.6 Behavioral Aspects of Grooming

In some species of carideans and stenopodideans, grooming behaviors take up a considerable portion of an individual's time budget. Little information on penaeoids and sergestoids has been reported on this aspect of their behavior. However, the lack of reports on grooming in these groups give the impression that, aside from grooming of antennae and anterior pereopods, grooming does not occupy them much. Little time appears to be devoted to GBG and none to gill cleaning, as the latter is done by a passive mechanism, the setiferous epipods (penaeoids) or not at all (sergestoids). In a few carideans, actual measures of the frequency and duration of various cleaning behaviors have been done (Bauer 1977; Palomar et al. 2005; Van Maurik and Wortham 2011, 2014). In *Heptacarpus sitchensis*, grooming occupies about 27% of its time budget. Palomar and colleagues found that *Alpheus macellarius* spent 37% of its time in grooming plus foraging, while Van Maurik and Wortham (2011, 2014) reported 27% and 19% time spent in cleaning behavior alone in *Macrobrachium grandimanus and M. rosenbergii,* respectively.

However, other more sedentary species appear to groom much less, which seems to be related to overall activity. For example, little GBG was observed from the relatively inactive filter feeder *Atya innocuous*, while a much more lively shrimp from the same streams, *Xiphocaris elongata,* engages in considerable GBG (Fryer 1977). One obvious explanation is that more active animals perform all behaviors more frequently than do more sluggish ones. On the other hand, mobile, lively shrimps encounter more diverse environmental situations more often than sedentary ones. This may expose them to more fouling, and their bodies and sensory sites have to remain cleaner to function efficiently. This is a hypothesis that needs to more thoroughly analyzed and tested.

Grooming behaviors such as antennular, antennal cleaning, and maxillipedal autogrooming are relatively stereotyped, frequent, and of short duration (Bauer 1977; Van Maurik and Wortham 2011, 2014). Bouts of antennular and antennal grooming are frequent but consist of brief movements than take from less than up to a few seconds. Mx3 autogrooming normally follows antennular, antennal, and pereopodal grooming and therefore is frequent as well. There is little variability in how these grooming movements are made. Antennules and other sensory appendages must remain constantly clean and capable of receiving stimuli so that the animal is consistently informed about its surroundings. However, GBG, gill and embryo cleaning are quite variable in form—a cleaning chela moves from here to there, nipping, picking, or brushing, changing its posture frequently. GBG and embryo cleaning by the fifth pereopods are similarly variable. This behavioral flexibility is necessary to deal with the greater variability, complexity, and area of body parts groomed. Bouts of these behaviors occur much less frequently but when they do happen, they are of greater duration, often several minutes. The shrimp can suffer a bit of short-term fouling as that will not greatly interfere with locomotion, gill function, or embryo care. In the long term, however, that fouling will have to be removed so that it does not develop further, i.e. settlement but not growth of a sessile organism, accumulation of sediment and debris. In the short term, this fouling can be endured until the next bout of grooming.

What stimulates a specific bout of cleaning behavior? The only grooming behavior that can be reliably stimulated is antennular grooming. Soluble compounds from food and conspecifics, especially receptive females, as well as novel environmental situations (new surroundings), can stimulate the rate of "sniffing" (antennular flicking, spinning). The accelerated rate of sniffing is invariably followed by increased antennular grooming by the third maxillipeds (e.g., Snow 1973; Monteclaro et al. 2010; pers. obs.). Stimuli which trigger other types of cleaning behavior are unknown. I have attempted to evoke general and gill cleaning by pipetting different materials on or around different species of shrimps, but to no avail. However, Ritchie and Hoeg (1981) found that rhizocephalan parasite larvae initially infect their anomuran crab hosts by attaching to their gills. When parasite larvae were introduced into a chamber with their porcellanid crab host, the crab soon vigorously cleaned their gills. Systematic experiments on stimuli which elicit grooming need to be done.

Grooming behaviors in general seem to be a low priority or "secondary" behavior which occurs when higher priority behaviors take place, such as feeding, mating, or other social interaction and escape from predators (Bauer 1989; Van Maurik and Wortham 2014). Perhaps the frequency and duration of most grooming behaviors are genetically programmed into the nervous system (Bauer 1989). Grooming may occur at fixed intervals and rates when not inhibited by higher priority behaviors.

Parasitism may affect grooming behavior. For example, McGrew and Hultgren (2011) showed that individuals of *Synalpheus elizabethae* (Alpheidae) infested by a bopyrid isopod parasite groomed about 80% less than did uninfested ones. A parasite may affect the behavior of its host so that shrimp does not attack and clean it off the body. I have observed thorid and palaemonid shrimps grooming large branchial or abdominal parasites (bopyrid isopods) that were infecting them (Fig. 4.19b). Perhaps the parasite mechanically or more likely physiologically "convinces" the host that it is part of the host's own body. For example, a large abdominal bopyrid seems to mimic a brood of embryos, which caridean and stenopodidean females groom and clean. When the bopyrid parasite dies, the shrimp perceives it as a foreign object and eventually picks and cleans it off the body. Similarly, the porcellanid crab *Petrolisthes cabrilloi* grooms the large female rhizocephalan parasite which resides on its abdomen (Ritchie and Hoeg 1981). In fact, the parasite cannot molt or continue breeding without grooming care from its host!. The relationships between parasite rejection or acceptance and grooming need to be explored both behaviorally and physiologically.

4.7 Other Antifouling Adaptations

There are many shrimp species which, although active, do not seem to engage in grooming behaviors other than antennular. Some shrimp taxa do not have appendages particularly specialized for general body grooming. Yet their body surfaces remain clean. In part, this may result from lack of sufficient observation or unnatural viewing conditions. However, if some species do not groom much or at all, what keeps them from being fouled? As shown in Fig. 4.1, there are alternate mechanisms that may be at work. Molting has already been mentioned; it completely replaces the old exoskeleton with a new clean one. But as we have seen in experiments, significant microfouling can build up in a short time. Female shrimps cannot molt while they are incubating embryos. Periods between molts may be long in slow-growing shrimps or during cold-water periods in temperate and high latitude species. Yet many of these species seem to remain free of body fouling.

Various causes for lack of fouling have been suggested for shrimps and other decapods (Bauer 1981, 2013; Becker and Wahl 1996). There is little evidence of secretion of antifouling compounds onto the surface of the cuticle via tegumental glands. Evidence for properties of cuticular surface or microtopography which might discourage larval settlement of both microbial and larger fouling organisms is scanty (Bers and Wahl 2004). However, there are more promising possibilities, especially behavioral strategies. A very simple tactic to deter algal fouling is to be active and exposed only at night. Glynn (1970) studied tropical sphaeromatid isopods that hide under rocks and within crevices during the day. He deprived one group of these shelters in the laboratory while others had access. The unprotected individuals quickly developed severe algal fouling which impaired their ability to move or feed, while those hiding in shelters during the day did not. Many, if not most, shrimps which live in lighted shallow-water habitats are nocturnal, which protects them from visually hunting predators. Unlighted habitats, such as the deep-sea or subterranean habitats, are in little danger of algal fouling.

Reports of specialized GBG morphology or behaviors are rare in penaeids or other penaeoids (Bauer 1981; Dall et al. 1990). Many of the benthic ones burrow shallowly into soft (sand, mud) bottoms during the day. The abrasion of sediment particles on the exoskeleton may keep epibiotic fouling low. Ólafsdóttir and Svavarsson (2002) found that protist fouling on deep-sea assellote isopod species was much higher on nonburrowing than burrowing species. Fouling pressures of some environments may be much lower than others. For example, strong currents may not allow settlement of larvae on hard surfaces like the crustacean cuticle (Wolff 1959). Coastal habitats, with their greater diversity and abundance of sessile organisms, may have higher fouling pressures than open-ocean and deep-sea environments. Most of the possibilities are hypotheses that need to be much better explored by systematic observations and experiments.

References

Abele LG, Martin JW (1989) American species of the deep-sea shrimp genus *Bythocaris* (crustacea, Decapoda, Hippolytidae). Bull Mar Sci 45:26–51

Alexander CG, Hindley JPR, Jones SG (1980) Structure and function of the third maxillipeds of the banana prawn, *Penaeus merguiensis*. Mar Biol 58:245–249

Alvarez F, Villalobos JL, Iliffe TM (2004) A new species of *Agostocaris* (Caridea: Agostocarididae) from Acklins Island, Bahamas. Proc Biol Soc Wash 117:382–390

Alvarez F, Iliffe TM, Villalobos JL (2006) Macromaxillocarididae, a new family of stenopodidean shrimp from an anchialine cave in the Bahamas, with the description of *Macromaxillocaris bahamaensis*, n. gen., n. sp. J Crustac Biol 26:366–378

Baeza JA, Liu X, Kostecka L, Wortham J (2019) Active parental care in the peppermint shrimp *Lysmata boggessi*: the effect of embryo age and circadian cycle. Mar Biol 166:132

Batang ZB, Suzuki H (2003) Gill cleaning mechanisms of the burrowing thalassinidean shrimps *Nihonotryphaea japonica* and *Upogebia major* (crustacea: Decapoda). J Zool 261:69–77

Bauer RT (1975) Grooming behaviour and morphology of the caridean shrimp *Pandalus danae* Stimpson (Decapoda: Natantia: Pandalidae). Zool J Linnean Soc 56:45–71

Bauer RT (1976) Antifouling adaptations of caridean shrimp: grooming morphology and behavior. Doctoral dissertation, Scripps Institution of Oceanography, University of California, San Diego, p 269

Bauer RT (1977) Antifouling adaptations of marine shrimp (crustacea: Decapoda: Caridea): functional morphology and adaptive significance of antennular preening by the third maxillipeds. Mar Biol 40:261–276

Bauer RT (1978) Antifouling adaptations of caridean shrimp: cleaning of the antennal flagellum and general body grooming. Mar Biol 49:69–82

Bauer RT (1979) Antifouling adaptations of marine shrimp (Decapoda: Caridea): gill cleaning mechanisms and grooming of brooded embryos. Zool J Linnean Soc 65:281–303

Bauer RT (1981) Grooming behavior and morphology in the decapod crustacea. J Crustac Biol 1: 153–173

Bauer RT (1984) Morphological trends in the genus *Heptacarpus* (Decapoda: Caridea) and their phylogenetic significance. J Crustac Biol 4:201–225

Bauer RT (1989) Decapod crustacean grooming: functional morphology, adaptive value, and phylogenetic significance. In: Felgenhauer BE, Watling L, Thistle AB (eds) Functional morphology of feeding and grooming in crustacea. Balkema, Rotterdam, pp 49–73

Bauer RT (1998) Gill-cleaning mechanisms of the crayfish *Procambarus clarkii* (Astacidea: Cambaridae): experimental testing of setobranch function. Invertebr Biol 177:129–143

Bauer RT (1999) Gill-cleaning mechanisms of a dendrobranchiate shrimp, *Rimapenaeus similis* (Decapoda: Penaeidae): description and experimental testing of function. J Morphol 242:125–139

Bauer RT (2002) The ineffectiveness of grooming in prevention of body fouling in the red swamp crawfish *Procambarus clarkii*. Aquaculture 208:39–49

Bauer RT (2004) Remarkable shrimps: adaptations and natural history of the carideans. University of Oklahoma Press, Norman

Bauer RT (2011) Chemical communication in decapod shrimps: the influence of mating and social systems on the relative importance of olfactory and contact pheromones. In: Breithaupt T, Thiel M (eds) Chemical communication in crustaceans. Springer Verlag, New York, pp 277–296

Bauer RT (2013) Adaptive modification of appendages for grooming (cleaning; antifouling) and reproduction in the crustacea. In: Thiel M, Watling L (eds) Functional morphology of crustacea. Oxford University Press, Oxford, England, pp 337–375

Becker K, Wahl M (1996) Behaviour patterns as natural antifouling mechanisms of tropical marine crabs. J Exp Mar Biol Ecol 203:245–258

Berkeley AA (1928) The musculature of *Pandalus danae*. Trans R Can Inst 16:181–231

Bers AV, Wahl M (2004) The influence of natural surface microtopographies on fouling. Biofouling 20:43–51

Breithaupt T, Thiel M (eds.) (2011) Chemical communication in crustaceans. In: Chemical communication in crustaceans. Springer Verlag, New York, pp 277–296

Brusca RC, Moore W, Shuster SM (2016) Invertebrates, 3rd edn. Sinauer Associates, Inc., Sunderland, Massachusetts

Burkenroad MD (1934) The Penaeidea of Louisiana with a discussion of their world relationships. Bull Am Mus Nat Hist 68:61–143

Butler TH (1980) Shrimps of the Pacific coast of Canada. Can Bull Fish Aquat Sci 202:1–280

Calman WT (1909) On a blind prawn from the sea of Galilee (*Typhlocaris Galilea* g. et sp. n.). Trans Linnean Soc London, Zool 11:93–97

Chace FA Jr (1975) Cave shrimps (Decapoda: Caridea) from the Dominican Republic. Proc Biol Soc Wash 88:29–44

Chace FA Jr (1976) Shrimps of the pasiphaeid genus *Leptochela* with descriptions of three new species (Crustacea: Decapoda: Caridea). Smithson Contrib Zool 222:51

Chace FA Jr (1984) The caridean shrimps (crustacea: Decapoda) of the albatross Philippine expedition, 1907-1910, part 2: families Glyphocrangonidae and Crangonidae. Smithson Contrib Zool 397:63

Chace FA Jr (1985) The caridean shrimps (crustacea: Decapoda) of the albatross Philippine expedition, 1907-1910, part 3: families Thalassocarididae and Pandalidae. Smithson Contrib Zool 411:143

Chace FA Jr (1986) The caridean shrimps (crustacea: Decapoda) of the albatross Philippine expedition, 1907-1910, part 4: families Oplophoridae and Nematocarcinidae. Smithson Contrib Zool 432:82

Chace FA Jr (1988) The caridean shrimps (crustacea: Decapoda) of the albatross Philippine expedition, 1907-1910, part 5: family Alpheidae. Smithson Contrib Zool 466:99

Chace FA Jr (1997) The caridean shrimps (crustacea: Decapoda) of the albatross Philippine expedition, 1907-1910, part 7: families Atyidae, Eugonatonotidae, Rhynchocinetidae, Bathypalaemonellidae, Processidae, and Hippolytidae. Smithson Contrib Zool 587:106

Chace FA Jr, Brown DE (1978) A new polychelate shrimp from the great barrier reef of Australia and its bearing on the family Bresiliidae (crustacea: Decapoda: Caridea). Proc Biol Soc Wash 91:756–766

Chace FA Jr, Bruce AC (1993) The caridean shrimps (crustacea: Decapoda) of the albatross Philippine expedition, 1907-1910, part 6: superfamily Palaemonoidea. Smithson Contrib Zool 543:152

Chace FA Jr, Holthuis LB (1978) *Psalidopus*: the scissor-foot shrimps (crustacea: Decapoda: Caridea). Smithson Contrib Zool 277:22

Chace FA Jr, Manning RB (1972) Two new caridean shrimps, one representing a new family, from marine pools on Ascension Island (crustacea: Decapoda: Natantia). Smithson Contrib Zool 131: 18

Cleva R (1990) Crustacea Decapoda: Les genres et les espèces indo-ouest pacifiques de Stylodactylidae. In: Crosnier, A. (ed.) Résultats des Campagnes MUSORSTOM, Volume 6. Mémoires du Muséum national d'Histoire naturelle, Paris. Nouvelle Série. Série A, Zoologie 145:71–136

Cleva R (2004) Stylodactylidae and Bathypalaemonellidae from Taiwan (crustacea: Decapoda: Caridea). Raffles Bull Zool 52:497–511

Cowles DL (1994) Swimming dynamics of the mesopelagic vertically migrating penaeid shrimp *Sergestes similis*: modes and speed of swimming. J Crustac Biol 14:247–257

d'Udekem d'Acoz C (1996) The genus *Hippolyte* leach, 1814 (crustacea: Decapoda: Caridea: Hippolytidae) in the East Atlantic Ocean and the Mediterranean Sea, with a checklist of all species in the genus. Zoologische Verhandelingen, Leiden 303:1–133

Dall W, Hill BC, Rothlisberg PC, Sharples DJ (1990) The biology of the Penaeidae. Adv Mar Biol 27:1–489

Dawson CE (1957) *Balanus* fouling of shrimp. Science 126:1068

De Grave S, Goulding LYD (2011) Comparative morphology of the pereiopod 1 carpo-propodal (P1-CP) antennal flagellar grooming brush in caridean shrimps (crustacea, Decapoda). Zool Anz 250:280–301

De Grave S, Arjun CP, Raghavan R (2018) The discovery of Euryrhynchidae (crustacea: Decapoda) in India, with the description of a new genus and species. Zootaxa 4462:367–378

Dounas C, Koukouras A (1989) *Odontozona minoica*, new species, from the eastern Mediterranean Sea (Decapoda: Stenopodidea). J Crustac Biol 9:341–348

d'Udekem d'Acoz C, De Grave S (2018) A new genus and species of large-bodied caridean shrimp from the Crozet Islands, Southern Ocean (crustacea, Decapoda, Lipkiidae) with a checklist of Antarctic and sub-Antarctic shrimps. Zootaxa 4392:201–240

Eldred B (1962) The attachment of the barnacle, *Balanus amphitrite niveus* Darwin, and other fouling organisms to the rock shrimp, *Sicyonia dorsalis* Kingsley. Crustaceana 3:203–206

Felgenhauer BE, Abele LG (1983) Ultrastructure and functional morphology of feeding and associated appendages in the tropical freshwater shrimp *Atya innocous* (Herbst) with notes on its ecology. J Crust Biol 3:336–363

Felgenhauer BE, Schram FR (1979) The functional morphology of the grooming appendages of *Palaemonetes kadiakensis* Rathbun 1902. Fieldiana Zool New Series 2:1–17

Fisher WS (1976) Relationships of epibiotic fouling and mortalities of eggs of the Dungeness crab *(Cancer magister)*. J Res Board Can 33:2849–2853

Fisher WS (1983a) Eggs of *Palaemon macrodactylus*. II. Association with aquatic bacteria. Biol Bull 164:201–213

Fisher WS (1983b) Eggs of *Palaemon macrodactylus*. III. Infection by the fungus, *Lagenidium callinectes*. Biol Bull 164:214–226

Fisher WS, Wickham DE (1976) Egg mortalities in wild populations of the Dungeness crab in central and North California. United States Marine Fishery Service 75:235–237

Förster C, Baeza JA (2001) Active brood care in the anomuran crab *Petrolisthes violaceus* (Decapoda: Anomura: Porcellanidae): grooming of brooded embryos by the fifth pereiopods. J Crustac Biol 21:606–615

Foxton P, Herring PJ (1970) Recent records of *Physetocaris microphthalma* Chace with notes on the male and description of the early larvae (Decapoda, Caridea). Crustaceana 18:93–104

Fryer G (1960) The feeding mechanism of some atyid prawns of the genus *Caridina*. Trans R Soc Edinb 64:217–244

Fryer G (1977) Studies of the functional morphology and ecology of the atyid prawns of Dominica. Philos Trans R Soc B: Biol Sci 277:57–129

Giri T, Wicksten MK (2001) Fouling of the caridean shrimp, *Lysmata wurdemanni* (Gibbes, 1850) by the barnacle, *Balanus improvisus* Darwin, 1854 and other epibionts. Crustaceana 74:1305–1314

Glynn PW (1970) Growth of algal epiphytes on a tropical marine isopod. J Exp Mar Biol Ecol 5:88–93

Goy JW (2010) Infraorder Stenopodidea Claus, 1872. In: Schram FR, von Vaupel Klein JC, Forest J, Charmantier-Daures M (eds) Treatise on Zoology – Anatomy, Taxonomy, Biology. The Crustacea, Decapoda.Vol. 9. Part A: Eucarida: Euphausiacea, Amphionidacea, and Decapoda (partim). Koninklijke Brill NV, Leiden, pp 215–265

Hackmann A, Delacave H, Robinson A, Labonte D, Federle W (2015) Functional morphology and efficiency of the antenna cleaner in *Camponotus rufifemur* ants. R Soc Open Sci 2:150129. https://doi.org/10.1098/rsos.150129

Hanamura Y, Masatsune Takeda T (1996) Establishment of a new genus *Bathystylodactylus* (crustacea: Decapoda: Stylodactylidae), with description of a new species from northwestern Pacific. Zool Sci 13:929–934

Hart CW Jr, Manning RB (1986) Two new shrimps (Procarididae and Agostocarididae, new family) from marine caves of the western North Atlantic. J Crustac Biol 6:408–416

Hindley JPR, Alexander CG (1978) Structure and function of the chelate pereiopods of the banana prawn, *Penaeus merguiensis*. Mar Biol 48:153–160

Hlavac TF (1975) Grooming systems of insects: structure, mechanics. Ann Entomol Soc Am 68: 823–826

Hobbs HH Jr, Hart CW Jr (1982) The shrimp genus *Atya* (Decapoda: Atyidae). Smithson Contrib Zool 364:143

Holthuis LB (1971) The Atlantic shrimps of the deep-sea genus *Glyphocrangon*. A. Milne Edwards, 1881. Bull Mar Sci 21:267–373

Huxley TH (1880) The crayfish. C. Kegan Paul and Co., London

Hyman LH (1940) The invertebrates: protozoa through Ctenophora. McGraw-Hill, New York

Hyman LH (1955) The invertebrates: Echinodermata. McGraw-Hill, New York

Ituarte RB, Spivak ED, Anger K (2005) Effects of salinity on embryonic development of *Palaemonetes argentinus* (crustacea: Decapoda: Palaemonidae) cultured *in vitro*. Invertebr Reprod Dev 47:213–223

Jander U (1966) Untersuchungen zur stammesgeschichte von putzbewegungen von Tracheaten. Z Tierpsychol 23:799–844

Judkins DC (1978) Pelagic shrimps of the *Sergestes edwardsii* species group (crustacea: Decapoda: Sergestidae). Smithson Contrib Zool 56:34

Kensley B (1983) New records of bresiliid shrimp from Australia, South Africa, Caribbean, and Gulf of Mexico (Decapoda: Natantia: Caridea). Smithson Contrib Zool 394:1–31

Kensley B, Williams D (1986) New shrimps (families Procarididae and Atyidae) from a submerged lava tube on Hawaii. J Crustac Biol 6:417–437

Koehl MAR (2011) Hydrodynamics of sniffing by crustaceans. In: Breithaupt T, Thiel M (eds) Chemical communication in crustaceans. Springer, New York, pp 85–102

Komai T (1999) A revision of the genus *Pandalus* (crustacea: Decapoda: Caridea: Pandalidae). J Nat Hist 33:1265–1372

Komai T, Kohtsuka H (2017) A new deep-sea species of the caridean genus *Bresilia* Calman, 1896 (crustacea: Decapoda: Bresiliidae) from Sagami Bay, Central Japan. Zootaxa 4299:405–414

Lane SJ, Tobalske BW, Moran AL, Caitlin CM, Shishido M, Woods HA (2018) Costs of epibionts on Antarctic Sea spiders. Mar Biol 165:137. https://doi.org/10.1007/s00227-018-3389-9

Manning RB, Chace FA Jr (1971) Shrimps of the family Processidae from the northwestern Atlantic Ocean (crustacea: Decapoda: Caridea). Smithson Contrib Zool 89:41

Manning RB, Hart CW Jr (1984) The status of the hippolytid shrimp genera *Barbouria* and *Ligur* (crustacea: Decapoda): a reevaluation. Proc Biol Soc Wash 97:655–665

Martin JW, Signorovitch J, Patel H (1998) Comparison of the carpal cleaning brush in two genera of hydrothermal vent shrimp (crustacea, Decapoda, Bresiliidae). J Morphol 235:31–39

Mayzel B, Haber M, Ilan M (2014) Chemical defense against fouling in the solitary ascidian *Phallusia nigra*. Biol Bull 227:232–241

McGrew M, Hultgren KM (2011) Bopyrid parasite infestation affects activity levels and morphology of the eusocial snapping shrimp *Synalpheus elizabethae*. Mar Ecol Prog Ser 431:195–204

McVean A (1982) Autotomy. In: Sandeman DC, Atwood HL (eds) Neural integration and behavior. The biology of crustacea, vol 4. Academic Press, New York, pp 107–132

Mejía-Ortíz LM, Yañez G, López-Mejía M (2017) Anchialocarididae, a new family of anchialine decapods and a new species of the genus *Agostocaris* from Cozumel Island, Mexico. Crustaceana 90:381–398

Monteclaro HM, Anraku K, Matsuoka T (2010) Response properties of crayfish antennules to hydrodynamic stimuli: functional differences in the lateral and medial flagella. J Exp Biol 213: 3683–3691

Okuno J (1996) *Cinetorhynchus manningi*, a new shrimp (crustacea: Decapoda: Caridea: Rhynchocinetidae) from the western Atlantic. Proc Biol Soc Wash 109:725–730

Okuno J (1997) A new shrimp of the genus *Rhynchocinetes* from the great Australian bight (crustacea: Decapoda: Rhynchocinetidae). Rec South Aust Museum 30:13–18

Ólafsdóttir SH, Svavarsson J (2002) Ciliate (protozoa) epibionts of deep-water asellote isopods (crustacea): pattern and diversity. J Crustac Biol 22:607–618

Palomar NE, Juinio-Meñez MA, Karplus I (2005) Behavior of the burrowing shrimp *Alpheus macellarius* in varying gravel substrate conditions. J Ethol 23:173–180

Pearson WH, Olla BL (1977) Chemoreception in the blue crab, *Callinectes sapidus*. Biol Bull 153: 346–354

Pérez Farfante I, Kensley B (1997) Penaeoid and sergestoid shrimps and prawns of the world. Keys and diagnoses for the families and genera. Mémoires du Muséum National d'Histoire Naturelle 175:1–233

Phillips C (1971) Incubation of the English prawn *Palaemon serratus*. J Mar Biol Assoc U K 5:43–48

Pohle G (1989) Gill and embryo grooming in lithodid crabs: comparative morphology based on *Lithodes maja*. In: Felgenhauer BE, Watling L, Thistle AB (eds) Functional morphology of feeding and grooming in crustacea. Balkema, Rotterdam, pp 75–94

Powell CB (1977) A revision of the African freshwater shrimp genus *Desmocaris* Sollaud, with ecological notes and description of a new species (Crustacea, Decapoda, Palaemonidae). Revue Zoologique Africaine 91:649–674

Read AT, McTeague JA, Govind CK (1991) Morphology and behavior of an unusually flexible limb in the snapping shrimp *Alpheus heterochelis*. Biol Bull 118:158–168

Ritchie L, Hoeg J (1981) The life history of *Lernaeodiscus porcellanae* (Cirripedia: Rhizocephala) and coevolution with its porcellanid host. J. Crust. Biol. 1:334–347

Ryland JS (1970) Bryozoans. Hutchinson, London

Sainte-Marie B, Bérubé I, Brillon S, Hazel F (2006) Observations on the growth of the sculptured shrimp *Sclerocrangon boreas* (Decapoda: Caridea). J Crustac Biol 26:55–62

Saito T, Komai T (2008) A review of species of the genera *Spongicola* de Haan, 1844 and *Paraspongicola* de Saint Laurent & Cleva, 1981 (crustacea, Decapoda, Stenopodidea, Spongicolidae). Zoosystema 30:87–147

Schöne H (1961) Complex behavior. In: Waterman TH (ed) The physiology of crustacea, 2nd edn. Academic Press, New York, pp 465–520

Snow PJ (1973) The antennular activities of the hermit crab *Pagurus alaskensis* (Benedict). J Exp Biol 58:745–765

Sokolov V (2000) Deep-sea shrimps of the genus *Bythocaris* G.O. Sars in the collections of Russian museums, with the description of a new species (crustacea: Decapoda: Hippolytidae). Zoologische Mededelingen 74:403–468

Suzuki H, McLay CL (1998) Gill-cleaning mechanisms of *Paratya curvirostris* (Caridea: Atyidae) and comparisons with seven species of Japanese atyid shrimps. J Crustac Biol 18:253–270

Thompson JR (1967) Comments on phylogeny of section Caridea (Decapoda Natantia) and the phylogenetic importance of the Oplophoroidea. Proceedings Symposium on Crustacea, Ernakulam, Marine Biological Association India 1:314–326

Torti MR, Boschi EE (1973) Nuevos aportos al conocimiento de los crustáceos Decápodos Caridea del género *Campylonotus* Bate 1988. Physis Sección A 32:65–84

Van Maurik LN, Wortham JL (2011) The grooming behaviors of the Hawaiian river shrimp, *Macrobrachium grandimanus*. J Crustac Biol 31:617–622

Van Maurik LN, Wortham JL (2014) Grooming as a secondary behavior in the shrimp *Macrobrachium rosenbergii* (crustacea, Decapoda, Caridea). ZooKeys 457:55–77

Vereshchaka AL, Kulagin DN, Lunina AA (2015) Phylogeny and new classification of hydrothermal vent and seep shrimps of the family Alvinocarididae (Decapoda). PLoS One 10(7): e0129975. https://doi.org/10.1371/journal.pone.0129975

Vereshchaka A, Kulagin D, Lunina AA (2021) New shrimp genus (Crustacea: Decapoda) from the deep Atlantic and an unusual cleaning mechanism of pelagic decapods. Diversity 13(11):536. https://doi.org/10.3390/d13110536

Vuillemin S (1967) La respiration chez crustacés décapodes. Année Biologique 6:47–82

Wahl M (1989) Marine epibiosis. I. Fouling and antifouling: some basic aspects. Mar Ecol Prog Ser 8:175–189

Wickham DE, Kuris AM (1985) The comparative ecology of nemertean egg predators. Am Zool 25:127–134

Wickler W, Seibt U (1970) Das Verhalten von *Hymenocera picta* Dana, einer Seesterne fressenden garnele (Decapod, Natantia, Gnathophyllidae). Z Tierpsychol 27:352–368

Wicksten MK (1989) *Encantada spinoculata*, a new genus and species of shrimp from the Galapagos Islands (Caridea: Bresiliidae). J Crustac Biol 9:667–671

Wilkins JL, Mcmahon BR (1972) Aspects of branchial irrigation in the lobster *Homarus americanus*. I. Functional analysis of scaphognathite beat, water pressures and currents. J Exp Biol 56:469–480

Williams AB, Chace FA Jr (1982) A new caridean shrimp of the family Bresiliidae from thermal vents of the Galapagos rift. J Crust Biol 2:136–147

Williamson DI (1973) *Amphionides reynaudii* (H. Milne Edwards), representative of a proposed new order of Eucaridan malacostraca. Crustaceana 25:35–50

Wilson EO (1971) The insect societies. Belknap Press of Harvard, Cambridge, Massachusetts
Wolff T (1959) Epifauna on certain decapod Crustacea. In: Hewer HR, Riley ND (eds) Proceedings: 15th international congress of zoology, London, 16–23 July 1958. International Congress of Zoology, London, pp 1060–1061
Yaldwyn JC (1960) Crustacea Decapoda Natantia from the Chatham rise; a deep water bottom fauna from New Zealand. N Z Dept Sci Industr Res Bull 139:13–53
Young JH (1959) Morphology of the white shrimp *Penaeus setiferus* (Linnaeus 1758). Fish Bull Fish Wildlife Service 59:1–168

Chapter 5
Coloration, Color Change, and Camouflage

5.1 Adaptive Value of Coloration

Animals such as birds, insects, amphibians, reptiles, and mammals often have bright and interesting colors and patterns. In some cases, the colors serve as signals or stimuli in mate attraction or social interactions among members of the same species. Coloration is a very important part of camouflage in many animals to conceal themselves from visually hunting predators. Shrimps too, especially those in shallower, more lighted regions of the sea, may also have beautiful and varied coloration which functions in camouflage but not in interspecific interactions like cephalopods and fishes. Coloration of shrimps can only be appreciated by viewing living individuals or photographs of recently captured ones because most shrimp color is leached out by the preservatives used in research and museum collections.

Some terrestrial animals (e.g., chameleons) are capable of changing their color very rapidly. Marine animals such as cephalopods (squids and octopi) and fishes are similarly adept in modifying color and color patterns for agonistic or mating displays, as well as for camouflage. Some shrimps are also able to alter or change color patterns in response to changes in their surroundings. Shrimps which inhabit neritic sand- or mud-bottoms have much drabber coloration, while mesopelagic and deep-sea (bathyal) shrimps display colors from pale to bright red. Mesopelagic shrimps and deep-sea shrimps may use bioluminescence to help conceal themselves or as an active defense against predators.

Color patterns are often taxonomically important in the detection of potentially new species. Some closely related sibling species are very similar or indistinguishable morphologically in museum collections but have been recognized initially by their color in life as potentially new species confirmed with molecular characters (Knowlton 1986). Although the investigation of color has been principally a basic research question in the past, more recently it has become of applied (commercial) importance for those involved in the ornamental shrimp trade. Marine aquarists prefer to buy and maintain brightly colored shrimps similar to those seen in the environment, but those kept in captivity often fade in color. Thus, knowledge about the chemical and morphological bases of color, its maintenance, mechanisms of color change, and adaptive value is one of the essential aspects of shrimp biology.

As I mentioned in the previous chapter, the fortuitous occurrence in the rocky intertidal pools near my graduate institution of an interesting and abundant shrimp species, *Heptacarpus sitchensis*, made possible the subject of my dissertation on shrimps. My initial attraction to this species were its color variations, with individuals with beautiful green, pink, white, and red-banded colors. One of my initial hypotheses about the shrimp was that the green color was due to the presence of symbiotic algal cells in their tissues. Another was that the different color forms were different species. These hypotheses were quickly rejected after some research and literature search. However, this led me to some exciting work on coloration that resulted in publications and stimulated work by others. This chapter will show that there is still much to be learned about coloration and its adaptive value in decapod shrimps.

5.2 Types and Sources of Color

A thin or membranous shrimp cuticle (exoskeleton) is transparent, devoid of pigmentation. A thicker and harder cuticle takes on an amber color which become darker with increase in thickness and degree of sclerotization. This latter process is a toughening of the cuticle by a cross-linking of cuticular proteins by a chemical process known as quinone tanning (Ghidalia 1985; Brusca et al. 2016). Although other processes such as light refraction and iridescence may be involved, most true color in shrimps comes from pigments, substances that selectively absorb wavelengths of incident (incoming) light and reflect others. Thus, a red pigment appears red because it absorbs most wavelengths of visible (white) light but reflects only the red wavelengths back to the eye of the perceiver. Substances that absorb all wavelengths, reflecting none, are black and those reflecting all white light, absorbing none, are white. Pigments in shrimps and other crustaceans are deposited within the exoskeleton or in cells below the cuticle termed chromatophores ("color-bearers"). Several chromatophores are usually associated together into a structure called the chromatosome (Ghidalia 1985; McNamara and Milograna 2015). In the older literature, the structure termed "chromatophore" usually refers to what is now more appropriately designated a chromatosome. These are the multirayed colored structures one sees upon close examination of shrimps (Plate 5.1 B-F). The individual chromatophores which compose them are highly branched cells with pigment granules are within. Most chromatophores contain but a single pigment but differently colored chromatophores are often aggregated into a di- or polychromatic chromatosome (Rao 1985; McNamara and Milograna 2015) (Table 5.1, Plate 5.1 B-E). Chromatosomes of a single color may be made up of several chromatophores, each with the same pigment type. Chromatosomes are usually located just below the cuticle in the epidermis (hypodermis) but are also found in the connective tissue

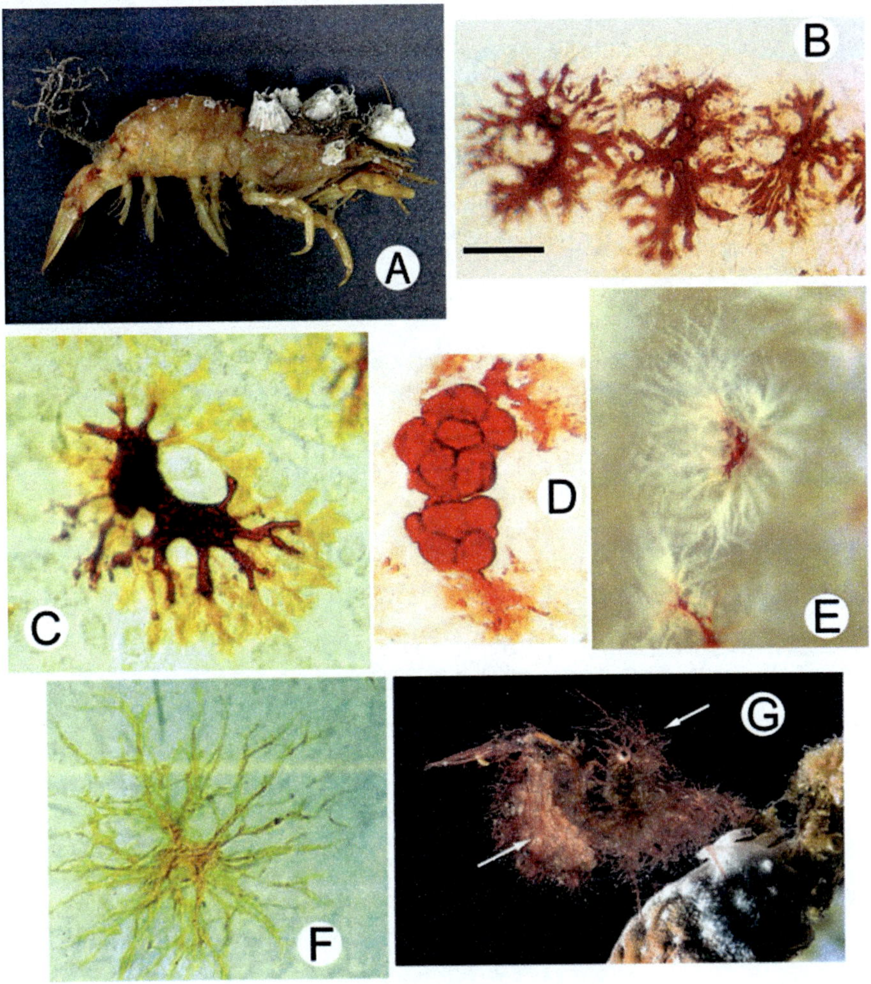

Plate 5.1 Macroscopic fouling; Chromatosomes (A) Severe fouling on *Sclerocrangon boreas*, with large white barnacles on the dorsal cephalothorax and a tuft of algal fouling on the fifth abdominal segment. (B-F) *Heptacarpus sitchensis* chromatosomes: (B) Red-yellow chromatosomes from a disruptive color leg band. (C) High magnification of a single red-yellow chromatosome with red pigment somewhat contracted. (D) Same type of chromatosome with red pigment concentrated. (E) Red-white chromatosome with red pigment concentrated and white reflecting material dispersed. (F) Yellow chromatosome from abdomen of green color morph. (G) *Phycocaris simulans* (Hippolytidae) female showing fascigerous ("hairy") condition, body with many long delicate setae resembling filamentous algae; upper right arrow points to anterior end of the shrimp, lower left to a mass of incubated embryos attached below the female's abdomen. Scale bar represents 38 μm in (B), 21 μm in (C), 11 μm in (D), 47 μm in (E), and 21 μm in (F). Courtesy of (A) Richard Larocque and Bernard Sainte-Marie from Sainte-Marie et al. (2006), (B-F) R.T. Bauer from Bauer (2004), (G) courtesy of Christian Gloor

Table 5.1 Chromatosome basis of color in the thorids *Heptacarpus sitchensis* and *H. paludicola* (from Bauer 1981)

Type of Chromatosome	Pigment Dispersion	Chromatosome Appearance	Chromatosome Density	Color of Body Part or Region
Red-white	Red: Low; white: High	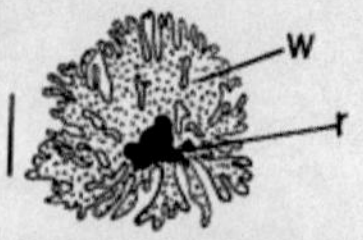	Abundant	White
Red-white	Red: Moderate to high; white: High		Abundant	Pink to magenta
Red	Red: Moderate to high		Moderate to abundant	Reddish brown to black
Yellow	Low to moderate	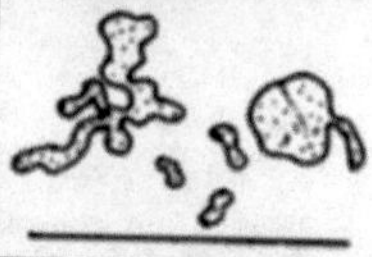	Low to moderate	Transparent to faint yellow
Yellow	High	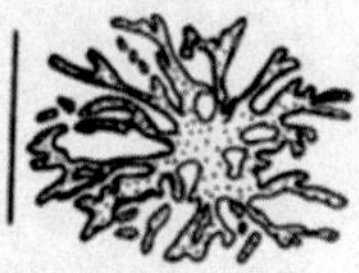	Abundant	Green
Red-yellow	Red: Low; yellow: High	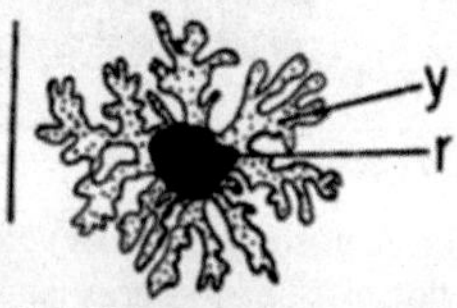	Abundant	Aquamarine
Red-yellow (red dense)	Red: Moderate to high; yellow: High		Abundant	Reddish brown to brown
Red-yellow (red low density)	Red and yellow both moderate to high	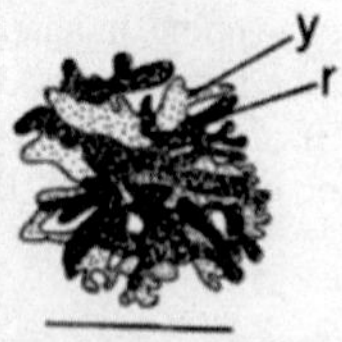	Abundant	Brown

Scale bar = 50 μm. *r*, *y* red and yellow pigments, respectively, *w* white crystalline substance membranes covering organs or tissues, e.g. the stomach, digestive gland (liver or hepatopancreas), hindgut, ventral nerve cord, and muscles. This is particularly true in shrimps with a thin, transparent cuticle. Both chromatosomes just below the cuticle and those deeper within the body may contribute to color. The color of a particular part or region of the shrimp's body depends on the pigment type within the exoskeleton (if any), the density and type of chromatosomes, and the dispersion of the pigments within the chromatosomes (Table 5.1).

The chemical makeup of crustacean pigments are of three main types (Ghidalia 1985; Noël 1994). Xanthophylls, like those coloring autumn leaves which have lost their chlorophyll green, are oxygenated carotenoids which make up the major crustacean pigments. Astaxanthin (bright red) is most common but canthaxanthin (orange-red) is also important. Lutein is another important xanthophyll which yields a yellow color. White color is produced not by pigments but pterins, heterocyclic compounds which are deposited as white crystals or beads. Ommochromes are a generally brown to black metabolite of the amino acid tryptophan. They are important as a screening pigment around the individual visual units of the compound eye, and thus are responsible for the black eye color of most crustaceans (Hallberg and Elofsson 1989). However, they are the pigments in black chromatophores in crangonids such as *Crangon crangon* (Noël 1994). Melanin, derived from the oxygenation of the amino acid tyrosine, gives rise to the dark (brown, black) pigmentation of many animals. However, it is not found in crustacean coloration per se; when present, it is associated with pathological conditions such as infections and wounds. A variety of tints, tones, and shades of color, ranging from black, violet, blue, green, brown, and red are formed when carotenoids are linked with different proteins, forming carotenoproteins (Wade et al. 2009). When shrimps and other crustaceans of varying coloration are cooked, they characteristically turn some variation of pink to red. This is because the carotenoproteins responsible for their color are denatured, and red color of the carotenoids, especially astaxanthin, is free to be expressed.

A transitory bluish color not associated with pigment granules can be found in chromatophores or throughout body tissues at certain times. It is most probably a carotenoprotein (Ghidalia 1985) but as yet its origin and composition are still unresolved. It often appears during color change, especially nocturnal color change. However, the more permanent blue color found in chromatophore pigment granules of shrimps is definitely a carotenoprotein. For example, Welton Lee (1966) chemically identified a bluish canthaxanthin-protein complex in the cuticle of the marine isopod *Pentidotea montereyensis*.

Chromatophores and chromatosomes in crustaceans are now generally described by the color of their coloring agents, e.g. red, red-white, white, yellow, red-yellow, black. However, in much of the older literature, they are described by other names such as leucophores (white), erythrophores (red), xanthophores (yellow), and melanophores (black). However, these terms are not commonly used today; the term "melanphore" is misleading, for as mentioned above, the black pigment granules found in black crangonid chromatophores are composed of ommochromes, not melanin. The chromatophore color is expressed when the pigment granules are dispersed from the center of the cell body into its multirayed branches (Plate 5.1 B-C, E-F). When the pigments aggregate toward the center of the cell (Plate 5.1 D), color is reduced. The micromechanics of the pigment movements is still not completely understood, but apparently is carried out by myosin motors that convey the pigment granules along actin filaments of the chromatophore cytoskeleton (McNamara and Milograna 2015).

5.3 Color and Camouflage

Different functions have been proposed for coloration in shrimps. One is that it is a protection against high-intensity solar radiation like that experienced in shallow, clear tropical waters. However, numerous shallow-water species are essentially transparent. Most shrimps live in deeper or more turbid waters in which light intensity decreases rapidly with depth. Deep-sea shrimps living in the aphotic bathyal zones are in no danger from solar radiation, and most have a reddish color of that is generally more intense with depth. In many animals (e.g., birds) males are much more brightly colored than females, an advertisement of a healthy mate with good genes. Such sexual dimorphism in color is rare in shrimps. In fact, in many species, females are larger and more brightly colored than the smaller more transparent males. There are certainly species in which males are large and more heavily armed than females, which helps such dominant males to fend off competing males and attract reproductive females (see Chap. 8, Mating Systems), but coloration is not a factor. A major reason for the unimportance of color in mate attraction in shrimps is that their vision is not very good. Image formation and resolution, as well as color perception is poor (e.g., Jaafar and Zeng 2012; Caves and Green 2018). Many shrimps are night active or live in dimly lit deeper or turbid waters. Shrimps in such habitats have superposition eyes in which the basic visual units (ommatidia) are not well isolated from each by other screening pigments. Superposition eyes are an adaptation which allows maximum use of low-intensity light but results in limited color vision and visual acuity and (Brusca et al. 2016).

The most obvious and documented evidence shows that color and color patterns of shrimps are a camouflage against visually hunting predators such as birds, fishes, cephalopods, and marine mammals, which have varying degrees of good to excellent visual acuity and color perception (e.g., Martin and Osorio 2008; Shuster et al. 2011; Stubbs and Stubbs 2016; Griebel 2002). The major types of color pattern protection in shrimps can be summarized as a general resemblance to background (mimicry; background blending), disruptive coloration (Cott 1957), motion-dazzle (Elias et al. 2019), and flicker-fusion camouflage (Umeton et al. 2017).

Shrimps must reduce or eliminate signals by which the predator recognizes them as prey. The body outline of a shrimp or some characteristic part of it is a key signal to visually hunting predators. In mimicry, the shrimp color pattern is a close match to the background on which it normally occurs. There are many examples of this among shrimps, e.g., the green coloration of some *Thor* and *Hippolyte* species that perch along sea grass blades and the close color resemblance of many symbiotic shrimps to that of their hosts (Plate 3.3C; Chap. 10, Symbioses). Crypsis (concealment) not only involves color but also shape, body structure, and behavior (Merilaita et al. 2017). For example, *Tozeuma* and *Hippolyte* seagrass species are elongate in shape and usually perch and cling parallel to the seagrass blade, often incorporating splotches of color which mimic encrusting epiphytic algae growing on the seagrass (Plate 3.3C). Some shrimp associates of sea urchins are slender and, with appropriate coloration, resemble urchin spines, especially as the shrimp orients parallel to the urchin spines (Chap. 10).

Individuals of some shrimp species are covered with plumose, branching, or long setae which resemble filamentous epiphytic algae, a condition described by the adjectives "pilose" or "fascigerous" (Plate 5.1 G, *Phycocaris*). Often the effectiveness of such adornments is enhanced by growth on them by microfouling organisms. Another way to hide the typical shrimp shape is transparency. Many shrimp species, especially smaller individuals, are essentially transparent except for the black color of their eyes and opaqueness of a full stomach or of other internal organs (Plate 5.3D). This "hidden in plain sight" strategy is common in many aquatic invertebrates (Johnsen 2001).

In disruptive coloration, the body outline of the shrimp is broken up by splotches, bands, or stripes of color. The predator sees these highly visible color marks which distract it from the shrimp's characteristic body outline or profile by which the predator recognizes it as a shrimp. Disruptive camouflage works especially well if the shrimp is motionless. Movement makes the shrimp more noticeable. This may be counteracted by motion-dazzle camouflage in which highly contrasting stripes of a fleeing shrimp disrupt motion perception by the predator so that it cannot accurately measure the shrimp's speed or trajectory (Hogan et al. 2016). Another related effect of movement is "flicker-fusion camouflage" (Umeton et al. 2017). Blurring of a prey's stripes during rapid motion may also alter the predator's perception of the prey's motion, as well as help blend it into its surroundings. The sudden change in the shrimp's distinct disruptive stripes in its resting state to a blurred image when in motion, and then back to the striped resting state may confuse or frighten the predator. Both dazzle and flicker-fusion camouflage in shrimps need experimental confirmation.

The degree of camouflage coloration differs widely among taxonomic groups, but some broad generalizations can be made. The coloration of pelagic shrimps varies with depth. Epipelagic species, which inhabit shallow lighted surface waters, are mostly transparent (e.g., Plate 3.2B). Mesopelagic species are partially red and partially translucent (e.g., Plate 3.2A), with degree of redness increasing with depth, while bathypelagic species are bright red (Vereshchaka et al. 2014) (Plate 3.5E-F). Aristeids, solenocerids, and benthesicymids are deep-sea penaeoid species that exhibit some tint or tone of red (Plate 3.1A-B, F), as do many other deeper-dwelling shrimps (e.g., the caridean *Bythocaris*, Plate 3.2G and *Eugonatonotus*, Plate 3.5A). Primarily nocturnal shrimps have reddish color patterns (e.g., *Cinetorhynchus, Rhynchocinetes* species, Plate 3.5C-D). The adaptive value of red coloration in low light and dark habitats will be explained later in the chapter (see Color and Depth, below).

Sicyoniids, which occur mainly in neritic and shallow shelf waters, vary from dull brown or gray, but some species have elements of brightly colored disruptive coloration (Plate 3.1D-E). The nearshore and neritic Penaeidae are generally a drably hued group as most occur over sand and mud bottoms to which their plain pigmentation is best camouflaged. For example, species of "white shrimps" (*Litopenaeus*) are a light blue-gray to white color (Plate 11.1 F, 11.2A). One penaeid exception is the "giant tiger prawn" *Penaeus monodon* which has bold white and black transverse stripes, a disruptive coloration (Plate 3.1C). Juvenile penaeoids that inhabit shallow seagrass and other vegetated habitats are often disruptively striped (Dall et al. 1990). Shallow-water carideans which occur over brightly or variably hued habitats such as algal flats, seagrass meadows, subtidal algal- or sessile invertebrate encrusted habitats show many color variations.

Color information on stenopodideans has been reviewed by Goy (2010). Most shallow-water Stenopodidae are white or translucent with red stripes and banding. An exception is the tropical and subtropical genus *Stenopus* whose species are brightly colored with species-specific color pattern mixes of red, white, orange, yellow, blues, and purple (e.g., Plate 3.8E). Free-living deep-sea Spongicolidae are reddish, typical of deep-water shrimps. However, many spongicolids which live as adults within deep-sea hexactinellid sponge hosts have little or no color (Plate 3.8H). As they are hidden and protected from visually hunting predators within their sponge hosts, there is no necessity to produce a color camouflage.

Cave shrimps, which live in total darkness, tend to lose most or all color as do many cavernicolous animals, with a whitish, yellowish, or pale red color (e.g., Hobbs 1973; Hobbs et al. 1977) (Plates 3.6B, 3.7E). Anchialine shrimps, which live in subterranean habitats connected on either end to sea and land, usually have some shade of red (e.g., Holthuis 1963, 1973; Chace and Manning 1972) (Plate 3.8D). This coloration is perhaps an adaptation to the dark of their subterranean habitats, similar to the preponderance of red coloration in deep-dwelling marine shrimps (see Color and Depth, below).

5.4 Mechanisms of Color Change

Many animals, such as cephalopods, fishes, amphibians, and reptiles, are capable of modifying their color or color patterns. Color change may be part of a camouflage strategy, i.e., the animal modifies color to blend into the background so as to be inconspicuous to other animals, e.g., a prey to visually hunting predators, or a predator to a potential prey. It may be part of a mating or agonistic display, particularly well known in squids and octopi, in which color change is nearly instantaneous. Such rapid change is possible because the chromatophores of squid and octopi are pigment-filled sacs that can be expanded by tiny muscles under nervous control to show color or contracted to hide it. However, in crustaceans, fishes, amphibians, and reptiles, color change is usually a slower, more gradual process. Many shrimps are capable of minor to major alteration in color to better resemble their background within a relatively short time. Others may completely replace one color pattern over a longer period of time in response to a change in their surroundings. Physiological (rapid) color changes take just a few minutes to an hour while morphological (slow) color modification occurs over days to weeks.

The basic color pattern, either realized or potential, is under control of genes, but there is considerable variability in the visible (phenotypic) expression of the color genotype depending on intrinsic factors such as sex, molt, and reproductive stage (Chassard-Bouchaud 1965), as well as temperature, light intensity, and stress (Siegenthaler et al. 2018). In some shrimps, color variations within and among populations are highly genetic. Populations of the anchialine shrimp *Halocaridina rubra* from different geographic areas of the Hawaiian Islands differ in color from nearly translucent to bright red, but individuals within the same population show much less variation. Vaught et al. (2014) showed that color variation among genetic lineages was due to differences in the number (density) of red chromatosomes. However, as will be seen below, the expression of the genetic color pattern in many species can be highly variable according to the background on which it occurs and the length of time it spends there.

5.5 Rapid Color Change

Rapid (physiological) response to a change in background has been recognized and studied for some time. Crangonid shrimps, like flatfish species from the same habitats, are masters at quick change that allow them to blend remarkably well into speckled sand, mud, and gravel substrates (Plate 5.2 A) (Koller 1927; Chassard-Bouchaud 1965; Siegenthaler et al. 2018). Experiments on species such as *Palaemon vulgaris* (Brown 1935), several other caridean species (Chassard-Bouchaud 1965), *Heptacarpus sitchensis* (Bauer 1982) and *Crangon crangon* (Siegenthaler et al. 2018) showed that individuals of many species will darken or increase expression of color on a black background but lighten considerably on white ones. These modifications can take place from minutes to an hour, and for shrimps, this is considered rapid color change. The overall result is a better match of the shrimp to the new color background, with the degree of match dependent on the chromatosomes and pigments that the shrimp possesses.

Rapid color change is physiologically mediated by hormones in crustaceans (Rao 1985; McNamara and Milograna 2015). Light stimuli received by the eyes are transmitted to endocrine organs, which release the pertinent hormones into the bloodstream. The hormones which stimulate pigment movements within shrimp chromatophores are secreted by the "X-organ, sinus gland" complex within the eyestalks. The nerve cells of this complex, together with those of some other nerve bodies, release neurosecretory peptide hormones (chromatophorotropins) into the

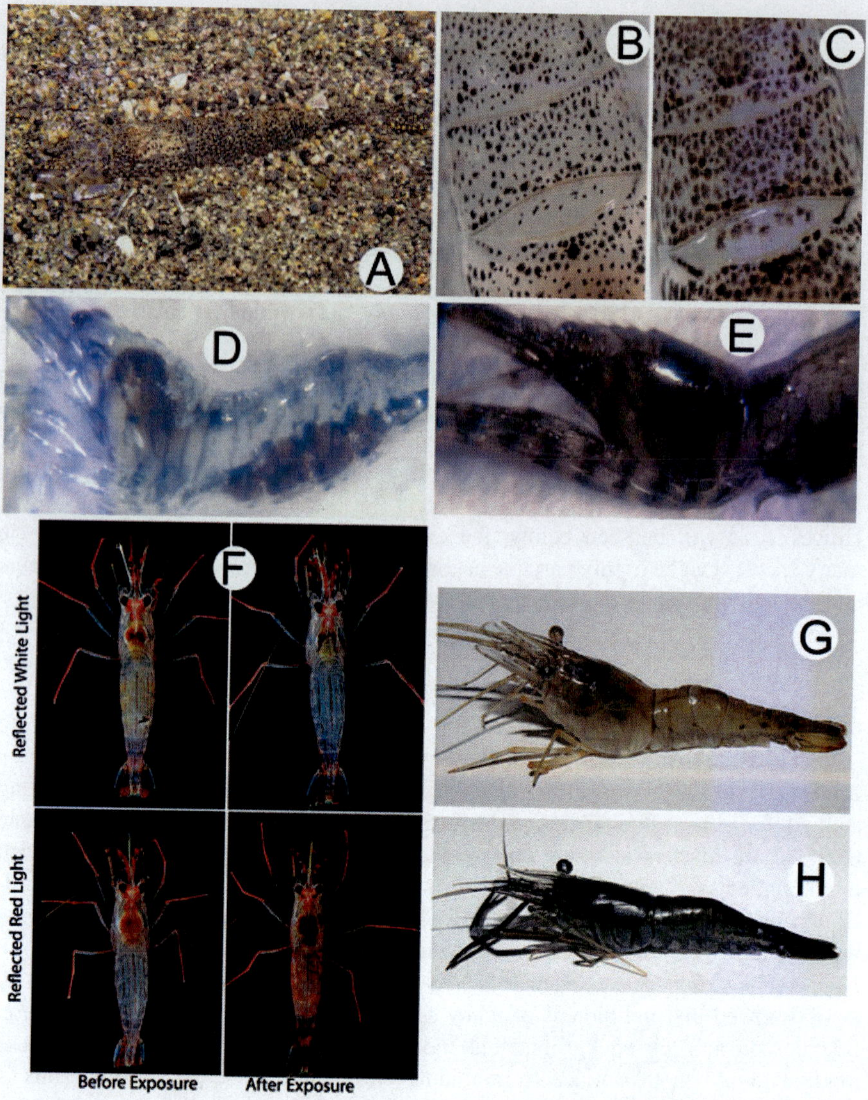

Plate 5.2 Rapid Color Change. (A) *Crangon stylirostris,* partially buried with chromatosomes matching the sediment background. (B,C). *Crangon crangon* abdominal segments with chromatosome pigment contracted (B) and dispersed (C) over white and black backgrounds, respectively. (D, E) *Heptacarpus sitchensis* banded color morph in bluish nighttime (nocturne) coloration (D) and then photographed again after 30 minutes exposure to light (E) with red pigment in chromatosomes dispersed (daytime condition). (F) Coloration of *Lysmata boggessi* before (upper left) and after (upper right) 15-day exposure to white reflected light; and before (lower left) and after (lower right) to 15-day exposure to red-reflected light. (G, H) Juveniles of *Macrobrachium tenellum* maintained for 60 days on a white background (G) and black background (H) before photographs were taken. (A) courtesy of Gregory Jensen from Jensen (2014); (B-C) modified from Siegenthaler et al. (2018) Creative Commons CC BY 4.0; (D–E) R.T. Bauer from Bauer (2004); (F) modified from Calvo et al. (2016) and (G–H) courtesy of Fernando Vega-Villasante from Aréchiga-Palomera et al. (2018)

bloodstream that then stimulate pigment dispersion or contraction in the chromatosomes (Rao 1985; McNamara and Milograna 2015).

Rapid color change is due to contraction and dispersion of pigments within individual chromatosomes (Plate 5.1 B-C). This can be quantified by selecting a group of chromatosomes from a specific area and then measuring the maximum diameter (dispersion) of a pigment sometime after background change. I did such an experiment with *Heptacarpus sitchensis* (Fig. 5.1a). After an hour on black backgrounds, the dispersion of red pigment of the red-yellow chromatosomes increased 30–45% but upon switching the shrimp to a white background, it contracted by about the same amount. This is similar to that found by earlier workers on other species of shrimps. In general, red or black pigments disperse more on darkly colored backgrounds and contract on lighter ones, but white pterins do just the opposite. However, the response varies among and within species, chromatophore pigment type, and body part. The response of chromatophores to light stimuli is based primarily on the wavelengths and intensity of light which are reflected back from the background.

Past studies on rapid color change used indices or direct measure of chromatophore dispersion and cell density (e.g., Brown 1935; Chassard-Bouchaud 1965; Bauer 1982) and described overall changes in color qualitatively. However, recent workers have evaluated color change more thoroughly using additional measures of color properties. Such properties include hue (the descriptive color, e.g., red, green, blue); its chroma (saturation), i.e., its concentration; and intensity (value, luminosity, brightness), i.e., the darkness or lightness of the color, ranging from white (total reflection of light) to black (total absorption of light). These properties can be captured photographically and analyzed with quantitative techniques using various color models (e.g., RGB: red, green, blue properties; HSI: hue, saturation, intensity) (Flores and Chien 2011; Vaught et al. 2014; Calvo et al. 2016; Aréchiga-Palomera et al. 2018; Siegenthaler et al. 2018). They are "noninvasive" in that there is less handling and trauma to the animal during observations and experiments than those measuring pigment dispersion of chromatophores. For example, Siegenthaler et al. (2018) showed rapid color change in *Crangon crangon* using coloration quantified as the percent area covered by dark (black, brown) pigmentation on uropods (Fig. 5.1b).

5.6 Day-Night Color Change

At the beginning of the twentieth century, F.W. Gamble and F.W. Keeble extensively documented the remarkable day-night color changes of *Hippolyte varians*, a small caridean that occurs in algal and seagrass habitats of shallow European waters. As the specific epithet "varians" suggests, there is considerable color diversity (e.g., green, brown, red) among individuals of this species. These color variations (morphs) were effusively described in the entertaining naturalist style of that period (Gamble and Keeble 1900, summarized in abstract form in Keeble and Gamble 1900). At night, all color morphs become a ghostly pale to dark blue, depending on

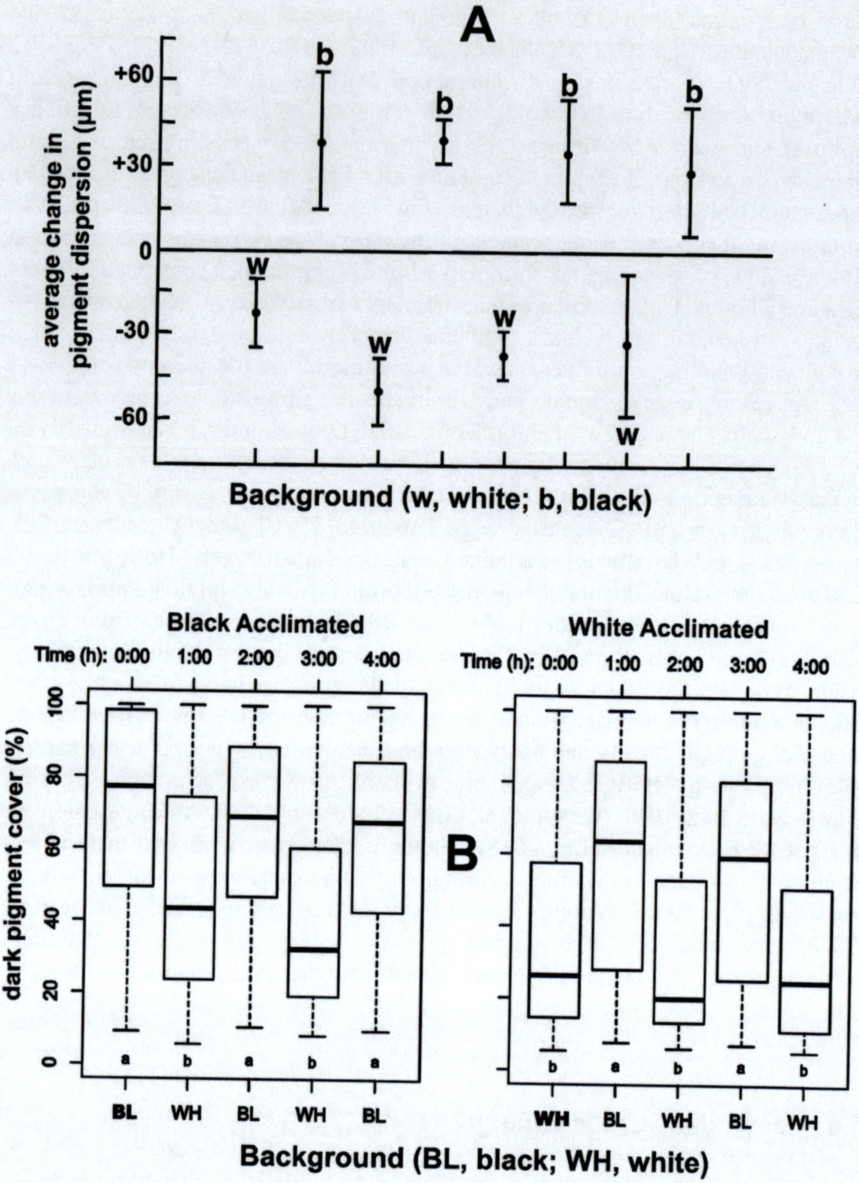

Fig. 5.1 Rapid color change: dispersion and contraction of chromatosome pigments on different backgrounds. (**a**) The mean change (± 95% confidence limits) in the maximum diameter of red pigment in red-yellow chromatosomes after one hour on white or black backgrounds in *Heptacarpus sitchensis*. The group of 5 shrimps began the experiment on a white background Bauer (1981). (**b**) Change in dark pigment cover during repeated shifts between black and white background in *Crangon crangon*. The experiment began after 24-hour acclimation to a black (left) or white (right) background. Box-and-whisker plots of dark pigment cover; different letters indicate groups that are statistically different ($P < 0.01$). Modified from (**a**) from Bauer (1981) and (**b**) Siegenthaler et al. (2018) Creative Commons CC BY 4.0

their daytime color patterns. The loss of incident and reflected light from the shrimp's surroundings trigger this event. Gamble and Keeble poetically termed this nighttime color as "nocturnes" and beautifully illustrated it with watercolor prints. This general loss of color is due to retraction of the red and yellow pigments from the chromatophore branches to the center of the cells. In the branches of red chromatophores, a pale blue color remains. This is the mysterious bluish color mentioned earlier which has not been clearly identified, but only hypothesized as a transitory carotenoprotein. In some species, e.g., *Palaemon vulgaris* and *H. varians*, the blue appears to diffuse out into surrounding tissues (Brown 1935; Gamble and Keeble 1900).

The thorid *Heptacarpus sitchensis* shows a dramatic nocturnal color change as well. As in *H. varians*, there are several color morphs. The banded morph, characterized by transverse dark bands composed of red-yellow chromatosomes, turns a pale aquamarine color at night (Plate 5.2 D). The red pigment is retracted but the yellow pigment remains dispersed. The transitory blue color appears or is unmasked by the retraction of red pigment and combines with the yellow pigment to produce the aquamarine color. Upon exposure to light, the daytime coloration quickly returns (in ~15 minutes in *H. sitchensis*) as retracted pigments disperse out again into the channels and branches of chromatophores (Plate 5.2 E). The blue color fades away, perhaps masked by daytime dispersion of red pigment, or perhaps it is photosensitive and is bleached out by light. Another color morph has a bright green abdomen during the day. The green color is produced primarily by abundant yellow chromatosomes combining with a blue color of unknown origin, perhaps a Tyndall blue from light scattering by abdominal tissues or a transitory dispersed carotenoprotein. At night, this yellow pigment is retracted or concentrated, its color is only slightly expressed, giving the abdomen a dull pale yellow color under white light. Nighttime color change has also been reported in a variety of carideans (Bruce 1975 for coral reef shrimps; several authors cited in Bauer 1981). However, a different nocturnal change was found in *Crangon crangon*; this shrimp shows a daily (circadian) rhythm of darkening at night with an overall lighter color during the day (Siegenthaler et al. 2018). In general, however, little about nocturnal color change has been studied in the last three decades, but research on it is certainly needed and a potentially productive area of investigation.

What might be the adaptive value of this general loss of color at night? It is hard to see how it might improve camouflage against visually hunting predators. However, in shallow-water habitats in which low-level illumination may occur, e.g., starlight and moonlight, the decrease in color pattern might be a mechanism for hiding the silhouette of shrimps swimming up off the bottom if viewed by predators below them. In *Crangon crangon,* the unusual darkening rather than the more usual lessening of color at night was hypothesized as adaptive by Siegenthaler et al. (2018). This shrimp is most active at night, leaving its daytime sediment cover at dusk, reburying at dawn. An increase in overall dark color may blend them in with the darkening sediment background when light is dim. For the majority of species that lose color at night, a metabolic explanation may be possible. As noted above, the dispersion of pigment granules through the chromatophore cytoskeleton is powered by myosin motors. Perhaps not only the movement but continued dispersion of pigment during the day is a significant metabolic cost which is saved by contraction of some or all pigment at night.

5.7 Morphological Color Change

Some shrimp species tend to remain on a single, more or less monochromatic color backdrop. If switched to another different set of surroundings, they are able to modify their color pattern, sometimes completely, to mimic that of their new environment. They do so by destruction and reabsorption of chromatophores and pigments and construction of new ones in the same or other areas of the body. This is termed morphological or "slow" color change because, unlike rapid color change, it takes from several days to weeks to complete. This phenomenon was first dramatically brought to light in the lengthy monograph of Gamble and Keeble (1900). They documented that individuals of *Hippolyte varians* came in several color variations (morphs), e.g., green, red, brown, which resembled the algal substrate on which they were found—more on that later. These workers found that when a color morph was switched from one background to that of a different color, with a few weeks the color of that individual could change to resemble the new background. Similar, although less flamboyant, slow color change was documented in *Palaemon vulgaris* by Brown (1935).

Morphological color change was most extensively studied by the French scientist Colette-Chassard-Bouchaud (1965) in a variety of species: the alpheid *Athanas nitescens*; the crangonid *Crangon crangon*; the palaemonids *Palaemon serratus* and *P. adspersus*; and the hippolytid *H. varians* of Gamble and Keeble fame. These shrimps were kept over black or white backgrounds for several weeks after which modifications in the number of different kinds of chromatosomes were measured in different body parts. *Athanas nitescens* is a small shrimp of green-blue to red-brown color with a mid-dorsal light disruptive stripe. It is found beneath rocks and algae over gravelly substrates. On a dark background, Chassard-Bouchaud measured a long-term increase in the number of red chromatosomes and a decrease in those with white pigment. On a white background, the change was the inverse. *C. crangon* has 4 types of chromatosomes with different pigments: black, white, red, and yellow. After a month on a black background, the number and dispersion of black chromatosomes increased on a black background but decreased on a white one; again, the change of white chromatophores was the inverse (Fig. 5.2). The red-yellow chromatophores showed no or comparatively little change. The degree of change varied somewhat between males and females, with the latter generally larger in body size. Similar, although less dramatic, changes were observed with the red-yellow and red-white chromatosomes of the two disruptively banded palaemonid species.

More recently, Siegenthaler et al. (2018), measuring whole body color with a photographic technique, similarly showed a darkening of color in *C. crangon* upon

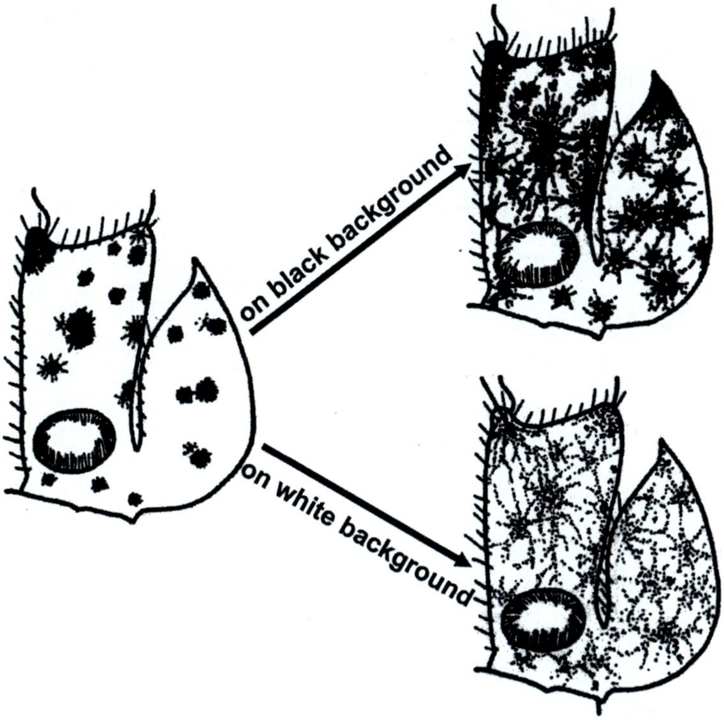

Fig. 5.2 Morphological color change in *Crangon crangon* chromatosomes experimentally shown by Chassard-Bouchaud (1965). Individuals taken from the natural habitat were maintained on either black or white backgrounds for 30 days. The change in pigmentation of the antennular peduncle and scale is shown. The solid black represents black pigment in chromatosomes, dotted lines represent white pigment. Note the increase in black pigment over black backgrounds and in white over white backgrounds. Modified from Chassard-Bouchaud (1965)

long-term exposure to a dark background (Fig. 5.3a). However, in none of the studies on rapid or morphological color change discussed above was the shrimp exposed to the often heterogeneously shaded mud, sand, and gravel backgrounds. *C. crangon* is a master of both rapid and long-term matching of its coloration to that type of background. How it manages to manipulate its various chromatosome types to natural backgrounds of composite shades is still a puzzle in need of resolution.

5.8 Applied Value of Color Change Studies

Ornamental shrimps are species with vivid coloration that are a favorite of millions of home aquarists as well as those viewing them in public aquaria. Unfortunately, many of these shrimps are taken from the wild. However, knowledge about cultivation of ornamental shrimps has progressed considerably in the last few decades

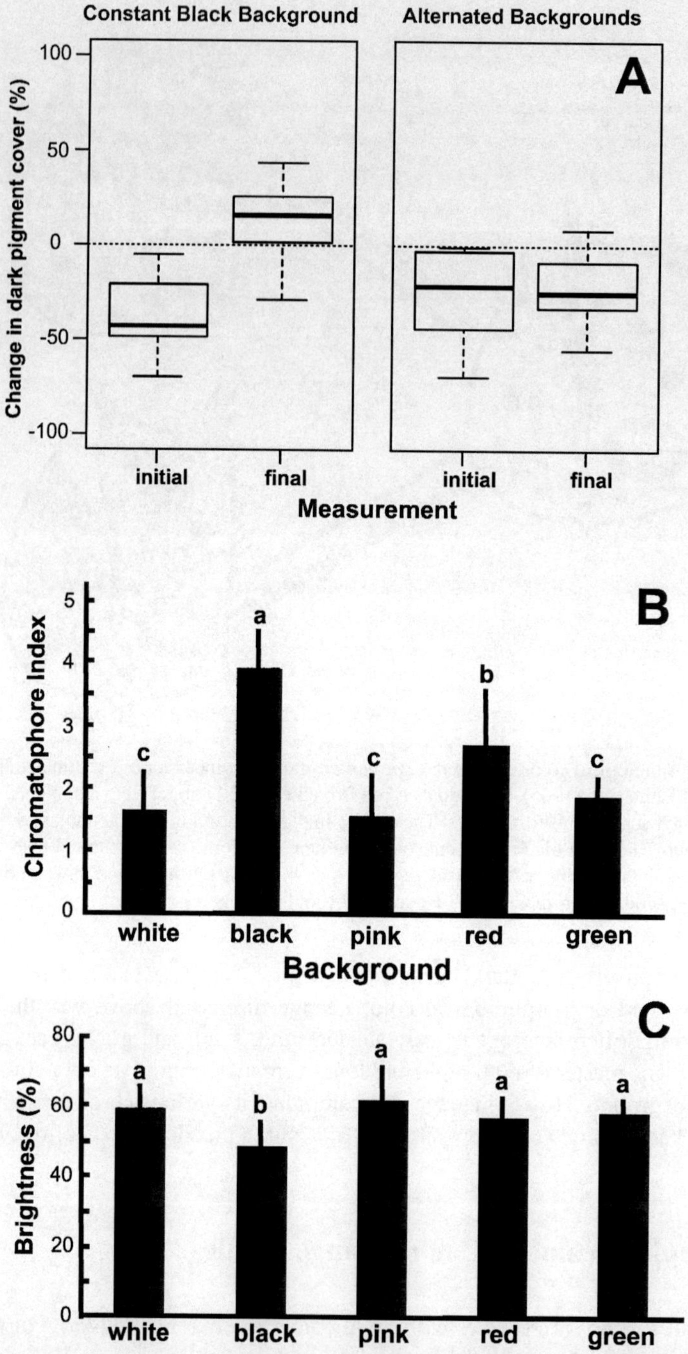

Fig. 5.3 Measurement of morphological color change in (**a**) *Crangon crangon* and (**b–c**) juveniles of *Macrobrachium tenellum*. (**a**) Shrimps were maintained initially on yellow sediment. One group (left side) was transferred to a black background for 21 days, while another (right side) was alternated daily between black and white backgrounds for the same period. The percent dark

(Calado 2008; see Chap. 11, Fisheries and Aquaculture). The trade of these shrimps has a value of 200–300 million US dollars per year (Calvo et al. 2016). Aquarists are attracted to the often dazzling colors of these shrimps. Thus, maintenance of these colors before sale and afterward in private homes or public aquaria is of great importance. However, shrimps in captivity often lose their natural color quickly, in spite of a nutritional diet, decreasing their attractiveness and commercial value.

Based on the basic science studies about color change cited above, researchers have recently investigated how the colors (and their properties) of the aquarium or culture environment may preserve and even enhance those of the shrimps. Calvo et al. (2016) showed this quite nicely, using a common reddish ornamental shrimp, *Lysmata boggessi*, as a model to test the effect of background color. They maintained two groups of shrimps in aquaria whose sides and bottoms were painted white or red so that the shrimp's eyes were receiving white- or red-reflected light, respectively. At the end of the experiment, the color of selected body parts was photographed and analyzed quantitatively using the RGB color model. The shrimps in the red aquaria preserved and intensified their overall red color while those from the white aquaria lost color (Plate 5.2 F). The increase in reddish appearance was due to an increase in the intensity (darkness) of blue and green hues, while that of the red hues did not change much. As the experimental aquaria were painted completely, sides and bottom, in one color, it is not clear if it was only the light reflected from the bottom or from all the surroundings that was most important. However, Siegenthaler et al. (2018) clearly showed that only bottom-reflected light was the important factor in color change and upkeep. This agrees with other studies using different backgrounds (bottoms).

Production of shrimps for food by aquaculture is a multi-billion dollar industry worldwide, as shown in Chap. 11. The color of shrimps in food markets is of course of great significance to the seafood industry. Some species of *Macrobrachium*, such as *M. rosenbergii* and *M. nipponense*, reach large body size and their cultivation is an important industry (Chap. 11). Diets designed to maintain their natural color for a good market value can be quite expensive. Less costly diets with good color upkeep by using a particular color background may be a cheaper alternative. Aréchiga-Palomera et al. (2018) maintained juveniles of *M. tenellum*, a species considered for aquaculture, in aquaria with variously colored gravel bottoms for 60 days. They measure chromatophore dispersion with the subjective Hogben and Slome (1931) index as well as the brightness (intensity, value) of the color on two parts of the body,

Fig. 5.3 (continued) pigment cover of a section on the tail fan was measured using a photographic (image analysis) technique, first at the beginning of the experiment and at the end. The data is summarized by box-and-whiskers plots which show the median (horizontal line), the interquartile range (box), and minimum and maximum observations (vertical line). (**b**) Juveniles were maintained on different color backgrounds for 60 days, after which the pigment dispersion in chromatosomes was measured using a subjective scale and (**c**) also analyzed with a photographic technique using the relative brightness of the second abdominal segment. Modified from (**a**) Siegenthaler et al. (2018) Creative Commons CC BY 4.0 and (**b**–**c**) Aréchiga-Palomera et al. (2018)

the lateral side of the cephalothorax and the second abdominal segment. The bottom colors chosen were white, black, pink, red, and green. On the backgrounds with the lowest brightness (black), the dispersion of chromatophores was significantly greater than that of shrimps on the other backgrounds (Fig. 5.3b). The overall color of shrimps, captured photographically and measured with the HSI model, became significantly darker on a black background than on white and colored backgrounds (Fig 5.3c, Plate 5.2 G-H). As shown by early workers, the greatest fading of color was over light (especially white) background, while the most intense (darkest) coloration is over a black one.

The practical use of background color manipulation in shrimps used for food marketing was shown by Parisenti et al. (2011) in the culture of the penaeid shrimp *Litopenaeus vannamei*, one of the most valuable aquaculture species. A number of individuals reared in white tanks were transferred to black tanks for 30 days, after which their coloration was compared to that of individuals from the same original group which remained in white tanks. Color differences were then analyzed at the end of the experiment. As might be expected, those from the black tanks had a darker color, more desirable in the marketplace, than those always reared in white tanks. After cooking, those shrimps from the black tanks had a more intense red color than those from white tanks. The more vivid red of cooked shrimps maintained in black tanks increased their culinary appeal and value. Interestingly, the total carotenoid content of individuals was not significantly different between the two groups. This suggests that individuals in the black tanks may not have developed more pigment, but that the difference between the two groups was due to contraction of pigments in the white-tank group. Thus, it seems obvious that, given a balanced nutritional diet, the simplest and more effective to bring out intense colors in shrimps is to maintain them over dark backgrounds.

5.9 Color Polymorphism and Camouflage

Phenotype, the characteristics of organisms that are expressed and can be measured or observed, are molded by the genotype as modified by the environment in which the organism develops. The degree of environmental influence in the expression of a particular gene or genes of a trait may vary greatly. Or, as it is often put, how much of the phenotype is "nature" (genetic) and how much is "nurture" (environmental)? In shrimps, the color phenotype encompasses the entire gamut from highly genetic, with little ambiental variation, to greatly dependent on the natural backgrounds or surroundings. Some shrimp species may only intensify or lighten their color pattern in response to their surroundings; in others coloration and pattern may be completely malleable. Some shrimp species may have a single color pattern, but others may be polymorphic in coloration (polychromatic) with different genetic color morphs in the population. As discussed earlier, the overwhelming adaptive value of coloration in shrimps is camouflage. Coloration works in concert with other traits or factors, e.g. morphology, sex, developmental stage, size, and behavior.

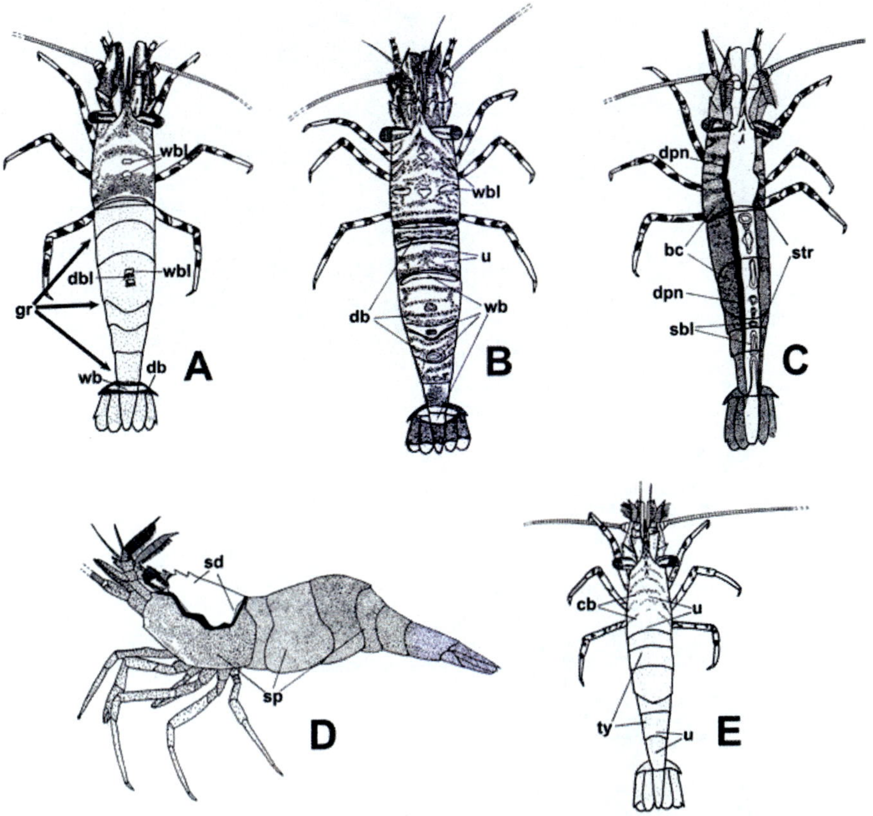

Fig. 5.4 Color morphs of *Heptacarpus sitchensis*: (**a**) green. (**b**) banded. (**c**) striped. (**d**) speckled saddleback. (**e**) transparent. *bc*, background color; *cb*, carapace band; *db*, dark band; *dbl,* dark blotch; *dpn,* darkly pigmented nerve cord; *gr,* area of green color; *sbl*, superficial (hypodermal) blotch; *sd* white saddle-shaped blotch; *sp,* speckled area; *str,* stripe; *ty*, dull translucent yellow area; *u,* unpigmented area; *wb,* white band; *wbl*, white blotch. From Bauer (1981)

Several species whose color-pattern and color-change ability has been well studied can be placed on this genetic-environmental continuum. At one extreme are those in which color is under tight genetic control, e.g., the anchialine shrimp *Halocaridina rubra* which was mentioned above. *Crangon crangon*, an inhabitant of heterogeneous mud to sand to gravel bottoms, has basically a single color morph, but one that rapid color change can quickly mold to match a particular substrate. Over time, it may darken or lighten overall by morphological color change. On the other hand, the thorid *Heptacarpus sitchensis* is highly polychromatic with 5 basic color morphs (Bauer 1981). This shrimp occurs in the rocky intertidal, with a multihued background of seagrasses: red, brown, and green algae; shell gravel and variously colored sessile invertebrates. The green morph (Fig. 5.4a, Plate 5.3A) has a green abdomen, very much like the color of the intensely green alga *Ulva lactuca* ("sea lettuce"). The bright green is due to an abundance of yellow chromatosomes

Plate 5.3 Color morphs; Bioluminescent emissions. (**a**) green, (**b**) striped, (**d**) transparent color morphs of *Heptacarpus sitchensis* (Thoridae) and (**c**) *H. taylori* saddleback morph. (**e**) Defensive bioluminescent secretion emitted from the mouth of *Heterocarpus sibogae* (Pandalidae). From (**a**–**d**) R.T. Bauer from Bauer (2004) and (**e**) courtesy of T.-Y. Chan and Crustaceana (Brill) from Chan et al. (2008)

enhanced with a sprinkling of red-yellow ones. The carapace, legs, and eyes have disruptive red-brown bands composed of red and red-yellow chromatosomes (Fig. 5.5). Between the bands are white to rose blotches and bands composed of red-white chromatosomes, with the exact tint depending on the degree of red pigment to white pterin dispersal. The banded morph has similar disruptive bands over the entire body (Fig. 5.4b). White splotches suggest bits of gravel from dead, bleached out coralline algae while the rose tints match the colors of living corallines common in the shrimp's habitat. The striped morph has a beautiful mid-dorsal stripe ranging from pink to red-purple, colors matching those of coralline algae in the tidepools (Fig. 5.4c, Plate 5.3B). The speckled morph combines the usual carapace and leg banding with a uniform speckled pattern, which varies in color from

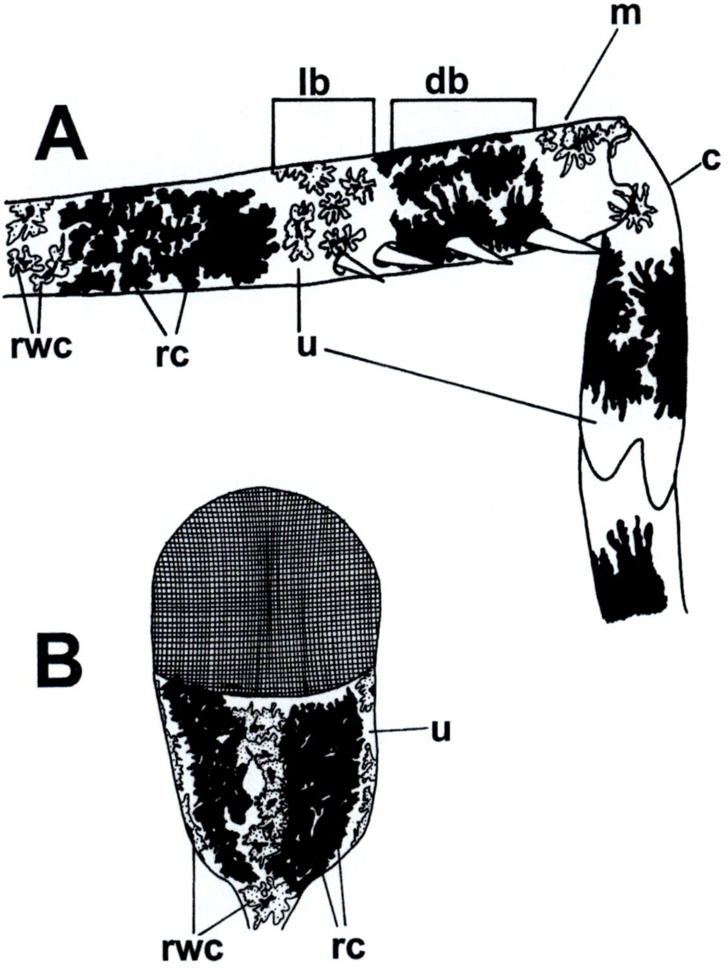

Fig. 5.5 Chromatosome composition of leg bands and eye stripes of *Heptacarpus sitchensis*. (**a**) Leg bands on meral and carpal segments of a walking leg. (**b**) eyestalk and eye. *c*, carpus; *db*, dark band; *lb*, light band; *m*, merus; *rc*, red chromatosome; *rwc*, red-white chromatosome; *u*, unpigmented area. Modified from Bauer (1981)

aquamarine to brown, depending on the relative dispersion of pigments of the numerous red-yellow chromatosomes. A variation of the green and speckled morph is the "saddleback" in which the dorsal cephalothorax is covered with a huge, dazzling white splotch (Fig. 5.4d); a similar but more spectacularly splotched morph is found in the related tidepool species *H. taylori* (Plate 5.3C). The transparent morph has little pigmentation (Fig. 5.4e, Plate 5.3D). Most males and juveniles fall into this latter category. They show weak carapace banding and the abdomen is sometimes a faint yellow from a relatively few yellow chromatosomes with poorly dispersed pigment.

The polymorphic color patterns of *H. sitchensis* are a disruptive camouflage. This is a highly mobile, active shrimp which walks, swims, and perches day and night on the heterogeneously hued rocky intertidal substrate. Each of the color patterns has elements that match similarly hued objects of the rocky intertidal, but the entire shrimp does not do so completely. The various elements of the color pattern distract the eyes of potential predators from the characteristic outline of the shrimp. Experiments with color change showed that this species does not pick a particular color background when given a choice, nor does it appear to select color backgrounds to which seem, to the human observer, best matched (Bauer 1982). When maintained in captivity over a monotonous color background, it gradually loses color, but close observation shows that the basic chromatosome pattern still exists but with much diminished pigment in the chromatosomes.

A closely related species, *Heptacarpus paludicola* lives in *Zostera* seagrass meadows over rather homogeneous sand and mud bottoms in bays and lagoons along the western Pacific coast of North America. The two major color forms are the green and transparent morphs, including large females, which are usually highly colored in *H. sitchensis.* However, in sampling the tidally flooded Argyle Lagoon on San Juan Island, Washington, USA, I found an enlightening situation. The lagoon shore was rimmed with beds of the green seagrass *Zostera* over a sandy mud bottom, good *H. paludicola* habitat. The slightly elevated lagoon is only flooded during the highest tides; at other times, it slowly drains back into the sea via a narrow and shallow tidal creek. Because of the water flow, the substrate is kept clean of fine sediment to reveal a gravel and rock bottom. This substrate is ideal for the growth of algae and sessile invertebrates, a multihued habitat similar to that of the rocky intertidal. In this creek, I found several color morphs similar to those of *H. sitchensis* (Bauer 1982). Upon close microscopic examination of the larger "transparent" individuals from the lagoon, I found that I could identify the pattern of the various color morphs found in the channel, but they were poorly expressed, i.e. the individual was nearly transparent macroscopically. In the Argyle Lagoon, the green morphs evidently could express their color pattern stimulated by the green of the *Zostera* seagrass background. Apparently, the other morphs lacked the multihued stimulus necessary for development of the potential pigmentation of their color pattern.

In summary, the color patterns of these two *Heptacarpus* species appear to be under tight genetic control, i.e., each morph is a specific genotype, but the expression of the color pattern depends on the intensity and heterogeneity of color in the substrate. Note that in these species, as in many others, the smaller individuals, i.e., juveniles and males in *Heptacarpus*, have relatively little expression of color. This is common in shrimp species. Transparency is a great camouflage, but it becomes less effective as size increases as it does in females. The correspondingly larger opaque internal organs (e.g., food-filled guts, digestive glands gonads, incubated embryos) and the darkly pigmented eyes become more apparent and can attract the eye of predator, especially upon movement of the shrimp. As size increases, the shrimp relies much more on camouflage coloration for protection from predators.

At the other extreme in phenotypic plasticity in coloration is *Hippolyte varians* which lives in shallow-water algal flats composed of various species. Building on the pioneering work of Gamble and Keeble (1900), Chassard-Bouchaud (1965) described in detail the pigmentation, color patterns, color change, and behavior involved in color camouflage of this species. There are a number of color morphs based on the basic body color (green, brown, red) and on the presence/absence of longitudinal striping (heterogenous or uniformly pigmented) (Fig. 5.6a-b). Feathery camouflage setae may or may not be present (smooth or "hairy"). All combinations of these various characters are possible, but five were most commonly expressed in Chassard-Bouchaud's observations: a uniformly colored (homogeneous) green, brown, or red (Fig. 5.6a); heterogenous morphs with a mainly transparent body except for longitudinal disruptive stripes of red or brown (Fig. 5.6b). These colors are produced by only a few types of pigments and chromatosomes. She described three basic pigment types: a yellow to green, a brown to red, and white. Various shades of green, brown, or red color are due to the relative dispersion of the yellow (greenish) pigment and contraction of red (brownish) pigment of the yellow-red chromatosomes. The white or red-white chromatosomes produce the white to brown to red mid-dorsal and lateral abdominal stripes of the heterogeneous morphs.

The green, brown, and red morphs are usually found on similarly hued algae: the green *Ulva lactuca,* brown *Dictyota dichotoma,* and red *Rhodymenia palmata,* respectively. Chassard (1956) did choice experiments in which she placed a morph into a tub with algae of different colors. In tests with the green, red, and brown morphs and appropriate pairs of algal species, the shrimp was successful most of the time in choosing the algae whose color it resembled the most. Homogeneous color morphs did not distinguish between algae of similar color but different morphology, e.g. brown algae with either laminate vs filamentous blades (see Fig. 5.5 in Bauer 2004). However, the heterogenous red and brown morphs are more common in nature on filamentous rather than laminate algae of their respective colors. This indicates that in heterogeneous morphs, both algal color and form are important. In all these choice experiments, shrimps displaced from their matching background were active and swimming about in search of their preferred color background during the day; once encountered, they remained clinging to their algal substrates.

The most remarkable demonstration of color plasticity in *H. varians* is revealed by its ability to change color and pattern completely when transferred from an alga of matching color to another with a different color. This is not an unnatural situation because shrimps may be dislocated from one plant to another by disturbance due to storms, algal death, attacks by predators, and movements through their habitat by larger animals. These color changes were first described by Gamble and Keeble (1900) but investigated with extensive experiments by Chassard-Bouchaud (1965). This was morphological color change. Chromatosomes and pigments of the initial color pattern were replaced by new pigments and chromatosomes in the correct body position for the new color pattern to match the new algal substrate. For example, Chassard transferred one type of morph, e.g. a heterogeneous morph from a filamentous alga of one color to a laminate green alga of another color for some weeks. The shrimps changed from a heterogeneous brown to a homogeneous green morph.

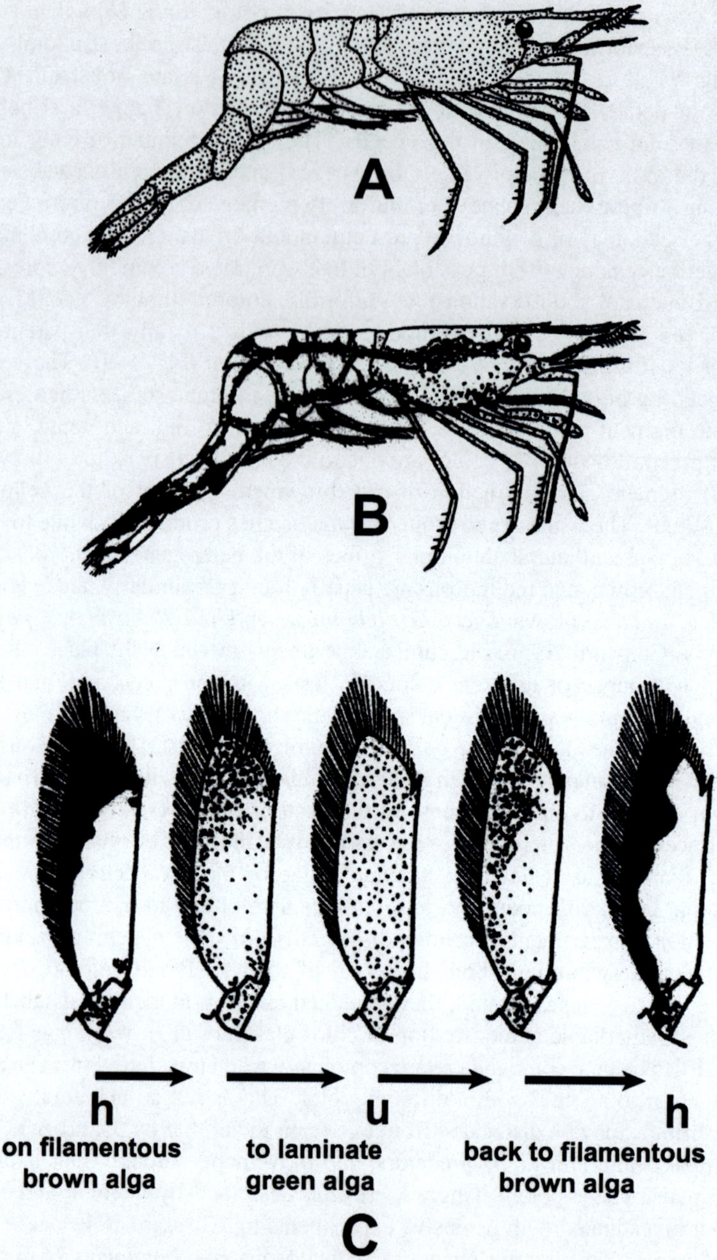

Fig. 5.6 Color morphs and morphological color change in *Hippolyte varians*. (**a**) The uniformly (homogeneously) pigmented morph (green, red, or brown). (**b**) Heterogeneous morph with brown or red longitudinal disruptive stripes on a transparent background color. (**c**) Change in color pattern of the antennal scale in a heterogeneous morph which was moved from the arborescent brown alga *Cystoseira* to the laminate green alga *Ulva* for a period of weeks and then back again to *Cystoseira*. The pattern changed from that of the heterogeneous brown morph (h) to the uniform (homogenous) green morph (u) and then back again. Modified from Chassard-Bouchaud (1965)

She then switched them back to the original algal type and the shrimps changed back to the initial morph type (Fig. 5.6c). Thus, color phenotype in this species appears to be totally malleable with little genotypic restriction. How the shrimp translates the visual information of the algal substrate (wavelengths of reflected light, morphological type) is still unknown.

Color and camouflage of two other species of *Hippolyte* have been investigated since Chassard-Bouchaud's work. *H. coerulescens* lives on the pelagic brown alga *Sargassum natans* in the Sargasso Sea. A fellow hippolytid, *Latreutes fucorum*, is common in the same habitat. Coloration, morphology, and behavior of these two species were investigated by Hacker and Madin (1991). The larger individuals of these two species, i.e., reproductive females, use a disruptive coloration as camouflage. It consists of a general brownish color, like that of the alga they cling to, but interrupted by bands, stripes, and spots of various color. Smaller individuals (juvenile females, males) tend to use transparency or a solid color similar to that of the alga to mask themselves from the several species of *Sargassum*-living fish predators. However, the two species differ in their overall shape, and this is related to mimicry of the plant part on which they are most commonly found. The body of *L. fucorum* is somewhat long and thin and perches on the elongate fronds that project from the main *Sargassum* stalk. This may explain why this species can also be so common in the shallow-water seagrass meadows (*Thalassia testudinum*) of the Western Atlantic (Bauer 1985). *H. coerulescens*, on the other hand, has a somewhat globular cephalothorax and abdomen which appear, both from side and dorsal view, similar to *Sargassum* vesicles (floats). The vesicles are located alongside the main stalk of the alga. This shrimp tends to cling perpendicular to the *Sargassum* main stalk so that its body shape, along with its coloration, mimics the *Sargassum* vesicles. This resemblance is enhanced by transparency of body parts and projections just anterior, between, and posterior to the main bulk of the cephalothorax and abdomen, enhancing their resemblance to two separate vesicles. Unlike *H. varians*, both of these Sargassum shrimps apparently lack the ability to change color.

Hippolyte obliquimanus, studied by Duarte et al. (2016) and Duarte and Flores (2017), is an interesting case in which color change ability is more than that of *Heptacarpus sitchensis* and *Hippolyte coerulescens* but perhaps less plastic than that of *H. varians. H. obliquimanus* occurs in mixed algal beds on rocky habitats along the north coast of São Paulo State, Brazil. The two dominant algal species are the brown alga *Sargassum furcatum* and red alga *Dichotomaria* (*Galaxaura*) *marginata*. The shrimp has two homogeneously colored morphs, one greenish-brown and the other pink, as well a heterogenous (striped) one with a generally transparent body interrupted by with stripes and bands. The homogeneous morphs resemble the color of, cling to, and are most abundant on the algal species of similar color. In choice experiments, similar to those done by Chassard-Bouchaud, they tend to select the alga with similar color (Fig. 5.7a). The striped morph is much more active and moves frequently between the two algal species. This color morph is more streamlined (elongate and thin) that the two more robust homogeneous morphs. However, the majority of striped morphs are males which are smaller in size than reproductive females. The difference in morphology and behavior between the striped morphs

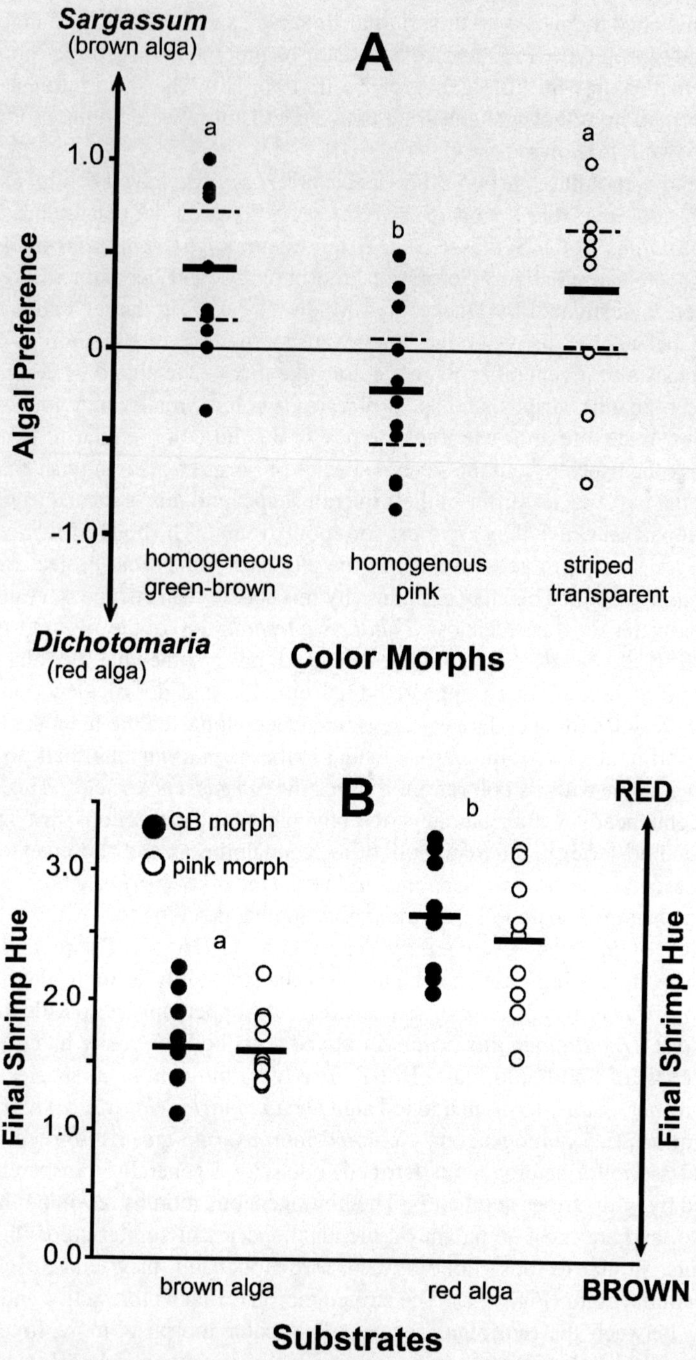

Fig. 5.7 Color preferences and color change in *Hippolyte obliquimanus*. (**a**) Individuals of three color morphs (homogeneous green-brown, homogeneous pink, striped transparent) were given a choice of the brown alga *Sargassum* and the red alga *Dichotomaria* (*Galaxaura*). Note that the green-brown and pink morphs chose algae similar to their own color pattern, while the disruptively

(male) and homogeneously colored morphs (female) is explained by sexual dimorphism. My interpretation is that the smaller males use disruptive transparency in order to mask their greater activity over differently colored substrates. As in many shrimp species, males are more active because of their "pure search" (promiscuous) mating system, in which males are frequently on the move in search of a receptive female (see Chap. 8, Mating Systems). Reproductive females of *H. obliquimanus*, especially those incubating embryos, are less streamlined and more stationary with their color mimicking that of the alga on which they perch.

However, there is color change capacity in the two homogeneous morphs of *H. obliquimanus* upon after displacement from their alga of color preference. Individuals of green-brown morphs from *Sargassum* and pink morphs from *Dichotomaria* were maintained by Duarte and colleagues in the lab for 5 days on both the same and the color-mismatched alga. Both morphs were capable of changing color toward that of the mismatched alga, i.e. the green-brown morph became more reddish on the red alga *Dichotomaria*, and individuals of the pink morphs switched to *Sargassum* became more brownish (Fig. 5.7b). On artificial substrates, however, the change was slight. Individuals which had been forced to modify color were able to change back to the color of their original substrate. Color change ability did vary between morphs, i.e. green-brown morphs became more like the red alga when switched than did the pink morphs toward brown when placed on *Sargassum*. The diversity of color modification was not so great as that demonstrated in *H. varians*, but on the other hand, the number of natural algal substrates was more limited. Additionally, the degree of color change was not so complete as in *H. varians*, and the length of time allowed by the experiment for change was limited to 5 days. Perhaps over a longer period, the degree of matching to a different algal substrate might be as great as it is in *H. varians*, which takes 12–20 days for complete change to a new substrate to take place (Chassard-Bouchaud 1965).

5.10 Warning Coloration

Many animals have bright colors and color patterns that make the animal quite apparent and visible against their background to an observer, just the opposite of camouflage coloration. The yellow and black banding of many bees and wasps, the outstanding colors of many other insects, the marine nudibranchs, and brightly colored frogs are just a few examples. These vivid colors and the unconcerned

Fig. 5.7 (continued) colored striped transparent morph showed no significant preference. (**b**) Green-brown (GB) and pink (P) morphs were placed on brown and red algal substrates and their change in hue was measured after 5 days. Note that the color of both morphs was browner on the brown alga *Sargassum* and pinker on the red alga *Dichotomaria*. Mean values are shown by solid black lines and different letters indicate significant statistical differences between groups. Modified from Duarte et al. (2016)

behaviors of the animals that display them are usually a sign that the organism is dangerous, toxic, or unpalatable. This warns the experienced predator to leave them alone. Such warning coloration is termed “aposematic.” There are certainly vibrantly colored shrimps but in most cases these colors are part of their camouflage, often accompanied by appropriate behavior and morphology. Cases of aposematic coloration in shrimps have rarely been demonstrated, but anecdotal observations by marine zoologists such as A.J. Bruce show that they may be more common than thought. Bruce (1975) noticed that strikingly colored shrimps such as *Hymenocera picta*, the clown or harlequin shrimp and *Gnathophyllum americanum*, the bumblebee shrimp, are left alone by predatory fishes in aquaria. Identification of other such species and the chemistry of the shrimp substances responsible for predator avoidance are fruitful areas for future research.

5.11 Color and Depth

Up to this point, I have been mainly discussing coloration in shallow-water benthic shrimps which live in relatively well-lighted waters. These shrimps detect light reflected from the bottom, and in many species this information is used to adjust coloration and pattern for the best camouflage. Benthic shrimps from deeper continental shelf and shallow slope depths receive radiation which is increasingly dim and monochromatic with depth. Epi- and mesopelagic shrimps swim above the bottom and receive light information only from above and diffuse scattered light from around them. Bathypelagic and bathyal benthic shrimps from the unlighted bathyal and abyssal zones receive no light information except that from occasional flashes or bursts of bioluminescent light produced by other deep-sea organisms. Coloration adaptations of these deeper-dwelling shrimps are quite different from those of shallow-water species.

The coloration of shrimps generally becomes less diverse and redder as light decreases in intensity and wavelength with depth. In the shallow, well-lighted waters of the epipelagic zone (~0–200 m in clear open oceanic waters), many organisms are transparent or nearly so. They visually blend into the clear waters around them, making them difficult for visual predators to see (Mcfall-Ngai 1990). The body outline of epipelagic luciferids, sergestids, and pasiphaeid shrimps is masked by their lack of coloration, translucent except for full digestive organs and black eyes. Several pasiphaeid species have common names of “glass shrimps” because of their transparency (McLaughlin et al. 2005).

Most mesopelagic shrimps migrate up toward or near the surface at night to feed on the abundant zoo- and phytoplankton concentrated there. They descend into the depths during the day to hide from visually hunting predators. As one goes deeper in the pelagic environment, shrimps become more reddish, termed the “half-reds” of the upper mesopelagic zone (~200 to 500–1000 m) (Foxton 1972; Herring and Roe 1988).These half-reds are characterized by a cephalothorax with dorsal reddish color produced by red chromatophores, and a pale to translucent abdomen. The dorsal reddish color makes them appear dark when viewed from above, blending them into the darkness of the deep water below them. Rapid color change in chromatophores of *Eusergestes arcticus* allows this shrimp, and probably similarly colored mesopelagic sergestids, to adjust their shade of red to that optimal for their depth as they migrate up and down in the water column throughout the day (Vestheim and Kaartved 2009). Deeper in the more dimly lit mesopelagic zone, shrimps become more completely scarlet to deep red, with much of their red pigmentation within the cuticle rather than in chromatophores. In the unlighted bathypelagic and bathyal benthic zones (> ~1000 m), shrimps and other crustaceans are almost always fully red in color, with most of their pigmentation in the cuticle. Fishes from the same depth ranges are usually black. Marine biologists Murray and Hjort (1912) referred to this bathyal assemblage as "the red crustacean/prawn, black fish" community in their classic book "The Depths of the Ocean."

The predominance of reddish coloration in deep-living crustaceans can be explained by two factors: (1) the quality (wavelength spectrum) of light with depth and (2) the physiological capacity of crustaceans to produce some pigments and not others. The different colors (wavelengths) composing incoming white light (sunlight) are differentially absorbed by water with depth from the longer (less energetic) to shorter (more energetic) wavelengths. Thus, the order of wavelength penetration from shallow water to deep is red, orange, yellow, green, blue, violet. The turbidity of the water affects this distribution somewhat. In clear oceanic water, blue wavelengths penetrate to the greatest depths, whereas green wavelengths reach greater depths in more turbid coastal waters. When diving in nearshore waters, one quickly sees that the bright colors of plants and animals wash out as one goes deeper until everything has a monochromatic green to bluish hue. At mesopelagic depths, the red pigmentation of the shrimp absorbs the ambient blue-green monochromatic light shining on it, so there is little or no light reflection back to the eye of a potential predator to reveal the shrimp. In the deep sea, the only light is the blue-green wavelength of bioluminescence (centered around 475 nm) produced by many organisms, including some predatory fishes and cephalopods. Deep-sea fishes are black, and so blend in with their lightless environment when lit up by bioluminescent light. Their black color is a result of the pigment melanin. However, crustaceans are physiologically unable to produce melanin, so instead they utilize red carotenoid pigments, especially astaxanthin, which they have in abundance. Shrimps with red color (reflected wavelength ~ 625–740 nm) appear black when viewed with bioluminescent wavelengths (~475 nm) which is absorbed and not reflected by the shrimp's pigments. Like the black deep-sea fishes, they are practically invisible when viewed against the dark of the unlit waters around them.

5.12 Bioluminescence and Photophores

Bioluminescence is light generated by living organisms. Although widespread in terrestrial insects and a few other invertebrates, it is nearly absent in freshwater organisms but common in marine animals, including many pelagic decapod shrimps (Haddock et al. 2010). It is generated by an energy-producing reaction between a luciferin molecule and oxygen which is catalyzed by an enzyme, luciferase. The energy is released not in the form of heat but rather as blue to blue-green light. There are different types of luciferins, and that found in decapods is termed coelenterazine, a molecule found in a number of other bioluminescing marine invertebrates. Coelenterazine is most likely synthesized by the shrimp and not obtained from its diet. This was shown by Thomson et al. (1995) in the oplophorid shrimp *Systellaspis debilis* by measurement of the concentration of coelenterazine in developing embryos. Although the female incubates and protects the embryos below the abdomen, the embryos are not connected physiologically to the female after spawning and thus receive no nutrition from her. However, Thomson and colleagues found that the concentration of coelenterazine increases as the embryos develop, showing that they must be synthesizing it by themselves.

Light is produced in most shrimps by structures termed photophores. Additionally, some shrimps emit a luminous fluid from the mouth fabricated within the hepatopancreas. Photophores may be cuticular (dermal) structures or they may be modified hepatic tubules. These hepatic organs are termed "Organs of Pesta" in sergestid shrimps and simply hepatic photophores in carideans. The known taxonomic distribution of bioluminescence in decapod shrimps is given in Table 5.2. It is probably not complete as many other species likely have bioluminescent structures which have not yet been observed or documented. Note that no completely benthic shrimps are known to have bioluminescence, which may explain its absence in the Stenopodidea (Table 5.2).

Cuticular photophores are found in a wide variety of mesopelagic shrimps, e.g., most of the *Sergia s.l.* species but not *Sergia s.s.*, some aristeids, solenocerids, possibly the benthesicymid *Gennadas*, three oplophorid genera, and a few pasiphaeid genera (Table 5.2). Photophores are tiny structures which can be arrayed on many of the body parts, e.g., the antennal scales, pereopods, carapace, abdominal pleura and sternites, and pleopods (Fig. 5.8). The photophores are oriented so that, at least when the shrimp is swimming, the light they produce is directed more or less ventrally. The photophore is covered by a cuticular window which is tinted by colors described as blue (often) and reddish to purple. These colors may serve as spectral filters to adjust the light produced by the photophore. The cuticular window may be thicker than the surrounding exoskeleton to serve as a lens. The photophore may contain many photocytes, the light-producing cells (Fig. 5.9a) or just a single photocyte (Fig. 5.9b). The main part of the photocyte is occupied by a large clear vacuole. Just above the photocyte is cap containing a white pigment (a pterin?) which serves as a reflector. Light from the photocytes is thus directed outward toward the exterior and not inward into the tissues. At the apical end of the

Table 5.2 Taxonomic distribution of bioluminescence reported in some species of decapod shrimp genera

Taxon	Cuticular (Dermal) Photophores	Hepatic Photophores	Hepatic Secretions Via the Mouth	Literature Source
Sergestoidea				
Sergestidae				
Sergestes s.l.	A	P (Organs of Pesta)	A	a, b
Sergia (s.l.) except *Sergia s.s.*	P	A	A	a, b
Penaeoidea				
Aristeidae				
Aristaeomorpha	P	A	A	c
Aristeus	P	A	A	c, d
Cerataspis coruscans	A	A	P	e
Benthesicymidae				
Gennadas	P	A	A	f
Solenoceridae				
Hadropenaeus	P	A	A	g
Hymenopenaeus	P	A	A	g
Mesopenaeus	P	A	A	g
Solenocera	P	A	A	c, h, i, j
Caridea				
Oplophoridae				
Oplophorus	P	A	P	i, j, k
Systellaspis	P	A	P	i, j, k
Janicella	P	A	P	i, j, k
Acanthephyridae				
Acanthephyra	A	A	P	i, j
Ephyrina	A	A	P	i, j
Hymenodora	A	A	P	i, j
Meningodora	A	A	P	i, j
Notostomus	A	A	P	i, j
Pandalidae				
Chlorotocoides	A	P	A	i, j
Heterocarpus	A	A	P	i, j, l
Plesionika	A	A	P	i, l
Stylopandalus	A	P	A	i, j
Thalassocaris	A	P	P	i, j
Pasiphaeidae				
Glyphus	A	A	P	i, j
Leptochela	?	?	?	i, j, m
Parapasiphae	?	?	?	i, n

P present, *A* absent or not yet observed; ?, bioluminescence observed, but body source unknown or unclear

(a) Judkins and Kensley (2008); (b) Vereshchaka (2017); (c) Crosnier (1978); (d) Chan et al. (2017); (e) Herring (1976); (f) Nowel et al. (2002); (g) Pérez Farfante (1977); (h) Yu and Chan (1986); (i) Herring (1985); (j) Herring and Gruner (2004); (k) Nowel et al. (1998); (l) Chan et al. (2008); (m) Chace (1940); (n) Widder et al. (1983)

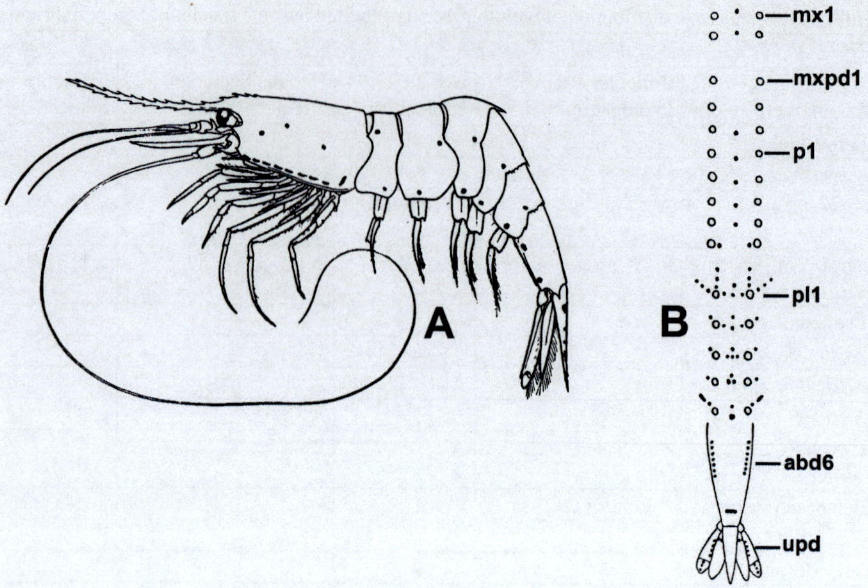

Fig. 5.8 Distribution of cuticular (dermal) photophores in two shrimps. (**a**) *Systellaspis debilis* (Oplophoridae). The black spots on the cephalothorax and abdomen show photophore positions. The ventral set on the abdomen is actually below the abdominal pleura and on the pleopod coxae. Not all photophores may be figured, i.e. may be more numerous than shown. (**b**) Schematic of photophore distribution (black spots and short lines) on the ventral surface of the sergestid *Robustosergia robusta.* Open circles indicate the insertions of appendages. *abd6,* sixth abdominal segment; *mx1,* maxillule; *mxpd1,* maxilliped 1; *p1,* pereopod 1; *pl1,* pleopod 1; *upd,* uropod. Modified from (**a**) Gruner (1992) and (**b**) Dennell (1955)

photocytes, just below the reflecting cap, are paracrystalline bodies which generate the bioluminescent light. There is a sheath of red-pigmented, light-absorbing cells which surround the sides of the photophore. They prevent light from escaping laterally. Thus, the white reflective cap and red-pigment screen guide and focus the light from the photocytes out through the cuticular window into the external environment (Nowel et al. 1998, 2002).

Hepatic photophores are found in all *Sergestes s.l.* species, in which they are termed the "Organs of Pesta" (Dennell 1955; Herring 1985; Herring and Gruner 2004). They are formed from modified terminal portions of the tubules which make up the digestive gland (hepatopancreas). Individuals have four major organs at each anterior and posterior corner of the hepatopancreas, and in some species there is a smaller additional pair a bit anterior or posterior of the mid-hepatopancreas (Foxton 1972) (Fig. 5.10a). Each organ (Fig. 5.10b) is composed of one to several tubules with the photophore portions surrounded on all sides with a screen composed of red-pigmented cells, except ventrally (the aperture). Dorsally in the luminous tubules is a layer of lipid microspheres which serve as a reflective cap. Light produced is directed by the red pigment screen and reflective cap only through the ventral

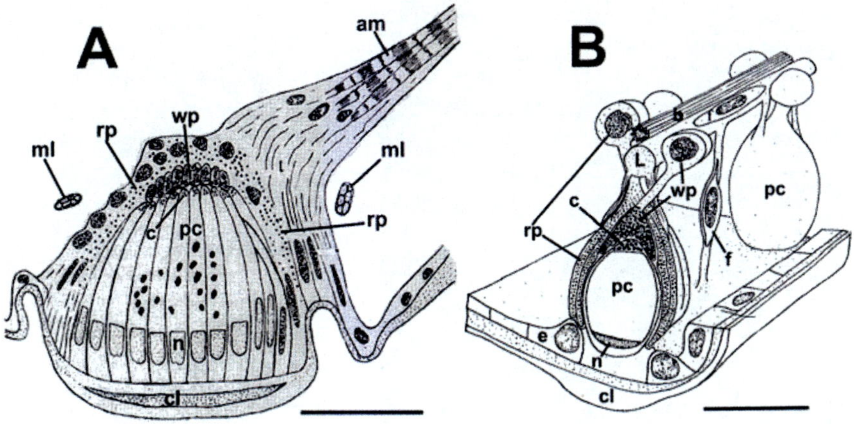

Fig. 5.9 Cuticular photophores of the caridean *Oplophorus spinosus*. (**a**) Section through a pleopod photophore with several photocytes (pc), a red pigment screening layer (rp), dorsal cap of white reflecting pigment (wp) and with the exoskeleton (cl, cuticle) thickened into a cuticular lens. Also shown are an accessory longitudinal muscle (am) and photophore muscle loops (ml), which are involved with tilting and rotation of the photophore. The light-producing paracrystalline bodies (c) are located just below the white reflective cap. (**b**) Section through a single (pc) photophore from the dactyl of the third maxilliped. A second photophore is shown linked to the other by a ligament (L) and both are anchored to the inner side of the cuticle by a fibroblast extension (f) of the ligament. The red pigment sheath (rp) and white pigment cap (wp) are supplied by cells whose basal portions are on either side of the ligament. As in (**a**), the light-producing paracrystalline bodies (c) are below the white reflective cap and above the large clear vacuole of the photocyte (pc). The cuticle (cl) below the photocyte is thickened into a clear lens-like structure. Also labeled are the photocyte nucleus (n), a blood vessel (b) above the ligament and epidermal cell (e). Scale bar represents 200 μm in (**a**) and 40 μm in (**b**). Modified from Nowel et al. (1998), courtesy of University of Chicago Press

cepthalothorax. Below the reflector in each tubule are the photocytes with paracrystalline inclusions which generate the blue-green fluorescent light of the photophores. The bluish tips of the tubules are made up of large cells with a clear homogeneous cytoplasm, perhaps serving as a spectral filter and lens.

Similarly located structures, simply called hepatic photophores, are found the pandalid carideans *Stylopandalus richardi*, *Chlorotocoides*, and *Thalassocaris* (Dennell 1955; Herring 1985; Herring and Gruner 2004) (Table 5.2). They differ from sergestid Organs of Pesta in that there are no paracrystalline structures in the photocytes, and the reflector cap is probably not made of lipid but perhaps a white (pterin) compound.

Sergestids and carideans with either cuticular or hepatic photophores mostly live in the dimly lit mesopelagic zone of the ocean. Some *Sergia s.l.* species with cuticular photophores are benthopelagic at slope depths (Vereshchaka et al. 2014) and so are exposed to dim light during the day as are mesopelagic species in the open ocean. These shrimps migrate up to or near the surface at night and then as dawn approaches, they move down to species-specific depths in the mesopelagic zone. As

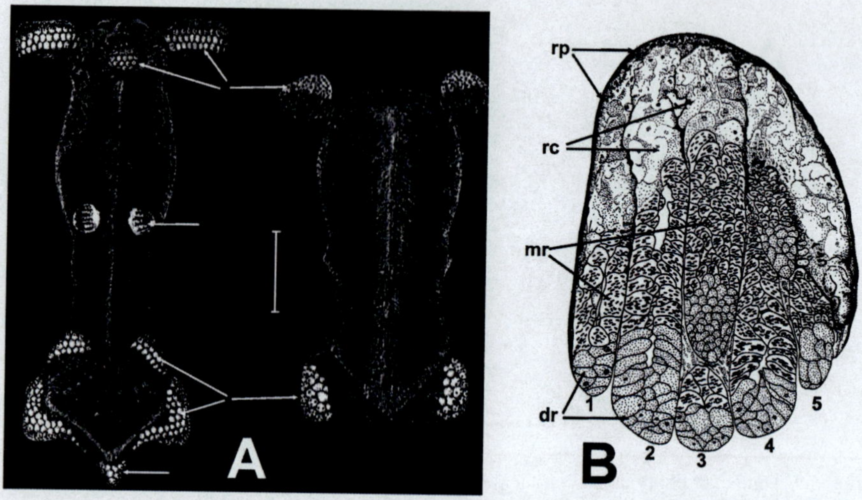

Fig. 5.10 Sergestid Organs of Pesta (OP). (**a**) Whole view of the hepatopancreas with the OP (bright areas indicated by arrows) in *Deosergestes henseni* (left) and *Allosergestes pectinatus* (right). (**b**) Section through five photophore tubules (numbered) of an *Eusergestes arcticus* OP. Labeled are the red pigment layer (rp), reflector cap (rc) of lipid droplets, middle region (mr) with light-generating paracrystalline inclusions, and bluish distal region (dr) composed of cells with homogeneous cytoplasm. Scale bar in (**a**) is 1.0 mm. Modified from Foxton (1972) courtesy of Crustaceana (Brill) and (**b**) from Gruner (1992) modified from Herring (1981)

explained above, their coloration helps them to blend into the dark depths below when viewed *from above*. But when viewed *from below* against the brighter surface waters, their general coloration and opaque internal organs show a silhouette with which predators can identify them. The photophores produce a light that matches, both in intensity and spectral composition, the downward-directed light coming from the surface, as shown by Warner et al. (1979) for *Eusergestes similis*. The shrimp's silhouette is hidden by replacing downwelling light from the surface which has been reflected or absorbed by the shrimp's body with equivalent bioluminescent light from the photophores. The shrimp can maintain this bioluminescence for long periods of time. It can adjust its light output to match temporally varying downwelling light, or when the shrimp changes depth.

Photophores of a horizontally swimming shrimp are located so that their light shines in a generally ventral direction for counterillumination. However, the position of the body may vary greatly as the shrimp twists and turns in its activities. To maintain photophore light in the correct orientation to maintain precise counterillumination, some types of photophores can rotate. This was shown elegantly by Latz and Case (1982), again with *E. similis*. Using captive animals in laboratory experiments, they demonstrated that the stalked posterior Organs of Pesta are capable of rotating up to 140° from the horizontal (pitch) plane with inclinations of the shrimp's body. These compensatory movements of the photophore organs maintain them horizontally so that light is directed ventrally, not at an angle, to match downwelling surface light. The shrimp's movable eyestalks similarly tilt upward to point toward the surface, i.e. the source of downwelling light. With ablation experiments, they showed that these compensatory movements were controlled by the organs of equilibrium and spatial orientation, i.e., the statocysts, located in the antennular peduncle (Chap. 2).

The cuticular photophores described in three genera of oplophorid carideans are also capable of limited tilt and/or rotation (Nowel et al. 1998). In these genera (Table 5.2), simple photophores on the flexor margin of the third-maxilliped dactylus are linked in series by a ligament and associated musculature. When the dactyl is extended and shrimp is horizontal, the linked photophores point downward, but as the tilt of the shrimp's body varies, the direction of photophore light can be adjusted 30–40° by dactylar flexion. The pleopod photophores have a different mechanism. The cuticular window of the photophore is somewhat flexible on three sides. Small muscles inserted on and around these articular areas can tilt the photophores from 40–50° from the horizontal. Thus, these shrimps can adjust their counterillumination with changes in body orientation, although the range of compensation is not as great as that of the sergestid Organs of Pesta. Many *Sergia s.l.* species have cuticular photophores, but they are not capable of tilt compensation (Nowel et al. 2002).

The three pandalid genera, *Stylopandalus*, *Thalassocaris*, and *Chlorotocoides*, have hepatic photophores which are connected to the main body of the hepatopancreas by narrow stalks of tissue. Thus, these organs are also capable of tilt to maintain their light directed ventrally. *Stylopandalus richardi* is mesopelagic (Omori 1974) and most *Thalassocaris* species are pelagic as well. In these groups, the tiltable hepatic organs may serve in counterillumination, as do the sergestid Organs of Pesta.

5.13 Defensive Bioluminescence

Photophores and bioluminescence may serve for other purposes besides counterillumination. One hypothesis is that nighttime bioluminescent flashing may serve as signals in some species for attracting a mate, as shown in marine ostracods by Morin and Cohen (1991) and suggested for the shallow-water pandalid *Thalassocaris crinita* (Marin and Chan 2011). However, in many species, bioluminescence is almost certainly a defensive mechanism. When attacked or otherwise disturbed, a number of meso-and bathypelagic shrimps, mainly carideans, are capable of emitting a luminous stream or cloud of blue to blue-green bioluminescence (~455–490 nm) (Herring 1985; Chan et al. 2008) (Table 5.2). This is wonderfully captured by the photograph of the pandalid *Heterocarpus sibogae* by T-Y Chan and colleagues (Plate 5.3E). When the hepatopancreas is dissected or probed in living shrimps, it may bioluminesce. The evidence to date indicates that the luminous secretions are produced by the hepatopancreas and then temporarily stored in the stomach before emission through the mouth (Herring 1976; Chan et al. 2008). As far as function is concerned, the most reasonable hypothesis is this is a defensive mechanism to startle and confuse an attacking predator, i.e., analogous to the ink clouds emitted by cephalopods as they swim away from a disturbance or predator. Observations from deep-sea submersibles and ROVs show that as the shrimp emits the luminous cloud, it performs the cardioid escape response, i.e., backward tail-flip swimming. It hides or obscures its getaway behind a bioluminescent screen, much like squid and octopi do with dark ink in lighted waters. A final comment: as it currently so difficult to collect, maintain, and observe mesopelagic shrimps, the chemistry and physiology of this defensive secretion is not well known and needs further investigation, as do many other topics about coloration and camouflage.

References

Aréchiga-Palomera MA, Vega-Villasante F, Montoya-Martínez C, Mendoza-González A, Badillo-Zapata D (2018) Background color effect on the pigmentation of prawn *Macrobrachium tenellum*. Lat Am J Aquat Res 46:610–614

Bauer RT (1981) Color patterns of the shrimps *Heptacarpus pictus* and *H. paludicola*. Mar Biol 46: 141–152

Bauer RT (1982) Polymorphism of colour pattern in the caridean shrimps *Heptacarpus pictus* and *H. paludicola*. Mar Behav Physiol 8:249–265

Bauer RT (1985) Diel and seasonal variation in species composition and abundance of caridean shrimp (crustacea, Decapoda) from seagrass meadows on the north coast of Puerto Rico. Bull Mar Sci 36:150–162

Bauer RT (2004) Remarkable shrimps: adaptations and natural history of the carideans. University of Oklahoma Press, Norman

Brown FA Jr (1935) Color changes in *Palaemonetes*. J Morphol 57:317–333

Bruce AJ (1975) Coral reef shrimps and their colour patterns. Endeavour 34:23–27

Brusca RC, Moore W, Shuster SM (2016) Invertebrates, 3rd edn. Sinauer Associates, Inc., Sunderland, Massachusetts

Calado R (2008) Marine ornamental shrimp: biology, aquaculture and conservation. Wiley-Blackwell, New York

Calvo NS, Roldan-Luna M, Argáez-Sosa JA, Martínez-Moreno GL, Mascaró M, Simões N (2016) Reflected-light influences the coloration of the peppermint shrimp, *Lysmata boggessi* (Decapoda: Caridea). J World Aquacult Soc 47:5. https://doi.org/10.1111/jwas.12314

Caves EM, Green PA, Johnsen S (2018) Mutual visual signaling between the cleaner shrimp *Ancylomenes pedersoni* and its client fish. Proc R Soc B 285(1881):20180800. https://doi.org/10.1098/rspb.2018.0800

Chace FA Jr (1940) Plankton of the Bermuda oceanographic expedition. IX. The bathypelagic caridean crustacea. Zoologica (NY) 25:117–209

Chace FA, Manning RB (1972) Two new caridean shrimps, one representing a new family, from marine pools on Ascension Island (crustacea: Decapoda: Natantia). Smithsonian Contrib Zool 131:1–18

Chan BKK, Lin I-C, Shih T-W, Chan T-Y (2008) Bioluminescent emissions of the deep-water pandalid shrimp, *Heterocarpus sibogae* De man, 1917 (Decapoda, Caridea, Pandalidae) under laboratory conditions. Crustaceana 81:341–350

Chan TY, Kumar AB, Yang CH (2017) Photophore counts in the deep-sea commercial shrimp *Aristeus alcocki* Ramadan, 1938 (crustacea: Decapoda: Aristeidae), with a revised key to the Indo-West Pacific species of the genus. Zootaxa 4329:392–400

Chassard C (1956) Polymorphisme des populations d'*Hippolyte varians* Leach et comportement en fonction de leur adaptation chromatique présente. Bull Soc Zool Fr 81:413–418

Chassard-Bouchaud C (1965) L'adaptation chromatique chez la Natantia (Crustacés Décapodes). Cah Biol Mar 6:469–576

Cott HB (1957) Adaptive coloration in animals. Methuen, London

Crosnier A (1978) Crustacés Décapodes Pénéides Aristeidae (Benthesicyminae, Aristeinae, Solenocerinae). Faune de Madagascar 46:1–197

Dall W, Hill BC, Rothlisberg PC, Sharples DJ (1990) The biology of the Penaeidae. Adv Mar Biol 27:1–489

Dennell R (1955) Observations on the luminescence of bathypelagic crustacea Decapoda of the Bermuda area. Zool J Linnean Soc London 42:393–406

Duarte RC, Flores AAV (2017) Morph-specific habitat and sex distribution in the caridean shrimp *Hippolyte obliquimanus*. J Mar Biol Assoc U K 97:235–242

Duarte RC, Steven M, Flores AAV (2016) Shape, colour plasticity, and habitat use indicate morph-specific camouflage strategies in a marine shrimp. BioMed Central Evol Biol 16:218. https://doi.org/10.1186/s12862-016-0796-8

Elias MAM, Anker A, Gawryszewski FM (2019) Microhabitat use and body size drive the evolution of colour patterns in snapping shrimps (Decapoda: Alpheidae: *Alpheus*). Biol J Linn Soc 128:806–816

Flores EE, Chien YH (2011) Chromatosomes in three phenotypes of *Neocaridina denticulata* Kemp, 1918: morphological and chromatic differences measured non-invasively. J Crustac Biol 31:590–597

Foxton P (1972) Further evidence of the taxonomic importance of the organs of Pesta in the genus *Sergestes* (Natantia, Penaeidea). Crustaceana 22:181–189

Gamble FW, Keeble FW (1900) *Hippolyte varians*: a study in color change. Q J Microsc Sci 43: 589–698

Ghidalia W (1985) Structural and biological aspects of pigments. In: Bliss DE, Mantel LH (eds) The biology of crustacea, Vol. 9, Integument, pigments, and hormonal processes. Academic Press Inc., New York, pp 301–393

Goy JW (2010) Infraorder Stenopodidea Claus, 1872. pp. 215–265 in F. R. Schram, J. C. von Vaupel Klein, J. Forest and M. Charmantier-Daures (eds.),Treatise on Zoology – Anatomy, Taxonomy, Biology. The Crustacea, Decapoda.Vol. 9. Part A: Eucarida: Euphausiacea, Amphionidacea, and Decapoda (partim). Koninklijke Brill NV, Leiden

Griebel U (2002) Color vision in marine mammals: a review. pp. 73–87 in M bright, P.C. Dworschak and M. Stachowitsch (eds.), A tribute to Jörg Ott. The Vienna School of Marine Biology: Facultas Universitätsverlag, Wien

Gruner HE (1992) Organes lumineux et luminescence. pp. 411–434 in P-P. Grassé (ed.) Traité de Zoologie. Tome VII (J. Forest, ed.). Crustacés. Morphologie, physiologie, reproduction, systématique. Masson, Paris

Hacker SD, Madin LP (1991) Why habitat architecture and color are important to shrimps living in pelagic *sargassum*: use of camouflage and plant-part mimicry. Mar Ecol Prog Ser 70:143–155

Haddock SHD, Moline MA, Case JF (2010) Bioluminescence in the sea. Annu Rev Mar Sci 2:443–493

Hallberg E, Elofsson R (1989) Construction of the pigment shield of the crustacean compound eye: a review. J Crustac Biol 9:359–372

Herring PJ (1976) Bioluminescence in decapod Crustacea. J Mar Biol Assoc U K 56:1029–1047

Herring PJ (1981) The comparative morphology of hepatic photophores in decapod crustacea. J Mar Biol Assoc UK 61:723–737

Herring PJ (1985) Bioluminescence in the crustacea. J Crustac Biol 5:557–573

Herring PJ, Gruner HE (2004) Luminous organs and luminescence. In: Forest J, von Vaupel Klein JC, Schram FR (eds) Treatise on zoology, anatomy, taxonomy, biology, Vol. 1. The Crustacea. E.J. Brill, Leiden, the Netherlands, pp 381–413

Herring PJ, Roe HSJ (1988) The photoecology of pelagic oceanic decapods. Symp Zool Soc London 59:263–290

Hobbs HH Jr (1973) Two new troglobitic shrimps (Decapoda: Alpheidae and Palaemonidae) from Oaxaca, Mexico. Bull Assoc Mexican Cave Stud 5:73–80
Hobbs HH Jr, Hobbs HH III, Daniel MA (1977) A review of the troglobitic decapod crustaceans of the Americas. Smithsonian Contributions to Zoology 244:183
Hogan BG, Scott-Samuel NE, Cuthill IC (2016) Contrast, contours and the confusion effect in dazzle camouflage. R Soc Open Sci 3:160180. https://doi.org/10.1098/rsos.160180
Hogben L, Slome D (1931) The pigmentary effector system. VI. The dual character of endocrine coordination in amphibian colour change. Proc R Soc Lond B 108:10–53
Holthuis LB (1963) On red coloured shrimps (Decapoda, Caridea) from tropical land-locked saltwater pools. Zoologische Mededelingen, Leiden 38:261–279
Holthuis LB (1973) Caridean shrimps found in land-locked pools at four Indo-West Pacific localities (Sinai peninsula, Funafuti atoll, Maui and Hawaii islands), with the description of one new genus and four new species. Zoologische Verhandelingen Rijksmuseum Natuurlijke Historie Leiden 128:1–48
Jaafar Z, Zeng Y (2012) Visual acuity of the goby-associated shrimp, *Alpheus rapax* fabricius, 1798 (Decapoda, Alpheidae). Crustaceana 85:1487–1497
Jensen GC (2014) Crabs and shrimps of the Pacific coast. A guide to shallow-water decapods from southeastern Alaska to the Mexican border. Mola Marine, Bremerton, Washington
Johnsen S (2001) Hidden in plain sight: the ecology and physiology of organismal transparency. Biol Bull 201:301–318
Judkins D, Kensley B (2008) New genera in the family Sergestidae (crustacea: Decapoda: Penaeidea). Proc Biol Soc Washington 121:72–84
Keeble FW, Gamble FW (1900) The colour-physiology of *Hippolyte varians*. Proc R Soc London B Biol Sci 65:461–468
Knowlton N (1986) Cryptic and sibling species among the decapod crustacea. J Crust Biol 6:356–363
Koller G (1927) Über chromatophoren-system, farbensinn und farbwechsel bei *Crangon vulgaris*. Z Vgl Physiol 5:191–246
Latz MI, Case JF (1982) Light organ and eyestalk compensation to body tilt in the luminescent midwater shrimp, *Sergestes similis*. J Exp Biol 98:83–104
Lee WL (1966) Color change and the ecology of the marine isopod *Idothea* (Pentidothea) *montereyensis* Maloney, 1933. Ecology 47:930–941
Marin I, Chan T-Y (2011) New records of the caridean shrimp family Thalassocarididae bate, 1888 (Decapoda, Caridea) from Asia. Crustaceana 84:243–249
Martin GR, Osorio D (2008) Vision in birds. In: Masland RH, Albright TD (eds) The senses, vol 1. Academic Press, London, pp 25–52
Mcfall-Ngai MJ (1990) Crypsis in the pelagic environment. Am Zool 30:175–188
McLaughlin PA, Camp DK et al (2005) Common and Scientific Names of Aquatic Invertebrates from the United States and Canada: Crustaceans (with CD-ROM). American Fisheries Society Special Publication, vol 31. American Fisheries Society, Bethesda, pp 1–533
McNamara JC, Milograna SB (2015) Adaptive color change and the molecular endocrinology of pigment translocation in crustacean chromatophores. In: Chang ES, Thiel M (eds) The natural history of the crustacea. Volume 4. Physiological regulation. Oxford University Press, Oxford, England, pp 68–101
Merilaita S, Scott-Samuel NE, I.C. (2017) How camouflage works. Philos Trans R Soc B 372: 20160341. https://doi.org/10.1098/rstb.2016.0341
Morin JG, Cohen AC (1991) Bioluminescent displays, courtship. and reproduction in ostracodes. In: Bauer RT, Martin JW (eds) Crustacean sexual biology. Columbia University Press, New York, pp 1–16
Murray J, Hjort J (1912) The depths of the ocean. Macmillan, London
Noël P, Chassard-Bouchaud C (1994) Chromatophores et pigmentation. In: Forest J (ed) Traité de Zoologie. Tome VII. Crustacés. Fascicule I. Masson, Paris, pp 91–105

Nowel M, Shelton PMJ, Herring PJ (1998) Cuticular photophores of two decapod crustaceans, *Oplophorus spinosus* and *Systellaspis debilis*. Biological Bulletin Woods Hole 195:290–307
Nowel MS, Shelton PMJ, Herring PJ, Gaten E (2002) Observations on the cuticular photophores of the sergestid shrimp *Sergia grandis* (Sund, 1920). Crustaceana 75:3–4. Special Issue for Prof. Jacques Forest): 551-566
Omori M (1974) The biology of pelagic shrimps in the ocean. Adv Mar Biol 12:233–324
Parisenti J, Beirão LH, Mouriño JK, Vierira FDN, Buglione CC, Maraschim M (2011) Effect of background color on shrimp pigmentation. Bol Inst Pesca 37:177–182
Pérez Farfante I (1977) American solenocerid shrimps of the genera *Hymenopenaeus, Haliporoides, Pleoticus, Hadropenaeus* new genus, and *Mesopenaeus* new genus. Fish Bull 75:261–346
Rao KR (1985) Pigmentary effectors. In: Bliss DE, Mantel L (eds) The biology of crustacea, Vol. 9, Integument, Pigments, and Hormonal Processes. Academic Press, New York, pp 395–462
Sainte-Marie B, Bérubé I, Brillon S, Hazel F (2006) Observations on the growth of the sculptured shrimp *Sclerocrangon boreas* (Decapoda: Caridea). J Crustac Biol 26:55–62
Shuster S, Machnik P, W. Schulze W. (2011) Behavioral assessment of the visual capabilities of fish. In: Farrell AP (ed) Encyclopedia of fish physiology: from genome to environment, Volume 1. Academic Press, San Diego, pp 143–149
Siegenthaler A, Mastin A, Dufaut C, Mondal D, Benvenuto C (2018) Background matching in the brown shrimp *Crangon crangon*: adaptive camouflage and behavioural plasticity. Sci Rep 8: 3292 l. https://doi.org/10.1038/s41598-018-21412-y
Stubbs AL, Stubbs CW (2016) Spectral discrimination in color blind animals via chromatic aberration and pupil shape. Proc Natl Acad Sci 113:8206–8211
Thomson CM, Herring PJ, Campbell AK (1995) Evidence for de novo biosynthesis of coelenterazine in the bioluminescent mid water shrimp, *Systellaspis debilis*. J Mar Biol Assoc UK 75:165–171
Umeton D, Read JCA, Rowe C (2017) Unravelling the illusion of flicker fusion. Biol Lett 13: 20160831. https://doi.org/10.1098/rsbl.2016.0831
Vaught RC, Havird JC, Santos SR (2014) Chromatosome variation in the anchialine shrimp *Halocaridina rubra* Holthuis, 1963 (Caridea: Atyidae). J Crustac Biol 34:647–657
Vereshchaka AL (2017) The shrimp superfamily Sergestoidea: a global phylogeny with definition of new families and an assessment of the pathways into principal biotopes. R Soc Open Sci 4: 170221. https://doi.org/10.1098/rsos.170221
Vereshchaka AL, Olesen J, Lunina AA (2014) Global diversity and phylogeny of pelagic shrimps of the former genera Sergestes and Sergia (crustacea, Dendrobranchiata, Sergestidae), with definition of eight new genera. PLoS One 9:e112057. https://doi.org/10.1371/journal.pone.0112057
Vestheim H, Kaartved S (2009) Vertical migration, feeding and colouration in the mesopelagic shrimp *Sergestes arcticus*. J Plankton Res 31:1427–1435
Wade NM, Tollenaere A, Hall MR, Degnan BM (2009) Evolution of a novel carotenoid-binding protein responsible for crustacean shell color. Mol Biol Evol 2:1851–1864
Warner JA, Latz MI, Case JF (1979) Cryptic bioluminescence in a midwater shrimp. Science, NY 203:1109–1110
Widder EA, Latz MI, Case JF (1983) Marine bioluminescence spectra measured with an optical multichannel detection system. Biol Bull 165:791–810
Yu HP, Chan T-Y (1986) The illustrated penaeoid prawns of Taiwan. Southern Materials Center, Inc, Tapei, The Republic of China

Chapter 6
Reproduction

6.1 Complexity of Shrimp Reproductive Morphology

The evolutionary goal of any bacterium, fungus, plant or animal species is to reproduce itself. All aspects of an organism's biology are ultimately related to reproduction. The basic structures and mechanisms which carry out this process are important to understand. Although there has been much study on reproduction of shrimps, there is still a lot to learn and many questions to answer. In Chap. 1, I briefly compared the overall variation in different biological aspects among the major shrimp groups, e.g. taxonomic diversity (number of species), ecological diversity (habitats and geographic distribution), and complexity of reproductive organs. It was stated that the carideans were by far more diverse in taxonomic and ecological diversity than the other major shrimp taxa. However, the dendrobranchiates, especially the penaeoid shrimps, are by far more complex and varied in their reproductive morphology than the carideans, stenopodideans, and procarididеans. In this chapter, the description of reproductive functional morphology will make this apparent, but keep in mind that the ultimate cause (selection pressure) and adaptive value of this dendrobranchiate reproductive diversity and complexity is still a mystery. Why do the penaeoids need such bewilderingly intricate genitalia, male reproductive tracts, and spermatophores to carry out insemination and fertilization compared to other shrimp taxa? This and other topics discussed in this chapter will lay the groundwork for the upcoming chapters on sexual systems, mating systems, and life histories.

6.1.1 *Female Gonads: Ovaries*

The organs which produce eggs and sperm are the gonads: ovaries in the female, testes in the male. The gonads are always paired in shrimps, although when mature and full of sex cells, they may appear clumped into a single mass. The ovaries are

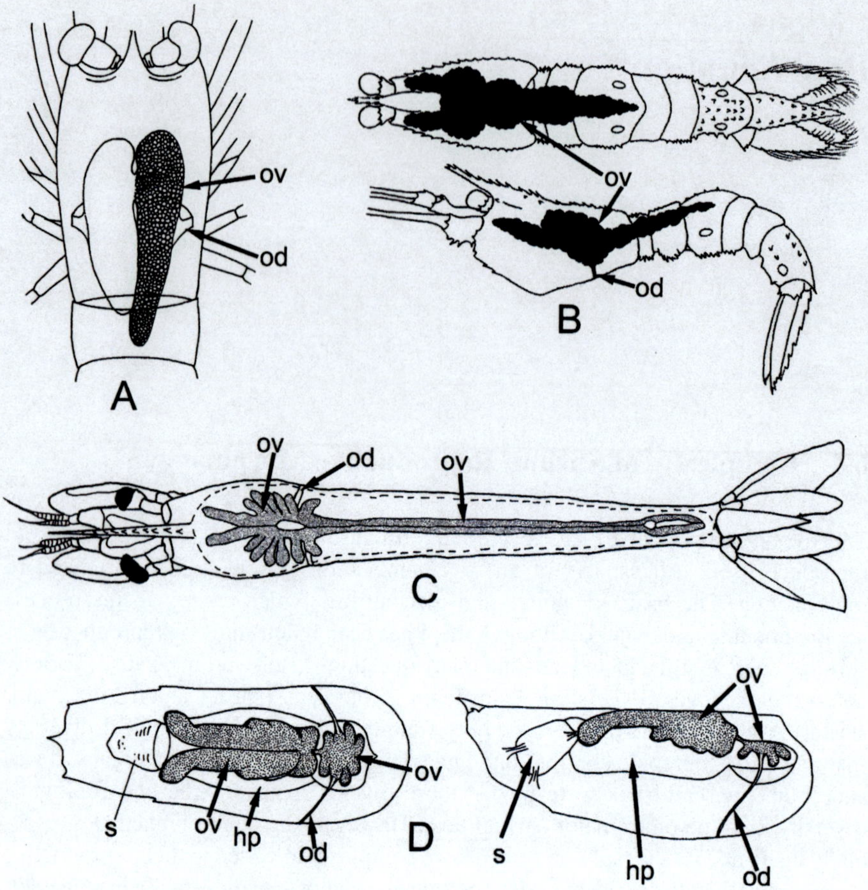

Fig. 6.1 Female reproductive systems. (**a**) *Processa edulis* (Caridea: Processidae) cephalothorax, dorsal view. (**b**) *Stenopus hispidus* (Stenopodidea: Stenopodidae): upper, dorsal view; lower, lateral view. (**c**) *Litopenaeus setiferus* (Penaeoidea: Penaeidae), dorsal view. (**d**) Cephalothorax of *Eusergestes similis* (Sergestoidea: Sergestidae): left, dorsal view; right, lateral view. hp, hepatopancreas; od., oviduct; ov, ovary; s, stomach; modified from (**a**) Noël (1976), (**b**) Gregati et al. (2014), (**c**) King (1948) and (**d**) Genthe (1969)

mainly located in the cephalothorax dorsal to the hepatopancreas and above the cardiac stomach when fully developed, i.e. full of yolk-filled (vitellogenic) egg cells (oocytes). Posteriorly, they dip below the heart, located just anterior to the cephalothoracic–abdominal junction. In carideans and stenopodideans, mature ("ripe") ovaries usually extend into the first two or three abdominal somites (Fig. 6.1a, b, Plate 6.1a, b) but even further in some genera, e.g. *Nematocarcinus* (Wenner 1979). In penaeoid and procaridid shrimps, the mature prespawning ovaries extend dorsally all the way back into the last posterior abdominal somite (Felgenhauer et al. 1988; Dall et al. 1990) (Plate 6.1c).

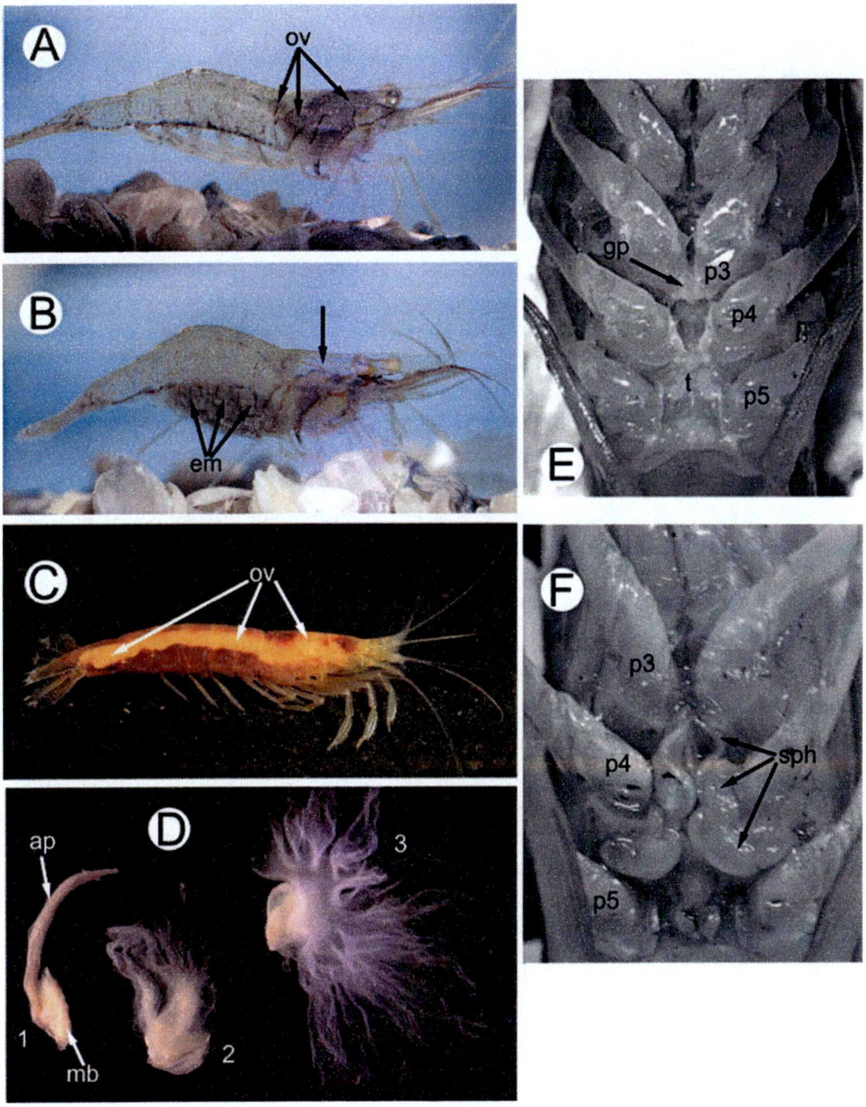

Plate 6.1 Gonads and Spermatophores. (**a**) Prespawning female *Palaemon (Palaemonetes) pugio* with mature ovaries. (**b**) Female *P. pugio* with recently spawned embryos attached below abdomen; space occupied by mature ovaries in (**a**) appears empty in (**b**), (unmarked arrow) because post-spawn ovaries are small and transparent. (**c**) Gravid female of procarididean *Vetericaris chaceorum,* with ovaries extending to end of abdomen. (**d**) Spermatophore reaction in penaeid shrimp *Farfantepenaeus duorarum*. From left to right: (1) newly emitted single spermatophore just exposed to seawater; (2) tightly rolled and pleated appendage begins to unfold; (3) appendage completely unfolded. The latter will cover and seal the aperture to the median seminal receptacle. (**e**, **f**) Underside of posteroventral thorax of a female *Solenocera melantho*, showing (**e**) thelycum area prior to insemination and (**f**) after insemination with twin spermatophores attached. ap, appendage; em, embryos; gp, gonopore; mb, main body; ov, ovaries; p3–p5, basal segments of pereopods 3–5; sph, spermatophore; t, center of thelycum. Courtesy of (**a**, **b**) R.T. Bauer, (**c**) Troy

In penaeoids, the cephalothoracic portion of each ovary is produced laterally into several lobes (Fig. 6.1c). The maturity or nearness to spawning of ovaries in a shrimp can conveniently be determined by observing the size and color of the ovaries located below the thin cuticle between the cephalothorax and first abdominal somite. In sergestids, the ovaries are restricted to the cephalothorax (Genthe 1969) (Fig. 6.1d). During spawning in all groups, the mature oocytes move through the oviducts, which extends from the ovaries to the gonopores located on the coxa of pereopod 3 (first walking leg) where the eggs are released to the outside. (Fig. 6.1).

6.1.2 *Male Gonads and Spermatophores*

In all decapod shrimps, the paired tubular male testes are similarly situated within the cephalothorax as are the female ovaries. Each testis is connected to its sperm duct (vas deferens) (Fig. 6.2a, b). Sperm produced by a testis move into the narrow proximal vas deferens (PVD), then pass into the often highly coiled median vas deferens (MVD), and then into the distal vas deferens (DVD), which terminates into a muscular ejaculatory duct (often termed "terminal ampoule" in dendrobranchiates) (Fig. 6.2b, c). The latter opens to the exterior through the male gonopore which opens on or close to the coxae of the last walking leg (fifth pereopod). During their passage through the vas deferens, the sperm are mixed with a mucoid gelatinous material to which other secreted layers may be added. In genera emitting simple spermatophores, the vasa deferentia are relatively uncomplicated tubes without regional differentiation. The resulting spermatophore is stored in the ejaculatory duct. The latter may have an evagination, the ejaculatory bulb, which holds additional spermatophore material, and both are surrounded by striated muscle bands which power ejaculation (Fig. 6.2c). In all shrimps, the androgenic gland, the source of male hormones, is situated on or just prior to the distal DVD (Charniaux-Cotton and Payen 1985; Sagi et al. 1997; Bortolini and Bauer 2016) (Fig. 6.2a, c).

6.1.3 *Male Gonadal System: Caridea, Stenopodidea, and Procarididea*

In many carideans and stenopodideans, the sperm mass is ejaculated as an amorphous spermatophore ("sperm bearer") which adheres to the female sternum between the bases of the last two walking legs (e.g., Nouvel 1939; Gregati et al. 2014). In the thorid *Heptacarpus sitchensis*, the spermatophore is ejected from each

---

Plate 6.1 (continued) Sakihara, Hawaii Department of Land and Natural Resources, (**d**) Bauer and Cash (1991), and (**e**, **f**) Jun Ohtomi

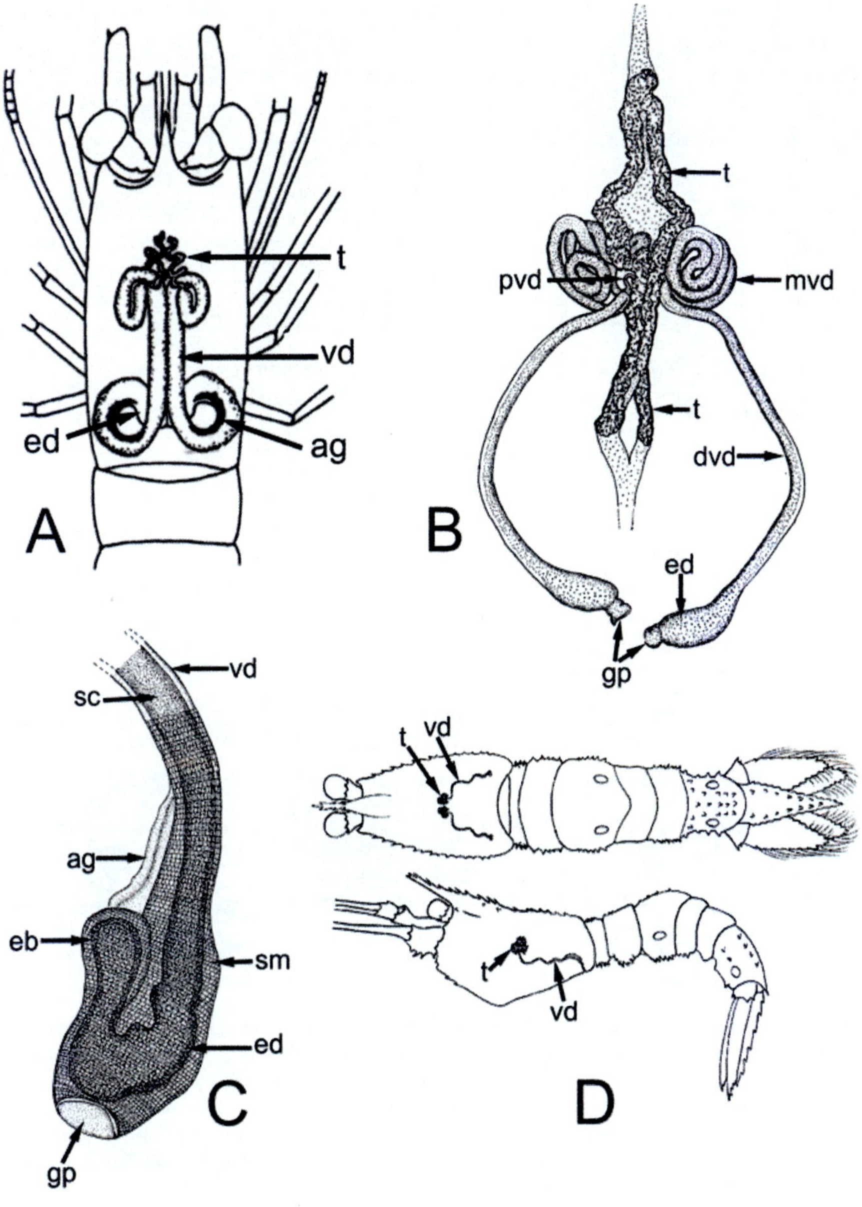

Fig. 6.2 Male reproductive system of carideans and stenopodideans. (**a**) Testes and associated ducts within the cephalothorax of the caridean *Processa edulis* (Processidae). (**b**) Testes and associated ducts in *Macrobrachium rosenbergii* (Palaemonidae). (**c**) Ejaculatory duct, bulb, and attached androgenic gland in *Heptacarpus sitchensis* (Thoridae). (**d**) Reduced male system in *Stenopus hispidus* (Stenopodidae). ag, androgenic gland; dvd, distal vas deferens; eb, ejaculatory bulb; ed; ejaculatory duct; mvd, median vas deferens; pvd, proximal vas deferens; sc, sperm cord; sm, striated muscle; t, testes; vd, vas deferens. Modified from (**a**) Noël (1976); (**b**) Chow et al. (1982); (**c**) Bauer (1976); (**d**) Gregati et al. (2014)

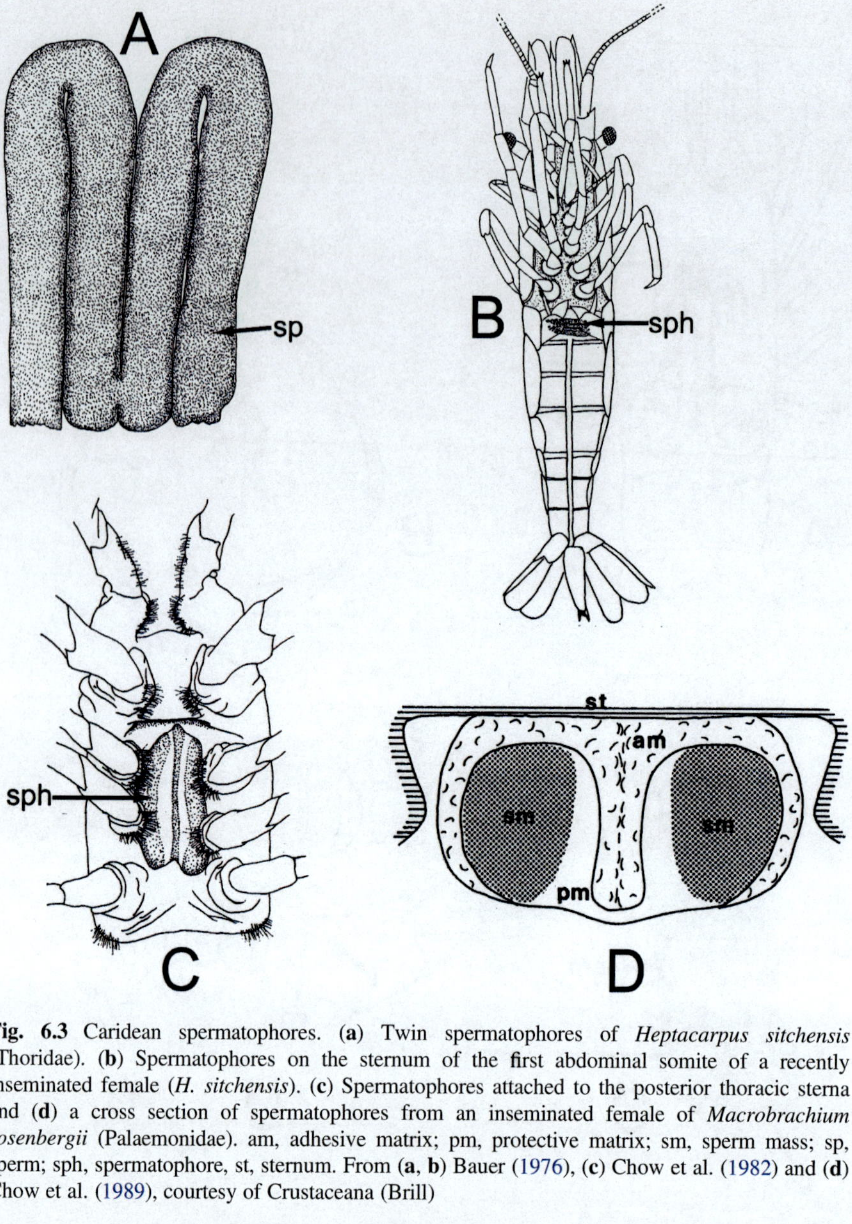

Fig. 6.3 Caridean spermatophores. (**a**) Twin spermatophores of *Heptacarpus sitchensis* (Thoridae). (**b**) Spermatophores on the sternum of the first abdominal somite of a recently inseminated female (*H. sitchensis*). (**c**) Spermatophores attached to the posterior thoracic sterna and (**d**) a cross section of spermatophores from an inseminated female of *Macrobrachium rosenbergii* (Palaemonidae). am, adhesive matrix; pm, protective matrix; sm, sperm mass; sp, sperm; sph, spermatophore, st, sternum. From (**a**, **b**) Bauer (1976), (**c**) Chow et al. (1982) and (**d**) Chow et al. (1989), courtesy of Crustaceana (Brill)

gonopore as a simple gelatinous sperm-filled tube which meets a similar one from the other side, forcing the two sperm masses to bend into a U-shape (Fig. 6.3a). This is deposited as a discrete structure on the underside of the first abdominal somite during copulation before spawning (Fig. 6.3b) (Bauer 1976). In *Crangon crangon*, the spermatophores may be inserted directly into the female gonopores, although there are observations to the contrary (Boddeke et al. 1991). In many carideans and stenopodideans, the spermatophores are deposited on the sterna of the last two or three thoracic somites.

However, in some palaemonids and the atyid *Potimirim potimirim*, spermatophores are more complex. In these species, the sperm mass is surrounded by a separate adhesive material for attachment to the female sterna, as well as a protective layer which prevents contact of the sperm with the outside environment until spawning (Fig. 6.3c, d). Different types of secretory epithelia produce and add these additional materials as the developing spermatophore passes through the proximal, medial, and distal portions of the vas deferens (Chow et al. 1982, 1989; Machado et al. 2021).

The only stenopodidean in which the male system has been studied is *Stenopus hispidus* (Gregati et al. 2014). The male reproductive tract of this shrimp is tiny compared to carideans and especially penaeoids (Fig. 6.2d). The very small testes connect to the gonopores by slender vasa deferentia. The small size of the male system may be related to its mating system, which is socially and sexually monogamous (see Chap. 8). The sticky ejaculated spermatophores are deposited on and adhere to the posterior two or three thoracic somites during copulation in *S. hispidus*. However, Saito and Koya (2001) found a spermatophore within the oviduct of *Spongicola japonica* and postulated that sperm from a single copulation might be used to inseminate more than one brood. As with the caridean *Crangon crangon*, I find this doubtful because the spermatophore would be swept out by eggs during spawning. Unlike some decapod crabs, there is no separate external seminal receptacle (spermatheca) or one confluent with the oviduct to store leftover sperm after spawning. However, there is a need for further research to determine if a hypothesis of sperm storage within the oviducts can be accepted in *Crangon* and *Spongicola.*

Although little studied, the male system of procarididеans appears similar to that of carideans (Felgenhauer et al. 1988).

6.1.4 Male Gonadal System: Penaeoidea and Sergestoidea

The highest diversity of male reproductive tracts is found in the dendrobranchiate shrimps and is directly related to the structure and function of their spermatophores. In those species in which the male system has been described, the more complex are those which produce an external spermatophore attached to the underside of the female (aristeids, solenocerids, some penaeids and benthesicymids). Perhaps the best group to begin with is *Penaeus* (*s.l.*) species, several of which have been intensively studied not only for their basic reproductive biology but also because of the commercial importance (see references in Bauer 1991; Bauer and Cash 1991; Feng et al. 2018a, b). In the genus *Litopenaeus* ("white shrimps"), the twin spermatophores are quite complex and are attached externally to the modified coxae and sterna of the last three thoracic somites, i.e., the thelycum (see Sect. 6.2, below). Each spermatophore consists of several components which help attach and adhere it to the female. For example, in *L. setiferus*, Pérez Farfante (1975) named a dozen different parts (e.g., flaps, flanges, blades, wings, plates, glutinous materials). The sperm mass of each spermatophore is contained a hardened sac which fuses with that of the other spermatophore to form a capsule termed the geminate body. The various other components attach the spermatophores to the female thelycum (Fig. 6.4A, B). As in all penaeoid males, there are paired testes, each with several lobes, vasa deferentia, and ejaculatory ducts (Fig. 6.4C). Each vas deferens is composed of a proximal, ascending and descending median portion leading to the distal vas deferens which terminates in the ejaculatory duct. Two longitudinal folds called typhlosoles project into the lumen of the vas and ejaculatory duct, dividing their spaces into sperm and accessory ducts. The epithelia of the accessory ducts secrete most of the various acellular components of the spermatophore which will surround and attach to the sperm sac in the sperm duct. Additionally, the ejaculatory duct of *Litopenaeus* has a multichambered adhesive gland (Fig. 6.4C) which secretes the glutinous substance added onto the spermatophore during ejaculation. As spawning occurs soon (within a few hours) after mating in *Litopenaeus* (Yano et al. 1988), the external sperm storage on the female is temporary.

In all the other genera in *Penaeus* (*s.l.*), e.g., *Farfantepenaeus*, males make simpler spermatophores which are stored on the female for long periods prior to spawning in a median external pocket (seminal receptacle or spermatheca) formed by a pair of evaginated sternal plates of the last thoracic somite (Fig. 6.4D). The spermatophores emitted during ejaculation consist of a main body or sperm sac which contains the sperm cells and surrounding secretion (Fig. 6.4E). Attached to it is a highly pleated appendage which, when exposed to seawater, unfolds into a beautifully delicate membrane which will cover and seal the aperture of the median seminal receptacle. (Bauer and Cash 1991; Feng et al. 2018a) (Plate 6.1d). The vasa deferentia have similar but less complex internal partitions than in *Litopenaeus* species. The ejaculatory ducts do not have an adhesive gland as in *Litopenaeus*. Instead, each is divided into two connecting chambers in which the appendage and main body of the spermatophore are stored until ejaculation (Fig. 6.4F).

Females of numerous other penaeid genera store spermatophores within twin seminal receptacles which are internalized cuticular pockets formed by invaginations into the posterior thoracic sterna (thelycum). A well-studied example is that of *Rimapenaeus* (formerly *Trachypenaeus*) *similis* (Plate 6.2a–c). The thelycum has a pair of lateral plates which cover a median sternal pocket that serves as an antechamber to the openings of the seminal receptacles. As in *Penaeus* (*s.l.*) species, the paired testes have several lobes, each composed of highly convoluted seminiferous tubules in which spermatogenesis and sperm production take place (Fig. 6.5A). A very thin proximal vas deferens (PVD) runs along the inner edges of the testicular lobes and collects sperm from the seminiferous tubules. The PVD joins the highly coiled internally undivided median vas deferens (MVD) which secretes a gelatinous material that mixes with and separates groups of sperms into small round sperm packets (spermatophores) of variable size (Plate 6.2d). These continue into and fill the distal vas deferens (DVD), a long simple tube that leads into and fills one part of the ejaculatory duct. The latter has an additional chamber which produces an

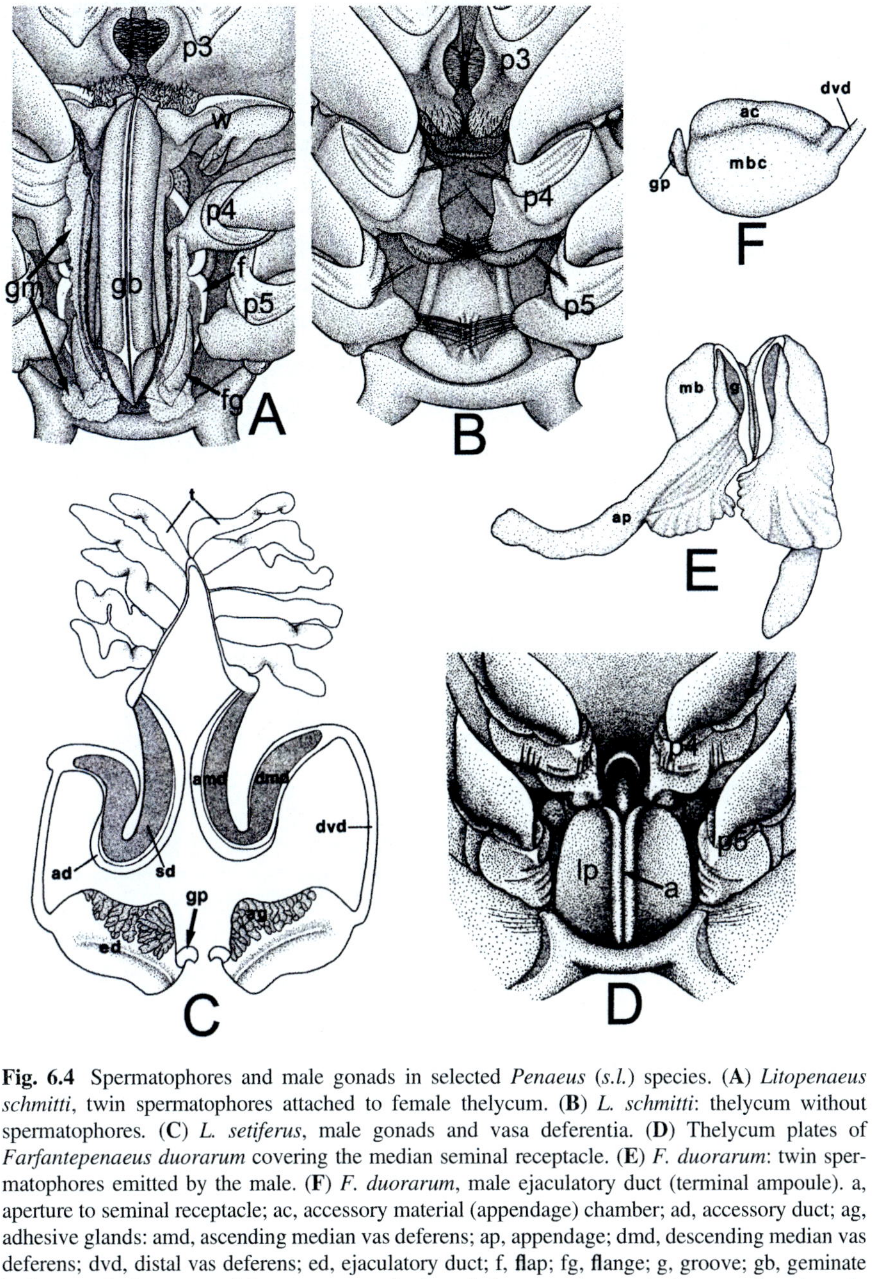

Fig. 6.4 Spermatophores and male gonads in selected *Penaeus* (*s.l.*) species. (**A**) *Litopenaeus schmitti*, twin spermatophores attached to female thelycum. (**B**) *L. schmitti*: thelycum without spermatophores. (**C**) *L. setiferus*, male gonads and vasa deferentia. (**D**) Thelycum plates of *Farfantepenaeus duorarum* covering the median seminal receptacle. (**E**) *F. duorarum*: twin spermatophores emitted by the male. (**F**) *F. duorarum*, male ejaculatory duct (terminal ampoule). a, aperture to seminal receptacle; ac, accessory material (appendage) chamber; ad, accessory duct; ag, adhesive glands: amd, ascending median vas deferens; ap, appendage; dmd, descending median vas deferens; dvd, distal vas deferens; ed, ejaculatory duct; f, flap; fg, flange; g, groove; gb, geminate body; gm, glutinous material; gp, gonopore; lp, lateral plate covering seminal receptacle; mb, main body of spermatophore; mbc, main body chamber; p3–5, basal segments of female pereopods 3–5; sd, sperm duct; t, testicular lobes; w, wing. From (**A**, **B**) Pérez Farfante (1975), (**C**, **E**, **F**) Bauer and Cash (1991) and (**D**) Pérez Farfante (1969)

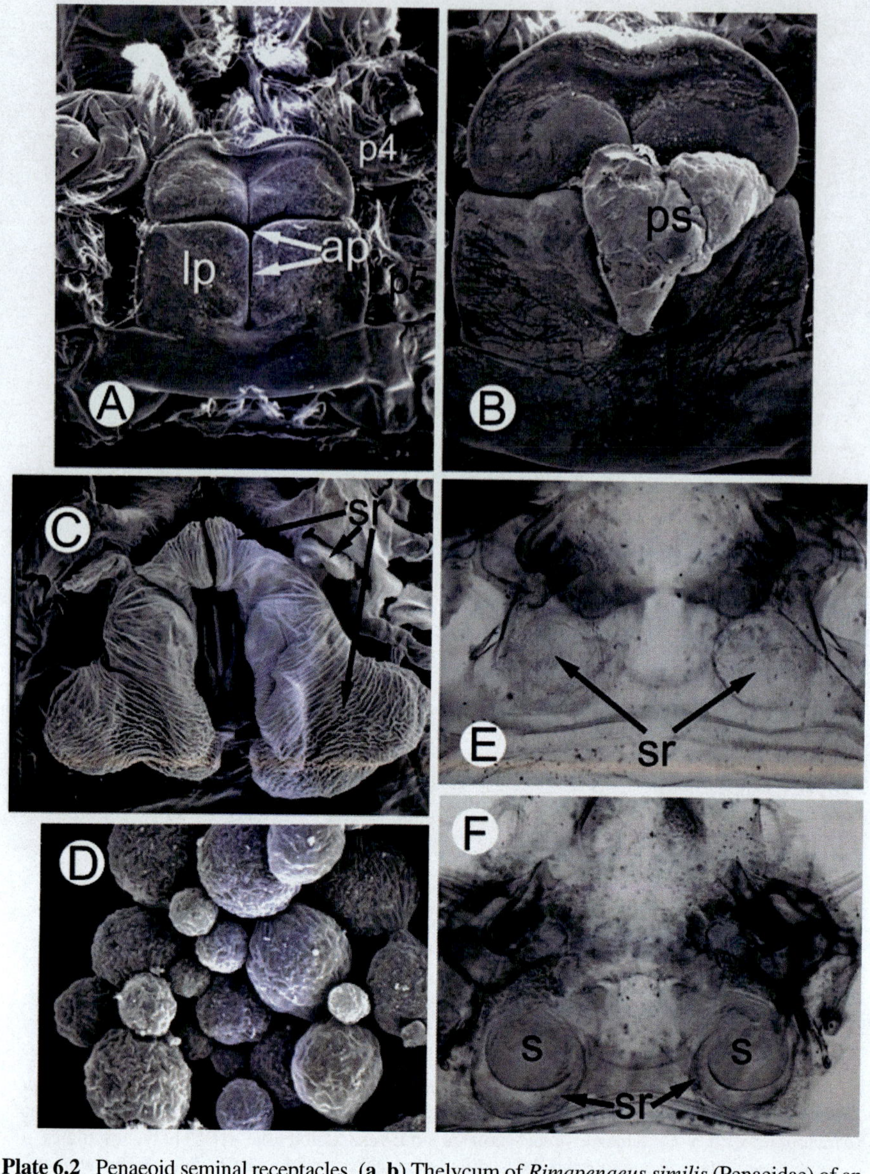

Plate 6.2 Penaeoid seminal receptacles. (**a**, **b**) Thelycum of *Rimapenaeus similis* (Penaeidae) of an (**a**) uninseminated and (**b**) inseminated female with male mating plug blocking the aperture to the median pocket, the antechamber to the seminal receptacles. (**c**) *R. similis:* Internal view from above of the seminal receptacles on the floor of the posterior cephalothorax. (**d**) *R. similis* spermatophores (sperm packets) before transfer from male ejaculatory duct. (**e**, **f**) Seminal receptacles of *Sicyonia dorsalis* (Sicyoniidae) of an (**e**) uninseminated female (receptacles empty) and a (**f**) fully inseminated female (receptacles full of sperm). ap, aperture to medial pocket of seminal receptacle; lp, lateral plate; ps, plug substance (mating plug); p4–p5, basal segments of pereopod 4–5; s, sperm mass; sr, seminal receptacle. From (**a**–**d**) Bauer and Lin (1993) and (**e**, **f**) Bauer (1991)

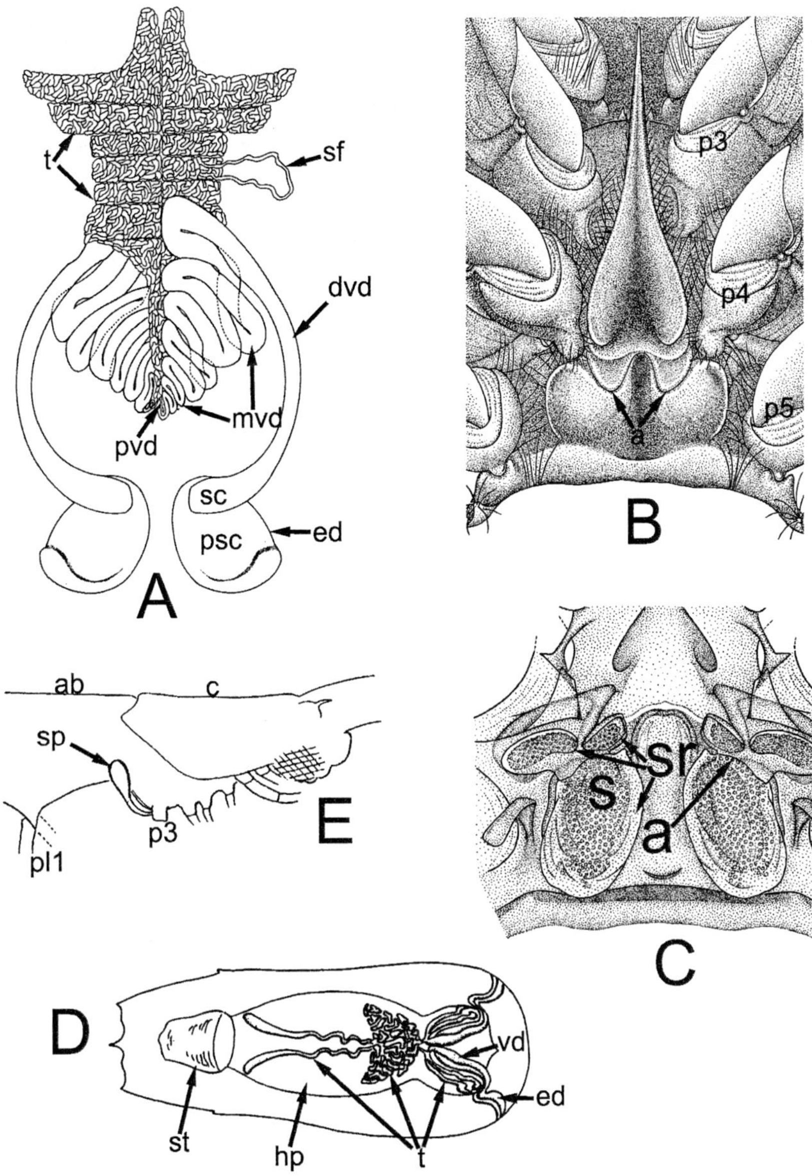

Fig. 6.5 Male gonads and female reproductive morphology in selected dendrobranchiate species. (**A**) Male reproductive tract of *Rimapenaeus similis* (Penaeidae). (**B**) Thelycum of *Sicyonia edwardsii* (Sicyoniidae) showing apertures to seminal receptacles. (**C**) Internal (dorsal) view of seminal receptacles full of sperm within ventral cephalothorax of female *S. edwardsii.* (**D**) Dorsal view (anterior to the left) of interior cephalothorax of *Eusergestes similis* showing male reproductive system. (**E**) Diagram of female *Lucifer penicillifer* cephalothorax and anterior abdomen, showing gourd-shaped spermatophore attached near female gonopore at base of third pereopod. a, apertures to seminal receptacle; ab, abdomen; c, cephalothorax; dvd, distal vas deferens; ed, ejaculatory duct; hp, hepatopancreas; mvd, median vas deferens; pl, pleopod 1; p3–5, basal portions of pereopods 3–5; psc, plug substance chamber; pvd, proximal vas deferens; s, sperm; sc, spermatophore chamber; sf, seminiferous tubule; sp, spermatophore; sr, seminal receptacle; st, stomach;

acellular plug substance. During insemination, the many sperm packets will be deposited within the seminal receptacles. The plug substance will fill the median pocket with the openings to the seminal receptacles and block the external aperture to the median pocket (Plate 6.2a, b). This substance serves as a mating plug that presumably prevents other males from inseminating the female ("paternity assurance device"). It also may protect the stored sperm from the outside environment until they are released to fertilize eggs during spawning. Similar male reproductive tracts, spermatheca and spermatophores have been described in the penaeids *Xiphopenaeus* and *Parapenaeus* (Burkenroad 1934).

In the Aristeidae, Solenoceridae, and some Benthesicymidae, the spermatophores are placed externally onto the female's thelycum. Well-described spermatophores have been found attached to females in the aristeids *Aristaeomorpha foliacea* and *Aristeus antennatus* (Heldt 1938; Tunesi 1987; Orsi Relini and Tunesi 1987). During the reproductive season, many females bear conjoined twin spermatophores (Heldt 1938; Demestre 1995), which indicates that spawning does not occur immediately after mating as in *Litopenaeus* and some solenocerids and perhaps some benthesicymids (see below). The testes, vasa deferentia, and ejaculatory ducts of aristeids are similar overall to those of penaeids. The vasa deferentia and ejaculatory duct are incompletely divided internally by a single typhlosole into two channels, one the sperm duct and the other an accessory duct, whose inner epithelia secrete various substances attached to the sperm mass. A continuous cord of sperm mass and accessory materials is found in the MVD, DVD, and into the ejaculatory duct. Upon ejaculation, sphincter muscles of the latter pinches off a length of this cord, forming one spermatophore. The spermatophore per se can be divided into a "wing" and a "sperm sac." The wings of the compound spermatophore face outward (ventrally) from the female and lodge it firmly within the thelycum. The sperm sac and adjacent materials rest within the depression of the thelycum. The accessory materials may protect the sperm from the outside environment until they are released during spawning.

In the Solenoceridae, spermatophores have been found on females on only four of the ten genera: *Pleoticus, Hadropenaeus, Mesopenaeus*, and *Solenocera* (Pérez Farfante 1977; Ohtomi and Yamamoto 2001) (Plate 6.1e, f). The twin spermatophores are externally placed on the female, and they are similar to those of *Litopenaeus* species in that there are a variety of wings, flanges, and plates attached to the pod-like sperm sacs (Plate 6.1f). These structures hook around the bases of the posterior pereopods and/or under protuberances of the female thelycum, holding the spermatophores firmly onto the female until spawning. In *Pleoticus* species, the spermatophores are commonly found on females with mature ovaries (Anderson and Lindner 1971), indicating that, as in aristeids, spawning does not occur immediately after mating, and sperm are stored on the female for some unknown period, perhaps

Fig. 6.5 (continued) t, testicular lobes; vd, vas deferens. From (**A**) Bauer and Lin (1993), (**B–C**) Pérez Farfante (1985), (**D**) Genthe (1969) and (**E**) Gordon (1956)

days to weeks. In contrast, reports of spermatophores on *Solenocera* females are uncommon, even when numerous reproductive females are examined (Heldt 1938; Ohtomi and Yamamoto 2001; Villalobos-Rojas and Wehrtmann 2014). These observations indicate that spawning must occur very soon after mating with spermatophore remnants soon discarded, as in *Litopenaeus*. This was confirmed in *S. melantho* by Ohtomi and Yamamoto (2001).

The small pelagic penaeoid family Benthesicymidae, with 12 genera, shows an interesting parallel evolution with the Penaeidae in spermatophore attachment and storage (Burkenroad 1936). In the genera *Altelatipes, Benthonectes*, and some species of *Benthesicymus*, the thelycum is open and the spermatophores, not yet observed, are apparently attached externally there. In *Bentheogennema, Gennadas*, and other species of *Benthesicymus*, the sperm is stored, either as an amorphous mass or in small packets (as in the penaeid *Rimapenaeus*) within twin seminal receptacles that vary in shape from that of sausages to a shallow depression in the thelycum. Burkenroad observed that males of open-thelycum benthesicymids deposit complex accessory substances around the sperm in the vasa deferentia, so that perhaps complex spermatophores are indeed attached to the open thelycum. If so, the females must spawn soon after mating with used spermatophores quickly discarded, as in *Litopenaeus* and *Solenocera* species. He also found accessory substances in the male genital tract of other benthesicymid species in which sperm in stored in female seminal receptacles. But he never observed these accessory substances on inseminated females, so that they must be discarded during insemination without sealing of seminal receptacles after mating as in some penaeids (e.g., *Farfantepenaeus, Rimapenaeus*).

The simplest penaeoid male reproductive tract is found in the sicyoniids (Bauer 1991). The vasa deferentia leading from the testes to the gonopores are like many carideans with simple spermatophores. The vasa are simple undivided tubes packed with spermatophoric mass terminating in a muscular ejaculatory duct. This mass is deposited and stored within internalized seminal receptacles for later spawning by the female (Fig. 6.5C; Plate 6.2e, f).

The penaeoid shrimp expert Martin Burkenroad always presented evolutionary hypotheses about his work. He proposed that the open thelycum with external spermatophores is the primitive condition in penaeoids. Internal sperm storage in seminal receptacles is the advanced (derived) condition (Burkenroad 1936). However, he observed that since the openings to the seminal receptacles of some benthesicymids are between the sixth and seventh thoracic somites of the female, but those of penaeids and sicyoniids are between the seventh and eighth, that evolution of the open thelyca to closed thelyca with seminal receptacle(s) must have occurred independently in the two groups (parallel evolution). Since Burkenroad's meticulous work on the benthesicymids, little has been done and much is still unknown about their reproductive biology. One interested in the reproductive and insemination biology of dendrobranchiates should begin with Burkenroad's major works (1934, 1936) on the subject as well as the more recent and amply illustrated Pérez Farfante and Kensley (1997). The brief but illuminating early comparison and summary of penaeoid reproductive morphology by Heldt (1938) is also recommended.

In sergestids and luciferids, the spermatophores are elliptical gourd-shaped structures from which projects a narrow curved tube or stalk, often termed the "wing" (Burkenroad 1934; Gordon 1956; Genthe 1969) (Fig. 6.5E). Interestingly, they are very similar in appearance and placement on the female as in euphausiacean shrimps, the eucarid sister group to the decapod shrimps (Spiridonov and Casanova 2010). Genthe (1969) described in detail the male reproductive morphology of *Eusergestes similis*. As in other shrimps, there are paired testes made up of seminiferous tubules, vasa deferentia, and ejaculatory ducts which open at the base of the last pair of pereopods (Fig. 6.5D). As in penaeoids, the internal structure of the sergestoid vas deferens is complex, with a sperm duct and accessory duct in which substances are added to the sperm mass. The ejaculatory duct has several glandular structures which add their secretions onto the spermatophore to complete it. The various secretions make up the sheath and tube of the spermatophore, as well as: (1) a plug substance to seal the openings to the female seminal receptacles after insemination; (2) a material that will connect to and form a bridge with the spermatophore emitted from the other side, forming a compound spermatophore; and (3) a glutinous material which will cover and temporarily stick the twin spermatophores to the female thelycum during insemination. The extreme lateral compression of the body in luciferids only allows a male to produce one spermatophore at a time, and thus males can only inseminate one of the two female receptacles during mating (Burkenroad 1934; Gordon 1956; Lee et al. 1992). Similar to some benthesicymids, the cuticular seminal receptacles of female sergestids and luciferids are formed by invaginations of the anterior margin of the female sixth thoracic somite, i.e., between somites 6 and 7. They open to the exterior very near the female gonopores.

6.2 Secondary Sexual Characters: Genitalia

Secondary sexual characters refer to differences between male and female external morphology related to reproduction. These include the genitalia, male and female structures involved with copulation, spermatophore transfer, and insemination. Male genitalia are found on the first and second pleopods, which are often termed gonopods, while those of the females may involve little to complex modifications of the posterior thoracic sternites.

6.2.1 Genitalia of the Procarididea, Stenopodidea, and Caridea

Description of the genitalia of the Procarididea is very simple; there are none! Thus, sexing of procarideans is difficult. The first description of this group was by Chace and Manning (1972) and they were unable to sex any of their specimens. The first two pleopods are not modified as possible sperm transfer devices (gonopods) as in carideans. However, females with developing ovaries can be recognized. Otherwise, one needs to search for the gonopores on the coxae of the third walking legs (sixth thoracic somite) of the female and the last walking leg (eighth thoracic somite) of the male. This can be difficult without the use of scanning electron microscopy, especially for the male gonopores (Felgenhauer et al. 1988). The Stenopodidea also have relatively little genital modifications. The first two male pleopods are not obviously modified for sperm transfer as in carideans and penaeoids (Goy 2010; Gregati et al. 2014).

The posterior sternites of female shrimps are broader than those males. The cephalothorax females must be more spacious in order to accommodate the ovaries which are quite voluminous near spawning when packed with vitellogenic oocytes. In addition, this in the area which will receive the externally placed spermatophores in some shrimp species or to accommodate seminal receptacles and associated structures in many dendrobranchiate species.

Long-term sperm storage in seminal receptacles is unknown in the procarididеans, stenopodideans, and carideans. A median pouch which has been described as a seminal receptacle is present on the last thoracic sternite of processid caridean females (Fig. 6.6a) (Noël 1976; Rasch and Bauer 2016) as well as in the atyid *Atyaephyra desmarestii* (Descouturelle 1971). Although sperm is deposited there just after mating, it is quickly used in the spawning which occurs within a few hours thereafter (Rasch and Bauer 2016).

In caridean males, each first pleopod usually has an appendix interna (AI) to link it together with the other, while the female does not. These AI can be more or less typical, i.e. a stalk tipped with cincinnuli (Fig. 6.6c, d; Plate 2.2c). However, in some genera, the AI is reduced to a sessile group of cincinnuli (Fig. 6.6b, e) near or at the tip of the endopod. Not all caridean taxa have an AI on pleopod 1, e.g. in *Thor* species, the pleopod-1 endopod lacks an obvious AI but does show a distinctive array of setae absent in the female (Bauer 1986). In the Campylonotidae, the medial edges of the large endopods bear cincinnuli which links the two endopods together (Fig. 6.6f), similar to the petasma of penaeoid males described below. Chace (1985) described the pleopod-1 endopod of thalassocaridid males as enlarged, uniquely convoluted, and heavily armed with setae and spines.

However, the most reliable way for sexing carideans is to look for the presence (males) or absence (females) of an appendix masculina (AM) on the medial edge of pleopod 2. The most typical type of AM, found in alpheids, hippolytoids, palaemonoids, and many other carideans, is a stalk or lobe of varying length usually tipped with long spines (Fig. 6.7c–f; Plate 2.2a, b). The AM of atyids is quite unusual

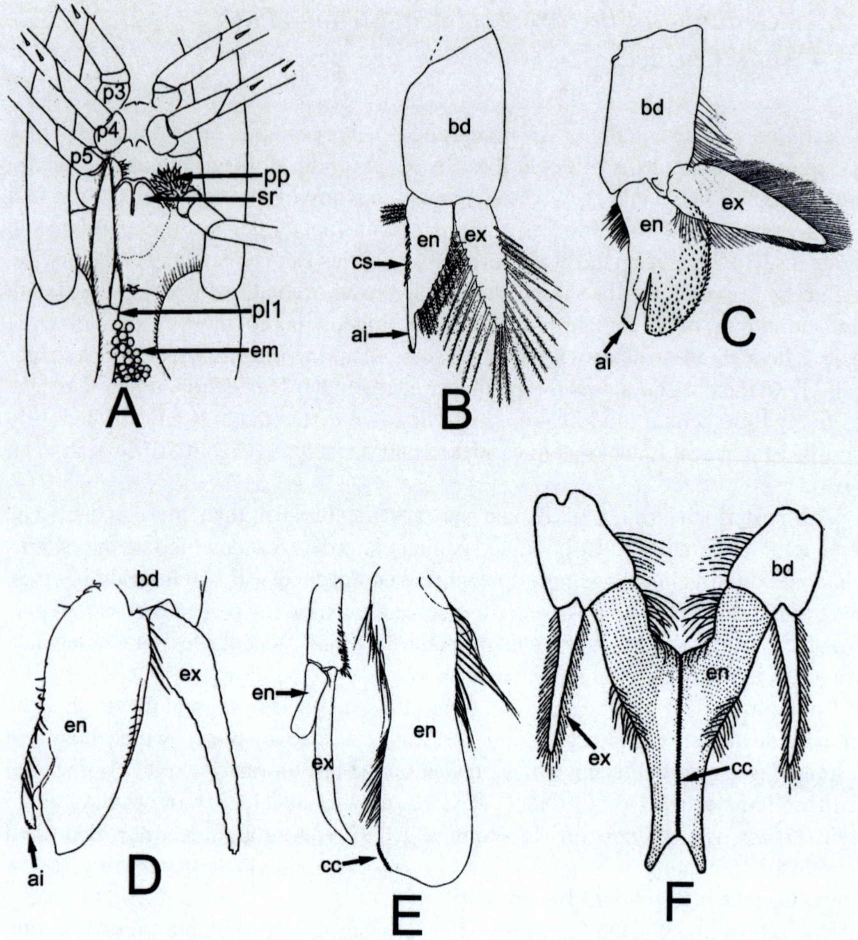

Fig. 6.6 Caridean reproductive morphology. (**a**) Seminal receptacle of *Processa edulis* (Processidae) on posterior thoracic sterna. (**b**) Appendix interna on pleopod 1 of male *Heptacarpus sitchensis* (Thoridae). (**c**) Stalked appendix interna on pleopod 1 of male *Atya innocuous* (Atyidae). (**d**) Stalked appendix interna on pleopod 1 of *Rhynchocinetes albatrossae* (Rhynchocinetidae). (**e**) Left: entire pleopod 2 *Psalidopus barbouri* (Psalidopodidae); Right: endopod showing unstalked appendix interna reduced to cincinnuli only. (**f**) First pleopods of *Campylonotus vagans* (Campylonotidae) linked to form a petasma-like gonopod. ai, appendix interna; bd, basipod; cc, cincinnuli; cs, curved setae; em, embryos; en, endopod; ex, exopod; p3–p5, basal segments of pereopods 3–5; pl1, pleopod 1; pp, coxal papillae; sr, seminal receptacle. From (**a**) Noël (1976), (**b**) Bauer (1976), (**c**) Hobbs and Hart (1982), (**d**) Chace (1997), (**e**) Chace and Holthuis (1978), and (**f**) Torti and Boschi (1973)

for carideans, dense with setae and quite large in relation to the endopod and AI (Fig. 6.7a, b). In two euryrhynchinid genera, the AM apparently has merged with the endopod into complex structure, or gonopod, with the loss of the usual AI (Powell 1976; Pachelle and Tavares 2018) (Fig. 6.7g). There are a few caridean taxa in which

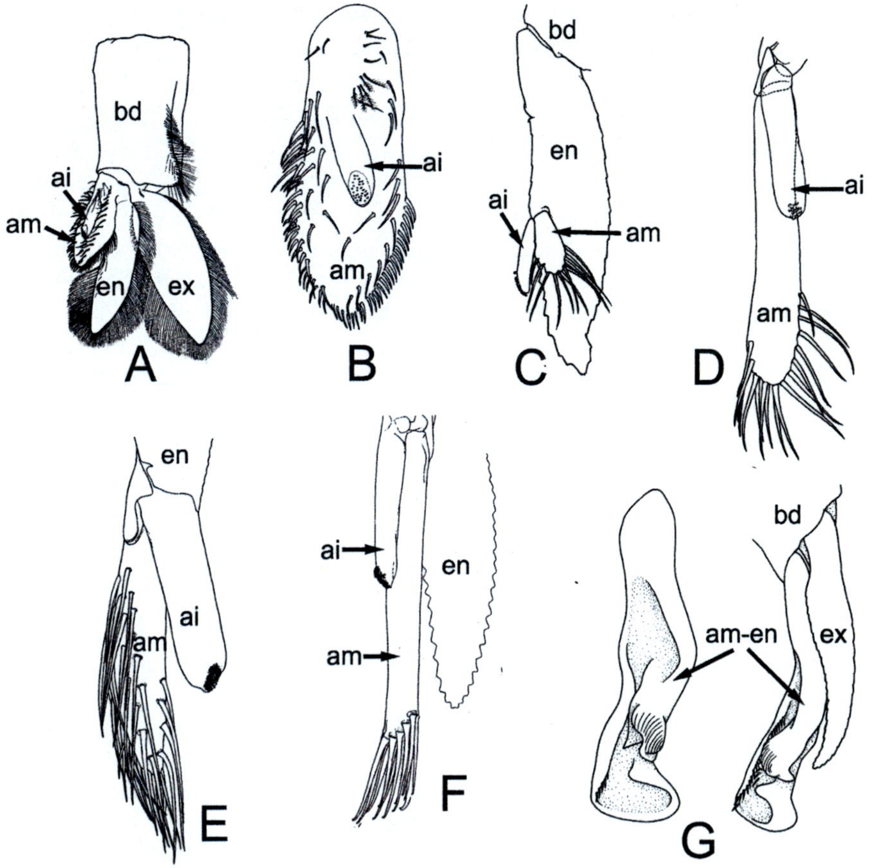

Fig. 6.7 Caridean appendices masculinae on the endopods of male second pleopods. (**a**) *Atya innocous* (Atyidae). (**b**) *A. dressleri*. (**c**) *Rhynchocinetes albatrossae* (Rhynchocinetidae). (**d**) *Meningodora vesca* (Acanthephyridae). (**e**) *Glyphocrangon unguiculata* (Glyphocrangonidae). (**f**) *Pontophilus spinosus* (Crangonidae). (**g**) *Euryrhynchoides holthuisi* (Euryrhynchidae). ai, appendix interna; am, appendix masculina; am-en, fused appendix masculina and endopod ("gonopod"); bd, basipod; en, endopod; ex, exopod. From (**a**, **b**) Hobbs and Hart (1982), (**c**) Chace (1997), (**d–f**) Chace (1984) and (**g**) Pachelle and Tavares (2018)

a male AM is lacking, e.g., the alpheid genus *Synalpheus* (Dardeau 1984; Tóth and Bauer 2007), the oplophorid genus *Janicella* (Chace 1986), and the pontoniine genus *Typton,* in which the AM is missing or vestigial, represented by only 2 stiff setae (Bruce 1972; De Grave 2010).

6.2.2 *Genitalia of Dendrobranchiata: Thelyca, Petasmata, and Appendices Masculinae*

Like their male reproductive tracts and spermatophores, the genitalia of dendrobranchiates species are remarkably variable and complex in both sexes. The female and male genitalia, termed "thelycum (thelyca)" and "petasma (petasmata)," respectively, were briefly introduced in Chap. 2. The term thelycum in dendrobranchiates shrimps refers to modifications of the female posterior sternites and/or pereopod coxae for attachment and/or storage of the male spermatophores (Bauer 1994). The genitalia of dendrobranchiate genera are beautifully illustrated and described in the classical monograph by Pérez Farfante and Kensley (1997). But in order to understand their role in copulation and insemination, some examples are required. Thelyca are categorized as "open" and "closed." In species with open thelyca, the twin spermatophores are attached externally; a depression or receptive space is formed by the various surrounding projections and protuberances of the posterior sternites and pereopodal coxae of the female. Open thelyca are found in the penaeoid families Aristeidae, Solenoceridae, some benthesicymids, and some Penaeidae (Figs. 6.4B and 6.8A; Plate 6.1e). Females which store spermatophores and sperm in invaginated cuticular pockets (seminal receptacles) are considered to have "closed" thelyca (some benthesicymids, many Penaeidae, all Sicyoniidae) (Figs. 6.4D and 6.8B, C, E). In some closed thelyca, evaginations (plates) of the posterior sternites form a median pocket or space which either form an externalized median seminal receptacle (as in all *Penaeus s.l.* except for *Litopenaeus*) or, as in the *Rimapenaeus, Xiphopenaeus*, and some other penaeid genera, cover the slit-like openings into the internalized seminal receptacles as described earlier (Fig. 6.8B; Plate 6.2a–c). The median external pocket is filled after insemination with a substance, the mating plug. In the closed thelyca of sicyoniid penaeoids, the openings into the seminal receptacles are not covered by median plates (Fig. 6.5B).

Among the many variations are the unusual thelyca of the penaeid *Metapenaeopsis* females (Fig. 6.8C) so well described by Pérez Farfante (1971). The two deep pits of the thelycum are the apertures to the tubular seminal receptacles (Fig. 6.8D). In inseminated females, the openings of the receptacles are stoppered by platelets formed by a hardened substance attached to the spermatophores—another example of a paternity assurance device deposited by the male in some penaeid species.

The thelyca of female sergestids and luciferids are much less complex than those of penaeoids and mainly restricted to the sternum of thoracic somite 6 (Burkenroad 1934; Genthe 1969; Pérez Farfante and Kensley 1997) (Fig. 6.8E). They are considered "closed" because sperm is stored in twin seminal receptacles (Fig. 6.8F). A shallow channel in the anterior edge of sternite 6 runs between the receptacle openings and to the female gonopores on each side. These channels and receptacle openings may be covered by coxal lamellae and a median sternal operculum.

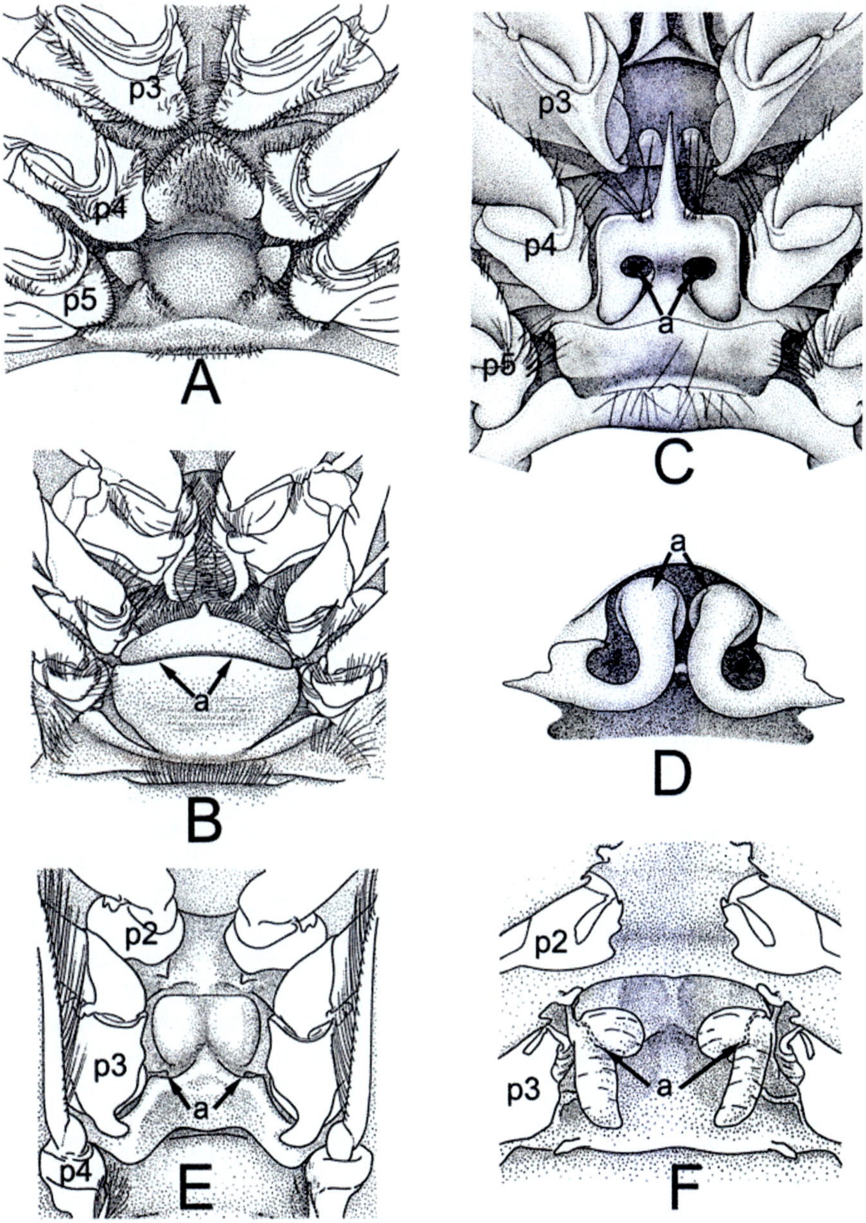

Fig. 6.8 Dendrobranchiate thelyca and seminal receptacles. (**A**) Open thelycum of *Aristeus antennatus* (Aristeidae). (**B**) Closed thelycum of *Xiphopenaeus kroyeri* (Penaeidae). (**C**) Closed thelycum of *Metapenaeopsis gerardoi* (Penaeidae). (**D**) Internal (dorsal) view of tubular seminal receptacles, *M. gerardoi*. (**E**) Closed thelycum of *Peisos petrunkevitchi* (Sergestidae). (**F**) Internal (dorsal) view of seminal receptacles of *P. petrunkevitchi*. a, apertures to seminal receptacles; p2–5 basal segments of pereopods 2–5. From (**A**, **B**; **E**, **F**) Pérez Farfante and Kensley (1997), courtesy of ©Publications Scientifiques du Muséum national d'Histoire naturelle, Paris and (**C**, **D**) Pérez Farfante (1971)

The male petasma consists of the linked endopods of the first pleopods (Chap. 2, Fig. 2.21J–L; Plate 2.2D, F). In juveniles, the endopods are separate, but in sexually mature males, they are hooked together with their medial coupling hooks (cincinnuli) into a single structure. Petasmata are characterized as "open, semi-open, semi-closed, and closed." These subjective categories are established on the degree of ventral flexure and closure of the median and lateral lobes of the petasma (Pérez Farfante and Kensley 1997). In the semi-open petasma of *Litopenaeus* species (Penaeidae), the lateral lobes fold toward each other but are very flexible and are able to spread out on about the same plane (Chap. 2, Fig. 2.21J). The lateral lobes of semi-closed petasmata similarly bend toward each other but are not so flexible and enclose a variably smaller space between them (Chap. 2, Fig. 2.21K). In closed petasmata, the lateral lobes are highly calcified and rigidly folded against the median lobes, enclosing very little space (Fig. 2.21L, Plate 2.2F). In the Aristeidae and Benthesicymidae as well as some solenocerid genera, the petasmata are considered open, as the lateral and median lobes of the petasma halves are spread apart on a similar plane (e.g., Fig. 6.9a). Semi-open petasmata are found in some solenocerid and penaeid genera, but the majority of penaeids have semi-closed petasmata. Completely closed petasmata are found only in the Sicyoniidae.

When males have open or semi-open petasmata, the thelyca of females are open, without seminal receptacles. Semi-closed or closed petasmata are always associated with closed thelyca and seminal receptacles. The byzantine petasma of *Metapenaeopsis* is asymmetrical, with the right and left endopods differently elaborated into various flaps, lobes, and projections and which terminates in a very complex tip equipped with peculiar and intricate twists, turns, blades, and spouts (Fig. 6.9b) (Pérez Farfante 1971). The generic name of *Macropetasma africana* is aptly chosen, given the long thin extensions of the lateral lobes of the petasmal endopods which reach nearly to the tip of the male cephalothorax (Fig. 6.9c)! Although penaeoid petasma are generally bizarre, that of *Trachypenaeopsis mobilispinis* is notably so (Fig. 6.9d). Although reasonable hypotheses on petasma function can be made in many penaeoid genera, it is particularly difficult to do so for these more exotic examples. Numerous carcinological careers could be devoted to this line of research.

The petasmata of male sergestids are similar to those of penaeoids in that the endopods of the first pleopods are joined together by coupling hooks, but otherwise are constructed quite differently. They are more similar to the petasmata of the decapod sister group, the euphausiacean shrimps (see Spiridonov and Casanova 2010). Each petasma half is composed of a number of finger-like processes and lobes equipped with various hooks, spines, and other setal types with a complex terminology important in sergestoid taxonomy (e.g., Burkenroad 1934; Genthe 1969; Judkins 1978, Pérez Farfante and Kensley 1997; Vereshchaka 2017) (Fig. 6.10A). Petasmata of luciferid males differ from those of sergestids in that the petasmal endopods are not joined together by cincinnuli as in sergestids. Each petasmal endopod is composed of a broad basal and distal part with lobes; the distal part is partially bent back or reflexed on the basal portion (Tirmizi and Qureshi 1971; Vereshchaka et al. 2016) (Fig. 6.10C, D). The latter authors believe each petasmal

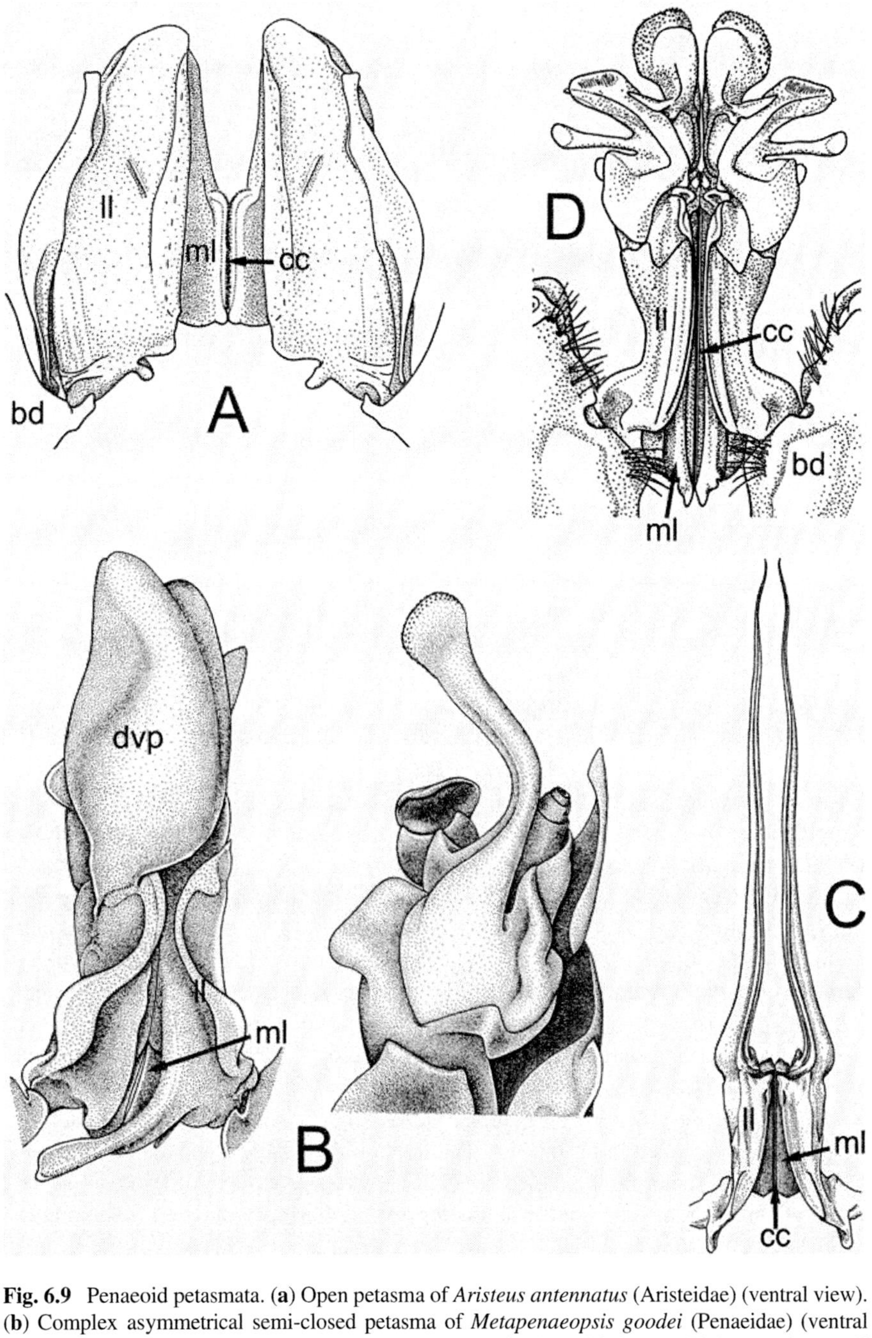

Fig. 6.9 Penaeoid petasmata. (**a**) Open petasma of *Aristeus antennatus* (Aristeidae) (ventral view). (**b**) Complex asymmetrical semi-closed petasma of *Metapenaeopsis goodei* (Penaeidae) (ventral view). Left, complete petasma, with large distoventral projection of right lateral lobe covering small left distoventral projection and distal part of the petasma; Right: Distal part of petasma with distoventral projection bent back to show its various spouts, lobes, projections. (**c**) Semi-closed petasma of *Macropetasma africana* with elongate distal extension of lateral lobes. (**d**) Semi-closed petasma of *Trachypenaeopsis mobilispinis,* with distal elaborations of the lateral lobes. bd, basipod

endopod acts independently during the transfer of the single spermatophore passed to the female during copulation (see below).

The males of all dendrobranchiates have an appendix masculina (AM) although it is very different from that of carideans. In penaeoids, the AM has a variety of shapes and sizes but is generally small and without long spines as in many carideans (Chap. 2, Fig. 2.21F–I; Plate 2.2E). In luciferids, the AM is described as a simple lamella or lobe with short setae (Tirmizi and Qureshi 1971; Judkins 1978; Pérez Farfante and Kensley 1997). The AM is usually presumed to have some role in copulation and insemination, albeit still poorly understood. Appendices internae are present in aristeid, solenocerid, and benthesicymid males on the second pleopods (Chap. 2, Fig. 2.21F, G) but they are quite small and lack cincinnuli.

6.3 Mechanics of Copulation and Insemination

After the above introduction to the equipment that males and females use to carry out spermatophore transfer and storage on the female, let us examine the mechanics of mating, copulation, and insemination. The modifications of the first and second pleopods of male shrimps are presumed to be genitalia (gonopods) because they are just posterior to the male gonopores and because these structures are opposed to the female genital area during copulation. Thus, it is reasonable to assume that they have some role in spermatophore transfer. This has been definitely shown experimentally in some species of brachyuran crabs, as well as in astacidean lobsters and crayfishes (Andrews 1911; Ryan 1967; Hartnoll 1969). Unlike shrimps, many crabs and crayfishes are large and the duration of copulation is usually lengthy. Therefore, manipulation experiments can and have been carried out without too much trouble that show the role of gonopods in insemination.

Shrimp biologists have generally hypothesized that direct spermatophore handling by the first and second pleopods does occur (e.g., Bauer 1976; Berg and Sandifer 1984). Other hypotheses have been proposed which are not mutually exclusive and may be concomitant with a spermatophore-handling function. One is that genitalia of some shrimp taxa serve as a lock-and-key mechanism for exact species identification between potential mating partners. The female genitalia are the "lock," and the male ones are the purported "key" of such a mechanical isolation mechanism. Another idea is that male genitalia may stimulate the female to accept the male during copulation. A third is that the seemingly exaggerated male gonopods of some species (dendrobranchiates especially) serve as courtship devices, in which the female may evaluate the male as a mating partner during copulation based on the

Fig. 6.9 (continued) of pleopod 1; cc, cincinnuli connecting median lobes; dvp, distoventral projection; ll, lateral lobe; ml, median lobe. From (**a**, **c**, **d**) Pérez Farfante and Kensley (1997), ©Publications Scientifiques du Muséum national d'Histoire naturelle, Paris and (**b**) Pérez Farfante (1971)

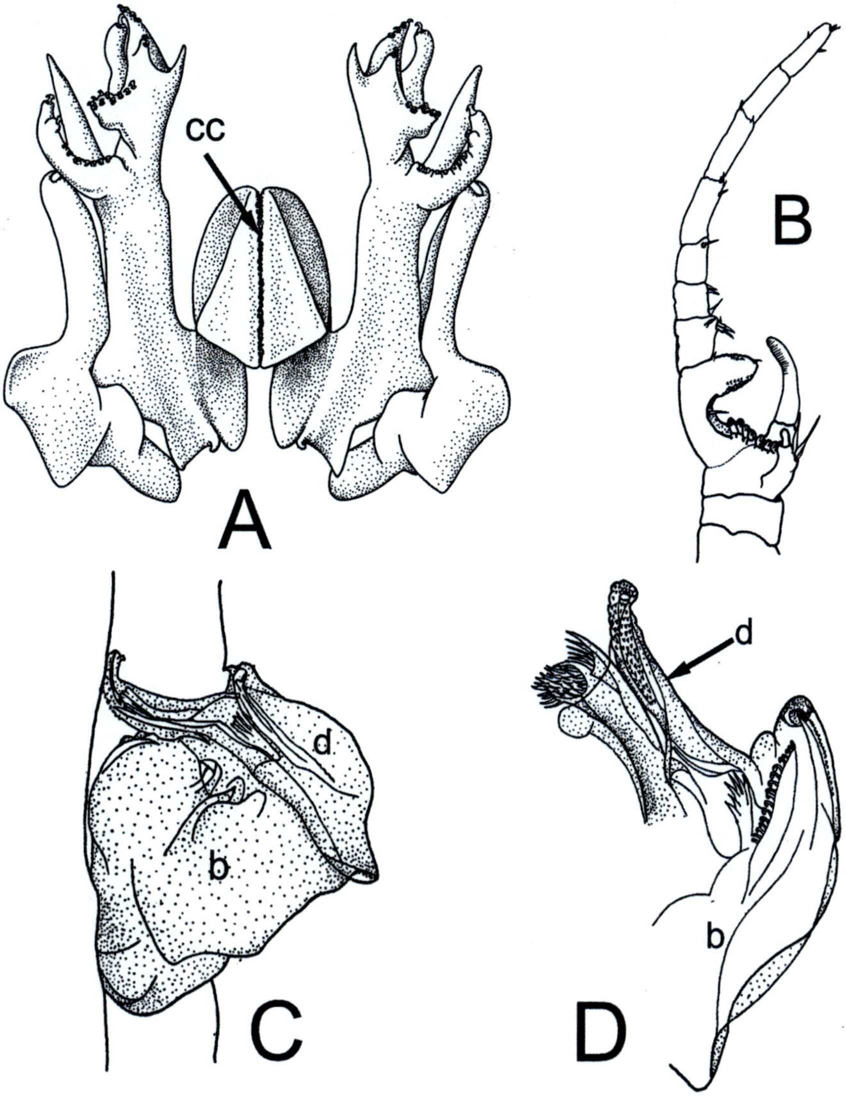

Fig. 6.10 Sergestoid petasmata. (**A**) Petasma of *Sergia robusta* (Sergestidae), with both petasmal endopods joined by cincinnuli (dorsal view). (**B**) Male antennular clasping organ of *Acetes marinus*. (**C**) Left petasmal endopod of *Lucifer penicillifer* (Luciferidae), medial view, showing basal ("palm") and distal ("finger") parts. (**D**) Same as (**C**), but with distal part reflexed to show complex projections, lobes, and spines. b, basal part; cc, cincinnuli; d, distal part. From (**A**) Pérez Farfante and Kensley (1997, ©Publications Scientifiques du Muséum national d'Histoire naturelle, Paris), (**B**) D'Incao and Silva Martins (2000) and (**C**, **D**) Tirmizi and Qureshi (1971), courtesy of Crustaceana (Brill)

quality of his gonopods (Eberhard 1985, 1991). The morphology of some penaeoid males suggests that the petasma participates in injection of sperm into female seminal receptacles. A more recent idea is that the petasma, together with other male structures, serves as an attachment device which links male to the female into an appropriate position for successful spermatophore transfer (Bauer 1996; Vereshchaka et al. 2017).

The copulatory position of shrimps is similar in all groups but the precopulatory or courtship behavior does vary both among and within the major shrimp groups (see Chap. 8). In most carideans and stenopodideans (*Stenopus hispidus*), a male identifies a receptive female, usually by contact or distance chemoreception. The male approaches, mounts, and grasps the female. He maneuvers his body so that he can dip below her, usually at an angle from 45 to 90° so that his genital orifices are near the area where the sperm mass or spermatophores will be deposited (Fig. 6.11a–c). The female usually lowers her anterior pleopods to allow him access (Bauer 1976; Berg and Sandifer 1984). The process of spermatophore transfer is usually brief, from a few to several seconds. In *S. hispidus,* the male's unspecialized first and second pleopods do not appear to be involved in this transfer. Instead, the spermatophore material seems to be applied directly to the female's unadorned genital area (Gregati et al. 2014).

In carideans, however, the anterior two pleopods do seem to be involved in sperm transfer. Observations on copulation and ablation experiments on *Heptacarpus sitchensis* suggested a model for this transfer (Fig. 6.12). When the spermatophores are emitted, the endopods of the first pleopods, linked together by their AI, swing forward to catch the twin spermatophores. These sticky structures do not become entangled on the male pleopods because they are held off of it by short, curved setae on the pleopod-1 endopods. The first pleopods move back toward the second pair, and then both pair swing forward toward the female underside. The spines of the AM push the spermatophores off the first pleopods and against the female. When the glutinous spermatophore strikes the female sternite, it preferentially adheres to it because there is a broad surface for its attachment. Similar observations and experiments were performed with *Palaemon (Palaemonetes) pugio* by Berg and Sandifer (1984) with comparable results and conclusions.

Can these findings be extrapolated to other carideans? Given the variation in the morphology of male pleopods 1 and 2 among and within caridean genera, the answer is "perhaps." Stenopodidean males inseminate females just fine without modifications of the anterior pleopods. Many caridean first pleopods have AI to link them together, but some do not. Most carideans have some sort of spiny AM but there is considerable variation as shown above. How do caridean males of *Synalpheus* species, which lack AM, transfer their spermatophores to the female? In the following chapter, I will discuss some carideans (*Lysmata* species) which are protandric simultaneous hermaphrodites. In these species, individuals develop first as males with attachment cincinnuli on the first and AM on the second pleopods, as in many carideans. But when these individuals change sex to female, they lose these appendices. Yet these female-phase hermaphrodites are perfectly capable of inseminating other female-phase individuals (Bauer and Holt 1998). Zhang and Lin (2004)

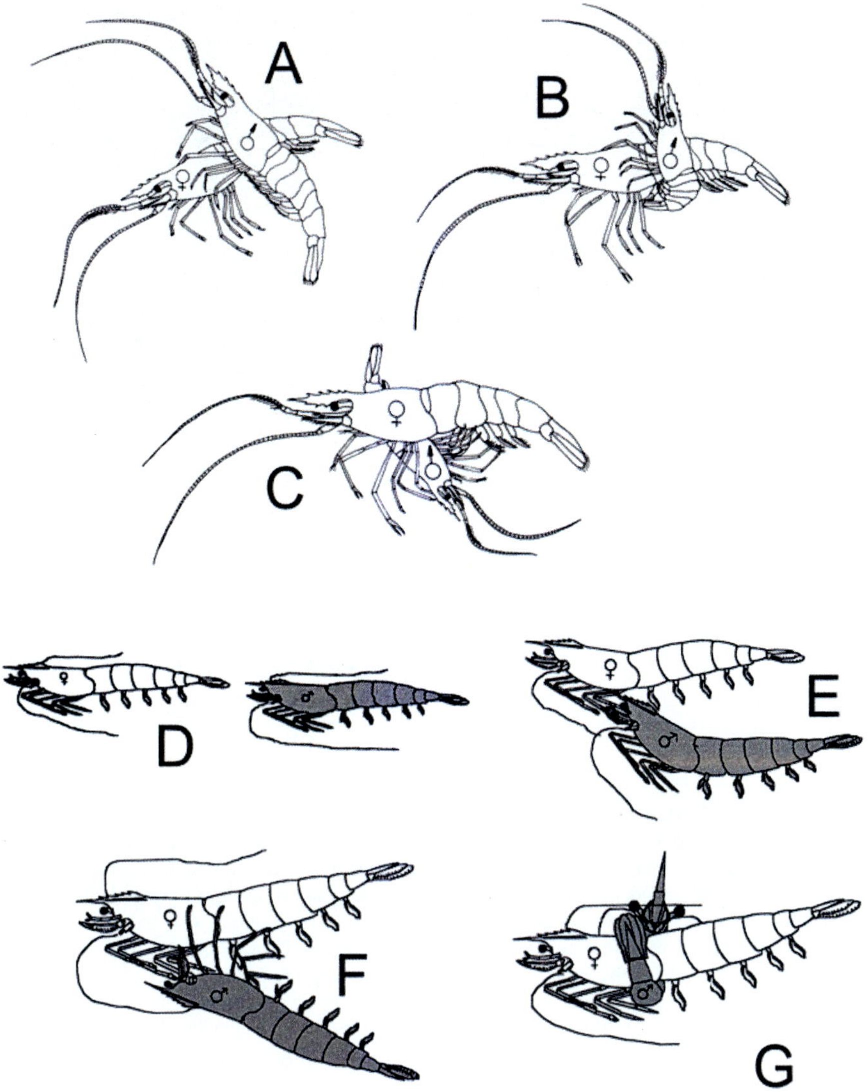

Fig. 6.11 Copulation in shrimps. (**a**–**c**) Caridean *Palaemon (Palaemonetes) pugio* (Palaemonidae): (**a**) male grasps and mounts female, (**b**) male dips below female, and (**c**) final oblique to perpendicular copulatory position in which male ejaculates and deposits spermatophores on underside of female's posterior thorax. (**d**–**g**) Penaeoid *Litopenaeus setiferus* (Penaeidae): (**d**) male (shaded) chases female from behind, (**e**) male probes and prods female genital area, (**f**) male embraces female from below and (**g**) male in perpendicular spermatophore transfer position below female. From (**a**–**c**) Berg and Sandifer (1984) and (**d**–**g**) Misamore and Browdy (1996). (**a**–**g**) courtesy of Journal of Crustacean Biology

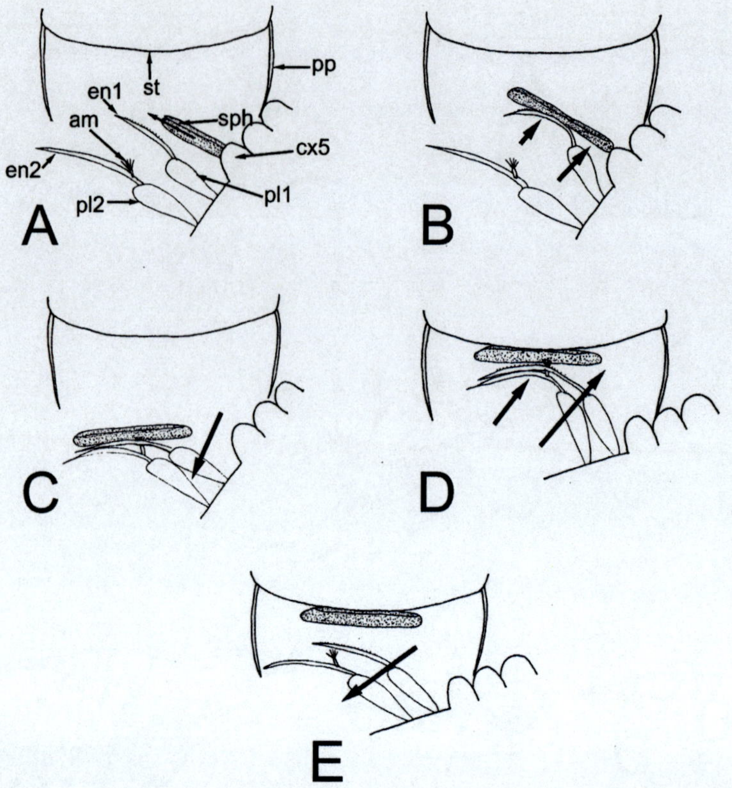

Fig. 6.12 Model of spermatophore transfer in the caridean *Heptacarpus sitchensis* (Thoridae). (**a**) Cross section of underside of female first abdominal segment at moment of spermatophore emission from the male gonopores and showing positions of male pleopods 1–2 on one side, lateral view. (**b**) First pleopod moves forward, bringing its endopod into contact with spermatophores, which adhere to its curved endopod spines. (**c**) Pleopod 1 moves backward against pleopod 2, whose appendix masculina spines push the spermatophores slightly off pleopod 1. (**d**) Pleopods move forward toward the female sternite with which spermatophores make broad area of contact. (**e**) As pleopods move backward, spermatophores preferentially adhere to the female sternite, with which they have a broad area of contact, rather than the appendix masculina spines, with which they have a small area of contact. am, appendix masculina; cx5, pereopod 5 coxal segment; en1,2, endopods of pleopods 1,2 (pl1, pl2); pp, pleural plate of first abdominal somite; sph, spermatophores; st, sternum. Modified from Bauer (1976)

showed that some, but not all, male-phase individuals of *L. wurdemanni* with pleopod 1 and 2 appendices removed were capable of inseminating females. I proposed in my studies on *H. sitchensis* that the copulatory appendices prevent the spermatophores from becoming fouled on the male pleopods. But if the ejaculation of spermatophores is forceful and the male gonopores are sufficiently near the female genital area, then the spermatophores have a good chance of sticking to the female even without the intervention of male appendices. In the *H. sitchensis* experiments, a good proportion of males without appendices (17%) were able to inseminate the female. Some males with *all* their equipment failed to inseminate. I believe that the male appendices do participate in spermatophore transfer; they make it more efficient when present than without. Additionally, as suggested above, they may help stimulate the female, locate the exact position of the female genital area before ejaculating, or serve as courtship devices by which females can evaluate male quality. The differences among caridean gonopods suggest that their role may vary among caridean taxa.

In penaeoid shrimps, there is good evidence that the male petasma is involved with spermatophore transfer to the female during copulation. In aristeids, benthesicymids, and some solenocerids, the petasma is a broad, sometimes ornate plate that apparently receives the twin spermatophores ejaculated by the male during copulation. The male might then press the spermatophores against the female thelycum to which it would adhere and attach. However, mating and copulation have not been observed in these pelagic/benthopelagic groups. The only direct observations of spermatophores being received by the petasma was reported in *Aristeus varidens* by Burukovsky (1978, cited in Orsi Relini and Tunesi 1987).

The mechanics of insemination in penaeoid species has been intensively scrutinized for many years with many questions remaining. Early workers made hypotheses that the petasma directly transfers spermatophores to the female thelycum and/or seminal receptacles. Their hypotheses imply that copulation position of male and female is head to head, tail to tail with ventral surfaces appressed. For the aristeids, solenocerids, and benthesicymids, we have no direct evidence. But mating has been observed in some penaeids and sicyoniids. In most of these species, e.g., *Penaeus monodon*, *Litopenaeus* and *Sicyonia* species (e.g., Bauer 1992; Primavera 1979; Misamore and Browdy 1996; Feng et al. 2018b), the male copulates from below the female, with his genital area appressed to that of the female, but at an angle, from about 20° to more usually near 90° (Fig. 6.11d–g). The report by Hudinaga (1942) of head-to-head copulation in *Melicertus japonicus* may have referred to the precopulatory "embrace" of some penaeids (Feng et al. 2018b) (as in Fig. 6.11f). This position prevents other males from copulating with the female until the embracing male can switch under the female to an oblique or perpendicular copulation position (Dall et al. 1990). The position of male to female during spermatophore transfer is important in understanding insemination mechanics.

In sicyoniid species, the most benthic of the penaeoids, mating occurs on the bottom. Male approaches females from behind, uses the rostrum and head region to prod her genital area, then tilt her abdomen up so that he can move below her and mate obliquely or at right angles, opposing his genital area with that of the female. If a receptive female is under the sediment surface, the male may first dig out her abdomen with his rostrum so that he can begin the prod, tilt, and copulation sequence (Bauer 1992) (Plate 6.3F–H).

Burkenroad (1934) proposed that the semi-open petasma of *Litopenaeus setiferus* was well adapted for receiving the complex twin spermatophores emitted by the male. The flexible nature of the petasma allows it to open widely and accept the spermatophores which fit into it perfectly. Then during copulation, it would be

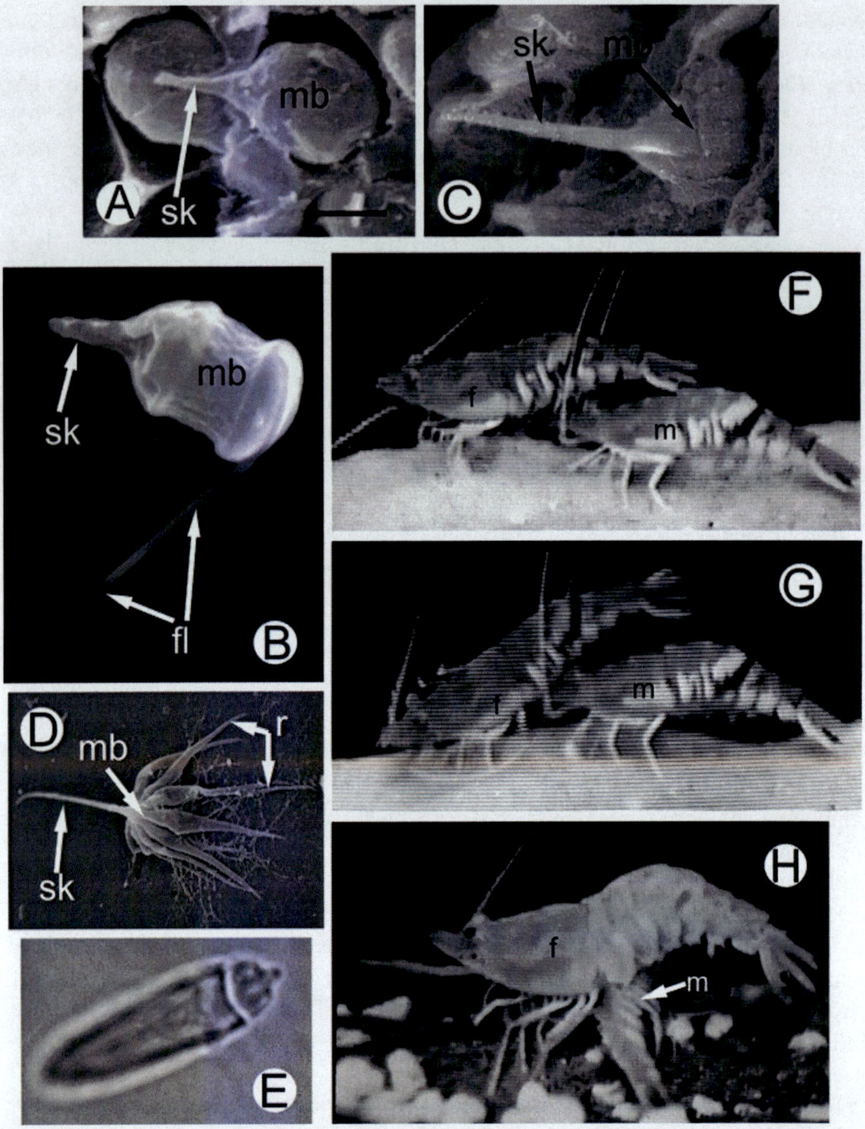

Plate 6.3 Sperm cells; copulatory behaviors. Sperm cells of (**A**) *Farfantepenaeus duorarum* (Penaeidae), (**B**) *Rimapenaeus similis* (Penaeidae), (**C**) *Lysmata wurdemanni* (Lysmatidae), (**D**) *Rhynchocinetes uritai* (Rhynchocinetidae), and (**E**) *Thor manningi* (Thoridae). (**F–H**) Mating in penaeoid *Sicyonia dorsalis*: (**F**) Courtship prodding of female by male; (**G**) male tilts female up to attain access to her underside and (**H**) Male in perpendicular copulatory position beneath female, male and female genitalia in contact (frames taken from video filmed with infrared light at night). f, female; fl, filament; m, male; mb, main body; r, main body rays or filaments; sk, spike. Scale bar represents 1.8 μm in (**A**), 1.7 μm in (**B**), 2.1 μm in (**C**), 5.5 μm in (**D**) and 6.0 μm in (**E**). From (**A**) Bauer and Cash (1991), (**B**) Bauer and Lin (1993), (**C**) Bauer and Holt (1998), (**D**) Bauer and Thiel (2011), (**E**) Bauer (1986) and (**F–H**) Bauer (1992) (**D**, **F–H** courtesy of Journal of Crustacean Biology)

pressed and fitted into the female thelycum. Something similar may occur in other penaeoids with open thelyca and externally attached spermatophores.

In female *Farfantepenaeus* and other *Penaeus s.l.* species with a median seminal receptacle, the lateral lobes of the semi-closed male petasma appear flexible enough to open and receive the twin spermatophores but are stiff enough to penetrate and open the apertures of the seminal receptacles to deposit the spermatophores (Burkenroad 1934; Bauer and Cash 1991). In these and other species with closed thelyca and seminal receptacles, mating occurs very soon after molting. The thelycum plates covering the median seminal receptacle are still soft and can be easily opened. As shown in *F. aztecus* and *F. duorarum* (Bauer and Cash 1991), only the sperm sacs are deposited within the median seminal receptacle. The spermatophore appendages spread apart and over the thelycum, sealing the aperture of the seminal receptacle. Feng et al. (2018b) used micro-computed tomography and scanning electron microscopy on spermatophores and genitalia of *Penaeus monodon* to make three-dimensional models which could be manipulated to test hypotheses about insemination mechanics. As in other *Penaeus s.l.* species, the male is curled around the female at an oblique or right angles during copulation, with genital surfaces in contact. The Feng et al. summary of the complex process of spermatophore transfer is that the petasma serves as a scoop to cradle and hold the twin spermatophores while they are pressed by the physically stiffened petasma and AM of the male into the aperture of the seminal receptacle softened by a recent molt. Feng and authors have an interesting video simulating the process.

Burkenroad (1934) proposed that the petasmata of other Penaeidae with stiffer, more tubular semi-closed or closed petasmata with channeled distal tips or closed spouts might function as syringe-like injection systems (e.g., *Rimapenaeus* and other *Trachypenaeus s.l.* genera, *Parapenaeopsis* and *Xiphopenaeus*), similar to those found in brachyuran crabs, astacidean lobsters and crayfishes. During ejaculation, the genital papillae bearing the gonopores become erect and inserted like penes into the open base of the petasma. Spermatophore material then flows, perhaps with the aid of the AM of the second pereopods, into the petasma interior and through the distal channels or spouts into the aperture of the median chamber leading to the seminal receptacles. The many small sperm packets of these genera then flow into the median chamber, followed by the copious plug substance, which pushes the sperm mass into the apertures of the internalized seminal receptacles. The plug material fills and hardens within the median chamber, extending to the outside, effectively sealing access to the seminal receptacle. In *Sicyonia* species, the tips of the completely closed petasma might be inserted directly into the apertures of the seminal receptacles which are not sealed afterward as in the other genera above.

The hypotheses suggested above by Burkenroad (1934) about sperm injection by *Rimapenaeus* and *Sicyonia* assume a head-to-head mating position. But, in *Sicyonia* species at least, the male definitely mates at right angles to the female. It only can inseminate one receptacle at a time so that a single male must copulate twice, once from each side (Bauer 1992) to inseminate both female receptacles. It is often unsuccessful in filling both receptacles and may copulate multiple times. Females sometimes mate with two or more different males before filling both receptacles.

One hypothesis involving insemination without direct involvement of the petasma has been suggested by studies on *Rimapenaeus similis* (Bauer and Lin 1993) and *Sicyonia dorsalis* (Bauer 1996). In this scenario, the complex petasma may simply help attach the male to the female while the male genital papilla of only one side is inserted into the opening leading to (*Rimapenaeus*) or directly into (*Sicyonia*) an aperture to the female seminal receptacle. Vereshchaka et al. (2017) have shown in *Gennadas* that if copulating partners are at right angles, one half of the petasma fits very exactly into the various processes of the female thelycum on one side. In open-thelycum penaeids such as *Litopenaeus*, the petasma may receive the twin spermatophores, but perhaps the posterior pereopods help manipulate the spermatophores into position on the female (King 1948). These various hypotheses need to be tested on mating experiments with living animals, which is challenging in most cases because the conditions conducive to mating are difficult to reproduce in the laboratory, and copulations are usually quite brief. Direct observation of what is really going on in the tangle of bodies and appendages is problematical to say the least. However, species that are being cultivated for aquaculture, in which mating can be observed, are a good place to start. Experiments involving manipulation or modification of the genitalia of mating partners may resolve some of these questions. The modeling techniques of Feng et al. (2018b) show considerable promise.

In the sergestoids, the function of the petasma in insemination is even less well known. Sergestid males have an antennular organ apparently for gripping the female during copulation (Fig. 6.10B) (Pérez Farfante and Kensley 1997; Vereshchaka et al. 2016). However, mating has not been observed, so we do not know if the copulatory position is head to head or at an angle, an important observation to have in planning and testing hypotheses on petasma function. One hypothesis, so often proposed for complex male genitalia, is that the thelycum and petasma serve as a lock-and-key mechanism that assures mating with a member of the same species. The complex genitalia of sergestid (and euphausiacean) genitalia might also be an adaptation to hold the male and female together during copulation in the rough and tumble of the often turbulent pelagic environment (Lunina and Vereshchaka 2017). Another hypothesis, not mutually exclusive with the others, is that the open petasma is a spermatophore transmission device in sergestoids. In sergestids, the petasma halves are linked by median coupling hooks as in penaeoids. The open petasma might accept the twin gourd-shaped spermatophores and insert them on the female thelycum so that their necks would be inserted into the seminal receptacle openings. The spermatophores hang on the female for an apparently short period, during which the material inside expands pushing the sperm mass into the receptacles (Burkenroad 1934; Genthe 1969).

Mating has not been seen in in luciferids either, but unlike sergestids, Vereshchaka et al. (2016) proposed that the female is grasped during mating by a holding device formed by processes and spines of the last abdominal somite and telson. In contrast to the sergestids, the petasma halves are not linked together, and this has led Vereshchaka et al. to propose that each half acts independently. It will be recalled that the luciferid male only has one spermatophore developed and ready for use at a time. Vereshchaka's group proposed that during ejaculation, the petasma half on one side grasps the single spermatophore ejaculated on the same side. The distal part of the petasma might be used as a finger, and a projection of the pleopod basipod as a thumb to enclose the spermatophore in the palm of the "hand" formed by the broad basal part of the petasmal endopod (Fig. 6.10C, D). Then it is passed in some way to one side of the female thelycum, with the elliptical spermatophore inserting into one of the female's twin seminal receptacles.

The role of the appendices masculinae in dendrobranchiates has been given little mention above. We saw in the carideans that there is good experimental evidence to show their role in copulation, at least in many species. But little is known about how these small appendices of dendrobranchiates function. It is probable that they do serve in some way, e.g., perhaps as a male sensory and/or stimulatory device, or perhaps in a mechanical role to support the petasma during copulation, as proposed by Bauer (1996) in *Sicyonia* species and Feng et al. (2018b) in *P. monodon*.

6.4 Eggs, Sperm, and Fertilization

6.4.1 Eggs

Just prior to spawning, a shrimp egg (oocyte) is composed mainly of yolk (vitellus), a substance rich in lipids and proteins that will sustain embryonic development. The egg includes the essential maternal nucleus undergoing meiosis and is surrounded by an outer egg membrane (vitelline envelope). The amount of yolk varies greatly depending on the type of embryonic development of the species between dendrobranchiates (sergestoids and penaeoids) and pleocyemates (carideans, stenopodideans, procarididеans). The dendrobranchiate egg is fertilized when spawned directly into the water, hatches within a short time (~day) and thereafter the larvae feed themselves and develop without any care from the female. In pleocyemates, however, the eggs (which are properly termed embryos after fertilization) are attached to the pleopods of the female, who cleans and aerates them until hatching them at a much more advanced larval stage. Pleocyemate embryos need much more yolk to sustain themselves during the longer embryonic development. As a result, pleocyemate eggs at spawning are generally larger, sometimes much more so, than dendrobranchiate eggs (Fig. 6.13) (Table 6.1).

Newly spawned caridean embryos vary in size from as small as 0.22 mm to as large as 4.5 mm in diameter, more than an order of magnitude difference. The overall average size is about 1.0 mm; values of about this size or smaller indicate an extended larval development in the plankton. The additional yolk required for abbreviated development produces a larger embryo, with just a few larval stages after hatching. In direct development, in which larval stages are eliminated, the embryo hatches as a postlarva or small juvenile (see Chap. 9) so that even more yolk increases embryo size. As Fig. 6.13 shows, the distribution of egg sizes in carideans is skewed to species with smaller eggs, indicating that more carideans have planktotrophic development than abbreviated or direct. Stenopodidean embryos

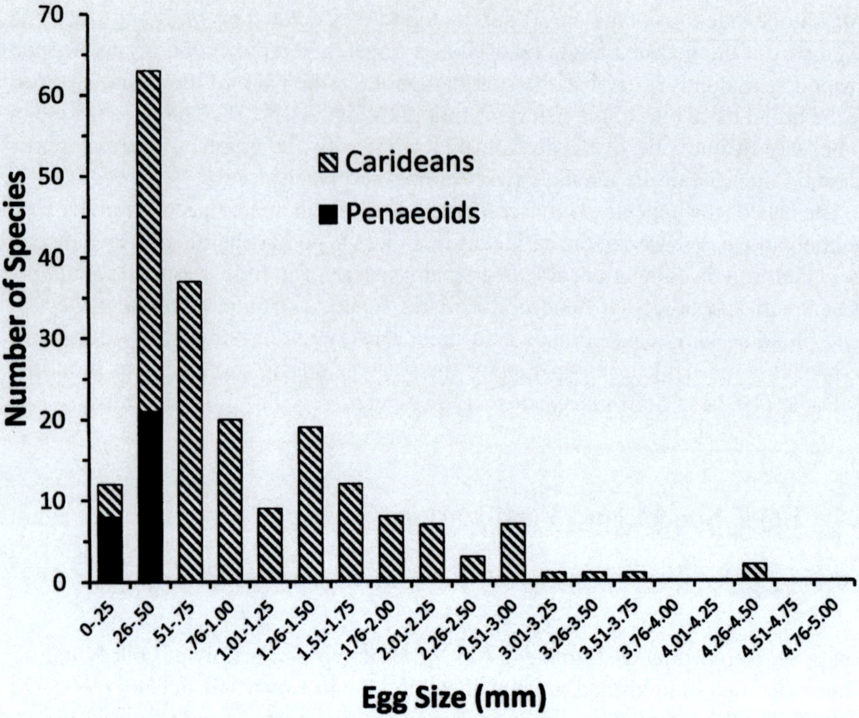

Fig. 6.13 Embryo or egg sizes (diameter) of 179 caridean and 29 penaeoid species, using early-stage caridean embryos and mature (prespawning) penaeoid oocytes. Where length and width were given, their average is given as the diameter. Caridean data is from many sources, while penaeoid data is mainly from Dall et al. (1990) as well as various other authors

Table 6.1 Summary comparison of egg (dendrobranchiates) or early-embryo size (pleocyemates) with spawn or brood size (number) in decapod shrimps. Egg or embryo size is given as a diameter in millimeters (mm); the pleocyemate embryo is usually oval in shape, and sizes are given in the literature as embryo length × width for many species. For those species, embryo diameter was calculated as the average or midpoint between embryo length and width; when egg volume was given, an approximation of diameter was calculated by trial and error substitution of values for radius or diameter in formulas for ellipsoids. Data is from various publications: for carideans those cited in Fig. 8.5 in Bauer (2004) plus numerous other more recent or discovered references; for penaeoids those cited in Dall et al. (1990) plus other additional publications

Taxon	Min-max egg/ embryo size	Average (mean) egg/ embryo size	Min-max spawn or brood size	Average (mean) spawn or brood size	*N* (number of species)
Penaeoidea	0.15–0.48	0.29	~5000–1,300,000	~290,000	34
Sergestoidea	0.19–0.28	0.23	~30–6800	~2800	7
Caridea	0.22–4.50	1.05	~10–250,000	~5000	183–219
Stenopodidea	0.40–2.25	0.90	~30–2400	~670	5–7

are similar to those of carideans: shallow-water species with planktotrophic larvae have small embryos, e.g., 0.40 mm and 0.55 mm in *Stenopus hispidus* and *S. goyi*, respectively (Zhang et al. 1998; Saito et al. 2009). In deep-sea spongicolid species with direct development, embryos are much larger: 1.78 mm for *Spongicola japonica* (Saito et al. 2009) and 2.25 mm in *Spongiocaris hexactinellicola* (Berggren 1993).

The penaeoids, on the other hand, produce many more very small eggs, varying among species reported from 0.15 to 0.48 mm in diameter, with an average size of 0.29 mm. This is because both embryonic and larval development take place in the plankton; the larva has to begin feeding very early in its development compared to that of carideans. The few values reported for sergestoids are similar to those of penaeoids but on the smaller side: 0.21 mm for the sergestid *Acetes japonicus* (Xiao and Greenwood 1993) and 0.19 mm for the luciferid *Belzebub faxoni* (Lee et al. 1992). In penaeids, the variation in egg size not is not related to amount of the yolk as in carideans, but rather to the amount of space between the egg proper and its outer membrane (perivitelline space). Penaeid species with somewhat larger eggs have a greater amount of perivitelline space which makes them buoyant compared to eggs of species with less (Dall et al. 1990). As a result, the larger eggs are more less dense and so pelagic, while smaller eggs are demersal, sinking to or toward the bottom, which has consequences for dispersal related to life history strategies (see Chap. 9).

6.4.2 Sperm

As in all decapod crustaceans, the sperm (spermatozoa) of shrimps are immobile cells, several micrometers in size and consist of the male haploid nucleus, usually an acrosome or acrosomal material, and a very small amount of cytoplasm (Felgenhauer and Abele 1991; Braga et al. 2013). In sicyoniids, solenocerids, and penaeids, the sperm cell consists of a spherical or tear-shaped main body composed of nucleus and spike-like appendage (Fig. 6.14A3–7; Plate 6.3A, B). The acrosomal complex is a structure important in binding the sperm to and penetration of the egg during fertilization. The acrosome is partly in the main body and continues into the spike (Fig. 6.14A3–7). In the penaeid *Rimapenaeus similis*, the main body is more cylindrical and a long, slender, and immotile filament, function unknown, protrudes from the top of the main body (Plate 6.3B) (Bauer and Lin 1993). Aristeid sperm lack a spike, as do those of the *Eusergestes arcticus* and *Peisos petrunkevitchi*, the only well-studied sergestoids (Braga et al. 2013) (Fig. 6.14A8–11). Some aristeids species lack acrosomal structures, as do the sergestid species.

Caridean sperm are similar in overall structure to that of typical dendrobranchiate sperm in that they usually have a main body and spike (Fig. 6.14A1–2, B–D; Plate 6.3C, D). However, there are some important differences. The main body of most species is cap-shaped so that the sperm resembles a thumbtack (Fig. 6.14B) (Plate 6.3C). There is not a distinct or organized acrosomal complex, as in most dendrobranchiates, but there is acrosomal-like material surrounding the sperm cap

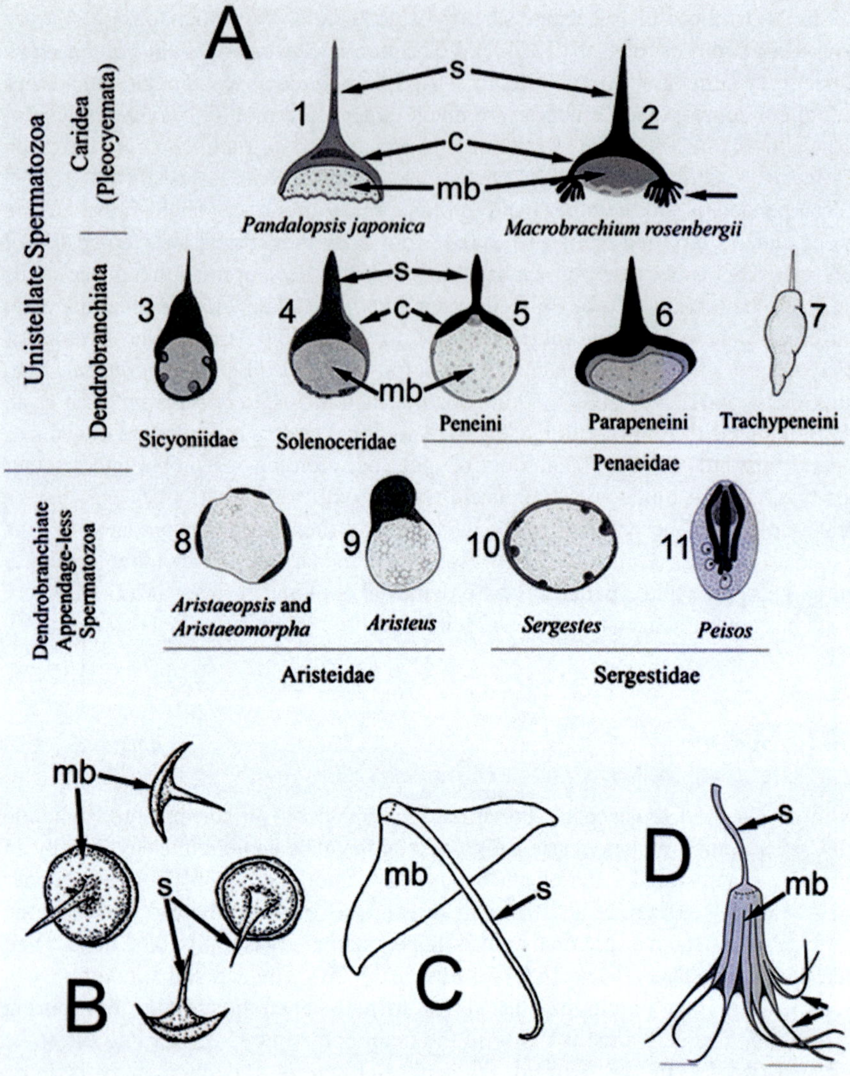

Fig. 6.14 Shrimp sperm cells. (**A**) Basic sperm cell morphology in two carideans (1–2) and different dendrobranchiate taxa (3–11). (**B–D**) Variation in caridean sperm cells: (**B**) thumb-tack shaped sperm of *Heptacarpus sitchensis* (Thoridae), (**C**) anchor-shaped sperm of *Campylonotus vagans* (Campylonotidae) and (**D**) multirayed sperm of *Rhynchocinetes typus*. c, cap; mb, main body; s, spike. Unmarked arrow in (**A**) indicates main body short rays or arms; unmarked arrows in (**D**) indicate main body rays in *R. typus*. Modified from (**A**, **D**) Braga et al. (2013), (**B**) Bauer (1976) and (**C**) Torti and Boschi (1973)

and extending into the spike (Felgenhauer and Abele 1991; Braga et al. 2013). Although the thumbtack shape is most common in caridean sperm, there are some notable variations. Three rays extend from the main body of *Pandalus* sperm

(Berkeley 1930). The sperm cells of *Campylonotus* species appear somewhat like a triangular anchor (Fig. 6.14C), while those of *Thor* species resemble, when exposed to seawater, a cartridge with basal case and distal projectile (Plate 6.3E). Extending from the main body of rhynchocinetid sperm are a series of connected radial arms that, when the sperm is exposed to seawater, unfold like an inverted umbrella (Fig. 6.14D) (Plate 6.3D) (Dupré and Barros 1983). The sperm cells of *Macrobrachium rosenbergii* (Fig 6.14A2) and *Potimirim potimirim* have similar but shorter radial arms extending from the main body (Lynn and Clark 1983a; Braga et al. 2013; Machado et al. 2021). Given the taxonomic diversity of carideans, it will not be surprising if other sperm morphologies will be found within the group.

Sperm cells of the procarididеans have, not surprisingly given their close relationship to carideans, a very similar thumbtack shape (Felgenhauer et al. 1988). Spermatozoa of stenopodideans studied to date have been described as unornamented flat discs, lacking a spike and distinct acrosomal structures (Felgenhauer and Abele 1991; Gregati et al. 2014).

6.4.3 Sperm Meets Egg: Fertilization

Wally (W.H.) Clark, Jr. and various colleagues pioneered studies on shrimp fertilization with their work on the penaeoid *Sicyonia ingentis* (Clark et al. 1986; Pillai et al. 1988; Griffin et al. 1988) and the caridean *Macrobrachium rosenbergii* (Palaemonidae) (Lynn and Clark 1983b). Enrique Dupré and Claudio Barros (2011) likewise have done important studies on fertilization with the caridean *Rhynchocinetes typus* (Rhynchocinetidae). Research on all these species has something in common: the shrimps were readily available and convenient to maintain in the laboratory. *S. ingentis* proved ideal for fertilization studies because Clark's students could tell by external inspection (presence of oocytes in female oviducts) that females were ready to spawn when exposed to darkness. When such a female is placed on top of a beaker of seawater after stimulating its gonopores with a probe, it goes into cataleptic paralysis, stays in position, and spawns directly into the beaker (Pillai et al. 1988). Oocytes come into contact with sperm emitted from the female's seminal receptacles, and researchers could collect such oocytes immediately and study fertilization right when it was occurring. Lynn and Clark (1983b) were able to study fertilization in *M. rosenbergii* because it is a major aquaculture species which can be readily obtained and easy to rear in the laboratory. *Rhynchocinetes typus* is a large shrimp abundant on the rock pilings around the Universidad Católica del Norte's marine laboratory at Coquimbo, Chile. Dupré and colleagues could collect it rather easily to do their studies. Females of these latter two species, as in all carideans, mate just or soon after molting, and premolt/prespawning females are easy to identify. Like sicyoniid species, these two carideans will readily mate and spawn in the laboratory. Thus, these three species were ideal models for fertilization studies. This shows the importance of finding and choosing the right species to work with for a particular type of research.

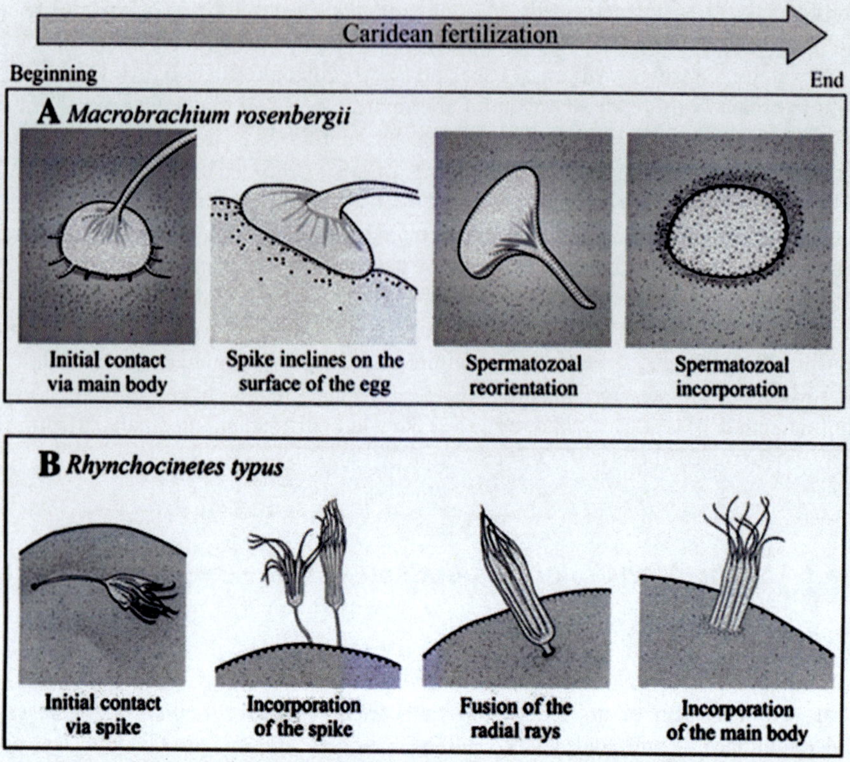

Fig. 6.15 Sperm-egg interaction during fertilization in two caridean shrimps. (**a**) *Macrobrachium rosenbergii* (sperm main body initially binds to egg, spike bends over to penetrate egg membrane) and (**b**) *Rhynchocinetes typus* (spike initially binds to and penetrates egg membrane). From Braga et al. (2013) based on work from (**a**) Lynn and Clark (1983b) and (**b**) Barros et al. (1986)

Thus, much is known about the fertilization process. When the oocyte is first spawned and contacts a sperm cell (usually several), several events take place so that only one sperm enters the egg. Its haploid nucleus will fuse with that of the female's egg nucleus to form the zygote to begin embryonic development. The shape of most shrimp sperm, with a long spike, suggests that the sperm might be simply pushed into the egg like a pin into a pin cushion. Of course, this oversimplifies the process, but it is the spike which penetrates the egg membrane (Fig. 6.15). However, it does so by biochemical and physiological means. In penaeoids, the sperm acrosome vesicle undergoes a dramatic chemical and morphological change (acrosome reaction) which penetrates the egg membrane and draws the sperm contents into the egg. These events also trigger an expansion of the egg membrane away from the main contents of the egg, keeping other sperm from entering the egg (polyspermy). The fertilization membrane will later develop into the hatching membrane of the embryo. In the caridean *M. rosenbergii*, however, it is the main body which first makes contact and fuses with the egg membrane. Then the spike slowly bends over to chemically bore into through the egg membrane (Fig. 6.15a). There is no dramatic acrosome reaction in caridean sperm, but similar processes involving sperm-egg binding substances to attach to and lytic enzymes to breakdown the egg membrane allow the sperm's genetic material to be drawn into the egg. Dupré and Barros (2011) show that, in *R. typus*, the radial arms of the sperm unfold when exposed to seawater prior to contact with an egg to form the characteristic inverted umbrella. However, it is the spike which makes contact and first penetrates the egg membrane (Fig. 6.15b). As the sperm cell is drawn into the egg, the radial arms are folded up again. The question naturally arises, what is the function, if any, of the unfolding then folding again of the sperm rays? In spite of these sophisticated and illuminating pioneering studies, there is still much to be learned about the biochemical and mechanical details of the fertilization process in shrimps.

6.5 Temporal Cycles of Molting, Mating, Spawning, and Incubation

Reproductive events such as vitellogenesis and ovarian maturation, mating, spawning and, in pleocyemate shrimps, embryo incubation are highly linked and coordinated within the molting cycle. In penaeoids with long-term sperm storage in seminal receptacles (closed thelyca species), mating occurs immediately or soon (up to a few days) after a reproductive female molt (e.g., Hudinaga 1942; de Saint-Brisson 1986; Primavera 1979; Bauer 1994; Sclezo 1991) (Fig. 6.16a). Molting females in these species have immature ovaries. Anderson et al. (1985) measured the number of sperm that female seminal receptacles contained in *Sicyonia ingentis*, and it was an order of magnitude greater than that needed to fertilize a single spawn. Females may undergo 2 or more cycles of ovarian maturation and spawning (up to 6 in *Artemesia longinaris*) using stored sperm from one mating (Anderson et al. 1985; Sclezo 1991). At the next molt, any remaining sperm are cast off with the molt skin (exuviae) of the seminal receptacles. Upon successful mating, the receptacles are filled once again for a new round of ovarian maturation and spawning during the reproductive season.

In aristeids such as *Aristeus antennatus* (Demestre 1995) and *Aristaeomorpha foliacea* (Tunesi 1987), and some solenocerids such as *Haliporoides, Pleoticus*, and *Hadropenaeus* (Pérez Farfante 1977), external spermatophores are commonly found during the reproductive season. In *A. antennatus*, females mate and receive spermatophores about midway during ovarian maturation (Fig. 6.16b). A few weeks later, spawning and release of sperm from the spermatophores take place. In this species, at least three cycles of ovarian maturation, molting, mating, and spawning occur during the reproductive season. Molting and mating take place midway during ovarian maturation and spawning. Unlike the closed-thelycum species, only one spawn per mating event occurs in this pattern (Demestre 1995). Unused spermatophores and sperm may either be groomed off or certainly cast off by the next molt.

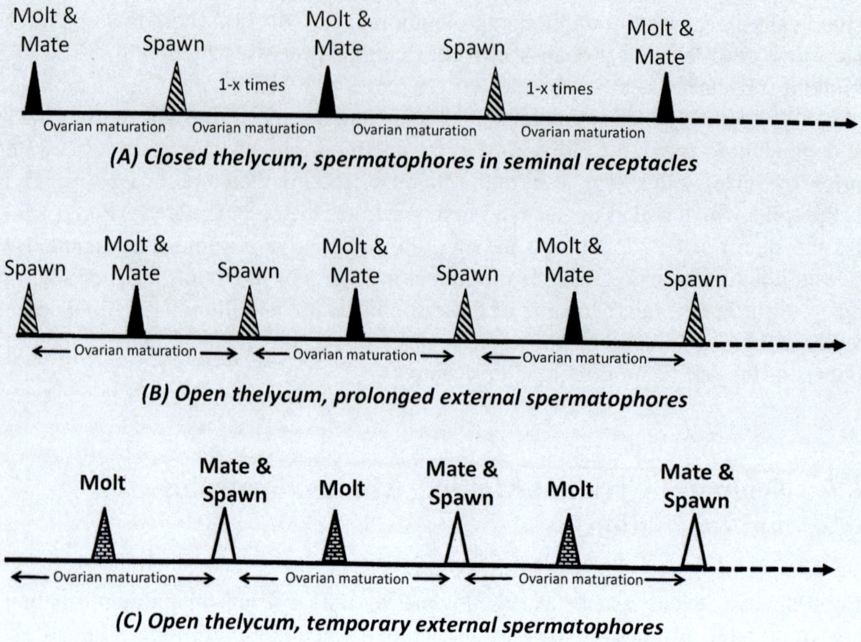

Fig. 6.16 Cycles of molting, mating, and spawning in penaeoid shrimps. (**a**) Closed-thelycum species (many penaeids and all sicyoniids). Spermatophores and sperm are stored in seminal receptacles. Mating occurs just after a female molt, followed by ovarian development and later spawning, which may be repeated one-several times before next molt and mating. (**b**) Open-thelycum species in which external spermatophores remain on the female for an extended period. Molting and mating occur midway during ovarian development of the female, spawning occurs later in aristeids which have been investigated, probable in several solenocerid species. (**c**) Open-thelycum species in which spermatophores are only briefly attached before spawning. Mating occurs in females with prespawning ovaries without female molting, soon followed by spawning (*Litopenaeus* and probably other open-thelycum penaeids; *Solenocera* and possibly some other solenocerids). Molts occur between episodes of mating and spawning

What might be the function of carrying external spermatophores for a month or more before spawning in these open-thelycum species? It may be that the sperm need further brief or longer maturation outside of the male duct before they are completely functional, a process known as capacitation, which occurs in many animals, including both open-thelycum *Litopenaeus* and closed-thelycum penaeid species (reviewed in Braga et al. 2013). Sperm taken directly from the male ejaculatory duct are much less effective in fertilization than those taken from female seminal receptacles or externally deposited spermatophores. Capacitation involves biochemical and physiological changes in the sperm so that the acrosome reaction only takes place when sperm contact spawned oocytes. Sperm capacitation has not yet been reported in carideans, perhaps because caridean sperm do not have organized acrosomes as do dendrobranchiate shrimps.

The external spermatophores of open-thelycum taxa such as *Litopenaeus* and *Solenocera* species are rarely found on females. This is because spawning takes place very soon (minutes, hours) after mating, as reported in *Litopenaeus* species (Misamore and Browdy 1996) and *Solenocera melantho* (Ohtomi and Yamamoto 2001). Mating and spawning are not linked to a molt; intermolt receptive females have mature ovaries ready for spawning (Fig. 6.16c).

Incubation of embryos is not an issue in penaeoids, as fertilized eggs are simply shed into the water and are on their own afterward. However, fertilized eggs are attached to the spawning female's third pereopods in the sergestoid *Belzebub faxoni* (Lee et al. 1992) and other luciferids, (e.g., Gordon 1956; note that Pérez Farfante and Kensley 1997 wrote "third pleopods" but this is undoubtedly an inadvertent error; Isa and Brian rarely made mistakes!). The embryos remain there during embryonic development to hatch as nauplii 24–40 h later. There is no information on any other care, e.g., grooming or aeration, of the embryos during their brief incubation. Spawning (oviposition) of *B. faxoni* occurs during intermolt periods, as does copulation with males. Thus, there is no close link of mating and molting in the luciferids studied. I have been unable to find any thorough studies on the molting and mating pattern of sergestids. In this group, there is no incubation of embryos: they are free-spawners like their penaeoid relatives. They do store sperm in seminal receptacles; their molt, mate, and later spawn pattern may be similar to that of luciferids or perhaps to closed-thelycum penaeoids. This is a subject badly in need of study, with best results likely from observations on a laboratory-friendly sergestid species.

In carideans, stenopodideans, and probably procarididеans, the pattern is distinct from those of dendrobranchiates. In the former taxa, hatching of a previous brood of embryos, molting, mating, and spawning are all closely coordinated and linked in time (e.g., Nouvel and Nouvel 1937; Höglund 1943; Bauer 1976; Noël 1976; Seibt and Wickler 1979; Berg and Sandifer 1984; Bauer 1989; Bauer and Abdalla 2001; Saito and Koya 2001; Gregati et al. 2010, 2014; Rasch and Bauer 2016) (Fig. 6.17a). After females with mature ovaries molt, they are immediately or soon receptive to males and readily mate. If females are carrying a previous brood, molting occurs a short time (up to a few days) after hatching of the embryos. Mating often takes place immediately or within minutes of molting provided that males are nearby (e.g., Bauer 1976; Bauer and Holt 1998; Bauer and Thiel 2011). Females deprived of male contact remain receptive to males until they spontaneously spawn several hours (Caskey and Bauer 2005) to 1–3 days after molting (e.g., Bauer 1976, 1989; Höglund 1943; Correa et al. 2000). Spawning may occur within minutes of successful mating but usually occurs 2–3 h afterward (e.g., Nouvel and Nouvel 1937; Höglund 1943; Bauer 1976; Bauer and Holt 1998; Bauer and Thiel 2011). *Rhynchocinetes typus* is somewhat unusual is that a male may deposit several spermatophores during a prolonged bouts of copulation and spawning; spawning begins immediately after placement of the first spermatophore.

The female molt soon before mating and spawning is often termed a parturial molt. Females of many carideans and stenopodideans have successive spawns during the reproductive season, i.e., when a female hatches the brood she is

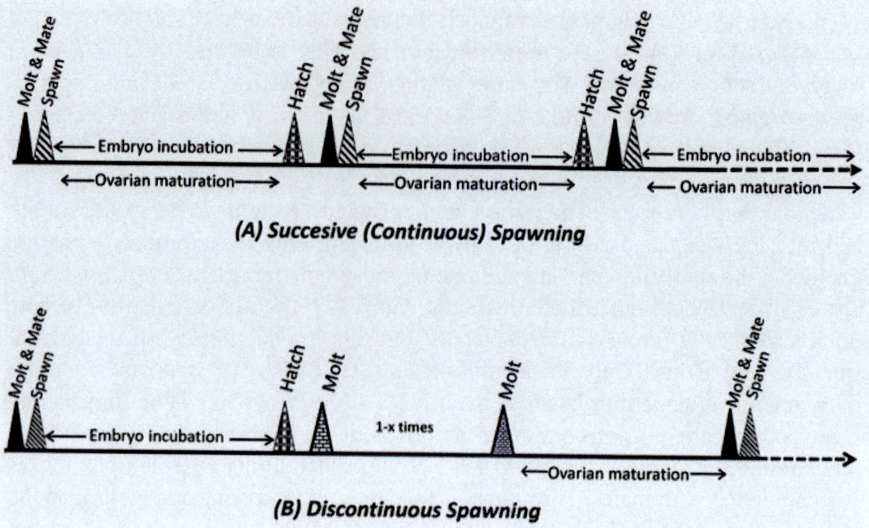

Fig. 6.17 Cycles of molting, mating, spawning, and hatching in caridean, stenopodidean, and possibly procarididean shrimps. (**a**) Successive spawning: while embryos from a previous spawn are being incubated, the ovaries simultaneously mature for the next spawning. After hatching of the current brood, females molt and then soon mate and spawn again. (**b**) Discontinuous spawning: while embryos from a previous spawn are developing, the ovaries do not mature for a new spawn. After embryos hatch, the female molts soon or some time afterward and may go through another molt or molts before ovarian maturation begins again for the next spawning

incubating from the previous spawning, the ovaries are mature and ready for a new spawn after the female molts and mates, which occurs shortly after hatching (continuous spawning, Fig. 6.17a). This pattern can be inferred from a population sample by recording both the degree of ovarian maturation of a female and the degree of development in the embryos in the brood she carries. Ovarian development can usually be easily observed through the translucent exoskeleton of small shrimps or in the area between the carapace and abdomen where the cuticle is thin in larger ones. A subjective scale of ovarian and embryonic development is normally sufficient (Bauer and VanHoy 1996; Bauer and Delahoussaye 2008; Rasch and Bauer 2015). In a population with such successive parturial molts, there will be a positive correlation between the two variables (Fig. 6.18). In other species, however, the female may delay a parturial molt until ovarian maturation is complete (Allen 1960; Bauer and Abdalla 2000). Another pattern (discontinuous spawning) is that the female may go through one or more nonreproductive molts before spawning again (e.g., Allen 1959; Bauer and Abdalla 2000) (Fig. 6.17b). These nonreproductive molts, both during and outside of the reproductive season, are opportunities for female growth. The resources and metabolic energy females use for vitellogenesis and ovarian maturation can instead be directed into growth.

Males of carideans which occur in socially monogamous male-female pairs may have molt cycles which are out of synchrony with the parturial molts of their female.

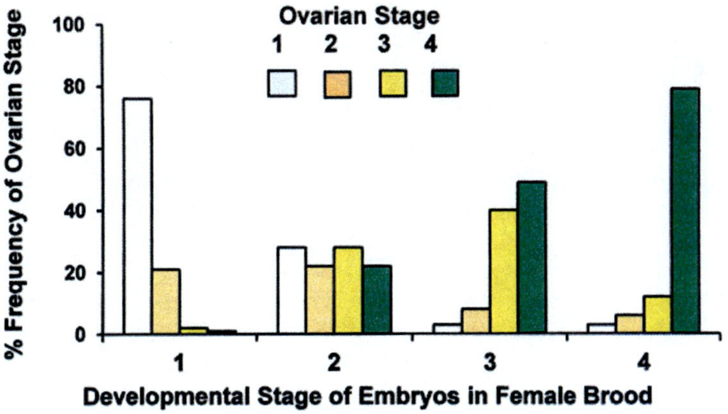

Fig. 6.18 Correlation between ovarian maturation and development of embryos incubated by caridean and stenopodidean females which have successive (continuous) spawnings. Ovarian stage ranges from 1 (little or no vitellogenic oocytes) to 4 (ovaries distended with vitellogenic oocytes, near to spawning); embryonic development ranges from 1 (recently spawned embryos, little development) to 4 (embryos with little yolk, large eyes, near to hatching (Allen 1966; Bauer 1986, courtesy of University of Chicago Press)

For example, males of *Stenopus hispidus* molt while their partner female is in intermolt. In this way, males have recovered from the molt (a vulnerable stage) and be in a hard-shell condition prepared to mate when the female molts. Furthermore, its hard-shell condition allows the male to properly defend its female from other males or predators while she recovers after a parturial molt (Gregati et al. 2014). Similarly, in the pair-living caridean simultaneous hermaphrodite *Lysmata amboinensis* (see next chapter), molts are staggered. When one member of the pair completes ovarian maturation, it molts, mates, and spawns while the other member, which serves as a male, is in intermolt; their roles are reversed in their next molt cycle (Fiedler 1998).

Why are the reproductive processes of ovarian maturation, molting, and spawning so closely linked in closed-thelycum penaeoids, carideans, and stenopodideans? In these penaeoids and also sergestoids, the seminal receptacles which store sperm are invaginations of the exoskeleton. Thus, their inner lining will be molted with the rest of exoskeleton along with any remaining sperm and associated seminal material. The external thelycum, the apertures to the seminal receptacles, and broken or fouled setae associated with them, will be renewed with clean and undamaged versions. The female will be ready for mating, storage of the spermatophores and subsequent spawning. Penaeoids with open thelyca also need to molt to remove old and fouled exoskeleton as well as any remains of previous external spermatophores. However, there are no seminal receptacles. Apparently molting between spawns is sufficient in these species. In carideans and stenopodideans, the areas that receive the external spermatophores and around the female gonopores need to be refurbished. In these pleocyemates, the underside of the abdomen, pleopod surfaces and the setae to which embryos will be attached for incubation must be replaced as well. These cuticular preparations, termed the "breeding dress," are discussed below.

6.6 Spawning, Parental Care, and Breeding Dress

Dendrobranchiates are termed "free-spawners" because inseminated females release eggs into the water column. Spawning behavior has been observed in some penaeoid shrimps in the laboratory and appears to follow a common general pattern (Pillai et al. 1988; Alfaro et al. 1993 and papers cited therein). Prespawning females are first quiescent on the bottom, then ascend into the water column, descending and ascending several times. Finally, up in the water column, the female releases a steady stream of eggs into the water while vigorously beating the pleopods, a behavior which disperses the eggs. The eggs are fertilized as they are released by a steady stream of sperm emitted from the exit slits of seminal receptacles in closed-thelycum species, or from sperm exposed by the breakup of external spermatophores in open-thelycum species. In dendrobranchiate species with seminal receptacles, the sperm might be forced out by contractions of the muscles attached to or around the seminal receptacles (e.g., Genthe 1969; Bauer and Lin 1993). However, the mechanism of sperm release is not well known. As sperm are released, they flow in sternal channels up to the nearby female gonopores where eggs are being released. The mechanism of sperm release from external spermatophores in open-thelycum species is unknown.

Discussion of parental care in dendrobranchiates will be short: there is none! Fertilized eggs are left on their own to develop, to hatch within a day or so, and then to undergo larval development in the sea (Schram 1986; Dall et al. 1990). As is often the case with sweeping generalizations, there is one exception—the luciferids. The females do attach a small brood of fertilized eggs to their third pereopods until hatching a day later (Lee et al. 1992). Whether the female actively does something to aid the brief embryonic development before hatching is not known. Female luciferids have no obvious morphological specializations for brood care, i.e. no breeding dress or discernable mechanisms for aerating or grooming or embryo aeration as in the pleocyemates.

The carideans, stenopodideans, and procarididеans are completely different from the dendrobranchiates in parental care. Embryos are attached to the pleopods of the female and are tended for until they hatch, a process known as incubation or brooding. This care consists of fanning behavior, in which the brooding female elevates the body a bit and beats the pleopods in place. This behavior circulates water around and among the embryos, providing them with oxygenated water and carrying away metabolic wastes, principally carbon dioxide and nitrogenous wastes (ammonia) (Phillips 1971). Sediment that has accumulated among the embryos may be dislodged as well. Females actively brush and groom the embryos, removing microbial growth, microfouling organisms, and particulate matter (Chap. 4). The length of the embryo incubation period closely coincides with the molting schedule of the female. If a brooding female molts before the embryos hatch, the latter will be cast off and left to the vagaries of the environment. Although embryos can physiologically survive without being attached to the female, they will soon succumb to scavengers or become smothered with particulate and microbial fouling organisms without female care (Phillips 1971; personal observations on *Heptacarpus sitchensis*).

Embryonic development, and thus incubation period, is highly influenced by water temperature. For example, in *Heptacarpus sitchensis* in southern California, embryo incubation is about 1 month during the winter but is only 2 weeks during the warmer summer period (personal observations). Thus, shrimps from higher latitudes have longer incubation periods, from a few to several months to almost a year in some Arctic and Antarctic species (e.g., Allen 1966; Bergström 2000; Sainte-Marie et al. 2006). Incubation may last less than a week in small tropical carideans which spawn repeatedly during their short lifetimes (Bauer 1989).

Female care of newly hatched larvae has been reported in a few caridean species. In crayfishes, this behavior is common—the newly hatched offspring remain several days and undergo a number of molts while still attached or associated with the female abdomen (Andrews 1907; Vogt and Tolle 2007). Embryos of the crangonid shrimp *Sclerocrangon boreas* are generously supplied with yolk and hatch as advanced larvae which are lecithotrophic, i.e., nonfeeding, subsisting on embryonic yolk remaining after hatching. These larvae cling to the pleopods of females, undergo one more larval molts and then another to a juvenile stage which also stays briefly on the mother (Guay et al. 2011). It is unlikely that the female gives any direct aid to brooded larvae and juveniles other than the protection provided by shelter among its pleopods. Interestingly, many of the broods also include a variable number of early-stage embryos and/or large "white eggs" which are either unfertilized or fertilized but aborted embryos. The intriguing hypothesis has been proposed that these serve as "trophic eggs" which the young juveniles feed on before leaving the mother. A similar case of extended incubation behavior has been described in the freshwater atyid *Dugastella valentina* (Cuesta et al. 2006). Females also allow the first two larval stages and the decapodid (pre-juvenile) stage to hang onto its pleopods before going off on their own in the same habitat as the adult. In this latter species, no possible trophic eggs were reported. Small white "chalky" eggs have been found among broods of a number of alpheid species (Felder 1982; Erdman and Blake 1987; Costa-Souza et al. 2014). However, the species in which they were reported are characterized by small embryo size with extended larval development, i.e. the larvae do not remain with the female after hatching. Several factors have been suggested to account for these small, infertile eggs, including parasitism, female immaturity, or unfavorable environmental conditions, e.g. development in waters cooler than optimal for these mainly tropical species (Felder 1982).

In carideans, morphological change occurs in the abdomen of prespawning females that prepares a spawning and incubation chamber for attachment and brooding of embryos. These adaptations are called the "breeding dress" of carideans. Although stenopodideans incubate and care for their embryos, no special breeding dress develops when female undergo a prespawning molt. It is likely that procaridideans, given the morphology of their females, are similar to carideans in this aspect. The breeding dress of carideans may be one of the key adaptations that are responsible for the biological success of the group.

Marine biologists C.F. Yonge (1955) and Hans Höglund (1943) synthesized knowledge about the breeding dress of carideans in their monographs on the reproductive biology of *Crangon crangon* (formerly *C. vulgaris*) and *Palaemon adspersus (*formerly *Leander squilla*), respectively. A female with mature ovaries prepares to spawn by first undergoing a molt at which the full breeding dress is fully attained or renewed. An immature female or nonbreeding female may begin to attain these features gradually through a series of molts, but at the first spawning molt, they emerge fully developed. The major characters are (1) the pleura of the first three abdominal segments become expanded and the sterna of these somites wider; (2) the protopods (peduncles) of the first 3–4 pleopods become noticeably elongated and more concave posteriorly; (3) the basipods of the pleopods develop a posterior flat extension (flange); (4) the underside of the posterior thorax and anterior pleopods gain additional setal groups; (5) long, stout and mostly naked "ovigerous" setae associated with embryo attachment are found mainly on the pleopods, as well as on the coxae of the last two pereopods (Fig. 6.19A–C). With these additional features, the female is ready for spawning, embryo attachment and then incubation.

The widening of the first three abdominal segments, increase in size of the anterior abdominal pleura, and the elongation and other changes in the pleopod peduncles create a much larger space under the abdomen to hold the eggs that will be spawned. In the more macruran-like families, such as Crangonidae and Alpheidae, these adaptations are not so well developed and the incubation chamber is therefore not so spacious (Yonge 1955). Some of the setal groups on the thorax, especially those around female gonopores, help to guide the eggs backward after they emerge from the gonopores. The pleopod flanges and pinnate setae on pleural borders and pleopod basipodites cover open spaces on the sides of the incubation chamber through which non-fertilized eggs might escape during the spawning process. The ovigerous setae are involved with attachment of the eggs to the pleopods.

Spawning has been observed in several caridean species but was particularly well documented by Höglund (1943) in *Palaemon adspersus*. I have been fortunate to observe spawning in the hippolytoids *Heptacarpus sitchensis* and *Lysmata wurdemanni* (Bauer 1976; Bauer and Holt 1998). After mating, the female spends some time vigorously grooming the underside of the abdomen and pleopods, interspersing bouts of grooming with vigorous beating of the pleopods. As the mating molt has recently occurred, the exoskeleton and its setae are new and clean. Höglund hypothesized that these activities help to renew and properly arrange the new setae so important to spawning and egg attachment. Molting cleans off any hatching membranes of embryos and associated debris from a previous brood, if present. As importantly, the cheliped picking at the underside of the inseminated female breaks apart the attached spermatophores and spreads sperm around the anterior part of the spawning chamber.

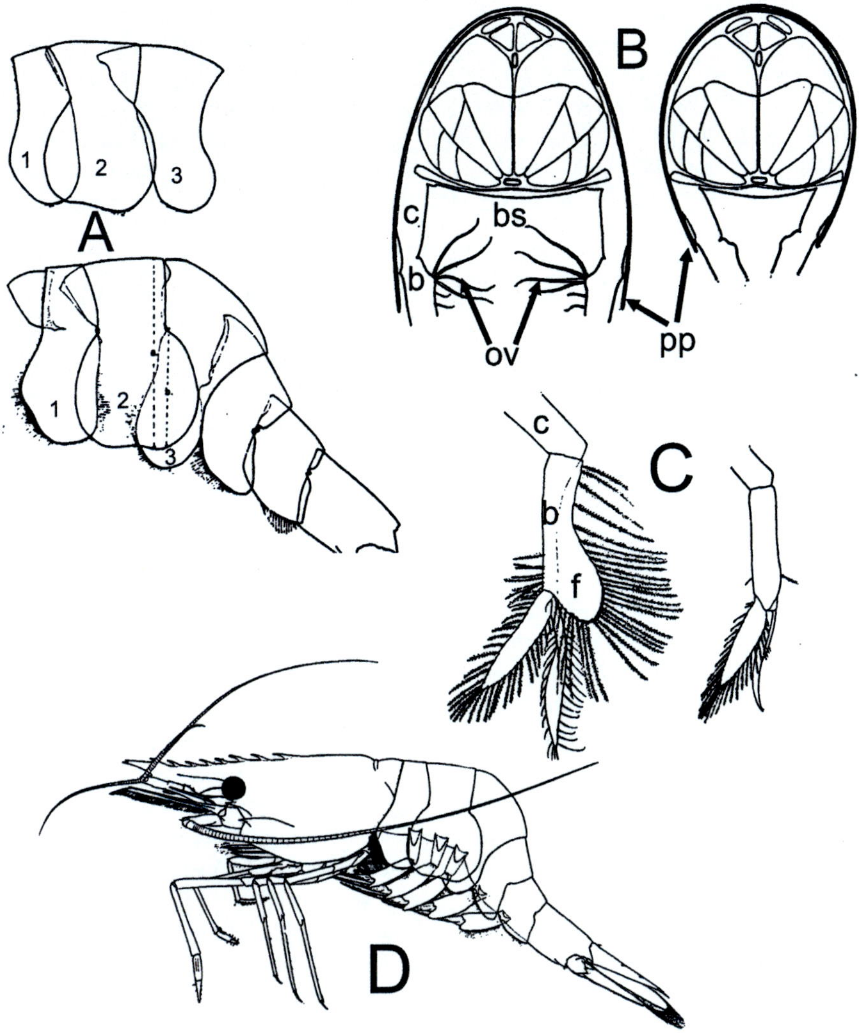

Fig. 6.19 Breeding dress of a caridean female, *Palaemon adspersus* (formerly *Leander squilla)* (Palaemonidae). (**A**) Anterior abdominal pleura of equally sized nonbreeding (above) and breeding (below) females. (**B**) Cross section of second abdominal somite of (left) a breeding (ovigerous) and (right) a nonbreeding female. (**C**) First pleopod of *Eualus* (*Spirontocaris*) *cranchii* in a (left) breeding and (right) nonbreeding female. (**D**) Hunched, somewhat S-shaped posture of a spawning female. 1–3, pleura of abdominal somites 1–3; b, basipodite; bs, brood space; c, coxopodite; f, flange; ov, ovigerous setae; pp, pleural plate. Modified from Höglund (1943)

When spawning begins, the female depresses her body and lowers the posterior few somites so that the body forms a somewhat S-shaped position (Fig. 6.19D). The anterior pleopods are extended forward so that the underside of the posterior thorax with gonopores is covered from below. As eggs are emitted from the gonopores, they move backward into the spawning/incubation chamber. Their movement is powered by a slow beat of the endopods of the first pleopods which powers a current within the chamber. The eggs are moved backward, swirling and bumping up against the roof (posterior thorax and anterior abdominal sternites) and floor (forward-extended pleopods) of the spawning/incubation chamber. This action causes the eggs to bump into sperm cells spread around the inner surfaces of the chamber so that fertilization takes place. The fertilized egg begins development and now is correctly termed an embryo.

The shrimps for which spawning was described above are typical hump-backed, laterally compressed epibenthic carideans with well-developed abdominal pleura. Carideans with a more depressed macruran-like body shape and short abdominal pleura may have a different spawning process, as in *Crangon crangon* (Höglund 1943). Spawning in this species is lobster-like, with the female lying on its side with the abdomen doubled against the thorax, a posture made necessary by a lack of a deep spawning/incubation chamber. Because of a shallower incubation chamber, the egg mass is not so voluminous, and the eggs are attached mainly to the proximal parts of the pleopods. In some shrimps, the egg mass may be so large that it extends well below the abdominal pleura with only the pleopod flanges partially covering its lower portion. This becomes more apparent as the developing embryos increase in size as they near hatching, which is usually the case.

The fertilized eggs attach to the pleopods and each other during spawning. As they churn backward into the spawning chamber, they are coated by an adhesive material secreted by large glands (mucilage packets) within the basipods of the pleopods (Yonge 1955; Fisher and Clark 1983). Gland ducts have not yet been found; Yonge believed the adhesive material ("cement") was emitted by the ovigerous setae during oviposition in *Crangon crangon*. However, Fisher and Clark (1983) presented good evidence in *Palaemon macrodactylus* that the substance is emitted onto the surface of the pleopods during the prespawning molt, not during spawning per se. Whatever the case, this material spreads by mixing and capillary action among the newly spawned eggs, coats them, and causes them to stick to each other and to appendages with ovigerous setae (pleopod peduncles and/or posterior pereopod coxae). The adhesive material forms simple or twisted connecting strands (funiculi) among the eggs and to the pleopods. The adhesive material binds with the egg membrane to form the outer coat or membrane of the embryo (along with other membranes later secreted by the embryo below the outer membrane) (Fisher and Clark 1983; Glas et al. 1997). In this way, the fertilized eggs form a cohesive mass that is firmly attached to the underside of the female until hatching. We will take up larval development and subsequent life history of the embryos in Chap. 9.

References

Alfaro J, Palacios JA, Aldave TM, Angulo RA (1993) Reproducción del camarón *Penaeus occidentalis* (Decapoda: Penaeidae) en el Golfo de Nicoya, Costa Rica. Rev Biol Trop 41: 563–572

Allen JA (1959) On the biology of *Pandalus borealis* Kroyer, with reference to a population off the Northumberland coast. J Mar Biol Assoc UK 38:189–220

Allen JA (1960) On the biology of *Crangon allmani* Kinahan in Northumberland waters. J Mar Biol Assoc UK 39:481–508

Allen JA (1966) The dynamics and interrelations of mixed population of Caridea found off the north-east coast of England. In: Barnes H (ed) Some contemporary studies in marine science. George Allen & Unwin, London, pp 45–66

Anderson WW, Lindner MJ (1971) Contributions to the biology of the royal red shrimp, *Hymenopenaeus robustus* Smith. Fish Bull 69:313–336

Anderson SLWH, Clark J, Chang ES (1985) Multiple spawning and molt synchrony in a free spawning shrimp (*Sicyonia ingentis*: Penaeoidea). Biol Bull 168:377–394

Andrews EA (1907) The attached young of the crayfish *Cambarus clarkii* and *Cambarus diogenes*. Am Nat 41:253–274

Andrews EA (1911) Male organs for sperm-transfer in the crayfish, *Cambarus affinis*: their structure and use. J Morphol 22:239–297

Barros C, Dupré E, Viveros L (1986) Sperm-egg interactions in the shrimp *Rhynchocinetes typus*. Gamete Res 14:171–180

Bauer RT (1976) Mating behaviour and spermatophore transfer in the *shrimp Heptacarpus pictus* (Stimpson) (Decapoda: Caridea: Hippolytidae). J Nat Hist 10(415):440

Bauer RT (1986) Sex change and life history pattern in the shrimp *Thor manningi* (Decapoda: Caridea): a novel case of protandric hermaphroditism. Biol Bull 170:11–31

Bauer RT (1989) Continuous reproduction and episodic recruitment in nine caridean shrimp species inhabiting a tropical seagrass meadow. J Exp Mar Biol Ecol 127:175–187

Bauer RT (1991) Sperm transfer and storage structures in penaeoid shrimps: a functional and phylogenetic perspective. In: Bauer RT, Martin JW (eds) Crustacean sexual biology. Columbia University Press, New York, pp 183–207

Bauer RT (1992) Repetitive copulation and variable success of insemination in the marine shrimp *Sicyonia dorsalis* (Decapoda, Penaeoidea). J Crust Biol 12:153–160

Bauer RT (1994) Usage of the terms thelycum and spermatheca in the reproductive morphology of the Decapoda, with special reference to the Penaeoidea. J Crust Biol 14:715–721

Bauer RT (1996) A test of hypotheses on male mating systems and female molting in decapod shrimp, using *Sicyonia dorsalis* (Decapoda: Penaeaoidea). J Crust Res 16:429–437

Bauer RT (2004) Remarkable shrimps: adaptations and natural history of the Carideans. University of Oklahoma Press, Norman

Bauer RT, Abdalla JA (2000) Patterns of brood production in the grass shrimp *Palaemonetes pugio* (Decapoda: Caridea). Invert Reprod Dev 38:107–113

Bauer RT, Abdalla JA (2001) Male mating tactics in the shrimp *Palaemonetes pugio* (Decapoda, Caridea): precopulatory mate guarding vs. pure searching. Ethology 107:185–199

Bauer RT, Cash CE (1991) Spermatophore structure and anatomy of the ejaculatory duct in *Penaeus setiferus, P. aztecus* and *P. duorarum* (Crustacea: Decapoda); homologies and functional significance. Trans Am Microsc Soc 110:144–162

Bauer RT, Delahoussaye J (2008) Life history migrations of the amphidromous river shrimp *Macrobrachium ohione* from a continental large river system. J Crust Biol 28:622–632

Bauer RT, Holt GJ (1998) Simultaneous hermaphroditism in the marine shrimp *Lysmata wurdemanni* (Caridea: Hippolytidae): an undescribed sexual system in the decapod Crustacea. Mar Biol 132:223–235

Bauer RT, Lin JM (1993) Spermatophores and plug substance of the marine shrimp *Trachypenaeus similis* (Crustacea: Decapoda: Penaeidae): formation in the male reproductive tract and disposition in the inseminated female. Biol Bull 185:174–185

Bauer RT, Thiel M (2011) First description of a pure-search mating system and protandry in the shrimp *Rhynchocinetes uritai* (Decapoda, Caridea). J Crust Biol 31:286–295

Bauer RT, VanHoy R (1996) Variation in sexual systems (protandry, gonochorism) and reproductive biology among three species of the shrimp genus *Thor* (Decopoda: Caridea). Bull Mar Sci 59:53–73

Berg AB, Sandifer PA (1984) Mating behavior of the grass shrimp *Palaemonetes pugio* Holthuis (Decapod: Caridea). J Crust Biol 4:417–424

Berggren M (1993) *Spongiocaris hexactinellicola*, a new species of stenopodidean shrimp (Decapoda: Stenopodidae) associated with hexactinellid sponges from Tartar Bank, Bahamas. J Crust Biol 16:784–792

Bergström B (2000) The biology of *Pandalus*. Adv Mar Biol 38:1–245

Berkeley AA (1930) The post-embryonic development of the common pandalids of British Columbia. Contrib Can Biol NS 6:79–163

Boddeke R, Bosschieter JR, Goudswaard PC (1991) Sex change, mating, and sperm transfer in *Crangon crangon* (L.). In: Bauer RT, Martin JW (eds) Crustacean sexual biology. Columbia University Press, New York, pp 164–182

Bortolini JL, Bauer RT (2016) Persistence of reduced androgenic glands after protandric sex change suggests a basis for simultaneous hermaphroditism in a caridean shrimp. Biol Bull 230:110–119

Braga A, Nakayama CL, Poersch L, Wasielesky W Jr (2013) Unistellate spermatozoa of decapods: comparative evaluation and evolution of the morphology. Zoomorphology 132:61–284

Bruce AJ (1972) Notes on some Indo-Pacific Pontoniinae, XXI. *Typton bawii* sp. nov., the first occurrence of the genus *Typton* Costa in the Indian Ocean (Decapoda Natantia, Palaemonidae). Crustaceana 23:243–254

Burkenroad MD (1934) The Penaeidea of Louisiana, with a discussion of their world relationships. Bull Am Mus Nat Hist 68:61–143

Burkenroad MD (1936) The Aristaeinae, Solenocerinae, and pelagic Penaeinae of the Binghan Oceanographic Collection. Materials for a revision of the oceanic Penaeidae. Bull Bingham Oceanogr Collect Yale Univ 5:1–151

Burukovsky RN (1978) Biology of the shrimp *Aristeus varidens*. Translated from Biologiya Morya 3:70–77

Caskey JL, Bauer RT (2005) Behavioral tests for a possible contact sex pheromone in the caridean shrimp *Palaemonetes pugio*. J Crust Biol 25:35–40

Chace FA Jr (1984) The caridean shrimps (Crustacea: Decapoda) of the Albatross Philippine Expedition, 1907-1910, Part 2: Families Glyphocrangonidae and Crangonidae. Smithsonian Contributions to Zoology, No. 397, 63 pp

Chace FA Jr (1985) The caridean shrimps (Crustacea: Decapoda) of the Albatross Philippine Expedition, 1907-1910, Part 3: Families Thalassocarididae and Pandalidae. Smithsonian Contributions to Zoology, No. 411, 143 pp

Chace FA Jr (1986) The caridean shrimps (Crustacea: Decapoda) of the Albatross Philippine Expedition, 1907-1910, Part 4: Families Oplophoridae and Nematocarcinidae. Smithsonian Contributions to Zoology, No. 432, 82 pp

Chace FA Jr, Holthuis LB (1978) *Psalidopus*: the scissor-foot shrimps (Crustacea: Decapoda: Caridea). Smithsonian Contributions to Zoology, No. 277, 22pp

Chace FA Jr (1997) The caridean shrimps (Crustacea: Decapoda) of the Albatross Philippine Expedition, 1907–1910, Part 7: Families Atyidae, Eugonatonotidae, Rhynchocinetidae, Bathypalaemonellidae, Processidae, and Hippolytidae. Smithson Contrib Zool 587:1–106

Chace FA, Manning RB (1972) Two new caridean shrimps, one representing a new family, from marine pools on Ascension Island (Crustacea: Decapoda: Natantia). Smithsonian Contributions to Zoology No. 131, 18pp

Charniaux-Cotton H, Payen G (1985) Sexual differentiation. In: Bliss DE, Mantel H (eds) The biology of Crustacea, Vol. 9, Integument, pigments, and hormonal processes. Academic Press, New York, pp 217–299

Chow S, Ogasawara Y, Taki Y (1982) Male reproductive system and fertilization of the palaemonid shrimp *Macrobrachium rosenbergii*. Bull Jpn Soc Sci Fish 48:177–183

Chow S, Taki Y, Ogasawara Y (1989) Homologous functional structure and origin of the spermatophores in six palaemonid shrimps (Decapoda, Caridea). Crustaceana 57:247–252

Clark WH Jr, Yudin AI, Shigekawa KJ, Griffin FJ (1986) Cellular events of fertilization in the Penaeidae, *Sicyonia ingentis*. In: Thompson MF, Sarojini R, Nagabhushanam R (eds) Indian ocean biology of benthic marine organisms. IBH, New Delhi, pp 143–149

Correa C, Baeza JA, Dupré E, Hinojosa IA, Thiel M (2000) Mating behavior and fertilization success of three ontogenetic stages of male rock shrimp *Rhynchocinetes typus* (Decapoda: Caridea). J Crust Biol 20:628–641

Costa-Souza AC, Schwarz da Rocha S, Arruda Bezerra LE, Oliveira Almeida A (2014) Breeding and heterosexual pairing in the snapping shrimp *Alpheus estuariensis* (Caridea: Alpheidae) in a tropical bay in northeastern Brazil. J Crust Biol 34:593–603

Cuesta JAE, Palacios-Theil PD, Rodríguez A (2006) A new rare case of parental care in decapods. Crustaceana 79:1401–1405

Dall W, Hill BC, Rothlisberg PC, Sharples DJ (1990) The biology of the Penaeidae. Adv Mar Biol 27:1–489

Dardeau MR (1984) *Synalpheus* shrimps (Crustacea: Decapoda: Alpheidae). I. The Gambarelloides Group, with a description of a new species. Mem Hourglass Cruises 7(2) 125 pp

De Grave S (2010) A new species of the genus *Typton* Costa (Decapoda, Palaemonidae, Pontoniinae) from Ascension Island. Stud Malacostraca 209–218

Demestre M (1995) Moult activity-related spawning success in the Mediterranean deep-water shrimp *Aristeus antennatus* (Decapoda: Dendrobranchiata). Mar Ecol Prog Ser 127:57–64

de Saint-Brisson SC (1986) The mating behavior of *Penaeus paulensis* Pérez-Farfante, 1967 (Decapoda, Penaeidea). Crustaceana 50:108–110

Descouturelle G (1971) Role des appendices sexuels males dan la copulations et d'un receptacle seminale femelle dans la ponte chez *Atyaephyra demaresti* Millet Crustacea-Decapoda-Caridea. Bull Acad Soc Lorraines Sci 10:10–14

D'Incao F, Silva Martins ST (2000) Brazilian species of the genera *Acetes* H. Milne Edwards and *Peisos* Burkenroad, 1945 (Decapoda, Sergestidae). J Crust Biol 20(special number 2):78–86

Dupré E, Barros C (1983) Fine structure of the mature spermatozoon of *Rhynchocinetes typus*, Crustacea Decapoda. Gamete Res 7:1–18

Dupré E, Barros C (2011) In vitro fertilization of the rock shrimp, *Rhynchocinetes typus* (Decapoda, Caridea): a review. Biol Res 44:125–133

Eberhard WG (1985) Sexual selection and animal genitalia. Harvard University Press, Cambridge, MA

Eberhard WG (1991) Copulatory courtship and cryptic female choice in insects. Biol Rev Camb Philos Soc 66:1–31

Erdman RB, Blake NJ (1987) Population dynamics of the sponge-dwelling alpheid *Synalpheus longicarpus*, with observations on *S. brooksi* and *S. pectiniger*, in shallow-water assemblages of the eastern Gulf of Mexico. J Crust Biol 7:328–337

Felder DF (1982) Reproduction of the snapping shrimps *Synalpheus fritzmuelleri* and *S. apioceros* (Crustacea: Decapoda: Alpheidae) on a sublittoral reef off Texas. J Crust Biol 2:535–543

Felgenhauer BE, Abele LG (1991) Morphological diversity of decapod spermatozoa. In: Bauer RT, Martin JW (eds) Crustacean sexual biology. Colombia University Press, New York, pp 322–341

Felgenhauer BE, Abele LG, Kim W (1988) Reproductive morphology of the anchialine shrimp *Procaris ascensionis*. J Crust Biol 8:333–349

Feng T, Paterson B, Johnston S (2018a) A morphological study of the male reproductive tract, post-testicular acrosome maturation and spermatophore formation in the black tiger prawn (*Penaeus monodon*). J Morphol 279:1290–1300

Feng T, Paterson B, Panagiotopoulou O, Green K, Mardon K, Johnston S (2018b) The functional anatomy of external genitalia in the black tiger prawn (*Penaeus monodon*) studied by micro-computed tomography and scanning electron microscopy. J Morphol 279:1346–1354

Fiedler GC (1998) Functional, simultaneous hermaphroditism in female-phase *Lysmata amboinensis* (Decapoda: Caridea: Hippolytidae). Pacific Science 52:161–169

Fisher WS, Clark WH Jr (1983) Eggs of *Palaemon macrodactylus*: I. Attachment to the pleopods and formation of the outer investment coat. Biol Bull 164:189–200

Genthe HC Jr (1969) The reproductive biology of *Sergestes similis* (Decapoda, Natantia). Mar Biol 2:203–217

Glas PS, Courtney LA, Rayburn JR, Fisher WS (1997) Embryonic coat of the grass shrimp *Palaemonetes pugio*. Biol Bull 192:231–242

Gordon I (1956) The Sergestidae of the great barrier reef expedition. Sci Rep Great Barrier Reef Expedition 6:323–333

Goy J (2010) Infraorder Stenopodidea Claus, 1872. In: Schram FR, Klein JCV (eds) Treatise on zoology: anatomy, taxonomy, and biology. The Crustacea. Vol. 9, Part A. Eucarida: Euphausiacea, Amphionidacea, and (partim). Koninklijke Brill NV, Leiden, pp 215–265

Gregati RA, Fransozo V, Lopez-Greco LS, Negreiros Fransozo ML (2010) Reproductive cycle and ovarian development of the marine ornamental shrimp *Stenopus hispidus* in captivity. Aquaculture 206:185–190

Gregati RA, Fransozo V, López-Greco LS, Negreiros-Fransozo ML, Bauer R (2014) Functional morphology of the reproductive system and sperm transfer in *Stenopus hispidus* (Crustacea: Decapoda: Stenopodidea), and their relation to the mating system. Invert Biol 133:381–393

Griffin FJ, Shigekawa K, Clark WH Jr (1988) Formation and structure of the acrosomal filament in the sperm of *Sicyonia ingentis*. J Exp Zool 246:94–102

Guay C, Sainte-Marie B, Brêthes JC (2011) Strong maternal effects and extreme heterogeneity of progeny development in the caridean shrimp *Sclerocrangon boreas* (Crangonidae). Mar Biol 158:2835–2845

Hartnoll RG (1969) Mating in the Brachyura. Crustaceana 16:161–181

Heldt JH (1938) De I'appareil genital des Penaeidae. Relations morphologique entre spermatophore. thelycum, el petasma. Travaux de la Station Zoologique Wimereux 13:349–358

Hobbs HH Jr., Hart CW Jr. (1982) The shrimp genus *Atya* (Decapoda: Atyidae). Smithsonian Contributions to Zoology, No. 364, 143 pp

Höglund H (1943) On the biology and larval development of *Leander squilla* (L.) forma typica De Man. Sven Hydrogr Biol Komm Skr Ny Ser Biol Band II, No. 6, 44 pp

Hudinaga M (1942) Reproduction, development, and rearing of *Penaeus japonicus*. Jpn J Zool 10: 305–393

Judkins DC (1978) Pelagic shrimps of the *Sergestes edwardsii* species group (Crustacea: Decapoda: Sergestidae). Smithson Contrib Zool 256:1–34

King JW (1948) A study of the reproductive organs of the common marine shrimp, *Penaeus setiferus* (Linnaeus). Biol Bull 94:244–262

Lee WY, Omori M, Peck RW (1992) Growth, reproduction and feeding behavior of the planktonic shrimp, *Lucifer faxoni* Borradaile, off the Texas coast. J Plankton Res 14:61–69

Lunina A, Vereshchaka A (2017) The role of the male copulatory organs in the colonization of the pelagic by shrimp-like eucarids. Deep Sea Res Part II Topical Stud Oceanogr 137:327–334

Lynn JW, Clark WH Jr (1983a) The fine structure of the mature sperm of the freshwater prawn, *Macrobrachium rosenbergii*. Biol Bull 164:459–470

Lynn JW, Clark WH Jr (1983b) A morphological examination of sperm-egg interaction in the freshwater prawn, *Macrobrachium rosenbergii*. Biol Bull 164:446–458

Machado M, Salti FC, Bertini G, Zara FJ, Negreiros-Fransozo ML (2021) Is *Potimirim potimirim* (Crustacea, Decapoda, Atyidae) a protandric hermaphrodite species? Behavioral and morphological aspects of the reproductive system. Arthropod Struct Dev 63:101060

Misamore MJ, Browdy CL (1996) Mating behavior in the white shrimps *Penaeus setiferus* and *P. vannamei*: a generalized model for mating in *Penaeus*. J Crustac Biol 16:61–70

Noël P (1976) L'évolution des caractères sexuels chez *Processa edulis* (Risso) (Décapode, Natantia). Vie et Milieu 26:65–104

Nouvel L (1939) Observation de l'accouplement chez une espéce de crevette *Crangon crangon*. C R Hebd Seances Acad Sci 209:639–641

Nouvel H, Nouvel L (1937) Recherches sur l'accouplement et la ponte chez les Crustaces Décapodes Natantia. Bull Soc Zool Fr 62:208–221

Ohtomi J, Yamamoto S (2001) The timing of mating related to ovarian maturation in the open-thelycum penaeoid shrimp *Solenocera*. Nippon Suisan Gakkaishi 69:467–474

Orsi Relini L, Tunesi L (1987) The structure of the spermatophore in *Aristeus antennatus* (Risso, 1816). Investigación Pesquera 51(suplemento 1):461–470

Pachelle PPG, Tavares M (2018) The freshwater shrimp family Euryrhynchidae Holthuis, 1950 (Crustacea: Decapoda: Caridea) revisited, with a taxonomic revision of the genus *Euryrhynchus* Miers, 1878. Zootaxa 4380:1–110

Pérez Farfante I (1969) Western Atlantic shrimps of the genus *Penaeus*. US Fish Wildl Serv Fish Bull 67:461–591

Pérez Farfante I (1971) Western Atlantic shrimps of the genus *Metapenaeopsis* (Crustacea, Decapoda, Penaeidae), with descriptions of three new species. Smithson Contrib Zool 79:1–37

Pérez Farfante I (1975) Spermatophores and thelyca of the American white shrimps, genus *Penaeus*, subgenus *Litopenaeus*. Fish Bull 73:463–486

Pérez Farfante I (1977) American solenocerid shrimps of the genera *Hymenopenaeus, Haliporoides, Pleoticus, Hadropenaeus* new genus, and *Mesopenaeus* new genus. Fish Bull 75:261–346

Pérez Farfante I (1985) The rock shrimp genus *Sicyonia* (Crustacea: Decapoda: Penaeoidea) in the Eastern Atlantic. Fish Bull 83:1–79

Pérez Farfante I, Kensley B (1997) Penaeoid and sergestoid shrimps and prawns of the world. Keys and diagnoses for the families and genera. Mémoires du Muséum national d'Histoire naturelle 175:1–233

Phillips G (1971) Incubation of the English prawn *Palaemon serratus*. J Mar Biol Assoc UK 51:43–48

Pillai MC, Griffin FJ, Clark WB Jr (1988) Induced spawning of the decapod crustacean *Sicyonia ingentis*. Biol Bull 174:181–185

Powell CB (1976) Two new freshwater shrimps from West Africa: the first euryrhynchids (Decapoda: Palaemonidae) reported from the Old World. Revue Zoologique Africaine 90: 884–902

Primavera JH (1979) Notes on the courtship and mating behavior in *Penaeus monodon* Fabricius (Decapoda, Natantia). Crustaceana 37:287–292

Rasch JA, Bauer RT (2015) Reproductive pattern and sexual system of the nocturnal seagrass shrimp *Ambidexter symmetricus* (Decapoda: Caridea: Processidae) in a Florida bay. Mar Freshw Res 67:1141–1152

Rasch JA, Bauer RT (2016) The functional morphology and role of the thelycum in insemination, and its relation to the mating system in the seagrass shrimp *Ambidexter symmetricus* (Decapoda: Processidae). Invert Biol 135:163–173

Ryan EP (1967) Structure and function of the reproductive system of the crab *Portunus sanguinolentus* (Herbst) (Brachyura, Portunidae). Proc Symp Crustacea Ernakulam India 2: 506–521

Sagi A, Snir E, Khalaila I (1997) Sexual differentiation in decapod crustaceans: role of the androgenic gland. Invert Reprod Dev 31:55–61

Sainte-Marie B, Bérubé I, Brillon S, Hazel F (2006) Observations on the growth of the sculptured shrimp *Sclerocrangon boreas* (Decapoda: Caridea). J Crust Biol 26:55–62

Saito T, Koya Y (2001) Gonadal maturation and embryonic development in the deep-sea sponge-associated shrimp, *Spongicola japonica* Kubo (Crustacea: Decapoda: Spongicolidae). Zool Sci 18:567–576

Saito T, Okuno J, Chan TY (2009) A new species of *Stenopus* (Crustacea: Decapoda: Stenopodidae) from the Indo-West Pacific, with a redefinition of the genus. Raffles Bull Zool supplement no 20:109–112
Schram FR (1986) Crustacea. Oxford University Press, Oxford
Sclezo M (1991) Control de la reproduccion en cautividad del camaron (*Artemesia longinaris*). Frente Maritimo 9, Section A:63–73
Seibt U, Wickler W (1979) The biological significance of the pair-bond in the shrimp *Hymenocera picta*. Z Tierpsychol 50:166–179
Spiridonov V, Casanova B (2010) Order Euphausiacea Dana, 1852. In: Schram FR, von Vaupel Klein JC (eds) Treatise on zoology—anatomy, taxonomy, biology. The Crustacea. Volume 9, Part A. Eucarida: Euphausiacea, Amphionidacea and Decapoda (partim). Koninklijke Brill NV, Leiden, pp 5–82
Tirmizi NM, Qureshi WM (1971) On the male genitalia of *Lucifer penicillifer* Hansen, 1919 (Decapoda, Sergestidae). Crustaceana 20:316–318
Torti MR, Boschi EE (1973) Nuevos aportos al conocimiento de los crustáceos Decápodos Caridea del género *Campylonotus* Bate 1988. Physis Sección A 32:65–84
Tóth E, Bauer RT (2007) Gonopore sexing technique allows determination of sex ratios and helper composition in eusocial shrimps. Mar Biol 151:1875–1886
Tunesi L (1987) La formation du spermatophore chez *Aristaeomorpha foliacea* (Risso, 1827). Investigación Pesquera 51(suplemento 1):471–476
Vereshchaka AL (2017) The shrimp superfamily Sergestoidea: a global phylogeny with definition of new families and an assessment of the pathways into principal biotopes. R Soc Open Sci 4. https://doi.org/10.1098/rsos.170221
Vereshchaka AL, Olesen J, Lunina AA (2016) A phylogeny-based revision of the family Luciferidae (Crustacea: Decapoda). Zool J Linn Soc 178:15–32
Vereshchaka AL, Lunina AA, Olesen J (2017) The genus *Gennadas* (Benthesicymidae: Decapoda): morphology of copulatory characters, phylogeny and coevolution of genital structures. R Soc Open Sci 4:171288. https://doi.org/10.1098/rsos.171288
Villalobos-Rojas F, Wehrtmann IS (2014) Secondary sexual characters and spermatophores of *Solenocera agassizii* (Decapoda: Solenoceridae), including a comparison with other solenocerid shrimp. Sex Early Dev Aquat Organ 1:45–55
Vogt G, Tolle L (2007) Brood care in freshwater crayfish and relationship with the offspring's sensory deficiencies. J Morphol 262:566–582
Wenner EL (1979) Distribution and reproduction of nematocarcinid shrimp (Decapoda: Caridea) from the Northwestern North Atlantic. Bull Mar Sci 29:380–393
Xiao Y, Greenwood JG (1993) The biology of *Acetes* (Crustacea, Sergestidae). Oceanogr Mar Biol Annu Rev 31:259–444
Yano I, Kanna RA, Oyama RN, Wyban J (1988) Mating behavior on the penaeid shrimp *Penaeus vannamei*. Mar Biol 97:171–175
Yonge CM (1955) Egg Attachment in *Crangon vulgaris* and other Caridea. Proc R Soc Edinb Sect B Biol Sci 65:369–400
Zhang D, Lin J (2004) Fertilization success without anterior pleopods in *Lysmata wurdemanni* (Decapoda: Caridea), a protandric simultaneous hermaphrodite. J Crust Biol 24:470–473
Zhang D, Lin J, Creswell RL (1998) Mating behavior and spawning of the banded coral shrimp *Stenopus hispidus* in the laboratory. J Crust Biol 18:511–518

Chapter 7
Sexual Systems

7.1 Sexual Systems and Hermaphroditism

Almost all organisms, particularly multicellular ones, reproduce sexually at some point in their life cycle. The advantages of sex are well-known and documented. Sexual reproduction produces new gene combinations (genetic diversity) for the environment to act upon. It also reduces the expression of and may promote the elimination of deleterious gene alleles. Organisms that reproduce sexually are able to fine-tune adaptations in stable environments and to adapt to new conditions in changing ones. The expression of male and female sex in a species is termed its sexual system, which also goes by the term "sex allocation." In many species, individuals are either male or female, i.e., sexes are separate (gonochory; dioecy). In others, individuals act as both male and female sometime during their lifetime (hermaphroditism; monoecy). Many different types of hermaphroditic sexual systems occur in multicellular plants and animals, aptly reviewed by David Policansky (1982).

Hermaphroditic organisms may function as male and female at the same time (simultaneous) or first as one sex and then another (sequential). Sequentially hermaphroditic organisms thus change ("reverse") their reproductive organs and function from one sex to the other. For example, in many fish species, individuals first mature and reproduce as females and then, when larger and older, as males (protogyny). Of special interest in this book are a number of shrimp species which mature and function first as males and then change to females (protandry). As we shall see, there are other species that mature first as male but, when older and larger, as females while retaining full male function (protandric simultaneous hermaphroditism). Besides defining these variations of hermaphroditism, I must also clarify some other terms and their usage. The term "intersex" has been used in crustacean literature in two different ways. Some authors refer to functionally hermaphroditic individuals as intersexes. Others use "intersex" to describe individuals which have both male and female structures but not function as a result of a genetic or developmental anomaly. They are not capable of reproduction both as male and female. Thus, in this chapter, "hermaphroditic" refers to functional hermaphrodites while "intersexes" are sexually abnormal individuals with both male and female characteristics (Martin and Scholtz 2012).

The first well-documented occurrence of hermaphroditism in shrimps was found in a European species, *Lysmata seticaudata* by Spitschakoff (1912). He correctly described the sex change from a male morphotype to that of functional female which produced and incubated viable embryos. Thus, the sex change was considered strictly protandrous. However, as will be seen later, although the direction of sex change was correct, his interpretation of the sexual system was somewhat in error. Almost three decades later, Alfreda Berkeley, first in a brief note (1929) and then the following year in an extensively documented study, definitely showed a protandrous (i.e., male to female) sex change in *Pandalus danae* (1930). From that time forward, numerous studies on hermaphroditic shrimps have been done (Table 7.1) including various reviews (Brook et al. 1994; Bauer 2000, 2004; Chiba 2007; Baeza 2018). The species counts of confirmed or probable hermaphroditism in shrimps have increased from approximately 30 (Brook et al. 1994; Bauer 2000) to 56 (Chiba 2007) to 90 (Table 7.1). Much of this increase is due to the recognition that all known lysmatid caridean species which have been studied (30; Table 7.1) are hermaphroditic. Prior to the initial recognition of the unusual lysmatid sexual system by Bauer and Holt (1998) and Fiedler (1998), only two *Lysmata* species were thought to be hermaphroditic (e.g., Spitschakoff 1912; Dohrn and Holthuis 1950; Charniaux-Cotton and Payen 1985). This stimulated other studies, especially by J. Antonio Baeza, a former doctoral student in my laboratory, who has verified hermaphroditism in a number of lysmatid and other caridean species (Table 7.1; References). Most studies on sexual biology and life history of shrimps indicate that gonochory is the most common sexual system in shrimps. Yet, sex change has many advantages for shrimps that are capable of doing so. As readers continue this chapter, they should keep in mind the question: "if sex change is so advantageous, why are not more shrimps hermaphroditic?"

7.2 The Direction of Sex Change and the Size Advantage Model

Shrimps, like other organisms, have limited resources for vital processes such as growth and reproduction. Males and females of gonochoric species can devote all their energetic resources available for reproduction into just one sexual function. Hermaphroditic organisms have the advantage of reproducing as both sexes but must divide up the total reproductive resources available among two sexual functions. Thus, hermaphroditic species evolve so as to allocate the resources to male and female reproduction in the most efficient way possible. According to the "Size Advantage Model," sequential hermaphroditism evolves when reproductive success

Table 7.1 Species with demonstrated or possible protandrous sexual systems

Family	Species	Sexual morphs	Supporting literature
Alpheidae			
	Arete dorsalis	PH, PM	Suzuki (1970)[a–d]; Nakashima (1987)[a–c]
	A. indicus	PH, PM, PF	Gherardi and Calloni (1993)[a–c]
	Salmoneus carvachoi	SH?	Oliveira et al. (2015)[a]
	Synalpheus paraneptunus	PH?	Toth and Bauer (2007, 2008)[b]
	Yagerocaris cozumeli	PH? SH?	Kensley (1988)[b, c]
Atyidae			
	Atyoida bisulcata	PH?	Radir (1930)[a, c]
	A. pilipes	PH, PF	Chace (1983)[a]
	Australatya striolata	PH, PF?	Smith and Williams (1982)[a, c]
	Caridina richtersi	PH?	Carpenter (1978)[a]; Holthuis (1954)[a]
	Paratya curvirostris	PH, PF	Carpenter (1978)[a–c]
	Potimirim mexicana	PH?	Alonso-Reyes et al. (2011)[a]; Bortolini and Álvarez (2012)[c]
Barbouriidae			
	Parhippolyte misticia	PSH	Onaga et al. (2012)[a–c, e]
	Pr. cf. uveae	PSH?	Baeza (2013, pers. obs.)
	Barbouria yanezi	PSH?	Baeza (2013, pers. obs.)
	Calliasmata nohochi	PSH?	Baeza (2013, pers. obs.)
Campylonotidae			
	Campylonotus rathbunae	PH	Yaldwyn (1960)[a, b]
	C. semistriatus	PH	Torti and Boschi (1973)[a, b]
	C. capensis	PH	Torti and Boschi (1973)[b]
	C. vagans	PH	Torti and Boschi (1973)[b]
Crangonidae			
	Argis dentata	PH, EMF?, PF?	Fréchette et al. (1970)[a,c]
	Crangon crangon	PH, PF	Boddeke et al. (1991)[b, c]; Schatte and Saborowski (2006)[d]
	Cr. franciscorum	PH, EMF?, PF	Gavio et al. (2006)[a–d]
Hippolytidae			
	Chorismus antarcticus	PH	Mascetti et al. (1997)[a, b]
	Hippolyte inermis	PH, PF	Zupo (1994)[a]; Zupo (2000)[d]; Zupo et al. (2008)[d]
	Nauticaris marionis	PH, PF	Pakhomov et al. (2000)[a, b]
Lysmatidae			
	Exhippolysmata ensirostris	PSH	Kagwade (1982)[a–c]; Sukumaran (1982)[a, c]

(continued)

Table 7.1 (continued)

Family	Species	Sexual morphs	Supporting literature
	E. oplophoroides	PSH	Laubenheimer and Rhyne (2008)[c]; Braga et al. (2009)[b, c]; Baeza et al. (2010)[a]
	Lysmata ankeri	PSH	Lin and Zhang (2001)[d, e] as *L. wurdemanni*; Baeza et al. (2009)[a, c–e, g]
	L. amboinensis	PSH	Fiedler (1998)[c, e]
	L. arvoredensis	PSH	Giraldes et al. (2018)[b, f]
	L. bahia	PSH	Baeza (2008)[a, c–e, g]
	L. boggessi	PSH	Baeza et al. (2009)[a, c–e, g]
	L. californica	PSH	Bauer and Newman (2004)[a–c, e]
	L. chica	PSH?	Baeza et al. (2009)[a, g]
	L. debelius	PSH	Fletcher et al. (1995)[f]; Simoes et al. (1998)[e]
	L. galapagensis	PSH	Baeza et al. (2009)[a, c–e, g]
	L. grabhami	PSH?	Wirtz (1997) ([c], based on pers. obs. R.T. Bauer)
	L. gracilirostris	PSH?	Baeza et al. (2009) (pers. obs., [f])
	L. hochi	PSH	Baeza and Anker (2008)[c–e, g]
	L. holthuisi	PSH	Anker et al. (2009)[a–e, g]
	L. intermedia	PSH	Baeza (2008)[a, c–e, g]
	L. kuekenthali	PSH?	Baeza et al. (2009)[a, g]
	L. moorei	PSH?	Baeza et al. (2009)[a, g]
	L. nayaritensis	PSH	Baeza et al. (2007)[a, c–e, g]
	L. olavoi	PSH?	Baeza et al. (2009)[f]
	L. nilita	PSH?	Dohrn and Holthuis (1950)[c]; d'Udekem d'Acoz (2003)[e]
	L. pederseni	PSH	Baeza (2010b)[a, c, g]
	L. rafa	PSH?	Rhyne and Anker (2007)[f]; Baeza (2009) ([f], pers. obs)
	L. rathbunae	PSH?	Baeza (2009)[a, g]
	L. seticaudata	PSH	Spitschakoff (1912)[c, g]; Dohrn (1950)[a–c]; d'Udekem d'Acoz (2003)[e]
	L. trisetacea	PSH?	Baeza et al. (2009)[a, g]
	L. udoi	PSH?	Baeza et al. (2009)[f]
	L. vittata	PSH?	Baeza et al. (2009)[a, g]; Alves et al. (2019)[e]
	L. wurdemanni	PSH	Bauer and Holt (1998)[a–c, e, g]
	Lysmatella prima	SH?	Murray et al. (2012)[d, e]
Merguiidae			
	Merguia rhizophorae	PH	Baeza (2010a)[a–c]; Alves et al. (2021)[a]
	M. oligodon	PH?	Malay and Rañises (2019)[a, b]
Pandalidae			
	Pandalus borealis	PPH, PF	Jägersten (1936)[a–c]; Allen (1959)[a–c]; Bergström (1997)[a]
	P. chani	PH	Komai (1999)[a, b]
	P. curvatus	PH?	Komai (1999)[f]
	P. danae	PH, EMF, PF	Berkeley (1930)[a–c]; Butler (1964)[a]; Marliave et al. (1993)[d]

(continued)

Table 7.1 (continued)

Family	Species	Sexual morphs	Supporting literature
	P. eous	PH, EMF	Berkeley (1930) as *P. borealis*[a–c]; Butler (1964)[a]
	P. formosanus	PH?	Komai (1999)[f]
	P. gracilis	PH	Oh et al. (2008)[a, b]
	P. goniurus	PH	Butler (1964)[a, b]
	P. gurneyi	PH?	Komai (1999)[f]
	P. hypsinotus	PH, EMF, PF	Berkeley (1930)[a–c]; Butler (1964)[a]
	P. ivanovi	PH	Komai and Eletskaya (2008)[a, b]
	P. jordani	PH,PF	Butler (1964)[a, b]
	P. latirostris	PH, EMF, PF	Onishi et al. (2001)[a–c]
	P. montagui	PH, PF, EMF	Jägersten (1936)[a–c]; Mistakidis (1957)[a–c]
	P. nipponensis	PH	Tamura (1950)[a, b]
	P. platyceros	PH	Berkeley (1930)[a–c]; Butler (1964)[a]
	P. prensor	PH	Hayashi (1988)[f]
	P. stenolepis	PH	Butler (1964)[a, b]
	P. teraoi	PH?	Komai (1999)[f]
	P. tridens	PH	Butler (1964)[a, b]
	Pandalopsis (now *Pandalus*) *dispar*	PH	Berkeley (1930)[a–c]; Butler (1964)[a]
	Pd. lamelligerus	PH?	Komai (1997b)[a? b?]
	Pd. longirostris	PH	Komai (1994)[b]
	Pd. japonica	PH	Kim et al. (2006)[b, c]
	Pd. gibba	PH?	Komai and Takeda (2002)[a, b]
	Pd. rubra	PH?	Komai (1994)[a? b]
Processidae			
	Processa edulis	PH?, PF	Noël (1973)[a]; Noël (1976)[b, c?]
Rhynchocinetidae			
	Rhynchocinetes uritai	PH	Bauer and Thiel (2011)[a–c]; Osawa et al. (2015)[a, b]
Thoridae			
	Thor amboinensis	PH	Baeza and Piantoni (2010)[a–c]
	T. manningi	PH, PM, PF	Bauer (1986)[a–c]

Sexual Morphs: EMF, protandrous early maturing females; PF, primary females; PH, protandric individuals; PM, primary males; PSH, protandric simultaneous hermaphrodites; SH, simultaneous hermaphrodites. ?, indicated but uncertain; Supporting Evidence: [a]Size-frequency distribution, degree of sexual dimorphism. [b]Transitional individuals (external morphology). [c]Gonadal evidence (transitional from testes to ovaries or ovotestes). [d]Direct observation of sexual development in laboratory-reared individuals. [e]For PSH species, mating experiments between female-phase simultaneous hermaphrodites. [f]Statement in the publication without specific supporting evidence. [g]For PSH species, sole observation of male gonopores in male phase, and both male and female gonopores in purported female-phase simultaneous hermaphrodites is correlated with larger size in one sex but not the other. Sequential hermaphrodites reproduce as one sex when younger and smaller but as the other sex when larger and older. Whether the sex change is male to female (protandry) or the reverse (protogyny) depends on the ecology and life history specifics for a particular species. If female reproductive ability is positively correlated with increasing size but that of male is not, protandry is favored. In the reverse situation, protogyny evolves. This model was initially proposed in a qualitative fashion by Michael Ghiselin (1969). It was formally tested quantitatively, using fishes and shrimps as examples, by Robert Warner (1975), then later with field studies on a variety of fish species throughout his career. The question of why and how a gonochoric species might evolve into a hermaphroditic one will be dealt with throughout this chapter.

In shrimps, hermaphroditism is always protandrous, with an individual reproducing first as a male and then later as a female or, in some hermaphroditic shrimps, first male and then both male and female simultaneously. Why male first? A basic biological fact is that the energetic costs of producing sperm are much less than those for making an egg cell. Many sperm can be produced for the same cost as an egg cell, which is abundantly rich in lipids and proteins needed to sustain development until hatching. This is particularly true in shrimps that incubate their embryos. As shown in the previous chapter, the size and energy content of a caridean or stenopodidean egg cell is greater than that of dendrobranchiate shrimps. Yet only a single sperm cell is needed to fertilize a single egg cell. A male shrimp does not have to use much energy to make more than enough sperm necessary to fertilize a brood of eggs of an individual reproducing as a female. However, to produce and incubate a brood of embryos, a female shrimp has to reach a certain size to have the energetic resources to do so. Fecundity in shrimps, as in many animals, is strongly correlated with female size. Larger females produce more offspring than smaller ones, as shown in Chap. 9. Small male size may be beneficial as long as male size is not correlated with male mating success. Small male size advantages may include avoidance of detection by predators or lower energetic needs for growth. Thus, in protandric sex changers, small (young) individuals can both grow and reproduce as males. However, in shrimps, individuals have to grow to a large enough size to have the resources to make large vitellogenic egg cells and to incubate the resultant embryos.

Considerations about the direction of sex change in a species must include the social organization, sexual dimorphism, and mating system of the species. There are many gonochoric shrimp species with small males and larger females, As stated above, this is due to the positive correlation with body size and female fecundity in shrimps. Many small male/large female species live at relatively high population abundances, i.e., aggregations or schools. As will be seen in the following chapter, males of such species do not defend females nor territories to attract them—mating is promiscuous, somewhat at random. The first male which contacts a receptive female usually is the one that mates with her ("pure search mating system"). There is little or no advantage to large male size in such a mating system. Males devote their energy to finding receptive females rather than to growth to a large size. It is from such small male/larger female taxa that protandry has evolved: mate as a male when small and a female when larger.

There are certainly a number of shrimp species in which males grow to a larger size than females (see next chapter). In these gonochoristic species, males gain matings by aggressively defending and guarding reproductive females from other males. Large size is important in male mating success; reproduction as a female when smaller would greatly diminish their ability to grow to the large male size needed in such species for reproductive success. Hermaphroditism has not evolved in any of these species: in such species, individuals do not reproduce first as females and then change to larger males (protogyny). Sexes remain separate. However, many fish species occur in social groups in which large dominant males do the mating (e.g., Warner 1984; Gemmell et al. 2019). In many of these species, individuals reproduce as females when smaller but large enough to produce eggs. If the dominant male of a social group dies, one of the females will change sex to male (protogyny) and become the dominant male. Some fish species are protandric, and in these species, the social organization and mating system is more like that of the small male/large shrimp species from which shrimp protandry probably evolved.

Thus, in the many small-male/large-female shrimp species in which male mating success is not correlated with large male size, how might protandry evolve? In such species, suppose that mutations occur that allow males near minimum female size to keep growing, to change sex and to reproduce as females ("male-first" hypothesis). Alternately, mutations might take place that permit juvenile females, in growing to the large size needed to produce eggs, to go through a male phase ("female-first hypothesis"). If a juvenile female can reproduce as a male just once before becoming a mature female, there is a tremendous increase in its reproductive fitness. Sperm production of the energetically inexpensive sperm should not detract much from growth. External differences between a juvenile female and similarly sized male are slight in carideans and stenopodideans, mainly some differences in pleopod 1–2 shape, hooks or spines and setation. In dendrobranchiates, the pleopod 1–2 differences are somewhat greater. Sex determination in shrimps is under control of endocrine organs, especially the androgenic gland, as will be discussed later, and is relatively flexible, so that two hypotheses mentioned above are not unreasonable. A major question that needs resolution is why hermaphroditism has evolved only in the Caridea and not in the dendrobranchiate (penaeoid, sergestoid) or stenopodidean shrimps. Another question is why protandry has not evolved more often in the Caridea, given the abundance of gonochoristic carideans with small males and larger females, the raw material for protandric species.

7.3 Types and Variations in Sexual Systems

The vast majority of shrimps have gonochoric sexual systems. However, about 90 species of carideans to date have been shown to have different variations of protandrous hermaphroditism. The range of protandrous sexual systems ranges from

(1) pure or partial protandry, i.e., all or some members of the population go through sex change to (2) protandric simultaneous hermaphroditism with first a male, then simultaneous hermaphrodite phase to (3) possibly pure simultaneous hermaphroditism, in which the male phase has been eliminated. The various sexual systems of shrimps are diagrammed in Fig. 7.1, with details of hermaphroditic species in Table 7.1.

7.3.1 Pure Protandry

All members of the population mature first as male (male phase: MP) before changing to female (female phase: FP). The gonads of juveniles develop as ovotestes but at first ovarian tissues are rudimentary and the testicular tissues develop and become functional (e.g., *Pandalus* species, Hoffman 1972; Charniaux-Cotton and Payen 1985). These individuals are functional MPs. With growth and age, the testicular tissue and male ducts (vasa deferentia, ejaculatory ducts) degenerate as the ovarian tissues develop (e.g., Hoffman 1972). The external male characters (endopod of first pleopod with cincinnuli; appendix masculina of second pleopod) change over some molts into those of breeding females (endopod of pleopod 1 changes shape, loses cincinnuli; appendix masculina is lost, breeding dress develops). These sex-changing individuals are called transitionals (Berkeley 1930; Butler 1964). Once the change is complete, the individuals function purely as reproductive females (FPs).

Given the variations of protandry described (Fig. 7.1), one might only have confidence that a particular population or the entire species is purely protandric when a fairly large sample has been examined, preferably from different locations. In some partially protandric species, most individuals of the populations are protandric but there may be a low and variable portion of the population that does not change sex However, it appears that pure protandry is fixed is some species, i.e., only there are only protandric genotypes. Some examples (Table 7.1) include the hippolytid *Chorismus antarcticus*; the pandalids *Pandalopsis dispar, Pandalus platyceros, P. tridens, P. goniurus, P. stenolepis* and *Rhynchocinetes uritai*. Although based on small sample sizes, campylonotid species, *Thor amboinensis* and *Merguia rhizophorae* are purely protandric (Table 7.1).

7.3.2 Partial Protandry with Primary and/or Early-Maturing Females

These are species in which protandry is not fixed. The majority of the population are protandric sex changers, but some variable proportion of the population develop only as females (primary females: PFs). Some smaller individuals in the same size

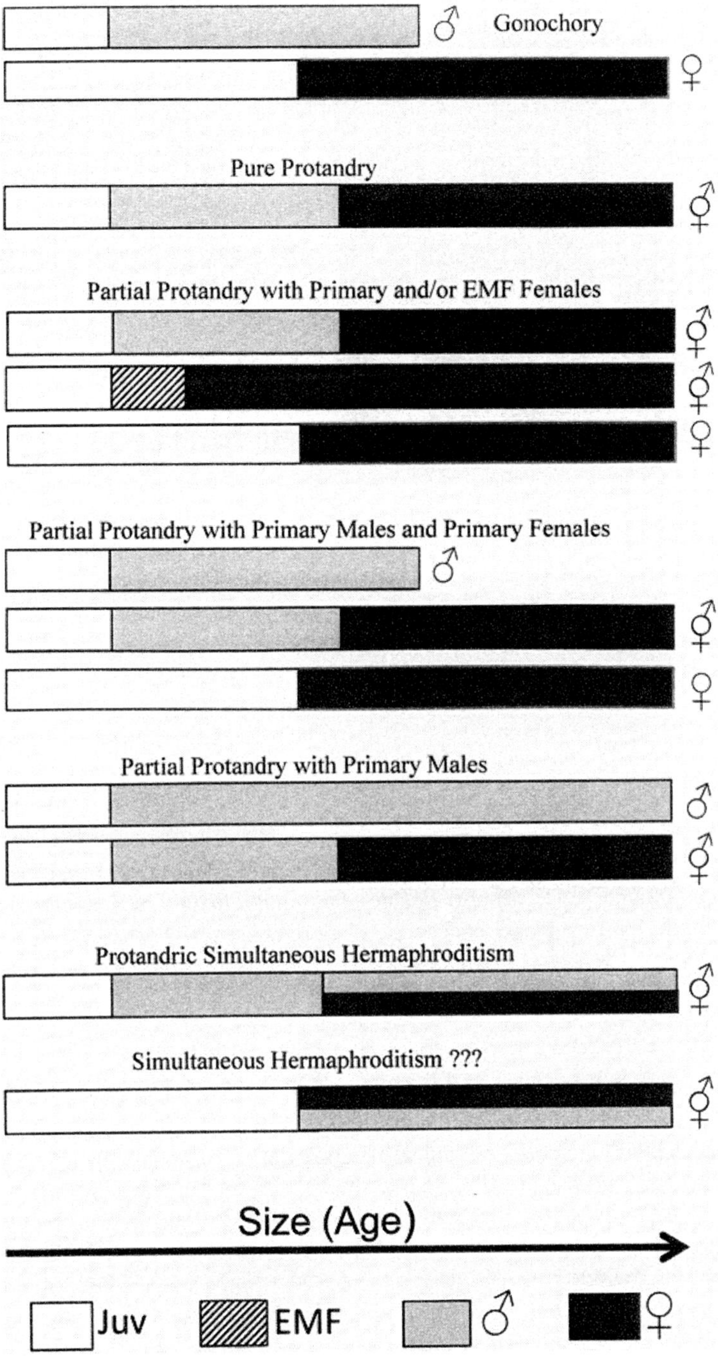

Fig. 7.1 Sexual systems of decapod shrimps. Boxes with no fill indicate juvenile (juv) phase; grey fill, male function; black fill, female function; hatched fill, aborted male function in early maturing females (EMF)

range as MPs show no sign of maleness (juvenile PFs). Good examples from Table 7.1 are *Paratya curvirostris*, *Crangon crangon*, *Nauticaris marionis, Pandalus jordani* and the Atlantic populations of *P. borealis* (the Pacific populations studied by Terry Butler are now recognized *P. eous*; Squires 1992; WoRMS 2022). Early-maturing females (EMFs) are individuals which begin to mature sexually first as males both in external morphology and gonads, but the process is cut short before they are functionally male. Instead, as the individual grows further, rudimentary male characteristics are replaced by functional female ones. The shrimp matures and breeds as a female. Such individuals were first recognized by studies on *Pandalus* by Jägersten (1936), Mistakidis (1957), and Butler (1964). Other species which have some mix of all three sexual types are *Argis dentata* and *Crangon franciscorum*.

7.3.3 Partial Protandry with Primary Males, Primary Females

A few species are a mix of protandric individuals (PH), primary (obligatory) females (PFs) and primary males (PMs), i.e., individuals which do not change into females. Some examples are the hippolytoid *Thor manningi* (Bauer 1986) and the alpheid *Arete indicus* (Gherardi and Calloni 1993), both discussed in detail below.

7.3.4 Partial Protandry with Primary Males

In this sexual system, all individuals start off as males. However, depending on the social interactions with other male-phase individuals, some may remain as males while others change sex to become female-phase individuals. This sexual system has been found in *Arete dorsalis*, as described below under the Alpheidae

7.3.5 Protandric Simultaneous Hermaphroditism

In this sexual system, individuals first mature with a typical male phenotype but later transition into the phenotype of typical breeding, embryo-producing females—but these "females" can also readily mate as males. The system was first recognized formally in *Lysmata amboinensis* by Fiedler (1998) and *Lysmata wurdemanni* by Bauer and Holt the same year. I proposed the term "Protandric Simultaneous Hermaphroditism" (PSH) for the system (Bauer 2000). It is unusual not only because individuals can mate both as male and female during the female/simultaneous hermaphroditic phase, but also because species in which it occurs may have very different social systems. Younger individuals with male characteristics on the first and second pleopods are termed "male-phase" or MPs as in other protandric systems. I originally proposed the term "female-phase" or FP for the simultaneous hermaphrodite as it has a typical female phenotype externally, except for the possession of male gonopores along with female ones. Other authors have used terms such as "euhermaphrodite" (EH) or SH or SHP ("simultaneous hermaphrodite") which are certainly adequate. However, I do not feel that they indicate sufficiently the fact that the individuals first go through a male phase. I have always preferred "FP" in my several publications on them. But I recognize that "FP" alone might be confused with the functionally female individual (after sex change) in protandric species. So, here I will use the somewhat unwieldy but more correct FPSH ("female-phase simultaneous hermaphrodite") to indicate individuals with an externally female phenotype but which have both male and female function. To date, PSH has only been found in hippolytoid shrimps in two closely related families, especially in the speciose genus *Lysmata*, in which all species examined to date have this system (Table 7.1). More details of PSH and its evolution will be discussed throughout the chapter.

7.3.6 *Pure Simultaneous Hermaphroditism?*

There is one lysmatid species (*Lysmatella prima*) which may have a PSH-type sexual system but in which the male-phase has been eliminated (Murray et al. 2012), making it a pure simultaneous hermaphrodite. However, this conclusion is based on preliminary work published only in a meeting abstract; it needs proper verification. *Salmoneus* species (Alpheidae) may also be true simultaneous hermaphrodites (Oliveira et al. 2015). This is also unverified at the moment but is a good possibility.

7.4 Identification of Protandrous Sexual Systems

Various kinds of evidence can be used to recognize and verify hermaphroditism in individual shrimps and their populations. Usually, more than one type of data is desirable and necessary to confirm and describe the sexual system for a species.

7.4.1 *Size-Frequency Distributions*

Often the first indication that a species is protandric comes from size-frequency distributions taken during studies on population biology. The frequency of males or females in different body-size classes, usually measured as carapace or total length, is plotted. Many gonochoric species of shrimps, especially those which are the target

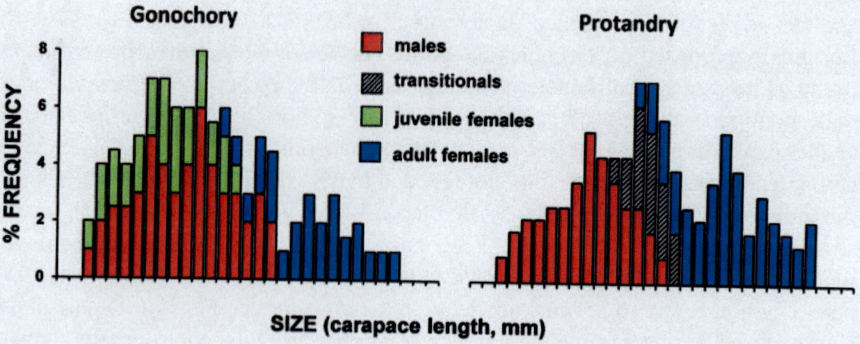

Fig. 7.2 Hypothetical size-frequency distributions of (left) a gonochoristic species (separate sexes) with small males and larger females and (right) a protandric species (sex-changer) with little overlap in size of male and female individuals, lack of juvenile females and the presence of transitional individuals. In gonochoric species, note the occurrence of juvenile females in the same size range of males, as well as the lack of transitional individuals. Modified from Bauer (2004)

of fisheries, have a sexual dimorphism in body size in which males are smaller than females. However, a protandric sexual system is suspected when there is a little or no overlap in male and female size classes and especially if few or no juvenile females are found in the smaller size classes (Fig. 7.2).

Caution: one can easily be deceived by size-frequency distributions, especially if they are from samples taken at only one season during the year. I learned this lesson while investigating the biology of the caridean *Heptacarpus sitchensis*, a tidepool/shallow-sublittoral species in southern California (Bauer 1976). The first samples taken in the spring (April) clearly showed complete separation of sex by body size. Having read studies on sex change in pandalids (Berkeley 1929, 1930; Butler 1964), I immediately suspected protandry. However, samples taken in the summer and autumn exhibited a pattern of recruitment of juvenile females in the same size range as adult males. The separation of male and female size classes in the spring was due to faster growth to larger size by females. The key piece of evidence, that of individuals transitioning from male to female, was not found. Some species initially reported to be protandrous have been based only on size-frequency evidence without other corroborating evidence. When size-frequency distributions from different seasons of the year show a significant separation of males and females, without juvenile females overlapping extensively with that of males, protandry is a good possibility. But it must be confirmed with other evidence, e.g., the presence of transitional individuals.

7.4.2 Transitional Characters

Transitionals are individuals from protandric species that show external or internal sexual characteristics intermediate between male and female. Common characters include a reduction and loss of the appendix interna or cincinnuli on the endopod of the first male pleopod, as well as a change in the shape of that endopod (Fig. 7.3A). Another key character is the gradual decrease in size and spines of the appendix masculina on the endopod of the male second pleopod (Fig. 7.3B, C). In *Thor manningi*, I found that the appendix masculina of primary males (PMs) and those of protandric males (MPs) were similar in relative size when both were small. With an increase in body size, the appendix masculina of PMs grew much larger while those of MPs, which changed little in absolute size, diminished in relative size compared to PMs (Fig. 7.11d, e; Fig. 6 in Bauer 1986). In MPs of protandric *Pandalus* species, the decrease and loss of male characters in the first two pleopods and change to female ones are well documented (e.g., Berkeley 1930; Butler 1964; Hoffman 1972) (Fig. 7.3A, B). However, such a gradual change may not occur during sex change, i.e., the male pleopod characters may be lost in a single molt (e.g., protandric crangonids). In a number of species, a vestige of the appendix masculina may persist on the second pleopod of breeding female-phase (FP) individuals e.g., in *Pandalus* species (e.g., Berreur-Bonnenfant and Charniaux-Cotton 1965; Hoffman 1972), *Arete indicus* (Gherardi and Calloni 1993), *Chorismus antarcticus* (Mascetti et al. 1997), *Campylonotus* species (Torti and Boschi 1973) as well as in female-phase simultaneous hermaphrodites (FPSHs), e.g., *Lysmata wurdemanni* (Bauer and Holt 1998) and *L. californica* (Bauer and Newman 2004).

In *Crangon crangon* and *C. franciscorum*, the male appendix masculina is small and disappears in a single molt during sex change in protandric individuals and so is of no use in identifying transitionals (Boddeke et al. 1991; Gavio et al. 2006). However, throughout the years, various investigators have found that changes in the shape and size of the pleopod 1 endopod has been useful in identifying transitional and protandric FPs ("secondary females") in *C. crangon* (Fig. 7.3D). Morphometric equations describing these changes have been developed, using measures of this character, to identify and distinguish primary from protandric FPs.

Although often inconspicuous and difficult to observe, the presence of both male and female gonopores in the same individual may indicate protandrous sex change. In some protandric species, the two sets of gonopores may be recognizable in both MPs and FPs. However, female gonopores are not open and functional in MPs nor are the male ones in FPs (*Pandalus* species: Berkeley 1929, 1930; Jägersten 1936; *Chorismus antarcticus*: Mascetti et al. 1997). However, in other protandric species, such as *Thor amboinensis* (Baeza and Piantoni 2010), both sets of gonopores have been found only in transitional individuals. In *Lysmata*, which are PSH species, male gonopores are present both in the male and simultaneous hermaphrodite (FPSH) female stages. However, the female gonopores only become visually obvious after change to the FPSH (Bauer and Holt 1998; Baeza 2009).

Other good transitional characters include the development of female breeding dress in older male individuals, e.g., an increase in the size (area) of the anterior abdominal pleurae, as well as the width of the basipod flanges of the anterior pleopods from MP to FP, as observed in *Thor manningi* (Fig. 7.11d–f), *Rhynchocinetes uritai* (Bauer and Thiel 2011; Osawa et al. 2015) (Plate 7.1C, D) and noted in the protandric atyid *Paratya curvirostris* (Carpenter 1978). In some

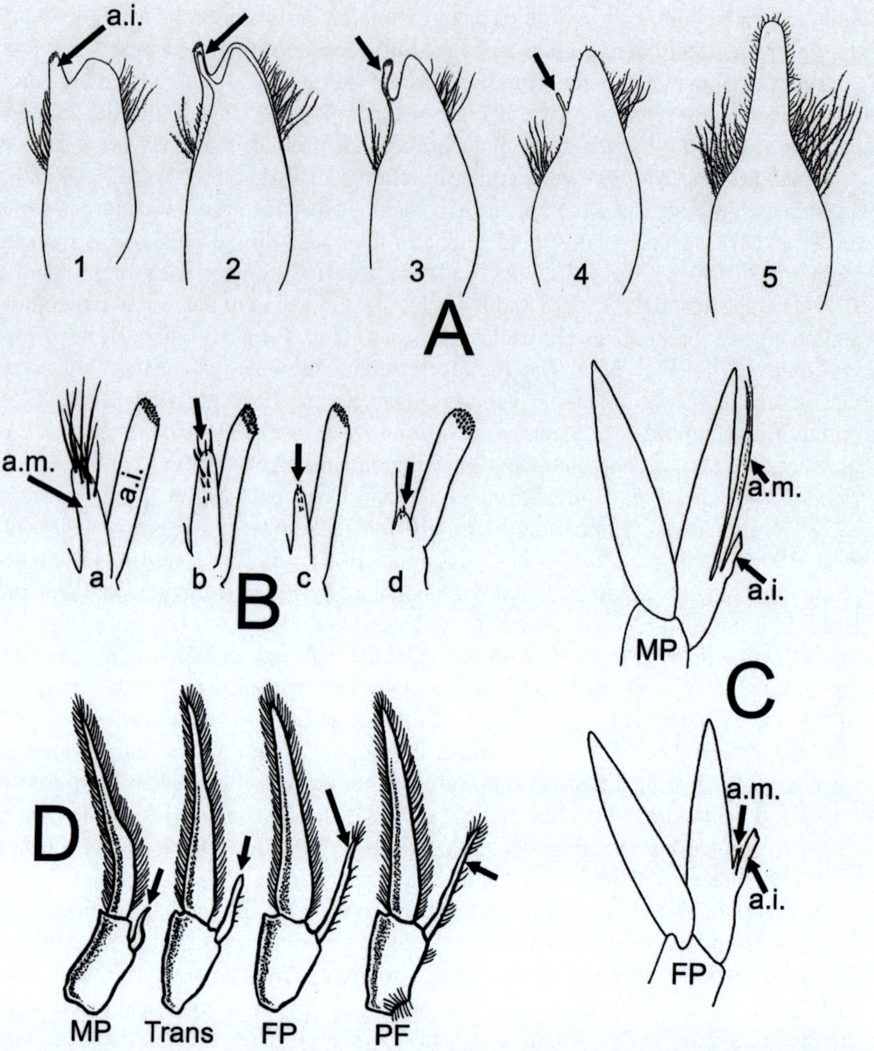

Fig. 7.3 Modifications in sexual characters of the endopods of anterior pleopods during sex change from male (MP, 1) to female phase (FP, 5) in protandric hermaphrodites. (**A**) Reduction and loss of pleopod 1 appendix interna (a.i., arrows) from MP to FP in *Pandalus borealis*. (**B**) Reduction and loss of appendix masculina (a.m., arrows) of pleopod 2 from MP (**a**) to FP (**d**) in *P. borealis*. (**C**) Fully developed a.m. of an MP of *Campylonotus capensis* (above) and its reduction to a vestigial a.m. in ovigerous females (below). (**D**) Increase in relative size of pleopod 1 endopod (arrows) from MP to FP, and comparison with that of primary (non-hermaphroditic) females (PF) in *Crangon crangon*. a.i., appendix interna; a.m. appendix masculina; FP, female-phase individual; MP, male-phase individual; PF, primary female; trans, individual in transition from male to female. Modified from (**A**, **B**) Jägersten (1936), (**C**) Torti and Boschi (1973), and (**D**) Boddeke et al. (1991)

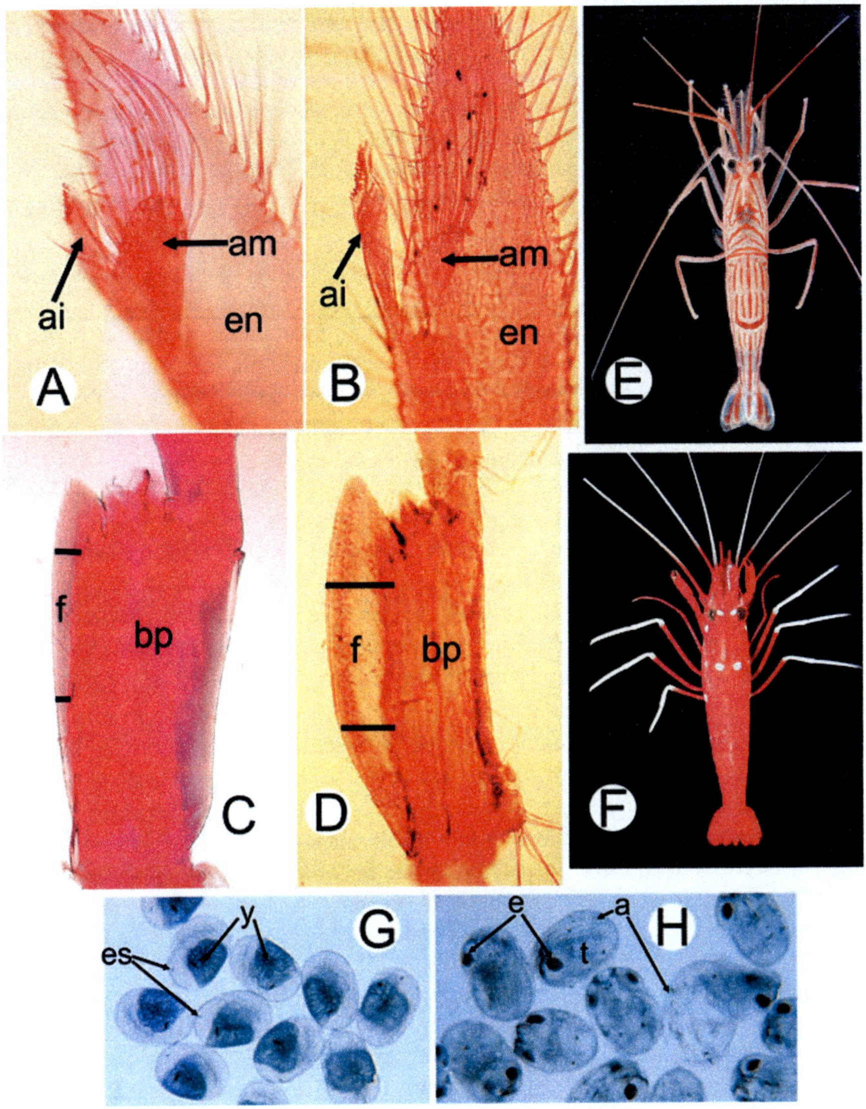

Plate 7.1 Some characteristics of hermaphroditic species; caridean embryos. (**A**–**D**) Comparison of pleopod 2 features in male-phase (MP) and transitional individuals during sex change of the protandric hermaphrodite *Rhynchocinetes uritai*. (**A**) Typical relative sizes of the appendix interna (a.i.) and appendix masculina (a.m.) in a *R. uritai* MP; compare with (**B**) in which the degenerating a.m. is reduced in size relative to the a.i.; the a.m. will disappear as the individual transitions to the female phase (FP). (**C**) Basipod flange of an *R. uritai* MP compared with (**D**) a transitional individual which will become an FP. (**E**, **F**) Comparison of color patterns of a "crowd" and "paired" species of *Lysmata*, a protandric simultaneous hermaphrodite genus. (**E**) *Lysmata ankeri* shows typical coloration of a "crowd" species while (**F**) *L. debelius* exhibits the showy coloration of a fish-cleaning "paired" species. (**G**, **H**) Embryos of *L. wurdemanni* at (**G**) pigment-eyespot stage and (**H**) near hatching. Note the just hatched (unfolded) embryo at the right. a, abdomen; ai, appendix interna; a.m., appendix masculina; bp, basipod of second pleopod; e, eye; en, endopod of second pleopod; es, pigment eyespot; f, basipod flange; y, yolk. Horizontal bars in (**C**, **D**) show the extent

protandric species, developing ovaries with large egg cells can be observed through the translucent carapace in larger MPs and transitionals (e.g., Carpenter 1978; Bauer 1986; Bauer and Thiel 2011).

The presence of both ovarian tissue (oogonia, oocytes) and testicular tissue (spermatogonia, sperm) in the same individual are compelling evidence for sex change. Histological demonstration of a change of gonadal tissue from testicular to ovarian has been demonstrated in a number of species (Table 7.1). However, simple dissections and macroscopic observation of gonads can often suffice to distinguish between testicular and ovarian tissues. Gonadal tissue undergoing vitellogenesis is quite obvious in carideans and stenopodideans, given the large size of embryos. Protandrous species have ovotestes, i.e., have both ovarian and testicular tissues in separate portions of the gonads. However, in young male-phase shrimps, the testicular tissue and male ducts are well developed while the ovarian portion and oviducts are rudimentary. After sex change occurs, vestiges of testicular tissue and male ducts may persist through one or more molts but usually later disappear (Fig. 7.4a, b). However, in PSH species, a small part of the gonads remains testicular with male ducts in the FPSH stage (Fig. 7.4c, d).

7.4.3 Direct Observation of Sex Change

Hermaphroditism in a species can be verified with near certainty given a good combination of the populational and morphological evidence discussed above. However, direct observation of sex change of living individuals maintained in the laboratory is very compelling evidence. Male-phase individuals of suspected protandrous species can be followed individually or in groups. Their sexual condition can be checked by direct inspection of the individual or by comparison of molt skins from successive molts. This has been done in several species amenable to laboratory observation, e.g., *Lysmata wurdemanni* (e.g., Bauer 2002), *L. boggessi* and *L. galapagensis* (Baeza 2009), *Pandalus danae* (Marliave et al. 1993), *Hippolyte inermis* (Zupo et al. 2008), *Crangon franciscorum* (Gavio et al. 2006), and *C. crangon* (Schatte and Saborowski 2006).

Plate 7.1 (continued) of the basipod flange, a female breeding dress character. Modified from (**A–D**) Bauer and Thiel (2011) courtesy of Journal of Crustacean Biology; (**E, F**) courtesy of Arthur Anker; (**G, H**) from Bauer (2004)

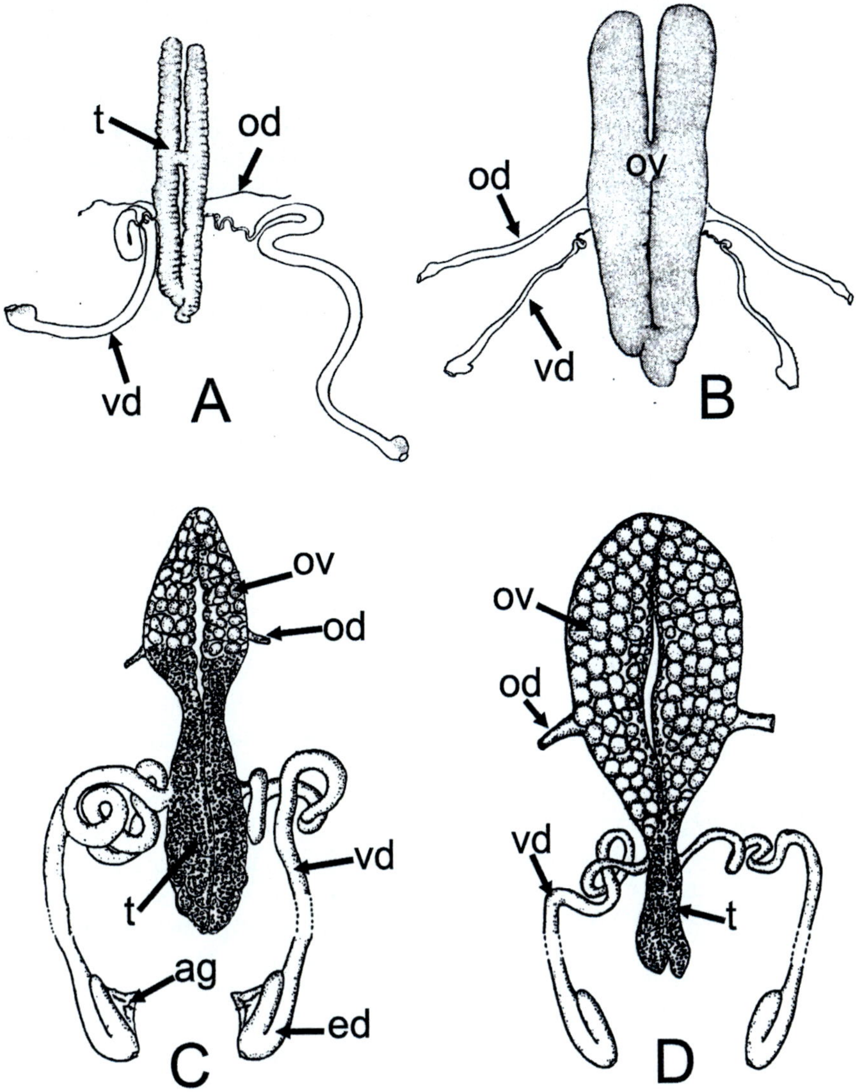

Fig. 7.4 Gonads of protandrous species. (**a**, **b**) *Pandalus borealis* protandric hermaphrodites: (**a**) Male-phase (MP) individual with functional testicular portions with well-developed male vasa deferentia but also with rudimentary oviducts. (**b**) Gonads transitioning to female-phase (FP) with well-developed ovarian portions, well-developed oviducts and deteriorating male ducts; the latter will disappear before or soon after attaining full breeding FP status. (**c**, **d**) Gonads of protandric simultaneous hermaphrodite *Lysmata seticaudata:* (**c**) Functional MP with rudimentary ovarian portion anteriorly but functional testicular tissue posteriorly, well developed male ducts, and fully formed androgenic gland. (**d**) Female-phase simultaneous hermaphrodite with mainly ovarian gonads anteriorly but retaining a small but functional testicular portion with male ducts, reduced androgenic gland not figured. ag, androgenic gland; ed, ejaculatory duct; od, oviduct; ov, ovaries; t, testes; vd, vas deferens. Modified from (**a**, **b**) Jägersten (1936) and (**c**, **d**) Charniaux-Cotton (1975)

7.5 Taxonomic Distribution of Hermaphroditism in Shrimps

As shown in Table 7.1, hermaphroditism in shrimps has only been verified in caridean shrimps, although it has been suspected in other shrimp taxa. Details about verified or suspected protandry in shrimps are presented below:

7.5.1 Dendrobranchiata (Sergestoidea and Penaeoidea)

Given the sexual dimorphism of dendrobranchiates, with males much to somewhat smaller than females, one might expect some cases of protandry. However, this is not the case. There are no verified examples of protandry in sergestoids and penaeoids. There were suggestions, based on size-frequency distributions, that *Solenocera membranacea, Melicertus (Penaeus) kerathurus*, and some other penaeids were or might be protandric (reviewed in Pérez-Farfante and Robertson 1992). Heegaard (1967) found little to no overlap between male and female sizes in *S. membranacea* and purported to see signs of transitional individuals based on pleopod 1 morphology. However, his samples were from a single summer sample. Froglia and Gramitto (1981, 1987) took samples of *S. membranacea* from different seasons of the year, as well with trawling gear with both a larger standard fishing mesh and a finer net end mesh. Nets with only the standard fishing mesh miss the smaller males and juveniles females of similar size. Samples from different seasons of the year indicated that the non-overlap of males and females was due to faster female growth. This was demonstrated by a sample from the autumn which showed a complete overlap between small male and juvenile female recruits (Fig. 7.5a). When these smaller individuals matured the next summer, the females had grown faster and larger than the males, resulting in the type of size-frequency distribution with low overlap in size between males and females that Heegaard had previously observed (Fig. 7.5b). Thus, the small overlap in size was not due to protandry but to differential growth. Froglia and Gramitto also showed that Heegaard's observations of transitional individuals based on pleopod 1 morphology were in error.

Other reports of hermaphroditism in penaeoids were cases of misinterpretation of intersexual individuals (sexual abnormalities). Pérez-Farfante and Robertson (1992) concluded that anomalies in the petasma and the presence of both male and female gonopores of a few individuals of *Litopenaeus vannamei* from an aquaculture pond were not good evidence of true hermaphroditism as there was no evidence of hermaphroditic gonads. Similar abnormalities in *Metapenaeus affinis, M. monoceros, and Penaeus semisulcatus* were also considered genetic or developmental abnormalities, i.e., "intersexes" as I defined them earlier for this chapter. I am unaware of any reports on hermaphroditism in sergestoids. All the confirmed cases of hermaphroditism to date are from the Caridea (Table 7.1) and are discussed below by family.

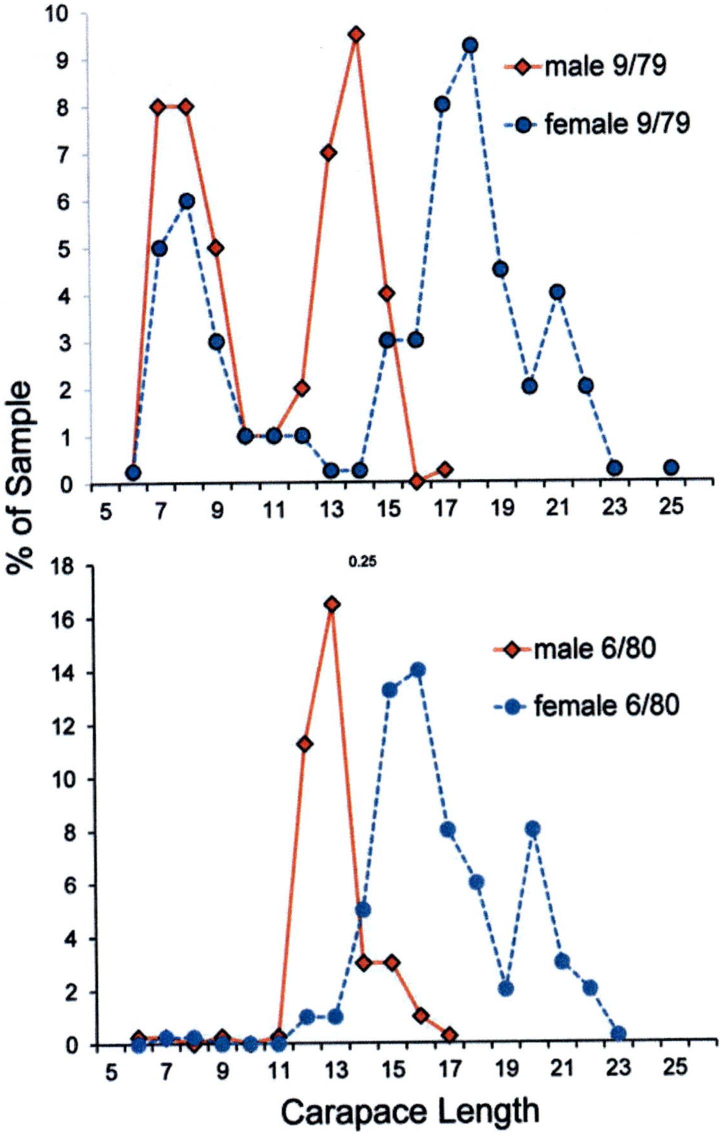

Fig. 7.5 Population structure of the penaeoid *Solenocera membranacea* from samples taken in (**a**) autumn (September 1979; $n = 358$) and (**b**) the following summer (June 1980; $n = 320$). The samples show the complete overlap of small male and female recruits in the autumn, with separation and low overlap in size in the summer due to faster growth of females. Modified from Froglia and Gramitto (1981)

7.5.2 *Alpheidae*

Although this is the second most speciose family of carideans, only two *Arete* species have been confirmed as protandric hermaphrodites (Table 7.1). This is not too surprising, given that sexual dimorphism in body size of alpheids is usually unlike that of most protandric species. Primary males are as large or larger than females in these two *Arete* species. Other alpheids studied are gonochoric, with males either slightly larger or smaller than females. Many alpheids are socially monogamous, i.e., live in male-female pairs, and the two sexes of most paired species in any shrimp family are similar in size, as will be seen in the next chapter.

The alpheid *Arete (Athanas) indicus,* a symbiont of sea urchins (*Echinometra mathei*), has a sexual system somewhat similar to *T. manningi*. There are primary males (PMs), protandric sex changers, and perhaps some primary females. Gherardi and Calloni (1993) demonstrated this with size-frequency distribution, morphological and histological evidence. PMs are not noticeably different in external appearance from protandric MPs. The differences arise as the protandric MPs transition from male into female breeding characters. Another difference is that PMs and protandric MPs may be as large as breeding females, i.e., MPs may not change sex until reaching a large size. It is also possible (my view) that the PMs of *A. indicus* are actually facultative sex changers, i.e., sexually labile individuals that may either stay male or change to female. The authors describe the mating system as promiscuous in the sense that mating encounters are quite brief, without male guarding of females prior to mating as in many alpheid species. (However, Correa and Thiel 2003, do not agree, considering this species as having a "search and attend" mating system). Thus, it may be advantageous for sexually labile individuals to delay change depending on the opportunities to mate as a male, e.g., the availability of potentially receptive females. PMs and MPs could conceivably be the same sexually labile genotype, with the physiological decision to change sex and at what size mediated by social and other environmental factors. Environmental control of sex change (ESD) will be discussed at greater length later in the chapter.

Arete dorsalis is also a sea urchin symbiont with the host *Heliocidaris (Anthocidaris) crassispina*. Suzuki (1970), collecting specimens in the Sagami Bay, Japan area, carefully described the sexual system as one in which individuals first develop as male when small, as females at intermediate size, and then change back to male when larger. The paper presents persuasive evidence of such a system in the form of size-frequency distributions, external morphology, and histology (Fig. 7.6). All individuals below 3 mm carapace length (CL) were juveniles and males (MPs), those between 3 and 6 mm CL females (FPs), and those larger than 6 mm were males. Thus, sex change, according to Suzuki, is protandric but with secondary change back to male (secondary MPs).

However, Nakashima (1987), sampling at a different location in Japan (Seto Marine Laboratory), contradicted the Suzuki study. In size-frequency distributions from year-round monthly samples, he found males in all size classes. He also reported, but did not present, histological evidence that contradicted that of Suzuki,

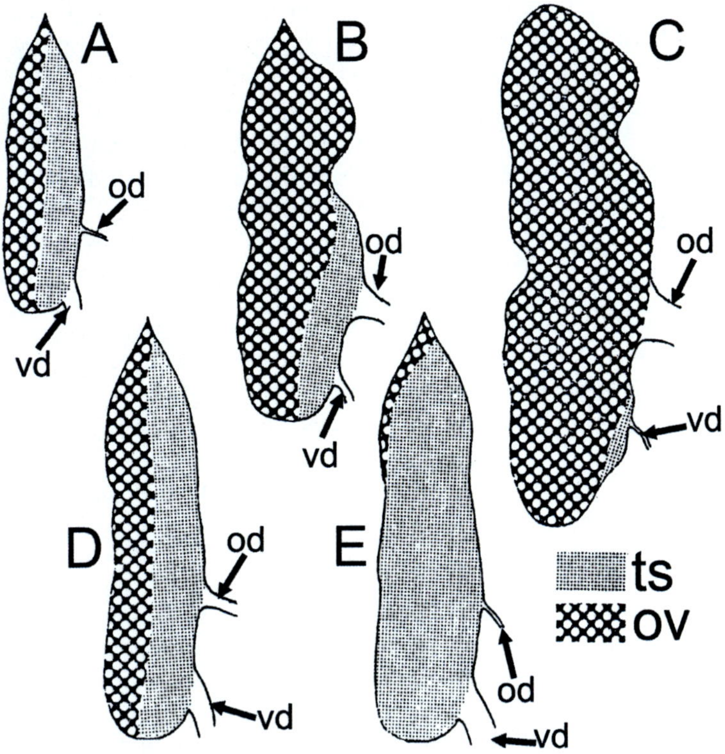

Fig. 7.6 Gonadal changes during transformation from male to female and then purportedly back to male phase in the alpheid *Athanas dorsalis*. (**a**) Juvenile. (**b**) Functional male-phase (MP). (**c**) Functional female phase (FP). (**d**) Transitional reversal from FP to secondary MP. (**e**) Functional secondary MP. od, proximal oviduct; ov, ovarian tissue; ts, testicular tissue; vd, proximal vas deferens. Modified from Suzuki (1970)

with no change from female back to male. He described the mating system, which is quite different than that of *A. indicus* described above. Reproductive females stay close to an urchin host while males, both small and large, search among urchins for such females. Once finding a reproductive female, the male stays and guards her until she undergoes the premating molt, mates with her and then is on his way to search for another female. Small males are perfectly capable of mating with females but larger males, with their outsized chela and better fighting ability, will displace them from a female before mating occurs (Nakashima 1987, 1995). Large male size has an advantage.

Nakashima concluded that all individuals are potentially protandric sex changers. Individuals settle from the plankton in the summer and first develop as males. In the following spring, most individuals are small males derived from the previous year's reproduction. They go from sea urchin to sea urchin looking for a female, but not many individuals have changed sex. When arriving at a sea urchin inhabited by larger male, fighting takes place and the major first cheliped of the smaller male may be lost. Such males are at a disadvantage in fighting over females, and so are stimulated to change into females which will assure their reproduction. The winning males of fights keep growing into the larger males of the population. They reproduce only as male as their large size gives them a male mating advantage. Thus, the males which never change sex (PMs) are actually facultative and not obligatory males. Given different circumstances when small, they could have developed into females. This sexual system is an example of social control of sex change. Nakashima admits the possibility that some larger females, if they survived into a second year, might be able to switch back to male function as the ovarian portion of the ovotestes never disappears completely in large males.

Kensley (1988) suggested that the anchialine cave species *Yagerocaris cozumeli* is hermaphroditic. All nine specimens examined by him had well-developed appendices masculinae but several of them were also ovigerous or showed gonadal development. This suggested simple protandry to him, but PSH is also a good possibility.

Salmoneus species are usually found in pairs but both individuals, which may or may not be brooding embryos, have appendices masculinae on their second pleopods. This is very strong but circumstantial evidence that they are simultaneous hermaphrodites (Anker 2011a, b; Oliveira et al. 2015). These shrimps are rare and difficult to collect, living in burrows or in rock or coral crevices. Once a researcher is able to find a population from which sufficient material can be collected for both morphological and experimental studies on live individuals, the resulting story will be indeed interesting!

Another unresolved alpheid sexual system is that of eusocial *Synalpheus* shrimps which live in colonies within sponges (Duffy and MacDonald 1999; Duffy 2007). All the breeding in a colony is done by one or two "queen" females. Mating males have not yet been identified, as *Synalpheus* is one of the few caridean genera in which males lack appendices masculinae. Besides the queen in eusocial *Synalpheus* species, colonies have several to numerous individuals ("helpers") of small to large size. To sex these individuals, Eva Tóth and I (2007, 2008) used scanning electron microscopy (SEM) to look for gonopores. Helpers of eusocial species are small, almost ant-like, and their gonopores are correspondingly quite tiny and very difficult to observe with light microscopy. With SEM, we found that in most species, all helpers, small to large, had either male or female gonopores, i.e., sexes were separate. In *S. paraneptunus*, however, all helpers and the single queen examined had both sets of gonopores, which we termed intersexes in that paper. Interestingly, the only queens of another species (*S. filidigitus*) that we had to examine also had both sets of gonopores. We concluded that the sexual system in *S. paraneptunus* might be some form of protandry or perhaps even simultaneous hermaphroditism. Because of the very small size of these shrimps and the form of preservation of samples examined, gonadal dissections were quite difficult. Histological work on specimens properly prepared for this procedure will be very important in resolving the sexual system of these colonial *Synalpheus* shrimps.

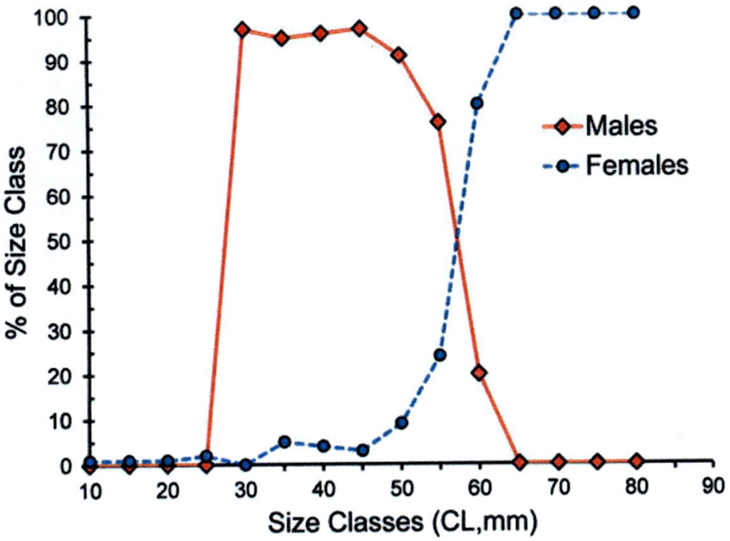

Fig. 7.7 Population structure of the atyid *Paratya curvirostris*, a protandric hermaphrodite studied by Carpenter (1978) from New Zealand streams. The percentage of male-phase (MP) and female-phase (FP) individuals is given for each size class from a total sample of ~4000 individuals accumulated from monthly collections over 1 year (1976). The size at which individuals breed as FPs is not indicated in Carpenter's paper but it is probable that at least those over 50–60 mm CL can do so. Modified from Carpenter (1978)

7.5.3 Atyidae

This freshwater family also has a large number of species, and many of them have the small-male, large-female sexual dimorphism that has given rise to protandry. Carpenter (1978) showed with good evidence from size-frequency distributions plus identification of transitional individuals that *Paratya curvirostris* from New Zealand streams is a protandric hermaphrodite. He did find a very low number of small immature females, so the species is not quite completely protandric (Fig. 7.7). He also reviewed evidence of other authors that *Caridina richtersii*, *Australatya striolata* and two species of *Atyoida* are protandric, perhaps with primary females. Chace (1983) gives some good data supporting protandry in the *Atyoida* species. The Hobbs and Hart (1982) monograph of *Atya* species showed that, unlike many atyid species, reproductive males tend to be larger than breeding females; thus, protandry appears unlikely in this genus. Although these authors were not specifically looking for hermaphroditism, their attention to sexual morphology would almost certainly have revealed protandry if present in the several *Atya* species examined.

Sexual dimorphism in body size is pronounced in the genus *Potimirim*, at first sight suggestive of protandry. Analysis of size-frequency distributions and histology of gonads led to a conclusion of protandry in *P. mexicana* (Alonso-Reyes et al. 2011; Bortolini and Álvarez 2012). Although this evidence is compelling, a more thorough analysis of transitional individuals and observations of sex change in captive specimens would help to accept a hypothesis of protandry in this species. On the other hand, Grilli et al. (2014) found no evidence of protandry in *P. brasiliana* and *P. potimirim* after observations and analyses of size-frequency distributions, male and female secondary sexual characters, and gonadal morphology. Based on a detailed analysis of secondary sexual characters and histology of gonads in both immature and mature individuals, Machado et al. (2021) found no evidence of any kind of indicating hermaphroditism in *P. potimirim*.

7.5.4 Campylonotidae

All five species of this deep-water genus from the Southern Hemisphere are considered protandric. Yaldwyn (1960) first suggested protandry on the basis of size-frequency data and individuals of intermediate size with transitional appendices masculinae and cincinnuli of the first pleopods. Some breeding females retained vestiges of the male appendix masculina. Torti and Boschi (1973) confirmed protandry in *Campylonotus* and illustrated the vestigial pleopod 2 appendix masculina in breeding females, showing that they were sex-changed males (Fig. 7.3C)

7.5.5 Crangonidae

A mix of protandrous individuals and primary females is known in 2 of the 23 genera of Crangonidae. Protandry was well documented in *Argis dentata* with various lines of evidence by Fréchette et al. (1970). It is a relatively long-lived species (5 years), with the first 2 years spent as male (MP) and the latter three as a female (FP). MPs were identified externally by the presence of cincinnuli ("crochets" = hooks) on their first pleopods and an "appendix interna" on the endopod of the second. However, as Komai (1997a) shows in his taxonomic monograph on this species, the latter structure is actually an appendix masculina. An externally recognizable transitional stage from male to female using pleopod characters was not observed by Fréchette et al., which they attributed to a rapid sex change, perhaps over a single molt. However, their histological analyses and macroscopic observations on gonads clearly show the change from male to female. Not all individuals are protandric; size-frequency distributions indicate that about a third of the population is primary females. Histology and gonadal structures also indicate that there may be EMFs (early-maturing females) in the population, i.e., individuals that begin to mature as males but then abort that developmental pathway to mature first as females.

There has been a fair amount of controversy about hermaphroditism in *Crangon crangon* (formerly *C. vulgaris*) as this species has been and still is an important fishery in European waters (Chap. 11). Boddeke et al. (1991) reviewed the history about sex change in this species and added their own evidence (size-frequency; gonadal morphology and histology) for partial protandry in this species. As in *A. dentata*, the small appendix masculina is lost in a single molt. However, using extensive morphological data, equations have been generated which allow investigators to distinguish protandric FPs ("secondary females") from primary females based on the structure of the endopod of the first pleopod (Fig. 7.3D). The proportion of protandric individuals and primary females may vary considerably from one population to another (Boddeke et al. 1991; Schatte and Saborowski 2006). The latter authors maintained 70 male individuals in the laboratory and recorded their sex based on successive molts. Only 1 of the 70 changed sex. *Crangon crangon* may be one of the species in which sex change is facultative, not genetically fixed, and which depends on either biotic or abiotic environmental conditions.

Gavio et al. (2006) conducted an exemplary study on possible protandry on the northwest Pacific coastal species *C. franciscorum*, utilizing several types of evidence (Table 7.1). They confirmed that sex change takes place in many individuals but that a considerable proportion of individuals are primary females. MPs may change sex at various sizes. The authors present a nice summary of the life cycle and the various sexual routes taken by different members of the population.

7.5.6 *Hippolytidae*

At least three species in this ecologically diverse family are protandric. Two of them come from the cold southern waters around the Antarctic. *Chorismus antarcticus* is a purely protandric species (Mascetti et al. 1997) (Fig. 7.8). Both male and female

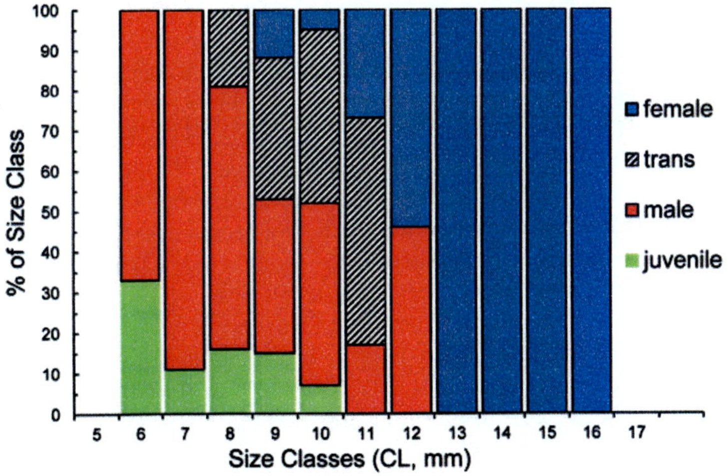

Fig. 7.8 Population structure of the hippolytid *Chorismus antarcticus*, a protandric hermaphrodite studied by Mascetti et al. (1997) from Antarctic shelf waters. Shown below for each size class is the percentage of male-phase, transitional, and sexually mature female-phase hermaphrodites, as well as juveniles (immature individuals which could not be sexed) from a sample of 125 individuals. From data in Mascetti et al. (1997)

gonopores are present in all individuals. In the smallest (youngest) individuals all gonopores are closed, but with increasing size the male appendix masculina develops and the male gonopores open: the shrimp is a functional male. In the transition to female, the male gonopores close, the appendix masculina is reduced in size, and the gonads begin to change to ovaries. With further increase in size, oocyte production in the gonads begins along with the opening of the female gonopores. However, these functional females retain a small vestige of the appendix masculina which shows their origin as males. In *Nauticaris marionis*, Pakhomov et al. (2000) used extensive size-frequency and sexual morphology observations to substantiate partial protandry in this species. Most individuals are protandric, but a small percentage of individuals show no evidence of male appendices or male gonads. Protandric FPs, however, keep vestiges of either of the latter two structures, indicating that they have changed sex.

A somewhat puzzling case of sex change is found in some populations of *Hippolyte inermis* which inhabits *Posidonia* seagrass meadows in the Mediterranean. Reverberi (1950) first hypothesized sex inversion in this species. As in many cases of suspected protandry, Reverberi's interest was stimulated by the dominance of males in the smaller size classes and females in the larger ones. His observations on gonadal condition were anecdotal but he noted the absence of female ovarian rudiments in the gonads of males, unusual in protandry. He concluded that some environmental factor (diet?) might be causing the sex change, rather than a genetically controlled protandry.

Valerio Zupo (1994) extensively sampled the population of *H. inermis* throughout some years in the Gulf of Naples. He identified two sexual types, α (alpha) and β (beta). Alpha individuals are protandric hermaphrodites, first developing as male, then changing to female. Betas develop and reproduce only as females, but their default sexual system is actually protandric. In order to understand the situation, we must trace out the life history of this shrimp as Zupo (1994) describes it. There are two major reproductive events, one in the spring (March-April) and one in the late summer/autumn (~September). The larvae of the September event become α individuals. They will develop into the protandric male phase over their first winter and mate as males in the spring reproduction. Those that survive the summer may also mate in the autumn again as males, after which they transform later in autumn (their second) into the α-females that participate in the spring reproduction of the following year. On average they live about 1–2 years. However, most of the offspring produced in the spring reproduction become β-females. But they are actually genetic protandrites whose male phase fails to develop because of their early diet. Leaves of *Posidonia* in the seagrass meadows sampled by Zupo developed heavy growths of the benthic diatom *Cocconeis* in the spring but not autumn. Newly settled postlarvae from the spring hatch that feed heavily on *Cocconeis* develop into β-females. Those from the fall hatch do not feed on *Cocconeis* as it is not common or available, and so these individuals develop into protandrites. Zupo and colleagues have shown experimentally that compounds in this diatom cause apoptosis (cell death) in the androgenic gland, whose hormones stimulate and control development of testes and other male characters. The destruction of the androgenic gland then prevents development of testes in protandric genotypes, resulting in β-females (Zupo 2000, 2001; Zupo and Messina 2007).

However, Cobos et al. (2005), working with Mediterranean populations of *H. inermis* from southern Spain, found no morphological evidence of sex change. It will be recalled that in several protandric species, older males often begin to lose their male characteristics in a series of molts while gaining female ones. Cobos and colleagues found no evidence of gradual loss of the male appendix masculina nor, with histological studies, of testicular change to ovaries. Thus, Zupo's conclusions of protandry were put into doubt. However, in addition to the dietary studies given above, Zupo et al. (2008) followed sexual development of individual postlarvae and juveniles in the laboratory. Several individuals did change sex, as substantiated by evidence from external morphology and histological study of gonads. They concluded that, unlike most caridean protandrites, that both external and gonad change occurs quickly, in one intermolt cycle.

So, what might cause the discrepancy between these two groups of investigators? As we will see later with *Pandalus* species, populations in different geographic areas or time periods in the same location may show different mixes of sexual types. The same may be occurring in *H. inermis*. The apoptotic diatom which prevents sex change in the Gulf of Naples population may not grow or be abundant in other *H. inermis* habitats. Another possibility is that two populations are sibling species whose morphological differences have not been yet recognized. Genomic studies on the two populations should be able to test this hypothesis.

The sexual system of *H. inermis* stimulated careful investigation of that in three other *Hippolyte* species, all of which have small-male and large-female size-frequency distributions. However, various studies, using some or all the criteria given above for recognizing protandry, have revealed only gonochory in these species (Espinoza-Fuenzalida et al. 2008; Terossi et al. 2008; Manjón-Cabeza et al. 2009). Thus, this is an example in which only one species of those examined in a speciose genus exhibits hermaphroditism, at least in one of its populations. All the *Hippolyte* species studied inhabit similar habitats, i.e., seagrass or algal meadows in tropical to temperate latitudes.

7.5.7 *Merguiidae*

The two species of this monogeneric, semiterrestrial family appear to be purely protandric. Baeza (2010a) demonstrated this in *Merguia rhizophorae* with observations on size-frequency and sexual morphology, supported by more extensive size-frequency data in Alves et al. (2021). Malay and Rañises (2019) study on *M. oligodon* indicates that it also is protandric.

7.5.8 Pandalidae

This family contains a large number of known protandrous species (Table 7.1). Yet protandry is restricted to only 2 of the 19 pandalid genera, *Pandalus* and *Pandalopsis*. These are large boreal shrimps of the North Atlantic and North Pacific which may live for several years. As a result of their large size, abundance, schooling behavior and therefore commercial (fishery) importance, they have been the focus of considerable study. Berkeley (1929) first summarized a protandrous life cycle in *Pandalus danae*, followed by detailed studies on life history and the anatomy of sex change in several North Pacific pandalid species (Berkeley 1930). Young individuals have ovotestes but in which the male (testicular) portion develops first. With increasing size, the testicular portion and ducts degenerate as the ovarian part develops and becomes functional (Fig. 7.4a, b). Corresponding changes occur in pleopods 1–2 (Fig. 7.3A, B). Appendices masculinae are gradually lost in a series of molts; and female pleopod and breeding characters develop. Size-frequency studies on population samples were concordant with the male to transitional to female sex change. Berkeley's studies were followed by others on several North Pacific pandalids by Butler (1964) and North Atlantic ones by Jägersten (1936), Mistakidis (1957), and Allen (1959).

Marliave et al. (1993) hatched *P. danae* in the laboratory, cultivated the larvae and reared many individuals from juvenile to male (MP) with sex change to female (FP) over the course of some years. Although previous morphological and populational evidence of protandry was already convincing, this demonstrated sex change without any doubt. All six species of *Pandalopsis*, a North Pacific genus, have either been shown to be purely protandric or are suspected of being so (Table 7.1)

The sexual composition of different populations and species often shows a mix of mainly protandric individuals along with some early-maturing females (EMFs) and primary females (PFs), summarized in Butler (1980) and Bergström (2000). Still unclear is whether EMFs and PFs are the same sexual genotype or are sexually labile genotypes in which the male phase is partially (EMF) or completely (PFs) repressed by environmental factors. The proportions of these sexual types among and within populations from year to year. Various hypotheses explaining these variations, including environmental sex determination (ESD) due to abiotic physical and/or demographic factors, or perhaps frequency-dependent selection on different sexual genotypes, will be discussed below along with those proposed for PSH lysmatid species.

7.5.9 Processidae

The genus *Processa*, with 49 species, comprises the bulk of the 66 species in the family. Noël (1973) proposed protandry (with some primary females) in the seagrass-dwelling *Processa edulis* from the eastern Atlantic (Roscoff) based on size-frequency information: small males and larger females, with only a small number of females overlapping in size with males. This led him to study sexual morphology in large samples from the Mediterranean (Banyuls-sur-Mer). His laboratory examined ~3500 individuals for evidence of transitional individuals based on pleopod and gonadal morphology (Noël 1976). Less than 1% could be considered transitional ("intersexes"). This is not very good morphological evidence for protandry. This proportion of intersexuals is not unusual for gonochoric species, such as *Homo sapiens* (e.g., Blackless et al. 2000). Noël proposed that the very low percentage of transitionals might be due to rapid sex change within a short time, e.g., perhaps within one intermolt. It will be recalled that this was also proposed with some support from direct observation of several individuals of *Hippolyte inermis* (Zupo et al. 2008). However, Labat and Noël (1987) analyzed size-frequency distributions in other populations of *P. edulis* which did not support protandry.

Extensive morphological and size-frequency studies on two tropical West Atlantic species (*P. bermudensis, P. riveroi*) by Bauer and Conner (2012) as well as the northeast Atlantic *P. nouveli* (Gros 1979) showed these species to be clearly gonochoric. In Table 7.1, I have indicated partial protandry (with primary females) for *P. edulis*, but given the controversial evidence presented, it should be considered a questionable case.

7.5.10 *Rhynchocinetidae*

Of the 26 described species of Rhynchocinetidae, sexual dimorphism and population structure have only been looked at carefully in *R. typus* (Correa and Thiel 2003), *R. uritai* (Bauer and Thiel 2011; Osawa et al. 2015) and two *Cinetorhynchus* species whose species status is yet to be clarified (Bauer et al. 2014; Junji Okuno, pers. comm.). There is a distinct dichotomy in sexual dimorphism in the family, with some species containing small males and large females and others with large dominant males bearing oversized chelipeds. *R. typus* and the two *Cinetorhynchus* species fit into the latter category. In *R. uritai*, there are only small males and larger females, with sexually transitional individuals of intermediate size, as first shown in a preliminary study done at the Shimoda Marine Research Center, Japan (Bauer and Thiel 2011). We found a few individuals of intermediate size with reduced appendices masculinae, enlarged female-like abdominal pleurae and pleopod flanges, and developing ovaries, all indicative of protandry (Plate 7.1A–D). A much more extensive study verified protandry (Osawa et al. 2015). First author Yumiko Osawa, a graduate student at the time, tirelessly took monthly samples which, when analyzed, convincingly showed the morphological transition from male to female (Fig. 7.9). We found that individuals recruit from the plankton in the autumn and then mature to and breed as males by the following spring to autumn. During their second winter, they make the transition to female by the next spring and produce broods continuously until the end of their life span sometime in the autumn

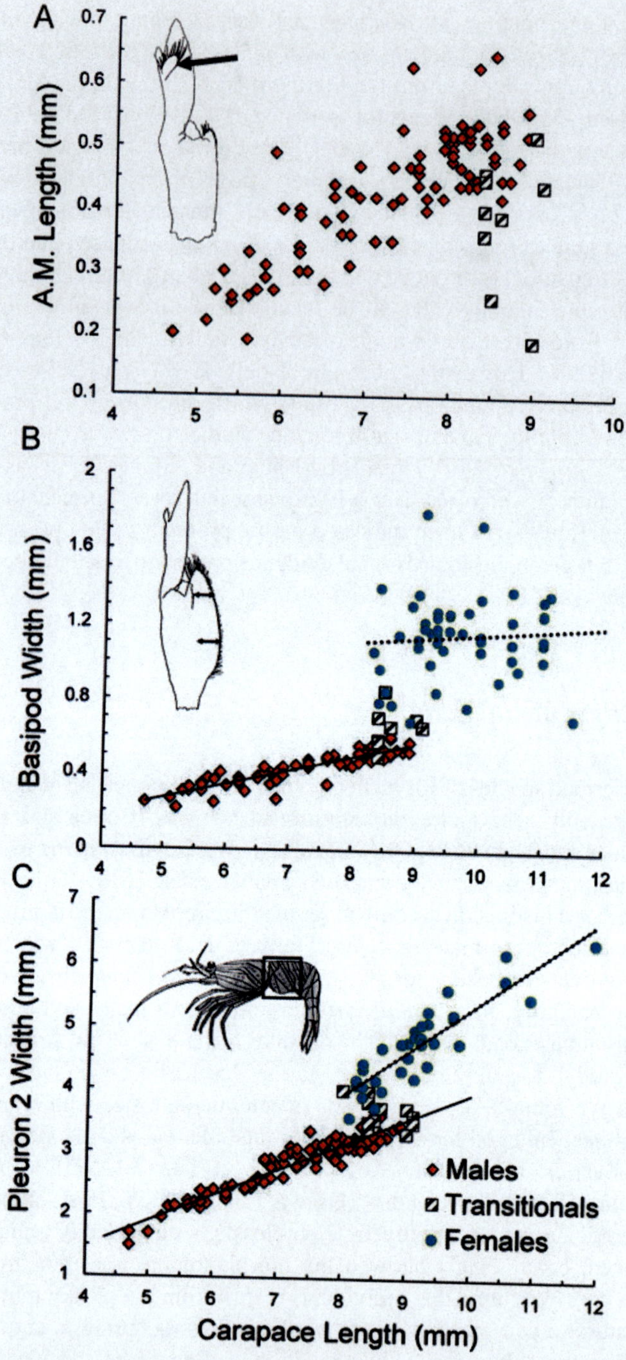

Fig. 7.9 Transitional secondary sexual characters in *Rhynchocinetes uritai*. (**a**) The size of the male stage (MP) appendix masculina (arrow) on pleopod 2 increases in size until they transition into the female stage (FP), when the appendix decreases in size and then is absent in the breeding FPs. (**b**) The width of the MP basipod flange (horizontal arrows) is narrow but increases in size as MPs

(Fig. 7.10). Other small male/large female rhynchocinetid species need to be investigated to determine if protandry is more widespread in the family.

7.5.11 *Thoridae*

Fenner Chace, Jr. (1972) described the small seagrass shrimp *Thor manningi* from a limited number of specimens. He noted that there were two kinds of males, one with small appendices masculinae and non-prehensile first pereopods, the other with large appendices masculinae and prehensile first pereopods (Fig. 7.11). He termed the first sexual type "nonfunctional" males. Given the size-frequency distribution, with smaller individuals with two different male morphotypes and larger individuals breeding females, he suspected some sort of "imperfect" protandry.

As a young assistant professor at the University of Puerto Rico, Río Piedras in the early 1980s, I began a year-long sampling program in seagrass beds to study the comparative reproductive ecology of tropical shrimps. Before leaving for Puerto Rico, Raymond Manning, my postdoctoral sponsor at the National Museum of Natural History (Smithsonian Institution), had advised me that the tropical seagrass habitat had a diverse and abundant shrimp fauna with which to do this project. Aided by an eager group of hard-working students, I was able to obtain and study relatively large samples over an entire year. Among the more abundant species was *T. manningi*. This ample material, not available to Chace (1972), allowed me to study in depth the sexual types and population structure of this species (Bauer 1986) (Figs. 7.11 and 7.12). It turned out that the "nonfunctional" males are male-phase protandric hermaphrodites (MPs), with a morphologically functional male phase which then changes to breeding female. The males with the large appendices masculinae and spinous prehensile third pereopods are obligate primary males (PMs), individuals which develop and reproduce only as males (Fig. 7.11). When the manuscript on this work was submitted to the journal The Biological Bulletin, Fenner was asked to review this manuscript. He was excited to see the mystery of his "nonfunctional" males resolved. It should be noted that in our samples we found a few small individuals which had no male characteristics, i.e., were juvenile primary females (PFs). In the Puerto Rico population, only about 1% of individuals were PFs, but in later studies on samples from Belize, the proportion of PFs was much higher (5–17%; Bauer and VanHoy 1996) (Fig. 7.12).

I originally proposed that the "supermale" features of PMs in *T. manningi* (prehensile third pereopods, massive appendix masculinae) evolved in an ancestral male genotype to grasp females for copulation *after* protandric genotypes arose in

Fig. 7.9 (continued) transition to breeding FPs in which the flange, a breeding dress character, is much wider than in MPs. (**c**) The increase in width of the anterior abdominal pleura, a female breeding-dress trait, from the MP to FP stages. Modified from figures prepared by and courtesy of Yumiko Osawa

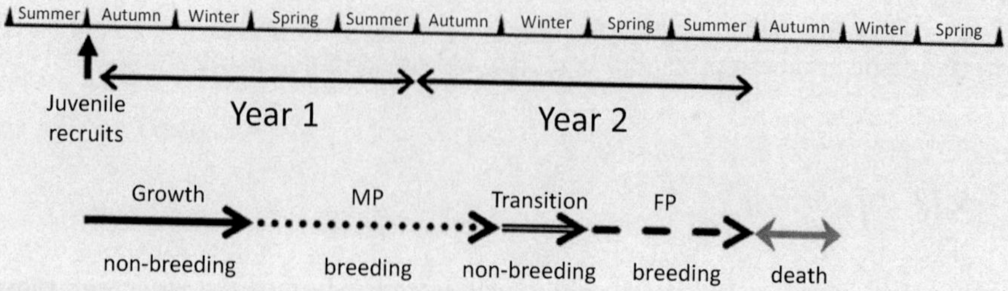

Fig. 7.10 Population cycle of *Rhynchocinetes uritai*: recruitment, growth, and sexual maturation of juveniles into the male phase (MP), with later transition into the female phase (FP), and then disappearance (death) from the population (from Osawa et al. 2015, courtesy of The Biological Bulletin)

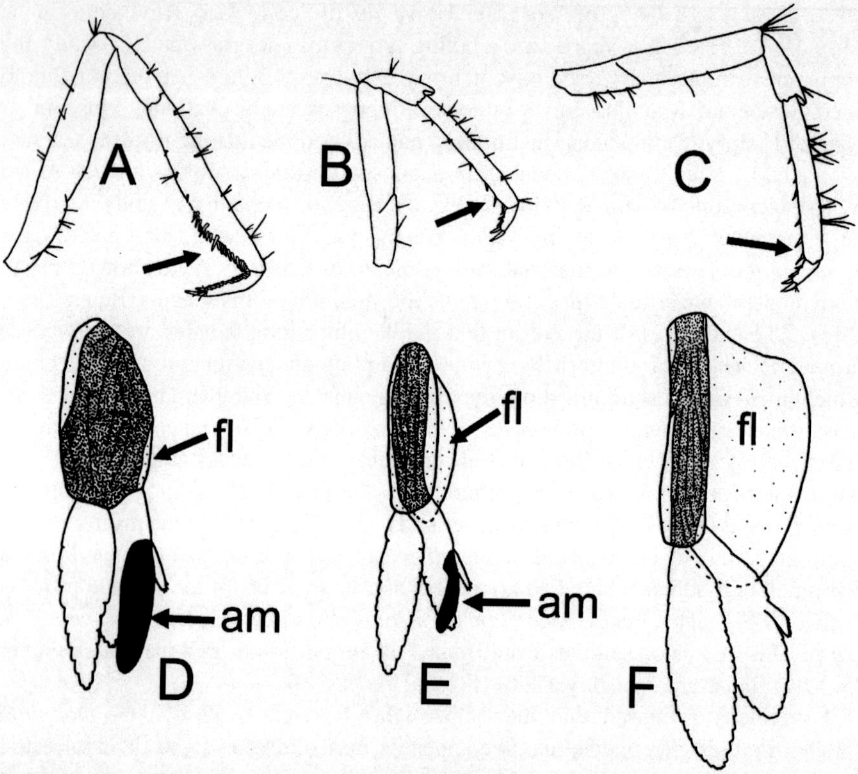

Fig. 7.11 Key characters of sexual morphs in *Thor manningi*. (**a**–**c**) Third pereopod of (**a**) primary males (PM), (**b**) male-phase (MP) and (**c**) female-phase (FP) hermaphrodites. Arrows indicate the distal two segments of the limb which are spinous and prehensile in PMs but lack these traits in MPs and FPs. (**d**–**f**) Comparison of the relative size of the basipod flange (a female breeding dress character) in (**d**) PMs, (**e**) MPs and (**f**) FPs. Note the massive size of the appendix masculina (shaded black) in PMs compared to that of MPs and its loss in FPs. am, appendix masculina; fl, basipod flange. Modified from Bauer (1986) courtesy of The Biological Bulletin

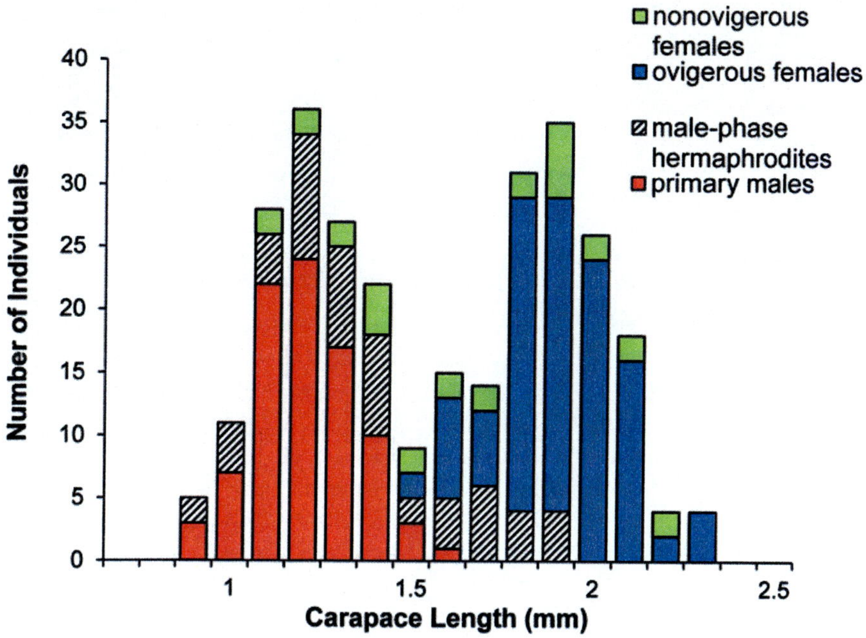

Fig. 7.12 Population structure of *Thor manningi*, a partially protandric species with protandric hermaphrodites, primary males, and some primary females (identifiable as the non-ovigerous females smaller than ~1.5 mm carapace length). Ovigerous females are a mix of mostly female-phase hermaphrodites and a small number of primary females, but the two sexual types cannot be distinguished after breeding begins. Shown is a December 1985 sample of 285 individuals from Cayo Enrique, southwest coast of Puerto Rico. From data in Bauer and VanHoy (1996) courtesy of Bulletin of Marine Science

the population. The protandric individuals, which reproduce both as male and female, would have a reproductive advantage over both primary males and females. But the protandric MPs lacked the superior mating equipment of the newly equipped primary males, which gave the latter a male mating advantage over protandric MPs. This would allow PMs superiority in inseminating females. Thus, the PM genotype has persisted alongside the protandrous one. However, analysis of extensive samples for two other *Thor* species (*T. dobkini, T. floridanus*) from very similar seagrass habitats rejected that scenario (Bauer and VanHoy 1996). Both these latter species are completely gonochoric—just primary males and females. But the males have the same sexual equipment as that of primary males of *T. manningi*. Therefore, a better interpretation of the situation might be that supermale characters were ancestral in a gonochoric *T. manningi* which allowed them to persist in the population when the new protandric genotype appeared,

Another *Thor* species (*T. amboinensis*) has a very different habitat than the three *Thor* species discussed above. This attractively colored circumtropical shrimp lives associated with sea anemones. Baeza and Piantoni (2010) carefully showed by several types of evidence that this species is purely protandric. They observed that these shrimps live in small groups with no apparent social structure. As is probable in the other *Thor* species above, the mating system is likely a promiscuous one like that of many gonochoric species with small males and larger females. The variation of sexual systems in *Thor* is quite remarkable, with gonochoric and both partially and purely protandric species.

7.5.12 Lysmatidae/Barbouriidae: Protandric Simultaneous Hermaphroditism

It will be recalled that PSH (protandric simultaneous hermaphroditism) is a sexual system in which individuals first mature sexually as males (male-phase: MP). When larger and older, the MP changes into an externally female phenotype (FP) which is actually a simultaneous hermaphrodite (FPSH), breeding equally well as male or female. The morphological evidence and size-frequency evidence for PSH in the common European species *Lysmata seticaudata* was shown by various studies over the years (Spitschakoff 1912; Dohrn 1950; Berreur-Bonnenfant and Charniaux-Cotton 1965; Charniaux-Cotton and Payen 1985). Younger individuals had male external characters. However, the gonads were ovotestes which were mainly testicular at first but with growth increasingly became ovarian. When the switch in external morphology to breeding female occurred, a small posterior portion of the gonads remained testicular with sperm production and vasa deferentia opening on male gonopores. The female system with separate gonopores was much better developed.

All these workers concluded that, because the female phase lacked male appendices masculinae on its second pleopods, *L. seticaudata* was strictly protandric. They believed that without male appendices, the shrimp could not function as a male. This mindset possibly prevented them from performing the simple experiment of trying to pair two of these female-phase individuals to see if one could act as a male while the other acted as female. Kagwade (1982) showed similar PSH morphological evidence for *Exhippolysmata ensirostris*. However, he was the first to clearly conclude that larger individuals of this species were simultaneous hermaphrodites. Interestingly, in a paper which immediately followed Kagwade's in the same journal edition, Sukumaran (1982) incorrectly rejected this hypothesis, again on the grounds that the female phase lacked appendices masculinae. However, Kagwade's hypothesis was later confirmed with the critical mating experiments on *E. oplophoroides* by Laubenheimer and Rhyne (2008) and further supported with histological studies by Braga et al. (2009).

In the early 1990s, I met G. Curt Fiedler, then a graduate student, at a scientific meeting and he told me that indeed *Lysmata amboinensis* "females" could mate as both male and female. I encouraged him to publish this exciting result as soon as possible. After a few years, I received a call from a recent master's student of Joan

Holt at the University of Texas (Austin) Marine Science Center in Port Aransas, Texas. He told me that successful mating among "females" of *L. wurdemanni*, common on nearby rock jetties, routinely occurred in their lab. Magazines for aquarium enthusiasts and other unpublished sources also reported this. I called Joan and she invited me to the marine laboratory and arranged for a student to show me where to collect them along the nearby rock jetties at night. I thus began work on the sexual morphology of the species, confirming the previous morphological work seen in *L. seticaudata*. Then I went forward, with a lot of student help, to take videos of matings between female-phase simultaneous hermaphrodites (FPSHs). This work resulted in Bauer and Holt (1998) demonstrating a PSH sexual system in *L. wurdemanni*; the same year, Curt Fiedler (1998) published his interesting work on *L. amboinensis*. These papers have generated a great deal of research and publication on this fascinating sexual system. It has been confirmed in all *Lysmata* species which have been examined, in *Exhippolysmata oplophoroides*, also a lysmatid, as well as *Parhippolyte misticia* of the related family Barbouriidae (Onaga et al. 2012), and is indicated in other species of that family (Table 7.1).

In most PSH species, the MP appears as a typical small caridean male, with a small appendix masculina on pleopod 2 and cincinnuli (reduced appendix interna) on the endopod of pleopod 1. In the FPSH stage, both features are lost but the male gonopores are retained and the female ones become visible. All this is the same in *Exhippolysmata* species, except pleopod 1 cincinnuli are rarely found in MPs or FPSHs (Kagwade 1982; Braga et al. 2009). After sex change, the female breeding dress is fully developed. In *Lysmata wurdemanni*, some individuals with male pleopod characters and observable ovarian development can be identified as transitionals. The transitionals also show developing elements of the female breeding dress, e.g., expansion of the anterior pleopod flanges. Transitionals occur over a relatively wide size range, showing that MPs may delay sex change until they reach a relatively large size (Fig. 7.13). As in *L. seticaudata*, the ovotestes of small *L. wurdemanni* MPs have an anterior undeveloped ovarian portion and a functional posterior testicular portion (Fig. 7.14a). In larger MPs, the ovarian portion expands relative to the testicular portion, the oviducts become evident, and oocytes begin vitellogenesis (Fig. 7.14b). After sex change, FPSHs show various stages of testicular reduction and ovarian maturation (Fig. 7.14c). However, even the largest FPSHs retain small but fully functional testicular portions and vasa deferentia, and readily mate as male as well as female.

Mating experiments are the final confirmation of functional simultaneous hermaphroditism in FPSHs. I paired embryo-incubating FPSHs of *L. wurdemanni* in small aquaria, with at least one carrying late-stage embryos and showing ovarian maturation. This indicated that it was near to hatching embryos after which a mating molt would soon take place. Using video surveillance hardware, I found that FPSHs could mate as males and inseminate other FPSHs. It was indeed thrilling to view this on the videotapes (Fig. 7.15). Further experiments and observations showed that FPSHs can mate as a male only an hour or so after mating as a female! Observations on isolated FPSHs showed that they are unable to self-fertilize, perhaps not too surprising in light of the potential negative consequences of such inbreeding. The

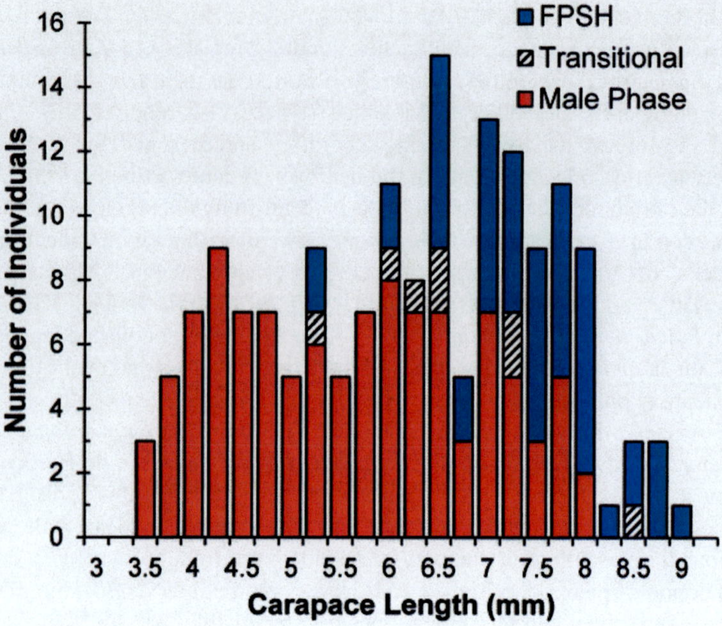

Fig. 7.13 Population structure of *Lysmata wurdemanni*, a protandric simultaneous hermaphrodite, from a June 1994 sample ($n = 165$) in south Texas, showing male-phase, transitional, and female-phase simultaneous hermaphrodites. From data in Bauer and Holt (1998)

morphological and experimental observations of *L. wurdemanni* (Bauer and Holt 1998) have now been done on many other PSH species (Table 7.1), most of them due to the efforts and publications of J. Antonio Baeza.

The comparative method has been and continues to be a powerful tool in the determination of possible causal (selective) factors of similar adaptations among a group of organisms. An association across species between a particular adaptive feature and some set of environmental factors may suggest the selection pressures responsible for the evolution of that adaptation. All lysmatid species, as well as some barbouriid ones, share a PSH sexual system. In Bauer (2000), I began a search for similarities in ecology, social organization, and mating systems of PSH species. None were or have yet been found which strongly indicates the selective factors involved.

All PSH species have small MPs and larger FPSHs, as one would expect in a protandrous sexual system. Some species live in aggregations, such as *L. wurdemanni, L. californica, and L. seticaudata*, which I termed "crowd" species as opposed to the fish-cleaning "paired" species (Bauer 2000). Crowd species appear to have typical pure-search, promiscuous mating systems. Some of them are opportunistic cleaners of fish (e.g., moray eels) and hard-shelled invertebrates, i.e., they will graze on fouling of other animals if allowed. However, at least three coral reef species (*L. grabhami, L. amboinensis, and L. debelius*) live in FPSH-FPSH pairs and

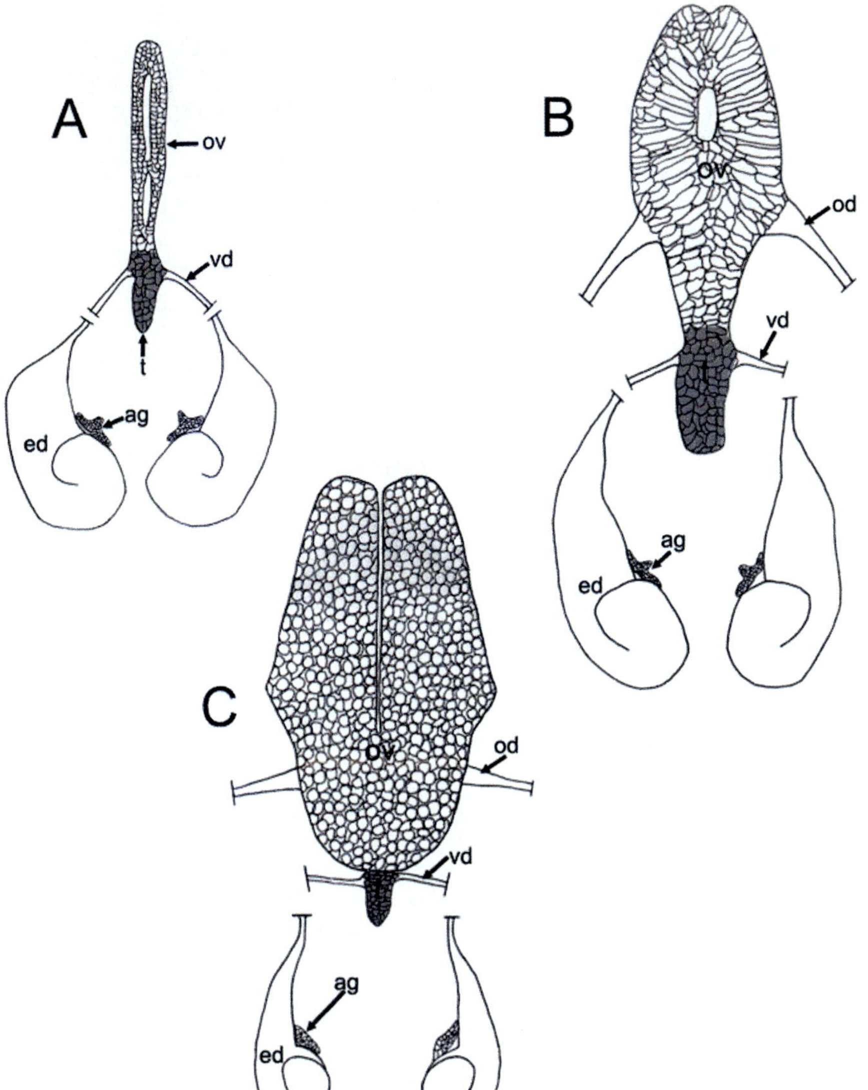

Fig. 7.14 Changes in the ovotestes of the protandric simultaneous hermaphrodite *Lysmata wurdemanni* from male-phase (MP) to female-phase simultaneous hermaphrodite (FPSH). (**a**) Small MP with anterior undifferentiated ovarian portion and posterior testicular portion. (**b**) Large MP with enlarging and maturing ovarian portion of the ovotestes with distinct oviduct and functional testicular portion. (**c**) Prespawning FPSH with ovarian portion of the ovotestes full of vitellogenic oocytes; note small but completely functional posterior testicular portion with vasa deferentia, ejaculatory ducts and a deteriorating androgenic gland. ag, androgenic gland; ed, ejaculatory duct; od, oviduct; ov, ovarian portion of ovotestes; t, testicular portion of ovotestes; vd, vas deferens. Shaded portions of the ovotestes are testicular, unshaded are ovarian. Modified from Bortolini and Bauer (2016) courtesy of The Biological Bulletin

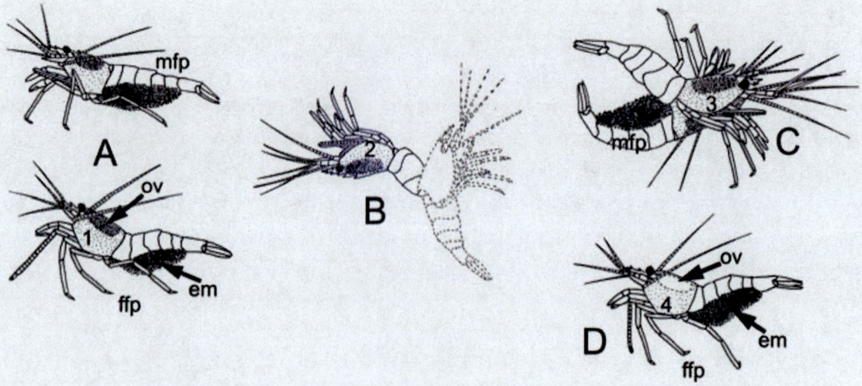

Fig. 7.15 Mating of two female-phase simultaneous hermaphrodites (FPSH) from videotape sequences. (**a**) One FPSH will mate as female (ffp; bottom). This individual can be identified through the (**a**–**d**) sequence using the numbers 1–4. Its potential male mating partner (mfp) is shown above. Note that both FPSHs have mature, oocyte-filled ovotestes and are carrying embryos. (**b**) After hatching its embryos, the ffp individual undergoes a prespawning molt. (**c**) The mfp seizes the newly molted ffp, dips below it, opposes its male gonopores just behind the ffp female gonopores, emits and attaches a sperm mass. (**d**) Within a few hours, the ffp has spawned (ovotestes emptied of oocytes) and attached a brood of fertilized oocytes (early embryos). em, brood of embryos; ffp, FPSH which will mate as a female; mfp; FPSH that will mate as a male; ov, ovarian portion of ovotestes; 1–4, labeled ffp from (**a**) through (**d**). Modified from Bauer and Holt (1998)

are socially monogamous (Limbaugh et al. 1961; Fiedler 1998; Wong and Michiels 2011). They are specialized fish-cleaners that maintain and guard cleaning stations on reefs to which they attract fishes that clean them (see Chap. 10). Baeza et al. (2009) also recognized an intermediate category of smaller aggregations termed "groups." The mating system of these species is unknown but probably promiscuous as well. Aggregated species are less brilliantly colored, usually some combination of reddish, pale, or uncolored stripes and blotches (Plate 7.1E). They are nocturnally active. Paired species are day-active and show striking coloration and behaviors to attract fish clients (Plate 7.1F). Habitats vary widely among PSH species: tropical to subtropical seagrass beds, coral and rocky reefs, and associations with moray eels, sea anemones, and sponges (Bauer 2000; Baeza 2009). The barbouriid PSH species are inhabitants of marine caves and anchialine habitats. Thus, we see that PSH species, with a very specialized sexual system, do not appear to share a common ecological or social factor trait might point to the origin of PSH except, of course, inheritance of a very advantageous sexual system from a common ancestor.

7.6 Role of the Androgenic Gland in Protandrous Sexual Systems

The androgenic gland (AG) is a small endocrine organ located on the distal end of the vasa deferentia in all male malacostracan crustaceans, including decapods (Fig. 7.14). It controls the expression of primary (gonadal) and secondary (external) sex characters in males (Charniaux-Cotton and Payen 1985; Sagi et al. 1997). Female secondary sex characters are controlled by hormones from the ovaries in the absence of a developed AG. Early studies on peracarids (amphipods, isopods) and decapods showed that removal of the AG from males caused their feminization and change to reproductive females. The reverse experiment, implantation of AG taken from males and implanted into females, induced the latter to become male. More recent and sophisticated studies by a group led by Amir Sagi (Ventura et al. 2011a, b; Ventura and Sagi 2012) have identified the hormone, analogous in function to testosterone in mammals, as a peptide, termed "IAG" (insulin-like androgenic gland hormone). They used various techniques to silence IAG expression in males of *Macrobrachium rosenbergii,* resulting in their switch to reproductive "neofemales." Conversely, genetic females can be treated with IAG to transform them into "neo-males," completely capable of producing sperm and mating successfully as males.

Thus, maleness is largely a function of the AG. Charniaux-Cotton and Payen (1985) reviewed studies that showed that an embryonic rudiment ("anlage") of the AG is present in both males and females of decapod species. In females, however, it never develops. In genetic males, it is triggered at an early stage and maleness develops. The AG develops normally in protandric males but later begins to deteriorate, bringing about external and internal morphological modifications resulting in sex inversion to female. Presumably, as the AG deteriorates, the IAG concentration is reduced to zero. In the sex-changed female, the AG is lost completely, as numerous studies on protandric species have shown (Carlisle 1959; Hoffman 1969; Fréchette et al. 1970; Gavio et al. 2006; Kim et al. 2006).

How is it possible that PSH individuals, after change to the female phase, retain some functional testicular tissue, vasa deferentia, male gonopores, sperm production, and male mating ability? In *Lysmata seticaudata*, Berreur-Bonnenfant and Charniaux-Cotton (1965) reported the complete loss of the AG in FPSH phase, reaffirmed in Charniaux-Cotton (1975). However, I hypothesized that perhaps an AG remnant remained after sex change, enough to produce sufficient IAG to maintain the reduced but functional masculinity of the FPSH. I collaborated with colleague José Luis Bortolini to investigate this hypothesis. We found (Bortolini and Bauer 2016) that a much reduced AG *is present* in the *L. wurdemanni* FPSHs. The relative size of the AG was only slightly larger in MPs than in FPSHs, but the AGs of MPs were healthy looking, packed with nucleated cells. In FPSHs, the AG was vacuolated with many fewer apparently active cells. The larger FPSHs had the most degenerated AGs; correspondingly, the relative size of the testicular part of the ovotestes was less than that of smaller FPSHs. These degenerate AGs apparently produce just enough AG hormone to keep the male system going in FPSHs. Mating studies (e.g., Bauer and Holt 1998; Bauer 2002; Baeza 2007a, b) show that FPSHs of any size are perfectly capable of mating as males and inseminating females. Our study on AG was subsequently followed by that of Zhang et al. (2017) who demonstrated the presence of IAG in both MPs and FPSHs of *L. wurdemanni*. As predicted by Bortolini and Bauer, the IAG concentration was indeed greater in MPs and less in the FPSHs, reflecting the gradual degeneration of AG in the latter. These findings probably apply to all PSH species.

7.7 Environmentally Determined Sex Change (ESD)

Phenotypic traits are based on genes, but their degree of development and final outcome is influenced by the surrounding biological and physical (abiotic) environment. In mammals and birds, sex is genetically determined by male and female sex chromosomes. However, in numerous organisms, sex is environmentally determined, e.g., by incubation temperature in turtles and crocodilians and larval diet in the shrimp *Hippolyte inermis* (Zupo et al. 2008). In protogynous fish species with male-dominated mating systems, sex is socially-mediated (e.g., Warner 1984; Ross 1990; Gemmell et al. 2019). When a dominant male dies, a female of its harem or group changes sex to female. This change is mediated by hormones triggered by social rather than physical factors.

Do individuals of hermaphroditic species of shrimps have similar sexual liability, i.e., can they be stimulated to change the course of sexual development because of changes in their environment? In purely protandric species, sex change is genetically fixed; all members of the population develop according to their genotype. However, in those species with a mix of sexual types, such as *Pandalus* species, the proportions of protandrites (PH), primary females (PF) and/or early-maturing females (EMF) may well vary in different physical or demographic environments. The question of environmental sex determination (ESD) was raised in shrimps by fishery catch data with species such as *P. jordani* (Charnov et al. 1978; Hannah and Jones 1991; Charnov and Hannah 2002), *P. borealis* (Charnov 1981) and other pandalids, nicely reviewed by Bergström (2000). Fishing pressure is highest on larger individuals because of their greater commercial value. In protandric shrimps, the larger individuals are females, either PF or EMF or completely PH females. The fishery data suggested that in heavily fished populations, the age and size of sex change to female phase was lower than in less-fished populations.

In his book and various papers, Charnov (1982) offered an explanation using sex allocation theory. As proposed by Fisher (1930) and Hamilton (1967), selection for the rarer sex maintains a 1:1 sex ratio in most sexually reproducing organisms. In populations with a high number of females, male births or development are favored because males will have abundant opportunities to mate with females. When males are in excess, females will be favored because males have more competition and less chance to inseminate a female. In terms of protandric shrimps, a lower proportion of adult females should stimulate protandric males to change sex early, resulting in a population composed of smaller protandric females, more EMFs and PFs.

However, how does a small juvenile shrimp measure the size and sexual composition of its population and then "decide" the course of its development? Charnov himself (1982) pondered "We can only wonder at what cues the animals may be using." Sexual development might be socially-mediated, i.e., by frequency and kind of contacts with other members of the population. *Pandalus* species occur in loose aggregations or in schools. Contact with surrounding conspecifics with their long antennae and/or via pheromones might allow developing juveniles to measure the size (age) and sexual composition of its population. This might be termed "socially-mediated sex change in different demographic environments" (Bauer 2002). However, the life cycle of protandric pandalids makes this unlikely. Protandric pandalids are cool-temperate and boreal species in which sex change occurs in the spring and early summer. Breeding takes place in autumn and winter (Butler 1964). As Bergström (1997) pointed out, the population structure when sex change takes place may not be the same as that in which breeding takes place, e.g., due to mortality, migration, and mixing with other populations. The potential sex changer would have to predict or forecast what the breeding population will be like some months in the future to make the correct physiological "decision" whether to change sex or not.

There is no good evidence that pandalid shrimps are able to do this or that it takes place at all. The large-scale laboratory experiment by Marliave et al. (1993) on *Pandalus danae* showed no effect on the sexual development of young shrimps reared under various population sex and size mixtures. These authors suggested that the size/age of sex change is a heritable trait and that variations among populations in time and space are due to the selection on different sexual genotypes and not by ESD. This view is supported by a 4-year study on a *P. borealis* population in a Swedish fjord by Bergström (1997) who found no correlation between population structure on the timing and type of sex change. He also proposed that the timing of sex change is genetically, not environmentally (demographically) controlled in pandalids. Bergström suggested that the population is composed of a mixture of genotypes programmed to develop directly into primary females or protandrites that change sex at different ages (PH, EMFs). Fluctuations in the age of sex change and proportions of sexual types may be a result of frequency-dependent selection for rarer sexual genotypes depending on a combination of mortality, recruitment, and growth (Bergström 1997, 2000). Thus, socially-mediated sex change in pandalids is supported by fishery/population data in some studies but not others. To date, there are no experimental (rearing) studies that substantiate ESD in pandalids.

On the other hand, there is ample evidence in *Lysmata* shrimps that the timing of sex change can be influenced by the mix of MPs and FHSPs in which a young individual (MP) is reared in the laboratory, as first indicated by Lin and Zhang (2001). Baeza and Bauer (2004) conducted laboratory experiments in which MPs large enough to change to the FPSHs (sex-change "candidates") were reared in different populations (treatments) which varied in size and sexual composition. The speed of MP sex change was inversely proportional to the abundance of FPSHs in

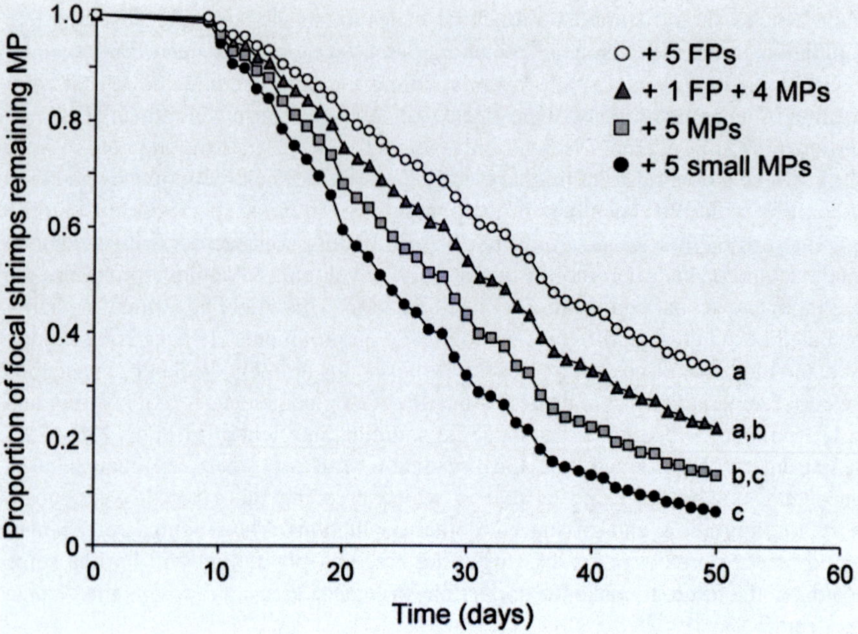

Fig. 7.16 Demographic (socially-mediated) sex change in *Lysmata wurdemanni*. Sex change was followed in four different groups composed of five sex change candidates (male-phase individuals large enough to change sex) plus different mixes of male-phase (MP) and female-phase simultaneous hermaphrodites (labeled FPs in this figure). Curves denote the proportion of focal MP candidates remaining MP, i.e., not changing to FP over time (days). Treatment curves labeled with different letters are significantly different statistically. From Baeza and Bauer (2004)

the group (Fig. 7.16). In groups with numerous FPSHs, an MP candidate has many opportunities to mate as a male, so it delays change. When reared alone, or with only MPs or a few FPSHs, an MP candidate changed more quickly, as opportunities to mate as male were zero to low, but after sex change, mating as a female was assured. Similar results of sex-ratio induction of sex change were found in *L. seticaudata* (Calado and Dinis 2007) and *L. amboinensis* (Tziouveli and Smith 2009).

However, is socially-mediated ESD in *Lysmata* well supported by field studies? In *L. wurdemanni*, Baldwin and Bauer (2003) found that sex change occurred earlier in the late spring to summer reproductive season (Fig. 7.17). In the autumn and winter months, sex change was delayed, and average MP size was considerably larger than earlier in the year. I hypothesized (Bauer 2002) that basic environmental factors such as temperature and photoperiod might be important in the age/size of sex change. In the laboratory, I maintained groups of MPs large enough to change sex under winter conditions (lower temperature, shorter daylength) but other groups under late/spring summer conditions (higher temperatures, longer day length) for a period of 14 weeks. Those reared under suboptimal winter conditions molted less and changed to FP more slowly; some never changed sex. Most MPs reared under optimal reproductive-season conditions changed to embryo-producing FPSHs

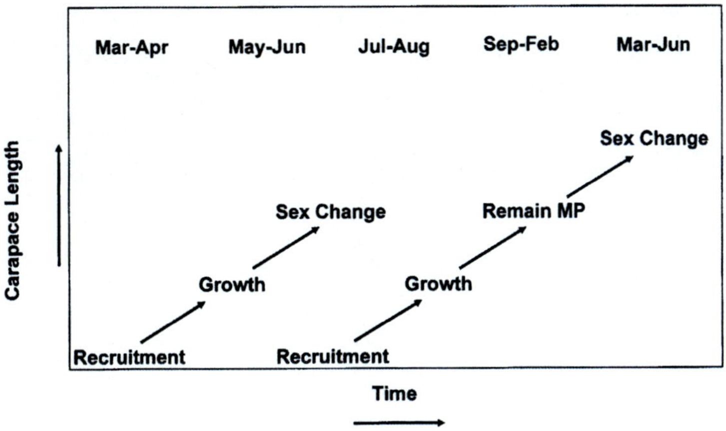

Fig. 7.17 Seasonal patterns of sex change in the protandric simultaneous hermaphrodite *Lysmata wurdemanni* in south Texas, USA. Spring recruits transition from male-phase to the female simultaneous hermaphrodite phase by summer, while those recruited later in the summer and early autumn delay sex change until early spring. These are the individuals which will produce the spring and early summer recruits. From Baldwin and Bauer (2003)

(Fig. 7.18a). At the end of 14 weeks, I changed temperature and photoperiod of the suboptimal-conditions group to optimal ones, and MPs quickly changed to FPSHs (Fig. 7.18b). Thus, individuals recruited in the late spring grow rapidly, changing sex quickly at small size to FPSH, with production of embryos during a favorable time of the year for larvae to feed and develop. On the other hand, MPs recruited in late summer and autumn delay sex change until the following spring, when these overwintering MPs change to FPSHs of large size with high fecundity. Instead of producing broods in the autumn or winter, time periods unfavorable for larval development, summer/autumn-recruited MPs put their energy into growth for successful female breeding at the start of the next year's breeding season. This is a pattern common in seasonally breeding gonochoristic shrimps; that is, spring-recruited females mature rapidly and breed all summer, while autumn-recruited ones delay reproduction until early in the next breeding season (Bauer 2004). Thus, sex change in *Lysmata* may be mediated by both the social/demographic environment (Baeza and Bauer 2004) and seasonally related abiotic factors acting on reproductive physiology (Bauer 2002, 2007).

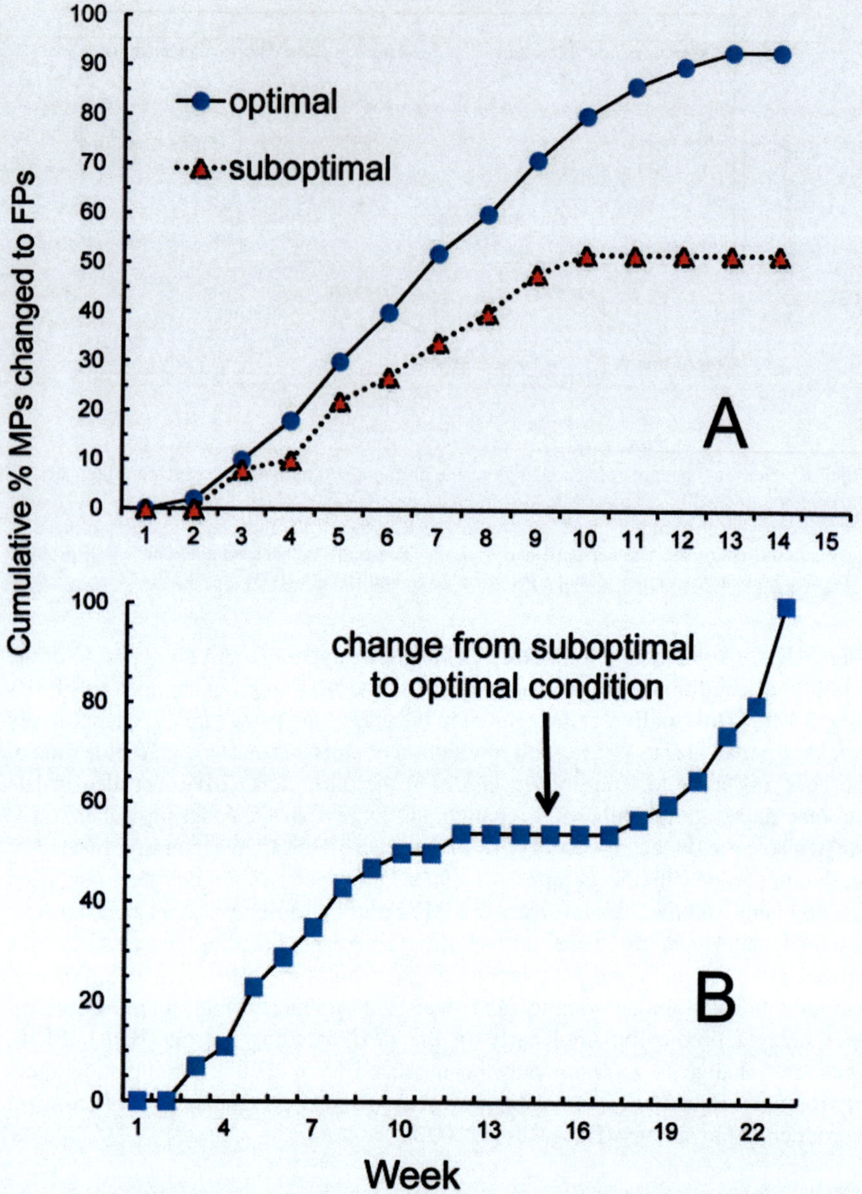

Fig. 7.18 Laboratory experiments on the effect of seasonal temperature and photoperiod on sex change in *Lysmata wurdemanni*. (**a**) Change of large male-phase (MP) individuals to female-phase simultaneous hermaphrodite (FPSH) maintained under optimal (summer) and suboptimal (winter) temperature and photoperiod for 14 weeks. (**b**) Change of MP to FPSH in MPs reared for 14 weeks under suboptimal (winter) temperature and photoperiod, and then switched (arrow) to optimal (summer) conditions for an additional 8 weeks. Modified from Bauer (2002) courtesy of The Biological Bulletin

7.8 Evolution of Protandrous Sexual Systems

7.8.1 *Female-First or Male-First Origin of Protandrous Sexual Systems?*

Although protandry will continue to be discovered in currently unstudied species, the vast literature on population structure and mating of many species confirm that gonochorism, or separate sexes, is the most common sexual system in shrimps. A basic question that must come to mind of the reader is "how did protandrous systems evolve from gonochoric ones?" Given that protandric species are scattered among a number of unrelated caridean families, it is likely that protandry has evolved independently several times. This is supported by ovotestes structure, which varies in at least three ways. In developing protandrous pandalids and atyids, small males have oogonia and oocytes within the ovotestes (medullar) while the outer part (cortical) is composed of spermatogonia and spermatocytes. During and after the change to female, the outer testicular tissue degenerates and is lost as the medullar ovarian cells increase and change the ovotestes to ovaries (Charniaux-Cotton and Payen 1985; Hoffman 1972; Carpenter 1978). In protandrous hippolytoids and crangonids, the most anterior part of the ovotestes is ovarian, the more posterior part is testicular. With change to a female phase, the ovarian part replaces testicular tissue from anterior to posterior, either completely (e.g., *Thor manningi*, Bauer 1986; crangonids; Boddeke et al. 1991; Gavio et al. 2006) or mostly (PSH lysmatid and barbouriid species, e.g., Berreur-Bonnenfant and Charniaux-Cotton 1965; Bauer and Holt 1998; Baeza 2009). Yet another pattern was described by Suzuki (1970), in which the ovarian (medial) and testicular tissues are arranged longitudinally in the ovotestes.

Given the Size Advantage Model discussed above, protandry must have evolved in small male/larger female species. One might ask, as did Charniaux-Cotton and Payen (1985), whether sex change evolved in males or in females of gonochoric species. Based on studies on the gonadal development of protandric pandalids as well as studies on androgenic glands (AG), they proposed that the males of gonochoristic species gave rise to protandric genotypes which replaced primary males. Suppose that mutations occurred that allowed males near minimum female size to keep growing but with AG reduction and loss, change sex and then reproduce as females ("male-first hypothesis") (Fig. 7.19). Recall that in decapods, all individuals have the embryonic rudiment ("anlage") of the androgenic glands (AG). The embryonic gonads may become either testes or ovaries, depending on the presence or absence of AG hormone. In the absence of a male gene or perhaps its inhibition by a female gene, the AG rudiment does not develop, and individuals are female, the "default" sex. In the male-first hypothesis, a mutation shut off or limited AG production, with the result that the individual changed sex to female. However. a male-first origin of protandry would entail added growth to the larger size needed by females to produce vitellogenic oocytes and an extension of the normally shorter male life span.

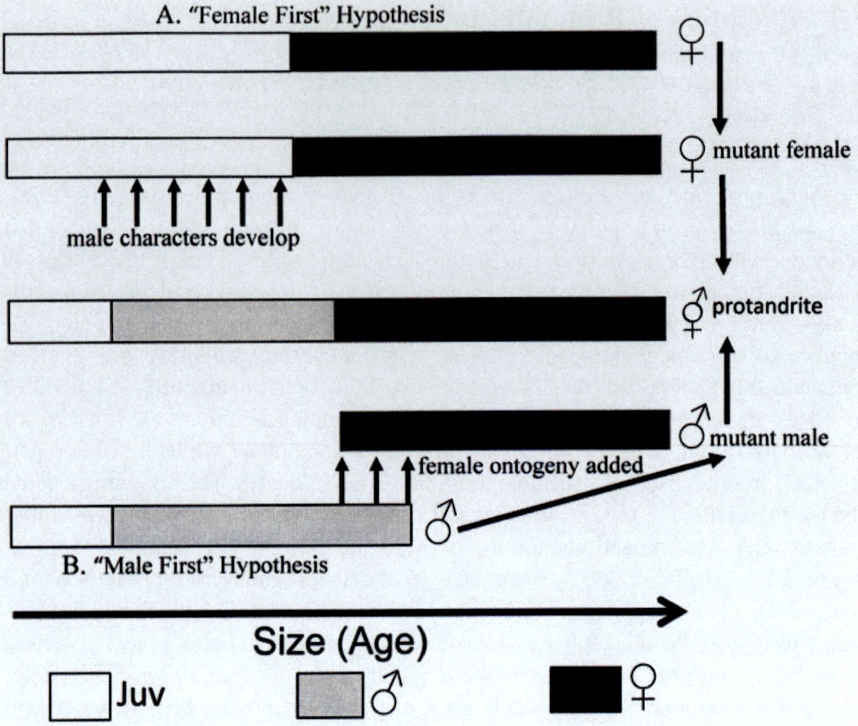

Fig. 7.19 Two hypotheses on the ancestral sex of protandric shrimps. (**a**) "Female first" hypothesis: androgenic glands develop transitorily in juvenile stages of mutant females, allowing the development of male morphology and function before later degeneration, leading to subsequent development of adult female morphology and function. (**b**) "Male first" hypothesis: androgenic glands of adult mutant males degenerate later in life, allowing sex change to female with subsequent extension of life span. Modified from Bauer (2006)

On the other hand, Hoffman (1972) cited histological evidence suggesting that females gave rise to protandry in *Pandalus*. In *P. platyceros*, oogenesis occurs before spermatogenesis in the juvenile shrimp. The primary spermatocytes of protandric individuals in this species are unusually large compared to those of gonochoristic carideans. He postulated that these spermatocytes were originally oocytes modified by secretions of the androgenic glands. Protandry might arise in an ancestor by a mutation triggering the transitory development of the AG in an originally genotypic female. Instead of passing through the normal juvenile female development, the individual would go through a male phase during the same size and growth period. This would be possible because of the apparently small energetic investment into male appendices and gamete production in males from "small-male" caridean species. Later degeneration of AG later would allow the change to female (Fig. 7.19).

In *Thor manningi,* which has primary males, I hypothesized a female-first evolution of protandric individuals which eventually replaced primary females in the population. Mutation allowing growth of the AG rudiment in juvenile females would trigger the development of transitory male sexual characters on the way to maturation as the larger embryo-producing FP. Given the small energetic cost of sperm and male secondary sexual appendices, even one insemination of another FP's brood of eggs by MPs would give protandric individuals a selective advantage over primary females. The production of neofemales from genetic males by ablation of the androgenic gland and more recently by molecular silencing of AG hormone in *Macrobrachium rosenbergii* (Ventura et al. 2011a, b) shows that this is a reasonable scenario.

In many organisms, sex chromosomes have been identified which might show the original sex of protandric individuals. Karyotyping techniques on shrimps have not yet identified such chromosomes in shrimps. However, progeny testing in which "neomales" (transformed genetic females) and "neofemales" (transformed males) mated with untransformed genetic females or males, respectively, can show which is the heterogametic sex. Using this technique with *M. rosenbergii,* males were shown to be the homogametic (ZZ) sex and females are the heterogametic one (ZW) (Ventura and Sagi 2012). Similar work with protandric species, e.g., crossing protandric MPs and FPs with neofemales and neomales might reveal the original sex of protandric species.

In conclusion, protandric species may have evolved either by a male-first vs a female-first trajectory. Protandry in unrelated species needs to be investigated on a case-by-case basis.

7.8.2 The Dendrobranchiate/Palaemonid Paradox

Protandry, by definition, describes the sexual system of species with small male-phase and larger female-phase individuals. It seems obvious that protandry must have evolved from gonochoristic species with this type of sexual dimorphism in body size. Males of penaeoid and sergestoid species are smaller than females. Although less diverse ecologically than carideans, these dendrobranchiates are a large group of species. Why are there no protandrous dendrobranchiates? This question occurred to me when finding that there are virtually no hermaphroditic species of insects nor other terrestrial arthropods (Policansky 1982; Gardner and Ross 2011). This is an enormous group of well over a million species. In terrestrial habitats, unlike aquatic ones, the fragile sperm and eggs must be protected from desiccation. Terrestrial animals, including arthropods, generally have internal fertilization and commit considerable resources to produce and maintain the complex genitalia and spermatophores needed to protect their gametes (Warner 1978; Policansky 1982). Dendrobranchiates also have much more complex genitalia (petasmata, thelyca, seminal receptacles), reproductive tracts and spermatophores than carideans. Although dendrobranchiate fertilization is external, sperm is protected before fertilization within complex external spermatophores or cuticular seminal receptacles. Male caridean genitalia are relatively small, simple appendices masculinae on the second pleopods and modest modifications of the first pleopods. Their reproductive tracts and spermatophores are nowhere nearly as complex as those of dendrobranchiates. Complex thelyca and seminal receptacles for sperm storage are unknown in carideans. Thus, the costs of sex change, i.e., the resources needed to construct complex reproductive structures, appear to be much higher in dendrobranchiates than in carideans. Perhaps this is the barrier that has prevented the evolution of protandrous sexual systems in dendrobranchiates which otherwise might benefit from it given their sexual dimorphism in body size.

Stenopodidean genitalia and spermatophores are even less complex than in carideans. However, males and females in this group live in socially monogamous pairs of more or less similar size. There are many pair-living carideans, especially among the Alpheidae and pontoniine palaemonids. Yet protandry is not expected in these species, given the Size-Advantage Model and their monogamous mating system. The "paired" species of *Lysmata* are one exception, and, if hermaphroditism in the alpheid *Salmoneus* is confirmed, that would be another. But I suggest that protandry has not been found or is not common in the Stenopodidea because of its sexual dimorphism and socially monogamous mating system.

Somewhat of a puzzler too is the apparent absence of protandrous systems in the caridean family Palaemonidae. This is a very large family with considerable variation in sexual dimorphism and mating system. There are species with large dominant males and smaller females (e.g., many *Macrobrachium* species) and there are pair-living species in which protandry is not particularly favored. Yet there are also numerous small male/larger female palaemonid species that seem preadapted for protandry. Yet none have been yet reported. I suspect that chance explains a large part of the distribution of protandrous sexual systems in carideans. Genotypic mutations, especially those related to the androgenic gland, begin the process, whether in a male-first or female-first scenario. Mutations are random DNA copying errors during the production of sexual cells. If and when they occur, and the selective environment of the shrimp is conducive, then such mutations resulting in protandry may take root in a species. Perhaps this simply has not occurred in the Palaemonidae—and perhaps it is an untestable hypothesis!

7.8.3 Evolution of Protandric Simultaneous Hermaphroditism (PSH)

In my 2004 book on carideans, I proposed that PSH was the "pinnacle" of protandrous sexual system in a series of evolutionary steps from gonochorism to simple protandry to PSH. The shift from protandry to PSH appears a somewhat minor physiological change compared to that of sex change itself. If the degeneration of the AG which leads to sex change is not complete, a reduced but very functional male system is retained, resulting in an individual with a largely female phenotype but which can successfully mate as a male as well as female. Of this there is no doubt; several studies cited previously have shown that FPSHs mate successfully as males even while carrying a brood of embryos or during maturation of the ovarian portions of the ovotestes (Bauer and Holt 1998; Bauer 2002; Zhang and Lin 2005; Baeza 2007a, b). To date, PSH is restricted to the two lysmatid genera and *Parhippolyte misticia* in the related family Barbouriidae (Table 7.1). It is surprising that PSH has not evolved more frequently in other families with protandric species, such as pandalids (*Pandalopsis and Pandalus*), given the small physiological and morphological changes involved and the reproductive advantage gained. However, PSH has only been recognized in the last two decades. This system may be shown to be more widespread as more species are scrutinized.

What are the possible barriers to the evolution of PSH? One such obstacle might be the physiological and behavioral costs of maleness to the FPSH individual. The cost of sperm production and male mating behavior seems small compared to egg production and brooding behavior in the FPSH individual. Baeza (2006) made various measurements and observations that indeed show that the cost of female function is much higher than that of male function. I looked at the cost of male mating behavior on FPSH fecundity by measuring brood production in FPSHs with various group combinations of FPSHs and MPs (Bauer 2005). Those FPSHs with the opportunity to mate as males did have a somewhat lower overall fecundity—so there is some cost after all. However, if the broods that an FPSH might fertilize when mating as a male are considered, its total reproductive output would still outweigh that of an FP of strictly protandric species. The relatively small cost of maleness in FPSHs is not enough to explain the apparently low incidence of PSH in caridean shrimps or other decapods.

One of the important selective pressures promoting the evolution of hermaphroditism is the availability of mating partners ("low density model"; Ghiselin 1969). Members of a species which are sessile, or mobile but living at low population density, or with limited mobility because of high predation pressure, have difficulty in finding a mating partner. In such a situation, hermaphroditism may evolve to ensure reproductive success. However, most lysmatid species live in either high-density aggregations ("crowds") or small groups (Bauer 2000; Baeza 2009). Furthermore, these shrimps are quite mobile so that encounters among individuals are frequent. In a protandric ancestor with a similar socioecology, FPs would have little problem in finding a male mating partner. There would seem to be little selective pressure in aggregated protandric species for evolution of PSH, considering also that there is some cost to maleness to FPSHs. Perhaps FPSHs are better at mating as a male than are MPs? This does not seem to be the case. Studies comparing the male-mating ability have either shown that MPs are similar or superior in inseminating FPSHs reproducing as female than FPSHs mating as males (Bauer 2002; Zhang and Lin 2005; Baeza 2007a, b).

Perhaps, as proposed for a number of evolutionary events, chance plays a role. Stephen J. Gould, in his 1989 book "Wonderful Life" presented a historical contingency explanation for a number of key evolutionary events. For example, the extinction of dinosaurs and the rise of mammals was due, in large part, to the chance asteroid strike at the end of the Cretaceous with the subsequent environmental

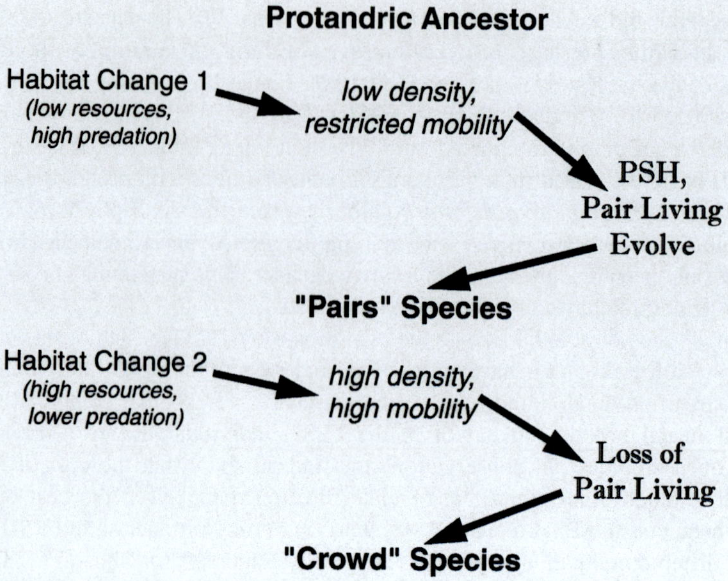

Fig. 7.20 Historical contingency hypothesis proposed for the evolution of protandric simultaneous hermaphroditism (PSH) and pair-living ("paired") species in *Lysmata* from a protandric ancestor, with the subsequent loss of pair living but not PSH in "crowd" species. From Bauer (2007)

upheavals and consequences. A unique adaptation such as PSH in shrimps might evolve under natural selection but its origin may also have stemmed from a cascade of unpredictable historical events, each dependent on a previous one (historical contingency, Gould 1989). In the case of PSH, the first step must be the mutations or gene combination responsible for the incomplete degeneration of the androgenic gland in a protandric species. This would allow the retention of reduced male morphology but functional male mating ability in female-phase individuals. Such an event would have to occur within the right selective environment for PSH to evolve. As pointed out previously, this mutant would have some cost of male function during the FPSH stage and also have abundant competition from protandric MPs for mating as a male in aggregated species. It is difficult to envision how PSH might have evolved in an aggregated protandric ancestor

I hypothesized (Bauer 2006, 2007) that a series of historical events occurred in lysmatids which might explain both PSH and the two extremes in social organization in lysmatid shrimps, "paired" and "crowd" (aggregated) species (Fig. 7.20). In a historical contingency hypothesis, the ancestral lysmatid (or related hippolytoid) ancestor was an aggregated species living at high density with a pure-search (promiscuous) mating system, with frequent contact among MPs and FPs. The first event necessary for PSH would have been a habitat change or invasion of a population into another habitat that resulted in low population size and/or mobility of individuals. This might come about with a reduction of resources (food, foraging space, shelter) and/or increased predation pressure. The result would be infrequent contact between potential mating partners. This situation would favor living in pairs, insuring a mating partner. Given the chance event of the mutations favoring partial rather than complete degeneration of the androgenic gland, PSH could evolve. PSH individuals would be able to mate both as male and female, and this genotype would eventually replace a purely protandric one. Thus, both pair living and PSH evolved together. Later, changes in some parts of the habitat became favorable again for high population density, and pair living was no longer favored, resulting in aggregated ("crowd") species with PSH. PSH is so favorable an adaptation that, once evolved, it is retained in aggregated species although originating in a different selective environment.

I suggested (Bauer 2000, 2006, 2007) that phylogenetic studies were necessary to test a PSH historical contingency hypothesis, i.e., to determine if paired species or aggregated species were ancestral in PSH species. This proposal was eagerly taken up in a series of studies using molecular (genomic) characters to generate phylogenies. The first of these, by former student Antonio Baeza et al. (2009), gave partial support to a paired-species origin of PSH but two subsequent studies (Fiedler et al. 2010; Baeza 2013) rejected that hypothesis. As with many molecular phylogenies, these studies varied in the genes utilized, sample sizes and species sampled. Although molecular phylogenies are often considered superior to others based on morphology or other traits, this is not always necessarily the case (Losos et al. 2012). Different molecular phylogenies testing the same evolutionary hypotheses often disagree in conclusions. Although support for the historical contingency origin of PSH is not now strong, it should be at least considered in the absence of a more convincing one, even as a "straw man" to encourage further studies on the subject.

As an alternative, Baeza (2013, 2018) proposed an evolutionary scenario based on sex allocation theory and his experimental work and observations on mating and brooding in *Lysmata wurdemanni*. His latest phylogenetic studies suggest that PSH arose in a hippolytoid group ancestral to the Lysmatidae. Given that, he suggested two factors that might be key to the PSH in that clade. One is that females of his lysmatid model species usually just copulates once with a single male-mating partner (monandry), resulting in reduced mating opportunities for MPs. The other is his observation that brooding FPSHs lose part of their brood due to limited brooding space below her abdomen as embryos mature and increase in size. In agreement with sex allocation theory, these conditions might provide the right selective environment for PSH given the necessary mutations for partial silencing of male function and morphology. However, complete or partial monandry is quite common in caridean shrimps in various gonochoric and protandric taxa (e.g., Bauer 1976; Bauer and Abdalla 2001; Noël 1976; Boddeke et al. 1991; Bauer and Thiel 2011), Likewise, embryo loss due to insufficient space in the brooding space below the abdomen is widespread and well documented in many caridean species (Corey and Reid 1991; see also Chap. 9). Thus, these two factors are not rare in the Caridea, but PSH has only evolved in the Lysmatidae and Barbouriidae to date. This is not convincing evidence in my view, although I urge to reader to read Baeza (2018) and pertinent papers within for a full explanation of his views and work.

Thus, the puzzle of PSH evolution remains unanswered and needs and is sure to stimulate further research on the biology of sexual systems in shrimps. The issue may become even more interesting given the preliminary observations of Murray et al. (2012). A lysmatid, *Lysmatella prima,* may have a form of PSH in which the male phase has been eliminated completely. Juveniles apparently mature sexually only as simultaneous hermaphrodites. If verified in *L. prima* or other carideans, this would be yet another evolutionary step in protandrous hermaphroditism in shrimps, replacing PSH as the "protandric pinnacle."

7.8.4 Final Thoughts

How common is hermaphroditism in shrimps? The total number of shrimp species is about 4400 (Chap. 3). At the beginning of this chapter, I indicated that numerous studies have been done on mating behavior, sexual biology, and especially population biology on species from the major groups of decapod shrimps. To tally the exact number of these studies would certainly be a time-consuming challenge which has not been done. But they must number in the several hundreds, if not thousands. Initial recognition of hermaphroditism comes from such studies. Yet only about 90 species to date are documented as possible or confirmed as protandrous (Table 7.1). In Bauer (2000), I reported possible or confirmed ~30 cases of protandry in carideans and predicted that more would be discovered. That certainly has been the case. But the evidence to date indicates that most shrimp species are gonochoric. The majority of sex-changing shrimps are found in just two unrelated caridean genera: *Pandalus* and *Lysmata*, in which all species studied to date are hermaphroditic. Other protandric species are scattered here and there in various unrelated caridean families. When one species of a genus is found to be hermaphroditic, it cannot be assumed that other closely related species will have the same sexual system, as we have seen. However, given the advantages of sex change, the question must be raised: "Why are not there more protandric or PSH shrimp species?" Some hypotheses have been raised in this chapter, but there is still no very convincing answer. Resolution of that question is a challenge that remains to be met.

References

Allen JA (1959) On biology of *Pandalus borealis* Kroyer, with reference to a population off the Northumberland coast. J Mar Biol Assoc UK 38:189–220

Alonso-Reyes MP, Bortolini-Rosales JL, Álvarez F (2011) Análisis discriminante aplicado a los grupos sexuales de *Potimirim mexicana*, camarón hermafrodita protándrico. Revista Mexicana de Biodiversidad 81:187–192

Alves DFR, López Greco LS, Barros-Alves SDP, Hirose GL (2019) Sexual system, reproductive cycle and embryonic development of the red-striped shrimp *Lysmata vittata*, an invader in the western Atlantic Ocean. PLoS One 14:0210723. https://doi.org/10.1371/journal.pone.0210723

Alves DFR, Barros-Alves SP, Almeida AC, Costa RC (2021) Sex change and reproductive output of the protandric shrimp *Merguia rhizophorae* (Rathbun, 1900) (Decapoda, Merguiidae). J Nat Hist 55:41–42
Anker A (2011a) Three new species of the alpheid shrimp genus *Salmoneus* Holthuis, 1955 (Crustacea, Decapoda) from the tropical Western Pacific. Zootaxa 2839:67–84
Anker A (2011b) Two new species of *Salmoneus* Holthuis, 1955 with a deep dorsal depression on the carapace (Crustacea, Decapoda, Alpheidae). Zootaxa 3041:39–50
Anker A, Baeza JA, De Grave S (2009) A new species of *Lysmata* (Crustacea, Decapoda, Hippolytidae) from the Pacific coast of Panama, with observations of its reproductive biology. Zool Stud 48:682–692
Baeza JA (2006) Testing three models on the adaptive significance of protandric simultaneous hermaphroditism in a marine shrimp. Evolution 60:1840–1850
Baeza JA (2007a) Sex allocation in a simultaneously hermaphroditic marine shrimp. Evolution 61: 2360–2373
Baeza JA (2007b) Male mating opportunities affect sex allocation in a protrandric-simultaneous hermaphroditic shrimp. Behav Ecol Sociobiol 61:365–370
Baeza JA (2008) Protandric simultaneous hermaphroditism in the shrimps *Lysmata bahia* and *Lysmata intermedia*. Invert Biol 127:181–188
Baeza JA (2009) Protandric simultaneous hermaphroditism is a conserved trait in *Lysmata* (Caridea: Lysmatidae): implications for the evolution of hermaphroditism in the genus. Smith Contrib Mar Sci 38:95–110
Baeza JA (2010a) Sexual system and natural history observations on semi-terrestrial shrimp *Merguia rhizophorae*. Invert Biol 129:266–276
Baeza JA (2010b) The symbiotic lifestyle and its evolutionary consequences: social monogamy and sex allocation in the hermaphroditic shrimp *Lysmata pederseni*. Naturwissenschaften 97:729–741
Baeza JA (2013) Multi-locus molecular phylogeny of broken-back shrimps (genus *Lysmata* and allies): a test of the "Tomlinson-Ghiselin hypothesis" explaining the evolution of simultaneous hermaphroditism. Mol Evol Phylogen 69:46–62
Baeza JA (2018) Sexual systems in shrimps (Infraorder Caridea Dana, 1852) with special reference to the historical and adaptive value of protandric simultaneous hermaphroditism. In: Leonard JL (ed) Transitions between sexual systems. Springer Nature, Cham, pp 269–310
Baeza JA, Anker A (2008) *Lysmata hochi* n. sp., a new hermaphroditic shrimp from the southwestern Caribbean Sea (Caridea: Hippolytidae). J Crust Biol 28:148–155
Baeza JA, Bauer RT (2004) Experimental test of socially mediated sex change in a protandric simultaneous hermaphrodite, the marine shrimp *Lysmata wurdemanni* (Caridea: Hippolytidae). Behav Ecol Sociobiol 5:544–550
Baeza JA, Piantoni C (2010) Sexual system, sex ratio, and group living in the shrimp *Thor amboinensis* (De Man): relevance to resource-monopolization and sex-allocation theories. Biol Bull 219:151–165
Baeza JA, Reitz JM, Collin R (2007) Protandric simultaneous hermaphroditism and sex ratio in *Lysmata nayaritensis* Wicksten, 2000 (Decapoda: Caridea). J Nat Hist 41:2843–2850
Baeza JA, Bolaños JA, Hernandez JE, López R (2009) A new species of *Lysmata* (Crustacea, Decapoda, Hippolytidae) from Venezuela, southeastern Caribbean Sea. Zootaxa 2240:60–68
Baeza JA, Braga AA, López-Greco LS, Perez E, Negreiros-Fransozo ML, Fransozo A (2010) Population dynamics, sex ratio and size at sex change in a protandric simultaneous hermaphrodite, the spiny shrimp *Exhippolysmata oplophoroides*. Mar Biol 157:2643–2653
Baldwin AP, Bauer RT (2003) Growth, survivorship, life span, and sex change in the hermaphroditic shrimp *Lysmata wurdemanni* (Decapoda: Caridea: Hippolytidae). Mar Biol 143:157–166
Bauer RT (1976) Mating behaviour and spermatophore transfer in the shrimp *Heptacarpus pictus* (Stimpson) (Decapoda: Caridea: Hippolytidae). J Nat Hist 10(415):440
Bauer RT (1986) Sex change and life history pattern in the shrimp *Thor manningi* (Decopoda: Caridea): a novel case of partial protandric hermaphroditism. Biol Bull 170:11–31

Bauer RT (2000) Simultaneous hermaphroditism in caridean shrimps: a unique and puzzling sexual system in the Decapoda. J Crust Biol 20(special no. 2):116–128
Bauer RT (2002) Test of hypotheses on the adaptive value of an extended male phase in the hermaphroditic shrimp *Lysmata wurdemanni* (Caridea: Hippolytidae). Biol Bull 203:347–357
Bauer RT (2004) Remarkable shrimps: adaptations and natural history of the Carideans. University of Oklahoma Press, Norman
Bauer RT (2005) Costs of maleness on brood production in the shrimp *Lysmata wurdemanni* (Decapoda: Hippolytidae: Caridea), a protandric simultaneous hermaphrodite. J Mar Biol Soc U K 85:101–106
Bauer RT (2006) Same sexual system but variable sociobiology: evolution of protandric simultaneous hermaphroditism in *Lysmata* shrimps. Integr Comp Biol 46:430–438
Bauer RT (2007) Hermaphroditism in caridean shrimps: mating systems, sociobiology, and evolution, with special reference to *Lysmata*. In: Duffy JE, Thiel M (eds) Evolutionary ecology of social and sexual systems: crustaceans as model organisms. Oxford University Press, New York, pp 232–248
Bauer RT, Abdalla JA (2001) Male mating tactics in the shrimp *Palaemonetes pugio* (Decapoda, Caridea): precopulatory mate guarding vs. pure searching. Ethology 107:185–199
Bauer RT, Conner SL (2012) Gonochoric sexual system in the caridean shrimps *Processa riveroi* and *P. bermudensis* (Decapoda: Processidae) inhabiting a tropical seagrass meadows. J Mar Biol Assoc U K 92:521–529
Bauer RT, Holt GJ (1998) Simultaneous hermaphroditism in the marine shrimp *Lysmata wurdemanni* (Caridea: Hippolytidae): an undescribed sexual system in the decapod Crustacea. Mar Biol 132:223–235
Bauer RT, Newman WA (2004) Protandric simultaneous hermaphroditism in the marine shrimp *Lysmata californica* (Caridea: Hippolytidae). J Crust Biol 24:131–139
Bauer RT, Thiel M (2011) First Description of a pure-search mating system and protandry in the shrimp *Rhynchocinetes uritai* (Decapoda, Caridea). J Crust Biol 31:286–295
Bauer RT, VanHoy R (1996) Variation in sexual systems (protandry, gonochorism) and reproductive biology among three species of the shrimp genus *Thor* (Decopoda: Caridea). Bull Mar Sci 59:53–73
Bauer RT, Okuno J, Thiel M (2014) Inferences on mating and sexual systems of two Pacific *Cinetorhynchus* shrimps (Decapoda, Rhynchocinetidae) based on sexual dimorphism in body size and cheliped weaponry. Zookeys 457:187–209
Bergström BI (1997) Do protandric pandalid shrimp have environmental sex determination? Mar Biol 128:397–407
Bergström B (2000) The biology of *Pandalus*. Adv Mar Biol 38:1–245
Berkeley AA (1929) Sex reversal in *Pandalus danae*. Am Nat 63:571–573
Berkeley AA (1930) The post-embryonic development of the common pandalids of British Columbia. Contrib Can Biol NS 6:79–163
Berreur-Bonnenfant J, Charniaux-Cotton H (1965) Hermaphrodisme protérandrique et fonctionnnement del a zone germinative chez la crevette *Pandalus borealis* Kröyer. Bull Soc Zool Fr 90:240–259
Blackless M, Charuvastra A, Derryck A, Fausto-Sterling A, Lauzanne L, Lee E (2000) How sexually dimorphic are we? Review and synthesis. Am J Hum Biol 12:151–166
Boddeke R, Bosschieter JR, Goudswaard PC (1991) Sex change, mating, and sperm transfer in *Crangon crangon* (L.). In: Bauer RT, Martin JW (eds) Crustacean sexual biology. Columbia University Press, New York, pp 164–182
Bortolini JL, Álvarez F (2012) Histología del ovotestis de *Potimirim mexicana*, un camarón hermafrodita protándrico. In: 18 Congreso Latinoamericano de macroinvertebrados de agua dulce: su importancia en ríos tropicales, 6–11 Feb 2012. Ciudad de la Investigación, Universidad de Costa Rica, San José, Costa Rica, pp 21–22
Bortolini JL, Bauer RT (2016) Persistence of reduced androgenic glands after protandric sex change suggests a basis for simultaneous hermaphroditism in a caridean shrimp. Biol Bull 230:110–119

Braga AA, López-Greco LS, Santos DC, Fransozo A (2009) Morphological evidence for protandric simultaneous hermaphroditism in the caridean *Exhippolysmata oplophoroides*. J Crust Biol 29: 34–41

Brook HJ, Rawlings TA, Davies RW (1994) Protogynous sex change in the intertidal isopod *Gnorimosphaeroma oregonense* (Crustacea: Isopoda). Biol Bull 187:99–111

Butler TH (1964) Growth, reproduction, and distribution of pandalid shrimps in British Columbia. J Fish Res Board Can 21:1403–1452

Butler TH (1980) Shrimps of the Pacific coast of Canada. Can Bull Fish Aquat Sci 202:1–280

Calado R, Dinis MT (2007) Minimization of precocious sexual phase change during culture of juvenile ornamental shrimps *Lysmata seticaudata* (Decapoda: Hippolytidae). Aquaculture 269: 299–305

Carlisle DB (1959) the sexual biology of *Pandalus borealis* (Crustacea Decapoda). III. The initiation of the female phase. J Mar Biol Assoc UK 38:493–506

Carpenter A (1978) Protandry in the freshwater shrimp, *Paratya curvirostris* (Heller, 1862) (Decapoda: Atyidae), with a review of the phenomenon and its significance in the Decapoda. J Roy Soc N Z 8(4):343–358

Chace FA Jr. (1972) The shrimps of the Smithsonian-Bredin Expeditions with a summary of the West Indian shallow-water species (Crustacea: Decapoda: Natantia). Smithson Contrib Zool, no. 98, 179 pp

Chace FA Jr. (1983) The *Atya*-like shrimps of the Indo-Pacific region (Decapoda: Atyidae). Smithson Contrib Zool, no. 384, 54 pp

Charniaux-Cotton H (1975) Hermaphroditism and gynandromorphism in malacostracan Crustacea. In: Reinboth R (ed) Intersexuality in the animal kingdom. Springer, New York, pp 91–105

Charniaux-Cotton H, Payen G (1985) Sexual differentiation. In: Bliss DE, Mantel LH (eds) The biology of Crustacea, vol 9. Integument, pigments, and hormonal processes. Academic, New York, pp 217–299

Charnov EL (1981) Sex reversal in *Pandalus borealis*: effect of a shrimp fishery? Mar Biol Lett 2: 53–57

Charnov EL (1982) The theory of sex allocation. Princeton University Press, Princeton, NJ

Charnov EL, Hannah RW (2002) Shrimp adjust their sex ratio to fluctuating age distributions. Evol Ecol Res 4:239–246

Charnov EL, Gotshall DW, Robinson JG (1978) Sex ratio: adaptive response to population fluctuations in pandalid shrimps. Science 200:201–206

Chiba S (2007) A review of ecological and evolutionary studies on hermaphroditic decapod crustaceans. Plankton Benthos Res 2:107–119

Cobos V, Diaz V, Raso JEG, Manjon-Cabeza ME (2005) Insights on the female reproductive system in *Hippolyte inermis* (Decapoda, Caridea): is this species really hermaphroditic? Invert Biol 128:310–320

Corey S, Reid DM (1991) Comparative fecundity of decapod crustaceans. I. The fecundity of thirty-three species of nine families of caridean shrimps. Crustaceana 60:270–294

Correa C, Thiel M (2003) Mating systems in caridean shrimp (Decapoda: Caridea) and their evolutionary consequences for sexual dimorphism and reproductive biology. Revista Chilena de Historia Natural 76:187–203

Dohrn PFR (1950) Studi sulla *Lysmata seticaudata* Risso (Hyppolytidae). I. Le condizioni normali della sessualita in natura. Pubblicazioni Stazione Zoologica di Napoli 22:257–272

Dohrn PFR, Holthuis LB (1950) *Lysmata nilita,* a newspecies of prawn (Crustacea Decapoda) from the Western Mediterrean. Pubblicazioni Stazione Zoologica di Napoli 22:340–347

d'Udekem d'Acoz C (2003) *Lysmata seticaudata* (Risso, 1816) and *L. nilita* Dohrn & Holthuis, 1950 are protandrous simultaneous hermaphrodites (Decapoda, Caridea, Hippolytidae). Crustaceana 75:1149–1152

Duffy JE (2007) Ecology and evolution of eusociality in sponge-dwelling shrimps. In: Duffy JE, Thiel M (eds) Evolutionary ecology of social and sexual systems: Crustaceans as model organisms. Oxford University Press, New. York, pp 387–412

Duffy JE, MacDonald KW (1999) Colony structure of the social snapping shrimp *Synalpheus filidigitus* in Belize. J Crust Biol 19:283–292

Espinoza-Fuenzalida NL, Thiel M, Dupre E, Baeza JA (2008) Is *Hippolyte williamsi* gonochoric or hermaphroditic? A multi-approach study and a review of sexual systems in *Hippolyte* shrimps. Mar Biol 155:623–635

Fiedler GC (1998) Functional, simultaneous hermaphroditism in female-phase *Lysmata amboinensis* (Decapoda: Caridea: Hippolytidae). Pac Sci 52:161–169

Fiedler GC, Rhyne AL, Segawa R, Aotsuka T, Schizas NV (2010) The evolution of euhermaphroditism in caridean shrimps: a molecular perspective of sexual systems and systematics. BMC Evol Biol 10:297. https://doi.org/10.1186/1471-2148-10-297

Fisher RA (1930) The genetical theory of natural selection. Oxford University Press, Oxford

Fletcher DJ, Kotter I, Wunsch M, Yasir I (1995) Preliminary observations on the reproductive biology of ornamental cleaner prawns. Int Zool Yb 34:73–77

Fréchette JG, Corrivault GW, Couture R (1970) Hermaphroditisme protérandrique chez une crevette de la famille des crangonidés *Argis dentata* Rathbun. Le Naturaliste Canadien 97: 805–822

Froglia C, Gramitto ME (1981) Critical remarks on the supposed protandric hermaphroditism in *Solenocera membranacea* (Risso) (Crustacea, Penaeidea). Rapports et Procés-verbaux des Réunions Commission Internationale pour L'exploration Scientifique de la Mer Méditerranée 27:211–214

Froglia C, Gramitto ME (1987) Notes on growth and biology of *Solenocera membranacea* (Risso, 1816) in the Central Adriatic Sea (Decapoda: Solenoceridae). Investigación Pesquera (suplemento 1) 51:189–199

Gardner A, Ross L (2011) The evolution of hermaphroditism by an infectious male-derived cell lineage: an inclusive-fitness analysis. Am Nat 178:191–201

Gavio MA, Orensanz JM, Amstrong D (2006) Evaluation of alternative life history hypotheses for the sand shrimp *Crangon franciscorum* (Decapoda: Caridea). J Crust Biol 26:295–307

Gemmell NJ, Todd EV, Goikoetxea A, Ortega-Recalde O, Hore TA (2019) Natural sex change in fish. Curr Top Dev Biol 134:71–116

Gherardi F, Calloni C (1993) Protandrous hermaphroditism in the tropical shrimp *Athanas indicus* (Decapoda: Caridea), a symbiont of sea urchins. J Crust Biol 13:675–689

Ghiselin MT (1969) The evolution of hermaphroditism among animals. Q Rev Biol 44:189–208

Giraldes BW, Macedo TP, Brandão MC, Baeza JA, Freire AS (2018) *Lysmata arvoredensis* nov. sp. a new species of shrimp from the south coast of Brazil with a key to species of *Lysmata* (Caridea: Lysmatidae) recorded in the southwestern Atlantic. PeerJ 6:e5561. https://doi.org/10.7717/peerj.5561

Gould SJ (1989) Wonderful life: the burgess shale and the nature of history. W.W. Norton, New York

Grilli NM, Terossi AM, Mantelatto FL (2014) Sexual system of the freshwater shrimps of the genus *Potimirim* Holthuis (Decapoda : Caridea : Atyidae): is there a pattern in this genus? Mar Freshw Res 65:759–765

Gros P (1979) Biologie de *Processa nouveli* Holthuisi, Al-Adhub et Williamson (Crustacea, Decapoda) dans le secteur nord du Golfe de Gascogne. Annals de l'Institut Océanographique Paris 55:49–70

Hamilton WD (1967) Extraordinary sex ratios. Science 156:477–488

Hannah RW, Jones SA (1991) Fishery induced changes in the population structure of pink shrimp (*Pandalus jordani*). Fish Bull 89:41–51

Hayashi KI (1988) Specific status of *Pandalus gracilis* (Decapoda, Caridea, Pandalidae). Nippon Suisan Gakkaishi 54:71–75

Heegaard P (1967) On behaviour, sex-ratio and growth of *Solenocera membranacea* (Risso) (Decapoda, Penaeidae). Crustaceana 13:227–237

Hobbs HH Jr., Hart CW Jr. (1982) The shrimp genus *Atya* (Decapoda: Atyidae). Smithson Contrib Zool, no. 364, 143 pp

Hoffman DL (1969) The development of the androgenic glands of a protandric shrimp. Biol Bull 137:286–297

Hoffman DL (1972) The development of the ovotestis and copulatory organs in a population of protandric shrimp, *Pandalus platyceros* Brandt from Lopez Sound, Washington. Biol Bull 142: 251–270

Holthuis LB (1954) On a collection of decapod Crustacea from the Republic of El Salvador (Central America). Zoologische Verhandelingen No. 23:1–43

Jägersten G (1936) Ŭber die Geschlechtsverhältnisse und das Wachstum bei *Pandalus*. Arkiv fűr Zoologi 28A(20):1–27

Kagwade PV (1982) The hermaphrodite prawn *Hippolysmata ensirostris* Kemp. Indian J Fish 28: 189–194

Kensley B (1988) New species and records of cave shrimps from the Yucatan peninsula (Decapoda: Agostocarididae and Hippolytidae). J Crust Biol 8:688–699

Kim DH, Choi JH, Kim JN, Cha HK, Oh TY, Kim DS, Han CH (2006) Gonad and androgenic gland development in relation to sexual morphology in *Pandalopsis japonica* Balss, 1914 (Decapoda, Pandalidae). Crustaceana 79:541–554

Komai T (1994) Deep-sea shrimps of the genus *Pandalopsis* (Decapoda: Caridea: Pandalidae) from the Pacific coast of eastern Hokkaido, Japan, with the description of two new species. J Crust Biol 14:538–559

Komai T (1997a) Revision of *Argis dentata* and related species (Decapoda: Caridea: Crangonidae), with description of a new species from the Okhotsk Sea. J Crust Biol 17:135–161

Komai T (1997b) Redescription of a little known pandalid shrimp, *Pandalopsis lamelligera* (Brandt) (Crustacea: Decapoda: Caridea) based upon topotypic material from Shantar Islands, northern Okhotsk Sea. J Nat Hist Mus Inst Chiba 4:139–145

Komai T (1999) A revision of the genus *Pandalus* (Crustacea: Decapoda: Caridea: Pandalidae). J Nat Hist 33:1265–1372

Komai T, Eletskaya M (2008) A new species of the pandalid shrimp genus *Pandalus* (Crustacea: Decapoda: Caridea) from the Sea of Okhotsk off eastern Sakhalin, Russian Far East. Zootaxa 1882:46–56

Komai T, Takeda M (2002) A new deep-water shrimp of the genus *Pandalopsis* (Decapoda, Caridea, Pandalidae) from Sagami Bay, Japan. Bull Natl Sci Mus 28:91–100

Labat JP, Noël PY (1987) Kinetical study of size structure and biological cycle of a Mediterranean population of *Processa edulis* (Decapoda, Caridea). Investigación Pesquera 51(suppl 1): 165–176

Laubenheimer H, Rhyne AL (2008) Experimental confirmation of protandric simultaneous hermaphroditism in a Caridean shrimp outside of the genus *Lysmata*. J Mar Biol Assoc U K 88: 301–305

Limbaugh C, Pederson H, Chace FA Jr (1961) Shrimps that clean fishes. Bull Mar Sci Gulf Caribb 11:237–257

Lin J, Zhang D (2001) Reproduction in a simultaneous hermaphroditic shrimp, *Lysmata wurdemanni*: any two will do? Mar Biol 139:919–922

Losos JB, Hillis DM, Greene HW (2012) Who speaks with forked tongue? Science 338:1428–1429

Machado M, Salti FC, Bertini G, Zara FJ, Negreiros-Fransozo ML (2021) Is *Potimirim potimirim* (Crustacea, Decapoda, Atyidae) a protandric hermaphrodite species? Behavioral and morphological aspects of the reproductive system. Arthropod Struct Dev 63:101060

Malay MCD, Rañises DML (2019) Observations on the habitat, color polymorphism, and sexual system of the semi-terrestrial shrimp *Merguia oligodon* (De Man, 1888) (Decapoda: Caridea: Merguiidae). J Crust Biol 39:764–769

Manjón-Cabeza ME, Cobos V, García Muñoz EJ, García Raso JE (2009) Structure and absolute growth of a population of *Hippolyte inermis* Leach 1815 (Decapoda: Caridea) from *Zostera marina* (L) meadows (Malaga, southern Spain). Sci Mar 73:377–386

Marliave JB, Gergits WF, Aota S (1993) F10 pandalid shrimp: sex determination; DNA and dopamine as indicators of domestication; and outcrossing for wild pigment pattern. Zoo Biol 12:435–451

Martin P, Scholtz G (2012) A case of intersexuality in the parthenogenetic Marmorkrebs (Decapoda: Astacida: Cambaridae). J Crust Biol 32:345–350

Mascetti P, de la Reguera RF, Albornoz L, Oyarzún S, Gorny M, Wehrtmann I (1997) Gonopore development and sex change in the Antarctic shrimp *Chorismus antarcticus* (Caridea: Hippolytidae). Polar Biol 17:384–388

Mistakidis MN (1957) The Biology of *Pandalus montagui* Leach. Fish Investig Ser II 21:1–52

Murray KB, Demko AM, Ward BK, Delillo CA, Balazadeh K, Bourque BD, Rhyne AL (2012) The absence of male features and functionality in the monotypic shrimp genus *Lysmatella*, a simultaneous hermaphrodite. Integr Comp Biol 52:300–E300

Nakashima Y (1987) Reproductive strategies in a partially protandric shrimp, *Athanas kominatoensis* (Decapoda: Alpheidae): Sex change as the best of a bad situation for subordinates. J Ethol 5:145–159

Nakashima Y (1995) Can small shrimps achieve copulation in the presence of larger ones? J Ethol 13:9–16

Noël P (1973) Cycle biologique et inversion sexuelle du Crustacé Décapode Natantia *Processa edulis*. Cahiers Biologie Marine 14:217–227

Noël P (1976) L'évolution des caractères sexuels chez *Processa edulis* (Risso) (Décapode, Natantia). Vie et Milieu 26:65–104

Oh CW, Byun JH, Choi JH, Kim HW (2008) Reproductive biology of *Pandalus gracilis* Stimpson, 1860 (Decapoda, Pandalidae) in the southeastern coastal waters of Korea. Crustaceana 81:797–811

Oliveira MV, Costa-Souza AC, Guimarães FJ, Almeida AO, Baeza JA (2015) Observations on the life history of a rare shrimp, *Salmoneus carvachoi* (Crustacea: Caridea: Alpheidae), a possible simultaneous hermaphrodite, inhabiting the tropical western Atlantic. Mar Biodivers Rec 8 (e141):1–8

Onaga T, Fiedler GC, Baeza JA (2012) Protandric simultaneous hermaphroditism in *Parhippolyte misticia* (Crustacea: Decapoda: Hippolytidae): implications for the evolution of mixed sexual systems in marine shrimps. J Crust Biol 32:383–394

Onishi Y, Chiba S, Goshima S (2001) Reproductive cycle of the protandric shrimp *Pandalus latirostris* in Saroma Lagoon. Benthos Res 56:9–20

Osawa Y, Aoki MN, Thiel M, Bauer RT (2015) Analysis of life-history traits in a sex-changing marine shrimp (Decapoda: Caridea: Rhynchocinetidae). Biol Bull 228:125–136

Pakhomov EA, Kuun P, McQuaid CD (2000) Biology of the bottom-dwelling shrimp *Nauticaris marionis* Bate, 1888 at the sub-Antarctic Prince Edward Islands. Polar Biol 23:522–530

Pérez-Farfante I, Robertson L (1992) Hermaphroditism in the penaeid shrimp *Penaeus vannamei* (Crustacea: Decapoda: Penaeidae). Aquaculture 103:367–376

Policansky D (1982) Sex change in plants and animals. Annu Rev Ecol Syst 13:417–495

Radir PL (1930) An interpretation of the taxonomic status of *Atya bisulcata* Randall and *Ortmannia henshawi* Rathbun. Ann Mag Nat Hist 5:351–354

Reverberi G (1950) La situazione sessuale di *Hippolyte viridis* e le condizioni che la reggono. Bollotino Zoolgico 4-6:91–94

Rhyne AL, Anker A (2007) *Lysmata rafa*, a new species of peppermint shrimp (Crustacea, Caridea, Hippolytidae) from the subtropical western Atlantic. Helgoland Mar Res 61:291–296

Ross RM (1990) The evolution of sex-change mechanisms in fishes. Environ Biol Fish 29:81–93

Sagi A, Snir E, Khalaila I (1997) Sexual differentiation in decapod crustaceans: role of the androgenic gland. Invert Reprod Dev 31:55–61

Schatte J, Saborowski R (2006) Change of external sexual characteristics during consecutive moults in *Crangon crangon* L. Helgoland Mar Res 60:70–73

Simoes F, Yasir I, Jones DA (1998) Reproductive biology of *Lysmata debelius* (Bruce 1983) and *L. amboinensis* (De Man), (Crustacea, Caridea), tropical marine cleaner shrimps important in the sea water trade. Abstract, World Aquaculture Society, Las Vegas, 3 pp

Smith MJ, Williams WD (1982) Taxonomic revision of Australian species of *Atyoida* Randall (Crustacea : Decapoda : Atyidae), with remarks on the taxonomy of the genera *Atyoida* and *Atya* Leach. Aust J Mar Freshwat Res 33:343–361

Spitschakoff T (1912) *Lysmata seticaudata* Risso als Beispiel eines echten Hermaphroditismus bei den Decapoden. Zeitschrift für wissenschaftliche Zoologie 100:190–209

Squires HJ (1992) Recognition of *Pandalus eous* Makarov, 1935, as a Pacific species not a variety of the Atlantic *Pandalus borealis* Krøyer, 1838 (Decapoda, Caridea). Crustaceana 63:257–262

Sukumaran KK (1982) On the gonad of the protandric prawn *Hippolysmata ensirostris* Kemp. Indian J Fish 28:195–198

Suzuki H (1970) Taxonomic review of four alpheid shrimps belonging to the genus *Athanas* with reference to their sexual phenomena. Sci Rep Yokohama Natl Univ Sect 2(17):1–37

Tamura T (1950) On the life history of *Pandalus nipponensis* Yokoya. Bull Jpn Soc Sci Fish 15: 721–724

Terossi M, López Greco LS, Mantelatto FL (2008) *Hippolyte obliquimanus* (Decapoda: Caridea: Hippolytidae): a gonochoric or hermaphroditic shrimp species? Mar Biol 154:127–135

Torti MR, Boschi EE (1973) Nuevos aportos al conociemento de los crustáceos Decápodos Caridea del género *Campylonotus* Bate 1988. Physis Sección A 32:65–84

Tóth E, Bauer RT (2007) Gonopore sexing technique allows determination of sex ratios and helper composition in eusocial shrimps. Mar Biol 151:1875–1886

Tóth E, Bauer RT (2008) *Synalpheus paraneptunus* (Crustacea: Decapoda: Caridea) populations with intersex gonopores: a sexual enigma among sponge-dwelling snapping shrimps. Invert Reprod Dev 51:49–59

Tziouveli V, Smith G (2009) Sexual maturity and environmental sex determination in the white-striped cleaner shrimp *Lysmata amboinensis*. Invert Reprod Dev 51:49–59

Ventura T, Sagi A (2012) The insulin-like androgenic gland hormone in crustaceans: from a single gene silencing to a wide array of sexual manipulation-based biotechnologies. Biotechnol Adv 30:1543–1550

Ventura T, Rosen O, Sagi A (2011a) From the discovery of the crustacean androgenic gland to the insulin-like hormone in six decades. Gen Comp Endocrinol 173:381–388

Ventura T, Manor R, Aflalo ED, Weil S, Khalaila I, Rosen O, Sagi A (2011b) Expression of an androgenic gland-specific insulin-like peptide during the course of prawn sexual and morphotypic differentiation. ISRN Endocrinol 2011:476283

Warner RR (1975) The adaptive significance of sequential hermaphroditism in animals. Am Nat 109:61–82

Warner RR (1978) The evolution of hermaphroditism and unisexuality in aquatic and terrestrial vertebrates. In: Reese ES, Lighter FJ (eds) Contrasts in behavior, vol 3. Wiley, New York, pp 77–101

Warner RR (1984) Mating behavior and hermaphroditism in coral reef fishes. Am Sci 72:128–136

Wirtz P (1997) Crustaceans symbionts of the sea anemone *Telmatactis cricoides* at Madeira and the Canary Islands. J Zool Lond 242:799–811

Wong JYW, Michiels NK (2011) Control of social monogamy through aggression in a hermaphroditic shrimp. Front Zool 8:30

WoRMS Editorial Board (2022) World Register of Marine Species. Available from https://www.marinespecies.org at VLIZ. Accessed 2022-03-05. https://doi.org/10.14284/170

Yaldwyn JC (1960) Crustacea Decapoda Natantia from the Chatham Rise; a deep water bottom fauna from New Zealand. N Z Dept Sci Ind Res Bull 139:13–53

Zhang D, Lin J (2005) Comparative mating success of smaller male-phase and larger male-role euhermaphrodite-phase shrimp, *Lysmata wurdemanni* (Caridea: Hippolytidae). Mar Biol 147: 1387–1392

Zhang D, Sun M, Liu X (2017) Phase-specific expression of an insulin-like androgenic gland factor in a marine shrimp *Lysmata wurdemanni*: Implication for maintaining protandric simultaneous hermaphroditism. PLoS One 12(2):e0172782. https://doi.org/10.1371/journal.pone.0172782

Zupo V (1994) Strategies of sexual inversion in *Hippolyte inermis* Leach (Crustacea, Decapoda) from a Mediterranean seagrass meadow. J Exp Mar Biol Ecol 178:131–145

Zupo V (2000) Effect of microalgal food on the sex reversal of *Hippolyte inermis* (Crustacea : Decapoda). Mar Ecol Prog Ser 201:251–259

Zupo V (2001) Influence of diet on sex differentiation of *Hippolyte inermis* Leach (Decapoda: Natantia) in the field. Hydrobiologia 449:131–140

Zupo V, Messina P (2007) How do dietary diatoms cause the sex reversal of the shrimp *Hippolyte inermis* Leach (Crustacea, Decapoda). Mar Biol 151:907–917

Zupo V, Messina P, Carcaterra A, Aflalo ED, Sagi A (2008) Experimental evidence of a sex reversal process in the shrimp *Hippolyte inermis*. Invert Reprod Dev 52:93–100

Chapter 8
Mating Systems

8.1 Sexual Tactics and Strategies

Males and females of shrimps and other animals use various behaviors to find a suitable mate, to inseminate or become inseminated, and to care for the offspring. A particular set of behavioral tactics and morphological traits compose adaptive reproductive strategies known as mating systems (Emlen and Oring 1977). These tactics include attraction and recognition of a mate, courtship behaviors, weaponry, mechanics of copulation, postmating and paternity assurance behaviors, and parental care of offspring (Emlen 2008; Shuker and Simmons 2014). Mating systems in shrimps, as in other animals, are usually based on the behavior of males (e.g., Correa and Thiel 2003). That is not to minimize the role of females; it is just that males are more often noticeably enthusiastic and eager to mate and to give up their sexual products to a female than the reverse. The male is generally the sex which seeks out the female, tries to convince her to pair up and mate using courtship behaviors. In many species, the male guards and defends the female from other males. On the other hand, females are generally the more reserved and choosy sex when it comes to mate selection. This is because females are the limiting resource in mating systems and males have to compete to obtain them (Darwin 1871). The underlying cause is the difference between sperm and eggs. Sperm cells are small, have little individual energy content, and are thus metabolically easy and cheap to produce. Males can produce many more sperm for the same energy cost of a single female egg cell. The latter must contain ample nutrients and energy-containing material to sustain embryonic development. In shrimps, this is especially true for caridean and stenopodidean shrimps. Females produce relatively large eggs with lots of yolk to support incubation of the embryo for weeks to months before the embryo can hatch out as a larva or juvenile. Furthermore, females of these pleocyemate groups spend considerable time in parental care, i.e., incubation of embryos. They must keep embryos aerated, clean and free from micropredators for long periods.

However, female dendrobranchiates (sergestoids and penaeoids) lack these energetic costs; they don't incubate their embryos (Chap. 6, Reproduction). Their eggs (oocytes) are fertilized by male sperm as the eggs are being spawned directly into the water. The dendrobranchiate female egg is nevertheless much larger than the male sperm cell. The fertilized eggs still need a period of embryonic development in the water before hatching out as a nauplius larva, albeit this period is much shorter (~1 day) than that of incubation in carideans and stenopodideans (weeks, months). Thus, the generalization that sperm can be produced more cheaply and abundantly by males than eggs by females also holds true for sergestoids and penaeoids. Carideans and stenopodidean males can copulate multiple times in a relatively short time period, e.g., a day. This is also true of sicyoniid shrimps (Bauer 1992, 1996a). Other penaeoid and sergestoid shrimp transfer sperm in complex spermatophores, which may take several days to replace (Chap. 6). However, if females of any decapod shrimp are not inseminated and must abort their spawn, it will take them weeks, months, or even 1–2 years to produce another batch, depending on the species (Chap. 9, Life Histories). Females have a lot more to lose than males from an unsuccessful mating.

Thus, females are usually the more selective sex and males are the suitors who must compete among themselves to obtain a mate. In most species, insemination of females, the limiting sexual resource, by males is almost assured, but the reverse is not true for males. The number of males ready to mate is usually greater than the number of receptive females, i.e., the "operational sex ratio" (OSR) is biased towards males. In most animal species, there is a considerable variability in the mating success of males; some or just a few males get most of the matings with females. This is termed "Bateman's Principle," first proposed by geneticist John Bateman's (1948) work on *Drosophila* fruit flies. Males usually have to compete to mate and inseminate females.

Males have evolved various adaptations for just this purpose, a process termed by Charles Darwin as "sexual selection." In his 1871 book on sexual selection, Darwin grouped such male adaptations into two basic categories, intrasexual competition and epigamy (attractiveness to females). In the first, individuals (usually males) battle among themselves, either directly or indirectly, for access to the mate which is the limiting reproductive resource (usually females). Rivalry among males may simply be a race or scramble to be the first to find and copulate with a receptive female. Male shrimps using this tactic are on average smaller or at least no larger than females and similar in relative size of weaponry. The small size may be adaptive in a number of ways: males can be swifter, less visible to predators and devote energy to searching behavior rather than growth to large size or heavy weaponry (e.g., Berglund 1981). Females are larger because fecundity is positively correlated with size: larger females produce more eggs. Alternately, in some shrimp species, males may defend and guard receptive or soon-to-be receptive females, fighting with or fending off other males until the female is ready for copulation. In these species, males are as large as or larger than females, with enlarged heavy chelipeds and/or third maxillipeds or other specialized weaponry such as snapping claws.

Darwin also pointed out that female choice for a suitable mate has led to the evolution of male epigamic characters, which refers to traits such as large size, showy morphology ("ornaments") and/or male courtship behaviors. Familiar examples include secondary sexual characters such as the showy plumage and courtship singing of many male birds compared to the rather dull (to human eyes) female dress; outsized horns of some male cervids (deer, elk, moose) and bovids (antelopes, sheep, cattle); lion manes; horns of rhinoceros beetles and so on. Such epigamic characters usually reflect the genetic quality of the male. In order to grow and produce such exaggerated and energetically expensive features, the individual must have favorable genes, e.g., for obtaining food, healthy growth, parasite resistance, and escape from predators ("good genes" hypothesis: Andersson 1986; Lorch et al. 2003; Emlen et al. 2012). In some species, epigamic traits are so exaggerated that they would seem to be deleterious rather than beneficial to the male's growth and survival: a handicap. Yet, in spite of these impediments, the male appears healthy and strong, of superior genetic quality ("handicap hypothesis," Zahavi 1975; Iwasa and Pomiankowski 1999). Epigamic traits thus serve as signals that females can use to choose a mate who will maintain or increase the genetic quality of its offspring (Hosken and House 2011). Female choice may be obvious or quite subtle, as will be seen later. It should be emphasized again that sex roles are sometimes reversed in various insect, amphibian, fish, and bird species (Darwin 1871; Eens and Pinxten 2000; Liker et al. 2013; Gwynne 2016). This is also true of certain shrimp species, in which females seek out dominant ("superior") males, as shall be seen below.

8.2 Mate Attraction and Recognition

8.2.1 Visual Cues

Before any mating can occur, males and females must be able to recognize when potential mates are presently or will soon be sexually receptive. Given that most mating systems are defined by male behavior, it is usually the male which is doing the searching and judging of receptivity of a potential mate. Vision is usually not very important in sex recognition in shrimps, although in *Alpheus* species, it plays a role in intersexual communication (Hughes 1996). The majority of shrimps occur in turbid waters, low-light or the unlighted habitats of the bathyal and abyssal zones, or are active at night. These shrimps have vision adapted mainly for detection of light and movement (superposition eyes; Brusca et al. 2016). Diurnal crustaceans living in shallow clear-water habitats often have apposition eyes which might allow sharp vision, but such visual or color acuity has not been demonstrated or at least is not common of decapod shrimps (see Chap. 5, Coloration: e.g., Jaafar and Zeng 2012; Caves et al. 2016). Behavioral tests for male or female detection of a receptive partner in a separate clear glass container are negative (e.g., Kamiguchi 1972; Díaz and Thiel 2004; Caskey and Bauer 2005). However, visual detection of movements and general form, for which superposition eyes are adapted, can be important in mate attraction in shrimps which live in aggregations. I have often observed in aquaria as well as in nature that when an individual of such species (e.g., *Heptacarpus sitchensis, H. paludicola; Palaemonetes pugio, Lysmata wurdemanni, Xiphocaris elongata,* various penaeids) finds a bit of food, its jerking and tearing feeding movements quickly attract nearby individuals. This same mode of detection may also be used to recognize a group of males attempting to copulate with a receptive female in the caridean *Rhynchocinetes typus* (Díaz and Thiel 2004). These authors termed this behavior "tumults" which may be a common and important visual cue for mate detection in many species.

8.2.2 Tactile Cues

In closed-thelycum penaeoids and all carideans and stenopodideans, a reproductive female is receptive just after it molts in preparation for mating. For a day or two, her cuticle is soft, which could be an indicator of her reproductive condition. However, in crustaceans it is difficult or impossible to separate touch from chemoreception, as both tactile and chemoreceptors are present on the same appendages (Hallberg and Skog 2011). Thus, the long antennal (and sometimes antennular) antennae, usually the first appendages to make contact with a potential mate, food, or other object, are termed "chemotactile." To try to determine the role of female texture after a molt in sex recognition in *Palaemonetes pugio*, former graduate student Jody Caskey and I (2005) devised some simple experiments in which males were allowed to contact prespawning postmolt (receptive) females, non-reproductive postmolt females, and newly molted males, all of which have soft cuticles. Test ("focal") males only showed interest in the prespawning females after a molt. It seems obvious from these observations that rather than a simple tactile cue, a chemical substance on the surface of the receptive female that is being "tasted" by the male, i.e., a contact pheromone, is involved. Some efforts have been made to identify such a contact sex pheromone which are discussed below. Verification of a receptive female may involve other tactile cues. Observations on some penaeid and sicyoniid shrimps reveal that males follow and prod the female genital area with the rostrum, with probable contact of antennular flagella (Hudinaga 1942; Misamore and Browdy 1996; Bauer 1996a). In the pandalid *Plesionika*, males use the multiarticulate second pereopods to check females for the presence of two large teeth on her 8th thoracic segment (Triay-Portella et al. 2018).

8.2.3 Sex Pheromones

As in many aquatic crustaceans, chemoreception of pheromones emitted by the opposite sex is the most important sense involved in sex attraction and recognition in shrimps (Bauer 2011). Pheromones are chemical compounds secreted by an individual which impact or change the behavior of another individual. They can be categorized as "olfactory" (smell; distance) or "gustatory" (taste, contact) pheromones. Olfactory pheromones are highly soluble and are emitted at very low concentrations into the water by a sender and perceived at a distance by a receiver. Gustatory pheromones are not very soluble and are detected at high concentrations so that the individual must have its receptor appendage very close to or in direct physical contact with the other individual of interest. In shrimps and other crustaceans, olfactory sex pheromones are perceived by the aesthetascs on the inner (medial) antennular flagella, as has been shown by simple ablation experiments (e.g., Ameyaw-Akumfi and Hazlett 1975; Obermeier and Schmitz 2003; Zhang and Lin 2006; Atema and Steinbach 2007) and physiological studies (Derby 1989; Hallberg and Skog 2011; Solari et al. 2017). Contact sex pheromones, on the other hand, are detected by bimodal chemotactile receptors on the long antennal flagella, as well as on other appendages such as the chelipeds and walking legs. The antennal flagella are of special importance because these lengthy and mobile structures are usually the first appendages to contact objects around the shrimp, including other individuals.

8.2.4 Olfactory Pheromones

Biologists suspect olfactory sex pheromones are at work by abrupt changes in male (usually) behavior in the presence of, but not in contact with, a prespawning (parturial) female which has recently molted. Such a female might be introduced into an aquarium with a male or males which then become agitated and begin typical searching behavior without initial physical contact. The male reaction is very much like that when a piece of food is introduced into aquaria with shrimps. They soon show a reaction of awareness (subtle alterations in behavior) followed by active searching behavior. This sort of distance response of males to newly molted parturial (receptive) females has been noted for several caridean species, e.g., *Pandalus danae* (Needler 1931), *P. borealis* (Carlisle 1959); *Heptacarpus sitchensis* and *H. paludicola* (Bauer 1976, 1979), *Lysmata wurdemanni* (Bauer and Holt 1998), *Processa edulis* (Noël 1976), *Crangon crangon* (Boddeke et al. 1991), *Palaemon elegans* (*Leander squilla*; Höglund 1943), *P. serratus* (Forster 1951), *P. pugio* (Bauer and Abdalla 2001), as well as closed-thelycum penaeoids, e.g., *Marsupenaeus japonicus* (Hudinaga 1942) and *Sicyonia dorsalis* (Bauer 1996a).

When water from an aquarium containing a newly molted, prespawning ("parturial") female is transferred into separate aquaria with only males, male searching behavior is stimulated in many species (e.g., Seibt 1973 with *Hymenocera picta*; Kamiguchi 1972 with *Palaemon paucidens*). Individuals of *H. picta* live in male-female pairs in specific shelters (beneath rock, coral rubble etc.) which they defend from other conspecifics. Seibt set up an aquarium with several pairs, each of which had a separate shelter. When one of the females underwent a prespawning (receptive) molt, all the males in the aquarium became active, left their pair partner, and searched for the receptive female (no sexual fidelity there!). On the other hand, Kamiguchi did more formal experiments with *P. paucidens* in which he exposed males to females in different molt and reproductive conditions. As expected, males introduced into a tank containing a parturial (postmolt prespawning) female all responded with searching, precopulatory behaviors, and copulation. Males also responded to postmolt females *without* mature ovaries (not receptive) with searching and a reduced amount of precopulatory behavior but no actual copulations. However, if Kamiguchi introduced at the same time water from a tank in which a parturial female had just molted, male response to non-receptive postmolt females was much greater, with a few copulations. When males were presented to premolt females with ripe ovaries (not yet receptive) there was no response *unless* Kamiguchi introduced "postmolt parturial female water" into the tank. Males searched but did not attempt copulation. The same responses took place using intermolt females carrying embryos. Thus, there is little doubt that some water-borne substance is given off by molting parturial females that stimulates males to sexual behaviors (i.e., a sex pheromone).

Unlike the examples above, in which the olfactory sex pheromone is released by the female at the time of molting, both males and females may release distance pheromones to attract a potential mate before the female molts and are ready to mate (see references in Bauer 1979, Dunham 1988 and Atema and Steinbach 2007). Males may attract premating females to guard and protect them before she is actually ready to mate (e.g., American lobster, *Homarus americanus*: Atema and Cowan 1986). Females may release their sex pheromones to attract a male mating partner which defends her from other males in return for mating privileges (e.g., blue crab *Callinectes sapidus*, Gleeson 1991). In shrimps which live in high densities, females may release a substance which alerts nearby males that the molt is near. In one such species, *Pandalus latirostris*, males can detect a distance pheromone from a prespawning female up to 12 hours before she undergoes the parturial molt (Chiba et al. 2003). In others, the time is shorter. In the penaeid *Marsupenaeus japonicus*, the male followed the female from 3-7 minutes before the mating molt (Hudinaga 1942). On the other hand, the penaeoid *Sicyonia dorsalis* followed both intermolt and premolt parturial females intermittently, touching them with the antennal flagella. However, the following behavior increased dramatically after the female molt. It appears that in this species, the brief following of any female is a testing of female molt status. After the female molts, an olfactory pheromone must be emitted because just the presence of a postmolt receptive female at a distance stimulates and excites the male to start searching.

In some species, males appear to be aware of parturial females only for a few minutes to an hour prior to her molt. Nearby males approach the female, occasionally moving forward to touch her. This vague male "awareness" has been reported in *P. pugio* (Bauer and Abdalla 2001), *Lysmata wurdemanni* (Bauer 2002) and *Rhynchocinetes uritai* (Bauer and Thiel 2011). Although a substance given off by the female is attractive to males, it may not always be adaptive for the female. As a mating molt grows close, the composition of female excretions given off in its urine and gills may inevitably change to a mixture that males have evolved to detect and recognize. The short-term attractiveness suggests that the female is actually avoiding the attraction of males as late as possible before the molt to prevent harassment and injuries from the abundant males around her in these high-density species. Thus, she tries to hide her upcoming molt for as long as possible. She does not need to advertise prior to the upcoming parturial molt. There are sufficient males around that soon one of them will soon detect and recognize her as a newly molted receptive female via olfactory and contact cues.

Mating is not associated with molting in some open-thelycum penaeoid species such as *Litopenaeus vannamei* as well as some solenocerids and benthesicymids (Chap. 6). In these species, females mate whenever ovaries are mature and ready to spawn. In *L. vannamei*, males are attracted to such females shortly before spawning and begin chasing and touching the female prior to mating (Yano et al. 1988; Misamore and Browdy 1996). Upon introduction of water from a tank containing a female near mating into a tank of males, the latter are stimulated to searching and chasing behavior. A water-borne pheromone is indicated by these observations.

How does one go about testing for the presence of an olfactory sex pheromone? As noted above, the detection of such a substance usually starts with anecdotal observations, e.g., males get excited at a distance in the presence of newly molted receptive females in carideans, stenopodideans and closed-thelycum penaeoids or open-thelycum penaeoid females near spawning. Choice experiments with some sort of "olfactometer" in which a male (or female), the "focal" individual, is allowed to choose between two or more treatment individuals, are most common. The usual apparatus is a Y-maze of different designs (e.g., Seibt 1973; Mathews 2003; Díaz and Thiel 2004) (Fig. 8.1 a, b). At upper end of the "Y" are two separate compartments, one with the test animal hypothesized to be attractive to the focal animal and the other with a control. The latter may a different type of individual, a neutral object or nothing (a "blank"). On the single end of the "Y" is another compartment with the focal subject, i.e., the potentially responsive individual. Water from a common tank is introduced or drawn equally through the test and control chambers; the stream from each meet in a common area at the junction of the two arms of the "Y." After allowing the focal animal to acclimate, she is released and is free to move up the stem of the "Y" to the common choice area. It can choose (or not) to enter or stay near to the entrance of the test or control areas. Positions of the test animal and control are varied at random. A simple statistic can analyze the hypothesis that the focal animal is attracted to the test shrimp or not. For example, one might test the hypothesis that males are attracted at a distance to a newly molted parturial female. The control in this case might be something unlikely to be attractive to the focal male, i.e., a newly molted male. intermolt female, an inanimate object or blank. In this case, an intermolt male would be the focal shrimp. Attraction of females to males can also be tested (i.e., the focal shrimp is a female, the test object is a male versus some control shrimp or inanimate object) as was done in Diaz and Thiel (2004) with *Rhynchocinetes typus*.

Other olfactometer designs have been used, depending on the species and circumstances of the investigator. Both anecdotal observations and formal tests for a

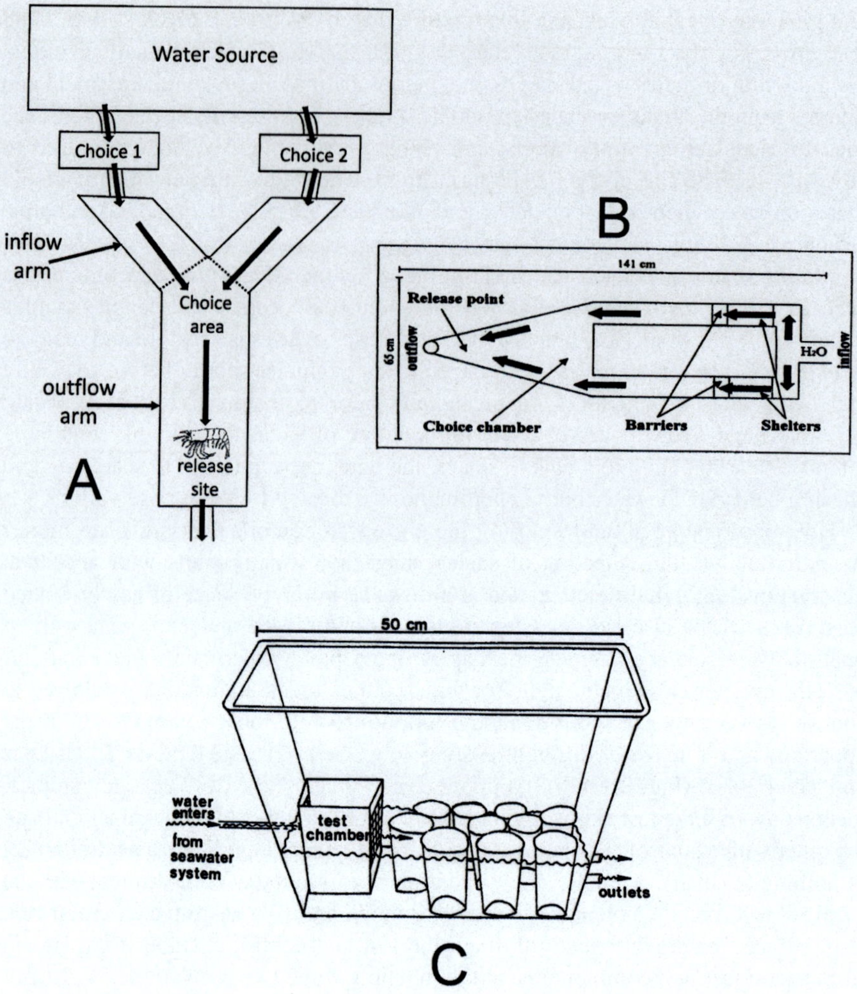

Fig. 8.1 Olfactory chemoreception choice-experiment apparatus. (**a**) Generic Y-maze design. (**b**) Y-maze design used in Díaz and Thiel (2004), courtesy of University of Chicago Press. (**c**) Olfactometer design used in Bauer (1979)

distance sex pheromone should be first attempted with objects or individuals known to be attractive, which can be used as controls. Food is a particularly good substance to use as most shrimps will respond to food introduced at a distance with noticeably increased searching behaviors. I made observations and did experiments on mating behavior and sex attraction in *Heptacarpus paludicola* (Bauer 1979). As in *H. sitchensis* (Bauer 1976), the males quickly begin to frantically search in an aquarium when food is placed in one end of the aquarium. Males of *H. sitchensis* and *H. paludicola* show no apparent reaction when a newly molted parturial (receptive) female is dropped in an aquarium. They seemed to have to touch the female with the antennal flagella to respond to them. Although this suggests a contact rather than an olfactory pheromone, I wanted to formally test for the latter.

Due to the high intrinsic activity levels of these shrimps, I did not think a Y-maze apparatus appropriate. My version of an olfactometer was a system with the test or control in a chamber with flow into and out of it to a larger chamber in which males were collectively housed in individual flow-through cups (Fig. 8.1c). I made repeated observations on activity of the males (swimming, walking) for a set period of time before and after the introduction of the test or control object. I first ran a series of experiments using a natural food item as the test subject and small natural rocks as control. The males downstream responded vigorously to the food but not the rock, as expected. I then ran an experiment using newly molted parturial females as the test and intermolt females with embryos as the control. A significantly greater number of males became active (increased swimming within their cups) with the receptive female but not when exposed to intermolt females. However, the male response was subtle, not frantic. This indicated to me that there was some perception of a substance given off by the sexually mature females. However, I should have also used a newly molted male as a control; perhaps the focal males were responding to a postmolt substance which was not sex-specific. This was pointed out by Dunham (1988) in his review on sex pheromones in crustaceans. He presented in the review the possible sorts of procedural errors that can occur in pheromone testing, such as proper novel stimulus controls, replication and "blind" procedures in which the observer recording responses does not know which stimulus (test individual or object) is being tested.

In nature, what type of locomotor response does a male shrimp use to find a female giving off a distance pheromone (or vice versa)? A simple chemotaxis, which is the following of an increasing concentration gradient of the pheromone directly to the female, probably does not occur in natural environments, at least for shrimps. This might be possible if water was flowing smoothly from the female (laminar flow). But water movement in aquatic habitats is usually turbulent, with swirls and eddies on various scales. The sort of reaction that shrimps seem to use is a chemo-klino-kinesis, like that used in flying insect males downwind of a female emitting sex pheromone (David et al. 1982; Kennedy 1983). The male responds to a chemical substance (the distance pheromone) by an increased rate of turning back and forth across the pheromone plume. When the scent decreases or lost, the male turns back and forth until the scent is picked up or is stronger. The same sort of behavior is probably that used by shrimps in combination with a positive rheotaxis, i.e., it simply orients and swims into the current, as the pheromone source (male or female) must be upstream. However, hydrodynamic analyses of sex pheromone perception have not been done for shrimps as in some other crustaceans (Webster and Weissburg 2009).

8.2.5 *Contact Sex Pheromones*

As noted above, physical contact of a postmolt parturial female by a male, usually with the chemotactile antennal flagella, induces its grasping response which begins copulatory behavior. Textural cues (soft postmolt cuticle) alone do not appear important (Caskey and Bauer 2005). Some substance on the postmolt cuticle stimulates the male to action. In high-density or aggregated ("schooling") species, this may be most important to prevent continual frenzied searching by males and/or a female adaptation to prevent harassment. Contact sex pheromones may also be important in individual recognition of the pair partner in socially monogamous species of carideans and stenopodideans (e.g., Seibt 1973; Johnson 1969, 1977).

8.2.6 *Chemical Composition of Sex Pheromones*

Given the importance of sex pheromones to crustacean mating systems, their evolutionary consequences, and possible applied value to aquaculture and fisheries, their chemical makeup is of great interest and importance (Bauer 2011; Karplus and Barki 2019). Such knowledge may be helpful in development of baits for fisheries, manipulation of matings in aquaculture and control of man-made compounds that might mimic and disrupt crustacean mating behavior (Bauer 2004). Insect pheromones were first identified in 1959, and now the composition of many such compounds is known (Yew and Chung 2015). This knowledge is very important in control of insect pests, where sex pheromones are routinely used to disrupt mating systems or to attract and then kill agricultural pests and disease vectors. Insect pheromones are often produced by specific glands in which the compounds of interest can be harvested much more easily than in aquatic crustaceans. Classic work on pheromones in lobsters, crayfishes, and other larger decapods shows that the primary source of their sex and other olfactory pheromones is the urine. Metabolites of various physiological processes, e.g., ovarian maturation and molting, are excreted in the urine, serving as possible chemical signals (e.g., Atema and Cowan 1986; Atema and Steinbach 2007; Hardege and Terschak 2011). The openings of the excretory organs (antennal glands) are situated in the path of the outgoing respiratory stream, so possible pheromones within the urine flow forward from the animal, including metabolites excreted through the gills (Herberholz and Schmitz 2001).

Olfactory pheromones are much diluted in the aquatic environment, which contains a bewildering number of chemical compounds, making extraction and identification of water-soluble (olfactory) pheromones quite difficult. In spite of these problems, there has been some progress with a few decapod species, including the shrimp *Lysmata wurdemanni* (Zhang et al. 2010a). Like many aggregated species, males of this species are stimulated to searching behaviors by molt water from newly molted parturial females. These investigators filtered and analyzed female molt water using high-pressure liquid chromatography (HPLC) and identified a small polar compound similar to uridine-di-phosphate (UDP), which has been confirmed as an olfactory sex pheromone in a number of crab species (Bublitz et al. 2008). In bioassays, males of *L. wurdemanni* reacted to filtrates containing the UDP-type compound but not to other filtrates without it or to control seawater. They did not respond to seawater solutions containing added UDP only, suggesting that perhaps a "pheromone blend" of UDP and other metabolites may be necessary to stimulate males.

Many contact sex pheromones (cuticular hydrocarbons) have been isolated and identified in insects (Howard and Blomquist 2005) but this is not the case in crustaceans. We have seen that males of many shrimp species become intensely stimulated to grasp and begin copulatory behavior with a postmolt parturial female upon touching her with his long chemotactile antennal flagella. Burkenroad (1947) made similar observations on *Palaemon* (*Palaemonetes*) *vulgaris* and suggested that males were using a "non-diffusible coating" on the female surface to recognize her. His observations on *P. vulgaris* and those of my lab on *P. pugio* showed that the male contact response to postmolt parturial females steadily declined with time after the molt; after 8 hours, the unmated female was no longer attractive (Caskey and Bauer 2005). In other shrimps, this period of attractiveness may last about 1 day (Bauer 1976; Díaz and Thiel 2004).

As the newly molted female cuticle hardens after the molt, simple textural cues (soft postmolt cuticle) are a possible signal to males that parturial females have just molted. However, Jody Caskey showed with *P. pugio* that such textural cues can be ruled out (Caskey and Bauer 2005). We (Caskey et al. 2009a) began work on a possible contact sex pheromone in this species by emulating insect biologists. We made whole-body extractions of postmolt parturial females as well as of postmolt and intermolt males with appropriate solvents. Gas chromatography-mass spectrometry (GC-MS) was used to analyze these extractions for cuticular hydrocarbons. Comparisons were made among postmolt parturial (sexually receptive) females, postmolt nonparturial females, postmolt males and intermolt females. We found a variety of hydrocarbons on the exoskeleton, as expected, as well as other compounds of possible interest. Significant differences among the various treatments were found, and a few hydrocarbons were unique to postmolt parturial females, which was encouraging. However, our bioassay (male response to sponge-tipped rod with extracts from various types of individuals) failed to evoke any response.

Zhang et al. (2011) used similar extraction techniques and treatments with an unrelated species, *Lysmata wurdemanni,* and also found differences in chemical composition according to sex and molt condition. But in this case, their bioassay was successful: males did respond by grasping polyethylene tubes coated with extracts of newly molted parturial female cuticles but not those of newly molted males. Furthermore, an important component of the cuticular hydrocarbon mix, Z-9-octadecenamide (Z-9-O), of newly molted parturial females by itself caused significant bioassay responses from males, as did Z-9-O blends which included other hydrocarbon components from the cuticle of such females.

Another possible candidate for a contact pheromone in shrimps was suggested by previous work on marine copepod crustaceans and other invertebrates in which surface glycoproteins serve in sex recognition (e.g., Kelly and Snell 1998). Carbohydrate components of such glycoproteins may act as signals in a contact pheromone. Chitin, the basic compound composing the exoskeleton of crustaceans and other arthropods, is the polysaccharide n-acetyl glucosamine. A derivative or associated residue present on the newly molted female exoskeleton, or perhaps secreted by the tegumental glands onto the exoskeleton, varying with sexual condition might operate as a contact sex pheromone. Jody Caskey and I, together with several colleagues, investigated this possibility in *P. pugio* (Caskey et al. 2009b). We used one possibility, glucosamine, as a test compound. In a copepod species, Kelly and Snell (1998) had found that lectins, a type of glycoprotein, and glucosamine (an exoskeletal component), both can bind to receptors on the long male antennules of copepods, blocking male receptors from perceiving female glycoprotein sex pheromone upon touching her. We tried this approach with mating experiments in which two males and a recently molted parturial female (< 4hr postmolt) were kept in aquaria to which either glucosamine or glucose (control) had been added, with a series of control runs with ordinary seawater. Behavior was recorded with video surveillance equipment until the parturial female molted. Frequency of copulations in the glucose-added and control seawater without any additions was the same. However, in the glucosamine-treated tanks, copulations were highly reduced. Our interpretation is that glucosamine bound with male chemoreceptors on its long antennal flagella and so blocked the males from perceiving the natural glucosamine-like substance (contact sex pheromone) on the female itself. Other observations were made using electrophysiological techniques that showed neurons in the excised male antennae responded to both glucosamine and cuticles of postmolt, parturial (sexually receptive) females, but not to glucose or to contact with cuticles of postmolt, nonparturial females. Our studies indicated that the compound associated with mate recognition in *P. pugio* is likely to be a glucosamine-containing glycoprotein.

Zhang et al. (2010b) again emulated our approach but with *L. wurdemanni*. Their results indicated that glycoproteins are not the contact sex pheromones involved in mating. Their results then stimulated their next work concluding that cuticular hydrocarbon, not glucosamine, are involved in mate recognition. The different results from studies on *P. pugio* and the unrelated *L. wurdemanni* might be due to interspecific differences in contact sex pheromone or different methodologies or techniques. Clearly, much more work on a variety of different species across shrimp taxa is required to resolve questions about shrimp contact sex pheromones.

8.3 Shrimp Mating Systems

Males and females use a variety of behaviors (tactics) to meet and successfully mate with a member of the other sex. The specific set or combination of behaviors can be grouped into a few basic strategies (mating systems). Wickler and Seibt (1981) proposed three common mating systems in Crustacea: (1) "pure searching" in which highly mobile males make frequent contact with females, seizing and mating with one if it is already receptive; the male then leaves to look for another female without guarding her after copulation; (2) Temporary mate (female) guarding, in which a male guards and defends the female for a brief period before mating. However, as we shall see below, either the male or female may seek out a mating partner; (3) monogamy (pair living) in which a male and female live together over at least two or to several mating and spawning events during a breeding season or longer.

The mating system which evolves in a species is a result of many factors, e.g., ancestry, distribution and abundance of individuals in the habitat, resource needs (i.e., food, shelter), mobility, ovarian and spawning cycles, and reproductive morphology. Although the details of these factors are important, Wickler and Seibt summarized the basic factors most simply as (1) the frequency of contact between males and females and (2) the "interspawn interval." These two variables determine the operational sex ratio (OSR), i.e., the number of males to the number of females receptive to mating. The frequency of contact (# females per day) depends on the abundance of males and females as well as their mobility. The latter is a combination of the locomotory ability, usually quite good in shrimps, but also the availability of receptive females and the predation pressure of the habitat. Shrimps may be quite swift swimmers and walkers but there may be lots of predators ready to gobble them up. They may be forced to hide much of time under and in crevices of rocky, coral, or algal substrates, venturing out only during limited periods of time, thus limiting contacts with females.

Receptivity to mating is mainly a female issue as they are the limiting resource in reproduction, as explained earlier. During the breeding season, sexually mature males are almost always capable and eager to mate except just before or after a molt. Wickler and Seibt's model assumes that female receptivity and mating occur just before spawning, thus their use of "interspawn interval." This is certainly the case in carideans and stenopodideans. Females are receptive to mating only within hours to a day or so after a parturial (prespawning) molt, and spawn shortly thereafter (hours to ~ a day in most species). In some open-thelycum penaeoids, females mate just before a spawn, when the ovaries are mature although a molt is not involved. The "interval between spawns" is thus a good estimator of female receptivity in these three groups. However, in closed-thelycum penaeoids, reproductive females are also interested in mating just after a molt, but it is not a prespawning molt. They spawn during the intermolt period, often more than once with the sperm stored from the last mating. In luciferid sergestoids and some other open-thelycum penaeoids, molting and mating occur midway between spawns.. As females of sergestid species have seminal receptacles and store sperm (Pérez Farfante and Kensley 1997), they may have a mate, molt, and intermolt spawn similar to luciferids or, alternately, their pattern may be similar to closed-thelycum penaeoids. The Wickler and Seibt model covers all shrimp groups if one replaces "interspawn interval" with "interval between periods of female receptivity."

As in all models, Wickler and Seibt made certain simplifying assumptions: (1) females have several successive breeding cycles at regular intervals, (2) females do not all become receptive at the same time (synchronously), (3) there is only one

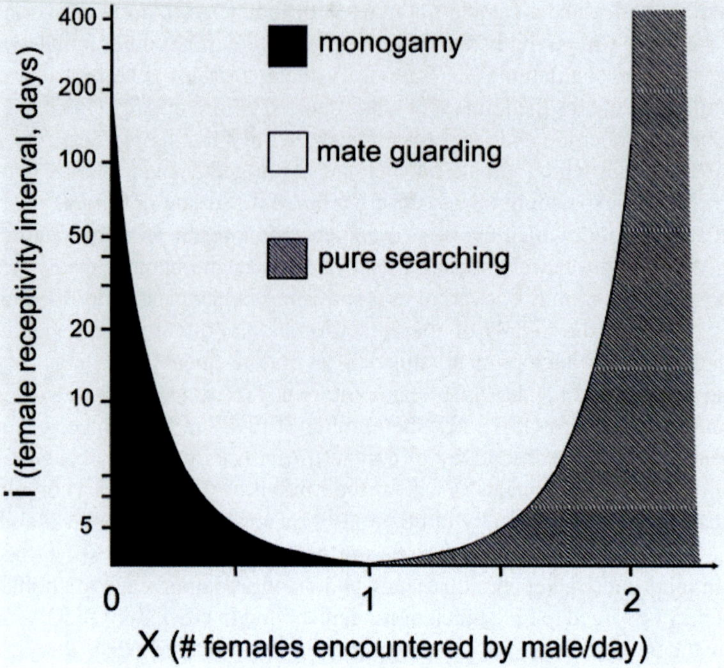

Fig. 8.2 The effects on mating system (pure searching, temporary mate guarding, monogamy) in decapod shrimps of (1) the frequency of male contacts with females per day and (2) female receptivity interval (days). Modified from Wickler and Seibt (1981)

copulation per spawn, and (4) all spawns have about the same number of eggs. The first assumption generally applies to most decapod shrimps, even those high latitude carideans which spawn only once or twice per year but have longer life spans than lower latitude species. Female receptivity does occur over regular intervals (6-12 months) rather than weeks to a few months usually found in mid- to lower latitude species. Assumption (2) is certainly a good one; I know of no shrimp species in which all or most females in a population molt and become receptive at about the same time. The single copulation assumption (3) is reasonable for males of many species, but does not always hold, as we will see later. The assumption (4) of a similar number of eggs per spawn for a species is reasonable but does vary with female size.

Using "*X*" (the frequency of daily male contacts with a female) and "*i*" (female receptivity interval) as the two variables, the Wickler and Seibt model makes good predictions about the mating tactics that will evolve in a shrimp species. Using their model, monogamy evolves when $X < 2/i$, mate guarding when $2/i < X < 2 - 2/i$" and pure searching when $X > 2-2/i$ (see their paper for the details). These predictions are displayed graphically in Fig. 8.2 and examples are shown in Table 8.1. It can be seen that pure searching is favored when male-female contacts (*X*) are as low as 1.6/ day when the female frequency interval "*i*" is short, e.g., 5 days. I use 5 days because

Table 8.1 Wickler and Seibt (1981) model predicting crustacean mating systems depending on frequency of male-female contacts and female receptivity interval. Below, I have calculated the predicted values of "X" (average number of females contacted by a male per day) given different values of "i" (female receptivity interval in days) for the three mating systems considered by Wickler and Seibt

i = female receptivity frequency	Monogamy: $X < 2/i$	Mate Guarding: $2/i < X < 2 - 2/I$	Pure searching: $X > 2 - 2/i$
5	0.4	0.4 < x < 1.6	>1.6
10	0.2	0.2 < x < 1.8	>1.8
15	0.133	0.133 < x < 1.867	>1.867
30	0.067	0.067 < x < 1.933	>1.933
180	0.011	0.011 < x < 1.989	>1.989
360	0.006	0.006 < x < 1.994	>1.994

this is the shortest interval that I have found for a decapod shrimp in the literature (the interspawn interval of the caridean, *Hippolyte obliquimanus,* formerly *H. curacaoensis*) (Bauer 1989). At this interval, the chance that any female is receptive is about 40% (2/*i*), so a male has a good chance of finding one in this condition by pure searching. In aggregated and schooling species, *X* (# male contacts with females per day) is likely much greater than 1.6. When population abundance is quite low and/or mobility of mating partners is limited by predation pressure, monogamy is favored, even at low "*i.*" For example, if *X* is 0.4 or less, it is better for a male to stay paired up even if "*i*" is as little as 5 days (a rare situation—it is usually longer). Unless cooperation between male and female in activities other than mating is involved, this kind of monogamy is just very extended mate guarding. A lengthy search for a mate is dangerous because it exposes the male to predation pressure. Temporary mate guarding, in which the male searches out a female, guards her until she is receptive, and then leaves to search for another female, is favored, according to the model, when *X* is between 0.4 and 1.6 when females are receptive frequently. At longer interspawn intervals, the lower *X* decreases and the higher one increases (Table 8.1).

My 2004 book on carideans accepted the three Wickler and Seibt categories of mating systems. I did not come across Correa and Thiel's 2003 paper on mating systems until after my book was in print. Their insightful analysis of mating systems revealed at least one other mating system, termed "neighborhoods of dominance." Both they and Asakura (2009) provide useful lists of species by mating category. However, the basic factors favoring one system or another, i.e., encounter rates between males and females and availability of receptive females, apply to all the mating systems. Sexual dimorphism in body size and weaponry are characteristics that co-evolve along with the mating system and are important clues for making a hypothesis about the mating system of a species. Other indicators are population abundance (density) and social organization. This sort of information can be obtained from field collections and observations. Just as important is to observe and study, when possible, is the behavior and interaction of individuals in a laboratory "mesocosm," i.e., an aquarium or water-table system large enough and

Table 8.2 Demographic, morphological, and behavioral characteristics of mating systems in major taxa of decapod shrimps

	Mating systems of decapod shrimp taxa			
Characteristics	Pure searching (Sergestoidea, Penaeoidea, Caridea)	Search and attend (Caridea)	Neighborhoods of dominance (Caridea)	Monogamy (Caridea, Stenopodidea)
Encounter Rates	High	Low	Moderate to high	Very low
Sexual Dimorphism in Body Size	High (female >> male)	Moderate (male > female)	Low (female >> male) to high (male >> female) according to male morphotype	Low
Sexual Dimorphism in Weaponry (major chelipeds or third maxillipeds)	Low	Moderate (male > female)	Low to high according to male morphotype	Low
Sex Which Searches For Opposite Sex	Male	Male	Small male for females; females for large dominant males	Male
Duration of Pair Bond	Very brief	Brief (hours to a few days)	Very brief or brief (hours to a few days) according to male morphotype	Continuous (duration of breeding season or longer)
Male Aggressiveness	Low	High with intruders, minimal with current female partner	Low or high according to male morphotype	High for intruders, minimal with female partner
Male Postcopulatory Guarding of Female	None	Brief (hours?)	None or brief (hours to a few days), according to male morphotype	Continuous

adequately equipped to simulate the natural environment of the species under study. Below, the basic mating systems proposed for decapod shrimps are presented and discussed. Characteristics of these mating systems are given in Table 8.2.

8.3.1 Pure Searching

A majority of decapod shrimps belong to this category, also termed "promiscuous" or polygamous because males might mate with multiply females (polygyny) but females mate with more than one male during a breeding season or even during one mating event (polyandry or polygynandry). Such species can be recognized by their (1) sexual dimorphism in body size, with males on average smaller, often much smaller, than females and (2) similarity in relative size and morphology of their weaponry (major chelipeds and/or third maxillipeds) between the sexes. No large males with oversized (hypertrophied) heavy weaponry in these species! Small male size is adaptive because males can devote less into growth and more energy to searching than females. Their smaller size and often less developed coloration are less noticeable to predators, and smaller size gives them more agility in escaping from them. Larger size is favored in females because fecundity (egg/embryo number) increases with size (Chap. 9). Many caridean and all dendrobranchiate (penaeoid and sergestoid) species possess these characteristics. Most protandric and protandric simultaneous hermaphrodites (PSH), by definition, have these traits as well (Chap. 7). A few PSH exceptions include *Lysmata wurdemanni*, in which some male-phase individuals may become as large as female-phase ones (Bauer 2002), and especially the few monogamous, pair-living *Lysmata* species (see below). But even in these exceptional PSH species, all individuals first develop in the male-phase (Chap. 7). As far as is known, there are no pure-searching species in the Stenopodidea, which are occur in male–female pairs (Goy 2010; Saito and Koya 2001) with the possible exception of the single species of Maxillocarididae, in which the few specimens collected by Alvarez et al. (2006) were all female.

Highly mobile benthic or benthopelagic species from homogeneous soft (mud/sand) bottoms often show the sexual dimorphism of pure-searching species. This group includes commercially important carideans (e.g., pandalids: Butler 1980; Bergström 2000; crangonids, Boddeke et al. 1991) and penaeoids (e.g., Omori 1974; Dall et al. 1990). Another homogeneous habitat is the pelagic realm, in which the shrimp species there have the same type of sexual dimorphism of pure searching ones, i.e., smaller to somewhat smaller male and larger female size, as well as relatively similar weaponry, e.g., sergestids (Omori 1974; Omori and Gluck 1979; Xiao and Greenwood 1993). Many pelagic shrimps live in swarms or shoals, sometimes permanent or temporary during the reproductive season (Omori 1974). The individuals of most aggregated species in these habitats are highly mobile and on the move. In addition, there is little or no structure in soft-bottom and pelagic habitats that might serve as shelters or refuges in which males could defend a female.

Although pure-searching species which occur on hard-bottom habitats are presented with plenty of potential shelters for a mating pair to take refuge, the high mobility and relatively high density of such species selects for the evolution of pure searching. Bathypelagic species, which typically occur in low densities, must use pure-searching to encounter a mate. However, some may form small temporary mating swarms (Omori 1974). Besides pheromone cues, bathypelagic and benthopelagic species equipped with photophores may use light cues to help with the search for a mate.

Laboratory observations help to confirm a hypothesis of pure searching in a species. Males and females wander and swim about in aquaria, or aggregate under natural or artificial shelters or within soft substrata provided (e.g., sand, small gravel). Male and female pairs do not form except for brief mating encounters;

males do not guard or defend females. Colleagues and I have made such observations on carideans, e.g., *Heptacarpus sitchensis* and *H. paludicola* (Bauer 1976, 1979), *Macrobrachium ohione* (pers. obs.), *Lysmata wurdemanni* (Bauer and Holt 1998) and *Rhynchocinetes uritai* (Bauer and Thiel 2011), as well penaeoids such as *Rimapenaeus similis*, *R. constrictus* (pers. obs. author) and *Sicyonia* species (Bauer 1992, 1996a). Similar observations by other authors include the carideans *Pandalus danae* (Needler 1931), *Processa edulis* (Noël 1976), *Crangon crangon* (Boddeke et al. 1991), *Plesionika* species (Triay-Portella et al. 2018), the penaeids *Litopenaeus vannamei* (Yano et al. 1988; Misamore and Browdy 1996) and *Marsupenaeus japonicus* (Hudinaga 1942). Other recent studies suggesting or confirming pure-searching caridean species include *Hippolyte williamsi* (Espinoza et al. 2008), *H. obliquimanus* (Duarte and Flores 2017), *Thor amboinensis* (Baeza and Piantoni 2010), *Macrobrachium pantanalense* (Hayd and Anger 2013), *Ambidexter symmetricus* (Rasch and Bauer 2016), *Acanthephyra pelagica* (Jorquera et al. 2016), and *Ancylomenes pedersoni* (Gilpin and Chadwick 2017).

Male competition for a receptive female depends on the male being the first to find and copulate with the female. One mating tactic involves male perception of a female distance pheromone which stimulates searching behavior. As noted earlier, distance pheromones are emitted by females of many species after a parturial molt (prespawning carideans and stenopodideans; closed-thelycum and/or seminal receptacle species, i.e., many penaeoids, probably sergestoids) or without a female molt in some open-thelycum penaeoids. Males are excited into searching until antennal or other contact chemoreception confirms identity of a receptive female. The first (quickest) male to arrive when the female is receptive usually obtains the mating, although there may be a scramble among nearby males, pushing and shoving into position. In others, in which a female may accept several copulations from different males, males take turns in precopulatory following and then copulating (*Marsupenaeus japonicus*, Hudinaga 1942; *Sicyonia dorsalis*, Bauer 1996a).

Chiba et al. (2003) tethered recently molted parturial females of *Pandalus latirostris* in their habitat (seagrass beds) and made snorkeling observations on them. Males frequently approached and attempted copulations with the tethered female. Other males did not approach while one male was mounting the female. Tethered females were eventually mated by several males. In laboratory competition experiments with males of varying sizes, males began to chase and mount premolt parturial females about 12 hours *before* the molt. Larger males mounted longer and often drove off other males attempting to mount the female. Females copulated with different males but larger males were most often first to copulate with her. Chiba et al. observed newly mated females quickly (after 15 minutes) rejected male advances, suggesting that unrestrained females might not mate with multiple males in a natural situation. However, larger size did make some difference in male reproductive success, so this might be the case in other pure-searching females. However, mounting behavior prior to mating was so brief that I would not compare it with the female-defense behavior of males in mate-guarding mating systems.

The western tropical Atlantic freshwater stream-dweller *Atya innocuous* is an aggregated species, often occurring in groups of several individuals on or under

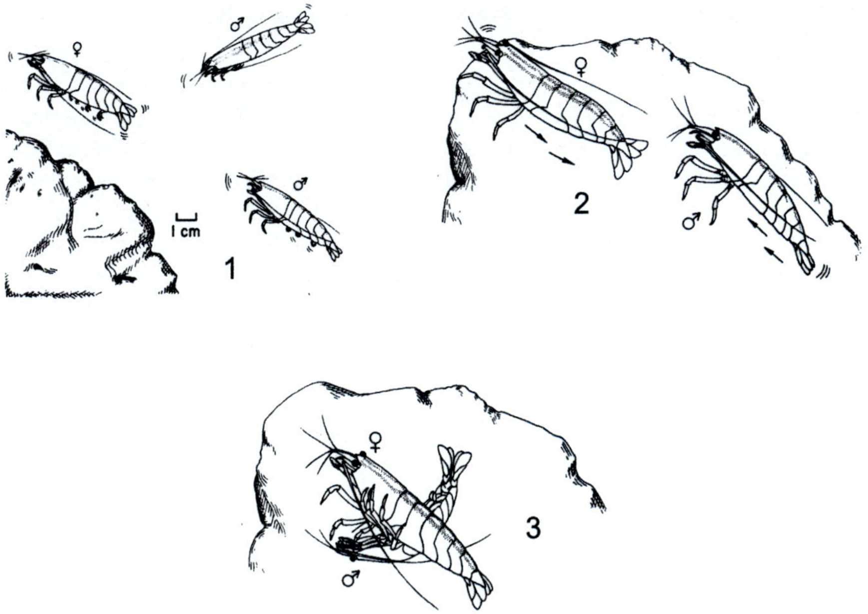

Fig. 8.3 Mating behavior in the pure-searching caridean *Atya innocuous*: 1. Two males pursue a receptive (newly molted, parturial) female; the female allows one of the males to approach her; he will contact and investigate her with both antennae and mutual pereopod touching. 2. When the male of choice is behind the female, she will move backward as he moves forward so that he can climb on top of her. 3. The male then dips below and perpendicular to the female, opposing their genital areas; sperm transfer takes place. Afterwards, the two will separate, and the female will reject any further male contacts. From Felgenhauer and Abele (1982)

rocks and large boulders (Chace Jr. and Hobbs Jr. 1969; Fryer 1977; pers. obs. in Puerto Rico). As the chelipeds are adapted for filter or brush-feeding (Chap. 3), the weaponry of both sexes is weak. Unlike most pure-searching species, *Atya* males are on average larger than females. Precopulatory behavior is somewhat elaborate and extended (Fig. 8.3) (Felgenhauer and Abele 1982). When a prespawning female molts, males in the same area quickly detect her and begin chasing her, up to 30 minutes or so (Fig. 8.3, part 1). Various males, one at a time, circle, touch and attempt to climb on her until one apparently satisfies her criteria, whatever they may be. After making antennal contact and mutual pereopod touching, the male moves behind her. She signals her assent for mating by slowing moving backwards as he moves forward (Fig. 8.3, part 2) so that she is beneath him. He then does the usual dip below her at right angles so that their genital areas meet and spermatophore transfer occurs (Fig. 8.3, part 3). After the copulation, she rejects all further male advances. Male-male competition and female choice are obvious in this species.

8.3.2 Temporary Mate (Female) Guarding Systems

In these mating systems, a male defends and protects a female which is either approaching a prespawning mating molt or one that has recently undergone that molt. Guarding of the female may last hours to days. Correa and Thiel (2003) recognized two variations of temporary mate guarding. In the first, termed a "search and attend" mating system, the male seeks out a parturial female soon before she molts and spawns. In the second "neighborhoods of dominance" system, it is the *female* who searches for a large dominant male equipped with hypertrophied weaponry. After pairing occurs, the male guards her soon before or just after her parturial molt. Smaller subordinate males may use simple pure-searching tactics to intercept and quickly mate with a newly molted female before or even when the dominant male is guarding her.

8.3.3 Search and Attend

This is a mating system characteristic of shrimp species in which reproductive females live in a refuge or shelter of some type. Individuals of such species do not live at high density, so contacts between individuals are not common ("low encounter rate"). Many of these species are symbiotic with larger invertebrates such as sea urchins, sea anemones and bivalves, which the shrimp uses as a shelter (see Chap. 10 Symbiosis). Although females are generally solitary, during the breeding season, they are often found paired with a reproductive male. The males hunt through the habitat for females, especially prespawning females near a mating molt. Their searching is probably aided by olfactory pheromones given off by the female. The guarding male shields the female from other searching males to ensure that he will be the one to inseminate her. After the female molts, he mates with her, and then soon abandons her to search for another premolt parturial female. Correlates (indicators) of the "search-and-attend" mating system have been nicely summarized in some of J. Antonio Baeza and colleagues' studies on symbiotic species (e.g., Baeza et al. 2011; Baeza and Díaz-Valdéz 2011) (Table 8.2). One is a positive correlation between host or refuge size and body size in females but not males. Females stay with the same host and grow with it or seek out larger refuges as they age and grow, but male partners pair with various female partners in different locations. The pairing of male and female doesn't last long. The result is a frequent mismatch between male and female size. A third correlate of this system is that there is often an overall sexual dimorphism in body size and weaponry. On average, reproductive males are somewhat larger than females and with better developed weaponry compared to females of similar size. However, as we will see below, not every species with this mating system exhibits all of these characteristics. As pointed out by Correa and Thiel (2003), search-and-attend mating systems have elements of both pure-searching (smaller males) and "neighborhoods of dominance" systems (see below) in which large males with hypertrophied weaponry are more successful in mating with and inseminating females.

A classic case of a shrimp with a search-and-attend mating system is the sea urchin associate, *Arete dorsalis* (formerly *Athanas kominatoensis*) (Nakashima 1987). This species is a protandric hermaphrodite (Chap. 7), so that all individuals develop first in a functional male-phase ("small males"). Some will go on to become females, others will grow larger and remain in the male-phase condition ("large males"). Most inhabited urchins have only a single shrimp but, during the breeding season, the number of hosts with male-female pairs increases. In most pairs, the female is in prespawning condition nearing a mating molt. Although small males can cohabit and mate with such a female, more large males do so. The evidence indicates that the large males, which have relatively more robust and toothy major chelae than females and small males, wander among urchins in search of a prespawning female. As males are paired with prespawning females but absent from urchins with postspawning (ovigerous) females, males must have some way of detecting female reproductive condition, probably via female distance and/or contact pheromones. If a male is already present, the intruding male will fight with and possibly evict the resident male, especially if the latter is smaller. The more effective weaponry and size of the larger males give them fighting superiority over smaller ones, as Nakashima's laboratory observations showed.

The mating system of a species may vary among populations of the same species. The sea urchin ectosymbiont *Gnathophylloides mineri* appears to have a search-and-attend mating system, (termed "sequential monogamy" by Maciá and Robinson 2012)) in Jamaica but is polygynous (pure searching) in the Virgin Islands. The authors proposed that the higher shrimp density on the host and smaller body size in the Virgin Island population select for pure searching, while the opposite has selected for a search-and-attend mating system in the Jamaican population. These results show that mating system may depend on a number of demographic and ecological factors, e.g., population density, shelter or host type and size, and predation among others (Baeza and Thiel 2007).

A more recent study on a search-and-attend" species is that on the pontoniine *Ascidonia flavomaculata*, a symbiont living in the branchial chamber of the ascidian ("sea squirt") *Ascidia mentula* (Baeza and Díaz-Valdéz 2011). Most individuals collected were solitary within the ascidian, but some hosts contained heterosexual pairs. Same-sex pairs were not encountered. Although there was no pronounced sexual dimorphism in body size or cheliped weaponry, the males which paired with females were larger than unpaired males, indicating greater competitive ability over smaller males. A positive correlation between female size and host (refuge) size was higher than that of males, but that between male and female size within pairs, a trait of monogamous species, was not significant--males were not long-term partners. This indicates that once a female finds an appropriate ascidian host, she stays and grows with her host, but males search from host to host looking for a reproductive female of any size, sequestering her from other males until she is receptive and mates with him, leaving soon after.

Another pontoniine, *Pontonia mexicana*, which lives within the pen-shell *Pinna carnea*, at first glance appeared to Baeza et al. (2011) to be monogamous—heterosexual pairs are more frequent than solitary individuals within these bivalves. However, a closer look shows that most females in the pairs had mature ovaries and thus were near to the mating-spawning molt. This suggests that males must switch hosts, abandoning the female after mating and searching for another, one near a mating molt. These are key traits of "search and attend." Sexual dimorphism was noted, i.e., relative cheliped and body size were slightly but significantly greater in males than females. Importantly though, as in the previous example, males paired with females were larger in body size than solitary males, indicative of large-male competitive superiority. Other traits indicating a "search and attend" were that body size of paired shrimps was poorly correlated and the relationship between host and shrimp body size was weak, all indicating that mating partners were not permanent and that males were moving among hosts-- perhaps females also.

As might be expected, given the variability of nature, there are species which do not fit well in one mating system or the other. In another sea urchin associate, *A. indicus*, Gherardi and Calloni (1993) found a mating system intermediate between a "pure searching" and a "search and attend." Unlike a pure-search system, males can become as large as females and fighting between individuals, especially between males, is fierce. On the other hand, urchins were never inhabited by a male–female pair, and mating encounters observed in the laboratory were brief, like those of pure-search species. Males search among urchins looking for a receptive female but do not stay with her unless she is ready to mate. If two males come across a receptive female at the same time, they may fight over her, with the larger male the winner.

8.3.4 Neighborhoods of Dominance (NOD)

In this mating system, proposed by Christy (1987) for brachyuran crabs and recognized by Correa and Thiel (2003) for shrimp species, males employ alternative mating tactics in the course of their ontogeny. Small males use pure-searching tactics for either finding unattached receptive females or sneak copulations from a female guarded by a distracted or harried dominant male. Large dominant males temporarily guard premating females within small territories which they control, e.g., a short-term mating nest on soft bottoms or under/around rock ledges and boulders in rocky habitats. In NOD species, prespawning females search for large dominant males to guard them prior to mating. Recall that prespawning carideans undergo a molt before mating and are vulnerable to harassment by males or even predation during the susceptible period right after the molt. Major cheliped and/or maxilliped 3 weaponry of females and smaller males are similar and relatively weak but hypertrophied and powerful in larger males (see Plates 3.3b, 3.5d, 3.6c–d, 8.1 a, b, i). These male transformations with growth may result in distinct male morphotypes, different not only in size and weaponry but also in color or other characters. Alternately, male changes with growth may simply transition gradually from smaller males to the

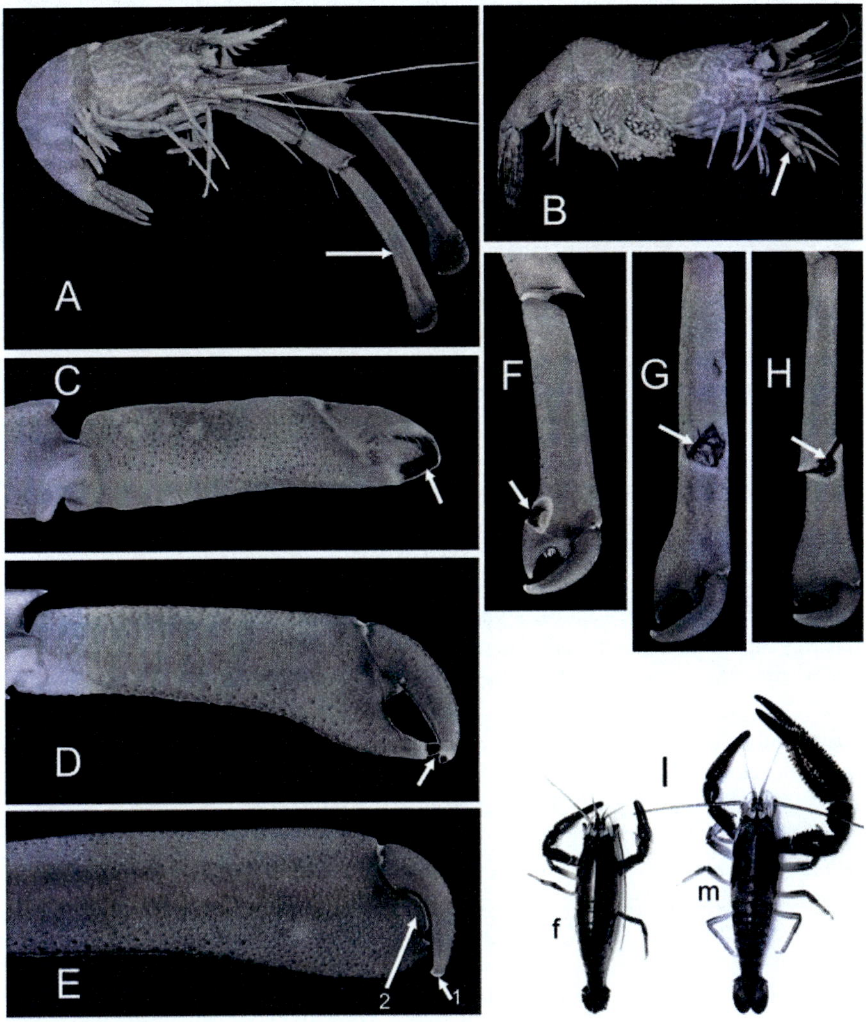

Plate 8.1 Sexual dimorphism of possible NOD (neighborhood of dominance) species. (**a–h**) *Cinetorhynchus cf. hendersoni* (Caridea, Rhynchocinetidae). (**a**) Large (dominant) male with hypertrophied wrench-like major chelae (arrow). (**b**) Reproductive female, major chelae not hypertrophied (arrow). (**c**) Female major chela with full set of stout black corneous scraping setae (arrow). (**d**) Major chela of intermediate-sized male with corneous setae reduced to just one on tip of each finger (arrow), chela more wrench-like. (**e**) Major chela of large dominant male; note absence of sclerotized setae (arrow 1); inner edge of propodal finger developed into a blade-like extension (arrow 2). (**f–g**) Major dominant male chelae with significant puncture wounds (arrows), while the chela in (**h**, arrow) was broken but later healed. Presumably, these injuries are due to male-male fighting contests. (**i**) Reproductive female (f, left) and large (dominant) male (m, right) of *Cryphiops caementarius* (Caridea, Palaemonidae); note difference in body and chela 1 size (only one of latter is usually hypertrophied in the male). From (**a–g**) Bauer et al. (2014) and (**i**) courtesy of María Morales Suazo and Cris Harris Toro, Crustacean Cultivation Laboratory of the Department of Aquaculture of the Universidad Católica del Norte, Chile

larger ones without distinct morphotypes. There is also a gradient in aggressiveness, with the greatest bellicosity displayed by the large dominant males.

One of best examples of a NOD mating system is *Macrobrachium rosenbergii*, which inhabits rivers and streams of Southeast Asia. This species is a significant component of shrimp aquaculture (Chap. 11) because females and especially the dominant males reach large sizes in a relatively short period (months to a year). The species is notable for its distinct male morphotypes with different mating strategies, as described in the excellent review of Karplus and Barki (2019). The morphotypes are sequential in the ontogeny of males. First to mature sexually are the small males (SM), with short, weak weaponry which use pure-searching tactics to mate. They do not guard females, grasping receptive females they encounter or detect, quickly copulating and then disengaging from the female without further interaction. Blue-claw (BC) males, on the other hand, are large in size and have hypertrophied second-cheliped weapons with blue coloration; these weapons can be nearly twice as long as the rest of their bodies. BC males make mating nests (shallow depressions on the bottom) and guard the space around them to await reproductive females nearing a mating molt. In this system, as in the American lobster *Homarus americanus* (Atema and Steinbach 2007), it is the female that seeks out a dominant (BC) male. The BC male guards one to several prespawning females (a harem). When a female molts, it mates with the BC, who continues guarding her for a day or two until her exoskeleton has hardened enough that she can fend off the advances of other males or attacks by other hungry conspecifics. The BC male also benefits from post-copulatory guarding because he prevents other males from depositing their sperm packets until the female is no longer receptive and/or has spawned. Intermediate is this sequence of male growth is the orange-claw (OC) male which is transitional in morphology and aggressive behavior. As its name implies, the chela (propodus and dactyl) is orange while the rest of the cheliped is bluish as in the BC male. OC males do not guard and use the same mating tactics as SM males. BC males are the dominant "alpha" males. Their presence, especially at higher population densities, can behaviorally inhibit SM and OC males from moving up the morphotype hierarchy.

BC males definitely inseminate more females than subordinate males—but the SM males get some too. This was shown by clever experiments done by Ra'anan and Sagi (1985). When a BC male is guarding a newly molted receptive female, SM and OC males slink around the edges of the BC's mating territory, dashing in to attempt copulation with the female while the BC male is fending off other males trying to do the same. Different combinations of SM and OC males were composed, together with a single BC male whose gonopores were plugged in some treatments and not in others. Reproductive females were introduced into the different treatments, and copulations by SM, OC, and BC males were recorded, as was the success of insemination (fertile broods produced or not). Although unstoppered dominant BC male were more successful in inseminating females, SM and OC males did fertilize guarded females in a number of trials. In one treatment, BC males with blocked gonopores guarded a harem against 5 SM males, which fertilized 34% of females in spite of BC guarding. In a comparison treatment, with the BC male fully functional,

84% of females were fertilized, so that presumably the BC male fertilized about 50% of females, with the other 34% divided up among 1 or more of the five "sneaker" SM males. Treatments with only OC males and a blocked BC male showed that OCs are less effective than SMs in fertilizing females.

The mate-guarding strategy of the BC male appears worth the high energy expenditure for growth to large body size with massive cheliped weapons, the time and effort spent in defending its mating territory and females and the period of reduced OC sexual activity during growth from SM to the BC. Small males do not pay these costs, but their pure-searching and "sneaking" mating tactics are less fruitful in fertilizing females.

Let us not forget the females: what do they gain by mating with a BC male? One obvious benefit is their survival and that of their offspring through the premating molt and spawning process. Mortality rates of unguarded females due to harassment by SM and OC males were about 20% in three *Macrobrachium* species due to harassment and cannibalism (Ra'anan and Sagi 1985; Karplus and Barki 2019). However, the preference for BC males by reproductive females is also a first-rate example of the "good genes" hypothesis. Males who survive to reach BC status in good physical condition should have, on average, a superior genotype (and some good luck as well). Carrying and dragging those oversized BC weapons are probably a handicap both in terms of energy expenditure for growth and predation risk. But those robust males that can overcome this handicap in good health likely have a high-quality genome.

The freshwater genus *Macrobrachium* has ~260 species. At least 24 of them have distinct sexually mature male morphotypes similar to *M. rosenbergii* (Karplus and Barki 2019). In addition, there are probably more species that fit into this category. It should be remembered too that there are *Macrobrachium* species with "reversed sexual dimorphism," i.e., with small males and larger females with proportionately similar weaponry, as noted above under "pure-searching" systems. Which type of sexual dimorphism and mating system might have had the ancestral *Macrobrachium*? Karplus and Barki found, based on a molecular phylogeny of *Macrobrachium* by Murphy and Austin (2005), that the most basal (ancestral) branches of the phylogeny were small male, larger female (pure searching") type species. Either NOD or pure-searching sexual dimorphism was found in all major clades and on all continents with members of the genus, indicating that NOD has evolved independently several times in the genus. However, the number of species sampled was small compared to the total number of species in the genus. Based on the high abundance of small-male, larger female species in the Caridea (and all dendrobranchiates), I propose that the *Macrobrachium* ancestor was such a pure-searching species. I suggest that molecular phylogeny colleagues test this in the future with larger sampling of species in the genus.

A similar mix of NOD and pure-searching species occurs in the family Rhynchocinetidae. The first studies on mating systems of this family was done by Martin Thiel and colleagues in a series of papers beginning with Correa et al. (2000) on *Rhynchocinetes typus*. This species inhabits cool-temperate waters along the west coast of South America in rocky subtidal habitats. As in *M. rosenbergii*, there are distinct males morphotypes termed: *typus* (smaller males with weaponry similar to females); *robustus* (dominant males larger than females, with hypertrophied chelipeds bearing big showy setal tufts of thin-walled setae and very large spear-like third maxillipeds) and *intermedius* (males transitional between *typus* and *robustus* in growth, size and weaponry).

However, there are several novel aspects about mating interactions, copulation and spawning in *R. typus* compared to *M. rosenbergii* and other *Macrobrachium* NOD species. Based on field observations and various experimental studies (e.g., Correa et al. 2000, 2003; Díaz and Thiel 2003, 2004; Thiel and Correa 2004), an overview of these interactions can be summarized as follows. Individuals forage over rocks and boulders at night, grazing on microbial films and small sessile organisms. During the day, they retire to shelters under rock ledges. Near to or just after a prespawning molt, reproductive females are attracted by male odors to shelters where the large dominant "robustus" males gather. Guarding by any male morphotype involves "caging" in which the male embraces and holds the female from above (Fig. 8.4a–1); guarding may last up to 3 hours rather than days as in *M. rosenbergii*. During this time, the male will periodically dip below the female and deposit packets of sperm (Fig. 8.4a-2). Unlike other carideans, there can be several episodes of spermatophore placement by the male, which in turn stimulates the female to begin spawning and fertilizing small groups of eggs. Finally, the male releases the female and goes on his way.

If a receptive (molted parturial) female is intercepted by one or more of the subordinate *typus* males before a *robustus* male, the former will cage and begin the copulatory sequence. If a *robustus* male comes along, it will dislodge and drive off the *typus* male (Fig. 8.4b–1) and begin caging. If two *robustus* males encounter a receptive female, they will display to each other by raising and waving their hypertrophied chelipeds at each other, thus perhaps showing their size and that of the cheliped setal tufts (Fig. 8.4b2–3). If one does not move away, fighting will begin. The combatants seize the major chelipeds of their opponent and attempt to pull it towards itself, pushing it into its large spine-tipped third maxillipeds, which act as spears. A common injury is damage or loss of an opponent's eye(s), and deaths do occur. Subordinate males are also attracted to the receptive female but the dominant male, if present, brushes them aside with sweeps of its large chelipeds. A fascinating aspect of this mating activity is that if a *typus* male succeeds in grasping the female while the robustus male is distracted (Fig. 8.4b-3), she may press down against the substrate so that he cannot dip below her to deposit spermatophores. If a subordinate *typus* male does succeed in placing spermatophores on a female and then a *robustus* male shows up, she may pick off the spermatophores of the *typus* male before mating with the *robustus*!! Clearly, the females prefer the sperm of a *robustus* male for fertilizing their eggs.

As in *Macrobrachium,* the two genera of Rhynchocinetidae (*Rhynchocinetes; Cinetorhynchus*) have both small-male, larger female (pure searching) species (e.g., *R. uritai,* Bauer and Thiel 2011; Osawa et al. 2015) and NOD-species with large dominant males equipped with hypertrophied weaponry (Bauer et al. 2014). The NOD mating system of *R. brucei* and *R. durbanensis* (Thiel et al. 2010; Prakash et al.

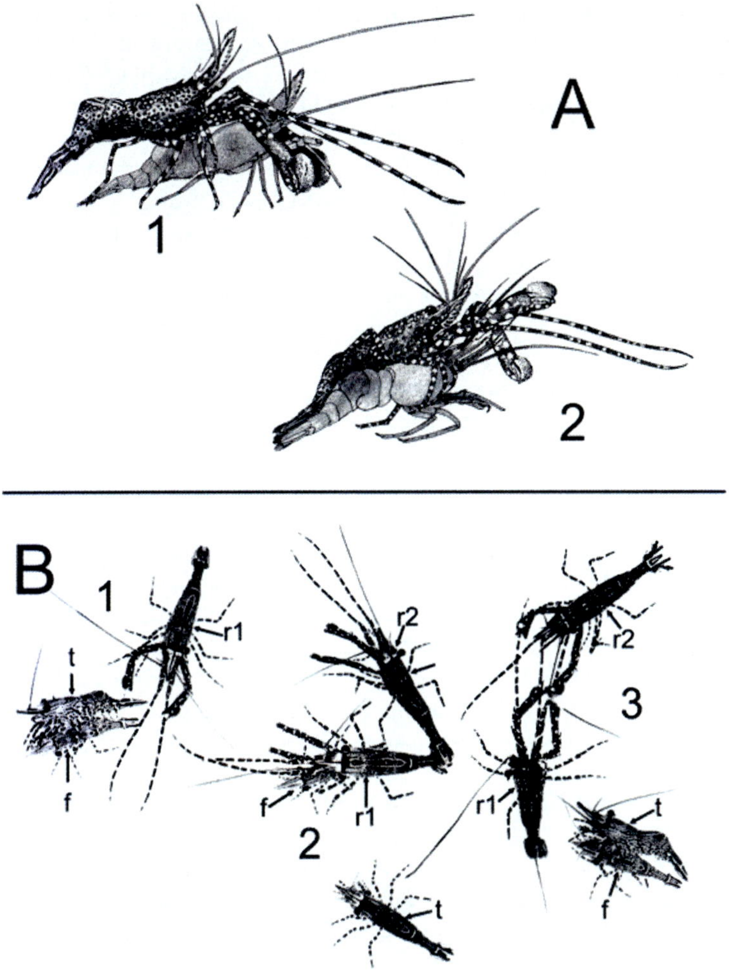

Fig. 8.4 Mating and copulation in *Rhynchocinetes typus*. (**a-1**). A *robustus* male (above) caging the female with its pereopods, and then (**a-2**) he copulates with her. Male somewhat below and at an oblique angle to the female so that their genital areas oppose during spermatophore transfer. (**b-1**). A *typus* male (t) is caging a female (f) but a *robustus* male r1 approaches and will dislodge the *typus* male; (**b-2**). *robustus* male r1 detects the approach of another *robustus* male (r2) and turns away from the female to confront and display to the intruder; (**b-3**). The two *robustus* males display and begin to fight while the *typus* male returns to the female and again attempts copulation. From (**a**) Correa et al. (2000) and (**b**) Correa et al., (2003) courtesy of Journal of Crustacean Biology

2016 respectively) is similar to that of *R. typus*. Although there are not yet any published studies on mating systems of *Cinetorhynchus* pure-searching type species, there is good morphological evidence of NOD-type species in the genus.

In 2011, my friend and colleague Martin Thiel collected a good number of specimens of two undescribed *Cinetorhynchus* species and sent them to me for a collaborative study with rhynchocinetid taxonomist Junji Okuno (Bauer et al. 2014). Truly amazing was the size and shape of the highly elongated major (first) chelipeds with wrench-like chelae of large males (Plate 8.1a). Unlike NOD *Rhynchocinetes* species, there is no sexual dimorphism in the third maxillipeds in NOD *Cinetorhynchus*. The major chelipeds of smaller males are similar to size and shape of females (Plate 8.1b), but as males grow larger, their first pair of chelipeds become elongated, and the movable chela finger becomes highly arched while the end of the propodal finger extends into a thin blade against which the dactylar can tightly close (Plate 8.1 c–e). The multiple corneous setae on the tips of chelae in smaller males and females (an adaptation for scraping food items from hard surfaces) are reduced and lost. The result is a massive wrench-like subchela, somewhat like that of a crangonid caridean. The loss of these feeding setae and the burden of carrying around these long clumsy appendages must indeed represent a handicap for the dominant males. However, it may demonstrate to receptive females that such a male has superior genes to overcome these handicaps. The major chelipeds of the large males often show serious punctures and crushing injuries (Plate 8.1 f–g). The form of the chelipeds and injuries on them suggest that pugnacious males may pinch and hold the chelipeds of opponents, perhaps flipping and shoving them aside them similar to the fighting behavior of rhinoceros beetles (McCullough et al. 2014). The details of mating behavior and contests between these males will be of great interest and a wonderful opportunity for research.

Other probable NOD species include *Cryphiops caementarius* which inhabit major rivers in Peru and Chile (Hartmann 1958). This species has NOD sexual dimorphism (Plate 8.1I) with two male morphotypes ("group 1 and 2") characterized by somewhat different coloration and, importantly, one hypertrophied major cheliped which is proportionately more massive in the larger (group 1) morphotype (Plate 3.6c) (Rojas et al. 2012). Although both male morphotypes show puncture wounds on their chelipeds, group 1 males had more than the smaller (group 2) morphotype. Anecdotal reports by river fishermen indicate that the group 1 males utilize the best shelters and are very territorial. All these observations indicate a NOD-like mating system. Based just on morphology (very elongated third maxillipeds and larger major chelae of males), certain hippolytids may have NOD-like mating systems (e.g., some *Heptacarpus* species, see Fig. 4.3 in Chap. 4; some *Saron* species (e.g., Plate 3.3a–b). As with the *Cinetorhynchus* and *Cryphiops* examples above, the mating behavior and presumed male agonistic contests of these species need study.

8.3.5 Monogamy

Like "search and attend" mating systems, population density and resultant encounters between individuals are fairly rare in species with this system (Table 8.2). A male and female live together through several breeding cycles during a breeding season or longer. They often live in a shelter or refuge which they defend from intruders, but from which they may leave to forage for food. There is usually little sexual dimorphism in body size and weaponry ("size-assortative pairing") which is thought to reduce the possibility of aggressive behavior between the male and female. Although monogamy implies sexual fidelity, this is not always the case, as will be seen. Unless sexual fidelity has been shown in a pairing species, the term "social monogamy" is often used. As Wickler and Seibt (1981) pointed out, social monogamy may be thought of an extension of temporary mate guarding ("search and attend"), i.e., prolonged mate guarding. Monogamy has certain advantages. Once paired, individuals have less exposure to predation risk and lower energetic expense incurred by mate searching. Inadvertent rather than direct cooperation between the male and female may occur when each independently defends the shared refuge. One partner may be allowed to feed on the unused or discarded food gathered and brought back to the shelter by the other for consumption.

In other monogamous animals, e.g., many birds and mammals, both parents may share in care of the young. However, this is not the case for caridean and stenopodidean shrimps, as only the females care for attached embryos. In sergestoids and penaeoids, there is no parental care at all—eggs are simply fertilized and spawned into the water, where they undergo embryonic development all by themselves.

How common is shrimp monogamy? There are no examples among the penaeoid and sergestoid shrimps; as noted previously, these groups have pure-searching mating systems. However, there are many observations in the literature, especially among symbiotic shrimps, of carideans and stenopodideans found in apparently monogamous pairs. However, careful study is required to determine if the pairings are temporary ("search and attend") or more permanent (monogamy).

One of the most thoroughly examined cases of shrimp monogamy is that of the caridean *Hymenocera picta* ("harlequin" or "clown" shrimp) by behavioral ecologists Wolfgang Wicker and Uta Seibt (Seibt 1973; Wickler 1973; Seibt and Wickler 1979). Unlike most shrimps, this a day-active species which preys on starfishes or scavenges on starfish remains. Its vivid coloration in the high-predation tropical reef environment, as well as the avoidance by fish predators in the laboratory, suggested to Bruce (1975) that *H. picta* may be protected from visual predators by its aposematic (warning) coloration. Males are slightly smaller than females but both have large plate-like cheliped weapons (Fig. 3.32 d). The various Wickler and Seibt studies found that this species occurs in pairs (~95% heterosexual, ~5% homosexual). The pairs were highly territorial and defensive of refugees. Males recognized and preferred to pair up again with their female partner when separated from them. However, males were not faithful; if a recently molted and unprotected female was available, the male would go out and mate with it but then return to its pair partner. The male does little for the female, although there is an inadvertent food sharing. In an aquarium situation at least, males were observed going out to pick up food and return to the pair's shelter. Shrimps are messy eaters, and the pair partner can feed on the left-over scraps. There is also no direct cooperative defense of the shelter, but defense by each partner independently certainly will contribute to the well-being of the other. However, Seibt and Wickler concluded that the male is the main beneficiary of the association. A female molts and produces a new batch of embryos about every 20 days. If the male stays with the same female and guards her from other males, he is assured of inseminating the female every time she molts. Given the low population abundance of the species, this is a much better strategy than constantly searching and competing with other males for newly molted females, which are scarce on any given day.

Many alpheid carideans live in socially monogamous male-female pairs as reproductive adults, e.g., most or all species of the very large genus *Alpheus*, as well as several species of *Betaeus, Synalpheus,* and *Aretopsis* (Hart 1964; Schein 1975; Anker and Baeza 2012; Soledade et al. 2018). Many species of *Alpheus* make burrows in sand or mud, often below rocks or other substrates, while several other species are symbiotic with invertebrate and fish hosts (Chap. 10, Symbioses). In the large behavioral literature of the genus are numerous detailed investigations on the social monogamy of the species, e.g., *A. heterochaelis* (Schein 1975; Rahman et al. 2003, 2004) and *A. angulosus* (formerly *A. angulatus*) (Mathews 2002a, 2002b, 2003, 2007). Some *Alpheus* species deviate a bit from strictly size-assortative pairing. Although males are similar in size to their pair partner, those searching for a partner prefer to pair with larger females as they produce more offspring than smaller ones (e.g., *A. heterochaelis;* Schein 1975; Rahman et al. 2004). They compete with other males for these larger females, which may explain their slightly larger snapping claws compared to a female of similar body size. Before pairing, interactions between a potential pair determine if an agonistic balance can be reached, resulting in size-assortative pairing. In *Alpheus angulosus,* Heuring and Hughes (2020) also showed that although males are partial to larger females, females have no such inclination with respect to male size. Bigger males compete more successfully than smaller males for the more desirable (higher fecundity) larger females. However, females of *A. angulosus* do not choose a mate based on size; they usually choose to remain with the same male during the reproductive season, although they may switch partners outside of the reproductive season. Heuring and Hughes suggested that these different types of male and female mate preferences account for the size-matched pairing rather than a predisposition for a large-sized mate in both sexes. Whatever slight sexual dimorphism in size and weaponry that may occur between monogamous pair partners, it is nowhere near the exaggerated sexual dimorphism of NOD and some search-and-attend species.

Lauren Mathews (2002a), who was doctoral student in my lab, found in *A. angulosus* that the male and female relationship includes inadvertent cooperation (division of labor). Although females did more of the "housework" (construction and maintenance of the shared burrow), males also participated. Both male and female went out of the burrow to forage for food and brought it back to the burrow, and the pair partner would take advantage and feed on the leftovers. The male and female participated separately in burrow defense, but males did a greater share; both individuals benefited from this joint protection of territory.

Nancy Knowlton (1980) studied two populations of *Alpheus* symbiotic with anemones (*Bartholomea annulata*) in adjacent areas of a tropical Caribbean habitat. Both populations usually occurred in male-female pairs, but one species, now recognized as *A. immaculatus*, occurred in an area where fish predation was high.

There was little sexual dimorphism in body size, weaponry (snapping claw) and coloration. This species was not only socially but also sexually monogamous. In a nearby area with a lower predation risk for shrimps moving about, males of a closely related species (*A. armatus*) were somewhat bigger than their female partners, had proportionately larger snapping claws, and had distinct black uropod spines. Nancy marked and followed the movements of males in both species. In the higher-risk area, *A. immaculatus* males stayed with their female partners during the study. On the other hand, in the lower-risk habitat, *A. armatus* males frequently abandoned their partner to seek out solitary females on other anemones, presumably to partner briefly and mate with them. The greater sexual dimorphism of *A. armatus* males might be explained by competition with other males attempting to pair and mate with the solitary females. Although shrimps do not have very acute vision, the distinctive black spines on the tail of *A. armatus* males, a possible indicator of male size and condition (genetic quality), might be detected and used by females as a male-selection tool. Thus, the mating system of *A. immaculatus* is monogamy with sexual fidelity, while that of *A. armatus* is somewhat between monogamy and search-and-attend. One can see how the two systems could evolve one from the other in either direction, depending on environmental conditions.

The alpheid genus *Betaeus* includes free-living species as well as pair-living symbiotic with other invertebrates. (Hart 1964; Anker and Baeza 2012). The lack of sexual dimorphism and high frequency of pairing in the symbiotic species indicates probable monogamy. As a graduate student in southern California, I frequently observed and collected two such species in sublittoral habitats in and around kelp beds. One species, *B. macginitieae*, occurred in pairs beneath sea urchins, with the pair close to the center of the urchin's oral disk. But there were often one to a few smaller (presumably juvenile) individuals just under the disk periphery. The male-female pair of the other species, *B. harfordi,* could often be seen peeking out from one side of the tiny head of an abalone (*Haliotus* species) from within the gastropod's mantle cavity. I could maintain both species with their host in the laboratory. When a solitary individual was introduced into the tank with a resident pair, one member of the pair would rush out, with its large major chelipeds cocked (a threat), and the two individuals would go back and forth threatening each other; full-blown fights frequently ensued. I think there is little doubt that the two species are at least socially monogamous, but a detailed study is needed to establish if sexual fidelity occurs. At night, the urchin host of *B. macginitieae* moves out from its home depression on the bottom to forage; it would be interesting and important to find out if the symbiotic pair moves with them, or forages out independently, either finding and returning to its specific host or moving in with any unoccupied host at daybreak.

Another *Betaeus* species, *B. longidactylus,* occurs in intertidal habitats from San Francisco Bay down into northern Baja California and Sonora, Mexico (Jensen 2014). Unlike *B. macginitieae* and *B. harfordi*, this is a free-living species in some habitats but symbiotic in others. In collecting in the rocky intertidal in southern California, I often found them unpaired in small to large groups under large stones or ledges, as has also been reported by Jensen (2014). Although similar in body size as females, larger males have hypertrophied major chelipeds, with a large gap between the fingers, with big tooth on the arched dactylar finger (Hart 1964) (Plate 3.2d; compare Plate 2c–d in Bauer 2004). The outsized chelipeds of larger males indicate that they may dominate in mating interactions, indicative of one of the two temporary female-guarding systems. However, MacGinitie (1937) also reported the same species in heterosexual pairs within burrows of urechid worms, as well as those of ghost and mud shrimps. His collections were from marine estuaries where shelter is scarce compared to that in the rocky intertidal. Perhaps the mating system is flexible in this species depending on the habitat, or perhaps morphologically similar sibling species have evolved in the two habitats. A species with comparable characteristics from the rocky intertidal of Chile, *B. lilianae*, has been studied more closely by Baeza et al. (2010). They found that the mating system is probably neither socially monogamous nor completely pure search. Perhaps a search-and-attend strategy is used by males. Clearly, there is a need for further field observations and laboratory experiments on the mating system of these two species.

As noted above, Antonio Baeza and colleagues have studied mating systems of several pontoniine species in which male-female pairs are commonly encountered. One of them which solidly falls into the monogamy category is *Paranchistus pycnodontae*, found within the mantle cavity of the winged pearl oyster *Pteria penguin* (Baeza et al. 2013). Although males are smaller on average than females, those paired with females were comparable in size and weaponry to their female partner, a classic trait of monogamy. As with *Hymenocera picta*, males were found with females regardless of their reproductive state, i.e., whether they were near a mating molt or not. Monogamy was similarly indicated in *Pontonia margarita*, which inhabits the pearl oyster *Pinctada mazatlanica* (Baeza 2008). A more complex situation was found in *Pontonia manningi*, a symbiotic shrimp which lives within the mantle cavity of the Atlantic winged oyster *Pteria colymbus*, itself symbiotic growing on *Pseudopterogorgia* gorgonians species in the Caribbean (Baeza et al. 2016). More heterosexual pairs than expected by chance were found, and females of those pairs were in various reproductive states, i.e., not necessarily near a mating molt, as in a search-and-attend system. A fair number of solitary females were brooding embryos, which might indicate that the males which inseminated them had abandoned them. The researchers concluded that the situation was complex, somewhat between promiscuous (pure searching) and monogamy

In the previous chapter, I stated that a few species of the protandric simultaneous hermaphroditic (PSH) genus *Lysmata,* reproductive adults are socially monogamous (Limbaugh et al. 1961; Fiedler 1998). They occur in pairs of female-phase simultaneous hermaphrodites (FPSH) which occupy a prominent area (station) on tropical reefs. The pair makes their living by cleaning fish clients of parasites, dead tissues and mucus (see Chap. 10, Symbioses). The best studied of these is the Indo-Pacific *Lysmata debelius* (Rufino and Jones 2001; Wong and Michiels 2011; Liu et al. 2019). When separated experimentally, the partners can recognize and prefer each other from other individuals for periods of at least 3 hours. In laboratory experiments, groups of 3-4 individuals are reduced to just a pair by aggressive behavior during molting of supernumerary individuals and strangers, apparently reducing competition for fish clients around a cleaning station. What would be important to know is what happens when one member of a resident established pair dies. How is a replacement chosen or accepted by the lone resident, and what form of competition takes place among possible candidates? In another monogamous species, *L. grabhami* from the tropical Atlantic, solitary or groups of individuals without cleaning stations can be observed elsewhere in the habitat (Wirtz 1997), and this is probably true of the other monogamous species, *L. amboinensis*. If all reef cleaning stations are occupied by a pair, what do these individuals do when they mature from the male-phase to the female-phase? Do they mate with another unattached hermaphrodite? A good field and laboratory study on these questions needs to be done!

Carideans certainly do not have exclusive rights to monogamy. In fact, some (perhaps all) species of Stenopodidae, especially *Stenopus* species, and all Spongicolidae, as far as is known, are socially monogamous (Goy 2010). The best-studied stenopodid is *Stenopus hispidus*, relatively common in shallow tropical reef environments around the world. Field observations and behavioral experiments by Limbaugh et al. (1961) and especially Johnson (1969, 1977) have shown much about the monogamous relationship. Males and females of similar sizes and weaponry form pairs that, during the day, aggressively defend shelters or small cleaning territories covering an area covered by their extended antennal flagella. Like the monogamous *Lysmata* species, *S. hispidus* is a fish cleaner during the day. At night, however, the male and female forage separately, scavenging over the bottom. Before dawn, they return to their original location and partner to resume attracting and cleaning fish clients. In nature, the male and female are rarely separated more than one night, but experiments in the lab show that pair partners may identify each other after nearly a month of separation. Encounters with strangers and intruders trigger aggressive behavior, often resulting in the death of one the combatants. If a single male meets an unattached female, interactions begin agonistically but gradually turn to courtship and pairing. According to present information, established pairs are permanent and probably sexually faithful until one of the pair dies, after which the survivor will try to acquire a new mate. Distance and particularly contact sex pheromones mediate recognition of a pair partner as well as the distinction between male and female strangers.

Except for the shallow-water genus *Microprosthema*, spongicolid stenopodideans are deep-sea associates of large sessile invertebrates, especially hexactinellid ("glass") sponges. The spongicolids live as adults in male-female pairs within the sponge cavity (atrium). The infrastructure of the sponge skeleton is a lattice of hard siliceous (glass) spicules. The enforced monogamous mating system and life history of these sponge associates are intimately linked to that of its hosts, as shown by several studies conducted by Tomomi Saito and various colleagues. The male-female pairs of these species live entrapped in the stiff mesh-like skeleton of the sponge host. The pairing of the male and female begins when they are still young and enter a developing sponge's skeleton when it is still flexible (Saito et al. 2002). Upon hardening with growth, the sponge's lattice-like skeleton imprisons the pair for life. The sexual dimorphism between male and female partner is what might be expected from a monogamous shrimp species. In *Spongicola japonica*, Saito found that the male was only slightly smaller than the female of the pair. Not surprisingly, there was no sexual dimorphism in the pereopod 3 cheliped, the major weapon of stenopodidean shrimps. Once ensconced within the hardened sponge skeleton, the male has nothing to fear from other males that might want to enter the sponge to pair with his female. Observation and collection of these deep-sea shrimps is difficult, and much work needs to be done on the many other spongicolid species.

8.4 Female Choice, Multiple Paternity and Sperm Competition

We have seen that the mating systems of shrimps, as in most animal groups, are more characterized by the behavior of the eager and aggressive males than on that of the more reserved and choosy females. However, female behavior is just as important in mating as that of males, although it may not always be so obvious. Females have a number of overt and cryptic tactics for determining the outcome of a mating and possible insemination encounter. First and foremost, the female can just reject the male aggressively or flee by backward tail flipping. Alternately, the male may seize a female, but during copulation she may not cooperate, e.g., by escaping his grasp or denying the male access to the area where he must deposit or insert his spermatophores. For example, she may hold the pleopods forward under her posterior thorax over her genital area so that if the male attempts insemination, it will fail, as in the carideans *Heptacarpus sitchensis* (Bauer 1976) and *Palaemonetes pugio* (Berg and Sandifer 1984). As noted earlier in *Rhynchocinetes typus*, a female may hunker down against the substrate so that a subordinate (less desirable) male cannot dip below her to place his spermatophores. In choice experiments in which a receptive female could choose between two tethered males, she usually chose the larger, more dominant male (Díaz and Thiel 2003). Presumably, the female would pick the better "good genes" dominant male to father her upcoming brood, as well as one that would protect her from harassment and injuries from other overeager males. As noted earlier, female of *Atya innocuous* actively chooses a male suitor with a specific backward movement (Felgenhauer and Abele 1982).

In sergestoid and penaeoid shrimps, the complex male genitalia (petasmata) or other male structures must attach or fit onto the female properly for effective insemination (e.g., Bauer 1996b; Lunina and Vereshchaka 2017; Vereshchaka et al. 2016, 2017; Feng et al. 2018). The male and female may go through all the steps of copulation, but the female may remain uninseminated or only partially so, as shown for *Sicyonia dorsalis* (Bauer 1992). The female sometimes rejects insemination in a yet unknown way, but a subtle shift in her body at a critical moment could result in a mismatch between her and the male's genitalia. Females of caridean, stenopodidean, open-thelycum penaeoids and sergestoids might pick off deposited spermatophores with their grooming limbs, although this has only been observed in *R. typus* females.

Although laboratory studies on variety of shrimps indicate that females often reject the advances of other males after a single copulation, there is evidence to the contrary in others. The evidence comes in two forms, direct behavioral observations and, more recently, microsatellite DNA parentage analyses from wild or captive shrimps. In carideans, these studies focus on the embryos carried by females. For example, studies on *Rhynchocinetes typus* by direct laboratory observations and experiments show that multiple males can mate with a female before she spawns. Fertilization of embryos in the same brood has been confirmed by microsatellite DNA data (Bailie et al. 2014). Of fifteen broods from females collected from a natural population, 75% were fertilized by more (2-4) than one male. Moreover, there was "reproductive skew," i.e., some of the males inseminating a female were the fathers of more embryos than others (Fig. 8.5a). In laboratory matings of *Palaemonetes pugio*, most females mated just once, but some females mated 2-5 times (Bauer and Abdalla 2001). Baragona et al. (2000) reported in a conference presentation that 8 of 10 female broods from wild-caught females of this species had multiple fathers.

Yue and Chang (2010) found in the atyid *Caridina ensifera* that all broods examined had been fertilized by more than one male (Fig. 8.5b). However, two different analyses were used, one more conservative estimating a maximum of five males inseminating one female (Fig. 8.5b), while the less conservative gave a maximum of 11 males per female brood! As with *R. typus*, there was considerable reproductive skew. Additionally, the males inseminating a female were not closely related to her, indicating that females avoid inbreeding. Similarly, Jorquera et al. (2016) discovered in the deep-sea penaeoid *Acanthephyra pelagica* that nearly 100% of female broods sampled were fertilized from between 2-4 different males. Reproductive skew again took place, with one male fertilizing the majority of embryos in a brood. However, these examples ultimately need to be verified by behavioral observations that show that, indeed, females do copulate with more than one male.

The caridean examples above were either pure-searching species or NOD species in which multiple mating is known. The multimale parentage found in broods of these species contrasts with that of a socially monogamous species, *Alpheus angulosus* (Mathews 2007). Her microsatellite data showed that the partner male was the only or primary parent in the broods of 53 pairs tested. In 17 of those pairs, there were some embryos which might have been inseminated by another male, but Mathews felt that that results may have been due to genotyping errors. Thus, this pair-living species is most likely both socially and sexually monogamous.

Multiple matings and multiparentage broods may occur in *Aristeus antennatus*, a penaeoid in which spermatophores are placed within an open-thelycum of the female. As in carideans, the external placement of the spermatophores make possible that insemination by more than one male may occur. However, at least in *A. antennatus*, the rate of multiple matings and insemination was lower compared to that of species cited above. Planella et al. (2019) found that 11.5% of females carried 2-3 sets of spermatophores rather than usual one pair (one mating).

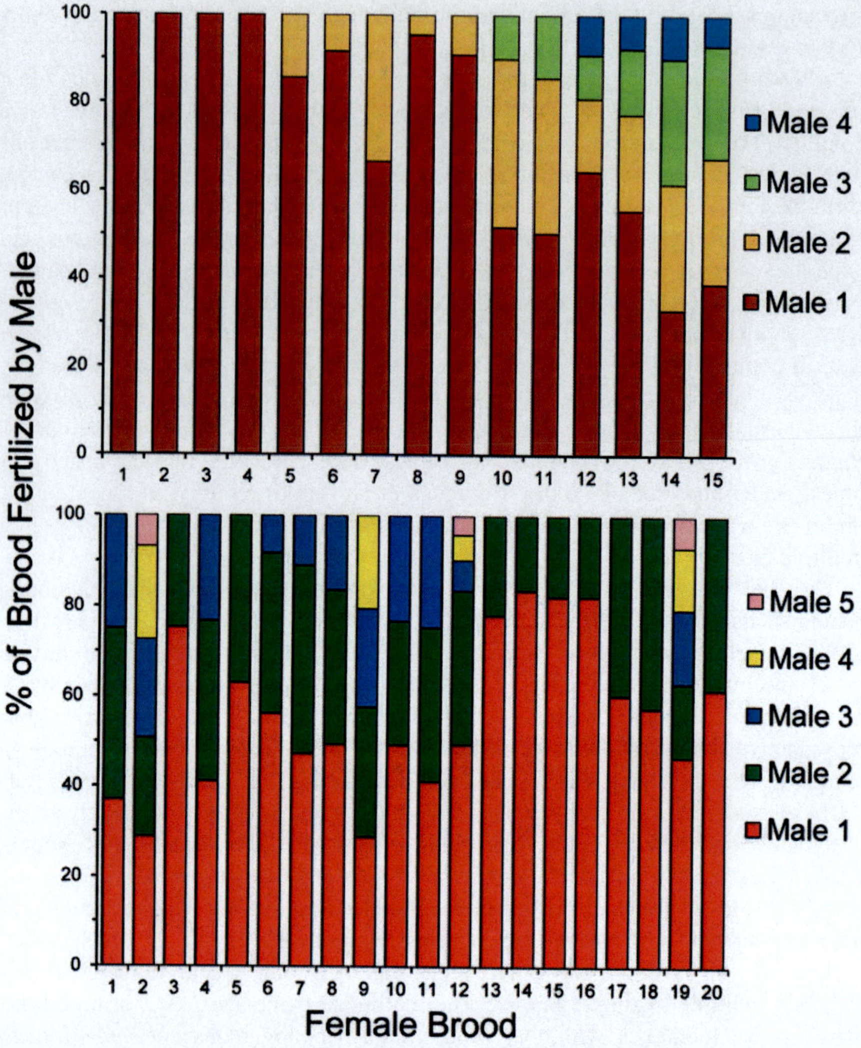

Fig. 8.5 Multimale paternity in caridean female broods showing reproductive skew: microsatellite DNA analyses using GERUD software indicates that individual female broods may be unequally fertilized by multiple males. **Upper graph:** A sample of 15 *Rhynchocinetes typus* females showing their broods variously fertilized by 4 different males. From Bailie et al. (2014) courtesy of Journal of Crustacean Biology. **Lower graph:** A sample of 20 *Caridina ensifera* female broods variously fertilized by 5 different males as estimated by the GERUD software. From Yue and Chang (2010), Creative Commons CC BY 4.0

Microsatellite DNA markers revealed that each set was from a different male. In one case, the same male had inseminated two different females. Their analyses further showed that, as in *Caridina ensifera*, females were selecting male mating partners that were not close relatives.

In closed-thelycum penaeids, males seal or plug the apertures to the sperm-storage structures of females (spermathecae; seminal receptacles) after mating (Chap. 6). Presumably, this prevents other males from inseminating the female while she is still soft after molting. One would predict that sperm from only one male would be found within the female spermathecae in these species. Fortunately, this has been tested by Knibb et al. (2014) with the penaeid *Fenneropenaeus merguiensis* maintained in an aquaculture facility. The microsatellite data supported the hypothesis that DNA from only one male was present in the median spermatheca of each female examined. At least in this species and presumably other similar closed-thelyca species, paternity assurance devices work well! It may be recalled that in sicyoniids, which have closed-thelyca with twin spermathecae, there are no mating plugs or seals. Females do mate with multiple males, at least in captivity, and copulating males may inseminate one, both or none of the spermathecae in a mating bout (Bauer 1992, 1996b). In some yet unidentified manner, females are cryptically choosing to accept or reject sperm from some copulating males. The probability of insemination by multiple males is high in *Sicyonia* species. I expect that some enterprising researcher will take up the challenge of testing this hypothesis using microsatellite data on sperm deposited within female spermathecae.

When multiple insemination of the female does occur in carideans, what determines the reproductive skew described above? Does the sperm of the first or last male inseminating the female predominate in fertilization? Or do males pick off some or all spermatophores placed by other males? Multiple spermatophores have been found on females of *Macrobrachium rosenbergii* in aquaculture ponds, indicating possible multiparentage of broods (Karplus and Barki 2019). In that species and *R. typus*, males rub the female sternum prior to depositing their spermatophores, perhaps to remove those deposited previously by other males (Correa et al. 2000).

Research on mating behavior and systems involving both field studies and laboratory observations and experiments are the realm of a branch of science termed "behavioral ecology." This type of research is a formalized branch of natural history which not only contributes to scientific knowledge but is also just good fun! Not only professional scientists but also lay naturalists are able to advance this field of study. Hopefully, the reader will be stimulated by this chapter to join in and advance our knowledge of mating systems as there are so many species to study and questions to be answered.

References

Alvarez F, Iliffe TM, Villalobos JL (2006) Macromaxillocarididae, a new family of stenopodidean shrimp from an anchialine cave in the Bahamas, with the description of *Macromaxillocaris bahamaensis*, n. gen., n. sp. J Crustac Biol 26:366–378

Ameyaw-Akumfi C, Hazlett BA (1975) Sex recognition in crayfish *Procambarus clarkii*. Science 190:1225–1226

Andersson M (1986) Evolution of condition-dependent sex ornaments and mating preferences: sexual selection based on viability differences. Evolution 40:804–816

Anker A, Baeza JA (2012) Molecular and morphological phylogeny of hooded shrimps, genera *Betaeus* and *Betaeopsis* (Decapoda, Alpheidae): Testing the center of origin biogeographic model and evolution of life history traits. Mol Phylogenet Evol 64:401–415

Asakura A (2009) The Evolution of Mating Systems in Decapod Crustaceans. In: Martin JW, Crandall KA, Felder DL (eds) Decapod Crustacean Phylogenetics. CRC Press, Boca Raton, FL, pp 121–182

Atema J, Cowan DF (1986) Sex-identifying urine and molt signals in lobster (*Homarus americanus*). J Chem Ecol 12:2065–2080

Atema J, Steinbach MA (2007) Chemical communication and social behavior of the lobster *Homarus americanus* and other decapod Crustacea. In: Duffy JE, Thiel M (eds) Evolutionary ecology of social and sexual systems: crustaceans as model organisms. Oxford University Press, Oxford, pp 115–146

Baeza JA (2008) Social monogamy in the shrimp *Pontonia margarita,* a symbiont of *Pinctada mazatlanica*, off the Pacific coast of Panama. Mar Biol 153:387–395

Baeza JA, Díaz-Valdéz M (2011) The symbiotic shrimp *Ascidonia flavomaculata* lives solitarily in the tunicate *Ascidia mentula*: implications for its mating system. Invertebr Biol 130:351–361

Baeza JA, Piantoni C (2010) Sexual system, sex ratio, and group living in the shrimp *Thor amboinensis* (De Man): relevance to resource-monopolization and sex-allocation theories. Biol Bull 219:151–165

Baeza JA, Thiel M (2007) The mating system of symbiotic crustaceans: a conceptual model based on optimality and ecological constraints. In: Duffy JE, Thiel M (eds) Evolutionary Ecology of Social and Sexual Systems: Crustaceans as Model Organisms. Oxford University Press, New York, pp 250–267

Baeza JA, Farías NE, Luppi TA, Spivak ED (2010) Refuge size, group living and symbiosis: testing the "resource economic monopolization" hypothesis with the shrimp *Betaeus lilianae* and description of its partnership with the crab *Platyxanthus crenulatus*. J Exp Mar Biol Ecol 389: 85–92

Baeza JA, Bolaños JA, Hernandez JE, Lira C, López R (2011) Monogamy does not last long in *Pontonia mexicana*, a symbiotic shrimp of the amber pen-shell *Pinna carnea* from the southeastern Caribbean Sea. J Exp Mar Biol Ecol 407:41–47

Baeza JA, Ritson-Williams R, Fuentes MS (2013) Sexual and mating system in a caridean shrimp symbiotic with the winged pearl oyster in the Coral Triangle. J Zool 289:172–181

Baeza JA, Simpson L, Ambrosio LJ, Guéron R, Mora N (2016) Monogamy in a hyper-symbiotic shrimp. PLoS One 11(3):e0149797

Bailie DA, Fitzpatrick S, Connolly M, Thiel M, Hynes R, Prodöhl PA (2014) Genetic assessment of parentage in the caridean rock shrimp *Rhynchocinetes typus* based on microsatellite markers. J Crustac Biol 34:658–662

Baragona MA, Haig-Ladewig A, Wang SY (2000) Multiple paternity in the grass shrimp *Palaemonetes pugio*. Am Zool 40:935

Bateman AJ (1948) Intra-sexual selection in *Drosophila*. Heredity 2:349–368

Bauer RT (1976) Mating behaviour and spermatophore transfer in the shrimp *Heptacarpus pictus* (Stimpson) (Decapoda: Caridea: Hippolytidae). J Nat Hist 10(415):440

Bauer RT (1979) Sex attraction and recognition in the shrimp *Heptacarpus paludicola* Holmes. Mar Behav Physiol 7(157):174

Bauer RT (1989) Continuous reproduction and episodic recruitment in nine caridean shrimp species inhabiting a tropical seagrass meadow. J Exp Mar Biol Ecol 127:175–187

Bauer RT (1992) Repetitive copulation and variable success of insemination in the marine shrimp *Sicyonia dorsalis* (Decapoda, Penaeoidea). J Crustac Biol 12:153–160

Bauer RT (1996a) A test of hypotheses on male mating systems and female molting in decapod shrimp, using *Sicyonia dorsalis* (Decapoda: Penaeoidea). J Crustac Biol 16:429–437

Bauer RT (1996b) Role of the petasma and appendices masculinae during copulation and insemination in the penaeoid shrimp, *Sicyonia dorsalis* (Crustacea: Decapoda: Dendrobranchiata). Invertebr Reprod Dev 29:173–184

Bauer RT (2002) Test of hypotheses on the adaptive value of an extended male phase in the hermaphroditic shrimp *Lysmata wurdemanni* (Caridea: Hippolytidae). Biol Bull 203:347–357

Bauer RT (2004) Remarkable Shrimps: Adaptations and Natural History of the Carideans. University of Oklahoma Press, Norman

Bauer RT (2011) Chemical communication in decapod shrimps: the influence of mating and social systems on the relative importance of olfactory and contact pheromones. In: Breithaupt T, Thiel M (eds) Chemical Communication in Crustaceans. Springer Verlag, New York, pp 277–296

Bauer RT, Abdalla JA (2001) Male mating tactics in the shrimp *Palaemonetes pugio* (Decapoda, Caridea): precopulatory mate guarding vs. pure searching. Ethology 107:185–199

Bauer RT, Holt GJ (1998) Simultaneous hermaphroditism in the marine shrimp *Lysmata wurdemanni* (Caridea: Hippolytidae): an undescribed sexual system in the decapod Crustacea. Mar Biol 132:223–235

Bauer RT, Thiel M (2011) First description of a pure-search mating system and protandry in the shrimp *Rhynchocinetes uritai* (Decapoda, Caridea). J Crustac Biol 31:286–295

Bauer RT, Okuno J, Thiel M (2014) Inferences on mating and sexual systems of two Pacific *Cinetorhynchus* shrimps (Decapoda, Rhynchocinetidae) based on sexual dimorphism in body size and cheliped weaponry. ZooKeys 457:187–209

Berg AB, Sandifer PA (1984) Mating behavior of the grass shrimp *Palaemonetes pugio* Holthuis (Decapod: Caridea). J Crustac Biol 4:417–424

Berglund A (1981) Sex dimorphism and skewed sex ratios in the prawn species *Palaemon* adspersus and *P. squilla*. Oikos 36:158–162

Bergström B (2000) The biology of *Pandalus*. Adv Mar Biol 38:1–245

Boddeke R, Bosschieter JR, Goudswaard PC (1991) Sex change, mating, and sperm transfer in *Crangon crangon* (L.). In: Bauer RT, Martin JW (eds) Crustacean sexual biology. Columbia University Press, New York, pp 164–182

Bruce AJ (1975) Coral reef shrimps and their colour patterns. Endeavour 34:23–27

Brusca RC, Moore W, Shuster SM, Haver N (2016) Invertebrates. Sinauer Associates, Inc., Sunderland, MA

Bublitz R, Sainte-Marie B, Newcomb-Hodgetts C, Fletcher N, Smith M, Hardege J (2008) Interspecific activity of the sex pheromone of the European shore crab (*Carcinus maenas*). Behaviour 145:1465–1478

Burkenroad MD (1947) Reproductive activities of decapod Crustacea. Am Nat 81:392–398

Butler TH (1980) Shrimps of the Pacific coast of Canada. Can Bull Fish Aquat Sci 202:1–280

Carlisle DB (1959) On the sexual biology of *Pandalus borealis* (Crustacea Decapoda). II. The termination of the male phase. J Mar Biolog Assoc UK 38:481–491

Caskey JL, Bauer RT (2005) Behavioral tests for a possible contact sex pheromone in the caridean shrimp *Palaemonetes pugio*. J Crustac Biol 25:35–40

Caskey JL, Hasenstein KH, Bauer RT (2009a) Studies on contact sex pheromones of the caridean shrimp *Palaemonetes pugio*: I. Cuticular hydrocarbons associated with mate recognition. Invertebr Reprod Dev 53:93–103

Caskey JL, Watson GM, Bauer RT (2009b) Studies on contact sex pheromones of the caridean shrimp *Palaemonetes pugio*. II. The role of glucosamine in mate recognition. Invertebr Reprod Dev 53:105–116

Caves EM, Frank TM, Johnsen S (2016) Spectral sensitivity, spatial resolution and temporal resolution and their implications for conspecific signalling in cleaner shrimp. J Exp Biol 219: 597–608

Chace FA Jr, Hobbs HH Jr (1969) The freshwater and terrestrial decapod crustaceans of the West Indies with special reference to Dominica. U.S. Natl Mus Bull 292:1–258

Chiba S, Goshima S, Shinomiya Y (2003) Male-male competition selects for delayed sex change in the protandrous shrimp *Pandalus latirostris*. Mar Biol 142:1153–1157

Christy JH (1987) Competitive mating, mate choice and mating associations of brachyuran crabs. Bull Mar Sci 41:177–191

Correa C, Thiel M (2003) Mating systems in caridean shrimp (Decapoda: Caridea) and their evolutionary consequences for sexual dimorphism and reproductive biology. Rev Chil Hist Nat 76:187–203

Correa C, Baeza JA, Dupré E, Hinojosa IA, Thiel M (2000) Mating behavior and fertilization success of three ontogenetic stages of male rock shrimp *Rhynchocinetes typus* (Decapoda: Caridea). J Crustac Biol 20:628–641

Correa C, Baeza JA, Hinojosa IA, Thiel M (2003) Male dominance hierarchy and mating tactics in the rock shrimp Rhynchocinetes typus (Decapoda: Caridea). J Crustac Biol 23:33–45

Dall W, Hill BC, Rothlisberg PC, Sharples DJ (1990) The biology of the Penaeidae. Adv Mar Biol 27:1–489

Darwin C (1871) The descent of man and selection in relation to sex. Murray, London

David CT, Kennedy JS, Ludlow AR, Perry JN, Wall C (1982) A reappraisal of insect flight towards a distant point source of wind-borne odor. J Chem Ecol 8:1207–1215

Derby CD (1989) Physiology of sensory neurons in morphologically identified cuticular sensilla of crustaceans. In: Felgenhauer BE, Watling L, Thistle AB (eds) Functional Morphology of Feeding and Grooming in Crustacea. A.A. Balkema, Rotterdam, pp 27–47

Díaz ER, Thiel M (2003) Female rock shrimp prefer dominant males. J Mar Biol Assoc UK 83:941–942

Díaz ER, Thiel M (2004) Chemical and visual communication during mate searching in rock shrimp. Biol Bull 206:134–143

Duarte RC, Flores AAV (2017) Morph-specific habitat and sex distribution in the caridean shrimp *Hippolyte obliquimanus*. J Mar Biol Assoc U K 97:235–242

Dunham PJ (1988) Pheromones and behavior in Crustacea. In: Laufer H, Downer RGH (eds) Endocrinology of Selected Invertebrate Types. Alan R. Liss, New York, pp 375–392

Eens M, Pinxten R (2000) Sex-role reversal in vertebrates: behavioural and endocrinological accounts. Behav Process 51:135–147

Emlen DJ (2008) The evolution of animal weapons. Annu Rev Ecol Evol Syst 39:387–413

Emlen ST, Oring LW (1977) Ecology, sexual selection, and the evolution of mating systems. Science 197:215–233

Emlen DJ, Warren IA, Johns A, Dworkin I, Lavine LC (2012) A mechanism of extreme growth and reliable signaling in sexually selected ornaments and weapons. Science 337:860–864

Espinoza NL, Thiel M, Dupre E, Baeza JA (2008) Is *Hippolyte williamsi* gonochoric or hermaphroditic? A multi-approach study and a review of sexual systems in Hippolyte shrimps. Mar Biol 155:623–635

Felgenhauer BE, Abele LG (1982) Aspects of mating behavior in the tropical freshwater shrimp *Atya innocous* (Herbst). Biotropica 14:296–300

Feng T, Paterson B, Panagiotopoulou O, Green K, Mardon K, Johnston S (2018) The functional anatomy of external genitalia in the black tiger prawn (*Penaeus monodon*) studied by micro-computed tomography and scanning electron microscopy. J Morphol 297:1346–1354

Fiedler GC (1998) Functional, simultaneous hermaphroditism in female-phase *Lysmata amboinensis* (Decapoda: Caridea: Hippolytidae). Pac Sci 52:161–169

Forster GR (1951) The biology of the common prawn, *Leander serratus* (Pennant). J Mar Biolog Soc UK 30:330–360

Fryer G (1977) Studies of the functional morphology and ecology of the atyid prawns of Dominica. Phil. Trans R Soc Lond B Biol Sci 277:57–129

Gherardi F, Calloni C (1993) Protandrous hermaphroditism in the tropical shrimp *Athanas indicus* (Decapoda: Caridea), a symbiont of sea urchins. J Crustac Biol 13:675–689

Gilpin JA, Chadwick NE (2017) Life-history traits and population structure of Pederson cleaner shrimps *Ancylomenes pedersoni*. Biol Bull 233:190–205

Gleeson RA (1991) Intrinsic factors mediating pheromone communication in the blue crab, *Callinectes sapidus*. In: Bauer RT, Martin JW (eds) Crustacean Sexual Biology. Columbia Univ. Press, New York, pp 17–32

Goy JW (2010) Infraorder Stenopodidea Claus, 1872. In: Schram FR, von Vaupel Klein JC, Forest J, Charmantier-Daures M (eds) Treatise on zoology – anatomy, taxonomy, biology. The Crustacea, Decapoda. Vol. 9. Part A: Eucarida: Euphausiacea, Amphionidacea, and Decapoda (partim). Koninklijke Brill NV, Leiden, pp 215–265

Gwynne DT (2016) Sexual selection: roles evolving. Curr Biol 26:R913–R936

Hallberg E, Skog M (2011) Chemosensory sensilla in crustaceans. In: Breithaupt T, Thiel M (eds) Chemical communication in crustaceans. Springer Verlag, New York, pp 103–121

Hardege JD, Terschak JA (2011) Identification of crustacean sex pheromones. In: Breithaupt T, Thiel M (eds) Chemical Communication in Crustaceans. Springer Verlag, New York, pp 373–392

Hart BFL (1964) Shrimps of the genus *Betaeus* on the Pacific Coast of North America with descriptions of three new species. Proc U S Natl Mus 115:431–466

Hartmann G (1958) Apuntes sobre la biología del camarón de río, *Cryphiops caementarius* (Molina) Palaemonidae, Decapoda. Pesca y Casa 8:15–28

Hayd L, Anger K (2013) Reproductive and morphometric traits of *Macrobrachium amazonicum* (Decapoda: Palaemonidae) from the Pantanal, Brazil, suggests initial speciation. Rev Biol Trop 61:39–57

Herberholz J, Schmitz B (2001) Signaling via water currents in behavioral interactions of snapping shrimp (*Alpheus heterochaelis*). Biol Bull 201:6–16

Heuring WL, Hughes M (2020) Continuously choosy males and seasonally faithful females: sex and season differences underlie size-assortative pairing. Anim Behav 160:91–98

Höglund H (1943) On the biology and larval development of *Leander squilla* (L.) forma typica De Man. Svenska Hydrografisk–Biologiska Kommissiones Skrifter Stockholm Ny Serie Band II 6: 44

Hosken DJ, House CM (2011) Sexual selection. Curr Biol 21:R62–R65

Howard RW, Blomquist GJ (2005) Ecological, behavioral, and biochemical aspects of insect hydrocarbons. Annu Rev Entomol 50:371–393

Hudinaga M (1942) Reproduction, development, and rearing of *Penaeus japonicus*. Jpn J Zool 10: 305–393

Hughes M (1996) The function of concurrent signals: visual and chemical communication in snapping shrimp. Anim Behav 52:247–257

Iwasa Y, Pomiankowski A (1999) Good parent and good genes models of handicap evolution. J Theor Biol 200:97–109

Jaafar Z, Zeng Y (2012) Visual acuity of the goby-associated shrimp, *Alpheus rapax fabricius*, 1798 (Decapoda, Alpheidae). Crustaceana 85:1487–1497

Jensen GC (2014) Crabs and Shrimps of the Pacific Coast. A Guide to Shallow-water Decapods from Southeastern Alaska to the Mexican Border. Mola Marine, Bremerton, WA

Johnson VR (1969) Behavior associated with pair formation in banded shrimp *Stenopus hispidus* (Olivier). Pac Sci 23:40–50

Johnson VR Jr (1977) Individual recognition in the banded shrimp *Stenopus hispidus* (Olivier). Anim Behav 25:418–428

Jorquera E, Kenchington E, Ruzzante DE (2016) High prevalence of multiple paternity in the deep-sea shrimp *Acanthephyra pelagica*. Mar Biol 163:89. https://doi.org/10.1007/s00227-016-2868-0

Kamiguchi Y (1972) Mating behavior in the freshwater prawn, *Palaemon paucidens*. A study of the sex pheromone and its effect on males. J Fac Sci Hokkaido Univ Ser VI Zool 18:347–355

Karplus I, Barki A (2019) Male morphotypes and alternative mating tactics in freshwater prawns of the genus *Macrobrachium:* a review. Rev Aquac 11:925–940

Kelly LS, Snell TW (1998) Role of surface glycoproteins in mate-guarding of the marine harpacticoid copepod *Tigriopus japonicus*. Mar Biol 130:605–612

Kennedy JS (1983) Zigzagging and casting as a programmed response to wind-borne odour: a review. Physiol Entomol 8:109–120

Knibb W, Quinna J, Lamont R, Whatmore P, Nguyen NH, Remilton C (2014) Reproductive behaviour of captive *Fenneropenaeus merguiensis*: Evidence for monogamy and high between family variances for offspring number. Aquaculture 426–427:60–65

Knowlton N (1980) Sexual selection and dimorphism in two demes of a symbiotic, pair-bonding snapping shrimp. Evolution 34:161–173

Liker A, Freckleton RP, Székely T (2013) The evolution of sex roles in birds is related to adult sex ratio. Nat Commun 4:1587. https://doi.org/10.1038/ncomms2600

Limbaugh C, Pederson H, Chace FA Jr (1961) Shrimps that clean fishes. Bull Mar Sci Gulf Caribb 11:237–257

Liu X, Zhang D, Lin T, Yan Z (2019) Chemical cues mediate social monogamy in a marine caridean shrimp, *Lysmata debelius*. J Ethol 38:41–49

Lorch PD, Proulx S, Rowe L, Day T (2003) Condition-dependent sexual selection can accelerate adaptation. Evol Ecol Res 5:867–881

Lunina A, Vereshchaka A (2017) The role of the male copulatory organs in the colonization of the pelagic by shrimp-like eucarids. Deep-Sea Res Part II Top Stud Oceanogr 137:327–334

MacGinitie GE (1937) Notes on the natural history of several marine Crustacea. Am Midl Nat 18: 1031–1037

Maciá S, Robinson MP (2012) Reproductive pattern in the caridean shrimp *Gnathophylloides mineri* Schmitt (Gnathophyllidae), a symbiont of sea urchins. J Crustac Biol 32:727–732

Mathews LM (2002a) Territorial cooperation and social monogamy: factors affecting intersexual behaviours in pair-living snapping shrimp. Anim Behav 63:767–777

Mathews LM (2002b) Tests of the mate-guarding hypothesis for social monogamy: Does population density, sex ratio, or female synchrony affect behavior of male snapping shrimp (*Alpheus angulatus*)? Behav Ecol Sociobiol 51:426–432

Mathews LM (2003) Tests of the mate-guarding hypothesis for social monogamy: male snapping shrimp prefer to associate with high-value females. Behav Ecol 14:63–67

Mathews LM (2007) Evidence for high rates of in-pair paternity in the socially monogamous snapping shrimp *Alpheus angulosus*. Aquat Biol 1:55–62

McCullough EL, Tobalske BW, Emlen DJ (2014) Structural adaptations to diverse fighting styles in sexually selected weapons. Proc Natl Acad Sci 111:14484–14488

Misamore MJ, Browdy CL (1996) Mating behavior in the white shrimps *Penaeus setiferus* and *P. vannamei*: a generalized model for mating in *Penaeus*. J Crustac Biol 16:61–70

Murphy NP, Austin CN (2005) Phylogenetic relationships of the globally distributed freshwater prawn genus *Macrobrachium* (Crustacea: Decapoda: Palaemonidae): biogeography, taxonomy and the convergent evolution of abbreviated larval development. Zool Scr 34:187–197

Nakashima Y (1987) Reproductive strategies in a partially protandric shrimp, *Athanas kominatoensis* (Decapoda: Alpheidae): Sex change as the best of a bad situation for subordinates. J Ethol 5:145–159

Needler AB (1931) Mating and oviposition in *Pandalus danae*. Can Field Nat 45:107–108

Noël P (1976) L'évolution des caractères sexuels chez *Processa edulis* (Risso) (Décapode, Natantia). Vie Milieu 26:65–104

Obermeier M, Schmitz B (2003) Recognition of dominance in the big-clawed shrimp (*Alpheus heterochaelis* Say 1818). Part II. Analysis of signal modality. Mar Freshw Behav Physiol 36: 17–29

Omori M (1974) The biology of pelagic shrimps in the ocean. Adv Mar Biol 12:233–324

Omori M, Gluck D (1979) Life history and vertical migration of the pelagic shrimp *Sergestes similis* off the southern California coast. Fish Bull 77:183–198

Osawa Y, Aoki MN, Bauer RT, Thiel M (2015) Numbers and sizes of shrimps *Rhynchocinetes uritai* Kubo (Decapoda: Caridea) caught in bait and refuge traps. J Crustac Biol 35:786–775

Pérez Farfante I, Kensley B (1997) Penaeoid and sergestoid shrimps and prawns of the world. Keys and diagnoses for the families and genera. Mémoires du Muséum national d'Histoire naturelle 175:1–233

Planella L, Vera M, García-Marín JL, Heras S, Roldán MI (2019) Mating structure of the blue and red shrimp, *Aristeus antennatus* (Risso, 1816) characterized by relatedness analysis. Sci Rep 9: 7227. https://doi.org/10.1038/s41598-019-43523-w

Prakash S, Ajithkumar TT, Bauer R, Thiel M, Subramoniam T (2016) Reproductive morphology and mating behavior in the hingebeak shrimp *Rhynchocinetes durbanensis* Gordon, 1936 (Decapoda: Caridea: Rhynchocinetidae) in India. J Mar Biol Assoc U K 96:1331–1340

Ra'anan Z, Sagi A (1985) Alternative mating strategies in male morphotypes of the freshwater prawn *Macrobrachium rosenbergii* (De Man). Biol Bull 169:592–601

Rahman N, Dunham DW, Govind CK (2003) Social monogamy in the big-clawed snapping shrimp, *Alpheus heterochelis*. Ethology 109:457–473

Rahman ND, Dunham W, Govind CK (2004) Mate choice in the big-clawed snapping shrimp, *Alpheus heterochaelis* Say, 1818. Crustaceana 77:95–11

Rasch JA, Bauer RT (2016) The functional morphology and role of the thelycum in insemination, and its relation to the mating system in the seagrass shrimp *Ambidexter symmetricus* (Decapoda: Processidae). Invertebr Biol 135:163–173

Rojas R, Morales MC, Rivadeneira MM, Thiel M (2012) Male morphotypes in the Andean river shrimp *Cryphiops caementarius* (Decapoda: Caridea): morphology, coloration and injuries. J Zool 288:21–32

Rufino MM, Jones DA (2001) Binary individual recognition in *Lysmata debelius* (Decapoda: Hippolytidae) under laboratory conditions. J Crustac Biol 21:388–392

Saito T, Koya Y (2001) Gonadal maturation and embryonic development in the deep-sea sponge-associated shrimp, *Spongicola japonica* Kubo (Crustacea: Decapoda: Spongicolidae). Zool Sci 18:567–576

Saito T, Uchida I, Takeda M (2002) Skeletal growth of the deep-sea hexactinellid sponge *Euplectella oweni*, and host selection by the symbiotic shrimp *Spongicola japonica* (Crustacea: Decapoda: Spongicolidae). J Zool 258:521–552

Schein H (1975) Aspects of the aggressive and sexual behaviour of *Alpheus heterochaelis* Say. Mar Freshw Behav Physiol 3:83–96

Seibt U (1973) Sense of smell and pair-bond in *Hymenocera picta*. Micronesica 9:231–236

Seibt U, Wickler W (1979) The biological significance of the pair-bond in the shrimp *Hymenocera picta*. Z Tierpsychol 50:166–179

Shuker DM, Simmons LW (eds) (2014) The Evolution of Insect Mating Systems. Oxford University Press, Oxford

Solari P, Sollai G, Masala C, Loy F, Palmas F, Sabatini A, Crnjar R (2017) Antennular morphology and contribution of aesthetascs in the detection of food-related compounds in the shrimp *Palaemon adspersus* Rathke, 1837 (Decapoda: Palaemonidae). Biol Bull 232:110–122

Soledade GO, Santos PS, Araujo MSLC, Mantelatto FL, Almeida AO (2018) Heterosexual pairing in three *Alpheus* (Crustacea: Alpheidae) snapping shrimps from northeastern Brazil. Vie et Milieu-Life and Environment 68:109–117

Thiel M, Correa C (2004) Female rock shrimp *Rhynchocinetes typus* mate in rapid succession up a male dominance hierarchy. Behav Ecol Sociobiol 57:62–68

Thiel M, Chak STC, Dumont CP (2010) Male morphotypes and mating behavior of the dancing shrimp *Rhynchocinetes brucei* (Decapoda: Caridea). J Crustac Biol 30:580–588

Triay-Portella R, González JA, Biscoitob M, Ruiz-Diaz R, Pajuelo JG (2018) Sex with the lights off: Can a morphological structure reveal the sex and functional sexual maturity in the genus *Plesionika*? Deep-Sea Res I 142:69–76

Vereshchaka AL, Olesen J, Lunina AA (2016) A phylogeny-based revision of the family Luciferidae (Crustacea: Decapoda). Zool J Linnean Soc 178:15–32

Vereshchaka AL, Lunina AA, Olesen J (2017) The genus *Gennadas* (Benthesicymidae: Decapoda): morphology of copulatory characters, phylogeny and coevolution of genital structures. R Soc Open Sci 4:171288. https://doi.org/10.1098/rsos.171288

Webster DR, Weissburg MJ (2009) The hydrodynamics of chemical cues among aquatic organisms. Annu Rev Fluid Mech 41:73–90

Wickler W (1973) Biology of *Hymenocera picta* Dana. Micronesica 9:225–230

Wickler W, Seibt U (1981) Monogamy in Crustacea and Man. Z Tierpsychol 57:215–234

Wirtz P (1997) Crustacean symbionts of the sea anemone *Telmatactis cricoides* at Madeira and the Canary Islands. J Zool (Lond) 242:799–811

Wong JWY, Michiels NK (2011) Control of social monogamy through aggression in a hermaphroditic shrimp. Front Zool 8:30

Xiao Y, Greenwood JG (1993) The biology of *Acetes* (Crustacea, Sergestidae). Oceanogr Mar Biol Annu Rev 31:259–444

Yano I, Kanna RA, Oyama RN, Wyban J (1988) Mating behavior on the penaeid shrimp *Penaeus vannamei*. Mar Biol 97:171–175

Yew JY, Chung H (2015) Insect pheromones: An overview of function, form, and discovery. Prog Lipid Res 59:88–105

Yue GH, Chang A (2010) Molecular evidence for high frequency of multiple paternity in a freshwater shrimp species *Caridina ensifera*. PLoS One 5(9):e12721. https://doi.org/10.1371/journal.pone.0012721

Zahavi A (1975) Mate selection-a selection for a handicap. J Theor Biol 53:205–214

Zhang D, Lin J (2006) Mate recognition in a simultaneous hermaphroditic shrimp, *Lysmata wurdemanni* (Caridea: Hippolytidae). Anim Behav 71:1191–1196

Zhang D, Lin J, Harley M, Hardege JD (2010a) Characterization of a sex pheromone in a simultaneous hermaphroditic shrimp, *Lysmata wurdemanni* (Caridea: Hippolytidae). Mar Biol 157:1–6

Zhang D, Lin J, Hardege JD (2010b) Surface glycoproteins are not the contact pheromones in the genus *Lysmata*. Mar Biol 157:171–176

Zhang D, Terschak JA, Harley MA, Lin J, Hardege JD (2011) Simultaneously hermaphroditic shrimp use lipophilic cuticular hydrocarbons as contact sex pheromones. PLoS One 6(4): e17720. https://doi.org/10.1371/journal.pone.0017720

Chapter 9
Life Histories

9.1 What Is a Life History?

The life history of an organism is an account of the characteristic events which occur from its birth until its death. These include development, growth, reproduction, and the shrimp's demise by senescence, accident, or, most often, predation. The recurring sequence of these events that produce the next generation of offspring is the life cycle. Important adaptations that contribute to successful survival and reproduction are termed "life-history traits." Examples are adaptations related to birth (hatching in shrimps), embryonic development, parental care or lack thereof, larval development, and dispersal prior to settlement in the adult habitat, followed by growth, reproduction, and mortality (life span). Many shrimps engage in migrations during one or more parts of their life cycle. The particular set of life-history traits of the life cycle is termed the "life-history strategy" of a species.

A description, or the "what," of a life-history trait naturally raises the question of the "how" and "why" of the life cycle. What are the physiological and morphological mechanisms (proximate causes) that result in the occurrence or nature of a particular life-history trait or pattern? What are the underlying selective pressures (ultimate causes) that explain the evolution of different life-history traits in various taxa (phylogenetic groups) of shrimps? The comparative method, as in other areas of biology, is a powerful tool for analyzing the "what," "how," and "why" of life histories. By comparison of life-history traits among diverse species from different habitats, common patterns emerge which are associated with environmental and biological factors (ultimate causes) which may explain these patterns. An understanding of shrimp life histories is important for the ecological well-being of aquatic communities, species conservation, management of fisheries, and the aquaculture of commercially important species.

9.2 Embryonic Development and Hatching

After an egg spawned by a female shrimp is fertilized, it goes through an embryonic development in which the basic body morphology of the larval form takes place (Fig. 9.1). The embryonic development of penaeoid and sergestoid (dendrobranchiate) shrimps is very different from that of the caridean and stenopodidean (pleocyemate) shrimps. Dendrobranchiate eggs are small with little yolk, which results in a hatching larval stage, the nauplius, with only three pairs of appendages (Fig. 9.2). This is the primitive crustacean larva which occurs in many other crustacean taxa. In penaeoid and sergestoid shrimps, females spawn fertilized eggs at night directly into the surrounding seawater. The fertilized egg is on its own in the sea because sergestoids and penaeoids do not incubate their embryos during development. However, in luciferids, fertilized eggs are attached briefly to the female but probably with little or no care by the female (Lee et al. 1992). Embryonic development takes anywhere from 8 to 24 hours in penaeoids (Dall et al. 1990). Spawning to hatching in sergestoids and luciferids is usually about 30 hours (Omori 1974; Lee et al. 1992), but developmental times vary with temperature (Omori 1979).

On the other hand, the oocytes spawned by carideans and stenopodideans are much larger and full of yolk, attached after fertilization below the abdomen and incubated (groomed, oxygenated) by the female until hatching. This allows a prolonged embryonic development with hatching as a more advanced larval stage, the zoea, with 8 pairs of appendages (Fig. 9.2). The period of embryonic development (spawning to hatching) is longer than in dendrobranchiates, from as short as 4–5 days in some small tropical carideans up to several months to 1–2 years in high-latitude and polar species.

Studies on shrimp embryology show that the basic dissimilarities between dendrobranchiate and pleocyemate development are primarily due to a large difference of embryo yolk content between the two groups as well as their phylogenetic history (Zilch 1978; Anderson 1982; Saito and Koya 2001; Müller et al. 2003; Hertzler 2005; Romero et al. 2018). The best general description of penaeoid embryonic development is that by Zilch (1978) on the penaeoid *Melicertus kerathurus* (as *Penaeus trisulcatus*) (see also Dall et al. 1990; Hertzler 2005). The cell division (cleavage) is holoblastic (cells divide completely), and early cell divisions form a blastula which is essentially a layer of embryonic tissue (blastoderm) surrounding a yolk mass, termed a germinal disk or blastodisk. Later, formation of a pocket or space into the blastula results in a typical invagination gastrula (Fig. 9.1a) with cells that will give rise to two of the basic cell layers: the endoderm (midgut; digestive system) and mesoderm (body segments and appendages). The developing naupliar body, its appendages, and the naupliar eye gradually take form within a germinal disk wrapped around the remaining yolk. This latter material will sustain the young shrimp for the remainder of embryonic development and the early part of larval development. The gut of sergestoid and penaeoid naupliar larvae is incomplete; they do not feed but subsist on remaining embryonic yolk until the first

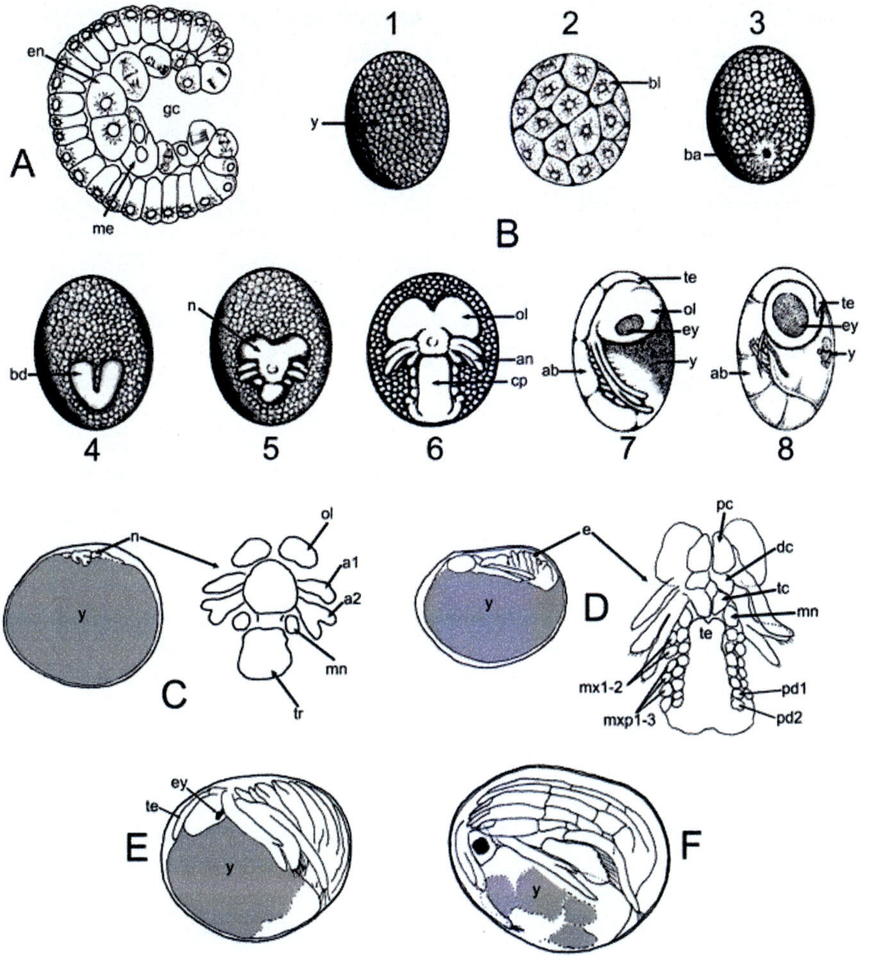

Fig. 9.1 Embryonic development. (**a**) Gastrula of penaeid *Melicertus kerathurus*. (**b**) Caridean *Macrobrachium olfersii*: 1, fertilized egg (embryo); 2, embryo with blastomeres; 3, with developing blastopore; 4, with developing blastodisk; 5, with embryonic nauplius; 6–7, developing embryo; 8, embryo just prior to hatching. (**c**–**f**). Stenopodidean *Spongicola japonica*: (**c**) embryonic naupliar stage; (**d**) embryo, further development, ventral view; (**e**) embryo at pigment eyespot stage; (**f**) embryo near hatching. a1, antennule; a2, antenna; ab, abdomen; an, antennae; ba, blastopore area; bd, blastodisc (germinal disk); bl, blastomere; cp, caudal papilla; e, developing embryo; en, endoderm; ey, eye; dc, deutocerebrum; gc, gastrocoel; me, mesoderm; mn, mandible; mx1,2, maxilla 1,2; mxp 1–3, maxillipeds 1–3; n, embryonic nauplius; ol, optic lobe; pc, protocerebrum; pd1,2, pereopods 1,2; tc, tritocerebrum; te, telson; tr, trunk; y, yolk. Modified from (**a**) from Zilch (1978), (**b**) Müller et al. (2003) courtesy of Journal of Crustacean Biology and (**c**–**f**) Saito and Koya (2001)

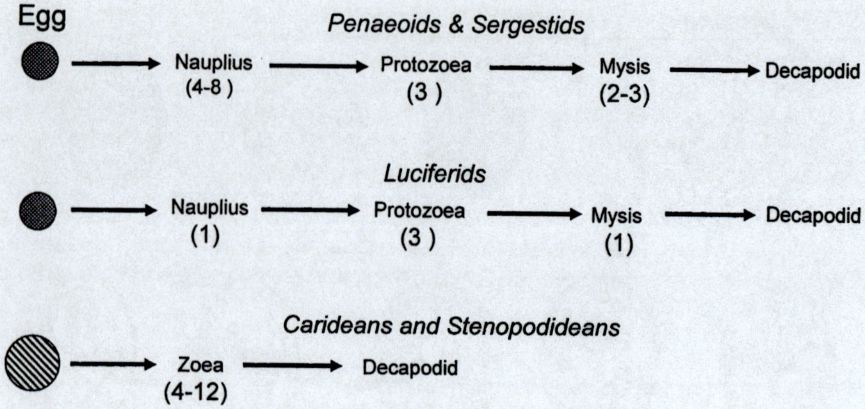

Fig. 9.2 Number and type of larval stages in shrimp species with extended (planktonic) development. Based primarily on Martin et al. (2014), Knight and Omori (1982), and Lee et al. (1992)

feeding stage, the protozoea (Omori 1974, 1979; Dall et al. 1990). When fully developed, the nauplius pushes against and punctures the embryonic membranes with its antennae and the spines at its tail end, breaking free as a hatched swimming larva. The pattern of holoblastic cleavage and an invagination gastrula are primitive crustacean embryological characters shared with the euphausiaceans, the sister group of decapod shrimps, in which hatching as a nauplius also occurs.

Caridean embryology differs from that of dendrobranchs in some important ways (Müller et al. 2003; Romero et al. 2018). After fertilization (Fig. 9.1b_1), the early cell divisions are complete (Fig. 9.1b_2) but later become superficial; i.e., the daughter cells are not completely separated. Cells from this equivalent of a blastula invade the yolk and end up at one pole of the embryo to form a blastopore (Fig. 9.1b_3). This ingression of cells is gastrulation in carideans, unlike the invagination gastrula of dendrobranchiates, but likewise will give rise to the endoderm and mesoderm. The area around the blastopore will generate a germinal disk (Fig. 9.1b_4) from which a transitory embryonic naupliar stage (Fig. 9.1b_5) can be recognized by its antennular, antennal, and mandibular rudiments and caudal (tail) papilla. The latter structure will produce the remaining body somites and appendages of the pre-zoea (Fig. 9.1b_{6-7}). A neurectoderm gives rise to the eyes, incipient brain, and exoskeleton. An ectodermal invagination, the stomodeum, forms the mouth and entrance to the midgut. The developing zoea is folded within the embryo, with abdomen against cephalothorax as it uses up much or most of the remaining yolk near hatching (Fig. 9.1b_{7-8}).

Relatively little is known about the embryology of the Stenopodidea, with observations by Brooks and Herrick (1891) on *Stenopus hispidus* and Saito and Koya (2001) on *Spongicola japonica*. In *S. hispidus*, development appears similar to that of carideans, i.e., superficial cleavage, formation of a blastoderm, gastrulation by ingression of cells, and development of a germinal disk. These similarities may have both a phylogenetic basis (both carideans and stenopodideans are pleocyemates) and a structural component (a relatively large amount of yolk, incubatory behavior, and hatching as a zoea). Saito and Koya's description of embryonic development of *Spongicola japonica* does not have much detail on the prenaupliar stages, but the naupliar and pre-zoeal stages are illustrated (Fig. 9.1 c–f). Development in this *Spongicola* species, unlike that of *S. hispidus*, is direct; i.e., there are no larval stages. The developing embryo passes through both naupliar and zoeal stages in the embryo, hatching as a benthic juvenile. Clearly, more studies are needed to fill in the gaps in knowledge about the embryology of stenopodideans.

9.3 Larval Development: Costs and Benefits

The life history of most shrimps includes larval stages in which further growth and development from the hatching stage, naupliar or zoeal, takes place. The growth stages, or instars, are accomplished by molting in which successive larval stages develop further toward the juvenile stage which will live in the adult habitat. Larval stages may be planktotrophic, i.e., feed on smaller organisms or particulate organic matter in the plankton. Most marine shrimps and many freshwater ones go through several planktonic larval stages after hatching—extended larval development (ELD). Others shortcut the larval development by hatching as an already advanced swimming lecithotrophic (non-feeding) larva. They derive their nutrition from yolk left over from their embryonic period and go through just a few further larval stages before becoming juveniles—abbreviated larval development (ALD). In some species, shortening of larval development goes even further. The shrimp hatches as a small juvenile (or nearly so) without a swimming larval stage—direct development (DD).

Each of these larval tactics has its costs and benefits (trade-offs). For example, ELD species have the advantage of dispersal far and wide via ocean currents. At least some larvae from a female's spawn will be carried to a suitable adult habitat or perhaps invade new favorable locales. However, many or most larvae will not find the right place to metamorphose and will die. ELD species need to produce lots of larvae to prevent this loss as well as that from the high mortality larvae suffered in the plankton. A favorable consequence of ELD is that high dispersal promotes genetic mixing among disparate populations; however, it may inhibit speciation and taxonomic diversity. On the other hand, the larvae of ALD species have limited dispersal, stay closer to the adult habitat, and suffer less predation in the water column because of their shorter stay there. The hatching juveniles of DD species are immediately released, stay in the adult habitat, and do not suffer predation in the plankton, but may compete with adults for food, space, or other resources. The limited dispersal of ALD and DD species may result in geographic and genetic isolation among populations and thus promote speciation and higher taxonomic diversity.

9.4 Extended Larval Development

Given that the larval life of shrimps and other crustaceans is such an important part of a species life history, considerable work has been done on larvae from the late nineteenth century to the present (e.g., Gurney 1942; Hardy 1956; Martin et al. 2014). Many publications on shrimps are morphological description of larval stages so that when larvae are collected, usually in plankton samples, they can be identified as closely as possible to species. Eggs and early larvae collected live from the plankton may be reared and then linked to the resultant postlarvae and/or juveniles of the parent species. Females of many shrimp species can be induced to spawn and hatch in the laboratory; the larvae are then reared and described. As a result, there is an abundant literature on larval development of many shrimp taxa. Such information is very useful for developmental and aquatic biologists, as well as fisheries scientists and aquaculturists. Excellent summaries of past studies, larval morphology, and terminology are found in the pertinent chapters of "Atlas of Crustacean Larvae," edited by Martin et al. (2014). Other key larval references include Anger (2001), which deals with many aspects of decapod larval development and physiology, and Williamson (1982). The general sequence of larval stages is compared in Fig. 9.2 among the major shrimp taxa with extended development. It should be noted that in the following summaries below, the details may vary within and among taxa.

9.4.1 Dendrobranchiata: Penaeoids and Sergestoids

The embryos of penaeoids and sergestoids hatch out as nauplii (Fig. 9.3a, d) (e.g., Pearson 1939; Knight and Omori 1982; Omori 1979; Choi and Hong 2001). As in other crustacean nauplii, the dendrobranchiate *nauplius* has an oval-shaped body with three pairs of appendages: uniramous antennules, biramous antennae, and mandibles consisting of a coxa and a palp. These limbs produce the jerky forward swimming characteristic of nauplii. A median naupliar eye is usually present (Fig. 9.3d). The naupliar stages of dendrobranchiates do not feed (Omori 1974, 1979; Dall et al. 1990), using leftover embryonic yolk for nutrition. In dendrobranchiate and other shrimps, development is anamorphic, i.e., gradual, not metamorphic. Appendages, somites, and other body parts are added from one instar to the next. In most dendrobranchiates, there are 4–8 naupliar stages. With each naupliar molt, the limb buds (rudiments) of the remaining head appendages take shape. The latter naupliar stages are often termed "metanauplii." The next major phase consists of three "*protozoeal*" stages (Fig. 9.3b, e). All head appendages are functional, and the antennules and antennae continue to participate in forward swimming. The first protozoea consists of a head with sessile (unstalked) compound eyes, a thorax partially covered by a carapace, small maxillipeds, and an abdomen. Body somites become distinct in the second and third protozoeal stages as do rudiments of the pereopods; the compound eyes become stalked. The bilobed telson

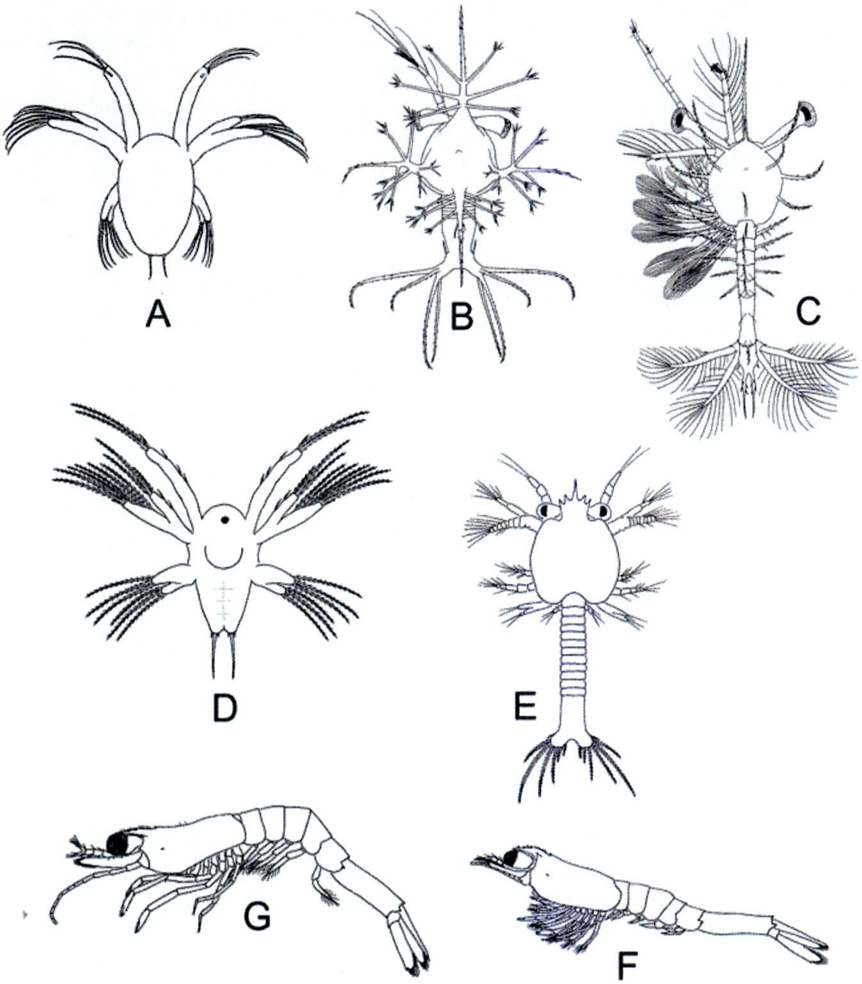

Fig. 9.3 Examples of dendrobranchiate larval stages. (**a–c**) *Eusergestes similis* (Sergestidae) (dorsal views). In (**c**), some appendages on right side of body are not illustrated for clarity. (**a**) Nauplius 1: note that naupliar stages lack a median naupliar eye in this species (cf. 9.3d). (**b**) Protozoea 2. (**c**) Mysis (Zoea) 1. (**d–g**) *Metapenaeopsis dalei* (Penaeidae): (**d**) Nauplius 4, ventral view; note median naupliar eye. (**e**) Protozoea 2, dorsal view. (**f**) Mysis 2, lateral view. (**g**) Decapodid (lateral view). From (**a–c**) Knight and Omori (1982) and (**d–g**) Choi and Hong (2001)

serves not only in locomotion but also as a sensory and grooming limb. In the last protozoeal stage, the uropods appear alongside the telson.

The next step in the series are 2–3 instars of "*mysis*" larvae, with a more shrimp-like body shape (Fig. 9.3f). Mysis larvae assume a vertical position in the water column to swim backward using the large natatory exopods on the now well-developed pereopods. This is a major change from forward swimming by the head appendages in the naupliar and protozoeal stages. Pleopod buds appear on the abdomen. The next distinct stage is the "*decapodid*" (sometime termed the megalopa), with the larva switching from primarily thoracic exopodal backward swimming to typical shrimp forward swimming with large setose pleopods (Fig. 9.3g). The shrimp larva assumes the horizontal body orientation of the juvenile and adult. The number of decapodid stages is variable. The decapodid is often called a "postlarva" by ecologists and fisheries and aquaculture biologists. However, as Anger (2001) has pointed out, the term postlarva should refer to the various instars between decapodid and juvenile, the latter defined by loss of all larval characters. In practical terms, for a benthic species, it is the settling stage: the postlarva transitions from a planktonic to a bottom way of life. In the luciferid sergestoid *Belzebub faxoni*, there is a single naupliar stage, 3 protozoeas, and a mysis before a "postlarva" (decapodid/juvenile) (Lee et al. 1992).

9.4.2 Carideans

The hatching stage in carideans with ELD is a basic *zoea*, a decapod larva with all head appendages, three pairs of maxillipeds bearing functional natatory exopods, and compound eyes (Fig. 9.4a, e). In ELD species, there are several zoeal stages, as many as 11 or more. The number varies among or even within species, depending both on taxon and on environmental conditions that the larva encounters in the plankton. The first three zoeal stages are easy to identify. Zoea 1 has sessile compound eyes (Fig. 9.4a, Plate 9.1a) which become stalked upon molting to Zoea 2 (Fig. 9.4b, Plate 9.1b) and remain so thereafter. In zoea 1 and 2, the uropods have not yet emerged from the last abdominal somite, but in Zoea 3, the uropods are found alongside the telson in their normal decapod shrimp position (Fig. 9.4c, Plate 9.1c). In subsequent zoeal stages, the pereopods develop, first as small limb buds which then become more complete with pereopodal exopods for swimming (Fig. 9.4d). All zoeal stages have a carapace covering the thorax. The zoeae of most caridean species swim upside down, tail facing forward (Fig. 9.4e). They can move about in any direction using their propeller-like thoracic exopods. The larva can also backflip, i.e., the abdominal-flexion escape response. Following the zoeal stages, the larva becomes a decapodid. It now has functional pleopods; the larva can now swim forward with these limbs as well as flip backward. The thoracic exopods are still functional, however, and also participate in swimming. In future molts, these exopods become reduced, and the transitional stages in which these and other larval characters are lost are termed "postlarvae" as in the dendrobranchiates. By this time, the individual has settled into the adult habitat as a juvenile. The period of larval development varies from a few weeks to several months, depending on the species and environmental conditions, especially temperature.

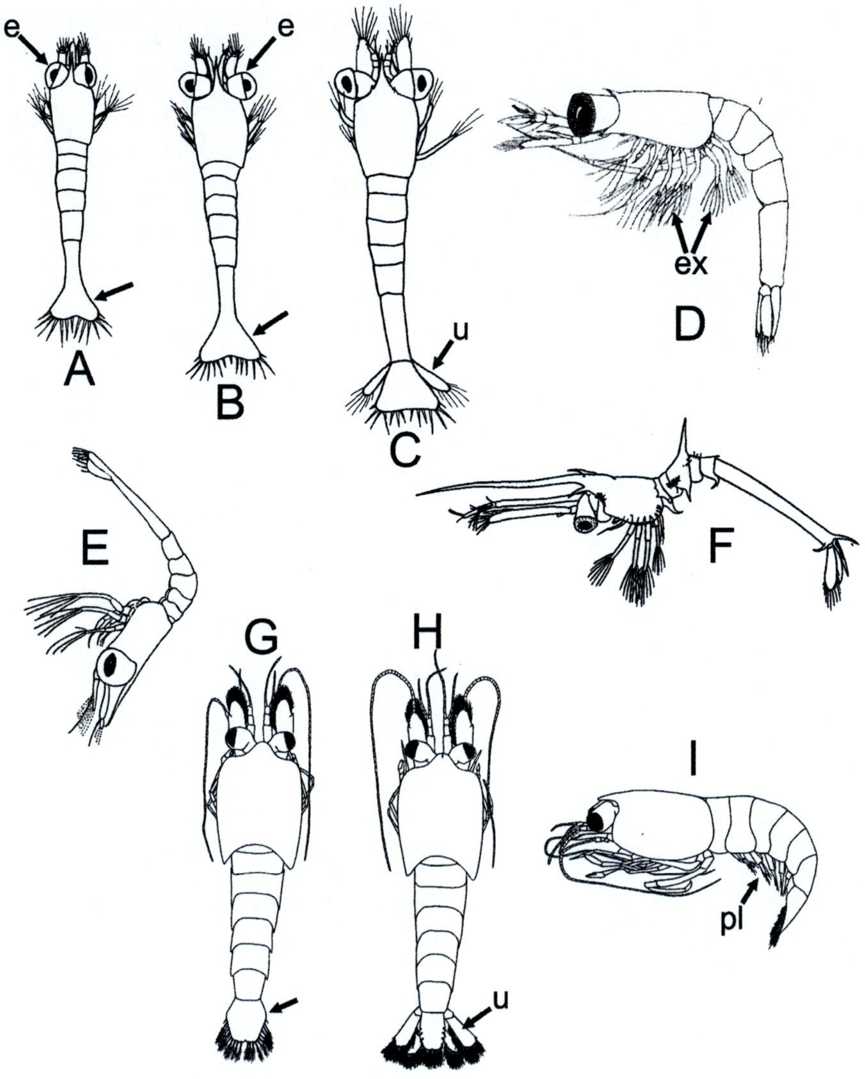

Fig. 9.4 Examples of caridean and stenopodidean larval stages. (**a–c**) *Paratya australiensis* (Atyidae): (**a**) Zoea 1. (**b**) Zoea 2. (**c**) Zoea 3. Note the change from sessile to stalked eyes from zoea 1 to 2 and emergence of uropods from zoea 2 to 3 (unmarked arrows). (**d**) Zoea 4 of *Alpheus estuariensis* (Alpheidae). (**e**) Typical vertical, upside-down swimming position of caridean zoeas (*Palaemon serratus*, Palaemonidae). (**f**) Zoea 4 of *Stenopus hispidus* (Stenopodidae). (**g–i**) *Lancaris singhalensis* with abbreviated but nearly direct development: (**g**) Stage-1 larva, dorsal view. (**h**) Stage-2 larva. (**i**) Stage-1 larva, lateral view. Note emergence of uropods in the stage-2 larva. e, eye; ex, pereopodal exopods; pl, pleopods; u, uropod; unmarked arrows indicate where uropods will be located when they emerge. Modified from (**a–c**) Walsh (1993), (**d**) Pires et al. (2008), Creative Commons BY 4.0, (**e**) from Bauer (2004) after Sollaud (1923), (**f**) Gurney (1942), and (**g–i**) Benzie and de Silva (1983). (**a–c, g–i**), courtesy of the Journal of Crustacean Biology

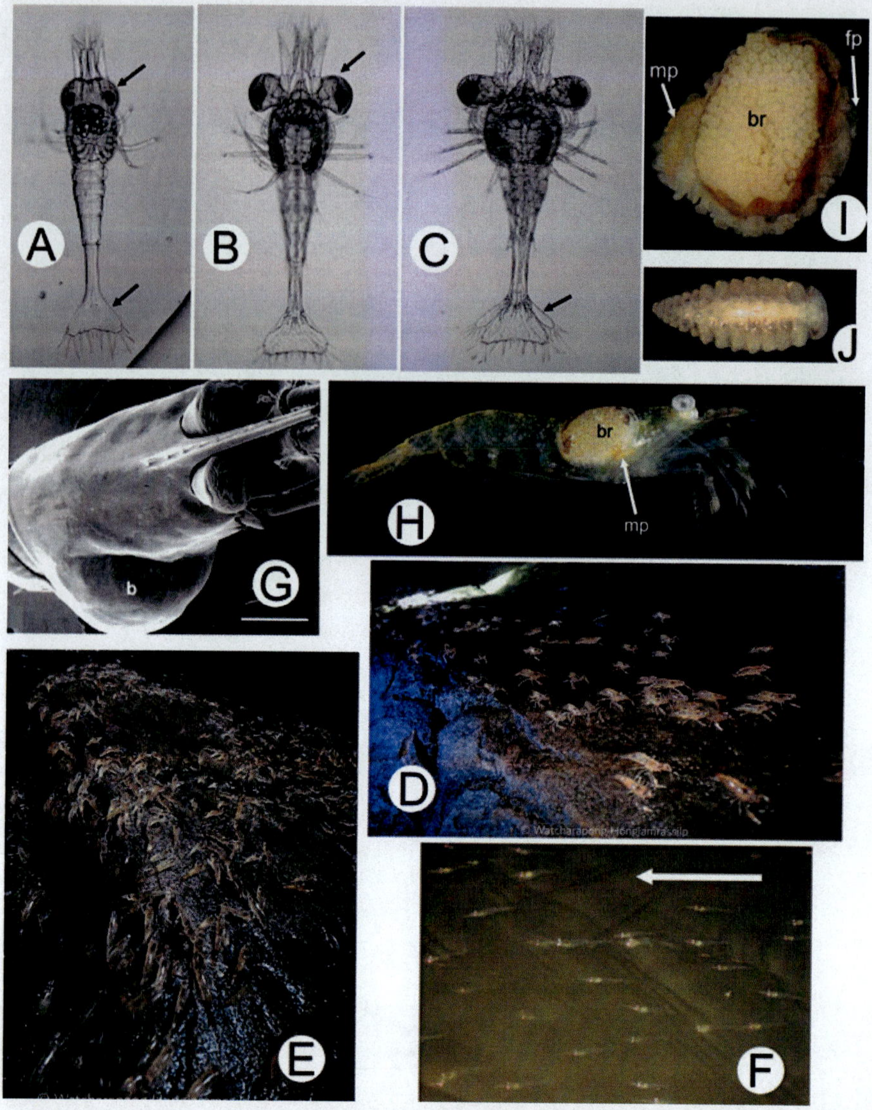

Plate 9.1 Larvae; mass movement of juveniles; bopyrid parasites. (**a–c**) Early larval stages of the caridean *Macrobrachium ohione*: (**a**) Zoea 1: upper arrow points to sessile (unstalked) eye; lower arrow indicates lack of uropods. (**b**) Zoea 2: arrow points to stalked eye. (**c**) Zoea 3: arrow points to newly emerged uropods. (**d**) Individuals (mostly juveniles) of *Macrobrachium dienbienphuense* crawling upstream, out of the water in the wet zone of the bank, to move around a stream barrier (e.g., rapids, waterfall). (**e**) Same as (**d**) but note individuals crawling up and over a vertical surface (left) to continue in the upstream direction. (**f**) Migrating juveniles of *M. ohione* swimming upstream (white arrow) along a riverbank. (**g**) Cephalothorax of a *Palaemonetes pugio* individual infected by a bopyrid parasite; note bulge on the right side of the carapace caused by the parasite pair within the gill chamber. (**h**) Processid *Ambidexter symmetricus* infested by the bopyrid parasite *Urobopyrus processae;* parasite female brood chamber filled with late-stage embryos. (**i**) Female of *U. processae* with dwarf male characteristically clinging to female abdomen; brood chamber filled with early-stage embryos. (**j**) Dwarf male of *U. processae*. b, bulge in carapace covering

9.4.3 *Stenopodideans*

The larval development of relatively few stenopodideans is known compared to dendrobranchiates and carideans. It is often difficult to obtain sufficient material to study larval development in this low diversity group. One reason for their scarcity is their monogamous mating system. Stenopodideans occur in male–female territorial pairs, which results in low abundance and density. Furthermore, most spongicolids are deep-sea species associated with glass (hexactinellid) sponges, so sampling of live material is technically difficult if not yet impossible in that group. Nonetheless, over the years, larval biologists and stenopodidean experts have been able to describe larval ontogeny in several species. Development of most Stenopodidae studied is extended and planktonic (ELD) while that of Spongicolidae is usually abbreviated or direct (ALD, DD; see below). The embryos of ELD species hatch out as a *zoea*, but besides the typical head appendages and three pairs of natatory maxillipeds of other decapod zoeae, the first pereopod is present and equipped with swimming exopods. Regarding the compound eyes (stalked or unstalked) and uropods (present or absent), the differences among the first three zoeal stages of ELD stenopodideans are the same as those described for the carideans. The number of zoeal stages in different species ranges from 4–5 to more usually 9–11. Some of the more distinctive features of stenopodidean zoeae are the massive mandibles accommodated by a notch in the anterior carapace. The rostrum is very long and thin (Fig. 9.4f). In later zoeal stages, the abdomen becomes sharply bent downward, sometimes at nearly right angles, between the 3rd and 4th abdominal somites. As in carideans, the *decapodid* acquires functional pleopods and forward swimming, and exopodal swimming by the thoracic appendages is reduced and then lost. The decapodid seeks out the bottom to settle and eventually lose all larval characters to become a juvenile, similar to the adult in morphology but still without reproductive function.

9.5 Abbreviated and Direct Larval Development

Numerous caridean and most spongicolid stenopodideans have evolved a shortened larval life, i.e., abbreviated development (ALD) in which there are only a few lecithotrophic larval stages. Other species have carried this further and eliminated all larval development. In these direct development (DD) species, the embryos hatch

Plate 9.1 (continued) branchial chamber indicates a parasite pair within; br, brood of embryos within female parasite's broods chamber; fp, head region of female parasite; mp, male parasite. From (**a–b, f**) Bauer and Delahoussaye (2008), (**c**) R.T. Bauer, (**d–e**) courtesy of Watcharapong Hongjamrassilp, (**g**) Cash and Bauer (1993), (**h–j**) Rasch and Bauer (2015), Creative Commons CC BY 4.0. (**a–c, f–g**) courtesy of Journal of Crustacean Biology

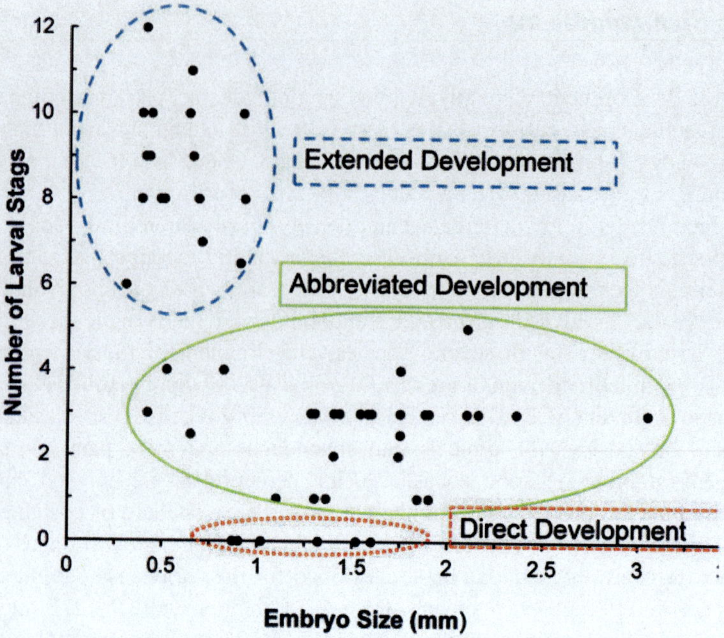

Fig. 9.5 Number of larval stages related to the size of early-stage embryos with extended (ELD), abbreviated (ALD), or direct development (DD) for 57 amphidromous or completely freshwater caridean species. Embryo size (diameter) defined as in Fig. 6.13, Chap. 6. Data from numerous sources (32 species from fig. 8.2 in Bauer 2004, plus 25 additional ones)

out as juveniles or nearly so; the distinction between ALD and DD is not always sharply dichotomous. These types of development usually evolve as a life-history tactic in habitats which are harsh or unfavorable for planktotrophic larvae or when a dispersal stage is disadvantageous in the life cycle.

In order to sustain the prolonged embryonic development needed to produce the more morphologically advanced hatching stage of ALD and DD species, a larger amount of nutritive yolk is packed into the egg cell. Consequently, embryos of ALD and DD species are large compared to ELD species (Figs. 9.5 and 9.6). Females of ELD caridean and stenopodid species incubate early (recently spawned) embryos which are usually less than 1 mm in diameter and as small as 0.22 (*Hippolyte obliquimanus*; Bauer 1991). Early-stage embryos of ALD and ELD species are usually between 1 and 3 mm in diameter, but as much as 4.5 mm (*Parapasiphae sulcatifrons*; Omori 1974).

The hatching larva of ALD species is always more advanced than the stage-1 larva of ELD species; its characters are a variable mix of early- and later-stage ELD zoeae. One example is *Thor floridanus,* a small marine hippolytid caridean from seagrass beds in south Florida. Its close relative, *T. dobkini*, is a small-embryo ELD species with at least 6 zoeal stages, while *T. floridanus* has larger embryos and ALD development with only 2 lecithotrophic zoeal stages and a decapodid. The first zoea

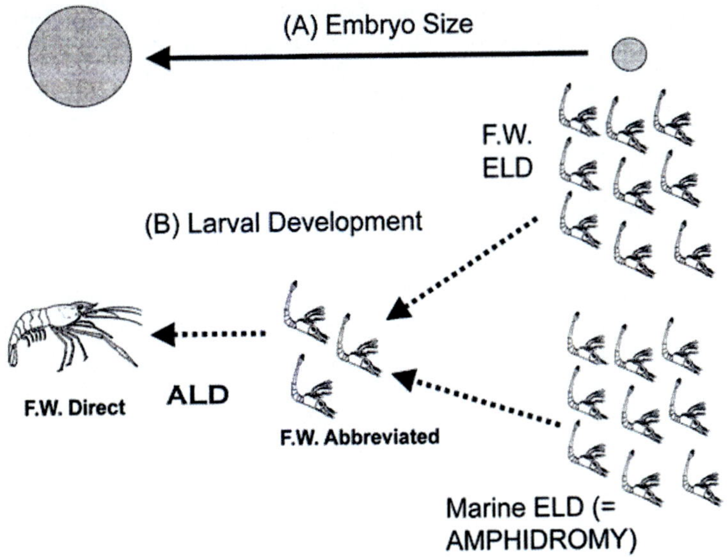

Fig. 9.6 The relationship among embryo size, larval development, and life history in freshwater shrimps. Species with small eggs and extended larval development (ELD) are generally amphidromous, although several ELD species are completely freshwater (F.W). Freshwater species with large embryos and abbreviated or direct larval development (ALD) are thought to be descended from ELD species. From Bauer (2013)

of *T. floridanus* is advanced in development in that, besides the sessile eyes, mouthparts, and maxillipeds with natatory exopods of typical ELD zoea 1, it already has nonfunctional pereopods and pleopods buds, characters that do not appear until later ELD zoeal stages. The cephalothorax contains a large amount of yolk, typical of non-feeding lecithotrophic larvae. The second zoeal stage is similar, but the eyes are stalked and free from the carapace, as in ELD species. The pereopods continue to develop but are not yet functional, and the uropods can be seen under the cuticle of the telson. With the next molt, the larva becomes a decapodid, the settling stage with functional pleopods and loss of natatory exopods on the thoracic appendages. The uropods become free on either side of the telson. Larval development takes only 2–4 days, depending on temperature compared to the weeks or months of ELD species (fig. 8.1 in Bauer 2004).

The completely freshwater *Lancaris* (*Caridina*) *singhalensis* (Atyidae) shows nearly direct development (DD). The hatching stage is much more precocious than that of *T. floridanus* and other ALD species, almost a juvenile, complete with all mouthparts, pereopods, and pleopods (Fig. 9.4g–i) (Benzie and de Silva 1983). It is benthic with functional walking legs and pleopods. Unlike zoeae of ELD species and ALD species like *T. floridanus*, it lacks natatory exopods on the thoracopods. The abdomen is already fully segmented. However, it is still considered a larva (but not a zoea) as it does not yet have uropods, i.e., a complete tail fan, and other appendage details are not fully formed. Larva 1 molts within a day of hatching to larva 2, at which it gains those uropods (Fig. 9.4h). At the next molt, it loses all larval features and is completely juvenile (Fig. 9.4i). Benzie and Silva found organic matter in the guts of larvae, so they apparently do feed before reaching the juvenile stage unlike most ALD species.

9.6 Developmental Strategies of Freshwater Species

Freshwater shrimps are all carideans, including many species of Palaemonidae (e.g., *Macrobrachium*), nearly all Atyidae (e.g., *Caridina, Atya*), Xiphocarididae, Euryrhynchidae, and Desmocarididae but just a few Alpheidae (Chap. 3, Families of Shrimps). Most live in tropical to warm-temperate habitats. Together, they compose about 25% of all shrimp species worldwide (De Grave et al. 2008). The consensus from a variety of paleontological, biogeographical, ecological, developmental, and phylogenetic studies is that freshwater shrimps are descended from marine ancestors (Bauer 2011a, 2011b). These ancestral species entered freshwater streams and rivers via estuaries, and from there into other freshwater habitats. There they found reduced competition, empty ecological niches, and reduced predation in some freshwater habitats, e.g., oceanic island streams with an impoverished fish fauna, or upstream of waterfalls and natural weirs difficult to access by fishes (McDowall 2007; Covich et al. 2009).

Most marine species have ELD, but in adaption to diverse freshwater habitats, various species found different solutions for successful larval development and recruitment. This process of adaptation by marine shrimps to freshwater, termed "freshwaterization" by Jalihal et al. (1993), has taken three principal paths (Figs. 9.5, 9.6): (1) production of fewer but larger embryos hatching to just a few advanced lecithotrophic larvae or juveniles (ALD, DD); (2) retention of broods with many small embryos and development in marine waters extended over several feeding larval stages (amphidromy), and, less frequently, (3) complete adaptation of ELD to freshwater habitats.

There are two major selection pressures determining whether a freshwater shrimp species has ELD or ALD/DD. One is whether larvae will encounter enough food in the plankton of their environment to sustain them through development and settlement as juveniles. The other has to do with the weak swimming ability of planktonic shrimp larvae. The majority of ALD and DD species are found in lentic bodies of freshwater such as ponds, lakes, seasonally inundated lowland forests and swamps, cave systems, and other freshwater habitats with limited or no access to the sea. In lotic habitats, i.e., rivers and streams, the weakly swimming planktonic larvae are carried away from their home habitat by rapid stream currents. ALD and DD are not favored in such habitats. In species living in rivers and streams with access to the sea, larvae are either released upstream to drift downstream or are carried by females to estuaries or the sea to hatch directly in brackish or fully marine waters. There the larvae go through extended planktonic development. These are the amphidromous species whose life cycle is described below. A more limited number of ELD freshwater shrimps live far from the sea in large permanent lakes or with access to seasonally inundated floodplains with a good supply of plankton for shrimp larvae to feed on. There is no need for the long trip to the sea and back again in these species.

9.6.1 Freshwater Abbreviated Larval Development and Direct Development Species

A number of freshwater species live in tropical and subtropical habitats without access to the sea. In these species, ALD or DD had to evolve, or the species would have become extinct. In some regions, seasonal flooding creates temporary pools or swamps deficient in plankton. Some examples are the Central and West African forest swamps and tidal grass mats inhabited by *Desmocaris* and euryrhynchid shrimps (Powell 1977, 1979). Similar habitats are found in some areas of South America's Amazonian regions. Studies by Ilse Walker (Kensley and Walker 1982; Walker 1992) and others clearly showed the selection pressures that result in ALD or ELD species. Walker found that palaemonid and euryrhynchid shrimps, ecologically dominant invertebrate groups of Amazonia's Rio Negro Basin, were ALD or DD species. Streams from the plateau leading into the basin and seasonally inundated, shady lowland forests ("igapó") are acidic from decaying vegetation and consequently poor in nutrients and plankton. ALD species, which develop through only a few lecithotrophic larval stages, and DD species, in which the hatching stage is benthic and juvenile, are the best adaptive solution in such plankton-deficient habitats.

9.6.2 Amphidromous Species

Many aquatic animals lead a dual (diadromous) life cycle. In anadromous species, such as Pacific salmon (*Oncorhynchus* species), the adults live in marine waters but migrate to freshwater rivers and streams to mate and spawn. After hatching and some growth, the young migrate back to the sea to mature (Hasler et al. 1978). Freshwater eels (*Anguilla* species) do just the opposite; the adults live in freshwater but migrate to the sea to reproduce. Larvae develop in the sea but are carried by ocean currents long distances back toward continents where they enter freshwater streams. They soon develop into juveniles which make the trip back up freshwater streams to grow and mature (catadromy), as shown in the seminal study by Schmidt (1922). The descendants of marine shrimps that invaded freshwater habitats evolved two different life-history strategies. As we have seen above, one group reduced or eliminated planktonic larval stages (ALD, DD), while the others retained larval development in the sea with evolution of a diadromous life cycle termed amphidromy (McDowall 1992, 2007, 2010; Bauer 2013). In amphidromy, juveniles and adults live upstream;

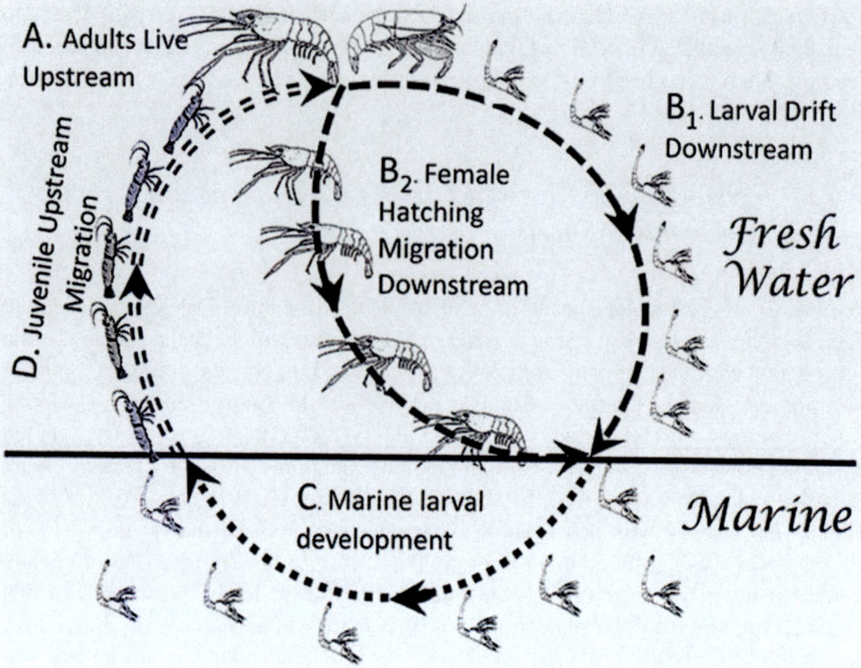

Fig. 9.7 Amphidromous life cycles in caridean shrimps: adults live and breed upstream in freshwater streams and rivers. (**a**) In many species, females hatch and release first-stage larvae upstream into the stream current to (**b**$_1$) drift down to marine waters. In other species (**b**$_2$), females migrate downstream and hatch their larvae directly in estuarine or marine waters. After larval development in the sea (**c**), the postlarvae enter a stream or river and migrate upstream as juveniles (D) to the adult freshwater habitat. From Bauer (2013)

mating and spawning occur there. In many species, upstream females hatch larvae directly in the stream current, which carries the larvae to the sea (Fig. 9.7a, b$_1$). In other species, females migrate downstream to the sea and hatch the larvae there (Fig. 9.7a, B$_2$). The larvae undergo extended development in estuaries or the open sea (9.7C). Postlarvae must find the mouths of rivers or streams and, as small juveniles, migrate upstream to the adult habitat (Fig. 9.7d). Amphidromy is not only found in shrimps but also in some fish and snail species. Amphidromous species are most diverse and numerous in streams and rivers of mountainous oceanic islands in lower latitudes. However, they are also found in continental rivers and streams in tropical and subtropical areas around the world (McDowall 2007, 2010). Given that there are no truly freshwater penaeoids, sergestoids, or stenopodideans, there is no amphidromy in these taxa.

How can freshwater shrimps living hundreds, sometimes thousands, of kilometers upstream release the weakly swimming zoea 1 into the current to drift to the sea, and then, after ELD, find a stream mouth, and migrate all that distance upstream to the adult habitat? As McDowall (2007) asked rhetorically, "why bother?" First of all, a basic change in development, from the ELD of marine ancestors to ALD or DD, is not a simple evolutionary step. However, shrimp populations that become landlocked or invade habitats unfavorable to planktonic larvae must evolve ALD/DD or go extinct. The change from marine to freshwater development in larvae is a difficult physiological challenge to meet. The ELD in the sea of most amphidromous species avoids this problem and has other important advantages. The hatching larvae do not have to struggle against stream currents to stay in the adult habitat. They just "go with the flow" to arrive in a relatively short time in estuarine and oceanic waters in plankton communities rich in larval food. Oceanic currents allow extensive dispersal to and invasion of other rivers and streams by the larvae, as well as recolonization of the home stream (Hunte 1978; McDowall 2007). Dispersal also promotes genetic interchange among populations of the same species from different locations, as shown by molecular studies, e.g., Page et al. (2008), Mashiko and Shy (2008), Dennenmoser et al. (2010), and Cook et al. (2012). On the downside, there is a great loss of larvae during the trip downstream to the sea, by predation in marine waters, and failure to find a freshwater river mouth as postlarvae. This loss is made possible by producing broods with many small embryos that will hatch into numerous ELD larvae.

Amphidromy in freshwater shrimps was initially recognized from description of their larval stages. Researchers trying to culture larvae of some tropical freshwater species found that the hatching stage (zoea 1) would not molt into zoea 2 unless they were transferred into brackish or full-strength seawater (e.g., Ling 1969; Hunte 1977, 1980). The high salt content of these waters is a physiological trigger for the larva's first molt. I tested this personally with two of the several amphidromous shrimps in the streams of Puerto Rico (*Atya innocuous* and *Xiphocaris elongata*) with similar results. Some of my colleagues were initially skeptical that shrimp larvae could go all the way to the sea to develop and then, after settling from the plankton as benthic juveniles, make their way back all the way upstream to the adult habitat. I was about to pursue this question when I was offered a promising position at the University of Louisiana, which then became my academic home for the remainder of my career. Fortunately, shortly before leaving, I met Alan Covich, a young freshwater ecologist working in Puerto Rico, who became quite interested. He went on to study and clarify, with numerous colleagues and students, many aspects of amphidromy. I did later get the opportunity to investigate an amphidromous species (*Macrobrachium ohione*) from the Mississippi River system (Bauer and Delahoussaye 2008 and later papers with graduate students cited below).

The hatching zoea-1 larva of amphidromous species travels downstream to the sea in two ways. Studies on the reproductive condition of females and plankton sampling of stream waters with plankton nets have shown that one mechanism is larval drift. The females simply stay upstream to hatch and release first-stage larvae into the stream current. The larvae will drift downstream to marine waters (Fig. 9.7b_1). Most amphidromous species live in streams and rivers in which distances from the adult habitat to the sea are relatively short, i.e., a few, several, or up to ~100–200 kilometers. This has been shown in streams in small mountainous islands in the Caribbean and Indo-Pacific, large islands (e.g., Japan, Taiwan), and continental locations relatively close to the sea, e.g., Central America (reviewed in

Bauer 2013). The current must be rapid enough so that the larvae can arrive to an estuary or coastal sea within a few days. First-stage larvae of amphidromous species do not feed (lecithotrophic); they retain sufficient yolk from the embryo to supply their nutritional needs for a few days (Plate 9.1a). If they do not molt within that time, their survival to a second instar decreases with every passing day; after a few days longer in freshwater, they die (e.g., *M. ohione*; Rome et al. 2009). However, some species, e.g., amphidromous populations of *M. amazonicum,* hatch with greater stores of yolk than most other amphidromous species (Anger and Hayd 2009). This allows them to remain lecithotrophic for a few more zoeal stages during longer distance drift to marine waters where planktonic food is available to begin feeding. Likewise, females of *Macrobrachium spinipes* release larvae as far as 400 km upstream. Their larvae may undergo the first few larval instars during drift downstream provided that they arrive in the estuary within 7 days (Novak et al. 2015).

Instead of relying on larval drift for transport of hatched larvae to the sea, the females of some amphidromous species travel downriver to release larvae near or directly into marine waters. In *M. rosenbergii* in southeast Asia (Ling 1969) and *Macrobrachium ohione* in the Mississippi and Atchafalaya River system (Bauer and Delahoussaye 2008), females migrate all the way downriver into the river estuaries where larvae hatch out. In coastal rivers of Peru and Chile, *Cryphiops caementarius* females travel hundreds of kilometers down from mountain streams and rivers to release larvae directly into the sea (Hartmann 1958). On the other hand, *M. malcolmsonii* females do not migrate all the way down to the River Godavari (India) estuary. Instead, during the monsoonal flood season, they travel downstream to about 80 km upstream to hatch and release their larvae which drift down to the estuary within a few days, depending on river flow (Ibrahim 1962). The movement of females downstream to release larvae has been shown by sampling at different points upstream to downstream along rivers. Migration downstream results in the decrease or disappearance of reproductive females from upstream sites, followed by their subsequent appearance in estuaries or far downstream near marine waters (e.g., Hartmann 1958; Ibrahim 1962; Bauer and Delahoussaye 2008). The technology of marking and tracking individuals of aquatic organisms such as shrimps has become more miniaturized and sophisticated (Arce and Moss 2002). However, this technology has not yet advanced sufficiently for practical use in marking and following the thousands of individual shrimps needed. Continued development of such techniques should be a goal of aquatic biologists interested in shrimp movements.

Most amphidromous species live in tropical and subtropical regions where temperature and food are usually sufficient for year-round reproduction. However, the reproductive seasons of amphidromous species usually occur or peak during the wet (rainy) seasons of the year when seasonal flooding takes place (Fig. 9.8). For successful larval drift, water levels must be high enough and currents rapid enough so stage-1 larvae can arrive in marine waters soon enough to molt to the first feeding stage (e.g., Hancock and Bunn 1997; Johnson et al. 1997; Bauer and Delahoussaye 2008; Novak et al. 2015). Downstream migrations of ovigerous females to or near marine waters do occur during seasonal river flooding (e.g., *Cryphiops*

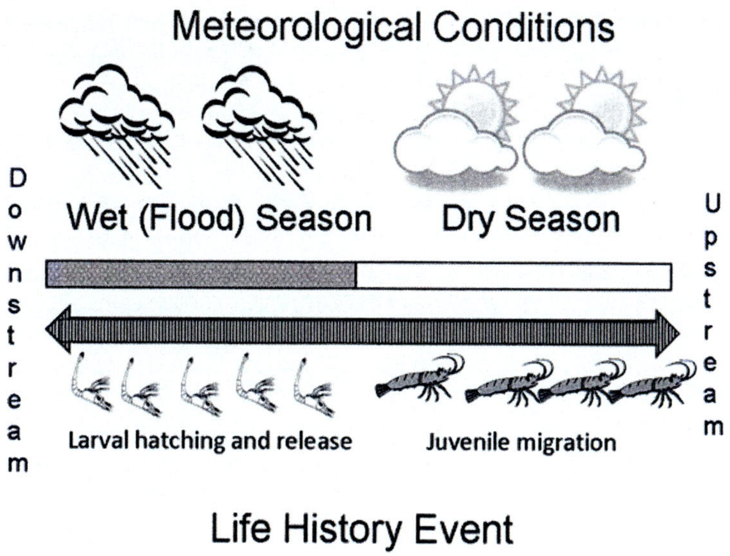

Fig. 9.8 Larval delivery to the sea by amphidromous females may occur by stream drift or female migration during the wet (rainy, flood) season of the year. The upriver juvenile migration after marine larval development generally takes place during the dry (low-flow) season. From Bauer (2013)

caementarius, Hartmann 1958; *Macrobrachium malcolmsonii*, Ibrahim 1962; *M. ohione*, Bauer and Delahoussaye 2008).

Among the most spectacular natural history phenomena are migrations, the mass movement of individuals from one location to a distant one, whether by insects, amphibians, fishes, birds, or mammals. Crustaceans have their share of migrations too (Bauer 2018). One such migration is the juvenile upstream trek of amphidromous ELD species. Once the stage-1 zoea has arrived in marine waters, it begins its series of larval stages to reach the decapodid/postlarval stages which settle from the plankton to the benthos. At this stage, the young shrimp has to find, by still unknown mechanisms, the mouth of a river or stream in order to move upstream, sometimes hundreds or, in a few species, even thousands of kilometers to the adult habitat in order to mature and breed, e.g., *Cryphiops caementarius* (Hartmann 1958), *M. malcolmsonii* (Ibrahim 1962), *M. rosenbergii* (Ling 1969), and *Macrobrachium ohione* (Bauer and Delahoussaye 2008; Olivier and Bauer 2011). The mass movement of thousands and thousands of small juvenile shrimps swimming or crawling upstream at night along the banks of streams and rivers is indeed impressive. However, not all mass upstream movements of shrimps are part of life-cycle migrations. They may also occur after sudden flooding or water-discharge events which displace a large number of shrimps downstream from their upstream habitat (Fièvet 1999; Hongjamrassilp et al. 2021).

The occurrence of juvenile life-cycle migrations of freshwater shrimps was suggested by or reported in publications starting in the 1960s (e.g., Ibrahim 1962;

Chace Jr. and Hobbs Jr. 1969; Hunte 1978; Hamano and Hayashi 1992). Study on this phenomenon has greatly increased in the last 25 years. In shallow rocky streams common in mountainous islands or coastal rivers, long lines and swarms of small shrimps swim or crawl in the shallows next to the riverbank (e.g., Benstead et al. 2000; Kikkert et al. 2009; Bauer and Delahoussaye 2008) (Plate 9.1 d–f). If water becomes too shallow or too high and turbulent (e.g., rapids, a waterfall), they can emerge onto the splash zone and crawl along the bank (Plate 9.1 d–e). Upon encountering a low natural dam, waterfall, or other stream obstacle, they may slowly crawl or hop up the obstacle, again staying within the wetted bank area next to the stream. In larger streams and rivers in which the water is deep along the bank, the small juveniles swim near the surface next to the shore (Plate 9.1f). As juveniles move upstream, they continue to molt and grow so that their average size increases with distance upstream from the river mouth (Bauer and Delahoussaye 2008).

The young shrimps migrate at night to avoid predation by visually hunting aquatic predators, e.g., fishes and birds. Juvenile migrators need to expend as little energy in migrating as possible, which explains two characteristics of the migration. The juveniles move right along the stream edge where flow is slowest due to friction of the water flow with the stream bank. Juvenile migrations usually take place when river velocity is low, so in many species they take place during the dry season or when seasonal floods are declining (Fig. 9.8) (Hartmann 1958; Ibrahim 1962; Bauer and Delahoussaye 2008; Novak et al. 2015). However, this is not always the case (e.g., Kikkert et al. 2009; Hoarau et al. 2018). There needs to be enough continuous flow for migration to take place. If streams are broken up into pools separated by dry streambed, then the migration will stop. Flow is an important directional cue; the shrimp orients into the current (positive rheotaxis) when migrating. Without flow, the shrimp becomes disoriented.

Human impacts on rivers and streams around the world are considerable. One such impact is the disruption of the migrations of aquatic animals, including amphidromous shrimps. Weirs dikes, dams, and levees can interrupt downstream female migrations, larval drift to the sea, and juvenile upstream migrations from the sea (Holmquist et al. 1998; March et al. 2003) (Fig. 9.9). These disruptions may cause considerable harm not only to amphidromous shrimps but also to the species diversity and ecological function of the affected streams. Some municipalities in tropical areas withdraw water from reservoirs behind dams for urban and agricultural use. If this is done at night, when larvae are drifting downstream, great numbers of larvae will be sucked out of the reservoir and lost. Although juvenile shrimps of amphidromous species can climb up and around obstacles in their path, they cannot make it over high dams. Such dams without any corrective measures eliminate amphidromous shrimp populations upstream. Migrating shrimps that are trapped downstream of the dam attract predators. However, dams with overflow spillways, fishways, or inclined ramps supplied with gentle flow ("shrimp ladders") can solve the problem (Bauer 2011b; Olivier et al. 2013). Another human impact that can seriously affect juvenile migrators is artificial light along the shore around bridges and in urban areas through which rivers and streams flow. The migrators swim away from the light and become disoriented. One solution that has been implemented in

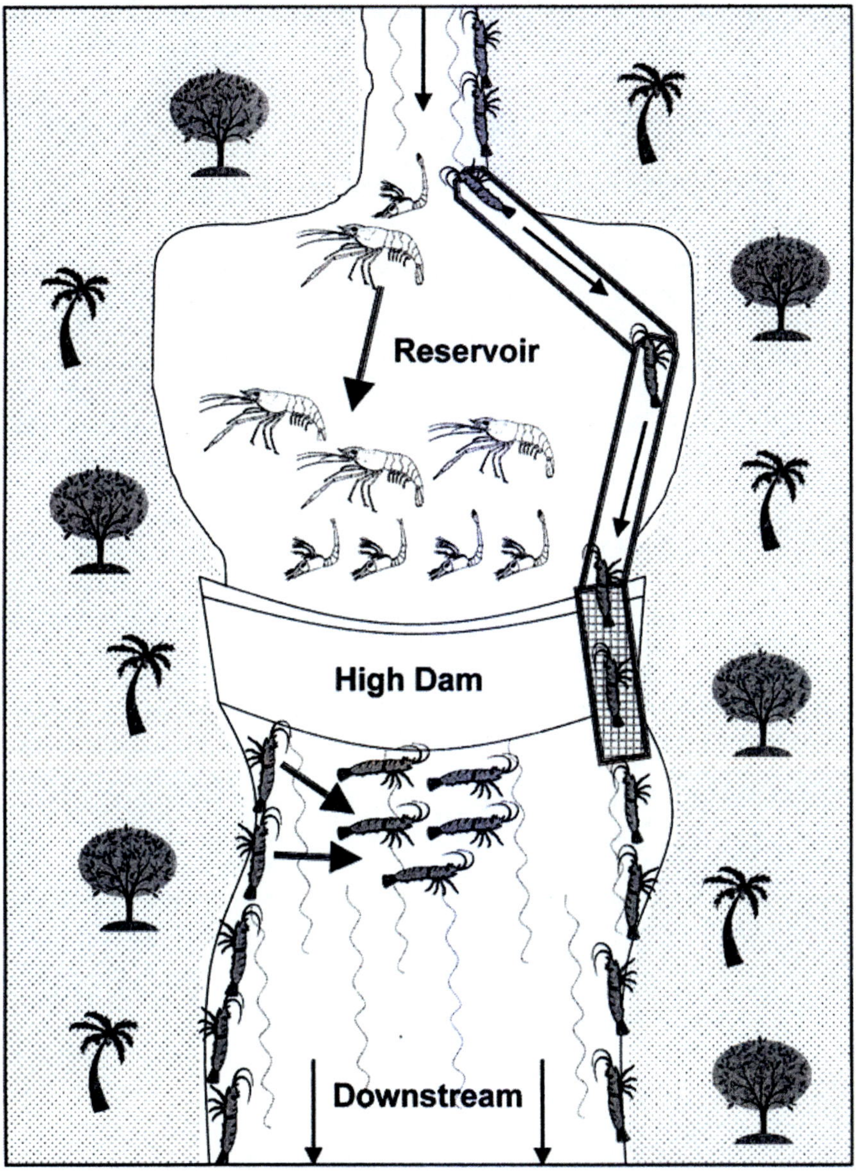

Fig. 9.9 Effect of dams on amphidromous migrations. Stage-1 larvae (upside-down swimmers) are released in upstream headwaters (top of figure) by females in one species. Adult females (unshaded, upright) of another species migrate downstream to hatch in coastal waters, but are trapped in the reservoir upstream of the high dam. Juveniles (shaded) of both species migrating upstream after larval development are blocked (left stream bank) and accumulate (arrows) downstream of the dam. On the right side of the stream, a shrimp ramp (rectangle with mesh fill) allows juveniles to climb up and over the dam. If a channel with directional flow (solid arrows) is provided, juveniles may be able to bypass the still waters of the reservoir and move into the headwaters upstream of the reservoir From Bauer (2011b)

Japan, at least, is lighting along only one bank of the stream which deflects the shrimp migrators to the opposite bank where they continue their upstream migration (Bauer 2011b).

9.6.3 Freshwater Extended Larval Development Species

Although most freshwater species with small eggs have extended larval development in the sea, there are some freshwater species which have taken a life-history route different from amphidromous species (Fig. 9.6). Besides studying ALD/DD species in headwater streams and seasonally flooded habitats in the Rio Negro Basin, Ilse Walker also investigated the shrimp fauna in an adjacent river system, the Solimões, an upper tributary of the Amazon River (Kensley and Walker 1982; Walker 1992). In its surrounding deforested and unshaded flood plains, seasonal flooding forms temporary lakes ("várzea"). These bodies of water, unlike those of the Rio Negro Basin, are not acidic, are high in nutrients, and are stable long enough to allow a plankton community to develop and flourish. No ALD species occur there. Because of the rich supply of rich larval food in the plankton, the far-upstream populations of *Macrobrachium amazonicum*, a species with ELD, can flourish (Magalhães 1985). Similarly, large and stable ancient lakes, such as Lakes Chad and Tanganyika in Africa, support thriving plankton communities that shrimp larvae can feed in. Lake Chad has at present no access to the sea and the distance from Lake Tanganyika to the Atlantic Ocean via (eventually) the Congo is several thousand kilometers away, too far even for amphidromous shrimps. However, these lakes have freshwater carideans with small eggs and presumably ELD (*Macrobrachium niloticum* in Lake Chad, and several species of *Limnocaridina* species, two *Caridella* species, and *M. moorei* in Lake Tanganyika (Mashiko et al. 1991).

Although most palaemonids from freshwater caves species have large eggs and ALD (Botello and Alvarez 2013), one cave species from the Yucatan Peninsula (Mexico), *Creaseria morleyi*, has extended planktonic development with at least 6 and possibly 8 larval stages. It completes its entire life cycle in the freshwater layer above the saline one characteristic of the anchialine caves it inhabits (Benítez et al. 2020). The larval series was pieced together from plankton samples in the cave's freshwater layer, showing that the first three stages have features like those of ELD species (e.g., sessile vs stalked eyes; uropods present or absent). This freshwater habitat contains little or no planktonic food, but all larval stages have some yolk. Apparently, this species has adapted to poor larval feeding conditions by doling out the embryonic yolk gradually until benthic feeding as a juvenile begins.

The transition from amphidromous ELD species to a completely freshwater life cycle of ALD species can be seen in progress today. In Australia, *Paratya australiensis* is a freshwater species with some populations at higher-elevation upriver streams and other lentic habitats. The type of larval development and female fecundity differs between headwater stream and downstream estuarine populations within the same river systems. Females of upstream habitats produce somewhat larger and fewer embryos that hatch to larger larvae than the estuarine "small-egg" populations. However, in all populations, development is planktonic with 8 zoeal stages, considered intermediate between that of ALD and completely amphidromous ELD atyids (Walsh 1993; Hancock and Bunn 1997; Hancock 1998). In populations of southeastern Australian upland rainforest streams, the entire life cycle is completed upstream. The populations studied by Hancock and Bunn are isolated from downstream ones by high dams and distance (190 km). The upstream habitats are plankton poor, but it appears that the larvae can feed on fine particulate matter derived from algal and microbial films of the pools (Walsh 1993). Interestingly, larvae from both upstream and downstream populations can develop in 15% salinity. Although it might be expected that larvae from the lentic upstream populations should be able develop in freshwater, this still needs testing. An upstream juvenile mass migration has not been observed in this species despite intensive field studies (Hancock and Bunn 1997) so the small-egg (estuarine) populations cannot yet be considered truly amphidromous.

The larger eggs of upstream populations may have evolved to provide a larger yolk supply for development in the plankton-poor upstream pools. The larger embryos hatch as large larvae, which has been a hypothesized to be an adaptation for resisting occasional high flow (flooding) and maintaining position to complete the life cycle upstream. These adaptations have not been selected for in small-egg estuarine populations which have access to the richer larval food supply found there. The large-egg upstream populations are evolving toward ALD, but a shortening of the larval series has not yet occurred. Genetic studies using molecular methods have shown that the various upstream and downstream *P. australiensis* populations in Australia likely represent a number of cryptic species in which some are transitioning from ELD to ALD (Cook et al. 2006; Rahman et al. 2020). Similar evolution of ELD in small-egg estuarine populations of *Macrobrachium nipponense* from Japan toward large-egg populations ALD-like ones in lacustrine habitats has been demonstrated by Mashiko and Numachi (2000) as well as in amphidromous and land-locked populations of *Caridina gracilipes* in Taiwan (Han et al. 2011). Within these two species, the various populations with diverging life-history strategies are likely to be distinct but morphologically similar (cryptic) species, or perhaps populations in the process of speciation.

9.7 Abbreviated and Direct Development in Marine Species

All penaeoids and sergestoids have extended larval development as do most carideans and stenopodids in the neritic waters of tropical and temperate seas, favorable habitats for ELD. However, ALD/DD is common in carideans from high-latitude to polar continental shelf waters as well as in a number of deep-sea carideans and all spongicolids in deep-sea waters. These latter habitats are not hospitable to planktonic larvae. Species living in these environments, like freshwater shrimps, have evolved various life-history tactics to deal with this situation: (1) to retain the ELD of ancestral species and adapt to the unfavorable consequences for larvae or (2) to avoid such conditions by evolving ALD or DD.

9.7.1 High-Latitude Species

A number of shallow-water caridean species from boreal, high austral, Arctic, and Antarctic waters have ALD or DD. These waters can be quite hostile to extended larval development. Water temperatures are very cold, close to freezing even in the summer which precludes rapid development. Perhaps more importantly, the summer is quite short, especially in polar waters, although plankton production (larval food) may be quite high. One problem for shrimps in these harsh waters is that the timing of larval hatching must be precisely timed to coincide with the brief window of elevated plankton productivity. However, it has been difficult for many species to evolve this degree of precision. Instead, they have evolved ALD or DD with hatching as lecithotrophic larvae more or less independent of planktonic larval food during development (Thorson 1950; Pearse et al. 1991).

There are several cold-water species which live in shallow water at high latitudes but are deep-sea species at lower latitudes, a phenomenon known as "equatorial submergence." These species have generally adopted the ALD/DD strategy which dodges both the problem of precise timing of larval release at high latitudes and the arduous trip to and from the bottom for planktonic larvae of deep-sea species. Examples include crangonid genera such as *Notocrangon, Sabinea, Argis, and Sclerocrangon* (Gurney 1942; Dobkin 1969; Komai 1997) as well as the thorids *Lebbeus polaris* and *L. groenlandicus* (Dobkin 1969; Bae and Oh 2014).

9.7.2 Deep-Sea Species

Some deep-sea carideans have ELD with its advantages of high female fecundity, dispersal, and favorable feeding in the plankton of upper surface waters. However, their larvae must make the long trip up to surface waters to find abundant larval food. Consequently, they will also suffer mortality from plankton predators. If they make it to the settling stage, the larvae then have to make the lengthy and arduous trip back down to the correct benthic or deep-pelagic habitat for adult life. This dangerous to-and-fro trip is analogous to that taken by amphidromous freshwater species. On the other hand, some species have evolved ALD or DD to avoid the hazards of ELD for deep-sea inhabitants. There is little larval food in the dark, cold deep sea. Females produce fewer but larger yolk-rich eggs that can nourish the extra development for hatching as advanced lecithotrophic larvae or directly as small juveniles. Although dispersal is limited, these young have the advantages of staying near their adult habitat to settle early. The perils of the journey from deep waters to the surface and back again are avoided by ALD/DD.

Excellent comparative studies showing ELD versus ALD/DD reproductive tactics were done by Elizabeth Wenner (1978, 1979) on the shrimp fauna from the continental slope of the Mid-Atlantic Bight (east coast of the USA). Females of two *Nematocarcinus* species spawned broods of several thousand small embryos indicative of an ELD strategy. In contrast, the two *Glyphocrangon* species and the crangonids *Sabinea hystrix* and *Metacrangon jacqueti* had small broods of large embryos, showing abbreviated development. Glyphocrangonids are bathyal and abyssal species, and all species for which embryo size has been reported have large ALD/DD embryos (Dobkin 1965, 1969; Holthuis 1971). The several bathyal benthic species of *Bythocaris* and *Cryptocheles pygmaea* (Bythocarididae) also produce broods of few large embryos (Dobkin 1969; Abele and Martin 1989). Beginning in the later 1970s, deep-sea hydrothermal vent and cold-vent habitats with abundant alvinocaridid shrimp populations were discovered (Grassle et al. 1979) and the reproduction of some species has been studied (e.g., Ramirez Llodra et al. 2000; Ramirez-Llodra and Segonzac 2006). The females of these species spawn many small embryos, indicating that they have retained the ancestral ELD option to colonize the widely spaced hydrothermal and other chemosynthetic habitats of the deep-sea.

Deep-mesopelagic and bathypelagic carideans make similar adaptive choices between low fecundity and ALD/DD versus high fecundity and ELD. Based on his observations on several species, Omori (1974) proposed that ALD is more prevalent than ELD in deep-pelagic species as proposed by Gurney (1942). Data on embryo size suggest that the females of many mesopelagic and bathypelagic species (Pasiphaeidae, Oplophoridae) incubate large embryos (> 1 mm), presumably with abbreviated development (Dobkin 1969; Omori 1974; Krygier and Pearcy 1981; Hopkins et al. 1989). However, deep species of the genus *Acanthephyra* have smaller embryos, and at least one was shown to have ELD (Gurney 1942). Hopkins et al. (1989) conducted a study on the biology of a group of mesopelagic and bathypelagic species from the eastern Gulf of Mexico. They found that some meso- and bathypelagic species brood small numerous embryos (ELD) while others took the option of fewer large embryos (ALD/DD).

Except for the shallow-water coral reef genus *Microprosthema,* spongicolids are associated (commensal) with deep-sea hexactinellid sponges. All these deep-water species have ALD or DD development (Goy 2010). Spongicolid females brood relatively few large embryos which hatch as decapodids, postlarvae, or juveniles (Saito and Konishi 1999; Saito et al. 2006).

9.7.3 ALD/DD in Shallow-Water Marine Species

Although most cases of ALD and DD in marine species occur in deep-sea and high-latitude species, there are some exceptional cases in shallow coastal seas. For example, in seagrass meadows of Florida (USA) waters, there are 3 species of *Thor.* Two of them, *T. dobkini* and *T. manningi*, have small stage-1 embryos while those of *T. floridanus* are about twice as large (Chace Jr. 1972). The small-egg species have ELD with at least 6 larval stages, while *T. floridanus* has ALD with only 3 larval stages (Dobkin 1968). Yet, they all appear to share the same type of habitat, so it is difficult to hypothesize what selective pressures account for ALD in *T. floridanus*. There are probably other similar cases from shallow marine waters, but they are rare.

An important exception to the statement above is a group of shallow-water sponge-dwelling shrimps in the genus *Synalpheus* (Alpheidae). Like eusocial insects such as termites, ants, and bees, these species live in colonies of several to hundreds of individuals in which most or all reproduction is done by a single female, the queen. The offspring, termed "helpers," stay in the sponge and participate in colony defense (Duffy 2002). If the queen's embryos hatched as planktonic or swimming lecithotrophic larvae, the eusocial (colonial) system could not have evolved or be maintained. As a result, direct development has evolved in these species. The hatching individual is a small crawling individual who will stay in the home sponge for most or all of its lifetime (see Chap. 10, Symbiosis, for further details).

9.8 Reproductive Investment, Output, and Fecundity

Perhaps, the most important life-history trait of an organism is the amount of energy that it devotes to producing offspring. The most common and measurable reproductive parameter of the costs involved is termed "reproductive output" (RO), which can be defined as the total mass of eggs spawned by a female divided by the body mass of the female (Clarke et al. 1991). A more common although perhaps less precise measure of RO is fecundity, i.e., the number or volume of spawned eggs (dendrobranchiates) or brooded embryos (carideans, stenopodideans) related to the size of the female. Comparisons of RO and fecundity within and among species from different habitats or geographic areas can suggest much about the selective pressures, both environmental and biological, which account for the RO of a female or species.

A laborious but precise estimate of RO can be obtained by dissecting out and carefully weighing the mature (prespawning) ovary full of vitellogenic oocytes (eggs) from a penaeoid or sergestoid shrimp female, followed by weighing the female with the ovaries removed. In caridean and stenopodidean shrimps, the brooded embryos can be simply stripped from the female and weighed instead of prespawning ovaries. In all groups, drying the mass of eggs and of the female before weighing to obtain "dry weight" is necessary because the water around and within the egg mass and female varies greatly, and water has no energetic value. When RO is measured by fecundity, the number of mature oocytes or brooded embryos is counted. In those carideans and stenopodideans in which brood number is a few hundred or less, counts can be done manually. However, in carideans and stenopodidean species with small eggs and/or large female size, the number of embryos per brood can be several thousand. In penaeoids and sergestoids, prespawning eggs number thousands to hundreds of thousands. Separating and counting that many eggs is simply unnecessary time and effort. An easier way to obtain a good estimate is to divide the total ovarian or embryo mass by the weight of a small subsample, and then multiply by the number of eggs or embryos in the subsample. The time-consuming labor involved probably accounts for the comparatively scarce data on fecundity values for dendrobranchiates compared to the more abundant data for carideans.

Another measure of fecundity preferred by many investigators is the total volume of the mass of eggs or embryos produced by a female. The average volume of a sample of egg or embryo is determined from the egg/embryo dimensions and then multiplied by the number per spawn or brood to give total volume. Measurements of mature eggs in dendrobranchiates or embryos in pleocyemates are made with a microscope equipped with an ocular micrometer or image analysis software. If the egg cell is more or less round, the formula for volume of a sphere can be used. Caridean and stenopodidean eggs are usually oval shaped in which case the length and width can be measured, and from those data the embryo volume, using the formula for that of an ellipsoid (e.g., Bauer 1991).

Measurement of egg number, weight, or volume is just one part of estimating reproductive output. Fecundity must be related or adjusted to female size in order to make intra- and interspecific comparisons. The easiest method is to simply use a linear dimension of the female, either total length or carapace length. The latter is preferable with shrimps because the carapace is a rigid structure whose length does not vary in handling. Precise measure of total length is more difficult as the bent or curved multisegmented abdomen of a shrimp is quite flexible. As a result, its measurement can be imprecise, depending on how the investigator is holding or pressing it down to make it straight. Once egg/embryo number or volume and female size are known, different methods of presenting the data for analysis are possible. The cube of carapace length is perhaps the best predictor of egg/embryo number or volume because it approximates the volume of the cephalothoracic space where eggs (oocytes) are produced and stored prior to spawning (Jensen 1958; Corey and Reid 1991). Although this may be so, a volume is difficult for most humans to visualize or understand intuitively. I think that is why many publications in the past and still today use egg/embryo number rather than volume to analyze and report. Perhaps a good compromise for such studies is to calculate and use volume for total egg volume (dendrobranchiates) and brood sizes (carideans, stenopodideans) but also to report the number and average egg or embryo size in length and width (millimeters or micrometers). In that way, the data can be compared with studies using either embryo volume or number.

9.9 Variation of Fecundity with Female and Egg (Embryo) Size

The two major variables that determine the number of spawned offspring produced by a female shrimp are her body size and the size of the eggs or embryos that she spawns. Females of larger size will produce more than smaller ones (Fig. 9.10). Females that spawn small eggs will produce more of them than females with large eggs, as demonstrated in numerous shrimp species (Fig. 9.11). Penaeoids and sergestoids (dendrobranchiates) spawn smaller but vastly more numerous eggs than do carideans and stenopodideans (Fig. 9.11 see also Chap. 6: Table 6.1, Fig. 6.13). Unlike the dendrobranchiates, the size of embryos in pleocyemates shows a great deal of variation due to the differences in the size and type of larvae that hatch from them (ELD vs ALD/DD).

Plots of egg number or volume versus female size are useful for analyzing life-history tactics of individual species. Do larger (older) females produce relatively more offspring for their size than smaller (younger) ones (positive allometry)? Or is it the reverse—do older females produce relatively fewer eggs for their size (negative allometry), indicating reproductive senescence? In many species, there is no significant variation in reproductive output with age (isometry). Allometric analyses of spawn or brood size utilize linear regression equations obtained from log transformation of the power curve $Y = (a)(X^b)$, where Y is the dependent variable (number of eggs or embryos); X is the independent variable (measure of female body size); "b," the slope of the regression line, is the allometric growth constant (relative growth rate); and "a" is the y-intercept (Gould, 1966; Hartnoll 1982). Logarithmic transformation gives the equation $\log Y = (b)(\log X) + \log a$. If X and Y are measures with the same dimension, e.g., egg mass volume (or egg/embryo number, which can considered a volume) and the cube of carapace length, then a "b" (slope) value not statistically different from 1 indicates isometry; significantly greater >1, positive allometry; and significantly less than <1, negative isometry (Fig. 9.12). However, most investigators just use the more easily visualized carapace length for X (female size), a linear measure, and in this case the "b" values for isometry, positive allometry, or negative allometry would be 3, >3, or < 3, respectively (Gould 1966).

Intra- and interspecific comparisons of spawn/brood size and female size can uncover possible selective forces, adaptations, and constraints acting on reproductive effort (Reaka 1979; Hines 1982). For example, I compared embryo production among 9 species of caridean shrimps from seagrass meadows in Puerto Rico (Bauer 1991). All these species are small, with average species carapace lengths of ovigerous females ranging from only 2 to 7 mm. The number of stage-1 embryos was isometric with female size in 7 species and positively allometric in two others: there was no evidence of reproductive senescence with size (age) in any of the species. As expected, females from larger-sized species incubated larger broods than smaller species. However, I asked the question: Do females of the small-sized species compensate for their smaller absolute body size by somehow increasing

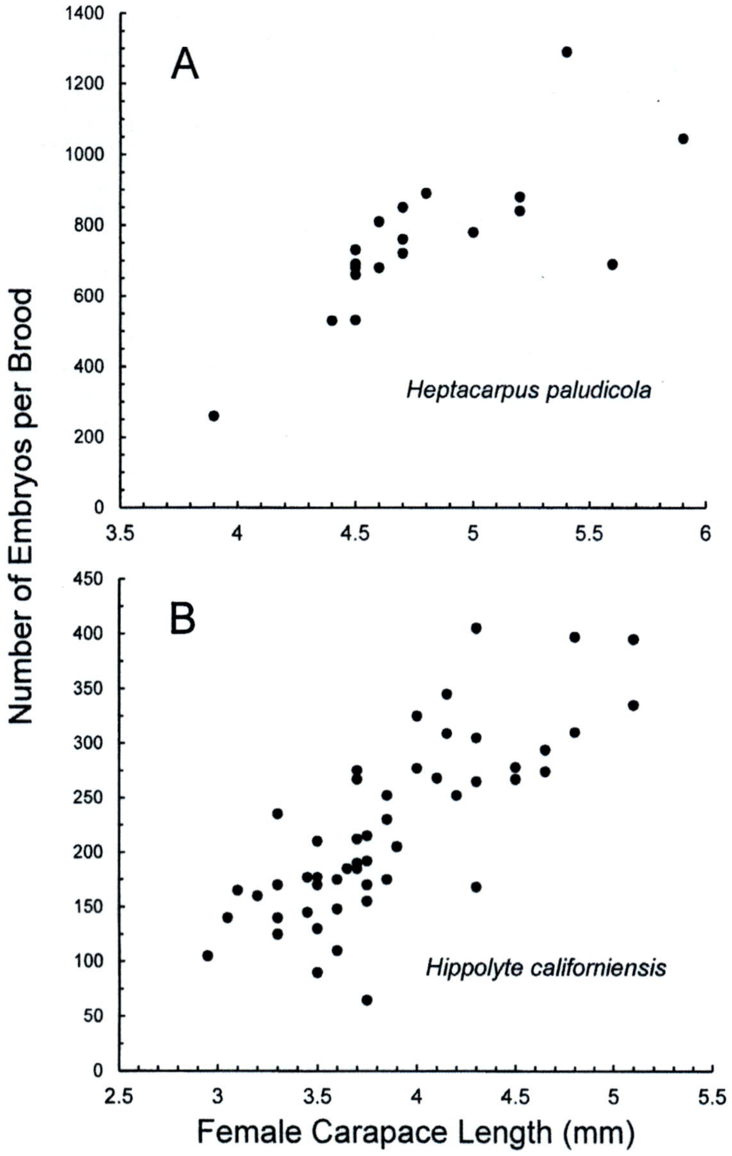

Fig. 9.10 Increase in fecundity with female size in the hippolytid carideans (**a**) *Heptacarpus paludicola* and (**b**) *Hippolyte californiensis.* Based on author collections from *Zostera marina* seagrass beds in (**a**) Argyle Lagoon, San Juan Island, Washington, and (**b**) Mission Bay (San Diego), California, USA

their reproductive output? The answer from the Bauer (1991) analysis was yes: when brood sizes were adjusted for differences in female body size, females of smaller species produced relatively greater numbers of embryos for their size than the larger

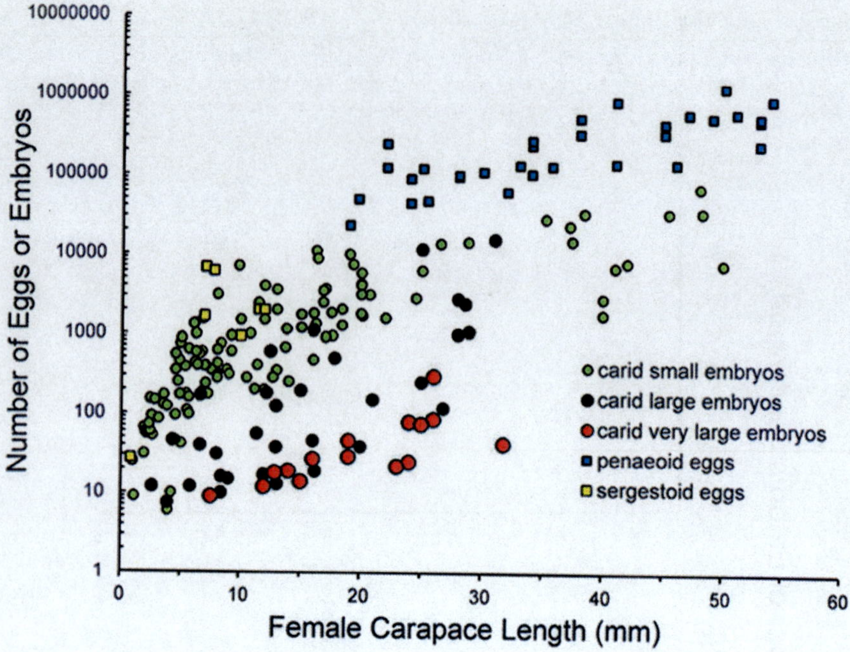

Fig. 9.11 Variation of fecundity with female size in major decapod shrimp taxa. Displayed are the number of mature prespawning (penaeoids, sergestids) or recently spawned eggs (luciferids) or early-stage embryos per brood (carideans, stenopodideans) related to female body size for 21 penaeoid species (30 observations), 6 sergestoid, 1 luciferid, and 171 caridean species. Mean or median size of reproductive females (prespawning penaeoid, sergestoid females and ovigerous caridean females) was used for most species and single observations for the remainder. For carideans, mean or median embryo size was used as in Fig. 6.13 in Chap. 6. From Bauer (2004) and other numerous sources

species. The amount of cephalothoracic space allotted for prespawning oocytes might be *relatively* greater than that of larger species. In addition, the eggs spawned by females of the smaller species were smaller in size than those of the larger species. Females could attach a greater number of embryos per unit of incubation space below the abdomen,. Additionally, the fact that smaller embryos contain less yolk meant that embryos developed and hatched sooner than those of larger species. Thus, females of these species made up for their smaller body size by a relatively higher reproductive output than larger-sized species.

9.10 Fecundity (Brood) Loss

As penaeoids and sergestoids spawn fertilized eggs directly into the sea without incubation, losses from spawning to hatching are difficult, if not impossible, to measure. But loss in fecundity before hatching likely does occur in these groups

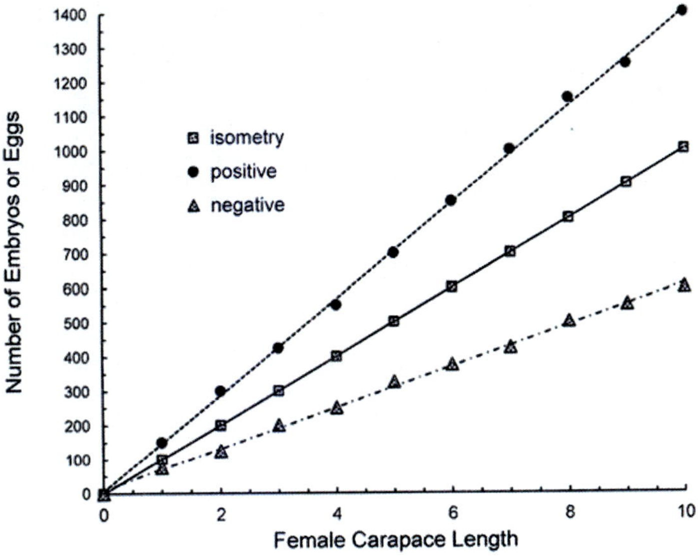

Fig. 9.12 A hypothetical example illustrating allometry (relative change) in shrimp fecundity (spawn or brood size) with female size. The relative fecundity may increase (positive allometry), decrease (negative allometry), or stay the same (isometry) with increasing female size. See the text for slope (b) values for each case

because of high mortality from planktonic predators and accidents of nature. However, loss of fecundity (brood loss) during embryo incubation by caridean and stenopodidean females can be measured, although I have not found such studies on the latter. Numerous caridean species have been examined, and Corey and Reid's (1991) work on 33 species from 9 caridean families should be recognized as seminal. This and other studies have shown that losses after embryo attachment below the female abdomen can range from 5 to 41% and perhaps more (e.g., *Philocheras trispinosus,* Oh and Hartnoll 1999; *Betaeus truncatus*, Lardies and Wehrtmann 2001; *Lysmata seticaudata*, Calado and Narciso 2003; *Hippolyte obliquimanus*, Terossi et al. 2010; *Plesionika izumiae*, Ahamed and Ohtomi 2011; *Crangon uritai*, Li et al. 2011; *Urocaris longicaudata*, Martínez and Romero 2017). However, significant brood loss does not occur in all species (e.g., Oh and Kim 2008; Puspitasari et al. 2013; Osawa et al. 2015).

The possible sources of embryo mortality within broods of pleocyemate (incubatory) decapods such as shrimps, crayfishes, lobsters, and crabs have been reviewed by Kuris (1991). The volume of caridean and other decapod embryos typically increases during development. Embryos commonly increase in size from 4 to 74% from spawning to hatching depending on the species (e.g., Corey and Reid 1991 and studies cited above). The swelling of the embryo is due to an uptake of water as the dense yolk is converted to the less dense tissues with higher aqueous content (Wehrtmann and Lardies 1999). The mass of embryos, tightly packed against the female's abdomen just after spawning, later may bulge out from beneath the abdominal incubatory as embryos develop and enlarge in volume. Especially in benthic species, the female may have to drag the brood mass against the substrate when moving along the bottom, hiding within a narrow space, or burrowing into the sediment. Mechanical abrasion against surfaces below may knock off embryos protruding out from below the female.

Another serious source of mortality in many pleocyemate decapods are embryo parasites and micropredators (bacteria, protists, small invertebrates such as nemertean egg predators). Sediment that accumulates among the mass of embryos during aeration of the brood may smother some of the embryos. However, this is not a big problem for carideans and presumably stenopodideans because of the careful and efficient grooming of the embryos by females in these two groups (Chap. 4, Shrimp Hygiene). Some embryos may simply abort due to some genetic defect or accident during spawning, fertilization, and attachment. Brood loss is measured by comparing the embryo number or volume in broods with early-stage (recently spawned) and late-stage (near hatching) embryos. Brood loss is indicated if slope of the regression equation for late-stage broods is lower than that of the early-stage ones. For example, in Fig. 9.12, if the number of early-stage embryos is represented by the positively allometric line, and the number of late-stage embryos by the isometric or negative allometry lines, this would show considerable brood loss. The slopes of the early-stage and late-stage embryo regression lines can be compared statistically to determine if any difference is significant or not.

9.11 Parasitic Castration by Bopyrid Isopods

Many parasites, both microbial and macroscopic, infest shrimps (e.g., Sinderman 1970; Overstreet 1978). Parasitism affects many aspects of an organism's biology, including female fecundity. One parasitic group in particular, the bopyrid isopods, have a dramatic effect on shrimp fecundity. Bopyrids are ectoparasites usually found within the gill chamber of marine shrimps. They occur in heterosexual pairs consisting of a large female and a very small, almost microscopic, dwarf male which clings most of the time to the female's reduced abdomen (Fig. 9.13, Plate 9.1 g–j). The bulk of the female parasite causes the gill cover of the carapace to bulge outward, which makes an infestation easy to identify (Plate 9.1g). Females and their dwarf males of one group of bopyrids, the hemiarthrines, attach below the shrimp's abdomens (Chap. 4, Fig. 4.19 b). Bopyrids mainly infest caridean shrimps; penaeoid and stenopodidean infestations are reported much less frequently in the literature (Dall et al. 1990; Calado et al. 2008; Goy 2010; Markham 2020). Sergestoid shrimps do not appear to suffer bopyrid infestations as none have yet been reported (Markham 2020). The question of why carideans seem to be more vulnerable to bopyrid attack compared to other shrimp groups is worthy of investigation.

The most dramatic effect of bopyrid infestations is the reduction of female shrimp fecundity to zero, commonly referred to by parasitologists as "parasitic castration" (O'Brien and Van Wyk 1985). Female ovaries remain immature, i.e., do not fill with

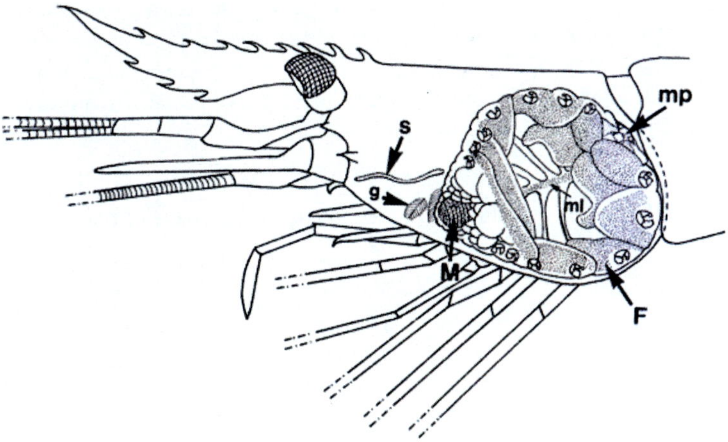

Fig. 9.13 *Palaemon (Palaemonetes) pugio* infested by the bopyrid parasite *Probopyrus pandalicola.* The parasite resides inside the shrimp's branchial chamber of shrimp. The small dwarf male clings to the underside of the abdomen of the much larger female parasite. The female parasite mouthparts face the inside of the shrimp's branchiostegite (gill cover) to suck the hemolymph (blood) of the host shrimp. F, female bopyrid with empty marsupium (brood chamber); g, anterior gills of shrimp; M, male bopyrid; ml, melanophores distributed along ventral midline of female; mp, mouthparts of female bopyrid; s, shrimp's scaphognathite (gill bailer). From Cash and Bauer (1993), courtesy of the Journal of Crustacean Biology

vitellogenic oocytes. Biomass and energy that female shrimps would use to produce a brood of embryos are hijacked by the female parasite to produce its own brood of embryos. The female shrimp is reproductively dead. Many examples have been reported and illustrated (e.g., O'Brien and Van Wyk 1985; Cash and Bauer 1993; Rasch and Bauer 2015; Lee et al. 2016; Markham 2020). Infestation rates range from low to quite high. As a graduate student in the 1970s, I studied a population of *Heptacarpus paludicola* (Hippolytidae) from *Zostera marina* seagrass beds in the Argyle Lagoon, San Juan Island, Washington, USA; 50% of females were infected, without embryos and with immature ovaries. Even worse, Beck (1979) reported 86–100% infection rates in various populations of *Palaemonetes paludosus* (Palaemonidae) from rivers and canals in Florida, USA! Infection rates of the relatively few penaeoid species parasitized by bopyrids are usually quite low, no more than 3% of the population (Dall et al. 1990; Ayub and Ahmed 2004). When penaeoid females are infected, their ovaries do not develop. Likewise, bopyridized *Stenopus hispidus* females are parasitically castrated as well (Calado et al. 2008).

There are always exceptions to the rule: in a few species of carideans, castration is not complete, although the female shrimp's fecundity is reduced. *Synalpheus yano* is a small pair-living alpheid which lives within sponge canals. Hernáez et al. (2010) reported that female *S. yano* infested by hemiarthrine (abdominal) bopyrids did produce and incubate some embryos. The female's brood shared the space available below her abdomen with the bopyrid parasite. However, fecundity of infested females was 41% lower than that of uninfested females. Likewise, Calado et al.

(2006) found that *Lysmata amboinensis* females with bopyrids produced a reduced number of embryos. In the related *L. seticaudata*, parasitic castration of females was complete (Calado et al. 2005). However, a significant reduction in fecundity, even if not total, is still a big blow to the reproductive output of the affected individuals.

9.12 Life Cycles, Seasonality, and Latitudinal Trends

The life histories of shrimps, as with other organisms, are centered around reproduction. Studies on reproductive patterns of numerous species have been done, along with related life-history traits such as recruitment pattern, growth, and life span. Generalizations about these life-history traits are a useful means of summarizing a large body of information about a species or taxon. Perhaps more importantly, generalizations serve as models or paradigms about which hypotheses can be made and predictions from them can be tested. As a result, such paradigms can be modified and improved, or simply discarded. Generalizations about latitudinal variation in annual reproductive patterns of freshwater and marine crustaceans such as shrimps have helped to identify the fundamental environmental factors responsible for observed trends (Sastry 1983; Bauer 1992).

One such generalization for freshwater and nearshore marine invertebrates is that reproduction occurs more or less continuously throughout the year in tropical habitats but becomes more restricted in time with increasing latitude (Orton 1920; Thorson 1950; Giese and Pearse 1974; Sastry 1983; Bauer 1992). Shallow tropical seas, with high water temperatures throughout the year, were considered by Orton to be a relatively stable and benign habitat for marine organisms which allow year-round reproduction. Temperature is a basic ("proximate") environmental factor that affects metabolic processes such as gametogenesis and breeding in aquatic habitats (Giese and Pearse 1974). Higher temperatures accelerate metabolic processes, while cooler ones depress them. Water temperature of aquatic environments is highly, although not exactly, correlated with latitude because of the unequal heating of the earth by the sun.

Another basic factor important to both freshwater and marine organisms with planktonic larvae is the seasonal pattern of plankton productivity (larval food). Thorson (1950) considered larval food supply to be the ultimate factor (actual selective pressure) that controls reproductive pattern in nearshore marine invertebrates. The basic factors important in plankton productivity are nutrient supply, temperature, and photoperiod, all of which vary seasonally with latitude. Plankton production tends to be low (oligotrophic) but relatively constant but throughout the year in tropical marine waters but increasing seasonal at higher latitudes. The continental shelves of mid-latitudes experience a rich spring plankton bloom made possible by the longer photoperiod, warmer water temperatures, and winter mixing of upper and deeper waters which bring up nutrient-rich waters from below. This is followed by a weaker plankton bloom in the autumn. In the Arctic and Southern

(Antarctic) oceans, a highly productive plankton bloom is limited to the very short summer at those latitudes.

In the following discussions, I will use various terms to describe the temperature regimes and associated correlates of different latitudinal areas. Tropical latitudes range from 23° S to 23° N; "warm temperate" refers to conditions from 23 to ~39° N or S; "cool temperate" from ~40 to 49° N or S; boreal and austral from ~50 to 65° N or S, respectively; and "polar" from 66 to 90° N or S. However, one must keep in mind that significant factors impacting reproduction such as temperature, salinity, nutrient supply to planktonic communities, and primary productivity are not always strictly correlated with latitude. For example, major oceanic gyres displace warm water poleward on the eastern sides of continents and cool water equatorward on their western sides. Regional currents, winds, and upwelling events may cause localized deviation from latitudinal trends in seasonality, as will be seen below. In freshwater shrimps, seasonality in precipitation (wet and dry seasons) may be important factors superimposed on temperature-related latitudinal variations.

Comparison among species from different areas and latitudes of the world is a good way to test for latitudinal trends in reproductive and related life-history traits (Bauer 1992). Does reproductive pattern vary with latitude, and thus with environmental variables such as temperature, plankton productivity and related factors in shrimp species? A useful approach for answering this question is to make comparisons and contrasts among unrelated species from similar latitudes and habitats: environmental factors are held constant, but the phylogenetic component of reproduction varies. On the other hand, related species from different latitudes can be compared. In this case, their ancestral reproductive biology is similar; differences in reproductive pattern might then be due to latitudinal differences in environmental factors. In the following sections, I describe patterns of breeding and related life-history traits such as growth, body size, and life span of shrimps.

9.12.1 Tropical and Warm-Temperate Shallow-Water Species: Dendrobranchiates

The breeding cycles of the benthic penaeoid and the pelagic sergestoid shrimps largely agree with the Ortonian and Thorsonian ideas of continuous reproduction in coastal tropical waters with more restricted but extended seasonality in warm-temperate regions. However, the details are not always so simple and clear cut. Let us start with a large group of neritic penaeoids, many of which are important fishery species (Chap. 11, Fisheries and Aquaculture) and thus have been intensively studied. These species have a variety of life cycles which have been conveniently summarized into four basic categories by Garcia and Le Reste (1981) and Dall et al. (1990) (Fig. 9.14). A relatively minor group of small-sized *Metapenaeus* species spend their entire life cycle in estuarine waters (Type 1, Fig. 9.14). Those penaeoids of major fishery and aquaculture interest are large-sized coastal species with a Type

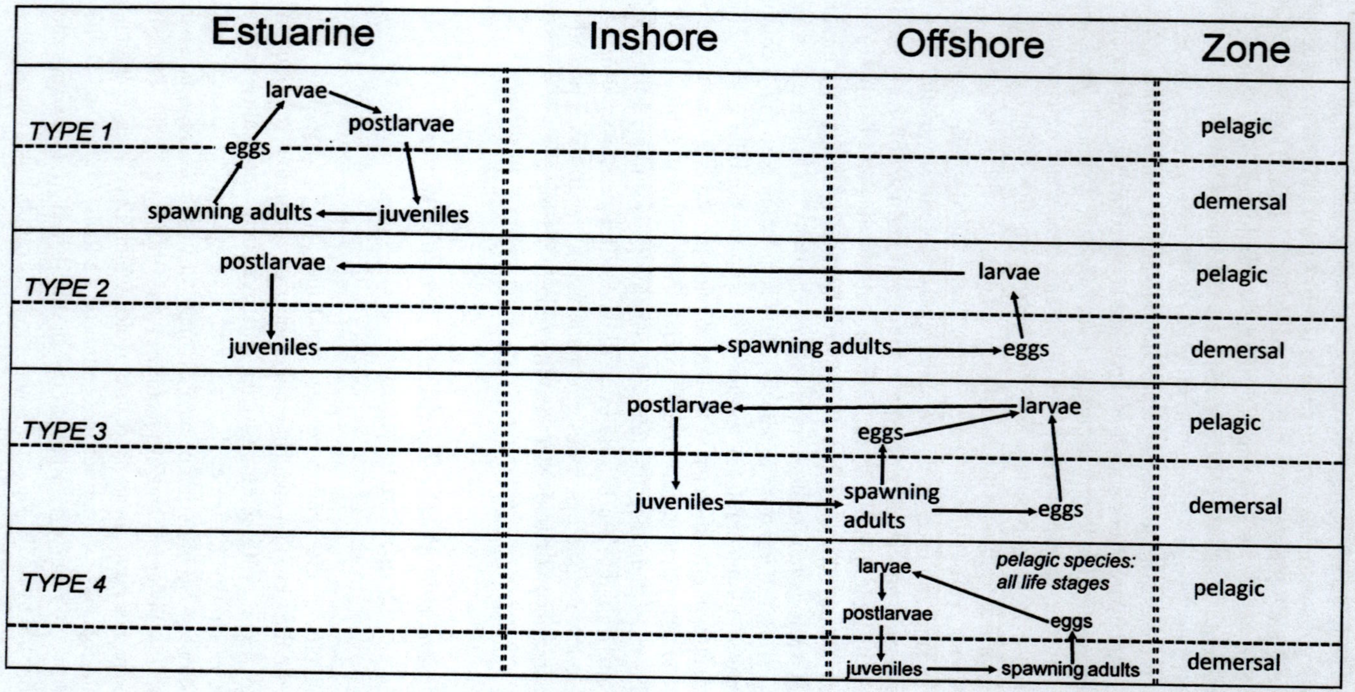

Fig. 9.14 Basic types of life-cycle patterns in penaeoid shrimps based on Garcia and Le Reste (1981) and Dall et al. (1990) showing the habitats utilized by various life stages. Demersal, on or near the sea floor; pelagic, up in the water column. Solid lines delineate life cycle types; dotted horizontal lines separate the pelagic and demersal (benthic or near bottom) part of the life cycle

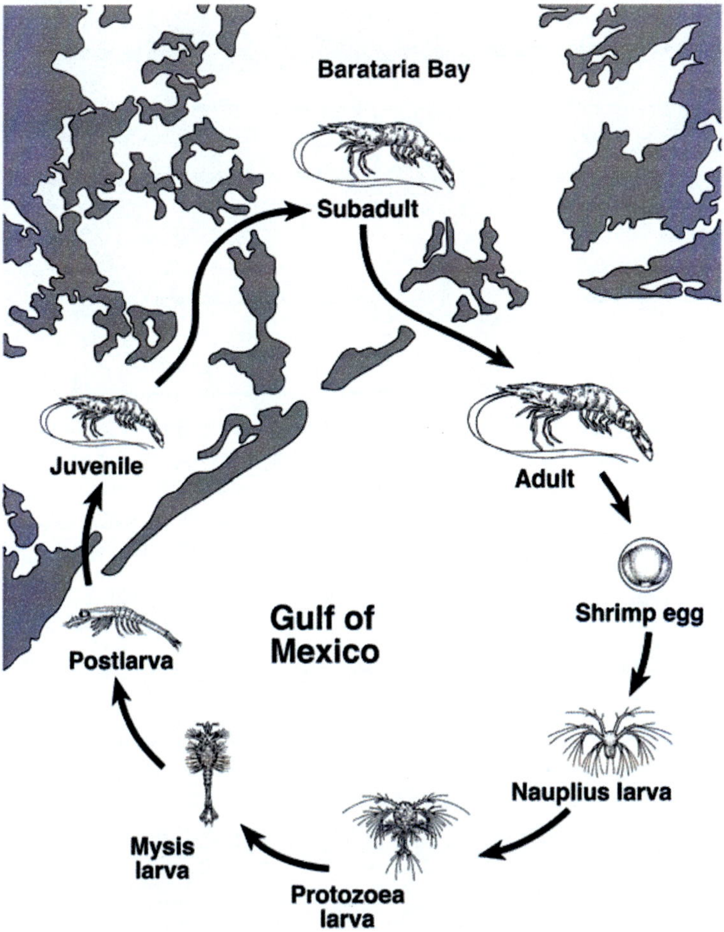

Fig. 9.15 Life cycle of the penaeoid shrimp *Litopenaeus setiferus* in the northcentral Gulf of Mexico. Courtesy of the NOAA Louisiana Sea Grant Program

2 (estuarine) life cycle, such as *Penaeus* (*s.s.*), *Farfantepenaeus*, and *Litopenaeus* species (Figs. 9.14 and 9.15). Adults live in fully marine nearshore waters where they mate and spawn. Larvae develop in the sea, and after development, postlarvae are carried by currents to or near estuaries in which they enter, using salinity and temperature cues to enter using high tides (Bauer 2018). Juvenile development to subadult takes place in the estuaries, which are very productive habitats, supplying the juveniles with abundant food and shelter. Other common coastal penaeoid species include the *Trachypenaeus* (*s.l.*) group which includes the well-studied genus *Rimapenaeus*, as well as *Trachypenaeus* (*s.s.*) and the important fisheries species, *Trachysalambria curvirostris*. These are Type-3 species (Fig. 9.14); all life stages live in fully marine waters; juveniles grow up in nearshore (inshore) waters, not in estuaries. Members of the genus *Sicyonia* share a similar Type-3 life history.

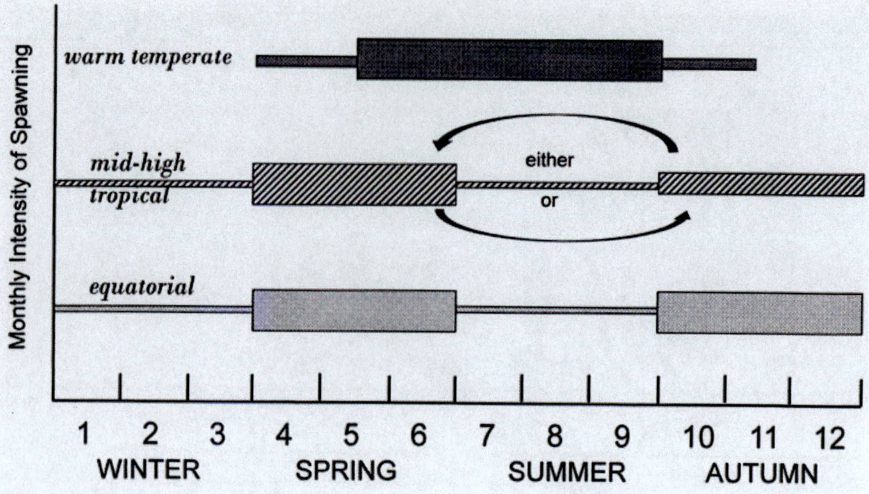

Fig. 9.16 Common seasonal patterns of reproduction (spawning) in tropical, mid-high tropical, and warm-temperate neritic penaeoid shrimps. The temporal pattern of spawning in the population is indicated by the horizontal extent of the spawning bars and its intensity by the vertical height of the bars. Note that spawning peaks are ~equally bimodal in the equatorial species; unequally bimodal in the mid-high tropical latitudes; and unimodal, restricted to spring-autumn in warm-temperate species

Although reproduction of coastal penaeoid populations living close to the equator may be continuous, there are often reproductive peaks at the population level. Near the equator, more or less equal spawning peaks of *Penaeus* (*s.l.*) species usually occur in the spring and autumn months between the summer and winter monsoonal seasons (various penaeoid species in Garcia and Le Reste 1981; Garcia 1988; Dall et al. 1990; *Farfantepenaeus subtilis* and *Litopenaeus schmitti*, Santos et al. 2020; *Solenocera crassicornis*, Sukumaran 1998). During inter-monsoonal periods, turbulence from high wind and current speeds is reduced, conditions more favorable for spawning and larval development. At somewhat higher latitudes near continental land masses, rising and falling temperatures can create the same conditions. Dall et al. (1990) termed these "spring" and "autumnal" peaks in tropical waters by analogy with temperate zones. In a strictly equatorial pattern, more females are spawning during these two peaks than between them (Fig. 9.16). Garcia (1988) proposed that the equatorial penaeoid pattern is one in which there were two generations per year. The major generation is born in the spring, but, given the high rapid growth and maturation of penaeiods, produces a minor reproduction in autumn at 6–7 months. This is followed by a major massive population spawning in the next spring when these individuals are one-year old. An autumnal generation, born in the autumn, has a minor spawning peak in the spring with their main spawning in the autumn. The result of the spawning interactions of these two generations, a major spring and minor autumnal, produces two population spawns of about equal size. At higher but still tropical latitudes, there are also spring and autumn peaks, but of unequal size (Fig. 9.16). For example, in the Gulf of Carpentaria (Australia), the percent of females spawning in the spring peak is greater than that in the autumnal peak [*Fenneropenaeus* (*Penaeus*) *merguiensis*, van der Velde et al. 2021]. Paradoxically, due to spring fishing pressures on this species, the females of the smaller autumnal peak produce, on the population level, the greatest proportion of eggs and larvae of the year.

In warm-temperate regions such as the northern Gulf of Mexico, with pronounced seasonality of warm spring to early autumn periods and cool to cold winters, there is usually a single extended spring-autumn reproduction in shallow-water penaeoids such as *Litopenaeus setiferus* (Linder and Anderson 1956), *Rimapenaeus similis, R. constrictus* (Bauer and Lin 1994), and the sicyoniid *Sicyonia brevirostris* (Kennedy et al. 1977) (Fig. 9.16). Penaeoids from other warm-temperate regions show a comparable pattern, e.g., *Melicertus kerathurus* (Conides et al. 2006), *Parapenaeus fissuroides* (Farhana and Ohtomi 2016), *Trachysalambria curvirostris* (Hossain and Ohtomi 2008), and *Solenocera melantho* (Ohtomi and Irieda 1997). In the Gulf of Mexico *Farfantepenaeus aztecus*, spawning was decidedly seasonal at shallower depths but became more continuous at greater depths in the study by Cook and Linder (1970). However, reproduction in *F. duorarum* in the warmer waters of the southeastern Gulf appears to be continuous based on juvenile recruitment patterns (Roberts 1986). At an extreme of northern warm-temperate waters (Yellow Sea, Korea), with cool-water temperatures most of the year but with summer high temperatures, *Metapenaeopsis dalei* and *Metapenaeus joyneri* have a restricted summer reproductive season (Choi et al. 2005; Cha et al. 2004). In the Gulf of California (25–30° N), *Farfantepenaeus californiensis* shows continuous reproduction during warmer years of equatorial temperatures. During years with somewhat lower water temperatures, it has two main reproductive peaks, a major one in the spring and a less intense one in the fall (Leal et al. 2001).

For the last few decades, a dedicated group of Brazilian scientists have studied the population dynamics and reproduction of the diverse penaeoid fauna of southeastern Brazil around 23° S latitude near the southeastern coastal town of Ubatuba. Adilson Fransozo and his scientific partner (and wife) María Lucía Negreiros began this effort, along with their then graduate students Fernando Mantelatto, Rogerio Costa, and Antonio Castilho who, now themselves professors, are mentoring numerous graduate students continuing this effort. This group and other associated scientists formed "NEBECC," the Portuguese acronym for the "Nucleus of Studies in Biology, Ecology, and Culture of Crustaceans." The coastal penaeoid fauna of southeastern Brazilian waters is very similar to that of the Gulf of Mexico, with species of *Litopenaeus, Farfantepenaeus, Rimapenaeus, Xiphopenaeus, Pleoticus,* and *Sicyonia,* as well as *Artemesia longinaris*, a strictly South Atlantic species. During much of the year, the waters around Ubatuba are typically tropical, with higher water temperatures and oligotrophic plankton productivity. However, during the spring and summer months, upwelling caused by changes in wind direction and coastal currents draws deep South Atlantic Cold Water (SACW) close to the coast (Valentin 2001). The nutrient-rich SACW stimulates an intense plankton bloom, supplying excellent food for shrimp larvae. The reproductive pattern of the penaeoid fauna in the area is generally continuous in that some females are reproducing and producing larvae throughout the year. However, there are other periods of high reproductive activity (e.g., Costa and Fransozo 2004; Castilho et al. 2007, 2008, 2015; Grabowski 2017), somewhat like the "midtropical" pattern shown in Fig. 9.16.

Many of the studies cited above show that neritic penaeids attain large size and sexual maturity in relatively little time, from 6 months to a year while attaining massive reproductive outputs. "Live fast and die young"—these species have relatively short life spans, 1–2 years compared to much longer ones in carideans from high latitudes and to penaeoids and carideans from deep-water habitats, as shown below.

Although they do not achieve the large size of their penaeoid relatives, sergestoids too have relatively short life spans. Basic life-history traits have been most extensively investigated in the nearshore neritic sergestoid genus *Acetes,* mainly because of its economic value in fisheries, especially *Acetes japonicus* of the tropical to warm-temperate Western Pacific and Indo-Pacific (Xiao and Greenwood 1993). Reproduction in *Acetes* species generally follows the classical paradigm of a continuous pattern in tropical waters with a more restricted spring-autumn reproductive season in warm-temperate waters (Xiao and Greenwood 1993; Amin et al. 2009a; Simões et al. 2013). Reproduction in *A. chinensis* from 35° N showed a pattern of summer and winter generations which is common in many warm-temperate carideans (Oh and Jeong 2003). Spawning in the spring produces a summer cohort of recruits which mature rapidly. Females spawn at the end of the summer and soon die. Their offspring will grow more slowly over the fall and winter to mature and spawn in the spring to produce the summer generation. Life span of the *A. chinensis* summer generation was estimated to be 3–4 months while that of the winter generation 11–12 months (Oh and Jeong 2003). Other warm-temperate populations of *Acetes* species may show this pattern or simply have one generation of about a year (Xiao and Greenwood 1993). *Acetes americanus* off the southeastern coast of Brazil (23° S) shows the continuous reproduction expected for a tropical species and a short life span of about 7 months (Simões et al. 2013). *Acetes indicus* from Malaysian waters (Malacca. ~2° N) also reproduces throughout the year but, paradoxically, has a 2-year life span (Amin et al. 2009b), quite long for a tropical member of this genus, a finding which needs further investigation. One other neritic sergestoid, *Peisos petrunkevitchi* from the cool waters off the Argentine coast, has a restricted reproductive season of only 3–4 months (Mallo and Boschi 1982).

Many other sergestids are open-ocean species in tropical to temperate seas, important components of meso- and bathypelagic habitats, which migrate near to the surface at night to feed. Despite their importance to open-ocean ecology, few studies have been done on their reproductive periodicity. The monograph on pelagic shrimps by Makoto Omori (1974) is still one of the best sources of the limited information available. *Lucensosergia lucens* from Suruga Bay, Japan (35° N), spawns from spring to fall, i.e., an extended seasonal breeding. On the other hand, *Eusergestes similis,* one of the best studied sergestoids, lives in cooler waters, from the North Pacific Drift (40–50° N) down to the southern cool California Current. In the latter location, Omori and Gluck (1979) found that reproduction, as in more northern Oregonian waters, was from winter to mid-spring, with some spring and summer pulses. Life span of females varied with location from 1 to 2 years.

Reproduction seasonality has been little studied in luciferid sergestoids, small delicate shrimps in which eggs are attached briefly before hatching (Lee et al. 1992; Teodoro et al. 2014). In *Belzebub (Lucifer) faxoni*, females with mature ovaries are found throughout the year at a tropical location (López 1966) although with summer-autumn peaks. Data in Lee et al. (1992) indicate that spawning was mostly restricted to the warmer spring to fall months in the warm-temperate Gulf of Mexico. Life span in these small (~8–10 mm) shrimps is short, 30–40 days.

9.12.2 Tropical and Warm-Temperate Shallow-Water Species: Pleocyemates

Unlike penaeoids and sergestoids, the carideans and stenopodideans incubate their embryos. Thus, besides ovarian maturity in females, one can use the presence/absence of embryos and their incubation periods as data for latitudinal comparisons of reproductive seasonality. Tropical neritic marine habitats have high water temperatures which speed up metabolic processes such as reproduction and growth rates throughout the year. Thus, life in the tropics is fast for carideans. Growth and sexual maturation are rapid, and duration of embryonic and larval development is relatively short. Reproductive effort is high. The price of this tempo of life in many carideans is a short life span and, unlike most tropical penaeoids, small body size. As latitude increases, the pace of life slows down, as will be seen first in warm-temperate and then in cool-temperate, boreal/austral, and polar latitudes, and this is reflected in life-history traits.

A good illustration of lifestyles in the tropics are life-history traits of carideans living in seagrass meadows of the Caribbean (e.g., Bauer 1985b, 1989, 1991; Martínez and Román 2009, 2013, 2014; Martínez and Romero 2017). These lush shallow-water habitats, dominated by the seagrass *Thalassia testudinum*, are host to abundant small caridean and penaeoid species. I studied this fauna as a young faculty member of the University of Puerto Rico, Río Piedras, in the early 1980s. Supported by an enthusiastic group of marine biology students, I made monthly collections in seagrass meadow on the north coast of Puerto Rico (18° N); over 13 months, we collected about 75,000 specimens of seagrass carideans. Night samples were by far the most productive, as these shrimps become active and move about more at night, and thus much more susceptible to being caught in the collecting nets (Fig. 9.17) (Bauer 1985b; De Grave et al. 2006; Briones et al. 2020). Analysis of all this material, i.e., separation of specimens, identifications, counting, measurement of body size, and embryo number and size, took about three years, again thanks to my efficient student's help.

In these seagrass meadows, we collected species from different families, such as Hippolytidae (*Latreutes fucorum*, *L. parvulus*; *Hippolyte obliquimanus*),

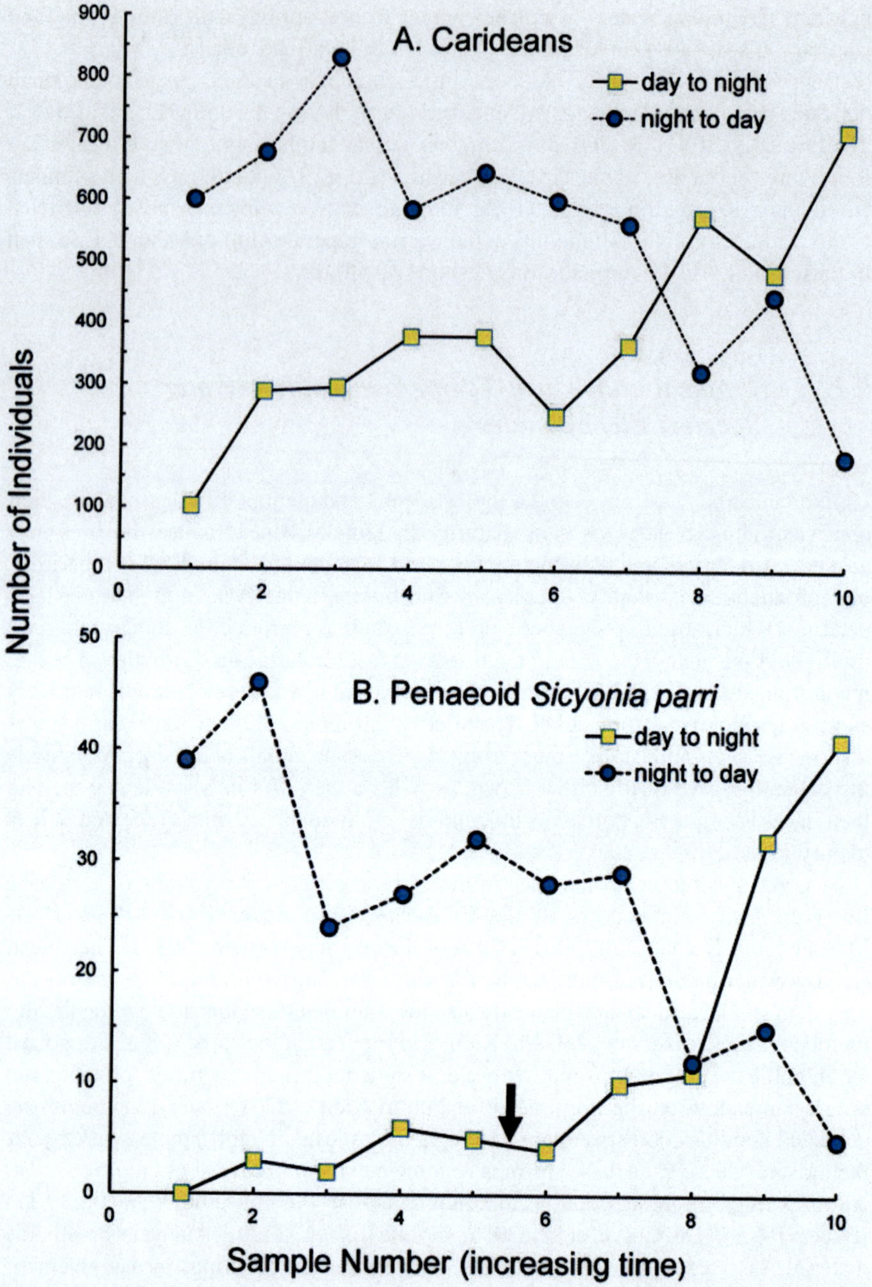

Fig. 9.17 Day–night comparisons in collectability (activity) of seagrass shrimps in Puerto Rico. Day to night: samples began just before sunset and terminated during the night. Night to day: samples began just before sunrise and terminated in early daylight hours. (**a**) Total caridean species. (**b**) Penaeoid *Sicyonia parri*. In (**b**), the day to night samples were begun a few hours before sunset (arrow) and ended a few hours after sunset and the onset of complete darkness. Based on data from Bauer (1985a), courtesy of Bulletin of Marine Science, and Bauer (1985b)

Processidae (e.g., *Processa riveroi* and *P. bermudensis*; Alpheidae (*Alpheus normanni*), Thoridae (*Thor manningi*), and Palaemonidae (*Cuapetes americanus; Leander tenuicornis*). They all have small body sizes, with reproductive females in the range of 2–7 mm CL (carapace length). Despite incubating small ELD embryos, these petite females can only carry average broods as low as 55–60 embryos in the three smallest species to about 600 in the largest. The small broods are due to low female body size; it is not for lack of reproductive effort! An analysis of brood production showed that the smaller of these species produced more embryos for their size than did larger ones, as pointed out earlier (Bauer 1991). Females produced successive broods frequently, one after the other; females with embryos near hatching usually had ovaries full of larger vitellogenic oocytes. This indicates that as soon as females hatched one brood, they molted, mated, and spawned another brood.

Incubated embryos develop and hatch rapidly in these species, thanks to the high and stable water temperatures. At a temperature range from 26 ° C in the winter to 31 ° C in the late summer, broods hatched in as little as 5 days in *Hippolyte obliquimanus* to a maximum of 10 days in *Alpheus normanni*. Unpublished cohort analysis of some of these species showed that the average life span was only about 6 months or less. Thus, these shrimps hatch out quickly, go through a larval period of perhaps a few weeks or a month, settle and sexually mature rapidly, and begin reproduction within a few months of settlement. Females pump out broods one after another for a few months more until their death, probably eaten by the abundant and diverse fish fauna and invertebrate predators of shallow tropical waters. Live fast and die young, indeed! Reproduction here can be characterized as continuous, year-round. Month to month intensity of reproduction varied in common among species but without a seasonal pattern, apparently due to episodically occurring factors affecting both adult reproduction and larval survival in the plankton (Bauer 1989). Recruitment of juveniles from the plankton followed a similar irregular pattern.

Outside of the tropics in latitudes up to about 35–40° N or S, factors affecting reproduction and recruitment are more seasonal, with a long period of warm water temperatures from late spring to early autumn followed by cool to cold winter temperatures. Phytoplankton production in coastal neritic waters is generally bimodal, with a spring to early summer burst of plankton productivity and a smaller peak in the early autumn, with lows in summer and especially winter. In these warm-temperate waters, caridean reproduction is no longer continuous but is generally compressed into a spring-autumn extended period of ~6–9 months. A good comparison of tropical and warm-temperate reproductive patterns of seagrass shrimps is that between my work in Puerto Rico and that of Kikuchi (1962) who investigated reproduction in the shrimp fauna of *Zostera marina* seagrass fauna at Tomioka Bay in southern Japan (32°N). Similar families and genera of carideans are found there (hippolytids, palaemonids, and processids) except for one species of Crangonidae, which is a cold-water group. Reproductive seasons, defined as periods when ovigerous females were found, were not continuous but seasonal and extended from 5 to 7.5 months. Most species were year-round residents except for the *Processa* species, which came onto the seagrass beds from surrounding habitats to mate, spawn, and brood embryos. Water temperatures varied from 14 ° C in mid-winter to 27 ° C in midsummer at this location—definitely seasonal. The average length of the late spring to autumn breeding season at this Japanese location was about 6 months compared to continuous reproduction in the Puerto Rican species. Reproductive seasons of the processid seagrass shrimps, *Ambidexter symmetricus*, from the northern Gulf of Mexico (29° N) and *Processa edulis* from the Mediterranean (42° N) were similar to those of the Japanese location (Rasch and Bauer 2015 and Noël 1976, respectively).

Not all species at the Japanese location began spawning in the warmest months of the year. *Heptacarpus* species began in winter and continued through the spring in the seagrass meadows studied by Kikuchi (1962). I found a similar pattern in *H. sitchensis* in La Jolla, California, also at 32° N but on the other side of the Pacific (Bauer 1976). Water temperatures are similar in the winter in both locations, but those of La Jolla, under the influence of the cool California Current, only reach to ~20 ° C in the summer. Perhaps those species which begin to breed in the winter are aiming to release larvae during the early part of a spring plankton bloom that typically occurs at the La Jolla location.

The pattern of an extended reproductive season, generally in the warmer more larva-friendly months of the year (spring-fall), is a common one in neritic warm-temperate latitudes, and there are many examples, e.g., *Palaemon* (*Palaemonetes) pugio* (Alon and Stancyk 1982), *Hippolyte inermis* (Zupo 1994), *Rhynchocinetes uritai* (Osawa et al. 2015), *Lysmata wurdemanni* (Bauer 2002), *Palaemon macrodactylus* (Omori and Chida 1988; Vázquez et al. 2012), *P. xiphias* (Guerao et al. 1994), and *L. seticaudata* (Dohrn 1950; Calado and Narciso 2003). These species are longer lived than the tropical seagrass species, with life spans of ~1–2 years. Average body sizes of these warm-temperate species are generally small, up to about 20 mm carapace length.

Two species from a transitional area between tropical and warm-temperate areas at 23°S, *Alpheus armillatus* (sand flat habitat) and *Exhippolysmata oplophoroides* (shallow-water soft bottoms), show continuous reproduction as in tropical species (Mossolin et al. 2006; Fransozo et al. 2005, respectively). However, in the latter species, there are late spring and summer peaks. *Alpheus armillatus* has a 1.3-year life span, while that of *E. oplophoroides* is 2 years (Pescinelli et al. 2018). It will be recalled that this area off southeastern Brazil has spring and summer upwelling which produces a plankton bloom in an otherwise oligotrophic area. This may account for the tendency toward seasonality, at least in *E. oplophoroides.* Two species of *Synalpheus* at 26° N (eastern Gulf of Mexico) had a similar pattern as *E. oplophoroides* with some ovigerous females all year, but the principal reproduction was in spring and summer. Interestingly, winter broods were smaller and many of the embryos were unviable (Felder 1982).

During the warm late spring to early fall months in warm-temperate zones, the pace of reproductive activity may approach that of tropical seas, with successive broods at short intervals. The rate of brood production depends, in part, on the incubation period of embryos, which is highly temperature dependent. In *Heptacarpus sitchensis*, which begins spawning in mid-winter in southern California, embryos in female broods may take 4–6 weeks to hatch, but in summer, that time is reduced to two weeks (Bauer 1976; pers. obs. on incubation periods). In the alpheid *Betaeus emarginatus*, incubation takes 69 days at 13 °C but only 23 days at 20 °C (Wehrtmann and López 2003). Short incubation periods have been reported for other warm-temperate species at higher water temperatures in the summer, e.g., two weeks at 27–28 ° C in *P. pugio* (Bauer and Abdalla 2000). *P. macrodactylus* spawned and incubated 6–9 broods in a 7-month reproductive season (Omori and Chida 1988).

In many of these warm-temperate caridean species, the life span and breeding pattern depend on the timing of that recruitment, centered around a nonreproductive season in the late autumn and winter. Some species, e.g., *Latreutes acicularis* and *Palaemon serrifer* (Kikuchi 1962). *P. pugio* (Alon and Stancyk 1982), and *Actinimenes (Periclimenes) ornatus* (Omori et al. 1994), have two overlapping generations, a short-lived "summer" generation and a longer-lived "winter" generation. Females of the winter generation are hatched in the previous late summer and early fall but do not become sexually mature until mid-winter or spring. They spawn through the spring and summer, producing the summer generation. They die off in the late autumn-early winter, with a life span of 1–1.5 years. Individuals of the summer generation mature and spawn the winter generation before dying off in the winter of their first year with a life span of ~6–8 months.

In other species, one can recognize a shorter and longer generation, whose reproductive pattern and life span depend on when their population cohort is recruited into the population. They are analogous to a summer and winter generation, but the overlap between them is more extensive. For example, in *Heptacarpus sitchensis* (southern California; Bauer 1976), samples in the autumn contained only juveniles and small nonbreeding adults hatched and recruited during the summer. Some of the females in this group mature and spawn by mid-winter, and by April, all females of this "short" generation are breeding. By the late summer, this population, recruited the previous year, disappears, with a life span of about 1 year, depending on the timing of their recruitment. The first recruits spawned by the short-generation females arrive in the spring. The individuals of this "long" generation mature and breed briefly in the summer, cease breeding in the autumn, and then begin to breed again the next mid-winter, with some surviving through the summer of their second year, with a life span of ~1.5 years. In this species, some individuals of both generations overwinter but differ somewhat in reproductive pattern and life span. A similar pattern, complicated by sex change, is shown by *Lysmata wurdemanni* from the western Gulf of Mexico (Baldwin and Bauer 2003).

Another pattern is shown by species such as *P. macrodactylus* from Japanese waters (38° N) (Omori and Chida 1988). This species lives on average for 2 years, with breeding females reaching the relatively large size for a warm-temperate species of 20 mm CL. The breeding period is seasonal, extending from mid-April through October. The average recruit arrives in the adult habitat in summer (Year 0), grows during the autumn and winter, and first breeds in the following spring-autumn (Year 1). It then goes into nonreproductive condition in the later autumn and winter and reproduces again in the spring and summer of Year 2 but disappears from the population soon after. Breeding females can spawn 6–9 broods during the breeding season, so that females spawn, incubate, and hatch embryos, and then mate and spawn again on average about every 3–4 weeks, a very high reproductive output, as in other warm-temperate carideans. A similar reproductive pattern is found in *Rhynchocinetes uritai* (Osawa et al. 2015). However, this latter species is a protandric hermaphrodite that reproduces as a male during its first breeding season and a female during its second (last) one (see Chap. 7, Sexual Systems: Fig. 7.10).

The Stenopodidae are tropical and warm-temperate neritic shrimps that brood embryos like the carideans. Unfortunately, there is little information on their reproductive seasonality due to their usual low abundances, a result of their territorial pair-living. Thus, the large sample sizes of individuals necessary for quantitative description of reproductive traits are lacking. Most of the work on breeding has been done in the laboratory on *Stenopus hispidus* which indicates that, in this tropical stenopodid at least, reproduction is continuous with successive spawns (Gregati et al. 2010).

9.12.3 Freshwater Carideans

Freshwater carideans are mainly a tropical and subtropical group, and it should be no surprise that the seasonality of breeding is comparable to their coastal marine caridean relatives. Tropical species usually have some breeding throughout the year (e.g., *Palaemon pandaliformis*, Paschoal et al. 2016). However, in many species, breeding is much more intense during the rainy (wet) season (e.g., *Atya lanipes,* Johnson et al. 1997; *Caridina* species from Hong Kong, Yam and Dudgeon 2006; *Xiphocaris elongata*, Cross et al. 2008; *Potimirim brasiliana*, Da Rocha et al. 2013; *Macrobrachium tenellum*, López-Uriarte et al. 2020). In contrast, the breeding season is confined to a 6-month period from spring to fall (the wet season) in the tropical species *Macrobrachium spinipes* in the Northern Territory, Australia (Novak et al. 2015).

Breeding seasons of warm-temperate freshwater carideans are seasonal but often extended. In the northeastern Gulf of Mexico. *Palaemon paludosus* has a long 9-month season from February to October, but in more tropical south Florida, breeding is continuous (Beck and Cowell 1976). In *Paratya australiensis* from southeastern Australia, Hancock and Bunn (1997) described a 7-month breeding season. In two lake species in South Korea, *Palaemon* (*Exopalaemon) modestus* and *P. paucidens,* breeding takes place for only 5 months during spring and summer (Oh et al. 2002; Puspitasari et al. 2013).

From these examples, it is apparent that temperature and hydrological factors are equally important in establishing a favorable environment for breeding of freshwater shrimps. High tropical temperatures with low annual variation favor continuous breeding in the tropics, while in warm-temperate areas, breeding is restricted to periods during spring to fall when water temperatures are high enough to stimulate and maintain breeding. In amphidromous species, in which larvae are released upstream or carried to the sea by females, there must be enough flow to carry out these activities. Thus, spawning is more intense during the rainy or wet season of the year or restricted to it. In non-amphidromous and land-locked species with either ALD or ELD, wet season conditions also favor breeding. As noted above, the various populations of *Paratya australiensis* in southeastern Australia are probably a complex of cryptic species. Populations at higher elevations are essentially land-locked by stream topography and hydrological conditions, as opposed to the amphidromous ("estuarine") ones at lower elevations. During the rainy season, water levels rise in pools and flooding of nearby creeks creates stable conditions for the larvae of upstream populations to feed and develop without being washed downstream. During the same period, the amphidromous downstream ("estuarine") populations of *P. australiensis* experience more favorable conditions for release and drift of their larvae to the sea. A similar situation has been shown by Han et al. (2011) for two *Caridina gracilipes* populations, one non-migratory from a land-locked lake and the other an amphidromous stream-dweller. As noted earlier, Walker (1992) showed, in a suite of freshwater species from Brazilian Amazonia, that breeding was dependent on wet-season flooding of habitats necessary for larval feeding and development.

Shrimps are difficult to age as they do not have structures with growth rings, like those of trees or otoliths (ear bones) of fishes. Their hard parts (exoskeleton) are shed at each molt and thus do not accumulate annual or season growth rings or bands. In shrimps, life spans are usually approximated from analyses of growth and population dynamics data. We saw above that such life span estimates of neritic marine species in tropical and warm-temperate habitats are relatively short, from several months to about 2 years. However, a suite of six freshwater *Macrobrachium* species from Costa Rica were estimated to have long maximum life spans of 4–11 years (Ibarra and Wehrtmann 2020)! These species attain rather large body sizes (~40–95 maximum CL). But even shrimps of much smaller body size (10–20 mm CL) in streams from Puerto Rico were shown by mark-recapture studies to have *minimum* life spans of 8 years (*Atya* species) and 5–11 years (*Xiphocaris elongata*) (Cross et al. 2008). The 5-year estimate for *X. elongata* came from populations in streams with predatory fishes, and the 11-year estimate was from population without these predators. On the other hand, shorter life spans have been estimated for small-sized *Macrobrachium* species (cited in Ibarra and Wehrtmann 2020), *Caridina* species from Hong Kong (~ 1 year) (Yam and Dudgeon 2006), and two *Palaemon* species (1.1–1.3 years) from lakes in Korea (Oh et al. 2002; Puspitasari et al. 2013). The biological and environmental factors that favor long life spans in some but not other freshwater carideans are an intriguing source for future investigation for population and freshwater biologists.

9.12.4 Cool-Temperate and Boreal/Austral Species

Marine coastal marine habitats at higher latitudes are populated only by caridean shrimps, as the dendrobranchiate and stenopodidean species are tropical and warm-temperate groups. Among the carideans, breeding seasonality, length of embryo incubation, and life span are what might be predicted from classical latitudinal paradigms about reproduction and life-history traits. Reproductive traits have evolved to favor major larval release in the highly seasonal major spring and the minor autumnal plankton blooms that provide food for the planktonic larvae of shrimps. One of the best comparative studies on life-history traits from these latitudes is J.A. Allen's (1966) comparative study on thirteen caridean species inhabiting coastal waters off the northeastern coast of England (Northumberland; 55° N). In this boreal locale, cold-water carideans such as crangonids, pandalids, and "spirontocarid" thorids predominate, with a few species from primarily warm-water families such as Processidae and Hippolytidae.

Allen studied this fauna over several years and found that the two major patterns of spawning, embryo incubation, and hatching of larvae are timed to coincide with the two major plankton blooms. Because of the colder water temperatures (5–12 ° C), embryo incubation periods are long, from 8 to 20 weeks. In *Crangon crangon (vulgaris)* and *C. allmanni*, incubation is 10–12 weeks in winter and 5–6 weeks in warmer nearshore waters in the summer. This contrasts with the 1–2 week incubation period of carideans in tropical waters and ~ 2–8 weeks in warm-temperate species. As a result, fewer broods per breeding season can be produced by the cold-water species. Those species targeting spawning for the spring bloom (*Pandalus montagui, P. borealis, Spirontocaris spinus, S. lilljeborgi, Pontophilus spinosus,* and *Caridion gordoni*) spawn in late November or December, incubate embryos for 10–20 weeks, and release larvae by April. Those females produce only one brood per year, except for *C. gordoni,* which may spawn two.

The spawning of the other species is timed to coincide with both spring and autumnal blooms. *Pandalina brevirostris, Crangon crangon, C. allmanni,* and *Eualus pusiolus* spawn variously from mid-autumn to spring, with usually 2 broods per year, one timed to hatch for the spring plankton bloom and the other for the autumnal one. Only two species, *Hippolyte varians* and *Processa canaliculata*, are "summer" breeders (April to October) which hatch two broods per year. In *H. varians*, the embryo incubation period is 10 weeks, so that larvae may enter the plankton just after either the spring or autumnal plankton bloom. Perhaps larvae of *H. varians* and *P. canaliculata* avoid competition with larvae of other species which are released during the major plankton blooms.

I examined reproductive information gathered by Butler (1980) from many sources on 24 species of cool-temperate crangonids, pandalids, and thorids from the Pacific coast of Canada (45–50° N). Ovigerous females were found variously from summer to mid-autumn, with some hatching in spring and others up to late summer. Incubation times and the number of broods were not reported except for the pandalids, which spawn in the autumn and hatch their only brood in the spring, in time for the spring plankton bloom. I suspect that a similar pattern of species, one timed to hatch in the spring, others for both spring and autumn blooms, and still others just after either bloom, is likely in this cool-temperate group of species. In the area covered by Butler, the only alpheids, a mainly tropical to warm-temperate group, were three species of *Betaeus*, which were ovigerous only 4–5 months during the late spring to early fall warm-water months. *Betaeus emarginatus,* an austral cool-temperate species, begins breeding in the autumn with hatching from mid-winter to spring (Lardies et al. 1998).

The life spans of species at Allen's boreal location are longer than those usually reported for species from lower latitudes: 2–3.5 years as opposed the several months to 2 years in the warm-water neritic species. In the boreal species, females brood embryos for longer periods compared to lower-latitude species, during which they cannot molt and grow. They may make up for this somewhat with growth during the nonbreeding season. Body sizes of boreal species overlap in size with warm-water species but on average are somewhat larger (~4–30 mm CL), especially the crangonid and pandalid species.

9.12.5 Polar Species

Polar waters, extending from 66 ° N and S to the poles, are home only to caridean shrimps. At these latitudes, the very cold-water temperatures have an inhibitory effect on growth and reproduction (Pearse et al. 1991). The combination of protracted ovarian maturation and lengthy incubation periods reduces the number of broods produced by a female to usually just one per year or one every two years, e.g., the arctic thorids *Lebbeus polaris* and *Eualus gaimardii* (Williams 1984), and the high Antarctic species *Chorismus antarcticus* (Hippolytidae), *Notocrangon antarcticus* (Crangonidae), and *Nematocarcinus lanceopes* (Gorny et al. 1992; Gorny and George 1997). The period of embryo incubation, from spawning to hatching, takes about 10 months in the latter three species. Compare that to less 5–10 days in tropical seagrass species studied by Bauer (1989)!

Seasonality is also extreme in polar habitats, with a long dark winter and a very short summer with a long daylight period. Water temperatures are near freezing during the winter and only a few degrees above that in summer, although somewhat higher in coastal areas of the Arctic. However, the long daylight period of the short summer along with high nutrient levels in the polar seas produces an intense plankton bloom providing an abundance of larval food for planktonic shrimp larvae. The difficulty for polar species is the timing of vitellogenesis, spawning, and incubation so that larvae hatch into the restricted period of the plankton bloom. Many species have not been able to evolve that adaptation, and thus there is a large proportion of ALD and DD species. In arctic polar regions, this includes the ALD species *Lebbeus groenlandicus* and *L. polaris* (Thoridae); *Sabinea septemcarinata* (Crangonidae); and the high Antarctic *Chorismus antarcticus* (Hippolytidae) and *Notocrangon antarcticus* (Dobkin 1969; Williams 1984; Węsławski 1987; Gorny et al. 1992). Females spawn large embryos rich in yolk that will sustain the 3 lecithotrophic larval stages until settlement. The adults of these and other polar species probably take advantage of the remains of the summer plankton bloom that falls to the benthos, providing food to make the large yolky embryos. The boreal-arctic genus *Sclerocrangon* has essentially direct development (Makarov 1968), with hatching of 1–2 benthic larval stages with extended maternal care (e.g., *S. boreas,* Guay et al. 2011).

Some species from polar regions have the other strategy; i.e., they appear to have adapted release of larvae to utilize the rich summer plankton bloom. The far-northern populations of *Pandalus* have planktotrophic larvae, as do the thorids *Eualus pusiolus, E. gaimardii, Spirontocaris lilljeborgi, S. spinus, and S. phippsii* (Węsławski 1987; Astthorsson and Gislason 1991).

The trend of slower growth giving rise to larger body size and longer life spans with latitude (colder water temperatures) is apparent in polar populations of Caridea. Life spans of polar species such as *Sabinea septemcarinata and Eualus gaimardii* are from 3–4 years, overlapping with those of boreal species. In the high Antarctic species *Chorismus antarcticus, Nematocarcinus lanceopes*, and *Notocrangon antarcticus*, maximum female life spans have been estimated at 7, 8–10, and 6 years, respectively (Gorny et al. 1993; Gorny and George 1997; Bluhm and Brey 2001).

9.13 Within Taxon Latitudinal Variation in Life-History Traits in Neritic Marine Species

Species in the same genus are closely related, sharing a recent common ancestor, and thus they have inherited a similar reproductive physiology and capacities. This makes possible testing of hypotheses on the effects on life-history traits due to latitudinal variation in environmental factors. Below, I will give various examples. Let us begin with some penaeoid genera and species. In the genus *Sicyonia,* a comparison of four species shows an increase in reproductive seasonality from tropical to cool-temperate areas (Bauer 1992) (Fig. 9.18). In *S. parri* (18° N, north coast of Puerto Rico), there was a low level of reproduction throughout the year with an episodic rather than seasonal pattern (Bauer and Rivera Vega 1992). At a southern hemisphere tropical location (23° S; southeastern Brazil), a high proportion of reproductive females (*S. dorsalis*) were found all year, but with peaks in the summer and winter months (Castilho et al. 2008). These peaks are undoubtedly related to the seasonal upwelling during the spring and summer months which produces high plankton productivity, i.e., larval food (Grabowski 2017). At 28° N off the Atlantic coast of Florida (USA), breeding occurred all year in *S. brevirostris* but with a major winter to early spring peak (Kennedy et al. 1977). In *S. ingentis* from the cool-water California Current at 34° N, reproduction was most seasonal with a marked summer-fall breeding season (Anderson et al. 1985). In these species, reproductive body size and estimated life span usually increased with latitude, from 3–9 mm CL (6-–8 months) in *S. parri*, to ~6–14 mm CL (1–2 years) in *S. dorsalis,* to 17–35 mm CL (20–22 months) in *S. brevirostris*, to 25–36 CL (> 24 months?) in *S. ingentis*.

Comparisons among populations of the same species are similarly instructive in understanding latitudinal patterns. The distribution of *Rimapenaeus constrictus* in

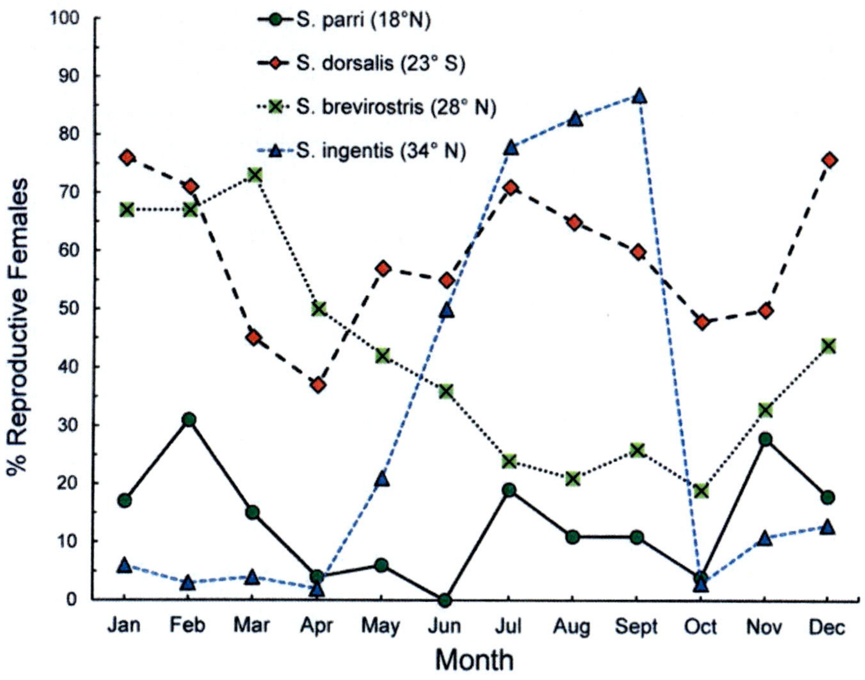

Fig. 9.18 Latitudinal variation in reproductive pattern (mature, prespawning females) in four *Sicyonia* species (Penaeoidea): *S. parri* (18° N, tropical), *S. dorsalis* (23° S, tropical/warm temperate transition), *S. brevirostris,* (28° N, warm temperate), and *S. ingentis* (34° N, warm/cool-temperate transition). Note that the *S. dorsalis* populations are from the southern hemisphere, where seasons are the inverse of the northern hemisphere locations. Based on data from Bauer (1992) and (*S. dorsalis*) Castilho et al. (2008)

coastal waters stretches in the Western Atlantic from Nova Scotia to Brazil. Reproductive studies on this species have been done in the northern Gulf of Mexico (30° N) (Bauer and Lin 1994) and on the other side of the equator around 23° in southeastern Brazil (Costa and Fransozo 2004; Garcia et al. 2016). In the more tropical location, reproductive females were found throughout the year with spring and summer peaks, while in the warm-temperate location, reproductive females only occurred from spring through early fall; breeding ceased in the colder temperatures of late fall and winter when conditions are not favorable for planktonic larvae. Castilho et al. (2007) investigated latitudinal variation of some life-history traits in the western South Atlantic penaeoid *Artemesia longinaris*. They compared populations from 23° southeastern Brazil with those from Mar del Plata (Argentina, 38° S). Female sexual maturity occurred at a larger body size at the higher latitude than in the tropical location, and females reached a larger maximum body size, indicating slower growth. These differences were attributed to lower water temperatures and increasing plankton productivity (larval food supply) at the higher latitude location.

The "classical" life-history paradigm for neritic shrimps is usually valid when comparing taxa across a broad stretch of latitude. However, this is not necessarily true when comparing related taxa or populations within a few degrees of latitude. Local variations in population dynamics and environmental conditions, such as temperature and plankton production, do not always correlate well with small changes in latitudes. Local upwelling, currents coming from higher or lower latitudes, and seasonal intrusion of oceanic water masses may weaken or negate expected correlations of life-history traits with environmental factors. Reproductive migrations from one area to another, as well as wide dispersal of larvae during the planktonic stage, may swamp latitudinal effects (Grabowski 2017). As with other generalizations, it is these deviations which often prove the rule by stimulating studies that explain why such digressions occur. A good example of such studies is that on intraspecific variation in life-history traits and seasonality of reproduction among different populations of *Sicyonia dorsalis*, *Rimapenaeus constrictus,* and *Pleoticus muelleri* across the narrow latitudinal range from 22 to 26° S latitude (Grabowski 2017).

Comparisons within species of neritic marine carideans mainly support the latitudinal model of life-history trends. Populations of deep-shelf to shallow-slope pandalid *Plesionika edwardsii* from the Cape Verde Islands (16° N), Canary Islands (28° N), and Madeira (33° N) showed expected latitudinal trends (González et al. 2016). From south to north, water temperatures decreased while maximum body size and size of female sexual maturity increased, accompanied by a slight increase in life span. In the boreal to subarctic *Pandalus borealis*, several life-history traits vary positively with latitude (reviewed in Haynes and Wigley 1969; and Bauer 2004). Females spawn just once a year with hatching of larvae from early to late spring. The length of the breeding season, from spawning to hatching, varies with water temperature from 6 months at 8 °C off Northumberland, England (55° N), to 12 months at nearly 0 °C off Spitzbergen, Norway (75–80° N). Water temperatures are not perfectly correlated with latitude because of current patterns and other factors. Thus, a southern population from the Gulf of Maine (42–43° N), under the influence of the arctic Labrador Current, has a longer breeding period than some of the more northern locations (Haynes and Wigley 1969). Incubation of a brood of embryos takes 3–5 months in boreal latitudes but 6 months at high Arctic latitudes. In the boreal populations of *P. borealis* studied by Allen (1966), life span was estimated at 3.5 years, but this increases to 7 years in high Arctic waters (Haynes and Wigley 1969). More recently, Nilssen and Aschan (2009) found that, in the far-northern population of *P. borealis* at Jan Mayen Island (77° N), females may not mature until 6–7 years, living as long as 11 years! Similar examples of latitudinal variation in life-history traits have been found in *Palaemon pugio* (reviewed in Bauer 2004) and *Betaeus truncatus* (Lardies and Wehrtmann 2001) as well as among Western Atlantic species of *Thor* (Bauer and VanHoy 1996). Like the penaeoids cited above, fecundity, reproductive output, and female body size in populations of the caridean *Exhippolysmata oplophoroides*, when compared across a narrow latitudinal range, showed mixed agreement with the latitudinal paradigm (Stanski et al. 2018).

In *Hippolyte obliquimanus*, trends in brood size and embryo size agreed with the latitudinal model, but body size did not (Terossi et al. 2010).

A different variation on the latitudinal paradigm occurs in *Crangon crangon*, a primarily cold-water species, but whose distribution ranges from subtropical (28° N) to subarctic waters of the North Atlantic and adjacent waters. In boreal waters, females of this species may produce two different kinds of "eggs" (embryos) depending on the season (Boddeke 1982). "Winter" eggs are significantly bigger in size than "summer" eggs; broods from summer eggs contain many more embryos than those of winter eggs. Burukovsky and Ivanov (2015) proposed that three basic latitudinal reproductive strategies have evolved in *C. crangon*. At the higher subarctic latitudes, females only spawn once during the short June–July (summer); embryos are the large "winter" ones. In boreal populations, females may spawn more than once, e.g., 2–3 times from November to September (e.g., Allen 1966). Boreal females spawn "winter" eggs during the colder months of the year and "summer" ones during warmer months of the year. Depending on the season (water temperature), incubation can vary from 4–5 weeks to 8–10 weeks. In the southernmost warm-temperate parts of the *C. crangon* distribution, females spawn only once, in the winter, but extrude the small "summer" eggs of boreal populations. Referring to the boreal populations of the North Sea, Bodekke hypothesized that the adaptive value of "winter" eggs had to do with low phytoplankton production in the boreal winter. Winter eggs hatch into larger-sized larvae than summer eggs; their larger size increases the chance of survival during the winter and early spring before the peak of the spring plankton bloom. The smaller larvae from the summer eggs hatch in other periods of the year with more favorable larval conditions. In the southernmost populations of *C. crangon*, the production of only small summer eggs occurs during the winter and is apparently timed so that the embryo-hatching period will coincide with the early spring plankton bloom. It is not clear how the summer spawning of winter eggs in subarctic populations is coordinated with favorable conditions in the plankton for successful larval development. Key factors to consider include the length of embryo incubation and larval development in the plankton, both of which vary from 3 weeks to 3 months each depending on water temperature (Campos and van der Veer 2008).

9.14 Deep-Sea Species

The effects of increasing depth on water temperature and light are somewhat like those experienced by an increase in latitude in marine species. Water temperatures decrease from neritic to bathyal to abyssal and hadal areas, with near freezing temperature in the latter environments. Light intensity decreases rapidly with depth, reaching only to ~800 m in the clearest oceanic waters. The lack of light prevents any photosynthesis, and so there is no basis for a trophic web from this source. Particulate matter (decomposing remains of organisms, feces, other detrital material) sinks down from the photosynthetic surface food web, thus providing some food for pelagic and benthic shrimps. Chemosynthetic food production takes place in deep-sea hydrothermal vents and cold seeps on the sea floor. Release of hydrogen sulfide and/or hydrocarbons at these sites supports thriving chemoautotrophic bacterial and associated animal communities which include caridean shrimps. Very near to hydrothermal vents, comfortable water temperatures occur. However, these chemosynthetic communities are only found in very limited areas of the deep sea. Thus, cold-water temperatures and impoverished food supplies limit and shape life-history traits of shrimps in the deep sea.

Seasonality of reproduction occurs to some extent in the dim to dark depths of the continental slope, especially its shallower parts. This may not be too surprising given that the slope begins at the edges of continental shelves, which do have seasonally varying food webs. Among the penaeoids, the solenocerids and aristeids are found at deep shelf and slope depths. They have a "Type-4" life cycle (Fig. 9.14), with all life stages taking place in deep-shelf and open-ocean habitats. Larvae must go to surface waters to feed and develop before returning to the bottom as postlarvae. *Aristeus antennatus (*Orsi Relini and Relini 1998; Kapiris and Thessalou-Legaki 2009) and *Aristaeomorpha foliacea* (Ragonese and Bianchini 1995), from warm- to cool-temperate areas in the Mediterranean, follow the common spring-summer breeding period of shrimps on the adjacent continental shelves. The solenocerids *Pleoticus* (*Hymenopenaeus) robustus* and *Haliporoides sibogae* from shallow slope depths in warm-temperate latitudes similarly show a restricted breeding pattern (fall or winter to spring; e.g., Anderson and Linder 1971; Ohtomi and Matsuoka 1998). Company and Sarda (1997) sampled a suite of 5 slope-dwelling *Plesionika* species (Caridea, Pandalidae) from the Mediterranean Sea over 3 years. Four of the five species showed seasonal breeding which became more restricted in time with species depth. The shallowest species, *P. heterocarpus*, from neritic to shallow slope depths, bred all year long, but the deepest-dwelling species, *P. acanthonotus,* reproduced only from late spring to early summer. Company and Sarda proposed a correlation between the fall of detrital material from surface plankton production and breeding pattern in these species. In the studied area of the Mediterranean, primary productivity occurs throughout the year, supported by nutrient inputs from nearby rivers, as well as by warmer water temperatures and greater light availability than in the deeper slope habitats. In the oceanic waters over the adjacent continental slope, the major plankton blooms take place from spring to autumn, generating the sinking phytodetrital material which supports the varying spring to fall breeding patterns of the deeper slope species.

Orton (1920) predicted that the cold temperatures, continual dark, and poor food supply of the deep sea would result in continuous, asynchronous reproduction; i.e., some females in the population might be reproducing at any given time throughout the year but not in seasonal synchrony. Unfortunately, it is very difficult to sample shrimps from the deep-slope, abyssal, and hadal benthic depths of the sea to test this hypothesis. However, Wenner (1978, 1979) was able to sample a group of glyphocrangonid, crangonid, and nematocarcinid carideans on the deep (200–3000 m) continental slopes and canyons off the eastern coast of the USA (36–37° N). This is a typical dark, cold (2–4 ° C) deep-sea habitat with no obvious seasonality in major environmental factors. Some females from these species were ovigerous or near spawning a new brood throughout the year but not at the same time (asynchronously). Some incubating females also had mature ovaries, indicating successive spawning, but some did not. In this group of species, Orton's hypothesis seems to hold true. However, as noted previously, different larval developmental strategies may be used by deep-sea species. Unlike the other carideans studied by Wenner, nematocarcinid females spawn many small embryos, indicating planktotrophic development (ELD). Their larvae need to go to surface waters to feed in seasonally productive waters. The glyphocrangonid and crangonid species produced relatively few large embryos with abbreviated or direct development, thus keeping their offspring on the bottom without the need of seasonal food supplies in surface waters. Both strategies are successful, indicating that phylogeny, i.e., the reproductive biology of ancestors, plays an important role in evolution of the reproductive traits in these deep-sea species.

A group of carideans which experience truly deep-sea conditions around hydrothermal vents are the alvinocaridids, e.g., *Rimicaris, Alvinocaris*, and *Mirocaris* species. Although difficult to sample because of the depths at which they occur, they are usually abundant around hydrothermal vents so that relatively good information on their reproduction has been obtained. Even though surrounded by the very cold deep-sea, water temperatures at the vents are extremely high, so that the shrimps can choose the temperature favorable for them from the surrounding heat gradient. Primary production by chemoautotrophic bacteria delivers a rich food source for these shrimps. Several studies show that females carry large broods of small-sized embryos, indicating planktotrophic ELD. Females carry multiple cohorts of oocytes showing that they spawn several broods in their lifetime. There is no evidence of synchrony among breeding females; i.e., reproduction is continuous and nonseasonal on a population level (Ramirez Llodra et al. 2000; Ramirez-Llodra and Segonzac 2006). *Lebbeus virentova*, a hippolytid caridean from a hydrothermal vent community, uses the other reproductive strategy found in Wenner's glyphocrangonid and crangonids shrimps. Females produce very large embryos indicating abbreviated development like their shallow-water, boreal-arctic relatives, *L. groenlandicus* and *L. polaris* (Nye and Copley 2014). *L. virentova* avoids the dangers facing the alvinocaridid planktotrophic larvae, but females produce many fewer offspring with limited dispersal potential. As with the alvinocaridids, reproduction was continuous at the population level, but the possibility of a seasonal peak was not ruled out.

A lack of seasonality is not necessarily the case for all deep-sea shrimps. Seasonality of breeding season and type of larval development might be influenced by seasonal pulses of sinking organic matter ("phytodetritus") from surface waters (Eckelbarger and Watling 1995; Young 2003). If seasonal surface phytoplankton and associated food web production is great enough, sufficient "leftovers" may sink to the bottom so that deep benthic adults can experience a seasonal windfall of food to support breeding (Tyler 1988). Generally, surface waters over oceanic deep-sea areas are oligotrophic and poor in productivity compared to coastal neritic waters. However, this is not so everywhere. Seasonal inputs of organic matter derived from plankton production, benthic plants, and terrestrial flooding can make their way into coastal waters down the continental shelf into the deep sea. Although the great expanses of mid- to low-latitude oceanic waters away from the continents are considered marine deserts with low primary productivity, there are oases of high plankton production (Sigman and Hain 2012). In equatorial waters on the eastern sides of major oceans (western sides of continents), there is a seasonally variable open-ocean upwelling due to divergence of equatorial surface waters. Convergence of northern and southern trade wind plus effects of the Coriolis force brings up nutrient-rich deep water which stimulates high primary production food webs (Flohn 1970). Particulate organic matter derived from decomposing microalgae, animal excretions, feces, and dead bodies eventually sink to the bottom, serving as a food source for deep-sea organisms. This seasonal food input might allow deep-dwelling female shrimps to assimilate energy for temporally synchronous reproduction. However, this remains to be demonstrated in benthic shrimps from the vast expanses of the deep sea. An asynchronous, nonseasonal reproductive pattern, continuous on the population level but not necessarily on an individual basis, appears to be the rule in deep-slope and abyssal benthic habitats.

Members of the stenopodidean family Spongicolidae, symbionts of deep-sea hexactinellid sponges, would be a good benthic group with which to test hypotheses on seasonality in the ocean depths. However, given their scarcity and the difficulty of sampling them, it is not surprising that there are almost no studies to report. Saito and Koya (2001) did manage to get small samples of females from different times of the year which did not show any obvious seasonality in reproduction of *Spongicola japonica*.

Similar trends in seasonality have been shown in several open-ocean, completely pelagic shrimps, particularly in the Caridea, on which most studies have concentrated. The pelagic sergestoids are acutely in need of study in this respect. The two major sources of information on temporal variation in reproduction of pelagic carideans are Omori's (1974) world review and the Hopkins et al. (1989) intensive investigation of 22 pelagic species from the eastern Gulf of Mexico. Both studies show that, as in bathyal and abyssal deep-sea shrimps, different strategies are used, with both environmental and phylogenetic factors at play. Omori's observations revealed that epipelagic and shallow mesopelagic shrimps tend to have seasonal breeding, depending on latitudinal plankton productivity, while deeper-dwelling species tend to be nonseasonal. As far as fecundity and larval development are concerned, Omori found that the number of deep-mesopelagic and bathypelagic species with large embryos and ALD is greater than that of epipelagic and shallow mesopelagic species, which were small-egg species and presumably with ELD. Although many pelagic shrimps are vertical migrators, rising to the surface at night to feed on the more abundant planktonic food there, the deep-mesopelagic and bathypelagic shrimps are usually weak migrators or do not move up toward the surface at night. Larvae of these species would have to migrate long distances to and from the surface every day, which is a long trip indeed for these small larvae. As a result, abbreviated or direct larval development in or near the deep adult habitat has evolved in many deep species to avoid this problem.

In contrast, Hopkins et al. (1989) showed that there were various patterns in the group of shrimps that they studied, which consisted mainly of oplophoroids but also two pandalid, a pasiphaeid, and a disciadidid species. Most of the shallower-dwelling species were seasonal breeders but the deeper ones nonseasonal. But there was no particular pattern with depth for embryo size and resulting larval development. In fact, of the group of seven deep-water oplophoroid species, six had small eggs showing probable ELD development. As with the benthic deep-sea species, both environmental factors and ancestry (phylogeny) must play a role in determining which life-history strategy has evolved.

In neritic species, the colder water temperatures with increasing latitude result in slower growth and a longer life span. The same tendency occurs with the colder waters of the deep sea. Omori (1974) estimated that mesopelagic sergestids and the penaeoid *Gennadas* species lived from 1 to 2 years, like their neritic relatives, but deeper-dwelling oplophoroids had a much longer life span of about 5 years, perhaps up to 8 years in the bathypelagic *Hymenodora* species. Likewise, benthic shrimps of the deep sea generally have longer life spans than neritic ones. Deep-shelf and shallow-slope penaeoids such as *Haliporoides sibogae and Parapenaeus longirostris* live 2–3 years (Ohtomi and Matsuoka 1998 and Sobrino et al. 2005, respectively) rather than 1–2 years of neritic penaeoids. Bathyal penaeoids such as *Pleoticus robustus* (5 years), *Aristeus antennatus* (3–8 years), and *Aristaeomorpha foliacea* (7–8 years) are long-lived (Anderson and Linder 1971; Sardà and Demestre 1987; Orsi Relini and Relini 1998; D'Onghia et al. 1998; Ragonese et al. 2012). The deep caridean pandalids *Heterocarpus woodmasoni* and *H. reedi* have longevity of 5 and 6 years, respectively (Rajasree et al. 2011; Roa and Ernst 1996).

9.15 The Life-History Continuum of Shrimps

Various kinds of models have been proposed which make different predictions about life-history traits given different broadly and sometimes ambiguously defined environmental conditions (e.g., MacArthur and Wilson 1967; Pianka 1970; Stearns 1976, 1992; Ramirez-Llodra 2002; Marshall et al. 2012). There has been a lot of discussion and disagreement about the validity of the different models. In my opinion, the trends and generalizations about life-history traits of decapod shrimps can best be understood by a continuum of "r" and "K" traits (MacArthur and Wilson 1967; Pianka 1970). In the logistic growth model, "r" refers to the species maximum rate of intrinsic increase, and "K" refers to the carrying capacity of the environment for a particular species, i.e., the number of individuals the habitat can hold without further increase or decrease. On one end of the "r-K continuum, "r-selected" species live in unpredictable or unstable environments, growing and reproducing rapidly when population density is low ("live fast, die young"). "r-selected" traits include rapid development and growth, early sexual maturity, high reproductive output per unit time, small body size, and short life span. At the other extreme, "K-selected" species live in more temporally stable, environmentally predictable habitats and

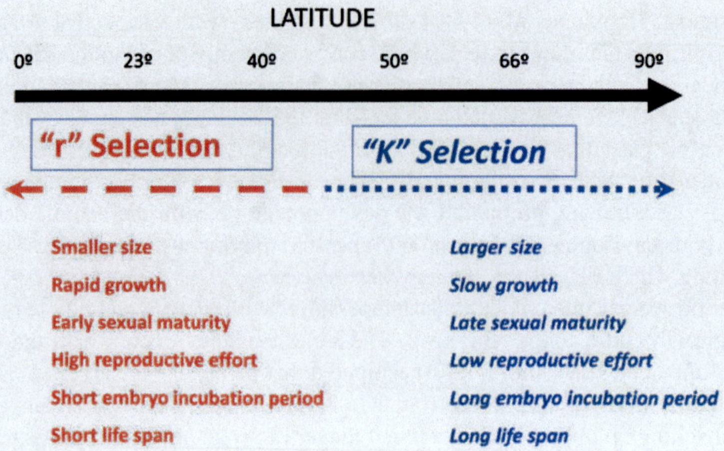

Fig. 9.19 Latitudinal variation in caridean life-history traits according to the "r-K" model of life-history patterns

devote more energy into growth and less into reproduction but live longer, so that over time their overall relative reproductive output is similar to that of "r-selected" species.

Where do the life-history traits of penaeoids, sergestoids, carideans, and stenopodideans fit along the r-k continuum in different major environments (Fig. 9.19)? The neritic penaeoids are tropical to warm-temperate species, and most have "r-selected" traits. The warm water temperatures in which they live allow them to grow rapidly to sexual maturity and to devote much energy to producing a few to several large spawns. However, some species of penaeoids reach a large body size, a "K" trait. This combination of high reproductive effort and, in some penaeoids, fast growth to a large body size is doomed to a relatively short life span. The neritic carideans show a better fit to the "r-K" continuum along a latitudinal gradient from equator to the poles (Fig. 9.19). At the "r" extreme, tropical shrimps mature rapidly within months of settlement and have several broods in succession before death at small body size at an early age. Much more energy is devoted to reproduction than to growth. Several broods can be spawned and incubated in a short time due to the warm water temperatures and adequate, albeit somewhat low, plankton productivity of tropical waters. At the "K" extreme, polar shrimps are constrained to slowly develop and grow to sexual maturity by the constant, very cold-water temperatures. The latter only allows extremely slow oogenesis and ovarian maturity which may take several months to two years in some species, allowing only one spawning per year or every two years. However, slow growth over an extended life span of several years accumulates, often to produce a large body size, as in many other organisms ("Bergmann's Rule" applied to marine organisms: Hirose et al. 2013; González et al. 2016). The reproductive output of high latitude shrimps is doled out gradually over several years.

The life-history trends with increasing depth in both penaeoids and carideans are similar to those exhibited by neritic carideans with increasing latitude. Deep-sea benthic and bathypelagic shrimps tend more toward the "K" end of the r-K life-history spectrum, which is especially apparent in their life spans of several years, as well as in larger body sizes indicating slow growth to sexual maturity. Shallower bathyal species show seasonality of reproduction like that of nearby continental shelf relatives. Given the lack of seasonal cues, very cold-water temperatures, and low food supplies in the totally dark deep-slope and abyssal/hadal depths, breeding and spawning show little temporal variation. Some species have retained the ELD of their ancestors, while others have avoided the problem of sending delicate larvae to the surface to feed by adopting the ALD/DD strategy. However, given the technical difficulty of sampling deep-sea shrimps, whose populations are not generally abundant, information about deep-sea shrimps is still scanty so that generalizations about them should be viewed with caution.

Most freshwater carideans occur in tropical to warm-temperate habitats. As might be expected, most show continuous reproduction with seasonal peaks, or seasonal periodicity in breeding, depending on latitude. However, superimposed on the usual environmental factors of temperature, photoperiod, and food availability is the meteorological pattern of the wet and dry seasons which affects reproduction as described earlier. Overall, breeding patterns, growth, and life span are r-selected in freshwater species, with short incubation periods and frequent broods, but recent studies have indicated K-selected traits such as surprisingly long life spans in a number of species, with quite large body sizes in some of the atyid, xiphocaridid, and *Macrobrachium* species.

Life-history traits of stenopodidean species are largely unknown because, as mentioned earlier, population sizes of these territorial monogamous species are too low to sample in large enough numbers for the necessary studies. The problem is difficult enough for the shallow-water Stenopodidae but doubly so for the deep-sea Spongicolidae, given the practical complications involved in sampling and observing deep-sea species hidden within sponge hosts!

We have seen that for any given major environment that shrimps inhabit, different life-history strategies may evolve within the same habitat for different taxa, instead of a single strategy common to most species. The reason for such variation might be explained by the thoughts expressed by Tyler (1995) and Young (2003) for deep-sea habitats, but which may apply to all major environments as follows. Any major habitat, such as benthic and pelagic of the neritic, bathyal, abyssal, and hadal zones of the sea, as well as lentic and lotic freshwater ones, is not really one homogeneous environment. Instead, it is a mosaic of habitats with different physical and biological features to which diverse species may adapted. It should not be a surprise that the reproductive and life-history strategies vary within any given habitat.

The various models and generalizations about life-history traits are usually analyzed and interpreted with the perspective of the ecological and environmental conditions in which modern species occur. However, it is important to recall that all taxa have an evolutionary history in which their life-history traits may have evolved under different environmental selection pressures than those experienced in recent times. Furthermore, the basic reproductive and developmental biology of organisms is highly conservative and not easily modified. The genetic makeup and its potential for mutation and change vary among taxa. Therefore, it is good to take the ancestry (phylogenetic history) of a shrimp taxon into account when interpreting its current life-history traits. For example, a deep-sea shrimp may be an ELD or ALD species because it has evolved its reproductive characteristics in response to the environmental conditions in which it now lives. Another possibility is that the reproductive traits inherited from a species' ancestors are genetically and developmentally too difficult to change under a different set of selective pressures. The challenge is to separate and determine the influence of these alternate hypotheses in any given species or taxon.

9.16 Global Warming and Climatic Effects on Life-History Patterns

There is little doubt that the earth's climate has changed throughout the history of the earth, as shown by numerous studies. Before human populations became abundant, the ups and downs in the temperature and chemical composition of earth's terrestrial and aquatic environments were correlated with natural variations in the concentration of the "greenhouse" gases in the atmosphere, primarily carbon dioxide. These temperature changes were responsible for large-scale changes in the earth's environments, such as the buildup of glaciers, their retreat, the expansion of tropical conditions to large parts of the earth, and large-scale rises and drops in sea level. These changes have occurred over long time scales, e.g., on the order of 100,000-year cycles during the last million years. In the late 1950s, Charles D. Keeling began long-term monitoring of carbon dioxide concentration in the atmosphere which showed its rapid rise compared to prehistoric time periods (Keeling 1997; Harris 2010). How might the current accelerated global warming and various accompanying changes in freshwater and marine habitats affect the life-history traits of shrimps and other aquatic organisms? As water temperatures increase, temperate-latitude shrimps might be able to survive and function physiologically by moving poleward, but their physical habitat, e.g., a river, lake, and rocky intertidal or subtidal reef, cannot pick up and move with them. Replacement habitats may not be available elsewhere. Plankton productivity in new habitats may be mismatched with a species reproductive cycle and larval development. Tropical habitats may become too warm for the diverse shrimp fauna there. High-latitude shrimps will be squeezed poleward into smaller and smaller areas of suitable living space. If a species is unable to move, seasonality of the physical and biological environment around them may change more rapidly than the species' ability to adapt their life-history traits to the new conditions (Varpe 2017). The human population and its governments of the last two centuries are responsible for the unnatural speed of climate change now experienced. However, it is in their (our) power to ameliorate the situation with decisive action.

The failure to do so will result in the deterioration or extinction of numerous shrimp species, many of which are so important to the ecology of aquatic habitats as well as to shrimp fisheries and aquaculture. Biologists interested in shrimps can contribute with studies analyzing the causative proximate and ultimate (evolutionary) factors, both abiotic and biotic, which mold the life-history traits of shrimp species.

References

Abele LG, Martin JW (1989) American species of the deep-sea shrimp genus *Bythocaris* (Crustacea, Decapoda, Hippolytidae). Bull Mar Sci 45:26–51

Ahamed F, Ohtomi J (2011) Reproductive biology of the pandalid shrimp *Plesionika izumiae* (Decapoda: Caridea). J Crustac Biol 31:441–449

Allen JA (1966) The dynamics and interrelationships of mixed populations of Caridea found off the north-east coast of England. In: Barnes H (ed) Some contemporary studies in marine science. Allen and Unwin, London, pp 45–66

Alon AC, Stancyk SE (1982) Variation in life-history patterns of the grass shrimp *Palaemonetes pugio* in two South Carolina estuarine systems. Mar Biol 68:265–276

Amin SMN, Arshad A, Siraj SS, Sidik BJ (2009a) Population structure, growth, mortality and yield per recruit of sergestid shrimp, *Acetes japonicus* (Decapod, Sergestidae) from the coastal waters of Malacca, Peninsular Malaysia. Indian J Mar Sci 38:57–68

Amin SMN, Arshad A, Bujang JS, Siraj SS, Goddard S (2009b) Reproductive biology of the sergestid shrimp *Acetes indicus* (Decapoda: Sergestidae) in coastal waters of Malacca, Peninsular Malaysia. Zool Stud 48:753–760

Anderson DT (1982) Embryology. In: Abele LG (ed) The biology of crustacea, Embryology, morphology, and genetics, vol 2. Academic Press, New York, pp 1–42

Anderson WW, Linder MJ (1971) Contributions to the biology of the royal red shrimp, *Hymenopenaeus robustus* Smith. Fish Bull 69:313–336

Anderson SLWH, Clark J, Chang ES (1985) Multiple spawning and molt synchrony in a free spawning shrimp (*Sicyonia ingentis*: Penaeoidea). Biol Bull 168:377–394

Anger K (2001) The biology of decapod crustacean larvae. A.A. Balkema, Lisse, p 419

Anger K, Hayd L (2009) From lecithotrophy to planktotrophy: ontogeny of larval feeding in the Amazon River prawn *Macrobrachium amazonicum*. Aquat Biol 7:19–30

Arce SM, Moss SM (2002) Tagging technology helps monitor shrimp performance in research, genetic improvement. Glob Aquacult Advocate 2002:42–43

Astthorsson OS, Gislason A (1991) Seasonal abundance and distribution of Caridea larvae in Ísafjord-deep, north-west Iceland. J Plankton Res 13:91–102

Ayub Z, Ahmed M (2004) Study on the host–parasite relationship of *Parapenaeopsis stylifera* (H. Milne Edwards) (Decapoda:Penaeidae) and *Parapenaeon japonica* (Thielemann) (Isopoda: Bopyridae). Hydrobiologia 523:225–228

Bae HJ, Oh C-W (2014) Reproduction and growth of the spiny lebbeid shrimp, *Lebbeus groenlandicus* (Fabricius, 1775) (Caridea, Hippolytidae) in the east sea of Korea. Crustaceana 87:1430–1446

Baldwin AP, Bauer RT (2003) Growth, survivorship, life span, and sex change in the hermaphroditic shrimp *Lysmata wurdemanni* (Decapoda: Caridea: Hippolytidae). Mar Biol 143:157–166

Bauer RT (1976) Mating behaviour and spermatophore transfer in the shrimp *Heptacarpus pictus* (Stimpson) (Decapoda: Caridea: Hippolytidae). J Nat Hist 10:415–440

Bauer RT (1985a) Penaeoid shrimp fauna from tropical seagrass meadows: species composition, diurnal and seasonal variation in abundance. Proc Biol Soc Wash 98:177–190

Bauer RT (1985b) Diel and seasonal variation in species composition and abundance of the caridean shrimps (Crustacea, Decapoda) from seagrass meadows in northern Puerto Rico. Bull Mar Sci 36:150–162

Bauer RT (1989) Continuous reproduction and episodic recruitment in nine caridean shrimp species inhabiting a tropical seagrass meadow. J Exp Mar Biol Ecol 127:175–187

Bauer RT (1991) Analysis of embryo production in a caridean shrimp guild inhabiting tropical seagrass meadows. In: Wenner A, Kuris A (eds) Crustacean egg production, Crustacean Issues, vol 7. CRC Press, Boca Raton, FL, pp 181–192

Bauer RT (1992) Testing generalizations about latitudinal variation in reproductive and recruitment patterns with caridean and sicyoniid shrimps. J Invertebr Reprod Dev 22:193–202

Bauer RT (2002) Reproductive ecology of a protandric simultaneous hermaphrodite, the shrimp *Lysmata wurdemanni* (Decapoda: Caridea: Hippolytidae). J Crustac Biol 22:742–749

Bauer RT (2004) Remarkable shrimps: adaptations and natural history of the carideans. University of Oklahoma Press, Norman

Bauer RT (2011a) Amphidromy and migrations of freshwater shrimps. I. Costs, benefits, evolutionary origins, and an unusual case of amphidromy. In: Asakura A (ed) New Frontiers in crustacean biology. Proceedings of the TCS summer meeting, Tokyo, 20–24 September 2009. Brill, Leiden, pp 145–156

Bauer RT (2011b) Amphidromy and migrations of freshwater shrimps. II. Delivery of hatching larvae to the sea, return juvenile upstream migration, and human impacts. In: Asakura A (ed) New Frontiers in crustacean biology. Proceedings of the TCS summer meeting, Tokyo, 20–24 September 2009. Brill, Leiden, pp 157–168

Bauer RT (2013) Amphidromy in shrimps: a life history pattern between rivers the sea. Lat Am J Aquat Res 41:633–650

Bauer RT (2018) Life-cycle and seasonal migrations. In: Thiel M, Watling L (eds) The natural history of the crustacea, Life histories, vol 5. Oxford University Press, Oxford, pp 203–229

Bauer RT, Abdalla JA (2000) Patterns of brood production in the grass shrimp *Palaemonetes pugio* (Decapoda: Caridea). Invertebr Reprod Dev 38:107–113

Bauer RT, Delahoussaye J (2008) Life history migrations of the amphidromous river shrimp *Macrobrachium ohione* from a continental large river system. J Crustac Biol 28:622–632

Bauer RT, Lin JM (1994) Temporal patterns of reproduction and recruitment in the marine shrimps *Trachypenaeus similis* (Smith) and *T. constrictus* (Stimpson) (Crustacea: Decapoda) from the Northcentral Gulf of Mexico. J Exp Mar Biol Ecol 182:205–222

Bauer RT, Rivera Vega L (1992) Pattern of reproduction and recruitment in two sicyoniid shrimp species (Decapoda, Penaeoidea) from a tropical seagrass habitat. J Exp Mar Biol Ecol 161:223–240

Bauer RT, VanHoy R (1996) Variation in sexual systems (protandry, gonochorism) and reproductive biology among three species of the shrimp genus *Thor* (Decopoda: Caridea). Bull Mar Sci 59:53–73

Beck JT (1979) Population interactions between a parasitic castrator, *Proboyrus pandalicola*, and one of its freshwater shrimp hosts, *Palaemonetes paludosus* (Decapoda: Caridea). Parasitology 79:431–449

Beck JT, Cowell BC (1976) Life history and ecology of the freshwater caridean shrimp, *Palaemonetes paludosus* (Gibbes). Am Midl Nat 96:52–65

Benítez SA, Iliffe TM, Martínez S, Ojeda JC, Villalobos JL, Alvarez F (2020) Larval development of the stygobitic shrimp *Creaseria morleyi* (Creaser, 1936) (Decapoda: Caridea: Palaemonidae) from the Yucatán Peninsula, Mexico. J Crustac Biol 40:221–229

Benstead JP, March JG, Pringle CM (2000) Estuarine larval development and upstream post-larval migration of freshwater shrimps in two tropical rivers of Puerto Rico. Biotropica 32:545–548

Benzie JAH, de Silva PK (1983) The abbreviated larval development of *Caridina singhalensis* Ortmann, 1894 (Decapoda, Atyidae). J Crustac Biol 3:117–126

Bluhm BA, Brey T (2001) Age determination in the Antarctic shrimp *Notocrangon antarcticus* (Crustacea: Decapoda), using the autofluorescent pigment lipofuscin. Mar Biol 138:247–257

Boddeke R (1982) The occurrence of "winter" and "summer" eggs in the brown shrimp (*Crangon crangon*) and the impact on recruitment. Neth J Sea Res 16:151–162

Botello A, Alvarez F (2013) Phylogenetic relationships among the freshwater genera of palaemonid shrimps (Crustacea: Decapoda) from Mexico: evidence of multiple invasions? Lat Am J Aquat Res 41:773–780

Briones P, Monroy LV, Estrada J, Lozano E (2020) Diversity of seagrass-associated decapod crustaceans in a tropical reef lagoon prior to large environmental changes: a baseline study. Diversity 2020(12):205. https://doi.org/10.3390/d12050205

Brooks WK, Herrick FH (1891) The embryology and metamorphosis of the Macrura. Mem Nat Acad Sci Wash 5:321–576

Burukovsky RN, Ivanov ES (2015) On the reproductive biology of the females of the shrimp *Crangon crangon* (L.) (Decapoda, Crangonidae). Oceanology 55:82–90

Butler TH (1980) Shrimps of the Pacific coast of Canada. Can Bull Fish Aquat Sci 202:1–280

Calado R, Narciso L (2003) Seasonal variation on embryo production and brood loss in the Monaco shrimp *Lysmata seticaudata* (Decapoda: Hippolytidae). J Mar Biol Assoc UK 83:959–962

Calado R, Bartilotti C, Narciso L (2005) Short report on the effect of a parasitic isopod on the reproductive performance of a shrimp. J Exp Mar Biol Ecol 321:13–18

Calado R, Vitorino A, Dinis MT (2006) Bopyrid isopods do not castrate the simultaneous hermaphroditic shrimp *Lysmata amboinensis* (Decapoda: Hippolytidae). Dis Aquat Org 73: 73–76

Calado R, Bartilotti C, Goy JW, Dinis MT (2008) Parasitic castration of the stenopodid shrimp *Stenopus hispidus* (Decapoda: Stenopodidae) induced by the bopyrid isopod *Argeiopsis inhacae* (Isopoda: Bopyridae). J Mar Biol Assoc U K 88:307–309

Campos J, van der Veer HW (2008) Autecology of *Crangon crangon* (L.) with an emphasis on latitudinal trends. Oceanogr Mar Biol Annu Rev 46:65–104

Cash CE, Bauer RT (1993) Adaptations of the branchial parasite *Probopyrus pandalicola* (Isopoda: Bopyridae) for survival and reproduction related to ecdysis of the host, *Palaemonetes pugio* (Caridea: Palaemonidae). J Crustac Biol 13:111–124

Castilho AL, Gavio MA, Costa RC, Boschi EE, Bauer RT, Fransozo A (2007) Latitudinal variation in population structure and reproductive pattern of the endemic South American shrimp *Artemesia longinaris* (Decapoda: Penaeoidea). J Crustac Biol 27:548–552

Castilho AL, Furlan M, Costa RC, Fransozo A (2008) Reproductive biology of the rock shrimp *Sicyonia dorsalis* (Decapoda: Penaeoidea) from the southeastern coast of Brazil. Invertebr Reprod Dev 52:59–68

Castilho AL, Bauer RT, Freire FAM, Fransozo V, Costa RC, Grabowski RC, Fransozo A (2015) Lifespan and reproductive dynamics of the commercially important sea bob shrimp *Xiphopenaeus kroyeri* (Penaeoidea): synthesis of a 5-year study. J Crustac Biol 35:30–40

Cha HK, Choi JH, Oh CW (2004) Reproductive biology and growth of the Shiba shrimp, *Metapenaeus joyneri* (Decapoda: Penaeidae), on the western coast of Korea. J Crustac Biol 24:93–100

Chace FA Jr (1972) The shrimps of the Smithsonian-Bredin expeditions with a summary of the West Indian shallow-water species (Crustacea: Decapoda: Natantia). Smithson Contrib Zool 98: 179

Chace FA Jr, Hobbs HH Jr (1969) The freshwater and terrestrial decapod crustaceans of the West Indies with special reference to Dominica. U.S. Natl Mus Bull 292:1–258

Choi JH, Hong SY (2001) Larval development of the kishi velvet shrimp, *Metapenaeopsis dalei* (Rathbun) (Decapoda: Penaeidae), reared in the laboratory. Fish Bull 99:275–291

Choi JH, Kim JN, Ma CW, Cha HK (2005) Growth and reproduction of the Kishi Velvet Shrimp, *Metapenaeopsis dalei* (Rathbun, 1902) (Decapoda, Penaeidae) in the western Sea of Korea. Crustaceana 78:947–963

Clarke A, Hopkins CCE, Nilssen EM (1991) Egg size and reproductive output in the deep-water prawn *Pandalus borealis* Kroyer, 1838. Funct Ecol 5:724–730

Company JB, Sarda F (1997) Reproductive patterns and population characteristics in five deep-water pandalid shrimps in the Western Mediterranean along a depth gradient (150-1100 m). Mar Ecol Prog Ser 148:49–58

Conides A, Glamuzina B, Jug-Dujakovic J, Papaconstantinou C, Kapiris K (2006) Age, growth, and mortality of the karamote shrimp, *Melicertus Kerathurus* (Forskål, 1775), in the east Ionian Sea (western Greece). Crustaceana 79:33–52

Cook HL, Linder MJ (1970) Synopsis of biological data on the brown shrimp *Penaeus aztecus aztecus* Ives, 1891. FAO Fish Rep 57:1417–1479

Cook BD, Baker AM, Page TJ, Grant SC, Fawcett JH, Hurwood DA, Hughes JM (2006) Biogeographic history of an Australian freshwater shrimp, *Paratya australiensis* (Atyidae): the role life history transition in phylogeographic diversification. Mol Ecol 15:1083–1093

Cook BD, Page TJ, Hughes JM (2012) Phylogeography of related diadromous species in continental and island settings, and a comparison of their potential and realized dispersal patterns. J Biogeogr 39:421–430

Corey S, Reid DM (1991) Comparative fecundity of decapod crustaceans. I. The fecundity of thirty-three species of nine families of caridean shrimps. Crustaceana 60:270–294

Costa RC, Fransozo A (2004) Reproductive biology of the shrimp *Rimapenaeus constrictus* (Decapoda, Penaeidae) in the Ubatuba region of Brazil. J Crustac Biol 24:274–281

Covich AP, Crowl TA, Hein CL, Townsend MJ, McDowall WH (2009) Predator–prey interactions in river networks: comparing shrimp spatial refugia in two drainage basins. Freshw Biol 54: 450–465

Cross WF, Covich AP, Crowl TA, Benstead JP, Ramirez A (2008) Secondary production, longevity, and consumption rates of freshwater shrimps in two tropical streams with contrasting geomorphology and food web structure. Freshw Biol 53:2504–2519

Da Rocha SS, Bueno SLDS, Shimizu RM, Mantelatto FL (2013) Reproductive biology and population structure of *Potimirim brasiliana* Villalobos, 1959 (Decapoda, Atyidae) from a littoral fast-flowing stream, Säo Paulo State, Brazil. Crustaceana 86:67–83

Dall W, Hill BC, Rothlisberg PC, Sharples DJ (1990) The biology of the Penaeidae. Adv Mar Biol 27:1–489

De Grave S, Livingston D, Speigh MR (2006) Diel variation in sea grass dwelling shrimp: when to sample at night? J Mar Biol Assoc UK 86:1421–1422

De Grave S, Cai S, Anker A (2008) Global diversity of shrimps (Decapoda: Caridea) in freshwater. Hydrobiologia 595:287–293

Dennenmoser S, Thiel M, Schubart CD (2010) High genetic variability with no apparent geographic structuring in the mtDNA of the amphidromous river shrimp *Cryphiops caementarius* (Decapoda: Palaemonidae) in northern-central Chile. J Crustac Biol 30:762–766

Dobkin S (1965) The early larval stages of *Glyphocrangon spinicauda* A. Milne Edwards. Bull Mar Sci 15:872–884

Dobkin S (1968) The larval development of a species of Thor (Decapoda: Caridea) from South Florida, U.S.A. Crustaceana Suppl 2:1–18

Dobkin S (1969) Abbreviated larval development in caridean shrimps and its significance in the artificial culture of these animals. FAO Fish Rep 57:935–946

Dohrn PFR (1950) Studi sulla *Lysmata seticaudata* Risso (Hyppolytidae). I. Le condizioni normali della sessualita in natura. Pubbl Stn Zool Napoli 22:257–272

D'Onghia G, Maiorano P, Matarrese A, Tursi A (1998) Distribution, biology, and population dynamics of *Aristaeomorpha foliacea* (Risso, 1827) (Decapoda, Natantia, Aristeidae) in the North-Western Ionian Sea (Mediterranean Sea). Crustaceana 71:518–544

Duffy JE (2002) The ecology and evolution of eusociality in sponge-dwelling shrimp. In: Kikuchi T (ed) Genes, behavior, and evolution in social insects. University of Hokkaido Press, Sapporo, pp 1–38

Eckelbarger KJ, Watling L (1995) Role of phylogenetic constraints in determining reproductive patterns in deep-sea. Invertebr Biol 114:256–269

Farhana Z, Ohtomi J (2016) Ovarian maturation, size at sexual maturity, and spawning season of *Parapenaeus Fissuroides* Crosnier, 1985 (Decapoda: Penaeidae). J Crustac Biol 36:815–822

Felder DF (1982) Reproduction of the snapping shrimps *Synalpheus fritzmuelleri* and *S. apioceros* (Crustacea: Decapoda: Alpheidae) on a sublittoral reef off Texas. J Crustac Biol 2:535–543

Fièvet É (1999) An experimental survey of freshwater shrimp upstream migration in an impounded stream at Guadeloupe Island, Lesser Antilles. Arch Hydrobiol 144:339–355

Flohn H (1970) Investigations of equatorial upwelling and its climatic role. In: Gordon AC (ed) Studies of physical oceanography, tribute to Georg Wust on his 80th birthday, vol 1. Gordon and Breach, New York, pp 93–10

Fransozo VRC, Costa GB, Cobo VJ (2005) Population biology of the spine shrimp *Exhippolysmata oplophoroides* (Holthuis) (Caridea: Hippolytidae) in a subtropical region (São Paulo, Brazil). Rev Bras Zool 22:1078–1084

Garcia S (1988) Tropical penaeid prawns. In: Gulland JA (ed) Fish population dynamics, 2nd edn. John Wiley, Chichester, pp 219–249

Garcia S, Le Reste L (1981) Life cycles, dynamics, exploitation and management of coastal penaeid shrimp stocks, FAO fish. Tech Pap 203:1–215

Garcia JR, Wolf MR, Costa RC, Castilho AL (2016) Growth and reproduction of the shrimp *Rimapenaeus constrictus* (Decapoda: Penaeidae) from the southeastern coast of Brazil. Reg Stud Mar Sci 6:1–9

Giese AC, Pearse JS (1974) Introduction: general principles. In: Giese AC, Pearse JS (eds) Reproduction of marine invertebrates. Acoelomate and pseudocoelomate metazoans. Academic Press, New York, pp 1–49

González JA, Pajuelo JG, Triay-Portella R, Ruiz-Díaz R, Delgado J, Góis AR, Martins A (2016) Latitudinal patterns in the life-history traits of three isolated Atlantic populations of the deep-water shrimp *Plesionika edwardsii* (Decapoda, Pandalidae). Deep-Sea Research I 117:2–38

Gorny M, George MR (1997) Oocyte development and gonad production of *Nematocarcinus lanceopes* (Decapoda: Caridea) in the eastern Weddell Sea, Antarctica. Polar Biol 17:191–198

Gorny M, Arntz WE, Clarke A, Gore DG (1992) Reproductive biology of caridean decapods from the Weddell Sea. Polar Biol 12:111–120

Gorny M, Brey T, Arntz W, Bruns T (1993) Growth, development and productivity of *Chorismus antarcticus* (Pfeffer) (Crustacea: Decapoda: Natantia) in the eastern Weddell Sea, Antarctica. J Exp Mar Biol Ecol 174:261–275

Gould SJ (1966) Allometry and size in ontogeny and phylogeny. Biol Rev 41:587–640

Goy JW (2010) Infraorder Stenopodidea Claus, 1872. In: Schram FR, von Vaupel Klein JC (eds) Treatise on zoology - anatomy, taxonomy, biology - the crustacea Volume 9 Part A. Koninklijke Brill NV, Leiden, pp 215–265

Grabowski R (2017) Geographic variation dendrobranchiate shrimps in the southeastern and southern regions of Brazil: test of hypotheses on the latitudinal paradigm. Postgraduate program thesis, in Biological Sciences (Zoology), State University Paulista "Juli de Mesquita Filho" (UNESP), Botucatu Biosciences Institute, Brazil.

Grassle JF, Berg CJ, Childress JJ, Hessler RR, Jannasch HJ, Karl DM, Lutz RA, Mickel TJ, Rhoads DC, Sanders HL, Smith KL, Somero GN, Turner RD, Tuttle JH, Walsh PJ, Williams AJ (1979) Galápagos 79: initial findings of a deep-sea biological quest. Oceanus 22:3–10

Gregati RA, Fransozo V, Lopez-Greco LS, Negreiros-Fransozo ML (2010) Reproductive cycle and ovarian development of the marine ornamental shrimp *Stenopus hispidus* in captivity. Aquaculture 206:185–190

Guay C, Sainte-Marie B, Brêthes J-C (2011) Strong maternal effects and extreme heterogeneity of progeny development in the caridean shrimp *Sclerocrangon boreas* (Crangonidae). Mar Biol 158:2835–2845

Guerao G, Pérez-Baquera J, Ribera C (1994) Growth and reproductive biology of *Palaemon xiphias* Risso, 1816 (Decapoda: Caridea: Palaemonidae). J Crustac Biol 14:80–88

Gurney R (1942) Larvae of decapod Crustacea. Ray Society, London

Hamano T, Hayashi KI (1992) Ecology of an atyid shrimp *Caridina japonica* (De Man, 1892) migrating to upstream habitats in the Shiwagi Rivulet, Tokushima Prefecture. Crustac Res 21:1–13

Han CC, Chang CS, Cheng IM, Fang LS, Tew KS (2011) Population dynamics of a landlocked and amphidromous freshwater shrimp, *Caridina gracilipes* (Decapoda: Caridea) in subtropical waters. J Crustac Biol 31:278–285

Hancock MA (1998) The relationship between egg size and embryonic and larval development in the freshwater atyid *Paratya australiensis* Kemp (Decapoda: Atyidae). Freshw Biol 39:715–723

Hancock MA, Bunn SE (1997) Population dynamics and life history of *Paratya australiensis* Kemp, 1917 (Decapoda: Atyidae) in upland rainforest streams, South-Eastern Queensland, Australia. Mar Freshw Res 48:361–369

Hardy AC (1956) The open sea. Its natural history (Part I) The world of Plankton. New naturalist No. 34, Collins, London

Harris DC (2010) Charles David Keeling and the story of atmospheric CO2 measurements. Anal Chem 82:7865–7870

Hartmann G (1958) Apuntes sobre la biología del camarón de río, *Cryphiops caementarius* (Molina) Palaemonidae, Decapoda. Pesca y Casa 8:15–28

Hartnoll RG (1982., Growth) In: Abele LG (ed) The biology of Crustacea, Embryology, morphology and genetics, vol 2. Academic Press, New York, pp 111–196

Hasler AD, Scholz AT, Horrall RM (1978) Olfactory imprinting and homing in salmon. Am Sci 66: 347–355

Haynes WB, Wigley RL (1969) Biology of the northern shrimp, *Pandalus borealis*, in the Gulf of Maine. Trans Am Fish Soc 98:60–76

Hernáez P, Martínez-Guerrero B, Anker A, Wehrtmann IS (2010) Fecundity and effects of bopyrid infestation on egg production in the Caribbean sponge-dwelling snapping shrimp *Synalpheus yano* (Decapoda: Alpheidae). J Mar Biol Assoc U K 90:691–698

Hertzler PL (2005) Cleavage and gastrulation in the shrimp *Penaeus (Litopenaeus) vannamei* (Malacostraca, Decapoda, Dendrobranchiata). Arthropod Struct Dev 34:455–469

Hines AH (1982) Allometric constraints and variables of reproductive effort in brachyuran crabs. Mar Biol 69:309–320

Hirose G, Fransozo V, Tropea C, López-Greco LS, Negreiros-Fransozo ML (2013) Comparison of body size, relative growth and size at onset sexual maturity of *Uca uruguayensis* (Crustacea: Decapoda: Ocypodidae) from different latitudes in the south-western Atlantic. J Mar Biol Assoc U K 93:781–788

Hoarau PE, Treilhes CRM, Valade PB (2018) Reproductive biology and recruitment in an amphidromous prawn *Macrobrachium australe* in Reunion Island. Invertebr Biol 137:231–239

Holmquist JG, Schmidt-Gengenbach JM, Buchanan-Yoshioka B (1998) High dams and marine-freshwater linkages: effects on native and introduced fauna in the Caribbean. Conserv Biol 12: 621–630

Holthuis LB (1971) The Atlantic shrimps of the deep-sea genus *Glyphocrangon.* A. Milne Edwards, 1881. Bull Mar Sci 21:267–373

Hongjamrassilp W, Maiphrom W, Blumstein DT (2021) Why do shrimp leave the water? Mechanisms and functions of parading behaviour in freshwater shrimp. J Zool 313:87–98

Hopkins TL, Garner JV Jr, Flock ME (1989) The caridean shrimp (Decapoda: Natantia) assemblage in the mesopelagic zone of the eastern Gulf of Mexico. Bull Mar Sci 45:1–14

Hossain MY, Ohtomi J (2008) Reproductive biology of the southern rough shrimp *Trachysalambria curvirostris* (Penaeidae) in Kagoshima Bay, southern Japan. J Crustac Biol 28:607–612

Hunte W (1977) Laboratory rearing of the atyid shrimps *Atya innocous* and Herbst and *Micratya poeyi* Guerin-Meneville (Decapoda, Atyidae). Aquaculture 11:373–378

Hunte W (1978) The distribution of freshwater shrimps (Atyidae and Palaemonidae) in Jamaica. Zool J Linnean Soc 64:135–150

Hunte W (1980) The laboratory rearing of larvae of the shrimp *Macrobrachium faustinum* (Decapoda, Palaemonidae). Caribb J Sci 16:57–60

Ibarra MA, Wehrtmann IS (2020) Estimates of growth and longevity of six species of freshwater shrimps (*Macrobrachium* species) (Decapoda: Caridea: Palaemonidae) from Costa Rica. J Crustac Biol 39:45–57

Ibrahim KH (1962) Observations on the fishery and biology of the freshwater prawn *Macrobrachium malcolmsonii* Milne Edwards in River Godavari. Indian J Fish 9:433–467

Jalihal DR, Sankolli KN, Shenoy S (1993) Evolution of larval developmental patterns and the process of freshwaterization in the prawn genus *Macrobrachium*. Crustaceana 65:365–376

Jensen JP (1958) The relation between body size and the number of eggs in marine malacostrakes. Meddel Dan Fisk Havunders Ny Ser II 19:25

Johnson SL, Covich AP, Crowl TA, Estrada-Pinto A, Bithorn J, Wurtsbaugh WA (1997) Do seasonality and disturbance influence reproduction in freshwater atyid shrimp in headwater steams, Puerto Rico? Verhandlungen Internationale Vereiningung für Theoretische und Angewandte Limnologie 26:2076–2081

Kapiris K, Thessalou-Legaki M (2009) Comparative reproduction aspects of the deep-water shrimps *Aristaeomorpha foliacea* and *Aristeus antennatus* (Decapoda, Aristeidae) in the Greek Ionian Sea (eastern Mediterranean). Int J Zool 2009:979512. https://doi.org/10.1155/2009/979512

Keeling CD (1997) Climate change and carbon dioxide: an introduction. Proc Natl Acad Sci U S A 94:8273–8274

Kennedy FS, Crane JJ, Schlieder RA, Barber DG (1977) Studies of the rock shrimp *Sicyonia brevirostris,* a new fishery resource on Florida's Atlantic shelf. Fla Mar Res Publ, Fla Dept Nat Res, Mar Res Lab 27:69

Kensley B, Walker I (1982) Palaemonid Shrimps from the Amazon Basin, Brazil (Crustacea: Decapoda: Natantia). Smithsonian Contrib Zool 362:28

Kikkert DA, Crowl TA, Covich AP (2009) Upstream migration of amphidromous shrimp in the Luquillo experimental Forest, Puerto Rico: temporal patterns and environmental cues. J N Am Benthol Soc 28:233–246

Kikuchi T (1962) An ecological study on animal community of *Zostera* belt, in Tomioka Bay, Amakusa, Kyushu (II). Community composition (2). Decapod crustaceans. Rec Ocean Works Jpn Spec Number 6:135–146

Knight M, Omori M (1982) Larval development of *Sergestes similis* Hansen (Crustacea, Decapoda, Sergestidae) reared in the laboratory. Fish Bull 2:217–243

Komai T (1997) Revision of *Argis dentata* and related species (Decapoda: Caridea: Crangonidae), with description of a new species from the Okhotsk Sea. J Crustac Biol 17:135–161

Krygier EE, Pearcy WG (1981) Vertical distribution and biology of pelagic decapod crustaceans off Oregon. J Crustac Biol 1:70–95

Kuris AM (1991) A review of patterns and causes of crustacean brood mortality. In: Wenner IA, Kuris A (eds) Crustacean egg production. Crustacean issues. Balkema Press, Boca Raton, FL, pp 117–141

Lardies MA, Wehrtmann IS (2001) Latitudinal variation in the reproductive biology of *Betaeus truncatus* (Decapoda: Alpheidae) along the Chilean coast. Ophelia 55:55–67

Lardies MA, Rojas JR, Wehrtmann IS (1998) Breeding biology of the snapping shrimp *Betaeus emarginatus* inhabiting a rock pool environment in central-southern Chile (Decapoda: Caridea: Alpheidae). Ophelia 49:221–231

Leal A, López J, Chávez EA, Hernández S, Méndez-Tenorio F (2001) Interannual variability of the reproductive period of the brown shrimp, *Farfantepenaeus californiensis* (Holmes, 1900) (Decapoda, Natantia). Crustaceana 74:839–851

Lee WY, Omori M, Peck RW (1992) Growth, reproduction and feeding behavior of the planktonic shrimp, *Lucifer faxoni* Borradaile, off the Texas coast. J Plankton Res 14:61–69

Lee IO, Bae HJ, Kim HG, Oh CW (2016) Effects of infestation by *Argeia pugettensis* Dana, 1853 (Isopoda, Bopyridae) on the growth and reproduction of the sand shrimp *Crangon hakodatei* Rathbun, 1902 (Decapoda, Crangonidae) in South Korea. Crustaceana 89:685–699

Li HY, Hong SY, Jin ZH (2011) Fecundity and brood loss of sand shrimp, *Crangon uritai* (Decapoda: Crangonidae). J Crustac Biol 31:34–40

Linder MJ, Anderson WW (1956) Growth, migrations, spawning and size distribution of shrimp, *Penaeus setiferus*. Fish Bull 56:555–645

Ling SW (1969) The general biology and development of *Macrobrachium rosenbergii* (de Man). FAO Fish Rep 57:589–606

López M (1966) Biologia de *Lucifer faxoni* Borradaile 1915, Cananéia, Brasil (Crustacea, Decapoda, Luciferidae). Bolm Inst Oceanogr S Paulo 15:47–54

López-Uriarte E, Vega F, Wehrtman IS, Cortés E, García MU, Badillo D, Chong O, Vargas MA (2020) Reproductive biology of the freshwater shrimp *Macrobrachium tenellum* (Smith, 1871) (Decapoda, Palaemonidae) in Mexico. J Crustac Biol 40:746–755

MacArthur RH, Wilson EO (1967) The theory of island biogeography. Princeton University Press, Princeton, NJ

Magalhães C (1985) Desenvolvimento larval obtido em laboratório de palaemonídeos da região Amazônica. I. *Macrobrachium amazonicum* (Heller, 1862) (Crustacea, Decapoda). Amazoniana 9:247–274

Makarov R (1968) On the larval development of the genus *Sclerocrangon* G. O. Sars (Caridea, Crangonidae). Crustaceana Suppl 1968(2):27–37

Mallo JC, Boschi EE (1982) Contribución al conocimiento del ciclo vital del camarón *Peiso petrunkevetchi* de la región de Mar Del Plata, Argentina (Crustacea, Decapoda, Sergestidae). Physis 41:85–98

March JG, Benstead JP, Pringle CM, Scatena FN (2003) Damming tropical island streams: problems, solutions, alternatives. Bioscience 53:1069–1078

Markham J (2020) Bibliographia Bopyridarum. Unpublished report (J Markham, jmarkham@seasurf.net)

Marshall DJ, Krug PJ, Krug K, Kupriyanova EK, Byrne M, Emlet RB (2012) The biogeography of marine invertebrate life histories. Annu Rev Ecol Evol Syst 43:97–114

Martin JW, Olesen J, Høeg JT (2014) Atlas of crustacean larvae. Johns Hopkins University Press, Baltimore, p 370

Martínez M, Román R (2009) Reproduction of *Periclimenes siankaanensis* (Decapoda, Caridea, Palaemonidae) in Bahía de la Ascensión, Quintana Roo, Mexico. Crustaceana 82:27–37

Martínez M, Román R (2013) Data on reproduction and fecundity of *Processa bermudensis* (Rankin, 1900) (Caridea, Processidae) from the southern coast of Quintana Roo, Mexico. Crustaceana 86:84–97

Martínez M, Román R (2014) Sexual maturity, fecundity, and embryo loss in the pontoniine shrimp *Cuapetes americanus* (Kingsley, 1878) (Decapoda: Palaemonidae) in Bahía De La Ascensión, Quintana Roo, Mexico. Crustaceana 82:27–37

Martínez M, Romero J (2017) Population structure and reproduction of the longtail grass shrimp *Urocaris longicaudata* (Decapoda, Caridea, Pontoniinae) in Laguna de Términos, SW Gulf of Mexico. Invertebr Reprod Dev 62:10–18

Mashiko K, Numachi KI (2000) Derivation of populations with different-sized eggs in the palaemonid prawn *Macrobrachium nipponense*. J Crustac Biol 20:118–127

Mashiko K, Shy JY (2008) Derivation of four morphologically affiliated species of *Macrobrachium* (Caridea, Palaemonidae) with divergent reproductive characteristics. J Crustac Biol 28:370–377

Mashiko K, Kawabata S, Okino T (1991) Reproductive and populational characteristics of a few caridean shrimps collected from Lake Tanganyika, East Africa. Arch Hydrobiol 122:69–78

McDowall RM (1992) Diadromy: origins and definitions of terminology. Copeia 1992:248–251

McDowall RM (2007) On amphidromy, a distinct form of diadromy in aquatic organisms. Fish 8:1–13

McDowall RM (2010) Why be amphidromous: expatrial dispersal and the place of source and sink population dynamics? Rev Fish Biol Fish 20:87–100

Mossolin EC, Shimizu RM, Bueno SLS (2006) Population structure of *Alpheus armillatus* (Decapoda, Alpheidae) in São Sebastião and Ilhabela, Southeastern Brazil. J Crustac Biol 26: 48–54

Müller YMR, Nazari EM, Costa MS (2003) Embryonic stages of the freshwater prawn *Macrobrachium olfersi* (Decapoda, Palaemonidae). J Crustac Biol 23:869–875

Nilssen EM, Aschan MM (2009) Catch, survey and life-history data for shrimp (*Pandalus borealis*) off Jan Mayen. Deep Sea Res II Top Stud Oceanogr 56:2023–2036

Noël P (1976) L'évolution des caractères sexuels chez *Processa edulis* (Risso) (Décapode, Natantia). Vie Milieu 26:65–104

Novak PA, Douglas MM, Garcia EA, Bayliss P, Pusey BJ (2015) A life-history account of *Macrobrachium spinipes* (Schenkel, 1902) (Cherabin) in a large tropical Australian River. Freshw Sci 34:620–633

Nye V, Copley JT (2014) Reproductive ecology of a hippolytid shrimp, *Lebbeus virentova* (Caridea: Hippolytidae) at the Von Damm Vent Field, Mid-Cayman Spreading Centre. Mar Biol 161:2371–2380

O'Brien J, Van Wyk PM (1985) Effects of crustacean parasitic castrators (epicaridean isopods and rhizocephalan barnacles) on growth of crustacean hosts. In: Wenner AM (ed) Crustacean issues 3: factors in adult growth. A.A. Balkema, Rotterdam, pp 191–218

Oh CW, Hartnoll RG (1999) Brood loss during incubation in *Philocheras trispinosus* (Decapoda) in Port Erin Bay, Isle of Man. J Crustac Biol 19:467–476

Oh CW, Jeong IJ (2003) Reproduction and population dynamics of *Acetes chinensis* (Decapoda: Sergestidae) on the western coast of Korea, Yellow Sea. J Crustac Biol 23:827–835

Oh CW, Kim JN (2008) Reproductive biology of *Exopalaemon carinicauda* (Decapoda, Palaemonidae) in the Hampyong Bay of Korea. Crustaceana 81:949–962

Oh CW, Suh HL, Park KY, Ma CW, Lim HS (2002) Growth and reproductive biology of the freshwater shrimp *Exopalaemon modestus* (Decapoda: Palaemonidae) in a lake of Korea. J Crustac Biol 22:357–366

Ohtomi J, Irieda S (1997) Growth of the deep-water mud shrimp *Solenocera melantho* De Man, 1907 (Decapoda, Penaeoidea, Solenoceridae) in Kagoshima Bay, Southern Japan. Crustaceana 70:45–58

Ohtomi J, Matsuoka T (1998) Reproduction and growth of Jack-knife shrimp, *Haliporoides sibogae*, off south-western Kyushu, Japan. Fish Res 38:271–281

Olivier TJ, Bauer RT (2011) Downstream migration of the river shrimp *Macrobrachium ohione* in the lower Mississippi River system. Am Midl Nat 166:379–393

Olivier TJ, Handy KQ, Bauer RT (2013) Impacts of river control structures on the juvenile migration of the amphidromous river shrimp M*acrobrachium ohione* (Smith): possible solutions for the restoration of upstream populations. Freshw Biol 58:1603–1613

Omori M (1974) The biology of pelagic shrimps in the ocean. Adv Mar Biol 12:233–324

Omori M (1979) Growth, feeding, and mortality of larval and early postlarval stages of the oceanic shrimp *Sergestes similis* Hansen. Limnol Oceanogr 24:273–288

Omori M, Chida Y (1988) Life history of a caridean shrimp *Palaemon macrodactylus* with special reference to the difference in reproductive features among ages. Nippon Suisan Gakkaishi 54: 365–375

Omori M, Gluck D (1979) Life history and vertical migration of the pelagic shrimp *Sergestes similis* off the southern California coast. Fish Bull 77:183–198

Omori K, Yanagisawa Y, Hori N (1994) Life history of the caridean shrimp *Periclimenes ornatus* Bruce associated with a sea anemone in southwest Japan. J Crustac Biol 14:132–145

Orsi Relini L, Relini G (1998) Seventeen instars of adult life in female *Aristeus antennatus* (crustacea: Decapoda: Aristeidae). A new interpretation of life span and growth. J Nat Hist 32:1719–1734

Orton JH (1920) Sea temperatures, breeding and distribution of marine animals. J Mar Biol Assoc UK 12:339–366

Osawa Y, Aoki MN, Thiel M, Bauer RT (2015) Analysis of life-history traits in a sex-changing marine shrimp (Decapoda: Caridea: Rhynchocinetidae). Biol Bull 228:125–136

Overstreet RM (1978) Marine Maladies? Worms, germs, and other symbionts from the Northern Gulf of Mexico. Mississippi-Alabama Sea Grant Consortium 78–021, 140 pp

Page TJ, Cook BD, Von Rintelen T, Von Rintelen K, Hughes JM (2008) Evolutionary relationships of atyid shrimp imply both ancient Caribbean radiations and common marine dispersals. J N Am Benthol Soc 27:68–83

Paschoal LRP, Guimarães FJ, Couto ECG (2016) Growth and reproductive biology of the amphidromous shrimp *Palaemon pandaliformis* (Decapoda: Caridea) in a neotropical river from northeastern Brazil. Fortschr Zool 33:1–14

Pearse JS, McLintock JB, Bosch I (1991) Reproduction of Antarctic benthic marine invertebrate tempos, modes, and timing. Am Zool 31:65–80

Pearson JC (1939) The early life histories of some American Penaeidae, chiefly the commercial shrimp *Penaeus setiferus* (Linn.). Bull Bureau Fish US Department of Commerce 49:1–73

Pescinelli RA, Davanso TM, Pantaleão JAF, Batista A, Bauer RT, Costa RC (2018) Population dynamics, relative growth and sex change of the protandric simultaneous hermaphrodite *Exhippolysmata oplophoroides* (Caridea: Lysmatidae) close to an upwelling area. J Mar Biol Assoc U K 98:727–734

Pianka E (1970) On r- and K-selection. Am Nat 104:592–597

Pires MAB, Abrunhosa FA, Maciel CR (2008) Early larval development in the labor elopment in the laboratory of *Alpheus estuariensis* (Crustacea: Caridea) from the Amazon Region. Rev Bras Zool 25:199–205

Powell CB (1977) A revision of the African freshwater shrimp genus *Desmocaris* Sollaud, with ecological notes and description of a new species (Crustacea, Decapoda, Palaemonidae). Rev Zool Afric 91:649–674

Powell CB (1979) Suppression of larval development in the African freshwater shrimp *Desmocaris trispinosa* (Aurivillius) (Decapoda Palaemonidae). Crustaceana Suppl 5:185–194

Puspitasari I, Oh CW, Lee WO (2013) Reproductive traits of the lake prawn *Palaemon paucidens* De Haan, 1844 (Decapoda, Palaemonidae), from Goesan Lake, South Korea. Crustaceana 86: 1191–1205

Ragonese S, Bianchini ML (1995) Size at sexual maturity in red shrimp females, *Aristaeomorpha foliacea,* from the Sicilian Channel (Mediterranean Sea). Crustaceana 68:73–82

Ragonese S, Vitale S, Dimechi M, De Santi A (2012) Growth discontinuity in males of the deep-water giant red shrimp *Aristaeomorpha foliacea* in the Mediterranean Sea. Mar Ecol 33:386–392

Rahman S, Schmidt D, Hughes JM (2020) Genetic structure of Australian glass shrimp, *Paratya australiensis,* in relation to altitude. PeerJ 8:e8139. https://doi.org/10.7717/peerj.8139

Rajasree R, Harikrishnan M, Kurup BM (2011) Stock characteristics and population dynamics of *Heterocarpus woodmasoni* Alcock. J Mar Biol Assoc India 53:101–107

Ramirez Llodra E, Tyler PA, Copley JTP (2000) Reproductive biology of three caridean shrimps, *Rimicaris exoculata, Chorocaris chacei*, and *Mirocaris fortunata* (Caridea: Decapoda), from hydrothermal vents. J Mar Biol Assoc UK 80:473–484

Ramirez-Llodra E (2002) Fecundity and life-history strategies in marine invertebrates. Adv Mar Biol 43:87–170

Ramirez-Llodra E, Segonzac M (2006) Reproductive biology of *Alvinocarls muricola* (Decapoda : Caridea : Alvinocarididae) from cold seeps in the Congo Basin. J Mar Biol Assoc U K 86:1347–1356

Rasch JA, Bauer RT (2015) Temporal variation in population structure of the isopod *Urobopyrus processae* Richardson, 1904 (Isopoda: Bopyridae) infesting the branchial chamber of the night shrimp *Ambidexter symmetricus* Manning and Chace, 1971 (Decapoda: Processidae). Nauplius 23:89–103

Reaka ML (1979) The evolutionary ecology of life history patterns in stomatopod Crustacea. In: Stancyk S (ed) Reproductive ecology of marine invertebrates. Univ. South Carolina Press, Columbia, pp 235–260

Roa R, Ernst B (1996) Age structure, annual growth, and variance of size-at-age of the shrimp *Heterocarpus reedi*. Mar Ecol Prog Ser 137:59–70

Roberts TW (1986) Abundance and distribution of pink shrimp in and around the Tortugas Sanctuary, 1981-1983. N Am J Fish Manag 6:311–327

Rome N, Conner SL, Bauer RT (2009) Delivery of hatching larvae to estuaries by an amphidromous river shrimp: tests of hypotheses based on larval moulting and distribution. Freshw Biol 54:1924–1932

Romero A, Turnbull MW, Baeza JA (2018) Embryonic development in the peppermint shrimp, *Lysmata boggessi* (Caridea: Lysmatidae). Biol Bull 234:165–179

Saito T, Konishi K (1999) Direct development in the sponge-associated deep-sea shrimp *Spongicola japonica* (Decapoda: Spongicolidae). J Crustac Biol 19:46–52

Saito T, Koya Y (2001) Gonad maturation and embryonic development in the deep-sea sponge-associated shrimp, *Spongicola japonica* Kubo (Crustacea: Decapoda: Spongicolidae)/. Zool Sci 18:567–576

Saito T, Tsuchida S, Yamamoto T (2006) *Spongicoloides iheyaensis*, a new species of deep-sea sponge-associated shrimp from the Iheya Ridge, Ryukyu Islands, southern Japan (Decapoda: Stenopodidea: Spongicolidae). J Crustac Biol 29:224–233

Santos RC, Perroca J, Costa R, Hirose G (2020) Population dynamics of *Farfantepenaeus subtilis* (Pérez-Farfante, 1967) and *Litopenaeus schmitti* (Burkenroad, 1936) (Decapoda: Penaeidae) and evidence of habitat partitioning in the northeast of Brazil. Reg Stud Mar Sci 35:101218. https://doi.org/10.1016/j.rsma.2020.101218

Sardà F, Demestre M (1987) Estudio biol6gico de la gamba *Aristeus antennatus* (Risso, 1816) en el Mar Catalan (NE de Espana). Investig Pesq 51(suplemento 1):213–232

Sastry AN (1983) Ecological aspects of reproduction. In: Vernberg FJ, Vernberg WB (eds) The biology of crustacea, Environmental adaptations, vol 8. Academic Press, New York, pp 179–270

Schmidt J (1922) The breeding places of the eel. Philos Trans R Soc B 211:179–208

Sigman DM, Hain MP (2012) The biological productivity of the ocean. Nat Educ Knowledge 3:21

Simões SM, D'Incao F, Fransozo A, Castilho AL, Costa RC (2013) Sex ratio, growth and recruitment of the pelagic shrimp *Acetes americanus* on the southeastern coast of Brazil. J Crustac Biol 33:1–9

Sinderman CJ (1970) Principal diseases of marine fish and shellfish. Academic Press, New York

Sobrino I, Silva C, Sbrana M, Kapiris K (2005) A review of the biology and fisheries of the deep water rose shrimp, *Parapenaeus longirostris,* in European Atlantic and Mediterranean waters (Decapoda, Dendrobranchiata, Penaeidae). Crustaceana 78:1153–1184

Sollaud E (1923) Le développement larvaire des "Palaemoninae." 1. Partie descriptive. La condensation progressive de l'ontogénèse. Bull Biol Fr Belg 77:510–603

Stanski G, Carvalho MM, Garcia JR, Goncalves GRL, Costa RC, Castilho AL (2018) Geographical variation and local environment effects in the reproductive output and fecundity of the shrimp *Exhippolysmata oplophoroides* (Decapoda: Caridea) in southeastern Brazil. Invertebr Reprod Dev 62:119–124

Stearns SC (1976) Life-history tactics: a review of the ideas. Q Rev Biol 51:3–47

Stearns SC (1992) The evolution of life histories. Oxford University Press

Sukumaran KK (1998) Studies on the fishery and biology of *Solenocera crassicornis* (H. Milne Edwards) from Bombay Waters. J Mar Biol Assoc India 20:32–39

Teodoro SAS, Pantaleão JAF, Fransozo MLN, Costa RC (2014) Ecological aspects and sexual maturity of a southwestern Atlantic population of the planktonic shrimp *Lucifer faxoni* (Decapoda: Sergestoidea). J Crustac Biol 34:422–430

Terossi M, Wehrtmann IS, Mantelatto FL (2010) Interpopulation comparison of reproduction of the Atlantic shrimp *Hippolyte obliquimanus* (Caridea: Hippolytidae). J Crustac Biol 30:571–579

Thorson G (1950) Reproductive and larval ecology of marine bottom invertebrates. Biol Rev Camb Philos Soc 25:1–45

Tyler PA (1988) Seasonality in the deep sea. Oceanogr Mar Biol Annu Rev 26:227–258

Tyler PA (1995) Conditions for the existence of life at the deep-sea floor: an update annual review. Oceanogr Mar Biol 33:221–244

Valentin JL (2001) The Cabo Frio upwelling system, Brazil. In: Seeliger U, Kjerfve B (eds) Coastal marine ecosystems of Latin America. Ecological studies (analysis and synthesis), vol 144. Springer, Berlin, Heidelberg, pp 97–105

van der Velde TD, Venables WN, Crocos PJ, Edgar S, Evans F, Rothlisberg PC (2021) Seasonal, interannual and spatial variability in the reproductive dynamics of *Penaeus merguiensis*. Mar Ecol Prog Ser 658:117–133

Varpe O (2017) Life history adaptations to seasonality. Integr Comp Biol 57:943–960

Vázquez MG, Bas CC, Spivak ED (2012) Life history traits of the invasive estuarine shrimp *Palaemon macrodactylus* (Caridea: Palaemonidae) in a marine environment (Mar del Plata, Argentina). Sci Mar 76:507–516

Walker I (1992) Life history traits of shrimps (Decapoda: Palaemonidae) of Amazonian inland waters and their phylogenetic interpretation. Stud Neotropical Fauna Environ 27:131–143

Walsh CJ (1993) Larval development of *Paratya australiensis* Kemp, 1917 (Decapoda: Caridea: Atyidae), reared in the laboratory, with comparisons of fecundity and egg and larval size between estuarine and riverine environments. J Crustac Biol 13:456–480

Wehrtmann IS, Lardies MA (1999) Egg production of *Austropandalus grayi* (Decapoda, Caridea, Pandalidae) from the Magellan region, South America. Sci Mar 63(Suppl. 1):325–331

Wehrtmann IS, López GA (2003) Effects of temperature on the embryonic development and hatchling size of *Betaeus emarginatus* (Decapoda: Caridea: Alpheidae). J Nat Hist 18:2165–2178

Wenner EL (1978) Comparative biology of four species of glyphocrangonid and crangonid shrimp from the continental slope of the middle Atlantic bight. Can J Zool 56:1052–1065

Wenner EL (1979) Distribution and reproduction of nematocarcinid shrimp (Decapoda: Caridea) from the northwestern North Atlantic. Bull Mar Sci 29:380–393

Węsławski JM (1987) Distribution of Decapoda (Crustacea) in South Spitsbergen coastal waters with remarks on their ecology and breeding biology. Pol Polar Res 8:121–134

Williams AB (1984) Shrimps, Lobsters, and Crabs of the Atlantic Coast of the Eastern United States, Maine to Florida. Smithsonian Institution Press, Washington, DC

Williamson DI (1982) Larval morphology and diversity. In: Abele LG (ed) The biology of crustacea, Embryology, morphology, and genetics, vol 2. Academic Press, New York, pp 43–110

Xiao Y, Greenwood JG (1993) The biology of *Acetes* (Crustacea, Sergestidae). Annu Rev Oceanogr Mar Biol 31:259–444

Yam RSW, Dudgeon D (2006) Production dynamics and growth of atyid shrimps (Decapoda: Caridina species) in 4 Hong Kong streams: the effects of site, season, and species. J N Am Benthol Soc 25:406–416

Young C (2003) Reproduction, development and life history traits. In: Tyler PA (ed) Ecosystems of the world, Ecosystems of the deep oceans, vol 28. Elsevier, Amsterdam, pp 381–426

Zilch R (1978) Ernbryologische untersuchungen an der holoblastischen ontogenese von *Penaeus trisulcatus* Leach (Crustacea, Decapoda). Zoomorphologie 90:67–100

Zupo V (1994) Strategies of sexual inversion in *Hippolyte inermis* leach (crustacea, Decapoda) from a Mediterranean seagrass meadow. J Exp Mar Biol Ecol 178:131–145

Chapter 10
Symbioses

10.1 Symbioses in Shrimps

In both the plant and animal world, there are extraordinary, sometimes amazing, associations and relationships between two very different species. The term "symbiosis," from the Greek words "the act of living together," describes this phenomenon very well. Symbioses can be more formally defined as intimate, usually prolonged ecological associations that have evolved between two different species. In most cases, the two species belong to very different taxonomic groups. One of the two species, usually the smaller in size, is termed the "symbiont," while the other usually the larger one is called the "host." Symbioses are grouped into three general categories: commensalism, mutualism, and parasitism (Fig. 10.1). In commensalism ("sharing the same table"), the symbiont benefits from the association while the host species is unaffected. In mutualism, both species gain from the relationship. A third category is the familiar parasitism ("one who eats at another's table") in which the symbiont benefits but the host is harmed. However, the actual relationship may be weakly or highly developed and may fit somewhere along a continuum between two of these categories.

Many decapod shrimps, especially marine species, have evolved symbiotic relationships with a wide variety of sessile benthic invertebrates such as sponges, cnidarians (e.g., sea anemones, corals, gorgonians), bivalve molluscs (e.g., clams), echinoderms (e.g., sea urchins, starfishes, feather stars), ascidians ("sea squirts"), and burrowing organisms such as echiurans ("spoon worms") and fellow crustaceans, principally callianassids ("ghost shrimps") and gebiids ("mud shrimps"). In a departure from the usual shrimp symbiont–invertebrate host symbioses, several alpheid shrimps have developed mutualistic relationships with gobiid fishes in which the shrimp (and its burrow) is the host. Several caridean shrimps clean (groom) fishes of parasites, excess mucus, and injured or dead tissues. This interaction has been shown or at least appears to be mutually beneficial. However, it does not quite fit the classical definition of symbioses in that the encounters between the

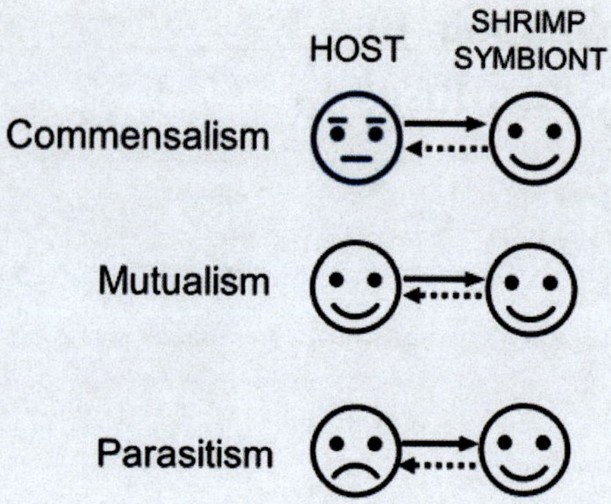

Fig. 10.1 Categories of symbiosis: the effects of a symbiotic shrimp on its host (dashed arrow) and that of host on the shrimp (solid arrow). Happy face = beneficial; neutral face = no effect; sad face, negative effect

two species are often brief although important for both. This interesting relationship may be considered a special case, a "cleaning symbiosis" which will be discussed later in depth. Finally, there are many potential cases of symbiosis in which the shrimp is often collected or observed together with another species, but the exact nature of the relationship is not yet known. In these instances, the shrimp species is usually called an "associate" of the host. When the biology of most shrimp–host associations is closely studied, they are commensalistic or mutualistic, although in some species, feeding on host tissues does occur, which qualifies as parasitic. However, the damage done often appears minor and barely apparent. Additionally, the hosts in which this occurs have considerable powers of regeneration, e.g., sponges, cnidarians such as corals and sea anemones, and echinoderms such as sea urchins. Perhaps commensalism verging on parasitism would be a more apt description of such a relationship.

Shrimps are termed "ectosymbionts" when they live on the exterior of the host, e.g., clinging to the spines of a sea urchin, or on the column or within the tentacles of a sea anemone. "Endosymbionts" reside within the host's body, e.g., within the bivalved shell or mantle cavity of a clam host or within the branchial chamber of an ascidian. Shrimps which, after the larval stage, must live and depend on the host species are "obligate" symbionts. "Facultative" symbionts may form a relationship with a characteristic host but can live free when that host is not available. Symbiotic species may be specialists, associating only with a single host, as *Alpheus armatus* does with the sea anemone *Bartholomea annulata* (McCammon et al. 2010). Another example is a suite of 4 pontoniine shrimps which occur only with a single species of scleractinian coral (*Galaxea fascicularis*) and apparently not with any of this coral genus' many other species (Bruce 1976a; Horká et al. 2016). Others shrimp species are symbiotic generalists, occurring with a few to several kinds of hosts, e.g., *Thor amboinensis*, which associates with very different cnidarian hosts such as several species of sea anemones, corals, alcyonaceans, and hydroids (Bruce 1976a).

10.2 Taxonomic and Geographic Distribution of Shrimp Symbioses

Although there are numerous shrimp symbionts and associates of other species, the dendrobranchiate shrimps (sergestoids and penaeoids) are not among them. This group of shrimps is overwhelmingly aggregated or schooling species that are either pelagic (sergestoids) or benthonektonic (penaeoids), the latter on soft sea bottoms (mud, grave, or sand substrates). The high mobility of dendrobranchiates and the lack of physical structure in their habitats are not favorable for prolonged interactions with host species with which they might evolve a symbiosis. Although there are some potential hosts in the pelagic realm, e.g., cnidarian jellyfishes and colonial tunicates, few symbioses or even associations with dendrobranchiates have evolved. The general lifestyle and availability of potential hosts are important factors predisposing evolution of symbioses in many benthic carideans and stenopodideans.

Among the Caridea, the families with the vast majority of symbiotic species are the Palaemonidae (especially the pontoniines) and Alpheidae, followed by the Hippolytidae, the gnathophyllid palaemonids, and a few genera from taxa such as the Pandalidae, Disciadidae, and palaemonid anchistioidids. Marine biologist and shrimp expert A.J. Bruce had a long and productive career in which shrimp symbioses have played a major part. His overview of shrimp symbioses, a long version (Bruce 1976a) and a shorter more widely available one (Bruce 1976b), are good introductions to the subject. As an example of the relative abundance of symbiotic species from different families, he gave a summary from collections at different sites in tropical Indo-West Pacific habitats (Fig. 10.2). In the Kenyan National Marine Park, 67 species were collected; 38 (57%) of these were considered symbionts of benthic invertebrate hosts, mostly pontoniine palaemonids along with a few alpheids and hippolytids. Results from other Indo-West Pacific sites were similar. In a collection of 50 pontoniine species from the Seychelles (western Indian Ocean), 74% were considered potential symbionts. The most important taxonomic groups which served as hosts were cnidarians, especially stony (reef-building) corals, as well as sea anemones and gorgonians; molluscs, especially bivalves; echinoderms such as comatulid crinoids ("feather stars"), sea urchins, and asteroids (starfishes); and poriferans (sponges). As we shall see, these groups are the most important of the many hosts of symbiotic shrimps. The range of hosts utilized by pontoniine and alpheid shrimp associates from the Indo-West Pacific is quite large, as shown in Fig. 10.2.

Many stenopodideans are symbiotic, summarized by Goy (2010). Among the family Stenopodidae, several species of *Odontozona* (and a single species of *Richardina*) are associates of various shallow-water and especially of deep-sea species of sessile invertebrates. However, the Spongicolidae, except for most species of *Microprosthema*, are symbiotic with deep-water sponges and a few other invertebrate groups.

Tropical waters are a hot spot for shrimp symbioses, especially the Indo-West Pacific. In his comparison of the incidence of shrimp associations among different

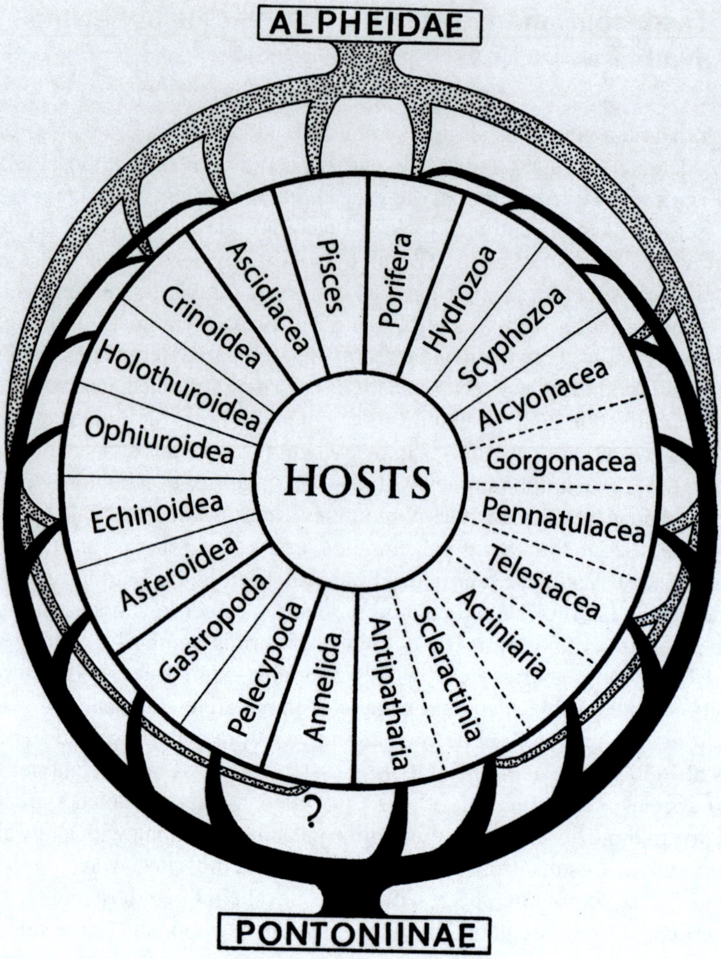

Fig. 10.2 Hosts of alpheid and pontoniine shrimp associates in the Indo-West Pacific. Modified from Bruce (1976b)

geographic regions, Bruce (1976b) showed that 59% of genera in the Indo-West Pacific had symbiotic species, followed by Eastern Pacific (50%) and Atlantic (41%). The overall species and generic diversity of potential shrimp associates and hosts are highest in the tropical Indo-West Pacific, with its extensive geographic area with coral reefs.

Most studies show that the incidence of shrimp symbioses is much lower in temperate and high latitude seas, with their lower overall species diversity of marine species. One factor is that the species abundance of pontoniine shrimps is very low in temperate and high latitudes, and this group has the highest percentage of symbiotic species overall. Just as important, the occurrence of shrimp symbioses may be underestimated because of insufficient sampling or lack of study on possible shrimp–host associations. Jonsson et al. (2001) investigated a group of pandalid and thorids in deeper waters (60–250 m) off the coast of Sweden using remote observation vehicles (ROV). These species, considered as facultative commensals, were variously found around and beneath the tentacles of two species of sea anemones. The investigators proposed that shrimp and other symbioses are not particularly rare in temperate waters but that the usual sampling methods may be inadequate to reveal them. Certainly, observation and sampling in the warm, shallow-, and clear-water habitats of coral reefs is easier and more pleasant than in the cold turbid waters often encountered at higher latitudes! Nonetheless, Jensen (2014), who has combined scientific studies with underwater photography, also describes several possible symbiotic relationships involving shrimps in the eastern North Pacific. The shrimp taxa involved in higher latitude symbioses are primarily hippolytoids (thorids, lysmatids), pandalids, and alpheids. The North and South Pacific temperate zone alpheid *Betaeus* has numerous commensal species (Hart 1964; Anker and Baeza 2012; Jensen 2014). However, as useful as Bruce's 1976 papers are, the considerable knowledge gained since then begs for a comprehensive worldwide analysis of the taxonomic and biogeographic distribution of symbiotic shrimps.

There are lots of freshwater shrimp species, mostly palaemonines and atyids—so where are the symbioses? Shrimp symbioses with freshwater invertebrates are, to my knowledge, restricted to just one so far, an atyid (*Caridina spongicola*) with a sponge (Zitzler and Cai 2006; von Rintelen et al. 2007), and perhaps one fish-cleaner shrimp (Garrone-Neto et al. 2014). Compared to the marine environment, the diversity of large sessile invertebrates which might host shrimp symbionts is low in freshwater, e.g., some sponges and bryozoans. The diversity of smaller invertebrates is certainly high, e.g., insects and their larvae and oligochaete worms, but their comparable size to shrimps makes them poor hosts. Although more cases of freshwater shrimp symbioses may be uncovered, there is little doubt that their abundance pales in comparison with that of marine ones.

10.3 Common Adaptations of Symbiotic Shrimps

For the symbiotic shrimp, the host is its substrate and habitat. Shrimps symbionts living with hosts of different taxonomic groups or on unrelated hosts with similar structural characteristics have often evolved comparable adaptations in morphology and coloration. Such characteristics may evolve independently among closely related species (parallel evolution) or among unrelated shrimp taxa (convergent evolution) or simply be inherited from a common symbiotic ancestor. A detailed description of these adaptations is presented in A.J. Bruce's (1976a) chapter on symbiotic shrimps. His observations were based on his vast experience in collecting and observing symbiotic shrimps, as well as describing these traits in numerous taxonomic papers. An understanding of these adaptations has been developed further since in his subsequent papers and those of many other authors.

Bruce used the free-living pontoniine *Periclimenes scriptus* (*elegans*) (Fig. 10.3a) to illustrate how the morphology of symbiotic species has become adapted to their

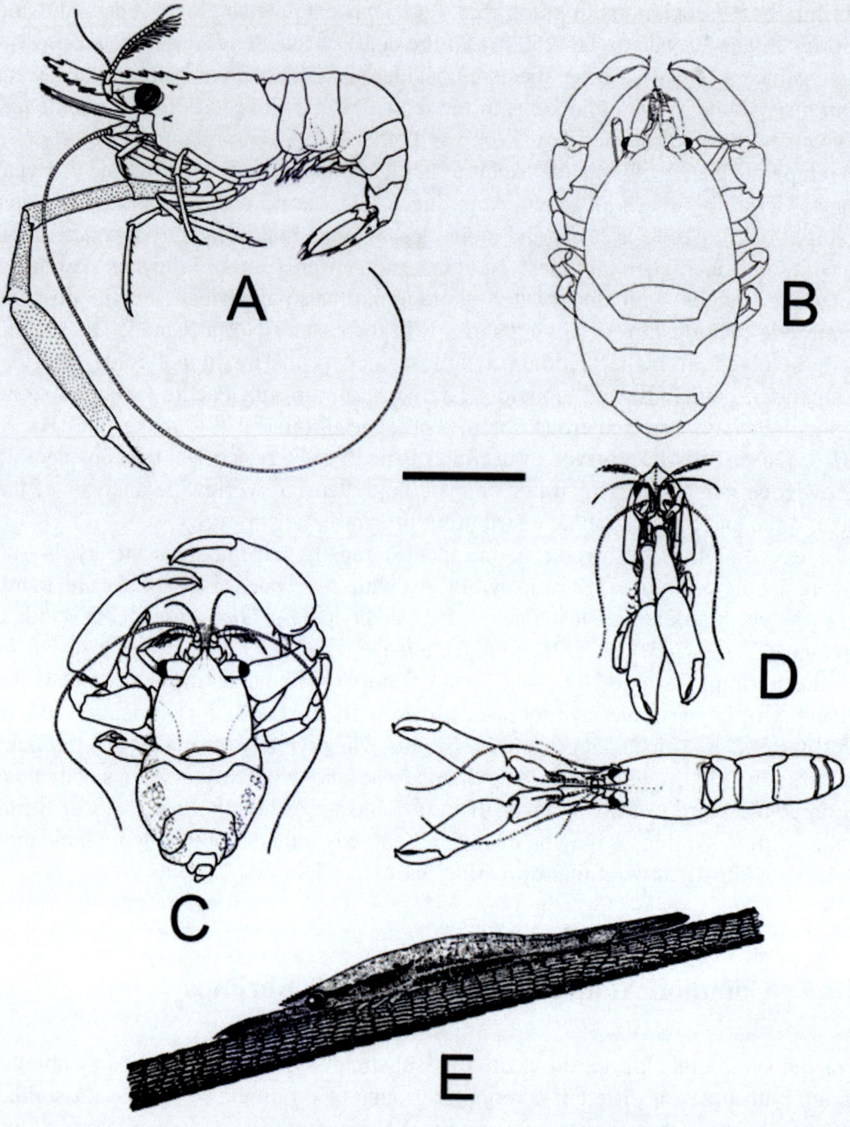

Fig. 10.3 Adaptive body shapes of symbiotic shrimps. (**a**) *Periclimenes scriptus* (*elegans*), with a typical body shape of a free-living epibenthic shrimp. (**b**) *Pontonia manningi* female, with the globular or swollen shape of endosymbionts symbiotic with bivalve molluscs. (**c**) *Platycaris latirostris*, a coral associate with a dorsoventrally compressed (depressed) body shape. (**d**) *Ischnopontonia lophos* with laterally compressed body (lateral, dorsal views). Both (**c**) and (**d**) are two different adaptations for living within narrow spaces among scleractinian corallites. (**e**) *Tuleariocaris neglecta*, with an elongated shape with short stout pereopods for gripping and living on or among sea urchin spines. Scale bar represents 5.2 mm in (**a**), 4.6 mm in (**b**), 3.6 mm in (**c**), 4.1 mm in (**d**) and 3.5 mm in (**e**). From (**a**, **c**–**d**) Bruce (1976a); (**b**) Fransen (2002); (**e**) Chace (1969), courtesy of Crustaceana (Brill)

habitat, i.e., a host species. *P. scriptus* is a an unspecialized epibenthic shrimp with typical natant characteristics: a somewhat laterally compressed body, well-developed toothed rostrum, various carapace spines, and slender elongate walking legs (pereopods 3–5) terminating in simple, long, slightly curved dactyls. Body shapes of symbiotic shrimps usually differ from this typical epibenthic caridean according to host structure. Endosymbiotic shrimps living within confined spaces, e.g., within bivalve molluscs or branchial chambers or siphons of ascidians, tend to be globular and swollen in shape, such as *Pontonia manningi* living in various bivalve species (Fransen 2002) (Fig. 10.3b) and *Paratypton siebenrocki* within coral cysts (Fig. 10.8). Females tend to be more distended than males, probably to provide more space under the abdomen for embryos. In order to move easily among the narrow spaces between individual corallites (zooids) of the scleractinian coral *Galaxea fascicularis*, bodies may be flattened (depressed) as in *Platycaris latirostris* (Fig. 10.3c) or laterally compressed, such as the pontoniines *Ischnopontonia lophos* (Fig. 10.3d) and *Anapontonia denticauda*, as well as the alpheid *Racilius compressus*. Some species of the pontoniines *Onycocaris* and *Typton* have evolved a wormlike shape, with short cephalothorax and long abdomen, to move easily within narrow sponge canals. Species of *Tuleariocaris* and *Stegopontonia*, which perch with their short stubby legs on long sea urchin spines, are slender and elongate (Fig. 10.3e; Plate 10.2D).

Many symbiotic shrimps have smooth bodies with reduced or few body projections. The obvious adaptive value, although rarely if ever tested, is probably to promote their easy movement on or within the host, or to prevent irritation of host tissues. Such species live nestled on or among hosts with soft external tissues, e.g., zooids of cnidarians such as corals, gorgonians, and zoantharians, or as endosymbionts within hosts such as sponges, bivalves, and ascidians. The external surface of the shrimp is made smooth by the reduction and loss of rostral, carapace, and telson spines. The angular projections of abdominal pleura of other shrimps become smoothly rounded, as do sharp projections of antennal scales (scaphocerites). On the other hand, some symbiotic species have expanded antennal scales, uropods, or telson with enlarged spines or projections. When a predator tries to pull them out of spaces on or within the host, the shrimp can use these sharp protuberances to wedge themselves within the host tissues (Bruce 1976a, b). The walking-leg dactyls of shrimps that live on or in soft-bodied hosts, e.g., externally on corals or within sponges or ascidians, are often adapted for digging into and clinging to host tissues. This may take the form of extra spines and hooks, or the dactylar spine may combine with a propodal protrusion or spine to form a rudimentary chela (Fig. 10.4).

As with other shrimps, the diet of symbiotic species is important to understand their natural history and morphology. Gut content analyses are one of the best ways to determine the diet of an animal, which can tell us a lot about their lifestyle. However, by the time food arrives in the stomachs of shrimps, it is so degraded by the mouthparts and structures within the gut that visual determination of food type is often very difficult. There are now many ways to examine gut contents, including stable isotope and DNA analyses, but very few have yet been done. In lieu of those, the structure of the mouthparts, especially the mandibles, often gives a clue to the

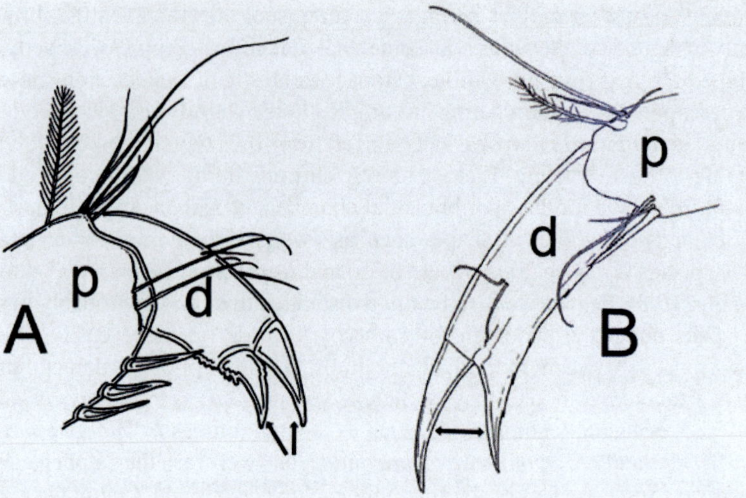

Fig. 10.4 Claw-like dactylar adaptations for clinging to soft tissues of invertebrate hosts. (**a**) Pereopod 4 dactyl with twin stout setal spines (arrows) (sponge shrimp *Spongiocaris hexactinellicola*). (**b**) Pereopod 5 dactyl with stout setal spine (left side of double arrow) and spine-like extension of dactyl per se (right side of double arrow) forming a rudimentary chela (coral associate *Periclimenes subcorallum*). d, dactyl; p, propodus. Modified from (**a**) Berggren (1993), courtesy of the Journal of Crustacean Biology and (**b**) Rauch et al. (2019) Creative Commons CC BY 4.0

diet. Bruce (1976a, b) pointed out that the incisor and molar mandibles of many symbiotic shrimps are highly reduced, indicating a diet of host mucus or other very soft material. Most marine animals, e.g., fishes and invertebrates (except crustaceans, covered with a hard, nonliving exoskeleton), have an external epidermis which secretes varying quantities of protective mucus, which is constantly produced and then sloughed off. This mucus is an energy-rich food often entangled with organic detritus and microbial organisms, all good food for shrimps (Arnal et al. 2001). It is soft food which requires little grinding or cutting, so reduction of mandibular processes is understandable in mucus-eating shrimps. Other symbiotic shrimps feed on a variety of food types. Some species are thought to be "freeloaders" which scavenge on the scraps from their host's food. Other species include host tissues in their diet, i.e., are at least partially parasitic. These various feeding habits are discussed below in the examples of symbionts of different host species.

Free-living and ectosymbiotic species of pontoniine palaemonids have well-developed eyes and presumably better vision, albeit poor, compared to endosymbiotic species (Dobson et al. 2014). The eyes of endosymbionts are reduced in various ways. The eyestalk is shorter, and the cornea has fewer facets of larger diameter with a somewhat disorganized arrangement compared to ectosymbiotic shrimps. The lower number of facets translates to poorer resolution of images, but the larger facet diameter may improve sensitivity to light in the very dark host spaces which endosymbionts inhabit. Ectosymbionts are more likely to have accessory eyes ("nebenaugen;" see Chap. 2), still of uncertain visual function, than endosymbionts. The reduction of endosymbiont eyes reflects their reduced lower visual needs in the murky confines of their host. Some visual function appears to be maintained as endosymbiotic species may need to leave their host to search for a mate or to return to their host if displaced from it.

As in free-living species, symbiotic shrimps show the usual bag of coloration tricks to disguise themselves from visually hunting predators, i.e., invisibility (transparency) (Plate 10.1A), imitation (mimicry of the host colors and/or structural features) (Plate 10.1B), interruption of body outline (disruptive coloration), or some combination thereof (Chap. 5, Coloration) (Bruce 1976a, b). The small size, cryptic habits, and transparent coloration of many coral symbionts help to protect them. Good examples are the two pontoniine *Fennera* species, which inhabit the spaces among branches of pocilloporids and poritid stony corals. They are perhaps the smallest of caridean species; females only attain 5–7 mm total length and can carry only up to 9 small-sized embryos (Bruce 1976a). They are translucent with only a few distracting red and yellow spots of color (Marin 2011). The freshwater sponge symbiont *Caridina spongicola* combines bold black transverse stripes on an otherwise transparent body to form its disruptive coloration. Urchin symbionts combine body shape and color pattern to achieve an effective crypsis. *Tuleariocaris* and *Stegopontonia* species, which live on long-spined urchins, combine an elongate body shape with a basal color like that of its host, interrupted with longitudinal white or light-colored stripes. This coloration gives the illusion of one or more spines instead of a shrimp body. Similarly, *Gnathophylloides mineri,* which usually lives on short-spined *Tripneustes* urchins, uses a similar trick, but given the short spines of the host, the shrimp has a short stubby body (Plate 10.1C). These and other examples of symbiotic shrimp coloration will be discussed below in the sections on symbionts of different host taxa.

Some symbiotic shrimps do not bother to conceal themselves; instead they use coloration termed "irrelevant" by Bruce (1976a, b). Their colors do not hide them from visual predators but instead are visually quite obvious, at least to the human eye. When I first saw the beautiful green color of *Coralliocaris graminea* (Plate 3.6G), I immediately thought that it must be a symbiont of a similarly colored host, e.g., a coral with a green zooxanthellae hue. On the contrary, the bright green or white colors of various *Coralliocaris* contrast strongly with those of their acroporid coral hosts. However, they live deep among the branches of the host which may keep them out of the reach of visual predators. The colors may have another function or perhaps are inherited traits adaptive in free-living ancestors which have not yet been lost. Other shrimps have bold color patterns but do not hide away unseen on their hosts; perhaps this type of conspicuous coloration is a warning to possible visual predators that the shrimps are toxic or distasteful (aposematic coloration), like that of many insects such as bees, wasps, and Monarch butterflies. Bruce (1976a, b) proposed this hypothesis based on his observation that in aquaria, fishes avoid these kinds of shrimps while usually quickly gobbling up other variously colored shrimps.

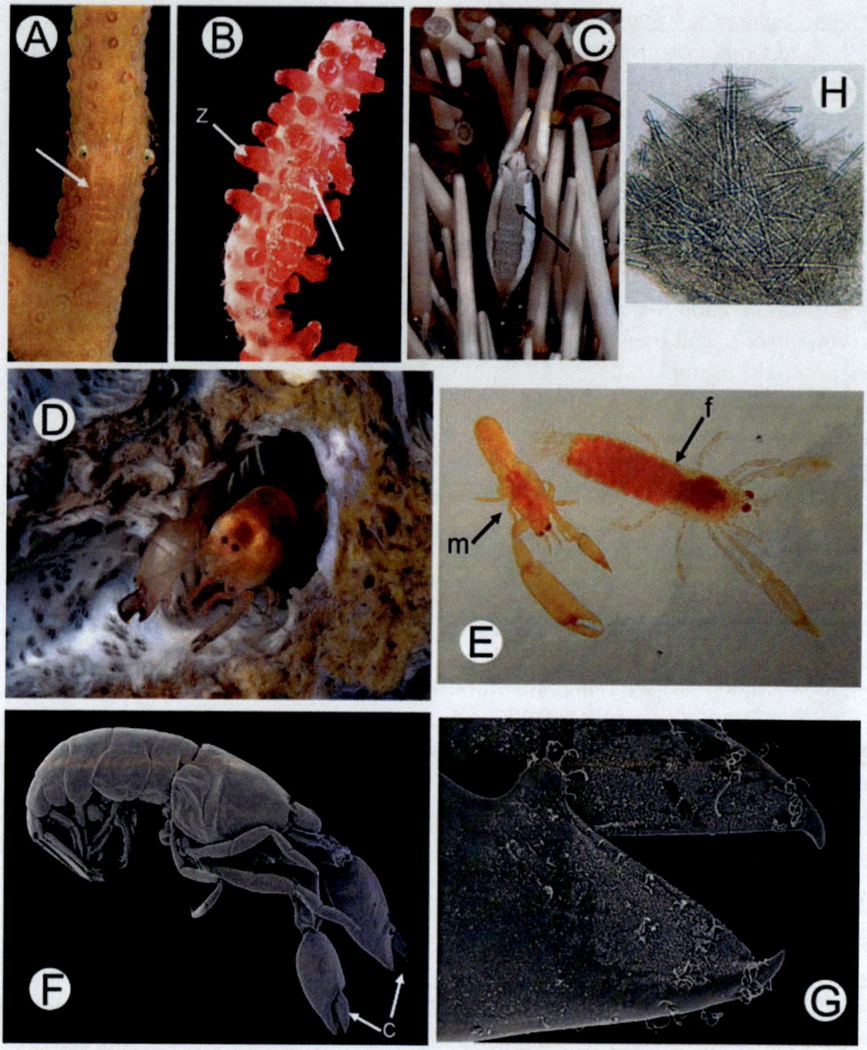

Plate 10.1 Symbiotic shrimps and their adaptations (1 of 2). (A) the transparent pontoniine *Harpiliopsis depressa* on a pocilloporid coral. (B) the pontoniine *Hamodactylus boschmai* with a similar color pattern as its pocilloporid coral host. (C) *Gnathophylloides mineri* (Palaemonidae *s.l.*) on a host urchin, *Tripneustes ventricosus*. (D) Individual of a *Synalpheus* species protruding from a sponge canal. (E) Male and female of the pontoniine *Typton carneus*, a micropredator of its fire sponge host, *Tedania* species. (F) SEM micrograph of a female *T. carneus*; note the shearing chelae; fingers magnified in (G). (H) Sponge tissues and spicules, common gut contents of *T. carneus*. Unmarked arrows in (A-C) point out the camouflaged shrimp. c, chelae; f, female; m, male; z, individual coral zooid. (A-D) courtesy of Zdeněk Ďuris, from collections sponsored by the Muséum national d'Histoire naturelle, Paris; (E-H) from Ďuris et al., 2011, courtesy of Zdeněk Ďuris and Creative Commons CC BY license

Endosymbiotic shrimps live hidden away in dark or dimly lit host microhabitats, e.g., within the tissues and shells of bivalve molluscs, the canals and spaces of a sponges, branchial chambers of ascidians, and the burrows of other crustaceans and vermiform invertebrates. Such species often have a pale white appearance due to a lack of coloration combined with opaque body tissues. Others are an orange or reddish color, or a combination of reddish stripes or splotches on an opaque body. This is very much like coloration of deep-sea species, both pelagic and benthic, and is likely due to similar selective pressures, i.e., a dark or dimly lit environment.

10.4 Sponge Dwellers

Sponges (phylum Porifera) provide a stable home for many species of shrimps. Sponges are sessile, often quite large, and estimates from growth-rate studies indicate that sponges may live for many years, from several years to decades to centuries (McGrath et al. 2018)! As sessile organisms, sponges cannot physically escape from predators and so have evolved a defensive skeleton which is variously composed of tough proteinaceous fibers and tiny sharp glasslike (siliceous) or calcareous spicules. Most sponges contain distasteful or toxic chemical compounds which makes them a disagreeable mouthful for potential predators. Sponges are full of canals and spaces of various sizes which are part of their unique "water-vascular" filtration system with which the sponge removes small organic particulate matter, microbes, and plankton for food (e.g., Brusca et al. 2016). Shrimps living within these canals and spaces are protected from most of their usual fish and invertebrate predators. They can feed on the material brought in by the water flow through the sponge, on the mucus which its host produces, and on sponge tissue itself. In the latter case, this would make them wholly or partially parasitic. However, such symbionts are in little danger of "eating themselves out of house and home" because sponges, like many lower organisms, have a remarkable ability for regeneration. Thus, it is not surprising that sponges are loaded with various symbionts, such as polychaete worms, ophiuroids, and crustaceans such as ostracods, amphipods, isopods, and shrimps (Ďuriš et al. 2011).

10.4.1 Caridean Sponge Associates

Sponge symbionts include several pontoniine genera such as *Periclimenaeus, Onycocaris, Typton, Thaumastocaris,* and *Periclimenes;* the palaemonoid *Anchistioides;* the alpheids *Alpheus* and especially *Synalpheus*; *Discias* (Disciadidae); and the hippolytids *Gelastocaris paronae* and a few *Lysmata species* (Bruce 1976b; Ďuriš et al. 2011). Most sponge shrimps are endosymbionts except for the *Periclimenes* and *Lysmata* species, such as *L. pederseni* (Baeza et al. 2016), which perch externally in the wider excurrent canals of the sponge. The

Stenopodidea has its share of sponge associates as well. There are only a few known in the Stenopodidae (one *Odontozona* and two *Richardina* species) but almost all the Spongicolidae, except for the mostly free-living, shallow-water genus *Microprosthema,* are associated with deep-water hexactinellid (glass) sponges (Saito and Takeda 2003; Goy 2010).

Shrimp symbionts may inhabit a single to several species of sponge hosts depending on their degree of specialization. The characteristics of a sponge species to which a shrimp associate adapts are mainly unknown. The shrimp has to overcome the diverse chemical defenses that sponges present, as well as other physical and ecological traits of the sponge. One factor that is important, however, is the space (diameter, volume) of the sponge canals. This was suggested as a feature involved in the host distribution of *Synalpheus* sponge associates (Duffy 1992; Hultgren and Duffy 2010) and *Paraclimenaeus* species (Cházaro-Olvera et al. 2013). In these studies, measurements of body size such as shrimp volume or size (height of the carapace) are highly correlated with sponge canal diameter or cross-sectional area of the sponge species. The shrimp fills and blocks the canal which allows it to defend its home from predatory intruders or shrimp competitors (Plate 10.1D). Many sponge shrimps, such as *Synalpheus, Paraclimenaeus*, and *Alpheus* species, have snapping claws. An intruder that suddenly finds a canal blocked by the cocked snapping claw of its aggressive defender might well be discouraged from trying to continue onward. Sponge shrimps, except for the eusocial ones discussed below, belong to taxa with "search and attend" or monogamous mating systems in which males defend their mating partner. A narrow canal would seem to be a favorable space to guard and protect a mate.

Various ideas about the diet of sponge endosymbionts, shut up within sponge canals, have been proposed throughout the years. What might the shrimps be eating? It was noted above that the water flow through the sponge promotes the growth of microbial growth within canals. Mucus secreted by the sponge as well as its tissues (albeit usually toxic or distasteful to most animals) might serve as a food source. The only way to really know is by examining the gut contents of shrimp associates, although visual identification of material is often difficult because of the mastication that occurs before arriving within the shrimp's stomach. However, sponge spicules are not digestible, and those taken in incidentally during feeding have been observed in the stomachs of some *Synalpheus* species examined by Duffy (2002). A thorough study on the pontoniine *Typton carneus*, a similarly tinted associate of the orange-red fire sponge *Tedania* (Plate 10.1E), showed that it feeds on sponge tissues. Nondigestible spicules and/or spongin fibers were identified in the stomachs of the 3 *Typton* species and other sponge endosymbionts (*Onycocaris, Periclimenaeus, Thaumastocaris*, and *Synalpheus*) (Ďuriš et al. 2011) (Plate 10.1H). One free-living species collected on sponges, *Cuapetes americanus*, was included for comparison. Seven of the 8 endosymbionts had sponge material in their guts, but the free-living (epibiotic) species did not.

The claw (chela) morphology of *Typton carneus* was closely examined in Ďuriš et al.'s study and found to be very well adapted for cutting and feeding on the tough spicule-ridden tissues of its host. The first chelae, as in all pontoniines, are adapted for picking up small particles and for grooming. In the female, both second chelae are equal in size, but in the male, one (major) is much larger than the other (minor) chela (Plate 10.1E). The outsized major claw of the male may be associated with female guarding in a "search and attend" type mating system. The minor second chela of the male, and both those of the female, resemble and probably function like a pair of shears (Plate 10.1 F–G). They are equipped with ridged cutting and rasping edges for chopping the sturdy tissue-containing spongin fibers and hard spicules of the host skeleton (Dŭriš et al. 2011). These authors suggested that other sponge endosymbionts have similar chela adaptations for feeding on the sponge host. Species of *Discias* (Disciadidae) are sponge associates and have unique first chelae with a disk-shaped dactylar finger that fits into a slot on the propodus of the claw (Fig. 3.18, Chap. 3). Brian Kensley (1983) suggested that this might be used as a scraper to remove and ingest the microbial film and mucus that usually coats sponge canals. More gut content and other dietary analyses of sponge symbionts are called for and are a wonderful opportunity for future research projects.

Although the spongin fibers and spicules might be difficult to digest, the many cells within that material would be nutritious food. Among the various sponge compounds ingested by a sponge-feeding shrimp are pigments that give some sponges their various typical colors. There is often a close correlation of color between that of an endosymbiont and its sponge host, e.g., the cherry-red color of *Typton carneus* and its similarly shaded host *Tedania.* A similar match of coloration between an ectosymbiont and its host usually suggests a camouflage function. That does not seem to be the case for shrimp endosymbionts, such as *T. carneus,* which are not normally exposed to visually hunting predators.

10.4.2 Spongicolid Sponge Associates

The other major group of sponge associates is the stenopodidean family Spongicolidae which, except for the shallow-water *Microprosthema,* usually occupy the central cavity of deep-water hexactinellid sponges. These vase-like sponges have a rigid lattice-like skeleton of fused siliceous spicules covered with a thin dermal membrane, hence their common name of "glass sponges." Collection of these delicate deep-sea sponges with their shrimp inhabitants intact is a difficult accomplishment. Notwithstanding, Tomioka Saito and colleagues have gathered considerable information about their shrimp inhabitants, e.g., the life cycle and mating system of the shrimp *Spongicola japonica* in relation to its host *Euplectella* (Saito et al. 2001, 2002). In these sponges, the still flexible skeleton of young sponges becomes rigidly stiffened with the growth and the development of longitudinal and circular bands of silicate (glass) bands of fused sponge spicules. The ~10–30 cm tall sponge is thus a rigid fine-mesh basket on which the living tissues of the sponge are supported (e.g., Brusca et al. 2016). Spongicolid pairs, consisting of a female with a somewhat smaller male, reside within this organic jail in an enforced monogamous relationship. They are securely protected from predators and provided with a constant flow of clean water that the sponge pumps through its body. The shrimps perhaps feed on fine particulate matter in sponge chambers and canals or possibly graze on sponge tissues. Thus, these sponge spongicolids may be commensals or parasites or something in between.

The shrimp pair is not always alone; a few to several juveniles and subadults may occur with them. Newly hatched and smaller juveniles can leave and reenter the sponge when they are smaller than the mesh of the sponge skeleton. They may disperse away from the home sponge or stay inside until their gonads begin to mature, at which point they leave the sponge, perhaps tearing through the skeletal mesh, and have a free-living period in which they search for a mating partner. While still immature, a male or female enters a young unoccupied sponge when its skeleton is still flexible and pliable enough for them to do so and are later joined by a partner of the appropriate sex. As the sponge skeleton hardens and becomes rigid, the pair are permanently enclosed in the sponge. The female goes through successive cycles of ovarian maturation, molting, mating with the male partner, spawning, and incubation of their embryos (Saito and Koya 2001). The pair will continue in this pattern (hopefully in harmony) for the rest of their life spans.

Spongicolids living within glass sponges have many reduced features compared to free-living relatives such as *Microprosthema*, as natural selection tends to eliminate useless or inadaptive features. Evolutionary trends in these and other spongicolid characters were nicely analyzed by Saito and Takeda (2003) and are summarized in Fig. 10.5. Spongicolids exoskeletons are uncalcified and membranous. The shrimps are protected in their sponge cages from predators and thus they can save the energy costs of calcifying their exoskeleton. Their eyes are generally reduced or degraded with relatively fewer corneal facets and little or no pigment (e.g., Fig. 3.43A, Chap. 3). They enjoy a nice clean environment in their sponge home because the sponge tissues filter out much potential fouling. As a result, their maxilliped 3 and pereopodal grooming brushes are sparse or absent. The exopods of the second and third maxillipeds are reduced or absent in these spongicolids, which might have something to do with reduced respiratory or gill-cleaning needs. The confines of their sponge domiciles do not require or allow much activity, and thus the sponge-living spongicolids have a much lower number of respiratory gills and epipods than the free-living *Microprosthema* spp. The extensive body spination characteristic of free-living stenopodideans is much diminished in deep-sea spongicolids within their predator-free shelter. Body spines may snag on host tissues as the shrimps move around within the sponge and thus are largely eliminated.

Sponge-dwelling spongicolid females produce relatively few large eggs (Fig. 10.6) with abbreviated or direct development compared to the numerous small eggs of their relatives, the free-living *Microprosthema* species, whose embryos hatch into planktonic larvae. The hatching as advanced larvae or juveniles from large spongicolid eggs is probably related to the patchy distribution of hexactinellid sponges in the deep sea. Planktonic larvae might disperse away from favorable areas of high concentration of the host species, but the offspring hatching from large spongicolid eggs remain in or near the areas where they can readily locate a host (Saito and Konishi 1999).

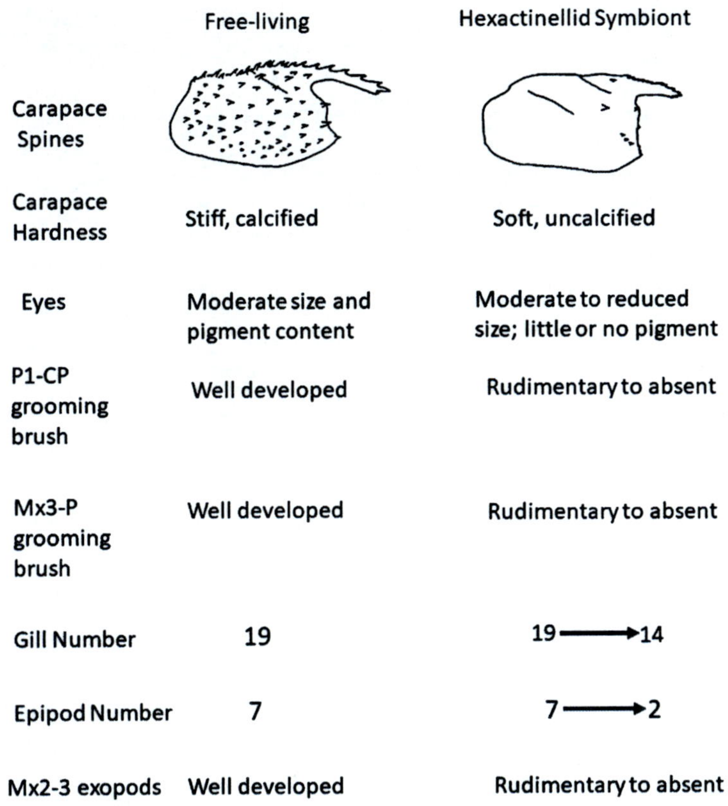

Fig. 10.5 Comparison of morphological characters between free-living spongicolids (*Microprosthema* species) with spongicolids modified by symbiosis with hexactinellid deep-sea sponges, based on Saito and Takeda (2003)

10.4.3 Social Organization in Synalpheus *Sponge Inhabitants*

The genus *Synalpheus* has over 160 species, and the majority are associates of sessile invertebrates such as crinoids, corals, and other cnidarians but especially tropical shallow-water sponges. The sponge-dwelling species show a great range of social systems. Some "asocial" pair-forming species live as a single heterosexual pair per sponge and are intolerant of other pairs (Fig. 10.7). Others are "communal," with several unrelated pairs inhabiting and defending their sponge home but relatively tolerant of each other. Some species, unique not only among shrimps but among any marine animal, are "eusocial." Several to thousands of closely related individuals live in colonies characterized by three attributes: (1) "reproductive skew" (only 1 or a few individuals of the colony produce offspring); (2) overlap of successive generations of offspring produced by the same female, the queen; and (3) a division of

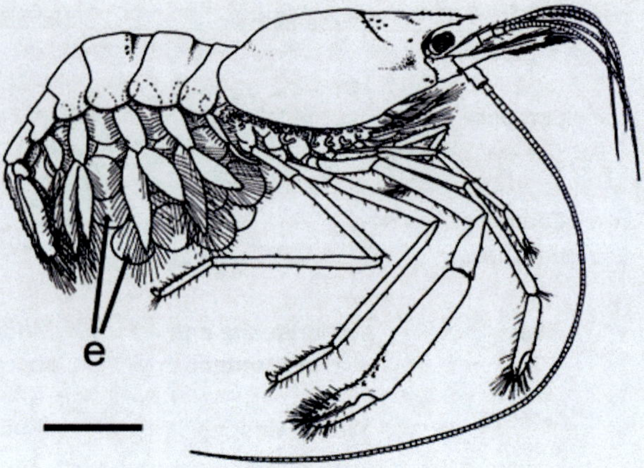

Fig. 10.6 *Spongiocaris hexactinellicola*, a spongicolid symbiotic with deep-sea hexactinellid sponges. Female with large embryos. e, embryos. Scale bar represents 5 mm. From Berggren (1993), courtesy of the Journal of Crustacean Biology

labor in activities such as defense, reproduction, and, to some degree, cooperative care of the young (Duffy 2002). Many hymenopteran insects (bees, wasps, ants), termites, some other insects (platyopodidid beetles, thrips, soldier aphids), and African mole rats are eusocial (Wilson 1975; Duffy 2002). To that list a number of *Synalpheus* sponge associates have been added by Emmett Duffy and his colleagues (e.g., Duffy 2002; Chak et al. 2017).

As a graduate student and postdoctoral fellow, Emmett was working with *Synalpheus* shrimps collected from tropical sponges. He often gathered from several to hundreds of individuals from the same sponge but curiously found only one female with embryos among these many individuals. I can only imagine the excitement he felt when it occurred to him that these might be eusocial species. These colonies have all the characteristics of eusociality, i.e., a single female (and usually a single male) that produces brood after brood of offspring which remain within the colony (overlapping generations). There is also division of labor in care of young and in colony defense. Emmett first showed, with allozyme analysis of genotypes, that colony members in one species, *S. regalis*, are siblings produced from a single female (the queen) and single male (the "king") (Duffy 1996). Subsequently, Duffy and his colleagues have uncovered considerable insight into *Synalpheus* sociality, especially with the West Atlantic and Caribbean "*gambarelloides*" species complex (e.g., Duffy 1992; Chak et al. 2017). However, observations on Indo-West Pacific sponge symbionts indicated that there may be numerous eusocial species in that diverse region as well (Duffy 2002; Didderen et al. 2006).

Synalpheus eusociality is most like that of termite societies, in which all individuals are diploid, and the mating system is serially monogamous, with a single reproductive queen and a single male mating partner. Most individuals, termed "helpers" or "workers," do not breed although they may have the capacity to do so, depending on the current social environment of the colony. However, the many nonreproductive individuals gain inclusive fitness, i.e., pass on their own genes, by staying in the colony and helping to protect other colony members, especially their mother the queen with whom they share 50% of genes.

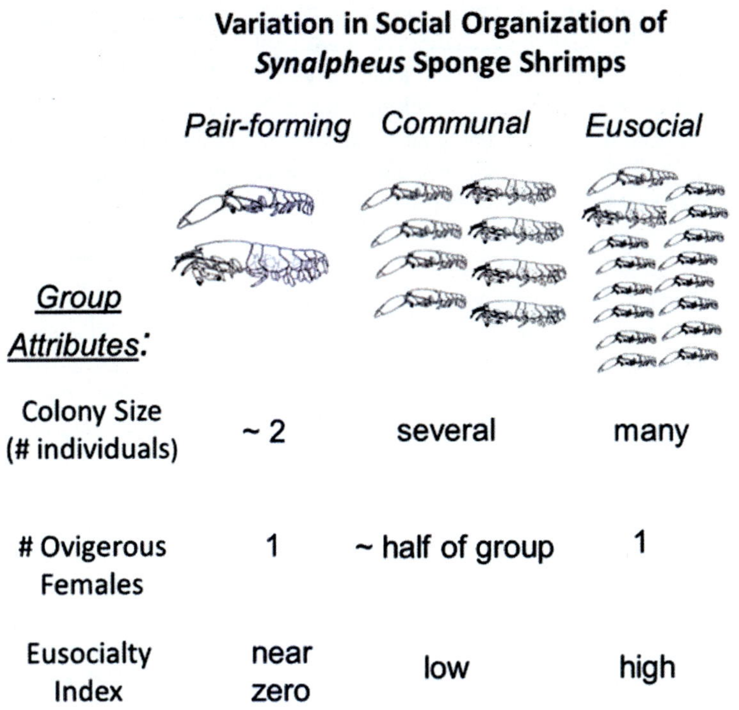

Fig. 10.7 *Synalpheus* sponge–shrimp social organization. Diagram shows colony (group) attributes of *Synalpheus* species that inhabit shallow-water Caribbean sponges, based on data in Chak et al. (2017). Comparisons are made of colony size, number of ovigerous females per colony, and a "eusociality index," a measure of reproductive skew (relative number of ovigerous females per colony or group)

Although not so strictly defined as in eusocial insects, there is division of tasks among colony members. All can defend the colony with their large snapping-claw weapon (Duffy 2002; Duffy et al. 2002). The larger (older) male helpers patrol around the colony with snapping claw cocked upon meeting another individual. They are often found near or around the exterior of the sponge, presumably to guard against intruders. Mass snapping events have been observed. A stimulus, such as an intruding congeneric *Synalpheus* or other possible competitor or predator, sets off snapping throughout the colony. Within the colony, there is little aggression among colony members. As they move around the canals of the sponge, individuals frequently meet, reciprocally touch antennules, and move around each other without discord. The normal activities of most individuals are grazing around the walls of the canals, autogrooming, and moving about within the colony. The gut content of a few species has been examined and contains, as in the *Typton* and other sponge associates discussed above, amorphous flocculent material and sponge spicules. This a mixture of sponge tissues and mucus indicates that their sponge is not only a home but also a food source, much as in wood-inhabiting termites. As they live in the excurrent canals in which the outflowing water has had plankton and organic detrital matter removed by the sponge, there may be little other food there for them to eat.

The larvae of eusocial insects are meticulously cared for by colony workers, and something similar but not so well developed occurs in eusocial *Synalpheus* colonies. The younger offspring tend to remain deeper (more protected) within the sponge and recently hatched offspring are found close to the queen. Genetic analyses of offspring and direct observations on captive colonies show that the queen is attended by a single large particularly aggressive male, probably the "king" of the colony. Thus, the individuals of eusocial *Synalpheus* species do belong to somewhat ill-defined and fluid castes marked by behavior and reproductive function.

The sexual system of eusocial *Synalpheus* was for some time difficult to characterize because of the problem of sexing individuals, except for reproductive females with developing ovaries and/or a brood of embryos. In eusocial colonies, the numerous young helpers as well as nonreproductive older ones are very difficult to sex. *Synalpheus* males lack the usual characteristic used for sexing males in carideans, i.e., the appendices masculinae on the second pleopod (Chap. 2). The shape of the first abdominal pleura (angular or hooked in the male) has been utilized, but it is somewhat subjective and difficult to use, depending on the age of the shrimp and the experience of the investigator. Another alternative is histology of gonads or simple dissections to observe the presence or absence of male ejaculatory ducts. Both are time-consuming, especially histology, and dissection of the often quite small helpers is challenging and often unreliable. Another alternative for sexing shrimps is the position of the gonopores which occur on different appendages in males and females (Chap. 2). Again, the small size of individuals and unremarkable appearance of eusocial *Synalphe*us gonopores make their observation very problematic, even with staining techniques when using the stereomicroscope standard for examination of morphology in shrimps and other small crustaceans.

Tóth and Bauer (2007) helped to resolve this problem by using scanning electron microscopy (SEM). The helpers are small enough so that several individuals can be mounted ventral size up and prepared on a single large observation stub, after which the position of the gonopores can be easily and clearly perceived by SEM. The several species we examined had a sex ratio of or close to 1:1, i.e., separate sexes (gonochoristic). In most of these species, there were a small number of individuals with both male and females gonopores, indicating that they had the potential to be either male or female. In one species, *S. filidigitus*, 3 queens examined had both sets of gonopores. In both species of the "*paraneptunus*" *Synalpheus* species group that we examined, all individuals had gonopores of both sexes (Tóth and Bauer 2008). Such individuals might have the potential to develop into male or female in a favorable demographic environment, or perhaps this condition indicates a protandric sexual system.

However, the great majority of individuals, except for the queen and king, will never reproduce ("high reproductive skew"). However, some of them may have the chance to do so by founding new colonies themselves. As in eusocial insects, new colonies are begun when possible to insure the survival of colony genotypes through time. The usual mode of establishing new colonies is by dispersion of potentially reproductive individuals. The mode of dispersal and invasion of new habitat in most marine invertebrates is by planktonic larvae. However, in eusocial *Synalpheus*, development is direct with hatching as immature juveniles, which guarantees colony cohesion and survival. Eusocial colony members are rarely found outside of the sponge. Solitary individuals wandering around on the sea floor exposed to predators would be very hard to observe and collect.

The question of dispersion to establish new colonies was examined by Eva Tóth with a field experiment reported preliminarily by Tóth and Bauer (2007). She paired up larger male and female helpers from different colonies in a depopulated sponge (Plate 10.2A). With time, the female helpers started developing ovaries and a few spawned some eggs, as did female helpers placed alone in a defaunated sponge. Eva concluded that in a normal colony, the queen somehow suppresses gonadal development in larger helpers. We hypothesized that some large helpers from established colonies, which have no or very little chance of reproducing, leave the colony in search of the rarely unoccupied sponge species preferred by the shrimp. If a male enters the sponge first, it may allow a potential female to enter and vice versa. Individuals of the same sex presumably will be prevented from entering. The members of the invading pair will be the queen and king of the new colony. However, this type of experiment need to be repeated and properly published to adequately test the hypothesis these initial findings raised.

Intriguingly, the rate of parasitism in obligate "*gambarelloides*" *Synalpheus* species by bopyrid isopods was found to be about seven times that of free-living *Synalpheus* (17% versus 2.5% (Duffy 1992). The higher incidence of infection may be due to the much greater population density of the sponge shrimps, but it is uncertain why. It would be understandable if parasite larvae remained in the colony and could easily infect shrimps within the colony. But bopyrid larvae are planktonic and have an intermediate planktonic copepod host. So, unless all these life cycle components are present within the sponge (unlikely), then something else is going on. However, high rates of parasitism occur in highly aggregated populations of free-living shrimps such as *Palaemon paludosus* (Beck, 1979) and *Heptacarpus paludicola* (pers. observation). As will be recalled from Chap. 9, these parasites have a devastating impact on shrimp fecundity. It is certainly an interesting question worthy of investigation.

What is the origin of eusociality in *Synalpheus* species? To investigate this question, Chak et al. (2017) examined the social organization and phylogenetic relationships of 39 species of the "*gambarelloides*" *Synalpheus* species complex. The group had 17 asocial pair-forming, 7 communal, and 6 eusocial species with 9 species variously intermediate between one social condition or other. The sociality of these species was described quantitatively with different colony attributes summarized in Fig. 10.7, including: (1) colony size (number of individuals in the sponge, e.g., two in asocial, several in communal, and many in eusocial species); (2) number of ovigerous females in the group or colony (just one in asocial, ~ half of individuals in communal; usually just 1 in eusocial species); and (3) a eusociality index, which is a measure of reproductive skew relating the number of ovigerous females relative to colony size. Combined with a molecular phylogeny, this information led to the

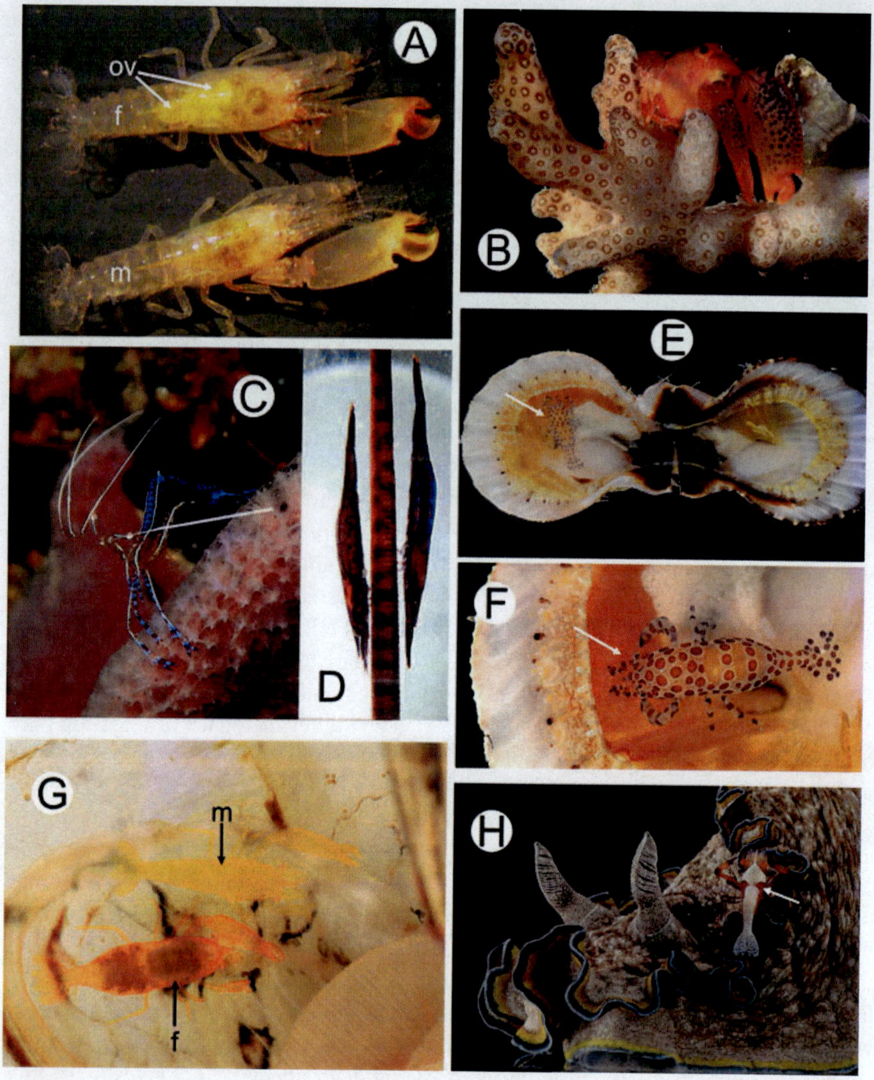

Plate 10.2 Symbiotic shrimps and their adaptations (2 of 2). (A) Sponge associate *Synalpheus rathbunae*; male and female (developing ovary) colonizing an unoccupied sponge host. (B) *Alpheus lottini*, a stony coral associate. (C) Cleaner shrimp *Ancylomenes pedersoni* (note bright white flagella of antennules and antennae, as well as white leg and body stripes). (D) Sea urchin associate *Tuleariocaris* species in typical position perched head-down on similarly colored spine of *Diadema* species. (E) Bivalve symbiont cf. *Anchistus pectinis* inside shell of living cf. *Bractechlamys vexillum*. (F) Close-up of cf. *Anchistus pectinis* inside host. (G) Male and female cf. *Neoanchistus* or *Anchistus* spp. inside bivalve host. (H) Nudibranch associate *Zenopontonia rex* on it host *Glossodoris* species. (A-B) are alpheids and (C-H) are pontoniine palaemonids. Unmarked arrows in (E-F, H) point out shrimp. f, female; m, male; ov, developing ovaries. (A) courtesy of Eva Tóth; (B-F, H) courtesy of Zdenek Ďuris, from collections sponsored by the Muséum national d'Histoire naturelle, Paris; (G) courtesy of Arthur Anker

conclusion that eusociality has evolved independently 4 times in the *gambarelloides* species complex. Communal living of several unrelated pairs is not a necessary intermediate step between asociality and eusociality in *Synalpheus*. Rather, both communal and eusocial species have evolved directly from asocial species. In the pair-forming ancestors of eusocial species, evolution of direct embryonic development with hatching as crawling juveniles allowed the retention of offspring. These individuals help with colony chores, but their reproductive function is suppressed. Thus, eusociality evolved from the “subsocial” route (extended family groups; offspring “stay at home to help”). The alternative would be the “parasocial” route in which several individuals form a common breeding group which becomes dominated by a single female, the queen, as in some wasp species (Krebs and Davis 1993; Queller and Strassmann 1998). This is not known in *Synalpheus*. The biology and natural history of *Synalpheus* and perhaps other eusocial shrimps is still relatively unexplored, as are other interesting social systems, especially in the Indo-West Pacific, where diversity of sponge and shrimp species is quite high.

10.5 Shrimps Associated with Cnidarians

The sessile cnidarians, e.g., corals and sea anemones, host a wide variety of shrimp associates, especially the pontoniines. The basic unit of a sessile cnidarian is the soft-bodied polyp which consists of a basal disk for attachment and a columnar body. The latter terminates in a flat oral disk with a central mouth surrounded by stinging tentacles. Many groups of sessile cnidarians are colonial, i.e., consist of an interconnected group of individuals (zooids) with a common skeleton, e.g., stony corals and sea fans. Others are solitary individuals, especially the anemones. These organisms make good hosts because, above all, they offer protection from predators such as fishes and other crustaceans. The host tentacles are loaded with stinging cells (cnidae) which discharge venomous and/or barbed threads (nematocysts). Shrimps are readily stung and eaten by cnidarians, but those which live among the tentacles have evolved mechanisms for acclimation to the host and vice versa. Cnidarian shrimp associates are mainly ectosymbionts as the main body space of the host is the digestive cavity, and that is not a very hospitable space for a shrimp to live!

The cnidarian may be a good feeding place for associated shrimps. Cnidarians constantly secrete mucus into which detritus, bacteria, and other particulate matter become entangled and need to be swept off the animal with its epidermal cilia. It is a form of self-cleaning as the cnidarian has no other appendages to do. The organic matter caught up in mucus are an excellent food source for shrimps. The cnidarian has a one-way gut, so that undigested prey is regurgitated back through the mouth along with the polyp’s feces. These are energy-rich foods for shrimps and other associates. The tissues of shallow-water cnidarians are full of intracellular algal cells (zooxanthellae) which have a mutualistic symbiosis with the host. There is some equivocal evidence that certain shrimp associates feed on host tissues (parasitism) which would be inadaptive for the coral or anemone if extensive.

Much of the voluminous literature on cnidarian shrimp associates is about their taxonomy (species descriptions) and systematics (classification and phylogeny). Species distributions on hosts have received considerable attention, e.g., Bruce (1972a, 1976a, b) and many other papers since), Marin and Chan (2014), Kou et al. (2015), Horká et al. (2016, 2018), and Rauch et al. (2019). There are certainly many more shrimp associates of colonial anthozoan hosts to be discovered and described. An updated and worldwide analysis of shrimp–host associations greatly needs a comprehensive analysis (any takers?).

10.5.1 Shrimp Associates of Corals and Related Colonial Anthozoans

The cnidarian class Anthozoa contains a number of related but often dissimilar taxa of colonial organisms whose common name includes "coral," such as the Scleractinia ("stony" reef-building corals), Alcyonacea ("soft" corals), Antipatharia ("black" corals), and Milleporidae ("fire corals"). Other colonial anthozoans are the gorgonians ("sea fans, "sea whips"), pennatulaceans ("sea pansies, sea pens"), zoantharians, and others. Sea anemones usually occur as single individuals and will be treated separately.

Relatively little is known about the natural history of shrimps associated with colonial anthozoans. The shrimps are difficult to observe hidden in their structurally complex host shelters. Their small size, usually nocturnal activity, and camouflage make observations on their activities in the field very difficult by snorkeling, SCUBA, submersibles, or ROVs. Laboratory setups of host and associates are an alternative. The maintenance and culture of corals and other colonial anthozoans is not easy, although techniques for doing so keep improving. However, many of these difficulties have been overcome for some species and have yielded some fascinating life history information.

It is already known or assumed with some confidence what the benefits for the shrimps in coral associations are shelter from predators and in many cases a food supply. What are the benefits and costs for the hosts? Shrimps, along with crab symbionts, can maintain coral health by helping to prevent particulate fouling and sedimentary suffocation of their hosts. Even in clear tropical coastal and oceanic waters, there are many fine particles, both of organic (microbial organisms, detritus) and inorganic origin (sediment). Corals do have cleaning mechanisms; they constantly secrete mucus in which fouling material accumulates. This aggregate is then sloughed off the coral polyp by the sweeping actions of their epidermal cilia. Polyps push off sediment whenever they move in and out of their zooid cups. However, wave action, storms, and sediment-laden runoff from nearby rivers often deposit suffocating material on coral reefs which may overcome their natural defenses. However, at least some coral associates assist their hosts in removing particulate matter. Two common associates of acroporid and pocilloporid corals in the Pacific are trapeziid crabs and the shrimp *Alpheus lottini* (Plate 10.2B), both of which occur in heterosexual pairs. The trapeziid crabs have been shown to increase coral survival by actively removing sediment by kicking it off with their back legs or shoving it away with their chelipeds (Stewart et al. 2006). The shrimp *A. lottini* is suspected of doing similar sediment removal, perhaps by picking or shoveling off sediment with its chelipeds. Coral colonies with both trapeziid crabs and *A. lottini* have better health and survival compared to those without (Stier et al. 2012). The role of these and other coral symbionts in sediment removal will become even more important with the intensified nearshore sedimentation from urban development and sewage discharge in tropical regions (Fabricius 2005).

Trapeziid crabs and *A. lottini* often co-occur on coral colonies in the Pacific and are aggressive not only to conspecifics but other species as well. They are very protective of their own space and their host. Coexistence of *A. lottini* with the crabs is made possible by a form of interspecific communication, termed "crabese" by Vannini (1985), by which they can be in close physical contact without any agonistic behavior. To occupy an area near a crab, the shrimp rubs its claws against those of the crab or just places its body in contact with that of the more dominant crab. If these signals of subordinance fail, the shrimp must go elsewhere as it is always the loser in an aggressive encounter with a crab.

However, *A. lottini* is not subordinate to some other symbionts inhabiting their coral homes. The narrow spaces between the many clumps of the aptly named "cauliflower coral" *(Pocillopora damicornis)* provide good shelter for the gobiid fishes *Paragobiodon echinocephalus* and *P. lacunicolus*, two *Trapezia* species, and the orange-red *Alpheus lottini* (Lassig 1977). In this system, the shrimp is appeased by *P. lacunicolus* but curiously has little interaction with *P. echinocephalus*, with which it often co-occurs in close physical proximity. The shrimp occasionally attacks *P. lacunicolus* but is usually placated by the fish when coming into physical proximity or contact. The fish shows subordinance by shivering when in contact with the shrimp, a behavior also observed in the fish's mating behavior. Peaceful coexistence among fellow symbionts allows more of them to inhabit the same coral head. As movements of the fish may increase sediment removal, host survival may be enhanced, and thus coexistence is mutually beneficial for both symbionts and host.

Alpheus lottini and trapeziid crabs also defend their coral host (Glynn 1976, 1980, 1987). The "crown-of-thorns" starfish *Acanthaster planci* is a major coral predator in the tropical Pacific. Although the starfish prefers non-branching corals, in some habitats such corals are rarer than branching pocilloporids inhabited by the two crustacean symbionts, both of which protect their host. When *A. planci* approaches a coral with an *A. lottini,* the shrimp moves to the edge of the colony and confronts the starfish by snapping at it and pinching its spines. This usually discourages the starfish, which moves away. The shrimp can smell the starfish at a distance and prepare to defend it if the predator approaches upcurrent of the coral and its symbiont. Another pair-living alpheid, *Synalpheus neomeris*, similarly defends its host, the soft coral *Dendronephthya hyalina* (Alcyonacea), from two species of nudibranch predators (Nakano and Fujii 2014). When the nudibranchs arrive on the coral, they begin to feed up to about one minute before the snapping shrimps attack and drive them away. The nudibranchs may make several attempts but finally give up after being repulsed each time by the shrimps. Many species of pontoniine shrimps are coral symbionts; they typically have large second cheliped weapons, some with snapping ability. It would not be surprising if some of these pontoniines also defend their coral hosts.

Several shrimp associates may share the same coral head or colony, but apparently avoid competition and aggression among themselves by utilizing different microhabitats within the coral (niche partitioning). Patton (1974) studied the assemblage of shrimps inhabiting *Pocillopora damicornis* around Heron Island, Australia. He classified the various species according to their body morphology, daily activity, and the area occupied on or within the coral.

One group, termed "agile" species, has the typical body form of an epibenthic shrimp like *Periclimenes scriptus* described by Bruce (1976a) (Fig. 10.3a). One of them, *Periclimenella spinifera,* is a free-living species that uses the coral as an occasional perch but is found on other kinds of substrates around the reef. *Cuapetes amymone* and *Periclimenes madreporae* are obligate associates of a variety of coral species (Bruce 1972a). They are rarely found on the coral at night, indicating they may leave the host at that time to feed. *Thor amboinensis* is an obligate symbiont but associates not only with corals but also sea anemones and other sessile invertebrates. Two other species, the pontoniine *Harpiliopsis beaupresii* and *A. lottini*, live among the coral branches and are obligate coral symbionts. The former has a depressed, flattened body form while *A. lottini* is laterally compressed, both of which are adaptations for living in narrow spaces. They were considered as "sedentary" species by Patton, especially the sluggish *H. beaupresii,* although *A. lottini* actively rushes out to attack starfish predators.

Some crustaceans symbiotic with corals live within pits or cysts in stony (scleractinian) corals. One family of crabs, the Cryptochiridae, with 52 species, can be fairly common, especially *Hapalocarcinus marsupialis* (Potts 1915a; Terrana et al. 2016). However, only one shrimp cyst dweller has been found among shrimp symbionts, the most specialized and peculiar coral associate of acroporid corals, *Paratypton siebenrocki,* first illustrated by Borradaile (1921) (Fig. 10.8). However, most of our knowledge on this species is from Bruce (1969a) with a few more details in Bruce and Trautwein (2007). The adult male and female are entrapped for a possibly long life within a bichambered cyst at the base of the coral's branches. Not surprisingly, their bodies are mostly transparent with little coloration, with the smooth exoskeleton largely free of spines typical of extreme endosymbionts. The cyst, formed by coral tissue growth, is difficult to discern on the exterior of the coral as it causes little noticeable swelling.

The male occupies the smaller upper chamber while the female is found in the larger lower chamber. The female shrimp has a narrow cephalothorax and a large broad abdomen which can accommodate numerous embryos. Her exoskeleton is membranous, indicating little calcification, and, with its globular and swollen appearance, shows a striking superficial resemblance to *H. marsupialis*, the cyst crab, suggesting convergent evolution. The smaller male is similar to the female in overall appearance, but cephalothorax and abdomen are more normally

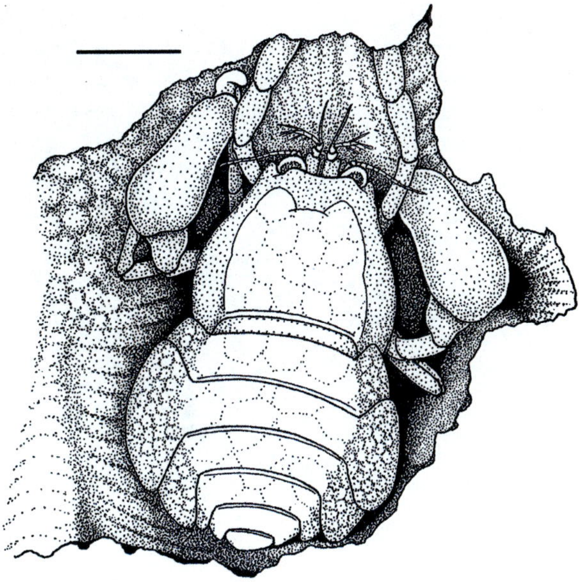

Fig. 10.8 *Paratypton siebenrocki* female exposed in the lower chamber of its coral cyst. Scale bar represents 2 mm. From Bruce (1977)

proportioned, the latter bearing broad enlarged pleopods. Unlike the female, which is sluggish and moves feebly when removed from the cyst, the male is quite capable of swimming. Bruce hypothesized that the male with its large pleopods produces a water current passing through both cyst chambers. Each chamber is equipped with a tiny pore (1.5 mm diameter) for entry and exit of the respiratory and food-bearing current. The small size and weak development of the mandible and its processes indicates feeding on small food particles (e.g., tiny plankters, detritus) brought in by the male-produced current. Most details about the natural history and life cycle await discovery, e.g., how the shrimp stimulates and controls formation of the cyst, which sex arrives first, and larval development. The female probably produces multiple broods, which means that the male would have to leave his room and enter the boudoir of his mate to do his duty. Fortunate indeed will be the marine biologist who discovers enough of these fascinating animals to study using a cleverly contrived laboratory mesocosm.

10.5.2 Anemone Shrimps

A number of shrimp species have found that association around or on sea anemones is a good place to find both refuge and food. A fish predator diving after a shrimp on an anemone is likely to encounter a mouthful of stinging tentacles. As with corals, the mucus, feces, undigested food scraps, and even tissues of a large anemone are a good food source. Several shrimp species have evolved symbiotic relationships with sea anemones, including varying degrees of commensalism, mutualism, and

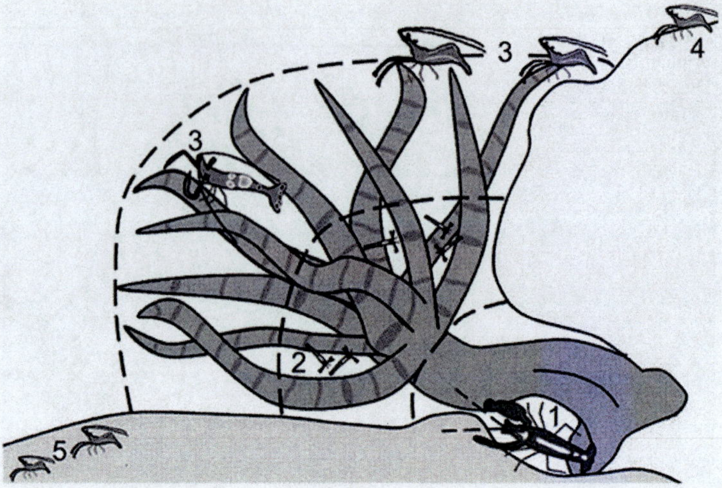

Fig. 10.9 Shrimp associates of *Bartholomea annulata*, a sea anemone which typically projects out from a crevice beneath rock or coral. The anemone and immediate surroundings may be divided into zones: (1) anemone column; (2) inner tentacle crown: (3) outer tentacle crown; (4) projecting rock or coral rubble forming crevice with soft substrate bottom; (5) soft substrate bottom next to anemone. Shrimp associates typically occupy different zones: *Alpheus armatus* beneath anemone column (zone 1); mysidan *Heteromysis actiniae* (zone 2); pontoniine *Periclimenes yucatanicus* (zone 3); and cleaner shrimp *Ancylomenes pedersoni* (zones 3,4,5, depending on position in social hierarchy). Modified from Gilpin and Chadwick (2020), courtesy of the Journal of Crustacean Biology

possibly parasitism. For example, in a coastal area of China, six species of shrimps routinely associate with 13 species of sea anemones in different combinations (Guo et al. 1996). In the tropical West Atlantic and Caribbean, various shrimp species are found on or around large common anemone species (e.g., Limbaugh et al. 1961; Gilpin and Chadwick 2020). In tropical waters, most are from pontoniine genera, especially *Periclimenes* and *Ancylomenes*, but also species from genera such as *Lysmata* and *Thor.* As we shall see later, many of them are fish cleaners. Although most anemone–shrimp associations have been best studied in shallow tropical waters, several have been observed in higher latitudes as well, especially involving thorid genera such as *Spirontocaris*, *Heptacarpus*, and *Lebbeus,* as well as *Pandalus* species (Jonsson et al. 2001; Jensen 2014).

An anemone host may have a single or multiple species of shrimps associated with it, and multiple individuals of one or more of those shrimp species may occur on a single host individual (Fig. 10.9). Some anemone shrimps are host specific while others may occur on a variety of anemone hosts (e.g., Guo et al. 1996; Nizinski 1989; Wirtz 1997; Jonsson et al. 2001; Khan et al. 2003; Gilpin and Chadwick 2020). These shrimp–anemone associates usually do not occur on all host individuals in an area; the factors which account for this patchy distribution are not well known. The location on the anemone by a shrimp species is not random. Different species are found on various areas of the anemone, e.g., among or just under the tentacles, on the column, or nearby on the substrate around the base of the anemone. For example, in the corkscrew anemone *Bartholomea annulata*, several individuals of *Ancylomenes pedersoni* may occupy various locations on or near the anemone depending on the social status of the shrimp individual (Gilpin and Chadwick 2020) (Fig. 10.9) (Plate 10.2C). This shrimp is a fish cleaner, and the most favorable spot for attracting fish clients is the outer tentacle crown, which is occupied by the largest individuals, usually reproductive females. Smaller individuals, such as nonreproductive females, males, and juveniles, are not on the anemone itself, but at varying distances around it on the substrate to which the anemone is attached. The smaller the individual, the farther from the anemone it is located. The social positions and locations are enforced by aggressive behaviors with larger individuals dominating the smaller ones (Mahnken 1972; Gilpin and Chadwick 2020). Distance from the protective shelter of the anemone tentacles presents more danger from fish predators. On the other hand, shrimps in these less favorable locations may leave at night to search for more favorable positions on other *B. annulata* in the vicinity. Another fish cleaner, *Periclimenes yucatanicus*, can also be found on the outer tentacle crown of *B. annulata*, whereas *Alpheus armatus* is found near the base of the anemone. Shrimp-like mysidaceans (Peracarida) can be found in the more basal part of the tentacle mass (Fig. 10.9).

The host occupied by a shrimp species and the natural history of the association may vary with locale. The studies by Nanette Chadwick and her colleagues on *A. pedersoni* were done in Caribbean waters (Virgin Islands). In Bermuda, Nizinski (1989) examined the natural history of a species identified as *Periclimenes anthophilus,* but which has been synonymized since with *A. pedersoni*. One might expect a very similar biology if they are from the same species. The anemone *B. annulata* was as common in the same reef habitat in Bermuda as *Condylactis gigantea*, but *A. pedersoni* ("*P. anthophilus*") was found exclusively on *C. gigantea*. Unlike *A. pedersoni* in the Virgin Islands, no fish-cleaning behaviors were observed in the "*P. anthophilus*" Bermuda population. One to several shrimps per anemone roamed freely among the tentacles of the host. No indication of territoriality, aggression, or subdivision of microhabitat on the anemone according to size or sex was observed in the Bermuda location. If the synonymy of "*P. anthophilus*" with *A. pedersoni* is correct, what factors might account for the differences in host choice and social organization observed between the Nizinski and Gilpin and Chadwick studies? Genetic and adaptive differences between the two populations in Bermuda and the Caribbean sites are possibilities. Disparities in social organization might be due to the morphology of the two anemone hosts. *C. gigantea* has large opaque tentacles, while *B. annulata* bears thin transparent ones. Nizinski hypothesized that it might be difficult and energetically expensive for "*P. anthophilus*" to visually survey their immediate surroundings among the thick tentacles of their host and thus exclude conspecifics, whereas the shrimps on *B. annulata* might be able to do so. Shrimp visual acuity is notoriously poor, even in clear tropical waters during the day, but it might be enough to identify a general shrimp-like shape.

What benefits, if any, does the host anemone receive from the association with shrimps? Shallow-water anemones, like corals and many other shallow-water cnidarians, have a mutualistic symbiosis with intracellular algae (zooxanthellae). The zooxanthellae provide additional food, via photosynthesis, to the anemone which is an important supplement to that taken by its feeding activities. Like all plants, zooxanthellae need nitrogen in order to grow, which is supplied by the anemone's cellular excretion as well as by uptake of nitrogen from the surrounding water. Most marine animals, including shrimps, excrete ammonia, an excellent source of nitrogen. In the Indo-West Pacific, many anemones host symbiotic fishes (*Amphiprion* and *Premnas*), which live among the anemone tentacles. The ammonia excreted by the fishes, which move among anemone tentacles, greatly enhances the growth of the host anemone and their zooxanthellae (Porat and Chadwick-Furman 2005). The small anemone shrimps, on the other hand, provide a relatively minor amount of nitrogen compared to that of the fishes. However, there are no anemone fishes in the West Atlantic and Caribbean areas where several studies on anemone–shrimp associations have been done. In this case, shrimp associates provide enough excess nitrogen above levels found in the surrounding water to enhance the growth of the anemone's symbiotic algae and thus tissue growth of the anemone itself (Spotte 1996). Some of these shrimps, especially *A. pedersoni*, are fish-cleaning species. Fish clients of anemone shrimps may excrete an even greater amount of ammonia in the area around the anemone while they are being cleaned, supplying vital nitrogen for the intracellular symbionts of the host anemone (Cantrell et al. 2015).

Another benefit that some anemone shrimps give to their hosts anemones is protection. A classic case is the relationship between the red and white *Alpheus armatus* and its host, the corkscrew anemone, *Bartholomea annulata* (Smith 1977). The anemone is usually attached by its basal (pedal) disk in a crevice below a rock or coral rubble above a sand bottom. *A. armatus*, which usually occurs as a heterosexual pair, resides near the base of the anemone. The shrimp actively defends the anemone from the conspicuously red fireworm *Hermodice carunculata,* a major predator of sessile cnidarians, such as corals, zoanthids, and sea anemones. The worm's body bristles with an array of toxic setae which is its defense against its own predators, i.e., fishes. When the worm attacks an undefended *B. annulata*, it begins by eating the anemone's tentacles and then, if unsatiated, goes on to eat other parts of its prey. The anemone tries to defend itself by contracting, which discharges tentacular stinging cells and gastric threads (acontia) through its mouth. But these defenses do not deter the polychaete predator very much. Unprotected anemones may survive and regenerate lost body parts, but a *Hermodice* attack is certainly a major blow to its chances of survival and reproduction.

The story is quite different when the anemone is defended by *A. armatus*. In a contest analogous to David versus Goliath, the 3-cm long shrimp rushes out to attack when the 15–30-cm worm approaches its host. *A. armatus* actively defends its host by rapidly snapping its major first cheliped and repeatedly pinching the worm with its other chelae Caribbean (Smith 1977; McCammon and Brooks 2014). Recall from Chap. 2 that the alpheid snapping claw emits a high energy jet of water that can stun, damage, and even kill an opponent or prey. Examination of *H. carunculata* assaulted by *A. armatus* in laboratory experiments showed a variety of tissue lesions and extruded internal organs. Some of the predators died soon thereafter or at least retreated as far away from the anemone and its defender as possible. Field observations showed that *B. caruncula*ta with *A. armatus* defenders have a much greater survival than those without.

Do the anemones act as a food source of hosted shrimps? Shrimps have been reported picking and nibbling at their host with their chelipeds. Usually, no harm seems to be done and they may simply be eating the mucus which is continually secreted by the host as a cleaning mechanism (e.g., Crawford 1992; Nizinski 1989). However, some anemone shrimps do eat host tentacles. Observations have been reported on tentacle clipping and consumption by *Ancylocaris brevicarpalis* on the host *Entacmaea quadricolor*, by *Ancylomenes holthuisi* on its preferred host *Stichodactyla haddoni*, and possibly by *Spirontocaris synderi* on *Urticina coriacea* (Fautin et al. 1995; Khan et al. 2003 and Jensen 2014, respectively). These observations were mainly done in the laboratory, where confined captive animals may not behave as they do in nature, and need to be confirmed with field studies. Widespread tentacle damage has not yet been verified in the field.

Crustaceans, including shrimps, are readily stung and eaten by sea anemones. How can anemone shrimps that occupy or roam through host tentacles not be affected by the host's stinging cells? Their exoskeleton per se is not a sufficient barrier to the nematocysts discharged by the anemone. Although it usually resides below the tentacle crown of its host anemone *Bolocera tuediae*, *Spirontocaris liljeborgii* sometimes blunders into and becomes entangled in its host's tentacles, receiving many stings in the process. However, the shrimp quickly jerks its way out of the tentacles without any apparent damage (Jonsson et al. 2001). However, in most other anemone shrimps, the shrimp and host must become acclimated if the shrimp is to stay or wander through the host tentacles. Protection of Indo-Pacific anemone fishes (*Amphiprion*) is due to their thick mucus coat; various compounds in the mucus inhibit the anemone nematocysts from discharging. If the mucus coat is interrupted by an injury or otherwise removed, the fish loses its immunity. It is not completely established if the important mucus components are innate (evolved) adaptations in these fishes or are in part acquired from the host mucus during acclimation (e.g., Fautin 1991; Elliott and Mariscal 1997). However, shrimps do not have a mucus coat, so how does a shrimp acquire immunity from a host anemone?

If unacclimated *Periclimenes rathbunae* and *Ancylomenes pedersoni* ("*P. anthophilus*") are introduced into a host anemone, they appear to be stung initially (Levine and Blanchard 1980). Individuals already associated with their preferred hosts are not stung if presented to different host individuals of the same species, but naive shrimps are susceptible. However, newly introduced shrimps gingerly limit their contact with the anemone and gradually become acclimated to it—or vice versa. To be accepted by an anemone host, the unacclimated shrimp endures a few hours of being stung and studded with host mucus. The investigators concluded that the shrimps might be covering their bodies with host mucus to prevent the anemone from recognizing them as foreign objects (chemical camouflage, as in anemone fishes). However, caridean shrimps, especially palaemonid shrimps, are often fastidious groomers. It seems improbable that they would foul themselves with mucus. Levine and Blanchard found the two shrimp species in their study were not able to acclimate to anemone species that they did not normally associate with.

In *A. pedersoni* (*P. anthophilus")* hosted by *Condylactis gigantea*, the "mucus" observed in the previous study was composed of discharged stinging cells of the host (Crawford 1992). He was able to eliminate acclimation simply by wiping the shrimp's exoskeleton with tissue paper. However, when an acclimated shrimp molted, it remained acclimated, but its molt skin (exuviae) did not. Crawford proposed that *P. anthophilus* and other anemone shrimps continually secrete a protective substance when exposed to an anemone. Newly molted shrimps have the new exoskeleton prepared and coated with the protective secretion before the old molt skin is shed. Can the substance purportedly secreted by anemone shrimps (through tegumental glands?) be identified and tested independently? If placed on nonsymbiotic shrimps or other objects, does the material prevent them from being stung by host anemones? Do some species of anemones serve as hosts because they are fooled by the chemical camouflage of the shrimp while other species are not? These questions are important to help explain the taxonomic distribution of anemone shrimps and their hosts, as well as the evolution of shrimp–anemone associations. I discussed the issue of acclimation in Bauer (2004), but unfortunately, not much work has been done since.

10.5.3 Jellyfish Shrimps

The medusa (jellyfish) is the pelagic life form of many cnidarians, except for the Anthozoa (corals, sea anemone, sea fans, etc.) which lack one in their life cycles. The sea abounds with a diverse array of jellyfishes, which might seem to be good places for shrimps to perch and associated with, as they do with the completely benthic anthozoans. The relatively few shrimp species found living with jellyfishes are mainly associated with the cnidarian class Scyphozoa, Order Rhizostomae. Jellyfishes of this group have long frilled extensions of their mouths, the oral arms, which are complexly fused to form a mass of mini-mouths for feeding. When preserved, many resemble a cauliflower or broccoli plant. One genus, *Cassiopea*, is more benthic than pelagic and is usually encountered in shallow tropical waters upside down, with its fused oral arms upward. In this position, they somewhat resemble large sea anemones, which led A.J. Bruce (1972b) to suggest why some pontoniine shrimps (e.g., *Ancylomenes holthuisi* and *Periclimenes tonga*) are occasionally found with them. More shrimps species found on rhizostomes and morphologically similar jellyfishes include *P. nomadophila*, *P. pavai*, *P. yucatanicus*, and *Ancylomenes pedersoni*, a few hippolytids (*Exhippolysmata ensirostris*, *Latreutes mucronatus*, and *L. anoplonyx*), and the pandalid *Chlorotocella gracilis* (Hayashi and Miyake 1968; Bruce 1972b; Filho et al.

2008). These are benthic shrimps and their presence on jellyfishes is possibly only occasional or accidental.

However, one exception is *P. pavai*, an obligate associate of rhizostomous or morphologically similar scyphozoan medusae (Filho et al. 2008; Gonçalves et al. 2016; Baeza et al. 2017). These shrimps occur as one or a few individuals per jellyfish and can be observed walking beneath the host's bell (subumbrellar surface) and over the oral arms where they manipulate and apparently feed on host mucus without harm to the host (commensalism). Males are smaller than females, and there is no sexual dimorphism in cheliped weaponry which, along with their distribution on hosts, indicates that they are nonterritorial with a pure-search (promiscuous) mating system (Baeza et al. 2017).

In the northeast Pacific, pandalid larvae have been found riding ("hitch-hiking") on small hydromedusae (Marliave and Mills 1993). The larvae choose hydromedusae with a bell diameter about that of their leg span. The hydromedusae are not physically harmed or eaten, but they must pulse more for longer periods, while resting and feeding less, than those without a shrimp larva riding on top of them. As a result of this energy expenditure and loss of feeding opportunity, the mortality of hosts with pandalid hitchhikers is greater than those without. The shrimp larva might thus be considered an energy parasite of the hydromedusae.

Large jellyfishes and similar gelatinous cnidarian pelagic relatives (e.g., siphonophores) are often abundant and highly aggregated in blooms in the pelagic environment. The few known shrimp associates are basically bottom dwellers that, for the most part, fortuitously or sporadically use them as a temporary substrate. Although pelagic shrimps such as some oplophorids, sergestoids, and benthesicymids do feed on jellyfishes (Moore et al. 1993; Arai 2005), they are predators, not associates. Pelagic shrimps are morphologically and behaviorally adapted for continuous swimming or drifting and do not need or normally use solid substrates. Perhaps this explains why associations between jellyfishes and pelagic shrimps have not evolved.

10.6 Shrimp Residents of Echinodermata

Although most shrimp associates live with cnidarians, the echinoderms also host a significant number of species. All echinoderms have a hard skeleton composed of calcareous plates (ossicles), but those with complex body structure, i.e., the sea urchins (Echinoidea) and feather stars (Crinoidea: Comatulida), attract more shrimp associates than other echinoderm groups. The bristling pointed spines of sea urchins and the array of flexible feathery arms of crinoids provide good cover for shrimps. Those echinoderms with a relatively barren upper (aboral) surface, i.e., the holothurians (sea cucumbers), asteroids (starfishes), and ophiuroids (brittle stars), host relatively few shrimp species. Shrimps which do live exposed on echinoderm hosts have evolved cryptic coloration to compensate somewhat for this exposure (Bruce 1982). However, some species hide beneath the lower (oral) side of the echinoderms, provided that the host is large enough. As with the cnidarians, echinoderm tissues and mucus are food sources, as are food scraps from the host feeding activities.

10.6.1 Echinoidea (Sea Urchin) Associates

Sea urchins are choice habitat for shrimp symbionts, providing both protection and feeding opportunities for their uninvited guests. Shrimps can live among or cling to the sharp spines of the host or may hide below the urchin. With their short stubby walking legs, the pontoniines *Tuleariocaris* and *Stegopontonia* grip the long thin spines of diadematid urchins (Fig. 10.3e) or swim among them with their anterior end pointing toward the urchin test (Plate 10.2D) (e.g., Chace 1969; Castro 1974; Berggren 1994). The circumtropical *Gnathophylloides mineri* lives among the short spines of various *Tripneustes* or other short-spined urchins (Castro 1971; Patton et al. 1985; Williamson et al. 2012) (Plate 10.1C). Urchin spine dwellers have an elongate (*Tuleariocaris, Stegopontonia*) or at least fusiform (*G. mineri, Levicaris mammillata*) body shape. The color patterns of the first three are both disruptive and mimetic so that the body outline is broken up by longitudinal lines and stripes which alternate between light or dark colors like those of the urchin spines and test. The effect is to break up the body form into a series of visual "pseudospines" that blend in with those of the host (Plate 10.1C, 10.2D). *L. mammillata* is completely mimetic, its reddish color blending in with the comparably colored thick spines of the red slate or pencil urchin, *Heterocentrotus mamillatus*.

Another characteristic of shrimps living on or among urchin spines is that they ingest epidermal tissues and mucus of the host. The rasping setae on the second maxillipeds are used by *G. mineri* and *L. mammillata* to feed on epidermal tissues covering the spines and test their hosts (Chap. 3). The feeding mechanism of *Tuleariocaris* and *Stegopontonia* has not been described, but the compounds (spinochromes) which give the shrimps their camouflage colors are the same as those found in the host urchin's pigments, as shown by chemical analyses of *T. holthuisi* on the urchin *Echinometra mathei* (Brasseur et al. 2018). Research on carbon and nitrogen stable isotopes showed that the shrimp's spinochromes are derived from feeding on host tissues. When deprived of its host, the shrimp loses color and dies within a few days. Although not yet studied, it is likely that *G. mineri* and *L. mammillata* obtain their mimetic colors from feeding on their host tissues as well. All these tissue feeders are thus ectoparasitic, although any overall damage to the host appears to be minor—the shrimps do not eat themselves out of house and home.

The shrimp *G. mineri* is well camouflaged on its usual host in the Caribbean, *Tripneustes ventricosus*. The urchin has a brown test and short (less than 1 cm long) white spines that tend to lay flat against the test. The alternating brown (or gray) and white longitudinal stripes of the small (~ 6 mm total length) shrimp mimic very well the spine and test pattern of the host. However, in Jamaican seagrass meadows that

Maciá and Robinson (2009) studied, both *T. ventricosus* and the green sea urchin, *Lytechinus variegatus*, were common and both hosted the shrimps. Unlike *T. ventricosus, L. variegatus* has much longer (a few centimeters) greenish spines that are held erect over a white test. The shrimp is much better camouflaged on *T. ventricosus*. However, *L. variegatus* covers itself with bits of algae, shell, and other debris held by the urchin's tube feet and spines. *T. ventricosus* does this as well but to a much lesser extent. Thus, the shrimp is much better hidden on *L. variegatus* and was three times more abundant than on the other urchin in the seagrass meadows studied. The shrimp can also move more freely among the longer erect spines of *L. variegatus* than among the short spines held close to the test in *T. ventricosus*. In laboratory choice experiments, the shrimp showed no preference between either of the urchin species with debris cover removed. However, they did choose *L. variegatus* with their normal heavy debris cover over *T. ventricosus* with its naturally low degree of covering. The shrimp may be adaptable enough to abandon its original host to which it is color-camouflaged when the opportunity to inhabit debris-covering urchins is available, assuming that the latter is a better antipredator defense. On the other hand, perhaps this population of *G. mineri* is in the process of adapting to a new preferred host. If so, will it lose or change its present camouflage coloration over evolutionary time?

The alpheid shrimp *Arete indicus* lives among bases of the long spines of its host *Echinometra mathei*. Unlike *Tuleariocaris holthuisi*, *A. indicus* faces outward toward the tips of the urchin spines (Fig. 10.10); it prefers to hide on the oral (lower) side of the urchin (Gherardi 1991). It is quite host specific. In the habitat studied by Gherardi, the shrimp never occurred on the long-spined *Diadema* or short-spined *Tripneustes*. Laboratory observations showed that the rapid spine movements of *Diadema* irritated the shrimp. If offered as a host, the *A. indicus* soon left it. It also did not stay long on *Tripneustes* whose short spines offered little protection to the shrimp. *A. indicus* has a somewhat similar mimetic coloration of stripes as *Tuleariocaris holthuisi* but is not so well camouflaged with its host (Brasseur et al. 2018). Gut contents of *A. indicus* showed no evidence of host tissues; their diet is like that of their hosts, i.e., bits and pieces of algae (Gherardi 1991). Brasseur and her colleagues did find some spinochromes among the pigments of *A. indicus,* but stable isotope studies did not indicate extensive feeding on host tissues. Unlike *T. holthuisi*, *A. indicus* did just fine in the laboratory when derived from its hosts and provided with food.

Other urchin shrimps include *Periclimenes colemani* which lives in male–female pairs on the short-spined *Asthenosoma intermedium* (Bruce 1982). Its white color with red spots closely matches the coloration of its host. It clears out a small bare area on the upper (aboral) surface of the urchin on which the pair reside, but more details are lacking. Its presence and activity on a specific area of the urchin test probably irritate the spines, which are then dropped. The alpheid *Betaeus macginitieae* occurs under *Strongylocentrotus* and *Centrostephanus* urchins in southern California (Hart 1964). I collected them for studies on grooming behavior and often observed the male–female pair hiding beneath the oral surface of the urchin. There were frequently 2–3 juveniles just beneath the periphery of the urchin.

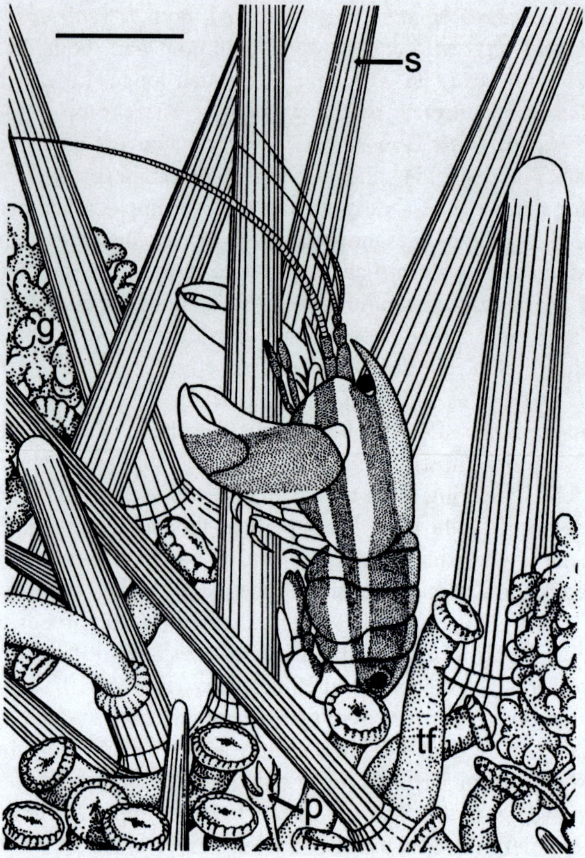

Fig. 10.10 *Arete indicus* (Alpheidae) in typical position facing toward spine tips of the sea urchin *Echinometra mathei*. g, dermal gill; p, pedicellaria; s, spine; tf, tube foot (podium). Scale bar represents 5 mm. Modified from Gherardi (1991), courtesy of Journal of Crustacean Biology

This shrimp is of a dark purple color like that of its host's spines. However, they do not cling to the spines; I kept them for long periods in the laboratory and fed them brine shrimp (*Artemia*). It may be recalled from Chap. 5 that colors such as dark purple can be due to carotenoid–protein complexes obtained from a general omnivore diet. It seems unlikely that they feed on host tissues; i.e., they have a commensal relationship with their hosts. Some shrimp species may be facultative associates with urchins. The western Pacific *Rhynchocinetes brucei* and *R. uritai* seek refuges such as crevices and small caves in rocky reefs during the daytime (Ory et al. 2013; pers. obs.). *R. brucei* prefers crevices occupied by sea anemones or alternately long-spined urchins which increase protection from predators. In these species, there is no color match to either of their hosts. Deep-sea urchins host shrimps as well. Four species of the deep-water pontoniine *Echinopericlimenes* inhabit venomous sea urchins living at shallow slope depths. As might be expected of shrimps living in the twilight environment of such depths, all have a deep red coloration interrupted in some of the species with disruptive white longitudinal lines that help conceal them among the urchin spines (Marin and Chan 2014)

10.6.2 Crinoid Shrimps

Crinoids are the only echinoderms in which the oral surface of the body faces upward. Their small body (calyx) has the basic pentamerous (5-part) symmetry of echinoderms so that the multibranched arms usually occur in multiples of five. Internal tissues of arms and calyx are enclosed by calcareous plates that make them unattractive prey to most marine predators. Their arms bear numerous side branches (pinnules) which give them a feathery appearance. The pinnules and arms bear grooves (ambulacra) from which project their small, soft papilla-like tube feet (podia). The podia secrete mucus which filters and entraps passing plankton and detrital material; podia and ciliary action pass the food bolus down the ambulacral grooves to the mouth. Comatulid crinoids ("feather stars") come in many variegated color patterns. All this complexity of morphology and color makes crinoids, especially comatulids, great hiding places with lots of microniches for shrimps to occupy. While protected by the numerous, color-confusing arms of the host, the shrimps can feast on the constant stream of food-laden mucus moving toward the feather star's mouth. As a result, feather stars are very appealing hosts to a variety of shrimp symbionts and are commonly infested with several shrimp species, including a few stenopodideans (*Odontozona, Microprosthema* species), alpheids (mostly *Synalpheus*), and numerous pontoniines (Potts 1915b; Bruce 1982; Huang et al. 2005a; Deheyn et al. 2006; Goy 2010).

Shrimp associates share their feather-star hosts with a diverse array of other symbionts such as galatheid and brachyuran crabs, clingfishes (Gobiesocidae), myzostomid annelids, gastropods, nematode worms, and ophiuroids (Huang et al. 2005a; Deheyn et al. 2006). *Synalpheus* species, with their large snapping claws, usually occupy the prime position on the central calyx except when clingfishes (Gobiesocidae) are present. The two symbionts are mutually exclusive, probably due to agonistic interactions. As Potts (1915b) pointed out, when disturbed, the host arms fold over the oral disk, forming a protective cage over any symbionts there. Pontoniine shrimps, such as *Palaemonella pottsi, Pontoniopsis comanthi*, and several *Periclimenes* (*s.l.*), species situate themselves on the arms and pinnules of the feather star. Pontoniines species may occupy distinct positions on the arms. In *Comanthina schlegeli* in the Indo-West Pacific, *Unguicaris* (*Periclimenes*) *pilipes* and *Palaemonella pottsi* wander over the oral surface of arms near the calyx, while *Periclimenes commensalis* scurry among the distal parts of the arms (Bruce and Zmarzly 1983).

Like many shrimp associates, comatulid shrimps have evolved camouflage coloration. As their hosts show bright and complexly variegated coloration, with lots of stripes and spots, their shrimp symbionts use a comparable coloration. One form is disruptive striping with inclusion of host colors in *Synalpheus stimpsoni*, *S. comatularum*, and *Pontoniopsis comanthi* on *Comanthus parvicirrus* (Fig. 10.11), while most pontoniine species on the same host are translucent tinted by host color and including some disruptive spots (Potts 1915b). In *S. stimpsonii* and other symbionts that occur in pairs, the male and female generally share the same

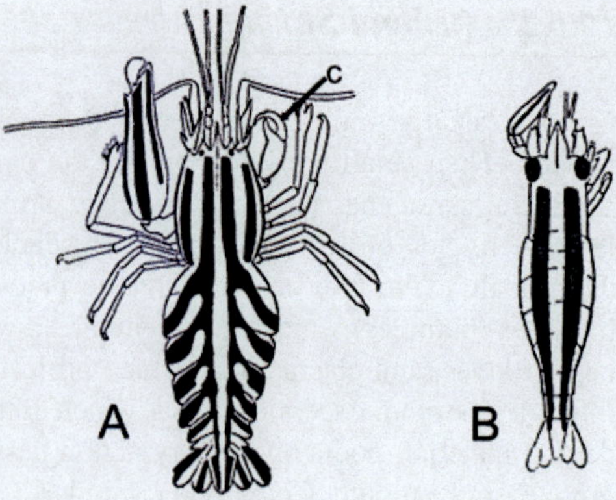

Fig. 10.11 Disruptive striped coloration of two symbionts on the comatulid crinoid (feather star) *Comaster schlegelii*. (**a**) *Synalpheus comatularum* (Alpheidae). (**b**) Pontoniine *Pontoniopsis comanthi* (Palaemonidae). c, minor chela of *S. comatularum* with hooked dactyl for digging into host tissue to maintain position. From Potts (1915b)

color pattern. However, in some pairs the color patterns differ. Potts hypothesized that, in these instances, either the male or female was a recent arrival to that individual host and would change its color pattern to match that of the host and its mate. *Comaster schlegelii* is extremely diverse in its color patterns and its pair-living obligate symbiont *Unguicaris pilipes* similarly varies its color according to that of its individual host (Bruce and Zmarzly 1983). These observations on covariation of symbiont color camouflage with color pattern of individual hosts show that the shrimps are using morphological color change (Chap. 5) to match up with their hosts.

10.6.3 Other Echinoderm–Shrimp Associations

As noted early, the much lower number of shrimp associates on the asteroid (starfish), holothuroid (sea cucumber), and ophiuroid (brittle star) echinoderms might be attributed to the lack of protective spines and simple morphology compared to sea urchins and crinoids. Many starfishes and holothuroids have evolved chemical defenses, such as the inclusion of distasteful or toxic saponins, which makes them unattractive to many predators. Ophiuroids are generally too small to support shrimp symbionts, but a few species do.

One of the few starfish symbionts that has received much scrutiny is *Zenopontonia (Periclimenes) soror*, a generalist which has been observed on at least 23 different species of starfishes (Antokhina and Britayev 2020). Over 50 individuals have been counted from a single starfish, although just a few individuals are more usual (Debelius 1999). Their preferred location on the host is near or beneath the oral (lower) surface of the starfish. Their color is similar (mimetic) to the host so that they presumably receive some protection when moving onto the exposed upper surface of the starfish (Olliff 2013). The shrimps prefer starfish hosts which have a more complex surface morphology, such as *Acanthaster planci*, a multiarmed species with many venomous spines (Antokhina and Britayev 2020). Starfish species which offer neither a structural nor chemical protective option to symbiotic shrimps, such as *Linckia laevigata*, are avoided. Starfishes with a large oral surface to hide below, such as *Culcita novaeguineae,* provide a good hiding space. The shrimps were observed picking at material on the starfish by Oliff, which might have been accumulated detritus or mucus, or, given the color mimicry to the host, might be obtained from host tissues (cf. Brasseur et al. 2018). Olliff did experiments in which the shrimps were maintained isolated and unfed, or isolated but fed a plankton extract, or not isolated (maintained with the host) and fed. The shrimps in the first two treatments died, but those in the third all survived, indicating that the shrimps were getting some alimentary supplement from the host. *C. novaeguineae* has several color morphs, and Oliff did experiments with shrimp–host color matching with the purple and yellow starfish morphs and their associates. He switched the dark purple shrimp color morphs from purple hosts to yellow hosts, and yellow shrimps to purple morphs, waited a week, and then measured shrimp color characteristics from photos evaluated with an image analysis program. The yellow shrimps transferred to a purple host turned purple. On the other hand, purple shrimps changed to yellow hosts did not change to yellow but rather to a pale shade of purple. Perhaps with more time the latter might have changed completely. The experiments indicate that morphological color change is involved in the symbiont color matching to host observed in *Z. soror*.

The gnathophyllid *Pycnocaris chagoae* (Plate 3.7D) usually occurs on holothurians (sea cucumbers) such as *Holothuria cinerascens* and *Labidodemas rugosum* (Prakash et al. 2011). The shrimp appears color adapted. *Zenopontonia rex* (formerly *Periclimenes imperator*) has also been found with holothurians (Bruce 1976a). Ophiuroids host few shrimp symbionts. One exception is the alpheid shrimp *Athanas ornithorhynchus* which occurs with the brittle star *Macrophiothrix longipeda* (Marin et al. 2005). The host lives on dead corals and muddy rocks, and the shrimp prefers to stay on the dorsal and lateral sides of the brittle star's arms, which are equipped with stout spines. The shrimps have a bold disruptive color pattern of red stripes on a white background which does not seem to match that of its drabber host; its possible role as camouflage is not clear. The shrimp might appear very different at depth, where red light does not penetrate, than it does when photographed in bright light. Little else is known except that the shrimp will cling tenaciously to its host. The only other Ophiuroidea reported to have shrimp symbionts are basket stars, ophiuroids in which the 5 arms are very highly branched, forming a web-like body that clings to rocks and sessile invertebrates; it is a filter feeder using its tube feet. A single basket star may host over 50 shrimp associates, still largely unstudied and unknown (Bruce 1982; Marin et al. 2005).

10.6.4 Host Recognition and Fidelity

Many sea urchins abrade out a home site in the rocky or coral substrate which they inhabit by day but leave and move about at night. Do urchin shrimps move along with their hosts or do they themselves leave the urchin to forage and, except for monogamous pairs, look for a mate? If so, do they come back to the same urchin every day? *Tuleariocaris neglecta* does leave its host urchin at night (Castro 1974), and it is likely that some urchin and other symbiotic shrimps do so as well. Shrimps may have to move from one host to another due to host death, predation pressure, or aggressive conspecific invaders (Antokhina and Britayev 2020). Juveniles need to find a suitable host after settling from the plankton or being evicted by conspecifics already entrenched on a host. They must do so quickly, especially during daytime, to avoid visually hunting predators.

Several studies have been done on echinoderm species regarding the sensory mechanisms used for location of hosts (e.g., sea urchins: Ache and Davenport 1972; Williamson et al. 2012; feather stars: Huang et al. 2005b; VandenSpiegel et al. 1998; starfishes: Olliff 2013; Antokhina and Britayev 2020). Chemoreceptive ability of shrimp species is usually tested with y-maze or similar choice experiments; visual capabilities are tested using various host species or geometric shapes simulating them. These studies point to chemoreception as the major sensory mechanism involved in host selection and fidelity, with vision much less so. This is perhaps predictable in view of the poor visual acuity of most shrimps as well as their generally nocturnal activity. Shrimp species can recognize and choose their original host species but apparently are not able to distinguish among specific host individuals (host imprinting: e.g., Ory et al. 2013; Brasseur et al. 2018; Antokhina and Britayev 2020).

10.7 Molluscan Hostels

The bivalve molluscs (Pelecypoda), e.g., clams, mussels, oysters, and scallops, are a wonderful habitat for symbiotic shrimps, which live protected from most predators within the pelecypod's shell. The shrimps are usually found within the mantle cavity of the host, a tissue space which houses the gills (Plate 10.2 E–F). These structures not only function in gaseous exchange but also have evolved into food collection mechanisms. The bivalve filters water from outside through the gills, and food particles become lodged within the mucus sheet that moves over the gills toward and then into its mouth. The mixture of mucus, plankton, and other organic matter is a great source of nutrition for the shrimp inhabitants. The water passing through the gills of the host also provides the shrimps with oxygenated water. The bivalve mantle cavity is a niche which has been exploited by a number of pontoniine shrimp species, especially from genera such *Conchodytes, Anchistus, Neoanchistus, Paranchistus, Platypontonia*, and *Pontonia*, among others (Bruce 1975, 1976a, b; Fransen 2002; Horká et al. 2016).

Bivalves that host shrimps are large in body size, as are their shrimp guests compared to other pontoniines (Bruce 1976b), up to 3 cm total length in the shrimp *Anchistus custos* (Johnson and Liang 1966) (Plate 10.2 E–G). Shrimp symbionts are mainly found in non-burrowing bivalves, such as *Tridacna* clams (Bruce 2000) and pen shells, the latter partially embedded in sand or other soft bottoms, but with their posterior ends extending above the substrate. Rates of infestation by a single species can be high (e.g., 30–60%, Johnson and Liang 1966; 57%, Aucoin and Himmelman 2010). Heterosexual pairs are common, with most shrimp symbionts having "search-and-attend" or monogamous mating systems (Plate 10.2G) (Hipeau-Jacquotte 1973a, b; Bruce 2000; Baeza 2008; Baeza et al. 2011, 2013, 2016)

The hidden habitat of these shrimps has prevented biologists from learning much about them. Fortunately, there are some extensive observations on a few species, like those of Johnson and Liang (1966) on *Anchistus custos* living within the pen shell *Pinna atropurpurea* in shallow waters around Singapore (Fig. 10.12a). In the laboratory, investigators were able to observe the activities of the shrimps by removing part of or the entire valve of one side. The shrimps cling to the edge of the host's gills with the walking legs while scraping the gills with setal brushes on the specialized first chelae (Fig. 10.12b–c). Particle-laden mucus accumulates in a setose food groove on the chela propodus. The chelae reach toward the maxillipeds, which remove the material and pass it into the shrimp's mouth. The gut contents of both shrimps and their host are composed of mucus and microscopic algae filtered from the plankton by the host's gills. No tissue damage or other harm to the host was observed, so that *A. custos,* like other shrimps living within bivalves, nicely fit the definition of a commensal. One clam can host several individuals of both sexes of *A. custos*, but aggressive or territorial behavior was not observed.

Hipeau-Jacquotte (1973a, b) studied territoriality and aggression among *Anchistus custos* and two other pontoniines, *Conchodytes biunguiculatus* and *Paranchistus ornatus*, in Madagascar. The former two species inhabit the pen shells *Pinna bicolor* and *Atrina vexillum*, while *P. ornatus* only lives with *A. vexillum*. Hipeau-Jacquotte observed that only a single male–female pair of just one species inhabits the mantle cavity of any host individual, although a juvenile or two may be present. Given the overlap among the three species in host use, there is ample opportunity for interspecific competition and conflict. Disputes are not resolved peacefully in these shrimps. When an individual of one shrimp species tries to enter the mantle cavity of an occupied host, fighting erupts, with the large second chelipeds of the shrimps inflicting injuries so severe that death may result. Among the three species, an intruding *C. biunguiculatus* regularly expels either of the other two species already established within a host, and *A. custos* is usually the winner over *P. ornatus*. The deciding factor in winning a battle is size. Adult body size is largest in *C. biunguiculatus*, while individuals of *A. custos* are generally bigger than those of *P. ornatus*.

Similarly, there is intraspecific fighting over possession of a host. Unlike Johnson and Liang, Hipeau-Jacquotte found heterosexual pairing, territoriality, and aggressive behavior among individuals of *Anchistus custos*. Intruding conspecifics cannot displace either member of a resident pair. When only a single individual occurs in the host, an intruder of the opposite sex may pair with it, but an intruder of the same sex is violently expelled, with the intruder suffering severe injuries or death in 50% of encounters. The host is an important limiting resource that must be protected by the symbiotic shrimps. That *A. custos* exhibits aggressive behavior in one clam species

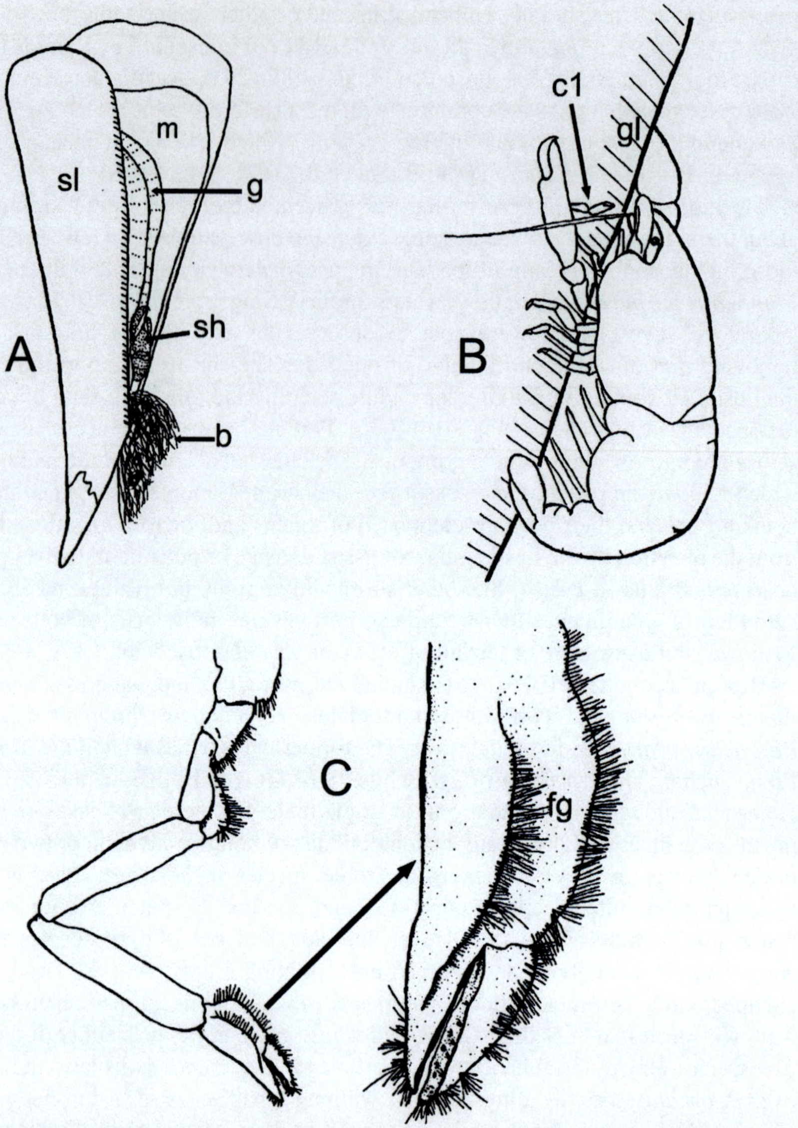

Fig. 10.12 The pontoniine *Anchistus custos*, a symbiont of the bivalve mollusc *Pinna atropurpurea.* (**a**) Shrimp in position on the gaping shell of the host. (**b**) Close-up of the shrimp clinging to the edge of a gill lamella, where it feeds by harvesting host mucus containing material filtered by the host. (**c**) Cheliped 1 (left) and close-up of chela 1 (right), showing setal scrapers surrounding the food groove. b, byssal threads (attachment organ); c1, cheliped 1; fg, food groove; g, gill; gl, edge of gill lamella; m, mantle; sh, shrimp *A. custos*; sl, shell valve. Modified from Johnson and Liang (1966)

(Hippeau-Jacquotte) but not another (Johnson and Liang) needs further study and explanation.

Within the same host individual, there may be some competition for the most favorable location. Almost all gravid females of *Pontonia domestica*, a commensal of the pen shell *Atrina rigida,* locate themselves among the area of gills near the host's posterior adductor muscle, the safest location within the host (Courtney and Couch 1981). The more motile males were more often found outside of the mantle cavity near the periphery of the shell. Non-gravid females were located equally between the two areas. Only 1–3 individuals were located per host: this and symbiont distribution by sex hint at a "search-and-attend" rather than a socially monogamous mating system.

The only other molluscan class that is known to harbor shrimp associates is the Gastropoda; even in this group, there are but a few examples. A single male–female pair of the alpheid *Betaeus harfordi* inhabits the mantle cavity of abalones (*Haliotis* spp.) along the Pacific west coast from central to Baja California (Hart 1964; Jensen 2014). When diving in kelp beds, I often observed a pair peering out of the anterior end of the abalone's mantle cavity on either side of the mollusc's reduced head. In the laboratory, they are quite aggressive to intraspecific intruders attempting to enter their host. Like most symbiotic shrimps, much needs to be learned about their natural history, i.e., is their abalone only a daytime host, as perhaps is the case with the urchin symbiont, *B. macginitieae*?

One pontoniine species with the regal name of *Zenopontonia rex* (formerly *Periclimenes imperator*) inhabits the dorsal surface of some shell-less gastropods, the brightly colored nudibranchs *Hexabranchus aureomarginatus, H. sanguineus*, and *Dendrodoris tuberculosa*, among others (Castro 1971; Shoup 1972; Debelius 1999). These large (2 cm) pontoniines have a white and red coloration like some body parts of their much larger hosts (Plate 10.2H). Castro found these in pairs on their hosts, but Shoup only found single individuals. The latter author made extensive natural history observations on this shrimp in the laboratory. He found that it never left the surface of the nudibranch. At night, the nudibranch *H. sanguineus* leaves the bottom and clumsily swims with violent undulations into the water column. To avoid being shaken off, the shrimp quits it usual position on the host's dorsal surface to attach more safely in the groove between the dorsal body of the host and its foot. During the day, it picks and plucks material from the host's dorsal surface and may go back to the host's posterior end to eat the host's feces as it is eliminated. It also hangs off the foot of the host in order to feed on items on the substrate around the host. However, if the host moves off, the shrimp will reluctantly let go of an attached food item to remain with the nudibranch (Debelius 1999). When Shoup introduced single *Z. rex* into the aquarium, it would try to mount any nudibranch nearby. But the resident shrimp aggressively drove it off. Shoup hypothesized that heterosexual pairing of such aggressive shrimps might occur during the coupling of their hermaphroditic nudibranch hosts, which come together for a prolonged mating. Nudibranchs contain toxic antipredator compounds that are advertised by their bright colors. The shrimp thus gains protection from its host, as it also gets free rides to different sites at which it may find food around the host.

10.8 Ascidian Homes

Members of the Ascidiacea (phylum Chordata) are commonly referred to ascidians or sea squirts; their only chordate feature as adults is a perforated pharynx. This is a large branchial chamber ("pharyngeal basket") which occupies most of the space of the animal. Ascidians are filter feeders; through an oral siphon, water flows into and through the branchial chamber and the inner mucus sheet which lines it. The mucus sheet is produced continually, carrying filtered plankton and other particulate matter eventually into the gut of the animal. The branchial chamber offers a protected home and a constant supply of plankton-enriched mucus as a food source. Numerous species of caridean genera, mainly pontoniine, such as *Pontonia, Odontonia, Dactylonia, Dasella, Ascidonia, Onycocaris, Typton*, and *Periclimenaeus*, are obligate associates of ascidians (e.g., Bruce 1976a, b; Fransen 2002, 2006; Dobson et al. 2014; Horká et al. 2016). The only alpheids reported as ascidian commensals are some species of *Synalpheus* (Bruce 1976b). A few shrimp species live outside the branchial chamber in the atrial chamber which surrounds it, e.g., *Odontonia sibogae* (Levitt-Barmats and Shenkar 2018). After flowing through the branchial chamber of the ascidian, water then flows out of the animal via the atrial siphon. Fecal pellets and remnants of mucus are deposited near the atrial siphon to be flushed out with the exhalatory flow, and shrimps living in the atrium can use this as a food source. Some ascidians are compound (colonial) in which several individuals share a common central atrial siphon. *Periclimenaeus* is a genus mainly associated with sponges, but some species live in compound ascidians whose communal exhalatory channels are architecturally similar to sponge canals (Horká et al. 2016). As with other endosymbionts such as molluscan shrimps, ascidian shrimps have relatively little coloration compared to the usually more colorful ectosymbiotic shrimps. The coloration of ascidian shrimps tends toward transparency, pale tints, and scattered chromatosomes.

The large, spacious barrel-like branchial chambers of solitary ascidians give endosymbiotic shrimps ample space to move around and feed in. Correspondingly, many pontoniine ascidian associates are large in body size for their group, a characteristic shared with molluscan shrimp associates, which as endosymbionts live in a relatively large enclosed space. Ascidian and molluscan shrimp associates also share other traits such as a globular or flattened body, smooth body surfaces with reduction or elimination of most spines, teeth, and other projections that could catch on host tissues, and pereopod dactyls to cling tightly to host tissues (e.g., Berggren 1999; Horká et al. 2016). The flattened body of *Notopontonia platycheles,* which inhabits the sea squirt *Herdmania momus*, appears to be an adaptation to the volume reduction of the host's branchial chamber when a disturbed host retracts (Berggren 1999).

Ascidian associates usually occur as single individuals or heterosexual pairs, rarely more considering that the space within the branchial chamber of even a large ascidian is limited. Although not truly colonial, solitary ascidians often occur close together in clumps or aggregations. This makes it possible for individual shrimps, especially males, to leave its host to look for females residing in other nearby ascidians. The information available from studies and scattered observations indicates that a "search-and-attend" mating system is common in ascidian shrimp associates (e.g., Berggren 1999; Baeza and Díaz-Valdéz 2011; Levitt-Barmats and Shenkar 2018). However, other mating systems are certainly possible in this group.

10.9 Shrimp Associations with Burrow Dwellers and Other Crustacea

Free-living alpheid shrimps prefer to live hidden under rocks, within crevices, constructed burrows, or other shelters. Burrow openings of many marine invertebrates are common in tidal mud and sand flats. In recent years, suction pumps, e.g., the yabby pump, have been used to extract the organisms making these burrows. In the process, many new species of shrimps have been added to the list of burrow-living crustaceans. Species of alpheid genera such as *Betaeus, Alpheus, Leptalpheus, Fenneralpheus, Salmoneus*, and *Athanopsis* (Fig. 10.13a–b) are often found in burrows made and inhabited by mud and ghost shrimps (Axiidea and Gebiidea, formerly grouped as "thalassanideans") such as *Neotrypaea, Lepidophthalmus*, and *Upogebia*, as well as echiurid and thalassematid echiuroid worms ("spoonworms, innkeeper worms") (MacGinitie 1937; Hart 1964; Anker et al. 2001; Anker and Baeza 2012). Not unexpectedly, shrimp associates living in the deep, dark host burrows usually have little color, variously described as colorless or transparent with scattered chromatosomes, or pale tinted by light shades of pink, red, or green. Like cave shrimps, as well as shrimp associates living in the murky microhabitats within

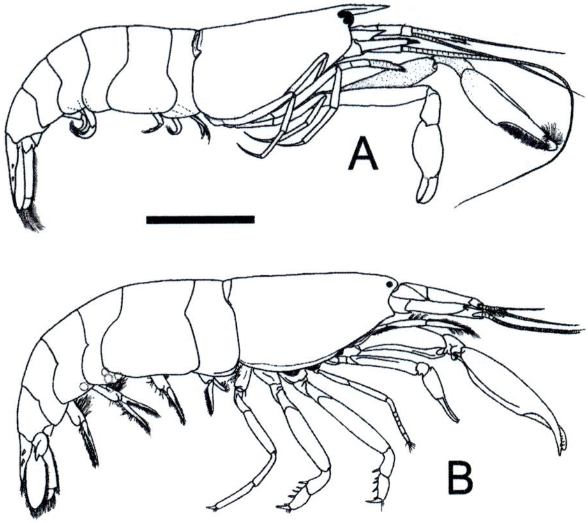

Fig. 10.13 Two alpheid shrimp associates of burrowing crustaceans. (**a**) *Athanas dentirostris* from *Upogebia* species burrows. (**b**) *Fenneralpheus chacei* from burrows of axiid ghost shrimps and/or stomatopods. Scale bar represents 3 mm in (**a**) and 9 mm in (**b**). From (**a**) Anker et al. (2001) and (**b**) Felder and Manning (1986), courtesy of Journal of Crustacean Biology

molluscan and ascidian hosts, little metabolic energy is expended in producing unnecessary camouflage coloration. Burrow-living shrimp associates often occur in pairs, but like most of their natural history and reproductive habits, little else is known beyond their taxonomy and host distribution. Innovative laboratory setups and field observation techniques need to be developed to reveal their activities and interactions, if any, within their hosts burrows.

A somewhat unusual shrimp–burrow dweller relationship has evolved between *Alpheus heterochaelis* and the xanthoid mud crab *Panopeus herbstii* (Silliman et al. 2003). The shrimp is a facultative or obligate symbiont, depending on the habitat. In tidal flats with soft mud bottom, the shrimp constructs its own burrows, but in tidal marshes, with a more difficult complex mud and marsh root substrate, it is unable to do so. In this habitat, it is only found living in the U-shaped burrows excavated and inhabited by *P. herbstii*. The mud crab primarily preys on marsh snails, but in the laboratory it will readily feed on small crabs and grass shrimps similar in size to *A. heterochaelis*. However, in the laboratory at least, the crab does not attack or feed on *A. heterochaelis* even when starved. The shrimp certainly enjoys benefits from burrow sharing with the crab: a living space giving protection from predators such as blue crabs (*Callinectes sapidus*) and fishes, burrow maintenance by the mud crab, and access to food scraps left by its predatory host. The association might not be one-sided given that snaps by this pistol shrimp may discourage potential predators of the host or at least warn the crab of their approach. The shrimp relationship with the host may best be described as commensalism bordering on mutualism.

Other unusual associations occur between shrimps and other crustaceans or even with other species of shrimps. The alpheid *Aretopsis amabilis* associates with hermit crabs of the genus *Dardanus* (Bruce 1969b). Bruce suggested that the shrimps might be commensals feeding on food scraps or feces produced by the hermit crab. The pontoniine *Periclimenes dardanicola* lives with *Dardanus* hermit crabs carrying sea anemones; when disturbed, it retreats into the hermit shell opening. However, another pontoniine (*Bathymenes poupini*) appears to be more closely associated with a sea anemone (*Calliactis* species) carried by a hermit crab (*Trizopagurus* species) than with the crab itself (Bruce and Okuno 2006).

Betaeid shrimps may be free-living or associate with a wide variety of invertebrates occupying shelters or burrows (Hart 1964; Anker and Baeza 2012). Some of the hosts are other crustaceans. In intertidal and shallow subtidal rocky areas of the Argentine coast, *B. lilianae* is found in aggregations in pools but is often associated with the brachyuran crab *Danielethus crenulatus*. Antonio Baeza et al. (2010) found that the shrimp choses to occupy shelters with the crab when possible. The relationship seems to be one-sided, with the shrimp receiving the benefits. In laboratory observations, the shrimp was found close below, between the claws, and on the dorsal carapace of the crab. On occasion, the crab acted aggressively toward the shrimp which is an indication that the shrimp may be an unwanted guest, but the shrimp is probably agile enough to stay out of the way of the crab's way when the latter becomes irritated. The shrimp receives protection from predators by the presence of its host and may feed on crab feces within its rock-crevice shelter. Another possibility is that the shrimp follows the crab which forages at night and then may feed on the crab's food scraps. Along the northwest Pacific coast, another betaeid, *B. setosus,* occurs solitarily underneath a heterosexual pair of the porcellanid crab *Pachycheles rudis* (Jensen 2014). It is interesting that the shrimp and the crab pair are size-matched, with shrimp size positively correlated with that of the crab pair. This suggests that the shrimp joins the crab pair when young and grows with it. In order to mate, one sex or the other of the shrimp, probably the male, must search for a receptive mate, and then return to its original crab pair, or at least one of similar size.

An association of one shrimp species (*Alpheus inca*) with another (*Alpheopsis chilensis*) was investigated by Boltana and Thiel (2001). The two species share the same burrow in over half of those examined but were also found living separately in other burrows. The pairing behavior and close correlation in size of the male and female in each species indicate social and probably sexual monogamy. Although *A. chilensis* pairs were smaller than *A. inca* pairs in the same burrow, there was a positive correlation in the size of one species with the other, indicating that the cohabiting pairs may have been together from an early age. Surprisingly, the females of each species were usually found in the same reproductive stage (with or without embryos; or with embryos at a similar stage of development). Alpheids are especially aggressive shrimps with heavy cheliped weaponry, and even short-term occupancy of a burrow by two species is unexpected. Many studies have shown that alpheid shrimps have a complex communication with conspecifics mediated by behavior and pheromones (e.g., Herberholz and Schmitz 2001). Boltana and Thiel proposed that these mechanisms might be employed by the two species to facilitate their joint occupancy of the same burrow. An obvious benefit for the shrimps is increased security and joint protection of their communal home from intruders and predators. A mesocosm study on their intraspecific cooperation would likely reveal interesting results.

10.10 Shrimps as Hosts: Goby Fish–Alpheid Shrimp Symbioses

In an unusual reversal in shrimp-symbiont roles, several alpheid shrimps have co-evolved a mutualistic relationship with many species of gobiid fishes—the shrimps are the host and the fishes the guests. The shrimp burrow is the shelter in which both host and guest share the night, and the fish pays its rent by serving as an alarm system for the shrimp. This is not an uncommon relationship in soft-bottom habitats of shallow tropical seas. In their excellent 2011 review, Karplus and Thompson list 120 gobiid fish species (20 genera) as guests, 30 of them *Amblyeleotris* and *Cryptocentrus* species; 14 described *Alpheus* species have been reported as hosts, with possibly three more. Most of these associations have been reported from the Indo-West Pacific with, in decreasing order, many fewer in the Red Sea, West Atlantic, and Hawaii. Indonesia is the center of diversity for goby–

shrimp associations. In most cases, the relationship is obligatory for fish guests and their hosts; only five of the 120 goby species reported by Karplus and Thompson are facultative symbionts.

Goby–shrimp partnerships are mainly found in sand-gravel habitats around coral reefs and seagrass meadows. During the day, the goby guest and the male–female pair of shrimp hosts can be observed near the entrance of the burrow constructed by the shrimps. The fish perches in front of the entrance feeding on small organisms in the sand, either by sorting through the sand or taking mouthfuls of it and filtering out small prey with the gills; it may also rise off the bottom and pluck out small plankters drifting by. The pair of shrimps actively forage around the entrance or take in loads of sand into the burrow to sort through food items at night. The shrimps are active during the day in burrow construction, maintenance, and feeding. The fishes, with their acute vision, serve as the eyes of the shrimps, which have poor visual acuity. The shrimp is in constant physical contact with the goby, using the long antennae to contact one of the fins of the fish. If the fish is in front, the shrimp touches its caudal fin; if to one side, a dorsal fin with one of the antennae; and if the fish is in back of the shrimp, the latter reaches back to touch one of the pectoral fins. The antennal flagella thus serve as the sensory structures that will detect warning signals from the fish when a predator approaches or other danger threatens. When potential peril is perceived by the fish, it transmits a low-intensity warning to the shrimp with flicks of its caudal fin sensed by the shrimp via its antennae. Such warnings can be triggered by the approach of any large fish, or a medium-sized fish which is either predatory or sediment-disturbing (e.g., goat fish), but not by a medium-sized coral-feeding or herbivorous fish, or by small-sized fishes. The fish symbiont can change its alert from low to greater intensity using the duration, amplitude, and frequency of the tail flicks. As the threat approaches and the warning strength increases, the shrimp may freeze or slowly retreat into its burrow. At some critical distance, the warning turns into a high-intensity alarm: the goby bends its body into a U in order to dive rapidly head-first into the burrow, which stimulates the shrimp into the backward tail flip, propelling it back into the burrow as well. A nice experimental study using both live predators and predator models manipulated by a clever pulley system allowed Lyons (2013) to observe and demonstrate many of the details of the warning-alarm system of the gobies *Nes longus* and *Ctenogobius saepepallens* with *A. floridanus* in the Bahamas.

Besides acting as sentinels for the shrimps, the gobies facilitate food acquisition by their hosts. The goby's feeding activities around the burrow entrance disturb the sediments, making food items more available to the shrimp. Goby feces deposited within or near the burrow entrance is an important dietary item for its hosts. This was shown by experiments on the shrimp *Alpheus bellulus* and its goby *Amblyeleotris japonica* done by Kohda et al. (2017). They maintained shrimps in three setups: (1) alone without food provided, (2) alone with goby feces offered as food, and (3) with a goby partner which was fed but the shrimp was not. In the first treatment, the shrimps lost weight as might be expected but weight loss was not significant in the latter two treatments, showing the importance of goby feces as food for this species. Anecdotal observations reported by Jaafar et al. (2014) indicated that two

Cryptocercus goby species bring algae, small bivalves, and other small items to their shrimp associates. These authors included a photo showing the goby *Amblyeleotris latifasciata* delivering a small crab it had captured near its partner's burrow entrance. It jumped rapidly up and down, stimulating the shrimp to come out of the burrow, remove the goby's prey, and then go back into the burrow, presumably to feed on the crab.

For its part, the shrimp–host constructs and maintains the burrows which are optimally built in a substrate with about 2/3 sand and 1/3 gravel (Karplus and Thompson 2011). The coral and shell gravel are needed to reinforce the easily collapsible sand burrow, usually made stronger by locating parts of it next to a coral or rock boulder. Burrow maintenance is a constant job for the shrimp, and Indo-West Pacific goby shrimps spend most of the morning in doing so, while spending their afternoons just outside the burrow entrance picking and searching for food particles. The position of the burrow opening is changed often to give the shrimp access to a new supply of substrate to forage in, to avoid conflicts with burrows of other alpheids that pop up nearby, and perhaps to confuse predators. One of the costs of the partnership to the shrimp host is that the occupancy of the burrow by the much larger goby causes a lot of burrow collapse which the shrimp must repair. Another is goby mating when the shrimp may have to endure two fish occupants for a time. As an added cost for the shrimp but a benefit for the gobies, one of the goby mating partners will brood its batch of developing embryos within the shrimps' burrow. Shrimps have also been observed picking and cleaning the body of their fish partners, presumably to remove dead skin, old mucus being shed, and ectoparasites, as do other fish cleaners (Jaafar et al. 2014). The shrimp *A. rapax* uses its second (grooming) chelipeds to clean its partner *Myersina macrostoma*. The goby shows the shrimp where to clean by nudging the shrimp with its head or rolling on its side to expose one part of its body or the other (Hou et al. 2013).

Goby shrimps may live without a goby partner, but their chances of survival are usually enhanced by a goby associate (Karplus and Thompson 2011; Jaafar and Hou 2012). Without their partner serving as a sentinel, shrimp hosts spend much more time within their burrows with less time for foraging and without other benefits bestowed by their partners. Obligatory goby associates are quickly killed by predators in the absence of a shrimp shelter (Thompson et al. 2005). It is the goby which initiates the partnership as it searches for a burrow. Goby associates are instinctively attracted to dim, dark places to hide in, such as a burrow. The goby tests a burrow for a shrimp partner by inserting its tail first into the burrow opening (Karplus and Thompson 2011). The shrimp probes the goby with its antennae and, if acceptable, will enlarge the burrow opening to allow the goby to enter. The gobies *Nes longus* and *Ctenogobius saepepallens* introduced into an aquarium with an established *A. floridanus* will pair up with a host within a few minutes (Lyons 2013). In nature, gobies might lose a shrimp partner and burrow when the shrimp dies, or the goby occupant is expelled by another goby. The ability to find a new partner is essential because obligate goby symbionts are very vulnerable to predators. Shrimp hosts and their partners are usually size-matched, i.e., larger gobies with larger shrimps and smaller gobies with smaller shrimps. This might come about in two ways: (1) the postlarval goby may pair up with a young shrimp big enough to make a burrow, and the partners grow together, or (2) a small goby with a large shrimp pair might be evicted by a bigger one looking for a larger burrow. Not any *Alpheus* species will do for goby symbionts, which have definite preferences for their hosts. *Myersina macrostoma* partners with *A. rapax.* In the same habitat *A. lobidens* and its burrows also occur, but the goby is not found with them. The goby, with it sharp vision, apparently distinguishes between the two species based on coloration and overall shape (Jaafar and Hou 2012).

The goby *Nes longus* is an obligatory associate of *A. floridanus*, while *Ctenogobius saepepallens* and *Bathygobius curacao* are facultative ones (Karplus 1992; Kramer et al. 2009). The symbiosis of these species has been the subject of several studies that together show how free-living gobies might co-evolve an obligatory relationship with an *Alpheus* species (Karplus 1992; Randall et al. 2005; Kramer et al. 2009; Lyons 2013). In the Florida Keys, the shrimp association with both *N. longus* and *B. curacao* was studied by Karplus (1992). The communication system with the shrimp of low-intensity caudal flick warnings and the high-alarm, head-first dive into the burrow by *N. longus* is sophisticated and well developed. With *B. curacao*, there is no low-intensity warning, only the high-alarm burrow dive, which *B. curacao* would probably do into any burrow, with or without shrimps, during an emergency. When with *N. longus,* the shrimp spends much more time outside of its burrow feeding than with *B. curacao*. Karplus occasionally observed an *A. floridanus* with two burrow openings, one stationed with a *N. longus* and one with *B. curacao*. The shrimp spent all its time at the opening with the obligate *N. longus*, switching from one burrow opening to the other when *N. longus* switched openings with the other goby. *A. floridanus* may also associate with another facultative symbiont, *C. saepepallens* (Randall et al. 2005; Kramer et al. 2009). Shrimp hosts with this goby showed similarly low activity outside the burrows as they did with *B. curacao*. *C. saepepallens* was always somewhere around a shrimp burrow but often not in contact with the shrimp. When the goby was close enough to the shrimp for it to make antennal contact, the latter stayed active outside the burrow. Unlike *B. curacao*, *C. saepepallens* makes the low-intensity caudal fin flicks characteristic of the obligate goby, *N. longus*. However, the shrimp did not react to them; sometimes the goby would make tail flicks in front of a burrow without a shrimp. *B. curacao* and *C. saepepallens* may represent a rudimentary stage in the evolution of shrimp–goby mutualism with some but not all its essential components.

Burrowing organisms often host more than one kind of symbiont in their homes. An interesting example of a three-way association between the host goby shrimp *A. bellulus*, an *Amblyeleotris* goby, and a porcellanid crab, *Enosteoides lobatus* in Indonesian waters has been described by Werding et al. (2016). Like other porcellanid crabs, *E. lobatus* is a filter feeder, extracting plankton and detrital particles from the water with its fan-like third maxillipeds. When the shrimp and goby are active outside the burrow, the crab sits at its entrance and filter feeds. If the shrimp and goby retreat into the burrow, the crab moves in and clings to the roof of burrow near the entrance. The crab, which also has relatively long antennal flagella, detects when the goby and shrimp move out again—danger over. The shrimp and goby do not act aggressively with the crab, which appears to be an unobtrusive commensal which receives a safe refuge to live and feed. Whether the crab is a host-specific facultative or obligatory symbiont still needs to be determined, as does its mating system and other aspects of its biology.

Biogeographical and molecular phylogenetic analyses of Indo-West Pacific shrimp–goby symbioses show that this mutualism has evolved independently twice in different groups (clades) of gobies nested within nonsymbiotic relatives (Thompson et al. 2005; Thacker et al. 2011). The small number of West Atlantic goby–shrimp symbioses represents yet another independent evolution of this relationship (Rüber et al. 2003). Perhaps indicative of the latter separate evolution of the fish–shrimp symbiosis, daily activity rhythms of the West Atlantic *A. floridanus* differ somewhat from that of Indo-West Pacific goby shrimps. Additionally, in the absence of a goby associate, *A. floridanus* stays active and keeps burrows open, while their Indo-Pacific counterparts close burrow entrances with only a small opening to allow a potential goby partner to insert its tail and establish a relationship with the shrimp (Karplus 1992; Karplus and Thompson 2011).

10.11 Shrimps That Clean Fishes: Cleaning Symbiosis

By the early 1950s, exploration of subtidal marine habitats with SCUBA (self-contained underwater diving apparatus) began in earnest, opening a new age in the direct observation and study of marine organisms. Among the early pioneers were Conrad Limbaugh, the first Diving Officer at the Scripps Institution of Oceanography, and Harry Pedersen, an underwater naturalist and photographer. Diving on coral reefs and other shallow-water habitats, they observed small shrimps mounting and picking at the bodies of live fishes, including predatory ones. They surmised that the shrimps were cleaning the fishes of ectoparasites, dead tissues around wounds, and excess mucus. Perhaps the best way to summarize their seminal field study is to quote directly from their 1961 paper with Fenner A. Chace, Jr., on shrimps that clean fishes: "Certain species of small, brightly colored shrimps have been observed to remove and eat parasites, injured tissue, and possibly undesirable food particles from a large variety of cooperating reef fishes. This symbiotic behavior is so well developed in some cases that the shrimps are invited to enter the mouths and gill cavities of their hosts, apparently with complete immunity from being eaten." Often criticized as anecdotal and speculative, e.g., Spotte (1998), this paper has stimulated numerous studies on the subject which have greatly expanded and corroborated Limbaugh et al.'s initial observations. There is little doubt that certain species of shrimps and fishes have evolved a mutualistic relationship characterized by communication between the symbiont cleaner shrimp and the fish, usually termed the "client."

This relationship is termed a "cleaning symbiosis" but does not quite fit the usual definition of a symbiosis as a prolonged intimate association between two species.

The contact between the two species in this case may be frequent, but it is not protracted—the two species spend most of their time apart, and cleaning bouts vary in duration from several seconds to several minutes. However, I suggested in Bauer (2004) that despite these caveats, the term is so well entrenched in the literature that it should be retained. The use of the term "cleaning symbiosis" to characterize this relationship was also justified by Vaughan et al. (2016) in their excellent review of the subject. They suggest a definition of cleaning symbiosis as "a cooperative interspecific behaviour where a cleaner removes and consumes materials that negatively impact a client and is preceded by their communication." Both the essential elements of actual cleaning with some sort of interspecific communication between the shrimp cleaner and the fish "client" are included in Limbaugh et al.'s quote above, although perhaps not so specifically. Several species of both fishes and shrimps specialize in cleaning fish clients.

In Chap. 4, we saw that autogrooming (self-cleaning) by shrimps is very important to the health and survival of shrimps and other crustaceans. Many species of birds, monkeys, apes, humans, and other social animals engage in allogrooming; i.e., one individual cleans another member of its own species of parasites, dead skin, etc.—a behavior which also enhances social bonds. Similar to cleaner fishes and shrimps, numerous species of birds (e.g., cattle egrets, oxpeckers) feed on ectoparasites, e.g., ticks and lice, on the surface of large mammals such as cattle, water buffalo, deer, hippopotami, and even alligators and crocodiles (e.g., Losey 1987; Poulin and Grutter 1996). Clearly, these cleaning services are beneficial for the clients which have no or only limited abilities to clean all their bodies by themselves. Fish are no exception; their fins are not adapted for autogrooming. They try to clean off ectoparasites and other irritations by rubbing themselves against hard surfaces ("chafing behavior"), body shaking, rapid acceleration, and coughing (Wisenden et al. 2009), none of them particularly efficient. Thus, the mutually beneficial associations have evolved between many fishes with shrimp and fish cleaners.

At least 52 shrimp species have been observed thus far in some sort of fish-cleaning behavior (supplementary table 2 from Vaughan et al. 2016). However, Huebner and Chadwick (2012) only considered that 6 palaemonid (*s.l.*) species are *proven* fish cleaners: 4 species of *Ancylomenes*, *Urocaridella degravei* ("sp. c"), and *Brachycarpus biunguiculatus*, i.e., species in which sufficient quantitative and field studies have been done to award this designation decisively. However, there is plenty of evidence on many other species in Vaughan et al.'s (2016) list to preliminarily designate them as fish cleaners. Note that on that list, there are no dendrobranchiate (penaeoid, sergestoid) species, probably for the reasons stated earlier for symbioses in general. Among the Caridea, cleaner shrimps are found mainly in the Lysmatidae and Palaemonidae (especially the pontoniines), a few species of Alpheidae, and one species each from the Barbouriidae (*Parhippolyte uveae*), Gnathophyllidae, Pandalidae, and Thoridae. Another thorid, *Lebbeus mundus*, should be added to the list (Jensen 2014). Additionally, 5 species of *Stenopus* (Stenopodidae) have been reported to clean fishes. The vast majority of reported shrimp cleaners are from tropical and warm-temperate waters, but there are certainly fish-cleaning shrimps from cool-temperate waters as well (e.g., Östlund-

Nilsson et al. 2005; Jensen 2014). The list of fish cleaners will certainly be updated and amended as time goes on.

Some shrimp cleaners have a lifestyle that is committed to attracting and cleaning fishes. They remain around specific areas or objects such as a large anemone or a specific coral head ("cleaning station"). The shrimps are active and visible during the day to attract fishes, which perceive them with their excellent vision. These specialized shrimps have bright flashy colors that stand out from their surroundings and use behaviors to attract the attention of potential fish clients. One such behavior is rocking of the body back and forth on their pereopods ("dancing"). The reader may recall that shrimps with setiferous pereopodal epipods use this behavior to clean the gills. But the palaemonid and stenopodidid shrimp cleaners lack these gill-cleaning epipods, so certainly this rocking behavior has evolved in cleaner shrimps to attract attention. These shrimps have long antennal and often antennular flagella, sometimes bright white, which are lashed and whipped back and forth. All these adaptations are related to a cleaning lifestyle; such shrimps have traditionally been termed "obligate" cleaners. Vaughan et al. (2016) proposed to substitute the term "dedicated" cleaners because "obligate" implies that most of their nutrition, contacts, and activities are tied to fish clients, which is probably not the case. I agree that "obligate" is best reserved for symbiotic species which fit into the "prolonged intimate contact" definition of symbiosis and will use "dedicated" instead of "obligate" for appropriate species of cleaner shrimps.

Some shrimp cleaners do so when the opportunity arises; their way of life does not revolve around cleaning. However, the fish client does cooperate or communicate with the shrimp by allowing it to crawl around and pick on its body, expose its gills, and open its mouth so that the shrimp can forage inside. These shrimp species are termed "facultative" cleaners. Good examples are *Lysmata californica* and *L. seticaudata*. They are nocturnally active, lack the flashy bright coloration, and display behaviors of dedicated cleaners. Both species hide out during the day, often in large groups in rocky caves inhabited by moray eels, which allow the shrimps to clean them (Limbaugh et al. 1961; Patzner 1982). If a crevice or cave is inhabited by spiny lobsters, *L. californica* will pick and nibble at the lobster's exoskeleton, or for that matter any organism which allows it to do so. I have entertained students and guests by letting them place their hand in an aquarium with *L. wurdemanni*, a Gulf of Mexico species. The shrimp quickly comes to it to pick and tug (clean) material from beneath the fingernails. However, this and several other *Lysmata* species are foraging omnivores that take a wide variety of food items. *L. seticaudata* can be abundant in Mediterranean seagrass meadows (Dohrn 1950) where opportunities to clean cooperating organisms are probably somewhat limited.

The following general description of a "typical" cleaning bout of a dedicated, diurnally active shrimp cleaner and its fish client is based on observations by numerous authors such as Limbaugh et al. (1961), Sargent and Wagenbach (1975) and others (Fig. 10.14). Fish clients begin the interaction by approaching the cleaning station upon which the shrimp displays by a rocking dance and/or antennal lashing. The fish stops near the station and poses in various ways: a stationary horizontal or head-down posture, fin extension, lifting of the gill operculum to

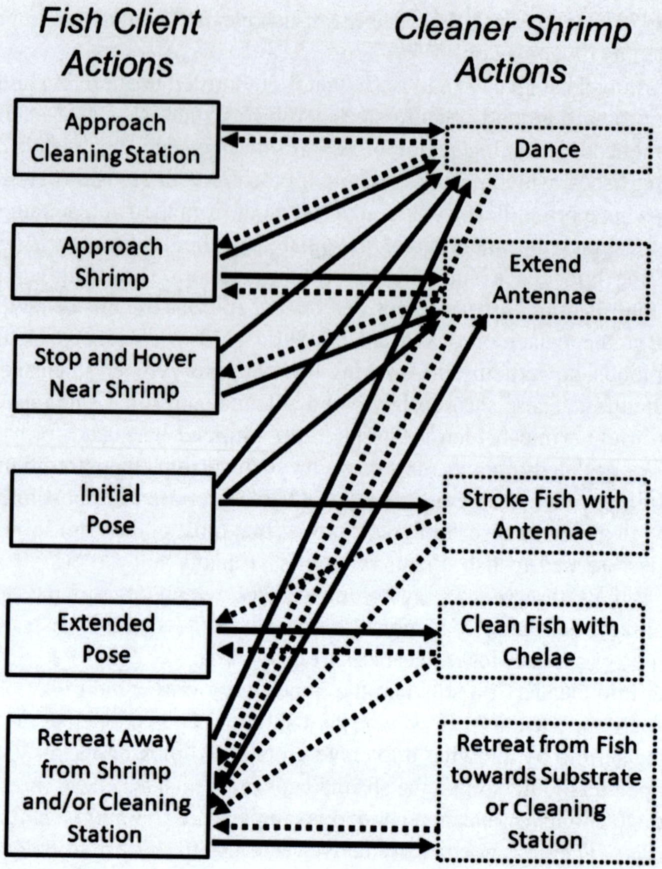

Fig. 10.14 Types of behavioral interactions between a fish client and the cleaner shrimp *Ancylomenes pedersoni* (as *Periclimenes anthophilus*) during typical bouts of cleaning behavior. Direction of arrow indicates the receiver of the action (e.g., fish approaches shrimp cleaning station, shrimp responds with dancing (rocking back and forth, waving both pairs of antennae)). Solid lines indicate fish actions; dashed lines indicate cleaner shrimp actions. Redrawn and modified from Sargent and Wagenbach (1975), courtesy of Bulletin of Marine Science

expose the gills, and/or mouth gaping to allow shrimp entry. Client fishes often change color, another possible invitation which the shrimp apparently perceives despite its relatively poor vision and color perception. The shrimp may respond by stroking the fish with antennal and antennular flagella and/or chelipeds. If not disturbed, it will hop onto the fish to begin its cleaning with one or both pairs of chelipeds (pereopods 1 and 2), It picks, nibbles, and tugs at damaged tissue, ectoparasites, and mucus on the various exposed fish surfaces. An especially stubborn item is often removed when the shrimp grips it with the chelipeds and forcefully tail flips backward. After cleaning, the shrimp retreats by hopping off the fish and returning to its cleaning station. The fish client usually leaves the cleaning station

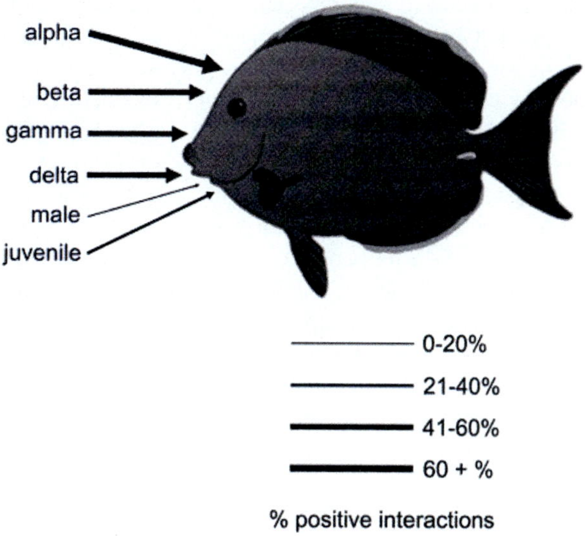

Fig. 10.15 Fish cleaning behavior by the cleaner shrimp *Ancylomenes pedersoni* with a client model in the laboratory, measured as percent positive interactions (signal, approach, clean) by different categories of individuals ranked by decreasing size and gender: alpha, beta, gamma, and delta females; males; juveniles. Modified from Gilpin and Chadwick (2020), courtesy of Journal of Crustacean Biology

and goes on about its normal activities or it may approach again to solicit another bout of cleaning behavior.

In *Ancylomenes pedersoni*, the hierarchy among females, males, and juveniles in occupying a favorable position on the host anemone may also apply to fish-cleaning opportunities. In a laboratory study, Gilpin and Chadwick (2020) found that positive fish-cleaning interactions (signaling, approaching, and cleaning behaviors) with a fish model in the laboratory were significantly greater in females than in males and juveniles. Larger-sized females had more positive interactions with the model than smaller ones (Fig. 10.15).

Fish cleaning by purely nocturnal cleaners such as the palaemonid *Brachycarpus biunguiculatus* occurs in which vision is obviously not a major factor. Adults of this species occur in pairs which hide out during the day but at night come out and perch on cleaning stations (rocks, boulders, or near urchins). They do not have the striking coloration of dedicated diurnal cleaners, behavioral dances, and antennal lashing displays (Corredor 1978; Criales and Corredor 1977). Perhaps guided mostly by chemoreception, a fish may approach the shrimp and sometimes touch them with the mouth while the shrimps stroke them with their antennae. To be cleaned, small fish lie on one side or the other, and the shrimp extends its larger third chelipeds to clip material from the fish and the smaller first and pair to pick up debris from cleaning. When one side is cleaned, the fish flips over so that the other side can be worked on. The shrimps do mount larger fishes such as morays. Unlike diurnal shrimp cleaners, *B. biunguilatus* does not clean inside the mouth or on the gills.

Stenopus hispidus is primarily a nighttime cleaner (Criales and Corredor 1977; Vaughan et al. 2018a). They also occur in male–female pairs with their small territory (crevice, cave, small clearing in a seagrass meadow) serving as a cleaning station. Unlike *B. biunguilatus,* they seize a small fish by its caudal peduncle using their massive third chelipeds and clean with the first two pair. Like the carideans, they mount larger fishes to clean them. Both *S. hispidus* and *S. scutellatus* have bright white antennae, but daytime antennal lashing to attract fishes seems to be a more frequent behavior in *S. scutellatus* (Limbaugh et al. 1961). However, it remains to be established if these two *Stenopus* species are dedicated or facultative cleaners in nature.

Do cleaner shrimps actually clean, and what exactly do they clean? It seemed apparent to investigators in earlier studies that the shrimps were feeding on something on the fish, probably ectoparasites, dead tissues, and mucus. However, this conclusion was criticized as being based only on anecdotal observations without any direct evidence, e.g., quantified removal of parasites or their presence in shrimp guts (Spotte 1998). Losey (1987) argued that the fish's reward was tactile stimulation (pleasurable sensation), common to all vertebrates, caused by the cleaner's walking, picking, and nipping at the fish surface—sort of a good human back scratch. However, in the last 20 years or so, several experimental studies and field observations have verified that, indeed, cleaner shrimps remove and ingest a wide variety of fish ectoparasites, especially the parasitic larvae of gnathiid isopods, monogenean flatworms (flukes), caligid copepods, ciliates, and hirudineans (leeches). Let us examine some of those studies.

One of the first studies showing a direct effect of shrimp cleaning on ectoparasites was done by Bunkley-Williams and Williams (1998). They introduced and parasitized common reef fishes with the newly settled juveniles of the parasitic cymothoid isopod *Anilocra haemuli* and then tested the ability of four reported shrimp cleaners, *Ancylomenes pedersoni*, *Lysmata grabhami*, *Stenopus hispidus,* and *S. scutellatus* to remove the parasite. Only *A. pedersoni* was able to do so and was very effective at it. Just after plucking the parasite from the fish, the shrimp drops to the bottom to eat it. The study shows that not all cleaner shrimps have the same abilities, relationships with clients, or prey preferences because the other three shrimp species have been shown or been reported as cleaners of other fishes with different parasites.

Gut content analyses of the cleaner shrimps *Urocaridella degravei* and *Periclimenes holthuisi* have shown the remnants of ectoparasites such as gnathiid isopods, copepods, and monogenean flukes (Becker and Grutter 2004). The chitinous, calcified exoskeletons of crustaceans, as well as the attachment hooks and organs of the flukes, are difficult for shrimps to digest. Consequently, body parts of these parasites can often be identified in shrimp guts. In laboratory experiments, these shrimps ate monogenean flatworms when offered to them but also significantly reduced them on parasitized captive fishes. *A. pedersoni* is also very effective in removing monogenean flukes (McCammon et al. 2010). Two temperate-water palaemonids, *Palaemon elegans* and *P. adspersus*, not usually considered cleaner shrimps, were very effective in removing and eating flukes and caligid copepod parasites from a flatfish, the plaice *Pleuronectes platessa* in the laboratory (Östlund-Nilsson et al. 2005). The shrimps were obviously attracted to the fishes which allowed them, as do all client fishes, to roam on their surface picking the ectoparasites.

Cleaner shrimps have varying abilities to clean ectoparasites. Their effectiveness may depend on their type of cleaning appendages (chelae), behavior, and alimentary preferences. This was nicely shown by a series of experiments with a grouper (*Epinephelus coioides)* involving three different types of ectoparasites and the cleaner shrimps *Lysmata amboinensis*, *L. vittata*, *Urocaridella antonbruunii*, and *Stenopus hispidus* (Vaughan et al. 2018a). The ectoparasites were the ciliate *Cryptocaryon irritans,* which causes a skin infection ("ich" or "white spot"), the fluke *Neobenedenia girellae*, and the hirudinean leech *Zeylanicobdella arugamensis.* The latter two suck blood and other body fluids. All are quite harmful to the grouper, especially in aquaculture where fish densities are high, and infections spread easily. Day–night differences in the behavior of both the fish host and cleaner shrimp, parasite susceptibility, and cleaner shrimp prey preferences may account for the variation in shrimp cleaner efficacy. The cleaner shrimps were good at cleaning up the parasite tomont stage which attaches to surfaces around the aquarium. It is this stage that asexually produces many trophonts (the infective stage). Some of the shrimps were effective at eating leech egg cases, an important component of reducing leech infections.

Among disputed ideas about shrimp cleaning is whether they clean and ingest injured or infected areas of damaged tissues from the surface of fishes as indicated by anecdotal field observations. However, laboratory experiments involving a serranid fish (*Pseudanthias squamipinnis*) and *Lysmata amboinensis*, a known cleaner shrimp, verified that this is indeed the case (Vaughan et al. 2018b). Anesthetized fishes were surgically inflicted with a standardized superficial wound which then developed the typical redness ("rubor") associated with healing and infected injuries. In the treatment group, fishes were allowed access to cleaner shrimps, while those in the control group were not. The shrimp's feeding on injured and/or infected tissues in the treatment group greatly accelerated wound healing compared to that of controls. Shrimp cleaning of wounds is a definite benefit for fish clients.

The mucus covering fish scales and skin plays various important roles in the biology of fishes. It allows the fish to swim through the dense medium of water (Hoyt 1975) more efficiently. Besides the hydrodynamic effect, fish mucus is a first defense against microbial infection and surface fouling (Dash et al. 2018). Fish constantly secrete mucus which later sloughs off the fish, carrying with it entrapped microbes, possible pathogens, and particles that accumulate in it. Shrimp cleaners take advantage of the high caloric and nutritional content of mucus by using it as a food source when the fish allows, and which may help the fish to rid the fish of old excess mucus. There are certainly many anecdotal observations of shrimps apparently picking and ingesting mucus off fishes (Titus et al. 2015).

Some cleaner fishes and shrimps "cheat" on the mutualistic nature of the cleaner–host relationship by snatching bits of mucus and fish tissue or otherwise inflicting minor damage during the course of a cleaning bout (Poulin and Vickery 1995; Soares et al. 2008; Chapuis and Bshary 2009; Vaughan et al. 2018b; Titus et al. 2019). Fish clients often punish cheating by chasing the cleaner, by mouthing the shrimp without hurting it, or by simply terminating the interaction and leaving the cleaning station. An incident of cheating is usually identified by "jolting" behavior, in which the fish client jerks away from the shrimp. Cheating is not always punished; a cleaning bout may continue after a jolt. Although the term "cheating" implies intent, i.e., a behavior with adaptive value, perhaps many or most incidents of cheating may simply be collateral damage of wound or ectoparasite removal. Obviously, the fish is irritated or experiences some pain when it responds to cleaning by jolting. Perhaps a comparison with human behavior is appropriate. When a health-care worker examines a wound or injects a patient, the latter often flinches or jerks away from the irritation or pain; the beneficial action has an unintended and nondeliberate cost.

In nature, the mutualistic cleaning symbiosis can only be maintained over evolutionary time if the cleaner shrimp does not cheat too frequently, and the fish client does not punish too severely. The benefits to the fish client of being cleaned have to outweigh the costs of any purposeful or accidental damage inflicted by the cleaner shrimp. The nutritional benefits to the shrimp from cleaning the client must exceed costs of punishment by the client fish, e.g., loss of food derived from cleaning and/or protection from injury and predation. In a perfect cleaning mutualism, the cleaner would never cheat (deliberately or incidentally) and the fish client would never punish. However, different degrees of precision in the mutualistic cleaning relationship have been shown by various studies. A good example is a comparison between the dedicated cleaner shrimps *Ancylomenes longicarpus* from the Indo-West Pacific and *A. pedersoni* in the Caribbean (Chapuis and Bshary 2009; Huebner and Chadwick 2012; Titus et al. 2019). *A. longicarpus* does frequently cheat on its clients but responds to client punishment by reducing the rate of cheating. It cheats less frequently on predatory fish clients than on non-predatory ones, perhaps because the risk of punishment by predators is greater. On the other hand, Huebner and Chadwick (2012) saw little cheating by *A. pedersoni* in the Virgin Islands, and no punishment when it did happen. Somewhat differently, Titus et al. (2019) showed frequent cheating by *A. pedersoni* in the Gulf of Honduras but little retribution by fish clients. Unlike *A. longicarpus*, *A. pedersoni* did not vary its cleaning services between predatory and non-predatory clients. The differences between cleaning symbiosis in the two *Ancylomenes* species may reflect the independent evolution of cleaning symbiosis in this genus in both the Indo-West Pacific and the Caribbean (Horká et al. 2018).

Cleaner shrimps appear protected from predation by clients, especially around cleaning stations or when cleaning. The evidence for this statement is mainly negative, i.e., a lack of observations showing predatory fish clients eating their cleaner shrimps in the vicinity of cleaning stations in nature or in laboratory studies, e.g., Limbaugh et al. (1961) and Vaughan et al. 2018a, b). Numerous observations have been made of cleaner shrimps foraging within the mouths of predatory fish clients. Predator clients may have an innate or perhaps learned tendency to leave cleaners, especially dedicated ones, alone. This is obviously beneficial to both species and is necessary for evolution of cleaning mutualism. However, this protection is not perfect, as shown by observations of predation on *Stenopus hispidus* by two fish species (Criales and Corredor 1977), as well as the presence of the facultative cleaner *Lysmata californica* in the guts of moray eels, known clients

(Limbaugh et al. 1961). Wirtz (1997) observed fishes eating *Lysmata grabhami* when it was displaced from its anemone host.

Some shrimp species, like some fish species (Côté and Cheney 2005), may mimic true cleaner species in morphology, color pattern, and behavior without actually being cleaners. The benefits of mimicry would be protection from fish predation and theft of mucus and tissues from duped fish clients. Most mentioned in this regard in shrimps is *Periclimenes yucatanicus*. It is an anemone shrimp and often occurs on the same anemone as the dedicated cleaner *Ancylomenes pedersoni*. It closely resembles *A. pedersoni* in general morphology, coloration, and even display behavior (Limbaugh et al. 1961; Mahnken 1972). These are all marks of a cleaner mimic, although in *P. yucatanicus* the adaptive value was only suggested to be protection from predation, not a free meal from defrauded fishes. However, several field investigations have failed to reveal good evidence of fish cleaning by *P. yucatanicus* (Huebner and Chadwick 2012). On the other hand, careful extensive field observations by Titus et al. (2017) show that it does clean fishes but only on a limited basis compared to *A. pedersoni* and other dedicated fish cleaners. Both Mahnken and Titus et al. suggested that *P. yucatanicus* represents a rudimentary stage in the evolution of true cleaning behavior, with the similarities to *A. pedersoni* due to convergent evolution.

10.12 Evolution of Symbioses in Shrimps

How did shrimps evolve symbiotic associations? The reconstruction of the evolutionary history of any adaptive feature is always difficult as we cannot go back in time to observe the steps directly. It is best approached with the comparative method, as did Charles Darwin for so many plant and animal adaptations, combined with construction of reliable phylogenies. In the case of symbioses, of course the most reasonable hypothesis is that symbioses evolved in a variety of free-living ancestors. As described earlier in the chapter for pontoniine shrimps associated with corals, free-living species show the least specialized body morphology and natural history (e.g., Bruce 1976a, b). Pontoniines associated with the coral *Pocillopora acuta* (*damicornis*) show a trend from free-living epibenthic species with a laterally compressed body, long pereopods, and simple dactyls to the highly modified morphology of symbiotic species for living among coral branches (Patton 1974). The free-living species use the coral as a daytime perch but are also found elsewhere around the reef. Within some genera with symbiotic species, one may find completely free-living species but others that have the tendency to associate with possible host species.

Some examples may show how obligate symbiosis can evolve from free-living ancestors. For example, the pontoniine *Cuapetes americanus* is one of the most abundant free-living species in tropical West Atlantic seagrass meadows that I studied (Bauer 1985), but it is also often found loosely associated with sea anemones which are hosts to related symbiotic species (Mahnken 1972). Similarly, *Thor manningi* is a common seagrass species, but I also found it likes to preferentially perch on sessile cnidarians such as fire coral (*Millepora* species) or sea anemones. Its congener *T. amboinensis* goes a big step further and is an obligate symbiont of sea anemones and corals. Like most alpheids, members of the cool-temperate genus *Betaeus* hide in some sort of shelter, at least when not out foraging or seeking a mate. This genus has both free-living and symbiotic species, the latter with living refuges such as beneath sea urchins, in the mantle cavity of abalones (Gastropoda), or in burrows of echiuroid worms or other crustaceans. Several species are gregarious and free-living. A few species may live free or in pairs when associated with a host, while others are socially monogamous obligate symbionts. A molecular and morphological phylogeny of the genus showed that the various symbioses evolved independently multiple times within the genus from free-living ancestors (Anker and Baeza 2012).

Free-living species were also the ancestors of symbiotic palaemonid shrimps according to a phylogenetic analysis by Horká et al. (2016). By looking at the distribution of hosts on the species genealogy, the authors concluded that cnidarians such as sea anemones and corals were the first palaemonid hosts. From there, some shrimp palaemonids began new lineages by switching to novel hosts such as molluscs, ascidians, sea urchins, and other invertebrates. The initial move from one host phylum to another must have involved specialized symbiotic species whose particular set of behavioral and morphological adaptations for living with cnidarians were preadaptations for life with a completely different host, e.g., endosymbiotic tendencies, reduction of spines, dorsal or lateral compression of the body, and feeding preferences. From one non-cnidarian group, such as an ascidian, another lineage might move into a similar type of host such as a bivalve mollusc, e.g., within an ascidian branchial chamber to the mantle cavity of a bivalve mollusc or vice versa. These phylum-level shifts probably occurred several times. A change back from a symbiotic to a free-living life is apparently rare.

Among the Stenopodidea, members of the family Stenopodidae are mostly tropical free-living, shallow-water species, while most spongicolids are deep-sea endosymbionts of hexactinellid sponges. The shallow-water, free-living spongicolid *Microprosthema* is an apparent transitional group between the two. What do phylogenetic analyses have to say about this? A strictly morphological analysis by Saito and Takeda (2003) suggested that, yes, a relatively unspecialized, shallow-water, free-living species gave rise to deep-sea endosymbiotic spongicolids. This hypothesis was tested independently by a detailed molecular phylogeny on numerous stenopodidean species by Chen et al. (2016). Their work seems to set Saito and Takeda's hypothesis upside down. The Chen study indicated that the earliest stenopodidean lineages were likely deep-water species that diverged into several lineages, a major one being the endosymbiotic sponge symbionts, while other lineages, including both ectosymbiotic and free-living spongicolids and stenopodids, invaded shallow tropical waters. However, those conclusions are not set in stone, and I am sure this is not the last word on the subject.

Fish cleaning by shrimps has certainly evolved independently several times. Taxa containing fish-cleaning species occur in distantly related caridean groups such as the Palaemonidae and Alpheoidea (Alpheidae, Lysmatidae, Thoridae). Within the Palaemonidae alone, dedicated cleaning symbioses may have evolved at least five times (clades containing *Urocaridella, Ancylomenes, Periclimenes, Palaemon, Brachycarpus*: Horká et al. 2018) In the Alpheoidea, fish cleaning includes species of *Alpheus* and *Lysmata* which probably represent independent evolutionary events. The steps in the evolution of fish cleaning might have been something like the following: many shrimp species are epibenthic foragers and grazers, picking over any surface in their search for food. If the shrimp hops on a sedentary or resting fish and begins to pick at its surface without injury, the fish may tolerate it just for the pleasurable tactile stimulation common to many vertebrates, including humans (Losey 1987). Shrimps would naturally tend to pick and eat food items such as ectoparasites and damaged and dead tissue around wounds and surface infections. This certainly has adaptive value for both fish and shrimp, leading to facultative fish cleaning as described earlier in *Lysmata californica* and *L. seticaudata*. That level of cleaning might then lead on to a more perfected mutualism as seen in dedicated shrimp cleaners today.

In the last 20 years, the number and sophistication of field and laboratory experiments on all types of symbioses has greatly improved our understanding of their evolution. However, many questions still need to be answered. What do symbiotic species do at night when field observations are difficult? Anecdotal evidence suggests that shrimp symbionts of sessile invertebrate hosts leave their hosts at night, when visual predators are not active, to forage for food or to search for mates. For example, fish-cleaning species such as *Ancylomenes pedersoni* live in socially structured groups on anemone hosts; individuals might leave the protection of their host anemone at night to search for another anemone where they might be able to increase their social status, which increases their fish-cleaning opportunities (Gilpin and Chadwick 2020). Some fish cleaners clean only or more efficiently at night (e.g., Corredor 1978; Vaughan et al. 2018a), so field observations day and night with underwater surveillance equipment trained on known cleaning stations would be helpful and avoid problems associated with SCUBA divers and their equipment. Twenty-four-hour video surveillance in the laboratory is much easier and practical than in the field, but the lab macrocosm has to be as natural as possible to get good observations. Questions about cheating and mimicry need further exploration, as well as the true adaptive value of shrimp cleaning to fish clients. Dietary analyses on all types of shrimp symbionts are badly needed as only a few good ones have been reported. These analyses can combine visual and chemical observations on gut contents as well as various stable isotopic and DNA methods to obtain information. Further comparative studies reconstructing the adaptive steps from free-living to obligate symbioses or dedicated fish cleaning, together with phylogenetic analyses, will help in fleshing out our understanding about shrimp symbioses.

References

Ache BW, Davenport D (1972) The sensory basis of host reception by symbiotic shrimps, genus *Betaeus*. Biol Bull 143:94–111

Anker A, Baeza JA (2012) Molecular and morphological phylogeny of hooded shrimps, genera *Betaeus* and *Betaeopsis* (Decapoda, Alpheidae): testing the center of origin biogeographic model and evolution of life history traits. Mol Phylogenet Evol 64:401–415

Anker A, Jeng MS, Chan TY (2001) Two unusual species of Alpheidae (Decapoda: Caridea) associated with upogebiid mud shrimps in the mudflats of Taiwan and Vietnam. J Crustac Biol 21:1049–1061

Antokhina TI, Britayev TA (2020) Host recognition behaviour and its specificity in pontoniine shrimp *Zenopontonia soror* (Nobili, 1904) (Decapoda: Caridea: Palaemonidae) associated with shallow-water sea stars. J Exp Mar Biol Ecol 524:151302

Arai MN (2005) Predation on pelagic coelenterates: a review. J Exp Mar Biol Ecol 85:523–536

Arnal I, Côté M, Morand S (2001) Why clean and be cleaned? The importance of client ectoparasites and parasites in a marine cleaning symbiosis. Behav Ecol Sociobiol 51:1–7

Aucoin S, Himmelman JH (2010) A first report on the shrimp *Pontonia* sp. and other potential symbionts in the mantle cavity of the penshell *Pinna carnea* in the Dominican Republic. Symbiosis 50:135–141

Baeza JA (2008) Social monogamy in the shrimp *Pontonia margarita*, a symbiont of *Pinctada mazatlanica,* off the Pacific coast of Panama. Mar Biol 153:387–395

Baeza JA, Díaz-Valdéz M (2011) The symbiotic shrimp *Ascidonia flavomaculata* lives solitarily in the tunicate *Ascidia mentula*: implications for its mating system. Invertebr Biol 130:351–361

Baeza JA, Farías NE, Luppi TA, Spivak ED (2010) Refuge size, group living and symbiosis: testing the "resource economic monopolization" hypothesis with the shrimp *Betaeus lilianae* and description of its partnership with the crab *Platyxanthus crenulatus*. J Exp Mar Biol Ecol 389: 85–92

Baeza JA, Bolaños JA, Hernandez JE, Lira C, López R (2011) Monogamy does not last long in *Pontonia mexicana*, a symbiotic shrimp of the amber pen-shell *Pinna carnea* from the southeastern Caribbean Sea. J Exp Mar Biol Ecol 407:41–47

Baeza JA, Ritson-Williams R, Fuentes MS (2013) Sexual and mating system in a caridean shrimp symbiotic with the winged pearl oyster in the Coral Triangle. J Zool 289:172–181

Baeza JA, Simpson L, Ambrosio LJ, Guéron R, Mora N (2016) Monogamy in a hyper-symbiotic shrimp. PLoS One 11(3):e0149797

Baeza JA, Barros-Alves SP, Lucena RA, Lima SFB, Alves DFR (2017) Host-use pattern of the shrimp *Periclimenes paivai* on the scyphozoan jellyfish *Lychnorhiza lucerna*: probing for territoriality and inferring its mating system. Helgol Mar Res 71:17. https://doi.org/10.1186/s10152-017-0497-8

Bauer RT (1985) Diel and seasonal variation in species composition and abundance of the caridean shrimps (Crustacea, Decapoda) from seagrass meadows in northern Puerto Rico. Bull Mar Sci 36:150–162

Bauer RT (2004) Remarkable shrimps: adaptations and natural history of the carideans. University of Oklahoma Press, Norman

Beck JT (1979) Population interactions between a parasitic castrator, *Proboyrus pandalicola*, and one of its freshwater shrimp hosts, *Palaemonetes paludosus* (Decapoda: Caridea). Parasitology 79:431–449

Becker JH, Grutter AS (2004) Cleaner shrimp do clean. Coral Reefs 23:515–520

Berggren M (1993) *Spongiocaris hexactinellicola*, a new species of stenopodidean shrimp (Decapoda: Stenopodidae) associated with hexactinellid sponges from Tartar Bank, Bahamas. J Crustac Biol 13:784–792

Berggren M (1994) *Periclimenes nomadophila* and *Tuleariocaris sarec*, two new species of pontoniine shrimps (Decapoda: Pontoniinae), from Inhaca Island, Moçambique. J Crustac Biol 14:782–802

Berggren M (1999) *Notopontonia platycheles*, a little-known shallow-water shrimp from western Australia (Decapoda: Pontoniinae). J Crustac Biol 19:180–187

Boltana S, Thiel M (2001) Associations between two species of snapping shrimp, *Alpheus inca* and *Alpheopsis chilensis* (Decapoda: Caridea: Alpheidae). J Marine Biol Assoc UK 81:633–638

Borradaile LA (1921) On the coral-gall prawn *Paratypton*. Mem Proc Manch Lit Philos Soc 65:1–11

Brasseur L, Caulier G, Lepoint G, Gerbaux P, Eeckhaut I (2018) *Echinometra mathaei* and its ectocommensal shrimps: the role of sea urchin spinochrome pigments in the symbiotic association. Sci Rep 8:17540. https://doi.org/10.1038/s41598-018-36079-8

Bruce AJ (1969a) Notes on some Indo-Pacific Pontoniinae. XIV. Observations on *Paratypton siebenrocki* Balss. Crustaceana 17:171–186

Bruce AJ (1969b) *Aretopsis amabilis* De Man, an alpheid shrimp commensal of pagurid crabs in the Seychelle Islands. J Mar Biol Assoc India 11:175–181

Bruce AJ (1972a) A review of information upon the coral hosts of commensal shrimps of the subfamily Pontoniinae, Kingsley, 1878 (Crustacea, Decapoda, Palaemonidae). Proceedings of the Symposium on Corals and Coral Reefs, 1969, Marine Biological Association of India: 399–418

Bruce AJ (1972b) An association between a pontoniinid shrimp and a rhizostomatous scyphozoan. Crustaceana 23:300–302

Bruce AJ (1975) Notes on Some Indo-Pacific Pontoniinae. XXVI. *Neoanchistus cardiodytes* gen. nov., sp.nov., a new mollusc-associated shrimp from Madagascar (Decapoda, Palaemonidae). Crustaceana 29:149–165

Bruce AJ (1976a) Shrimps and prawns of coral reefs, with special reference to commensalism. In: Jones O, Endean R (eds) Biology and geology of coral reefs, Vol. III, Biology 2. Academic Press, New York, pp 38–94

Bruce AJ (1976b) Coral reef Caridea and commensalism. Micronesica 12:83–98

Bruce AJ (1977) Pontoniine shrimps of Heron Island. Atoll Res Bull 245:133

Bruce AJ (1982) The shrimps associated with Indo-West Pacific echinoderms, with the description of a news species in the genus *Periclimenes* Costa, 1844 (Crustacea: Pontoniinae). Mem Aust Mus 18:191–216

Bruce AJ (2000) Biological observations on the commensal shrimp *Paranchistus armatus* (H. Milne Edwards) (Crustacea: Decapoda: Pontoniinae) The Beagle: Records of the Museums and Art Galleries of the Northern Territory 16:91–96

Bruce AJ, Okuno J (2006) *Periclimenes dardanicola* n. sp., a new species of hermit crab associated shrimp (Crustacea, Decapoda, Palaemonidae) from the western Pacific. Zoosystema 28:367–377

Bruce AJ, Trautwein SE (2007) The coral gall shrimp, *Paratypton siebenrocki* Balss, 1914 (Crustacea: Decapoda: Pontoniinae), occurrence in French Polynesia, with possible abbreviated larval development. Cah Biol Mar 48:225–228

Bruce AJ, Zmarzly DL (1983) *Periclimenes pilipes*, new species, a crinoid associate from Enewetak Atoll, Marshall Islands (Crustacea: Decapoda: Pontoniinae). J Crustac Biol 3:644–654

Brusca RC, Moore W, Shuster SM (2016) Invertebrates, 3rd edn. Sinauer Associates, Sunderland

Bunkley-Williams L, Williams EH Jr (1998) Ability of Pederson cleaner shrimp to remove juveniles of the parasitic cymothoid isopod, *Anilocra haemuli*, from the host. Crustaceana 71: 862–869

Cantrell CE, Henry RP, Chadwick NE (2015) Nitrogen transfer in a Caribbean mutualistic network. Mar Biol 162:2327–2338

Castro P (1974) A new host and notes on the behavior of *Tuleariocaris neglecta* Chace, 1969 (Decapoda, Palaemonidae, Pontoniinae), a symbiont of diadematid sea urchins. Crustaceana 26: 318–320

Castro R (1971) The natantian shrimps (Crustacea, Decapoda) associated with invertebrates in Hawaii. Pac Sci 25:395–403

Chace FA Jr (1969) A new genus and five new species of shrimps (Decapoda, Palaemonidae, Pontoniinae) from the Western Atlantic. Crustaceana 16:251–272

Chak STC, Duffy JE, Hultgren KM, Rubenstein DR (2017) Evolutionary transitions towards eusociality in snapping shrimps. Nat Ecol Evol 1:0096. https://doi.org/10.1038/s41559-017-0096

Chapuis L, Bshary R (2009) Strategic adjustment of service quality to client identity in the cleaner shrimp, *Periclimenes longicarpus*. Anim Behav 78:455–459

Cházaro-Olvera S, Winfield I, Barcena-Cisneros M, Ortiz M (2013) Species of the genus *Periclimenaeus* (Decapoda, Caridea, Palaemonidae) associated with sponges from the Veracruz Coral Reef System National Park, SW Gulf Of Mexico. Crustaceana 86:641–650

Chen CL, Goy JW, Bracken-Grissom HD, Felder DL, Tsang LM, Chan T-Y (2016) Phylogeny of Stenopodidea (Crustacea: Decapoda) shrimps inferred from nuclear and mitochondrial genes reveals non-monophyly of the families Spongicolidae and Stenopodidae and most of their composite genera. Invertebr Syst 30:479–490

Corredor L (1978) Notes on the behavior and ecology of the new fish cleaner shrimp *Brachycarpus biunguiculatus* (Lucas) (Decapoda, Natantia, Palaemonidae). Crustaceana 35:35–40

Côté I, Cheney K (2005) Choosing when to be a cleaner-fish mimic. Nature 433:211–212

Courtney LA, Couch JA (1981) Aspects of the host-commensal relationship between a palaemonid shrimp (*Pontonia domestica*) and the pen shell (*Atrina rigida*). Northeast Gulf Science 5:49–54

Crawford JA (1992) Acclimation of the shrimp, *Periclimenes anthophilus*, to the giant sea anemone, *Condylactis gigantea*. Bull Mar Sci 50:331–341

Criales MM, Corredor L (1977) Aspectos etológicos y ecológicos de camarones limpiadores de peces (Natantia: Palaemonidae, Hippolytidae, Stenopodidae). Anales Del Instituto de Investigaciones Marinas de Punta de Betín 9:141–156

Dash S, Das SK, Samal J, Thatoi HN (2018) Epidermal mucus, a major determinant in fish health: a review. Iranian J Vet Res 19:72–81

Debelius H (1999) Crustacea Guide of the World. IKAN, Frankfurt

Deheyn DD, Lyskin S, Eeckhaut I (2006) Assemblages of symbionts in tropical shallow-water crinoids and assessment of symbionts' host-specificity. Symbiosis 42:161–168

Didderen K, Fransen CHJM, de Voogd NG (2006) Observations on sponge-dwelling colonies of *Synalpheus* (Decapoda, Alpheidae) of Sulawesi, Indonesia. Crustaceana 79:961–975

Dobson NC, De Grave S, Johnson ML (2014) Linking eye design with host symbiont relationships in pontoniine shrimps (Crustacea, Decapoda, Palaemonidae). PLoS One 9:99505. https://doi.org/10.1371/journal.pone.0099505

Dohrn PFR (1950) Studi sulla *Lysmata seticaudata* Risso (Hyppolytidae). I. Le condizioni normali della sessualita in natura. Pubblicazioni della Stazione Zoologica di Napoli 22:257–272

Duffy JE (1992) Host use patterns and demography in a guild of tropical sponge-dwelling shrimps. Mar Ecol Prog Ser 90:127–138

Duffy JE (1996) Eusociality in a coral-reef shrimp. Nature 381:512–514

Duffy JE (2002) The ecology and evolution of eusociality in sponge-dwelling shrimp. In: Kikuchi T (ed) Genes, behavior, and evolution in social insects. University of Hokkaido Press, Sapporo, Japan, pp 1–38

Duffy JE, Morrison CL, Macdonald KS (2002) Colony defense and behavioral differentiation in the eusocial shrimp *Synalpheus regalis*. Behav Ecol Sociobiol 51:488–495

Ďuriš Z, Horká I, Juračka PJ, Petrusek A, Sandford F (2011) These squatters are not innocent: the evidence of parasitism in sponge-inhabiting shrimps. PLoS One 6:e21987. https://doi.org/10.1371/journal.pone.0021987

Elliott JK, Mariscal RN (1997) Acclimation or innate protection of anemonefishes from sea anemones? Copeia 97:284–289

Fabricius KE (2005) Effects of terrestrial runoff on the ecology of corals and coral reefs: review and synthesis. Mar Pollut Bull 50:125–146

Fautin DG (1991) The anemonefish symbiosis: what is known and what is not. Symbiosis 10:23–46

Fautin DG, Guo C-C, Hwang H-S (1995) Costs and benefits of the symbiosis between the anemoneshrimp *Periclimenes brevicarpalis* and its host *Entacmaea quadricolor*. Mar Ecol Prog Ser 129:77–84

Felder DL, Manning RB (1986) New genus and two new species of alpheid shrimps (Decapoda: Caridea) from south Florida. J Crustac Biol 6:497–508

Filho JEM, Stampar SN, Morandini AC, Mossolin EC (2008) Cleaner shrimp (Caridea: Palaemonidae) associated with scyphozoan jellyfish. Vie et Milieu Life and Environment 58: 133–140

Fransen CHJM (2002) Taxonomy, phylogeny, historical biogeography, and historical ecology of the genus *Pontonia* (Crustacea: Decapoda: Caridea: Palaemonidae). Zoologische verhandelingen Leiden 336:1–433

Fransen CHJM (2006) On Pontoniinae (Crustacea, Decapoda, Palaemonidae) collected from ascidians. Zoosystema 28:713–746

Garrone-Neto D, Gadig OBF, Zuanon J, Carvalho LN (2014) Cleaning interactions between shrimps (Palaemonidae) and freshwater stingrays (Potamotrygonidae) in the Parana River, Southeastern Brazil. Ichthyol Explor Freshw 24:379–384

Gherardi F (1991) Eco-ethological aspects of the symbiosis between the shrimp *Athanas indicus* (Coutière 1903) and the sea urchin *Echinometra mathaei* (de Blainville 1825). Trop Zool 4:107–128

Gilpin JA, Chadwick NE (2020) Social behavior of the Pederson cleaner shrimp *Ancylomenes pedersoni* (Chace, 1958) (Decapoda: Caridea: Palemonidae) in the Caribbean Sea: size- and gender-based dominance hierarchy. J Crustac Biol 40:24–36

Glynn PW (1976) Some physical and biological determinants of coral community structure in the Eastern Pacific. Ecol Monogr 46:431–456

Glynn PW (1980) Defense by symbiotic crustacea of host corals elicited by chemical cues from predator. Oecologia 3:287–290

Glynn PW (1987) Some ecological consequences of coral-crustacean guard mutualisms in the Indian and Pacific oceans. Symbiosis 4:301–324

Gonçalves GRL, Wolf MR, da Costa RC, Castilho AL (2016) Decapod crustacean associations with scyphozoan jellyfish (Rhizostomeae: Pelagiidae) in the southeastern Brazilian coast. Symbiosis 69:193–198

Goy JW (2010) Infraorder Stenopodidea Claus, 1872. In: Schram FR, von Vaupel Klein JC, Forest J, Charmantier-Daures M (eds) Treatise on zoology – anatomy, taxonomy, biology. The Crustacea, Decapoda. Vol. 9. Part A: Eucarida: Euphausiacea, Amphionidacea, and Decapoda (partim). Koninklijke Brill NV, Leiden, pp 215–265

Guo C-C, Hwang J-S, Fautin DG (1996) Host selection by shrimps symbiotic with sea anemones: a field survey and experimental laboratory analysis. J Exp Mar Biol Ecol 202:165–176

Hart JFL (1964) Shrimps of the genus *Betaeus* on the Pacific coast of North America with description of three new species. Proc US Natl Mus 115:431–466

Hayashi K, Miyake S (1968) Three caridean shrimps associated with a medusa from Tanabe Bay, Japan. Publ Seto Mar Biol Lab 16:11–19

Herberholz J, Schmitz B (2001) Signaling via water currents in behavioral interactions of snapping shrimp (*Alpheus heterochaelis*). Biol Bull 201:6–16

Hipeau-Jacquotte R (1973a) Manifestation d'un comportement territorial chez les crevettes Pontoniinae associées aux mollusques Pinnidae a Tuléar (Madagascar). J Exp Mar Biol Ecol 13:63–71

Hipeau-Jacquotte R (1973b) Étude des crevette Pontoniinae (Palaemonidae) associées aux mollusques Pinnidae a Tuléar (Madagascar). 6. Comportement sexual. Tethys 5:403–408

Horká I, De Grave S, Fransen CHJM, Petrusek A, Duris Z (2016) Multiple host switching events shape the evolution of symbiotic palaemonid shrimps (Crustacea: Decapoda). Sci Rep 6:26486. https://doi.org/10.1038/srep26486

Horká I, De Grave S, Fransen CHJM, Petrusek A, Ďuriš Z (2018) Multiple origins and strong phenotypic convergence in fish-cleaning palaemonid shrimp lineages. Mol Phylogenet Evol 124:71–81
Hou Z, Liew J, Jaafar Z (2013) Cleaning symbiosis in an obligate goby–shrimp association. Mar Biol 160:2775–2779
Hoyt JW (1975) Hydrodynamic drag reduction due to fish slimes. In: Wu TY-T, Brokaw CJ, Brennen C (eds) Swimming and flying in nature. Springer Science+Business Media, New York, pp 653–672
Huang HD, Rittschof D, Jeng M-S (2005a) Multispecies associations of macrosymbionts on the comatulid crinoid *Comanthina schlegeli* (Carpenter) in southern Taiwan. Symbiosis 39:47–51
Huang H-D, Rittschof D, Jeng M-S (2005b) Visual orientation of the symbiotic snapping shrimp *Synalpheus demani*. J Exp Mar Biol Ecol 326:56–66
Huebner LK, Chadwick NE (2012) Patterns of cleaning behaviour on coral reef fish by the anemoneshrimp *Ancylomenes pedersoni*. J Mar Biol Assoc U K 92:1557–1562
Hultgren KM, Duffy JE (2010) Sponge host characteristics shape the community structure of their shrimp associates. Mar Ecol Prog Ser 407:1–12
Jaafar Z, Hou Z (2012) Partner choice in gobiid fish *Myersina macrostoma* living in association with the alpheid shrimp *Alpheus rapax*. Symbiosis 56:121–127
Jaafar Z, Tan W, Dexiang C (2014) Goby and shrimp associations: more than meets the eye. Coral Reefs 33:864
Jensen GC (2014) Crabs and shrimps of the Pacific coast. A guide to shallow-water decapods from Southeastern Alaska to the Mexican Border. MolaMarine, Bremerton, Washington
Johnson DS, Liang M (1966) On the biology of the Watchman prawn, *Anchistus custos* (Crustacea; Decapoda; Palaemonidae), an Indo-West Pacific commensal of the bivalve *Pinna*. J Zool (Lond) 150:433–455
Jonsson LG, Lundälv T, Johannesson K (2001) Symbiotic associations between anthozoans and crustaceans in a temperate coastal area. Mar Biol Prog Ser 209:189–195
Karplus I (1992) Obligatory and facultative goby-shrimp partnerships in the Western Tropical Atlantic. Symbiosis 12:275–291
Karplus I, Thompson AR (2011) The partnership between gobiid fishes and burrowing alpheid shrimps. In: Patzner RA, Van Tassell JL, Kovačić M, Kapoor BG (eds) The biology of gobies. Science Publishers (CRC Press), Boca Raton, pp 559–607
Kensley B (1983) New records of bresiliid shrimp from Australia, South Africa, Caribbean, and Gulf of Mexico (Decapoda: Natantia: Caridea). Smithson Contrib Zool 394:31
Khan RN, Becker JHA, Crowther AL, Lawn ID (2003) Sea anemone host selection by the symbiotic saddled cleaner shrimp *Periclimenes holthuisi*. Mar Freshw Res 54:653–656
Kohda M, Yamanouchi H, Hirata T, Satoh S, Ota K (2017) A novel aspect of goby–shrimp symbiosis: gobies provide droppings in their burrows as vital food for their partner shrimps. Mar Biol 164:22. https://doi.org/10.1007/s00227-016-3060-2
Kou Q, Li XZ, Chan TY, Chu KH (2015) Divergent evolutionary pathways and host shifts among the commensal pontoniine shrimps: a preliminary analysis based on selected Indo-Pacific species. Org Divers Evol 15:369–377
Kramer A, Van Tassell JL, Patzner RA (2009) A comparative study of two goby shrimp associations in the Caribbean Sea. Symbiosis 49:137–141
Krebs JR, Davis NB (1993) An introduction to behavioural ecology. Blackwell Publishing, Malden
Lassig BR (1977) Communication and coexistence in a coral community. Mar Biol 42:85–92
Levine DM, Blanchard OJ Jr (1980) Acclimation of two shrimps of the genus *Periclimenes* to sea anemones. Bull Mar Sci 30:460–466
Levitt-Barmats Y, Shenkar N (2018) Observations on the symbiotic relationship between the caridean shrimp *Odontonia sibogae* (Bruce, 1972) and its ascidian host *Herdmania momus* (Savigny, 1816). PLoS One 13:0192045. https://doi.org/10.1371/journal.pone.0192045
Limbaugh C, Pederson H, Chace FA Jr (1961) Shrimps that clean fishes. Bull Mar Sci Gulf Caribb 11:237–257

Losey GS (1987) Cleaning symbiosis. Symbiosis 4:22–258

Lyons PJ (2013) The benefit of obligate versus facultative strategies in a shrimp–goby mutualism. Behav Ecol Sociobiol 67:737–745

MacGinitie GE (1937) Notes on the natural history of several marine Crustacea. Am Midl Nat 18: 1031–1037

Maciá S, Robinson MP (2009) Why be cryptic? Choice of host urchin is not based on camouflage in the caridean shrimp *Gnathophylloides mineri*. Acta Ethologica 12, Article number: 105

Mahnken C (1972) Observations on cleaner shrimps of the genus *Periclimenes*. Bull Natl Hist Mus Los Angeles County 14:71–83

Marin I (2011) *Fennera holthuisi* sp. nov., a new coral-associated pontoniine shrimp (Decapoda: Palaemonidae: Pontoniinae) from Moorea, French Polynesia. Zootaxa 2890:29–39

Marin IM, Chan T-Y (2014) Deep water echinoid-associated pontoniine shrimp "*Periclimenes hertwigi* Balss, 1913" species group (Crustacea: Decapoda: Caridea: Palaemonidae): species review, description of a new genus and species from Philippines. Zootaxa 3835:301–324

Marin IN, Anker A, Britayev TA, Palmer AR (2005) Symbiosis between the alpheid shrimp, *Athanas ornithorhynchus* Banner and Banner, 1973 (Crustacea: Decapoda), and the brittle star, *Macrophiothrix longipeda* (Lamarck, 1816) (Echinodermata: Ophiuroidea). Zool Stud 44:234–241

Marliave JB, Mills CE (1993) Piggyback riding by pandalid shrimp larvae on hydromedusae. Can J Zool 71:257–263

McCammon A, Sikkel PC, Nemeth D (2010) Effects of three Caribbean cleaner shrimps on ectoparasitic monogeneans in a semi-natural environment. Coral Reefs 29:419–426

McCammon AM, Brooks WR (2014) Protection of host anemones by snapping shrimps: a case for symbiotic mutualism? Symbiosis 63:71–78

McGrath EC, Woods L, Jompa J, Haris A, Bell JJ (2018) Growth and longevity in giant barrel sponges: redwoods of the reef or pines in the Indo-Pacific? Sci Rep 8:15317. https://doi.org/10.1038/s41598-018-33294-1

Moore PG, Rainbow PS, Larson RJ (1993) The mesopelagic *Notostomus robustus* Smith (Decapoda: Oplophoridae) observed in situ feeding on the medusan *Atolla wyvillei* Haeckel in the northwest Atlantic, with notes on gut contents and mouthpart morphology. J Crustac Biol 13:690–696

Nakano R, Fujii T (2014) The soft-coral associated pistol shrimp *Synalpheus neomeris* (De Man) (Decapoda: Alpheidae) defends its host against nudibranchs in Okinawa, Japan. Raffles Bull Zool 62:759–763

Nizinski MS (1989) Ecological distribution, demography and behavioral observations on *Periclimenes anthophilus*, an atypical symbiotic cleaner shrimp. Bull Mar Sci 45:174–188

Olliff ERR (2013) Symbiosis of the sea star shrimp, *Periclimenes soror* Nobili, 1904 (Decapoda, Palaemonidae), and cushion star, *Culcita novaeguineae* Müller & Troschel, 1842 (Echinodermata, Asteroidea, Oreasteridae): host finding and benefits. Crustaceana 86:564–577

Ory NC, Dudgeon, Thiel M (2013) Host-use patterns and factors influencing the choice between anemone and urchin hosts by a caridean shrimp. J Exp Mar Biol Ecol 449:85–92

Östlund-Nilsson S, Becker JHA, Nilsson GE (2005) Shrimps remove ectoparasites from fishes in temperate waters. Biol Lett 1:454–456

Patton WK (1974) Community structure among the animals inhabiting the coral *Pocillopora damicornis* at Heron Island, Australia. In: Vernberg WB (ed) Symbiosis in the Sea. University of South Carolina Press, Columbia, pp 219–243

Patton WK, Patton RJ, Barnes A (1985) On the biology of *Gnathophylloides mineri*, a shrimp inhabiting the sea urchin *Tripneustes ventricosus*. J Crustac Biol 5:616–626

Patzner RA (1982) *Lysmata seticaudata* (Decapoda, Crustacea) als Putzergarnele im Mittelmeer. Helgoländer Meeresun 35:227–230

Porat D, Chadwick-Furman NE (2005) Effects of anemonefish on giant sea anemones: ammonium uptake, zooxanthella content and tissue regeneration. Mar Freshw Behav Physiol 38:43–51

Potts FA (1915a) *Hapalocarcinus,* the gall-forming crab, with some notes on the related genus *Cryptochirus*. Pap Dept Mar Biol Carnegie Inst Wash 8:33–69

Potts FA (1915b) The fauna associated with the crinoids of a tropical coral reef: with especial reference to its colour variations. Carnegie Institution of Washington. Papers Dept Marine Biol 8:71–97.

Poulin R, Grutter AS (1996) Cleaning symbioses: proximate and adaptive explanations. Bioscience 46:512–517

Poulin R, Vickery WL (1995) Cleaning symbiosis as an evolutionary game: to cheat or not to cheat? J Theor Biol 175:63–70

Prakash SI, Babu M, Gopi T, Thangappanpillai AK, Balasubramanian T (2011) Discovery of the shrimp *Pycnocaris chagoae* Bruce, 1972 (Decapoda: Caridea: Gnathophyllidae) in the Lakshadweep Archipelago, India. Zootaxa 2998:66–68

Queller DC, Strassmann JE (1998) Kin selection and social insects. Bioscience 48:165–175

Randall JE, Lobel PS, Kennedy CW (2005) Comparative ecology of the gobies *Nes longus* and *Ctenogobius* saepepallens, both symbiotic with the snapping shrimp *Alpheus floridanus*. Environ Biol Fish 74:119–127

Rauch CBW, Hoeksema B, Hermanto B, Fransen CHJM (2019) Shrimps of the genus *Periclimenes* (Crustacea, Decapoda, Palaemonidae) associated with mushroom corals (Scleractinia, Fungiidae): linking DNA barcodes to morphology. Contrib Zool 88:201–235

Rüber L, Van Tassell JL, Zardoya R (2003) Rapid speciation and ecological divergence in the American seven-spined gobies (Gobiidae, Gobiosomatini) inferred from a molecular phylogeny. Evolution 57:1584–1598

Saito T, Konishi K (1999) Direct development in the sponge-associated deep-sea shrimp *Spongicola japonica* (Decapoda: Spongicolidae). J Crustac Biol 19:46–52

Saito T, Koya Y (2001) Gonad maturation and embryonic development in the deep-sea sponge associated shrimp, *Spongicola japonica* Kubo (Crustacea: Decapoda: Spongicolidae). Zool Sci 18:567–576

Saito T, Uchida I, Takeda M (2001) Pair formation in *Spongicola japonica* (Crustacea: Stenopodidea: Spongicolidae), a shrimp associated with deep-sea hexactinellid sponges. J Mar Biol Assoc UK 81:789–797

Saito T, Uchida I, Takeda M (2002) Skeletal growth of the deep-sea hexactinellid sponge *Euplectella oweni*, and host selection by the symbiotic shrimp *Spongicola japonica* (Crustacea: Decapoda: Spongicolidae). J Zool 258:521–529

Saito TP, Takeda M (2003) Phylogeny of the family Spongicolidae (Crustacea: Stenopodidea): evolutionary trend from shallow-water free-living to deep-water sponge-associated habitat. J Mar Biol Assoc UK 83:119–131

Sargent RC, Wagenbach GE (1975) Cleaning behavior of the shrimp, *Periclimenes anthophilus* Holthuis and Eibl-Eibesfeldt (Crustacea: Decapoda: Natantia). Bull Mar Sci 25:466–472

Shoup JB (1972) Notes on the behavior of *Periclimenes imperator* Bruce, an ectocommensal on the dorid nudibranch *Hexabranchus marginatus* Quoy & Gaimard (Decapoda, Palaemonidae). Crustaceana 23:109–111

Silliman BR, Layman CA, Altieri AH (2003) Symbiosis between an alpheid shrimp and a xanthoid crab in salt marshes of mid-Atlantic states, USA. J Crustac Biol 23:876–887

Smith WL (1977) Beneficial behavior of a symbiotic shrimp to its host anemone. Bull Mar Sci 27: 343–346

Soares MC, Bshary R, Cardoso SC, Côté IM (2008) The meaning of jolts by fish clients of cleaning gobies. Ethology 114:209–214

Spotte S (1996) Supply of regenerated nitrogen to sea anemones by their symbiotic shrimp. J Exp Mar Biol Ecol 198:27–36

Spotte S (1998) "Cleaner" shrimps? Helgoländer Meeresun 52:59–64

Stewart HL, Holbrook SJ, Schmitt RJ, Brooks AJ (2006) Symbiotic crabs maintain coral health by clearing sediments. Coral Reefs 25:609–615

Stier AC, Gil MA, McKeon CS, Lemer S, Leray M et al (2012) Housekeeping mutualisms: do more symbionts facilitate host performance? PLoS One 7(4):e32079. https://doi.org/10.1371/journal.pone.0032079

Terrana L, Caulier G, Todinanahary G, Lepoint G, Eeckhaut I (2016) Characteristics of the infestation of *Seriatopora* corals by the coral gall crab *Hapalocarcinus marsupialis* Stimpson, 1859 on the great reef of Toliara, Madagascar. Symbiosis 69:113–122

Thacker CE, Thompson AR, Roje DM (2011) Phylogeny and evolution of Indo-Pacific shrimp-associated gobies (Gobiiformes: Gobiidae). Mol Phylogenet Evol 59:168–176

Thompson AR, Thacker CE, Shaw EY (2005) Phylogeography of marine mutualists: parallel patterns of genetic structure between obligate goby and shrimp partners. Mol Ecol 14:3557–3572

Titus BM, Daly M, Exton DA (2015) Temporal patterns of Pederson shrimp (*Ancylomenes pedersoni* Chace 1958) cleaning interactions on Caribbean coral reefs. Mar Biol 162:1651–1664

Titus BM, Vondriska C, Daly M (2017) Comparative behavioural observations demonstrate the 'cleaner' shrimp *Periclimenes yucatanicus* engages in true symbiotic cleaning interactions. R Soc Open Sci 4:170078

Titus BM, Daly M, Vondriska C, Hamilton I, Exton DA (2019) Lack of strategic service provisioning by Pederson's cleaner shrimp (*Ancylomenes pedersoni*) highlights independent evolution of cleaning behaviors between ocean basins. Sci Rep 9:629. https://doi.org/10.1038/s41598-018-37418-5

Tóth E, Bauer RT (2007) Gonopore sexing technique allows determination of sex ratios and helper composition in eusocial shrimps. Mar Biol 151:1875–1886

Tóth E, Bauer RT (2008) *Synalpheus paraneptunus* (Crustacea: Decapoda: Caridea) populations with intersex gonopores: a sexual enigma among sponge-dwelling snapping shrimps. Invertebr Reprod Dev 51:49–59

VandenSpiegel D, Eeckhaut I, Jangoux M (1998) Host selection by *Synalpheus stimpsoni* (De Man), an ectosymbiotic shrimp of comatulid crinoids, inferred by a field survey and laboratory experiments. J Exp Mar Biol Ecol 225:185–196

Vannini M (1985) A shrimp that speaks crab-ese. J Crustac Biol 5:160–167

Vaughan DB, Grutter AS, Costello MJ, Hutson KS (2016) Cleaner fishes and shrimp diversity and a re-evaluation of cleaning symbioses. Fish 2016:1–19. https://doi.org/10.1111/faf.12198

Vaughan DB, Grutter AS, Ferguson HW, Jones R, Hutson KS (2018a) Cleaner shrimp are true cleaners of injured fish. Mar Biol 165. https://doi.org/10.1007/s00227-018-3379-y

Vaughan DB, Grutter AS, Hutson KS (2018b) Cleaner shrimp are a sustainable option to treat parasitic disease in farmed fish. Sci Rep 8:13959. https://doi.org/10.1038/s41598-018-32293-6

von Rintelen K, von Rintelen T, Meixner M, Carsten Lüter Y, Cai Y, Glaubrecht M (2007) Freshwater shrimp–sponge association from an ancient lake. Biol Lett 3:262–264

Werding B, Christensen B, Hiller A (2016) Three way symbiosis between a goby, a shrimp, and a crab. Mar Biodivers. https://doi.org/10.1007/s12526-016-0453-x

Williamson JE, Gleeson C, Bell JE, Vaïtilingon D (2012) The role of visual and chemical cues in host detection by the symbiotic shrimp *Gnathophylloides mineri*. J Exp Mar Biol Ecol 414–415: 38–43

Wilson EO (1975) Sociobiology. The new synthesis. The Belknap Press of Harvard University Press, Cambridge

Wirtz P (1997) Crustacean symbionts of the sea anemone *Telmatactis cricoides* at Madeira and the Canary Islands. J Zool 242:799–811

Wisenden BD, Goater CP, James CT (2009) Behavioral defenses against parasites and pathogens. In: Zaccone C, Perriere A, Mathis A, Kapoor G (eds) Fish defenses, vol 2. Science Publishers, Enfield, pp 151–168

Zitzler K, Cai Y (2006) *Caridina spongicola,* new species, a freshwater shrimp (Crustacea: Decapoda: Atyidae) from the ancient Malili lake system of Sulawesi, Indonesia. Raffles Bull Zool 54:271–276

Chapter 11
Fisheries and Aquaculture

11.1 A Desirable Food Product

Shrimps are a tasty, nutritious, and valuable commodity. The tail (abdomen) is a large chunk of muscle, almost all protein which is very low in fat, albeit high in dietary cholesterol. Shrimps are a highly popular and healthy food product around the world whether boiled, fried, grilled, or eaten raw in ceviche and sushi. Dried, salted, and pickled shrimps are important ingredients in many Asian dishes, sauces, and dips (Omori 1975). Shrimp paste, a fermented product, is an essential component of South Chinese and Southeast Asian foods enjoyed around the world. Although food for human consumption is the principal use of fished shrimps, they are often locally important as bait for subsistence and recreational fisheries. Shrimp fisheries and aquaculture are both culturally and economically significant in many countries, especially in developing countries, giving employment to many throughout the world.

11.2 Facts and Figures

In this chapter, I present a number of tables, graphs, and other illustrations about shrimp fisheries and aquaculture. Where does the data come from? In addition to citing other specific sources, much of the information comes from an agency of the United Nations, the Food and Agriculture Organization of the United Nations (FAO). Among its aims are alleviation of hunger, improved nutrition and food security around the world. The agency produces many publications, such as annual summaries of fisheries and aquaculture statistics (FAO Yearbook. Fishery and Aquaculture Statistics, "FAS") and a summary analysis of these statistics (The State of World Fisheries and Aquaculture; "SOFIA"). The actual yearly data for production by species, country, region, and other categories was obtained from an

FAO website in searchable form (FIGIS: Fisheries Global Information System) for both fisheries and aquaculture from 1950 onwards. Specific data is given for 187 fishery species and 33 aquaculture species (including generic and familial species groups) (FIGIS 2021a, b). Because of the massive time-consuming labor in collecting these statistics, the information in all these sources is 2–3 years behind the data of publication. FAO is very conservative about scientific names and does not easily change those used in its data compilation. This is understandable, as the frequent but necessary scientific name changes in taxonomy and systematics would cause confusion and difficulty in presenting data over a long time period. I will use the most current taxonomic names in this chapter but will follow FAO common names in the text below, perhaps a good compromise to avoid confusion for the reader.

How are these data collected and how accurate are they? Garibaldi (2012) presents the process by which data are solicited from all countries, supplemented by any information that can be gleaned from published reports, publications, and similar sources. Data from many countries is considered unsatisfactory and must be evaluated and corrected when possible. Some countries do not submit actual data but instead target (goal) estimates. Subsistence and small-scale fisheries are often unreported or underrepresented in formal catch statistics (Vance and Rothlisberg 2020). Some countries fudge the data for political expediency. In some cases, FAO will make their own estimates for a particular year based on past or other information. For example, in the 2020 SOFIA, FAO did a significant revision of data from past years which improved the previous data set. In spite of these caveats, the massive size and scale of data gathered considerably improve the reliability of the overall data set. Although reported numbers may not be absolutely correct, the data is very good for looking at trends and making comparisons among years, countries, species, and other components of the data. It is the best source of information on fisheries and aquaculture available, and FAO, as with other UN agencies, is to be commended for providing the service to scientists and the general public.

All researchers love a good review article or book to help them get information for their work. In addition to specific citations given throughout the chapter, the following review publications have been particularly useful: from "Natural History of the Crustacea" volume 9 "Fisheries and Aquaculture (2020): the chapters by Bauer on shrimp fisheries, Boyd and Jescovitch on penaeid shrimp aquaculture and Valenti and Flickinger on freshwater shrimp culture. The book by New et al. (2010) on the biology and aquaculture of freshwater prawns (*Macrobrachium* species) was very useful as well. See Chap. 3 (Families of Shrimps) for more information and illustrations about individual species, genera, or families discussed below.

11.3 Shrimp Fisheries and Aquaculture Production: Economic Value

First, let's put the quantity (metric tonnes "t") and value (in $US) of shrimp fisheries and aquaculture in perspective by comparison with world fisheries and aquaculture overall. FAO gives the monetary value for the first ("ex vessel') sale by the fisherman to a buyer (e.g., directly to consumer, or vendor or distributor). In 2018, the total biomass taken in all animal fisheries was 96.4×10^6 (= million) t (all aquatic animals excluding mammals) (Fig. 11.1) with a first-sale value of $\$150.2 \times 10^9$ (= billion) for an average price of $1565/t (FAO-FAS 2020). For all shrimp fisheries (marine, brackish, freshwater), 3.7×10^6 t were taken worldwide in 2018 (Fig. 11.1), with an average value of $4500/t worth $\$16.8 \times 10^9$. Thus, shrimps make up only 3.8% of the total global fisheries catch by weight—but 11.2% of the global fisheries in monetary value. Shrimps are among the most valuable of fishery products, along with tunas, cephalopods (squid, octopus), and lobsters (FAO SOFIA 2020). The outsized importance of shrimps in the world's seafood economy is even much more so in animal aquaculture. FAO statistics for 2019 show that the global production of cultivated shrimps was 7.1×10^6 t with a first-sale value of $\$45.1 \times 10^9$ compared with a total animal aquaculture production of 85.4×10^6 t with an overall value of $\$259.8 \times 10^9$. Thus, shrimps made up 8.3% of aquaculture products by weight but 17.4% in first-sale value. The ultimate economic (market value) of both fisheries and aquaculture is much greater than the first-sale price. Besides its nutritional, monetary, and cultural value, shrimp fishing and aquaculture gives jobs and livelihood,

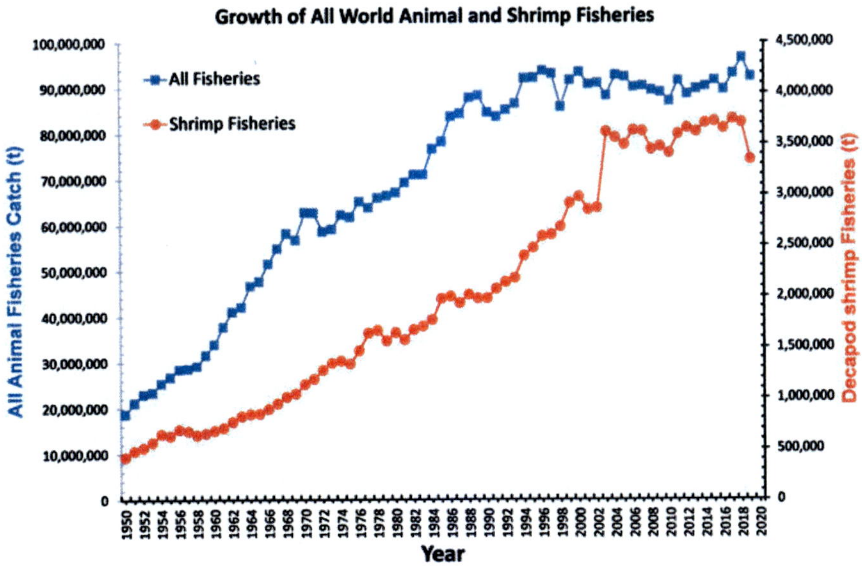

Fig. 11.1 Catch (t, metric tonnes) compared between all animal fisheries and decapod shrimp fisheries, 1950 to 2019 (data from FIGIS 2021a)

directly or indirectly to millions of persons around the world. No doubt about it—shrimps are a very valuable resource for the world's food supply and economy!

11.4 Composition of the World's Shrimp Fisheries

Shrimp species that are abundant and either form schools or are at least moderately aggregated are good candidates for catching them in mass by various kinds of trawls and nets. This leaves out the stenopodideans, which are never abundant enough to make their fishing worthwhile except in small numbers for the ornamental shrimp trade (see below). Additionally, their small body size and relatively small abdomen do not make them desirable as a food item. However, several species of penaeoid, sergestid, and caridean species have the features necessary to support fisheries. With the exception of the pelagic sergestids, the most important species are fished over extensive soft bottoms of the continental shelves accessible and favorable for trawling, the principal method for fishing shrimps. Shelf marine habitats have a much larger fishable area than freshwater rivers, streams, and lakes. This is one reason that a relatively small fraction of shrimp fisheries production come from freshwater environments (e.g., 7.1% in 2019).

Marine shrimp fisheries are often categorized as either "warm-water," i.e., those in tropical and subtropical oceans, and "cold-water," those in temperate and higher latitudes. The taxonomic make-up of fisheries in these two regions is quite distinct. Warm-water fisheries are overwhelmingly dominated by penaeoid shrimps, especially by species of Penaeidae (Figs. 11.2, 11.3, 11.4). Very important too, however, are fisheries of sergestid shrimps, in particularly the Akiami paste shrimp, *Acetes japonicus*. Cold-water species are principally pandalids, although the crangonid species *Crangon crangon* has been and still is significant in northeast Atlantic and Mediterranean fisheries. On a global basis, the dominance of dendrobranchiate (penaeoid and sergestid) species has been constant since the 1950s, when FAO begin to publish fisheries statistics (Figs. 11.3 and 11.4). Shrimps fished commercially in freshwater habitats are all palaemonid carideans. The temperate latitude *Macrobrachium nipponense* and *Exopalaemon modestus* are the most productive of these species.

A closer look at the shrimp fisheries of the United States of America (USA) is instructive because of its geographic location which includes both subtropical and cool-temperate waters. Thus, USA shrimp fisheries include both warm-water (penaeoid) fisheries along the South Atlantic (southeast) coast and in the Gulf of Mexico, as well as cold-water (caridean) fisheries in New England (northeast coast) and USA west coast (northern California, Oregon, Washington, Alaska) (Fig. 11.5a). In the south Atlantic and Gulf of Mexico, commercial shrimp fisheries are completely penaeoid: mainly the Northern white shrimp *Litopenaeus setiferus*, the Northern brown shrimp *Farfantepenaeus aztecus*, and Northern pink shrimp, *F. duorarum*, with minor contributions by the Rock shrimp *Sicyonia brevirostris* and Atlantic seabob, *Xiphopenaeus kroyeri.* The cold-water fisheries are

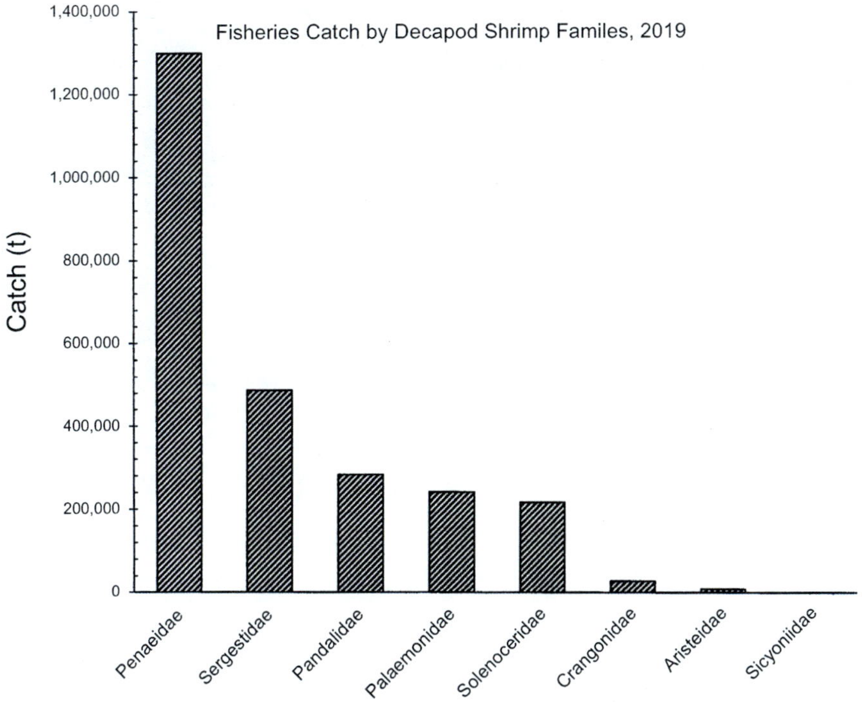

Fig. 11.2 Composition of world shrimp fisheries catch by weight and family for 2019 (data from FIGIS 2021a). 2019 Sicyoniidae catch (199 t) is too small to show on the graph. Total shrimp catch not identified to taxon (FAO "Natantia nei,") was 780,000 t in 2019

overwhelmingly pandalid: *Pandalus borealis* (New England) and *P. eous* on the USA west coast (grouped with *P. borealis* by FAO as Northern prawn), *P. jordani* (Ocean shrimp), *P. platyceros* (spot shrimp) and other pandalids, with a minor input from the Pacific rock shrimp, *Sicyonia ingentis*. The preponderance of the warm-water penaeoid species, which account for 82% of the USA fisheries catch (2018 values) versus 18% cold-water caridean species has been as consistent over time as in the world fisheries. FAO FIGIS does not give data on monetary value for individual shrimp species, but data from USA fisheries reports shows that, at least for USA fishers, the warm–warm penaeoids are much more valuable than the cold-water caridean ones (Fig. 11.5b). Shrimps were the fifth most valuable fished species in the USA, ranking behind lobsters, crabs, salmon, and scallops in 2018 (NMFS 2020). However, on a dollar per weight basis, penaeoid shrimps are worth more than double ($3780/t) than the average of ($1300/t) of all USA fished animal species.

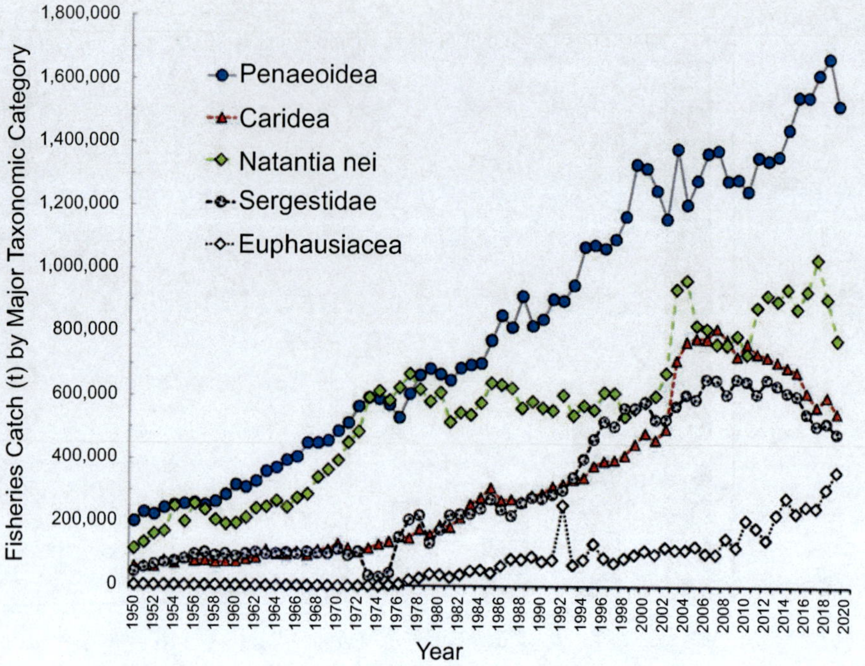

Fig. 11.3 Annual contribution of major shrimp groups to global shrimp fisheries catch from 1950–2019. Natantia nei is the FAO designation for shrimps not identified or specified to taxonomic group. The catch of krill (Euphausiacea), the taxonomic sister group of decapod shrimps, is included for comparison (data from FIGIS 2021a)

11.5 Life-History Traits and Shrimp Productivity

Caridean shrimps are much more diverse than penaeoids and sergestoids with about eight times as many species as the dendrobranchiate species. Marine carideans occur in all latitudes, depths, and major habitats of marine waters. The dendrobranchiates are largely restricted to subtropical and tropical waters. Numerous carideans species occur in temperate and especially tropical freshwater lakes, rivers, and streams. However, as shown above, it is the penaeoids, along with a few sergestids, that dominate decapod shrimp fisheries both in terms of production (weight) and value (in $US) both globally and in the USA, a major fishing country with both warm-water and cold-water fisheries. The same relationship holds (even more so) for the shrimp aquaculture industry, as will be shown in later sections. The causal factors underlying this disparity are essential knowledge both in terms of basic and applied science, and thus to world nutrition, economy and culture. In short, differences in life-history traits among species explain why some species are so productive in fisheries and aquaculture, and most others are not.

In Chap. 9 (Life Histories; Fig. 9.19), I proposed that the "r-" and "K-" selection model, based on the intrinsic rate of natural increase ("r") and the logistic growth

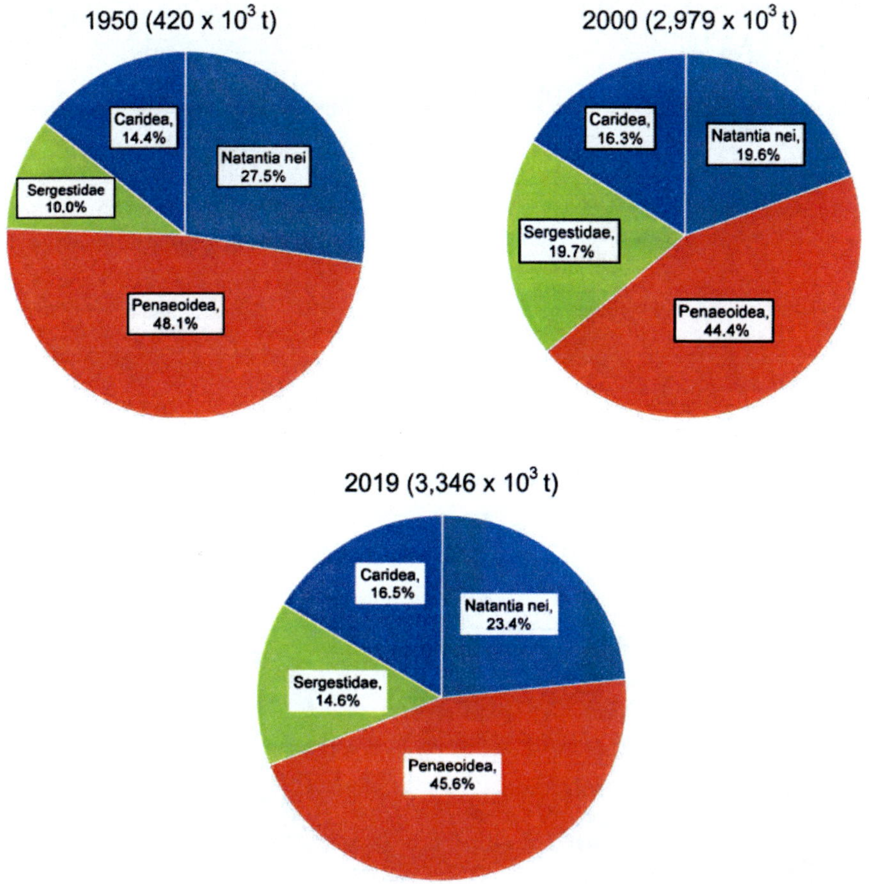

Fig. 11.4 Percentage of the global shrimp fisheries catch from sergestid, penaeoid, caridean and unidentified (Natantia nei) shrimps by weight in 1950, 2000, and 2019 demonstrates that the taxonomic composition of the decapod shrimp fisheries has changed little throughout the years. Data from FIGIS (2021a)

curve, with early rapid growth to variation about a long-term plateau or asymptote ("K") in populations size, is most helpful in understanding life-history traits in decapod shrimps, as in many types of animals (Pianka 1970). Penaeid shrimps contributing most to fisheries and aquaculture are definitely on the "r" end of the r-K selection continuum. These species mainly have r-selected traits, i.e., rapid development and growth, early sexual maturity, high reproductive effort, and a short life span—life in the fast lane with rapid burnout. Many carideans, especially the subtropical and tropical ones, have these traits but have not evolved the genetic and physiological capacity to grow to a large body size because of various selective pressures. In the shrimp fisheries, commercial penaeoids usually have all the above r-selected traits, but unlike most carideans, have evolved phenomenal growth rates in

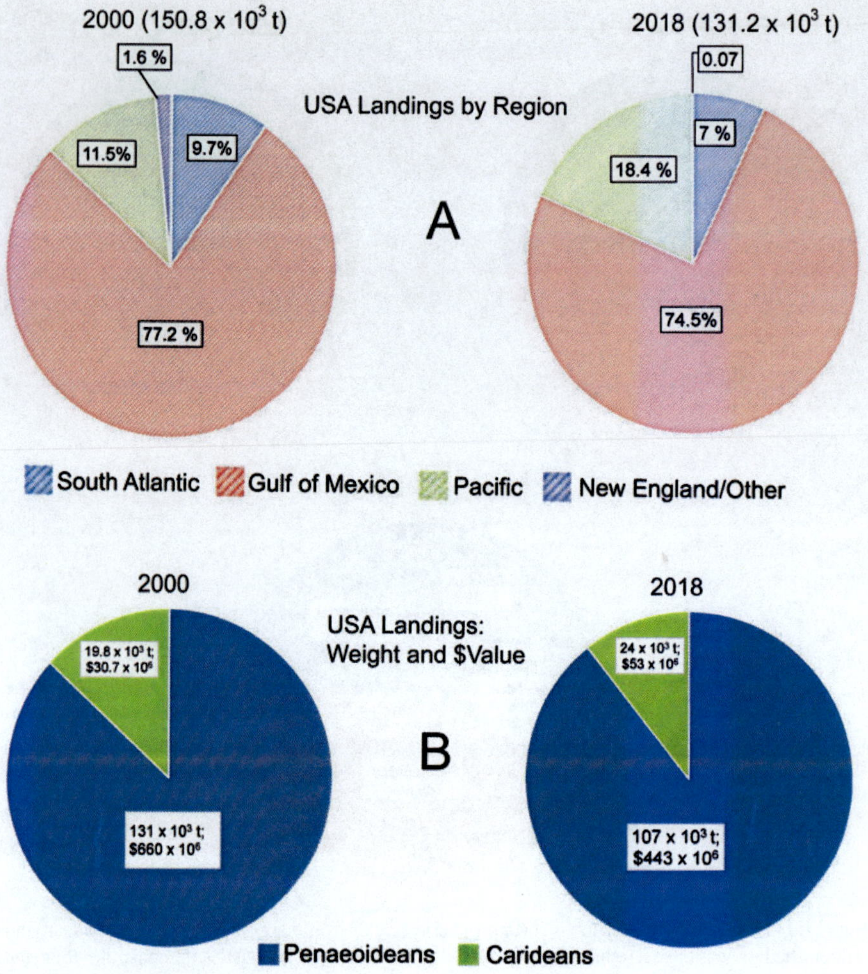

Fig. 11.5 USA shrimp fishery information compared between 2000 and 2018 (data from NMFS 2020). (**a**) Catch by weight and region. (**b**) Penaeoid (warm-water fisheries) and caridean (cold-water fisheries) production in the USA by weight and monetary value

the warm waters in which they occur. In contrast, the most important carideans which compose the cold-water fisheries are the K-selected pandalids and a crangonid from temperate and cold-water seas. They attain individual and population sizes worthy of fishing on a commercial scale. However, growth is slow, but life span is long compared to penaeoids. Some examples will help to illustrate these generalizations and the few exceptions to them (Table 11.1). Note that in Table 11.1 the data applies to females, as they are the most important sex in terms of life-history traits. In addition, the females are larger than the males in fished shrimp species and cultured penaeoid species. An exception to this generalization is cultured *Macrobrachium* species, in which dominant males are larger than females.

Table 11.1 Comparison of "r-" vs. "K-"selected life-history traits of *females* of commercially important dendrobranchiate (penaeoid, sergestid) and caridean species in nature

	Penaeoids and Sergestids					Carideans			
Life-History Trait	*Penaeus monodon* (F, A) (PN)	*Litopenaeus vannamei* (F, A) (PN)	*Litopenaeus setiferus* (F, A) (PN)	*Pleoticus muelleri* (F, A) (SL)	*Acetes japonicus* (F) (SG)	*Pandalus borealis* (F) (PD)	*Crangon crangon* (F) (CG)	*Macrobrachium rosenbergii* (F, A) (PL)	*Macro nipponense* (F, A) (PL)
Maximum size (tl) and weight	34 cm; 260 g	23 cm; ~90 g	20 cm; 90 g	19 cm; ~60 g	3 cm; 0.1 g	18 cm; 13 g	9.6 cm; 4 g	25 cm; ~50 g	7.5 cm, 6 gm
Age of first reproduction	~6 months	6–7 months	~6–8 months	~5–6 months	3–10 months	18 months-7 years	1–2 years	≥ 4 months	1 year
Spawns/year	1–2	1–2 (?)	1–2	2	1	≤ 1	1–2	several	1 (?)
Egg size (mm)	0.28	0.22	0.28 mm	0.26	0.21	1.1 mm	0.40	0.65 mm	0.45
Fecundity (# eggs/ embryos)	250,000–800,000	80,000–450,000	500,000–1000,000	100,000–1000,000	700–7000	600–4900	1000–14,000	5000–100,000	500–5000
Larval Development	2–3 weeks	2 weeks	2–3 weeks	3–4 weeks	6 weeks	4–6 months	5–8 weeks	3–4 weeks	2–3 weeks
Growth rate	Rapid	Rapid	Rapid	Rapid	Moderate	Slow	Slow	Rapid	Moderate
Natural Life Span	1.5–2 years	~1 year	1–1.5 years	~20 months	3–10 months	2.5–11 years	2–4 years	1–2 years	~2.5 years

(F) Fishery species; (A) Aquaculture species; CG, Crangonidae; PN, Penaeidae; SL, Solenoceridae; SG, Sergestidae; PD, Pandalidae; PL, Palaemonidae; g, grams; tl, total length. Data from Allen (1966), Amin et al. (2009), Ar et al. (2016), Aranceta-Garza et al. (2016), Aye (2018), Bauer (2004), Bergström (2000), Boschi (1989), Crespi and New (2009), FAO Cultured Aquatic Species Information Program (2021), *Penaeus vannamei;* Holthuis (1980), Iversen et al. (1993), Kutty and Weimin (2010), Linder and Anderson (1956), Lloyd and Yonge (1947), Macchi et al. (1998), MacLean and Brown (1991), Mashiko (1990), Motoh (1981), Nilssen and Aschan (2009), Ogawa et al. (1991), Scelzo (2016), Shumway et al. (1985), Temming and Damm (2002)

The penaeids *Penaeus monodon, L. vannamei* and *L. setiferus* are typical of fished penaeids. Their reproductive capacity is enormous; females spawn hundreds of thousands of small eggs directly into the water, with no incubation, followed by just of few weeks of larval development in the plankton. After settling as postlarvae only several millimeters in length and milligrams in weight, within just 4–6 months they grow to large shrimps ~100 mm or more weighing tens of grams! *Penaeus monodon* and especially *L. vannamei* are the mainstays of marine (penaeid) shrimp aquaculture because of these r-selected traits. *P. monodon* is a really large shrimp, reaching a maximum count of 6 per kilo (2–3/lb); *L. vannamei* and *L. setiferus* are quite large as well, with maximum counts of about 11/kg (5/lb). Although most fished adults do not attain these maximum values, there is no doubt that these penaeid species have very rapid growth rates in the warm waters in which they occur.

The solenocerid penaeoid *Pleoticus muelleri* supports a substantial fishery (Table 11.2). It lives in transitional waters between subtropical and cool-temperate waters along the southeastern coasts of South America penaeoid important in fisheries with mostly r-selected traits is (Boschi 1989) (Table 11.1). One productive penaeid that is somewhat more K-selected is *Parapenaeus longirostris,* which lives in cooler waters of the deep shelf and shallow slope. It lives a bit longer (2–3 years) than warm-water penaeids and its females have a somewhat lower fecundity (~90,000 eggs/spawn) (Sobrino et al. 2005). Surprisingly, a rather small-sized (maximum 0.1 g), subtropical temperate water sergestid, *Acetes japonicus*, sustains the most productive shrimp fishery worldwide. It has been on or near the top of list of most productive fished shrimp species from the 1950s to present (Table 11.2). What it lacks in size is made up for by the tremendous number of individuals in its massive schools. For its size, it has comparable fecundity to penaeoids but moderate growth to sexual maturity. Overall, it is more r-than K-selected (Table 11.1).

Only a few carideans contribute substantially to global fisheries as shown by fishery statistics from 1950 to present: pandalids, especially *Pandalus borealis, P. jordani* and the crangonid *Crangon crangon*. The fished pandalids are large-sized species for marine carideans, although not anywhere near as large as the penaeids. Compare the weight of a very large *P. borealis*, 13 g (counts of 77 shrimps/kg; 35/lb) compared to the 90–100 gm weight of very large penaeids such as *P. monodon* or *L. setiferus*. However, depending on the latitude (and corresponding water temperatures), a female *P. borealis* may take anywhere from 1.5–7 years to reach sexual maturity. After spawning, incubation of the embryos to hatching make take 4–5 months or more so that many females may only breed once in their lifetime (Bauer 2004). Fecundity is low compared to most penaeids, with brood sizes in the low to several thousands compared to the hundreds of thousands of commercial penaeoids (Table 11.1). Larval development lasts several months compared to the 2–3 weeks of warm-water penaeids.

Similarly, very large *C. crangon* are only ~4 g which translates to 250 individuals/kilo (~114/lb) compared to much lower counts of shrimp per unit weight for the penaeoids. These shrimps are northern hemisphere species from temperate to high latitudes—cold-water species. These smaller carideans have similar life-history traits to pandalids, slow growth but to an even smaller size, longer larval developmental

Table 11.2 Principal species in global marine and brackish-water shrimp fisheries ranked by capture production (metric tonnes, t) and listed by decadal intervals from 1950 to present

Species	Common Name (FAO)	2019	2010	2000	1990	1980	1970	1960	1950
Acetes japonicus (3 cm) (SG) (*e*)	Akiami paste shrimp	1 (402,061)	1	1	2	2	2	2	4
Trachysalambria curvirostris (10 cm) (PN)	Southern rough shrimp	2 (243,076)	3	3	5	14	21	–	–
Penaeus monodon (34 cm) (PN) (*e*)	Giant tiger prawn	3 (229,144)	4	4	4	16	13	10	10
Pleoticus muelleri (19 cm) (SL)	Argentine red shrimp	4 (216,442)	9	12	17	37	37	24	13
Fenneropenaeus chinensis (18 cm) (PN)	Fleshy prawn	5 (215,550)	6	7	9	10	10	–	–
*Pandalus borealis** (18 cm) (PD)	Northern prawn	6 (197,100)	2	2	3	3	5	7	8
Penaeus (*s.l.*) *spp.* (~15–30 cm) (PN)	Penaeus shrimps nei	7 (191,786)	5	5	1	1	1	1	1
Metapenaeus spp. (~10–20 cm) (PN) (*e*)	Metapenaeus shrimps nei	8 (90,058)	8	8	10	6	8	11	11
Fenneropenaeus merguiensis (24 cm) (PN) (*e*)	Banana prawn	9 (87,344)	7	6	7	5	7	8	9
Sergestidae (SG) (~2–4 cm)	Sergestid shrimps nei	10 (80,794)	8	14	8	7	11	–	–
Pandalus spp. (PD)	*Pandalus* shrimps nei	11 (58,464)	34	13	13	17	16	–	–
Xiphopenaeus kroyeri (14 cm) (PN)	Atlantic seabob	12 (35,597)	13	11	16	25	14	15	20
Parapenaeus longirostris (19 cm) (PN)	Deep-water rose shrimp	13 (29,223)	16	17	15	15	17	18	6

(continued)

Table 11.2 (continued)

Species	Common Name (FAO)	2019	2010	2000	1990	1980	1970	1960	1950
Farfantepenaeus californiensis (PN)	Yellowleg shrimp	14 (29,171)	23	51	42	38	29	26	23
Crangon crangon (9 cm) (CG) (*e*)	Common shrimp	15 (27,852)	12	15	14	9	4	4	2
Litopenaeus stylirostris (23 cm) (PN) (*e*)	Blue shrimp	16 (27,267)	18	–	–	–	–	–	–
Farfantepenaeus notialis (19 cm) (PN) (*e*)	Southern pink shrimp	17 (22,288)	20	16	19	18	20	19	18
Pandalus jordani (16 cm) (PD)	Ocean shrimp	18 (17,580)	15	18	–	–	–	–	–
Penaeus semisulcatus (23 cm) (PN)	Green tiger prawn	19 (13,348)	19	22	31	26	25	21	17
Farfantepeneus aztecus (24 cm) (PN) (*e*)	Northern brown shrimp	20 (12,094)	14	9	6	4	3	3	3

Unidentified decapod shrimps ("Natantia nei") are not included; data on freshwater species are given in the text. Maximum size (total length) is given for each species. The capture weight for 2019 is given in parentheses below the species rank. Data on capture production and species names are from FIGIS (2021a); other information is from Holthuis (1980), Shumway et al. (1985) (*P. borealis*), and Bauer (2018). (*e*), estuary-dependent species; nei, (FAO "not elsewhere identified", i.e., not identified to species or other taxon); *s.l., sensu lato*; – no catch reported; * FAO (2020) includes the Pacific *Pandalus eous* with the Atlantic *P. borealis*. CG, Crangonidae; PD, Pandalidae; PN, Penaeidae; SG, Sergestidae; SL, Solenoceridae. Generic names of penaeoids follow Pérez Farfante and Kensley (1997) times and low fecundity compared to the much larger and fecund penaeoids. The overwhelming effect of cool to cold waters has selected for a life-history strategy incorporating a longer time to first reproduction, slow incubation of embryos to hatching, extended larval development and longer life spans to a relatively small body size compared to commercial penaeids, all of which results in lower biomass productivity. However, these K-selected characteristics are adaptive in that populations are more stable over time, parceling out their reproductive output more slowly. The r-selected penaeoids tend to have more boom-and-bust population cycles on a yearly basis.

In shrimp aquaculture, only r-selected species attaining large body sizes within several months are profitable. Life-history traits for four of the most valuable aquaculture species are given in Table 11.1. Not surprisingly, two of them are penaeids which are the mainstay of shrimp culture today, *L. vannamei* and *P. monodon*, both important in world fisheries, especially the latter. Somewhat surprisingly in terms of life-history strategies, the two other significant aquaculture species are freshwater carideans, *Macrobrachium rosenbergii* and *M. nipponense* (Palaemonidae). They overwhelmingly dominate freshwater shrimp culture, the former species native to rivers in tropical southeast Asia and the latter to subtropical temperate lakes and rivers from northern China to Vietnam. Appropriately given the common name of Giant river prawn, the dominant blue-claw males of *M. rosenbergii* can reach very large lobster-like sizes, e.g., maximum of 320 mm total length and 400 g, nearly a pound in weight. Although most are not that large, especially in culture, they are still impressive shrimps. Females are considerably smaller, especially those harvested in culture, but still in the tens of grams range. Growth is extremely rapid in their warm-water environment, with females gaining sexual maturity is a little as 4 months. Fecundity is quite high for a caridean, with several thousand to 100,000 small eggs per brood—and several broods can be produced per year. Incubation time is short as is larval development (about 2–4 weeks each). The natural life span is 1–2 years for females, at least, although for males, it may be much longer in nature. While *M. rosenbergii* is at the extreme r-end of the shrimp r–k continuum, the life history of the Oriental river prawn, *M. nipponense*, is somewhat more towards the K-end of the spectrum. Females are an order of magnitude smaller, less fecund, with a much more moderate growth rate and longer natural life span than those of *M. rosenbergii* (Table 11.1). However, they do very well in culture and are preferred as a food item over *M. rosenbergii in China*, where their production and consumption are greatest, mainly due to the public's familiarity with and culinary preference for them (Kutty and Weimin 2010).

11.6 Globally Important Fishery Species

11.6.1 Sergestidae

Members of this family, especially the Akiami paste shrimp *A. japonicus*, are a significant component of Asian cooking and seasoning. They are prepared for use in a variety of ways: dried, salted, pickled, or fermented in the forms of shrimp paste and sauces (Xiao and Greenwood 1993). They have topped the list of most productive species for decades. They are fished in shallow nearshore and estuarine waters from Korea to Indonesia to the west coast of India. This epipelagic species is captured particularly during their juvenile phase when massive swarms hundreds of meters long and tens of meters wide form near river mouths and over shallow seagrass meadows. Fishers often locate these swarms at night by their greenish bioluminescent glow, although it is not certain whether it is produced by the shrimps themselves or perhaps by their disturbance of bioluminescent plankton. Artisanal and commercial fishing are done with a variety of gear, such as push nets; seines; and stow, bag, and lift nets according to depth (Omori 1975).

11.6.2 Penaeoidea

Large-bodied penaeoid shrimps, with their massive meaty abdomens, are highly prized by consumers and are the most productive segment of the commercial shrimp fisheries (Table 11.2, Figs. 11.2 and 11.3). Several of these are estuary-dependent species, i.e., those with the Type-2 penaeoid life cycle described in in Chap. 9 (see Dall et al. 1990; Bauer 2018; Vance and Rothlisberg 2020). Individuals are spawned and go through larval development at sea. The small postlarvae settle to the bottom and seek out, aided by currents and salinity cues, brackish-water estuarine habitats. These environments, enriched by the high primary productivity of salt marshes, mangroves and seagrass beds, provide shelter and abundant food for the rapidly growing juveniles. Upon reaching a large (fishable) size as subadults, they go out to sea and continue growing, become sexually mature and reproduce. Top-producing estuarine-dependent species include the Indo-West Pacific *Penaeus monodon*, *Fenneropenaeus* (*F.*) *merguiensis*, *Metapenaeus* species, *Litopenaeus stylirostris* and, from eastern and western tropical Atlantic coasts. *Farfantepenaeus* (*Ft.*) *notialis.* Estuary-dependent species significant in USA fisheries are the southeastern Atlantic and Gulf of Mexico *L. setiferus*, *Ft. duorarum* and *Ft. aztecus*. The eastern Pacific White leg shrimp *L. vannamei,* so important in aquaculture, also supports a significant fishery along coasts from the Gulf of California to Peru.

In contrast, some commercial penaeoids live and reproduce offshore over the outer continental shelves. Juveniles of these Type-3 species move into and grow in nearshore waters but not in estuaries. Currently, the most important of these are the Indo-Pacific penaeids *Trachysalambria* (*Trachypenaeus*) *curvirostris, F. chinensis,*

Gulf of California *Ft. californiensis,* Western Atlantic *Xiphopenaeus kroyeri,* and the solenocerid *Pleoticus muelleri*, the latter fished along the coasts of southeastern Brazil, Uruguay, and Argentina.

Type-4 species, in which all life stages take place offshore, are represented in Table 11.2 only by the Deep-water rose shrimp *Parapenaeus longirostris*, which is the basis of a major fishery in the Mediterranean Sea. Also significant in Mediterranean fisheries are two deep-water Type 4 aristeids, the Blue and red shrimp, *Aristeus antennatus* and the Giant red shrimp *Aristaeomorpha foliacea*. The subtropical West Atlantic *Sicyonia brevirostris* and East Pacific *S. ingentis* support local fisheries producing a few to several hundreds of tonnes per year.

11.6.3 Caridea

The most productive caridean fisheries are those from cool-temperate and cold-water latitudes, especially those of the Northern Hemisphere, which has more extensive coastal areas than the Southern Hemisphere. Penaeoids are practically absent from these latitudes. As noted earlier, the most important fished carideans are the Northern prawn *Pandalus borealis s.s.*, *P. eous* and, more recently, the Ocean shrimp *P. jordani* along with several North Pacific pandalids (Table 11.2). *P. borealis s.s.* has a North Atlantic distribution in coastal seas from Massachusetts USA, along northeastern Canada, Greenland, Iceland, and high latitude European countries. *P. eous* is fished from Oregon USA around the North Pacific coast as far south as Japan. Together (listed by FAO as *P. borealis*), they have been top producers in global fisheries for a long time, as shown by FAO statistics from 1950 to present (Table 11.2). Pandalids are Type-3 species, with all life history stages occurring in fully marine waters. The protandric sexual system and other aspects of *Pandalus* life history have been discussed in Chapters 7 and 9. A few species of deep-water Type-4 pandalid genera *Heterocarpus* and *Plesionika* support small fisheries in different locales around the world (FIGIS 2021a).

Another top species is the much smaller-sized North Atlantic *Crangon crangon,* fished in the Mediterranean and Black Seas, Atlantic and northern coasts of Europe, with most production by the Netherlands, followed by Germany and Denmark. Unlike the pandalids, *C. crangon* is a Type-2 estuary-dependent species. The adults reproduce in coastal marine waters, but the larvae enter productive coastal estuaries and mudflats to grow as juveniles, later migrating back out to sea to mature as adults (e.g., Hufnagl et al. 2014). This species was near the top of shrimp capture fisheries prior to the 1970s but declined in significance as other fisheries developed worldwide (Table 11.2). However, it is still an important producer in terms of biomass and commerce, especially in European waters.

Although a few species of palaemonids such as *Palaemon adspersus, P. serratus and P. longirostris* from coastal eastern Atlantic and adjacent seas produce a few to several hundred tons annually (FIGIS 2021a), their fisheries are minor on a global scale. However, two species of tropical nearshore palaemonids, *Nematopalaemon schmitti* (West Atlantic; South America) and *N. hastatus* (East Atlantic: Africa), have shown some fairly substantial catches from about the year 2000 on. The latter species has been listed among the top 20 marine fishery species for some years (e.g., 17th in 2010). Various aspects of their life history and fisheries can be found in Pereira et al. (2017) and Enin et al. (1996), respectively.

Freshwater fisheries make up a less than 10%. of world fisheries catch compared to marine ones. One factor is simply the relatively small amount of area of fishable freshwater lakes, rivers, and streams compared to the extensive marine areas of the continental shelves and shallow slopes where most shrimp fisheries are located. All significant freshwater fisheries are composed of palaemonid shrimps, especially the Oriental prawn *Macrobrachium nipponense* and the somewhat smaller Siberian prawn *Exopalaemon modestus* (maximum 60 mm total length) from temperate to subtropical East Asia. The total catch of these two species together is divided between them in FAO statistics, as the major reporting country, China (The People's Republic of China), has not recorded each species separately. In addition, reporting on the two species only began in 2003. Nonetheless, in 2019, the two species each are reported at 98,000 t each which, when compared to the marine species (Table 11.2), puts them as the ninth and tenth rank in global shrimp fishery production. *Macrobrachium rosenbergii*, so important in shrimp aquaculture, produces far less (17,000 t in 2019) but is still significant (just below *Pandalus jordani* in 2019). Atyid shrimps from tropical and subtropical rivers and streams are often the target of small-scale artisanal fisheries. Three pasiphaeid species produced only a few tens of tonnes in 2019 and even less before that.

11.7 Top Shrimp Fishing Countries

The distribution of globally important fishery species can be understood on the basis of the higher productivity of the warm-water penaeoid and cold-water caridean species (Table 11.3). Top producers are countries with extensive subtropical and tropical coastal areas in East and Southeast Asia (Indo-West Pacific), especially China, India, and Indonesia. Fisheries in these countries are primarily based on penaeoid and sergestid (*Acetes japonicus*) species. Countries of the Americas, such as the USA (Gulf of Mexico, s.e. Atlantic coast), Mexico, northern Peru, Brazil, and Argentina with warm marine waters host significant penaeoid fisheries. However, countries with temperate and cold waters such as Greenland, Canada, the Russian Federation, Norway, and the northwest USA have sizable pandalid caridean fisheries. Two tropical African countries have fisheries currently on the top 20 list. These are Nigeria, based mainly on *Farfantepenaeus notialis*, and Cameroon, fishing both *Penaeus* species and the warm-water marine palaemonid *Nematopalaemon hastatus*. However, no production of the latter species has been reported since 2016, hopefully due to a lack of reporting to FAO rather than fishery collapse due to overexploitation or other factors.

Table 11.3 Major fishing countries of marine shrimps ranked by production (metric tonnes) and listed by decadal intervals from 1950 to present (from FIGIS 2021a)

Country	2019	2010	2000	1990	1980	1970	1960	1950
China	1 (1,049,126)	1	1	1	2	3	2	5
India	2 (437,152)	2	2	2	1	2	3	2
Indonesia	3 (288,040)	3	3	4	5	8	6	13
Argentina	4 (215,727)	9	14	26	63	60	48	25
United States of America	5 (130,914)	7	4	3	3	1	1	1
Malaysia	6 (106,517)	8	7	6	6	9	10	8
Greenland	7 (100,105)	6	8	7	13	23	35	45
Mexico	8 (86,240)	10	11	10	7	4	4	3
Canada	9 (61,155)	4	5	15	22	41	47	46
Thailand	10 (60,955)	11	9	5	4	5	14	12
Peru	11 (50,197)	20	66	33	62	70	69	–
Nigeria	12 (45,195)	19	24	47	51	50	51	44
Russian Federation*	13 (44,399)	31	15	19	21	–	69	–
Philippines	14 (43,452)	12	12	14	14	6	12	14
Brazil	15 (33,300)	13	13	11	10	10	8	6
Cameroon	16 (32,628)	25	86	67	74	54	–	–
Vietnam	17 (31,880)	5	6	16	12	12	11	15
Norway	18 (28,035)	16	10	8	11	25	16	20
South Korea	19 (27,662)	14	16	9	11	19	15	7
Myanmar	20 (25,674)	15	21	32	–	–	–	–

Capture weights (t) are given for 2019 below country rank. Freshwater shrimp fisheries are discussed in the text. –, no catch reported. *, prior to 1988, the Russian Federation was reported by its former larger entity, the Union of Soviet Socialist Republics

Although the countries above were the top 20 in 2019, Table 11.3 shows that their ranking has varied throughout the last several decades with the vagaries of the fishing economy and fleets, national and international politics and, of course the fluctuations in shrimp populations themselves. These species oscillations can be due to several factors, e.g., variation in fishing pressure and environmental factors affecting mortality of different life stages. Those interested in changes in shrimp capture over time in countries not listed in Table 11.3 should visit the FAO FIGIS website for capture fisheries.

11.8 Fishing Techniques

11.8.1 Catchability and Behavior

Nets of various types are the main method of fishing shrimps. The ease in which these relatively small animals can be caught is termed their catchability. This characteristic depends primarily on their degree of aggregation and their general behavior. Important shrimp fishery species congregate in groups of moderate to high densities (schools). The behavior of the shrimps also affects their availability to fishers. J.W. Penn (1984) described three categories of catchability based on these two factors, degree of aggregation and day-night behavioral pattern, for some major penaeid-fishery species. The first two groups (types 1 and 2) are shrimps best fished only at night when they are out and about. Type-1 (e.g., *Ft. duorarum; Melicertus latisulcatus*) are primarily nocturnal although often not active and/or buried at night, but always buried during the day. Type-2 species (e.g., *Ft. aztecus; Penaeus esculentus*) are always active at night, buried during the day but occasionally emerging at that time also. In contrast, the Gulf of Mexico *Litopenaeus setiferus* and Indo-West Pacific *Fenneropenaeus merguiensis* are day-active and never burrow. The latter species forms schools so dense that they stir up the sediment and make the water highly turbid, forming visible areas termed "mud boils." The industrial shrimp fisheries off North Australia use light aircraft to spot mud boils and thus target the schools for fishing vessels (Vance and Rothlisberg 2020).

11.8.2 Otter and Beam Trawls

Industrial shrimp fisheries began with beam trawls dragged by sailing vessels in the early nineteenth century. Invented in Britain in 1860, the lighter otter trawl has replaced the much heavier and bulkier beam trawls. Together with the development of more powerful and larger motorized fishing vessels in the late nineteenth century, the otter trawl is largely responsible for the accelerated growth of industrial shrimp fisheries (Iversen et al. 1993; Gillett 2008). This trawl is a long, bag-like net of variable size dragged over the bottom and equipped with boards on either side on the

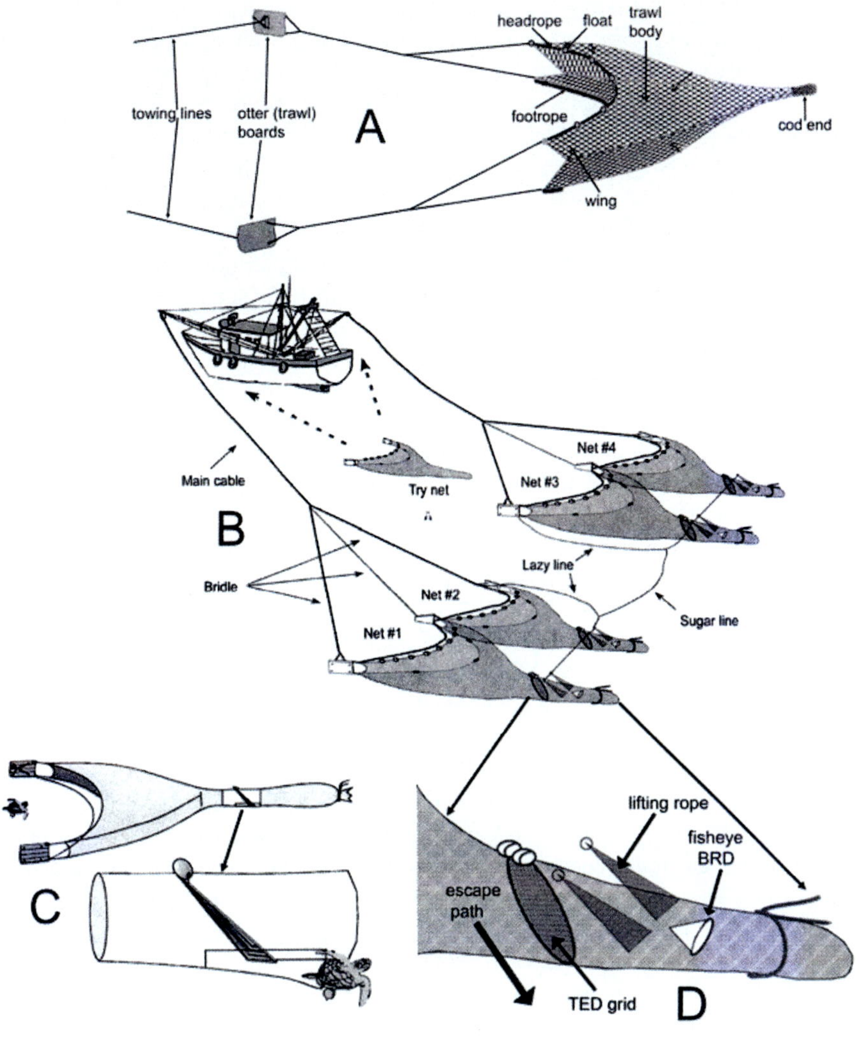

Fig. 11.6 Shrimp otter trawling and bycatch exclusion devices (BRD). (**a**) Diagram of an otter trawl. (**b**) A Gulf of Mexico shrimp trawler with twin outriggers, each hauling two otter trawls equipped with turtle exclusion devices (TED, bottom opening) and a "fisheye BRD"; the "try net" takes a preliminary trawl to determine if there is sufficient shrimp abundance to set the otter trawls. (**c**) Trawl with TED inserted (above) and a turtle escaping from the TED (below). (**d**) Closer view of the TED and fisheye BRD; arrow shows the escape path of a turtle or other large animal. From (**a**) McConnaughey et al. (2019). (**b**, **d**) Scott-Denton et al. (2012), courtesy of National Oceanic and Atmospheric Administration (NOAA). (**c**) Mitchell et al. (1995) courtesy of NOAA

front of the net (Fig. 11.6a–b). Most of the bag is a relatively large mesh, e.g., ~80 mm (~3 in), but the mesh is small enough (~20–60 mm, i.e., ~1–2.5 in) at the end of the net ("cod end") to retain shrimps being fished. The front end of the net stretches out on each side as wings, each of which is attached to a rectangular trawl ("otter") board rounded on one end. The otter boards are attached by chain bridles and wire cables to the main towing cable. The latter is connected to a pulley and winch system which raises and lowers the net. The upper edge of the net mouth (headrope, float line) is equipped with floats and the bottom edge (footrope, lead line) has small weights attached. Their combined action keeps the net mouth open vertically during towing. The mouth of the net stretches wide horizontally by action of the trawl (otter) boards which are forced in opposing directions by water resistance during towing (Dumont and Sundstrom 1961). As the net moves forward, shrimps on or in the bottom jump up and do a backward tail flip (retrograde escape response) which directs them into the net. The top of the net mouth is made so as to project forward like a hood which helps to herd the shrimps back into the net. "Tickler" chains are usually attached to the bottom of the net mouth to disturb the first few cm of bottom sediment and startle shrimps to jump up into the path of the net opening. Otter trawls have been a vital part of many collecting trips for research I have done. I often wondered about the origin of the term "otter" boards and trawl. One possibility is based on the herding of fishes by pairs of hunting river otters in Ireland (Gillett 2008, footnote, p. 9). If true, the name is certainly appropriate.

After towing for a variable time (e.g., 1–2 h), the net is brought up to the surface. A pulley and winch system hoists the cod end above the deck. A "lazy line" is attached around the rear of the net to maneuver it above a sorting table where the cod end is opened to release the. Most commercial trawlers are "double rigged," that is, they tow one to two otter trawls from each side attached to outrigger booms (Fig. 11.6b; Plate 11.1a), a technique developed by Gulf of Mexico shrimp fishers (Gillett 2008).

While otter trawls are the net of choice over relatively smooth soft bottoms (sand, muds), the beam trawl is still the better option over rough terrain with rocks and other obstacles that will rip, tear, and snag otter trawls. The net bag of a beam trawl is attached to a rigid pole (beam) while the bottom to heavy metal runners or skids (Fig. 11.7a). This arrangement keeps the mouth open, and the runners facilitate movement of the trawl over rough bottoms while keeping the mouth of the net bag open. The beam trawl is weighty and cumbersome, restricting its useful size. Beam trawls are used for fishing pandalids in coastal areas of the North Pacific (Roberts 2005), in Australia for *Fenneropenaeus merguiensis* (Vance and Rothlisberg 2020) and in the North Sea for *Crangon crangon* (Gillett 2008).

11.8.3 Skimmer and Butterfly Nets

Skimmer and butterfly nets are used by smaller vessels to shrimp in shallower waters of estuaries, bays, and tidal passes. Both types have net bag mouths attached to an L-shaped or rectangular frame (Fig. 11.7b, Plate 11.1b–c). They are lowered from one or both sides of the boat and pulled or pushed through the water. In keeping with their simplicity and size, all the net does not need to be removed from the water after a trawl; only the cod end is lifted and emptied on board. In the somewhat larger and

Plate 11.1 Shrimp fishery techniques, catch and bycatch. (**a**) Otter trawler equipped with outriggers, twin trawls and a try net. (**b**) Skimmer trawl with twin skimmer nets set in the water. (**c**) Skimmer trawl being lowered into the water. (**d**) Catch from an otter trawl, northern Gulf of Mexico, consisting of penaeid shrimps mixed with extensive fish bycatch. (**e**) Otter trawl penaeid shrimp catch on sorting table with much of bycatch removed. (**f**) Cleaned otter trawl catch with a mix of penaeids *Litopenaeus setiferus* (Northern white shrimp) and *Farfantepenaeus aztecus* (Northern brown shrimp). (**g**) Turtle exclusion device (TED) with Nordmøre grid. (**h**) View of the Nordmøre grid through the escape opening of a TED (covering flap removed to view grid). ep, escape path of turtle or other large animal in front of the grid; g, Nordmøre grid; L-f, L-shaped frame of skimmer net; n, net; ob, otter (trawl) boards; or, outrigger; tn, try net; s, skid of skimmer net; sh, arrow showing path of shrimps through the grid into the cod end of the net. Courtesy of (**a**–**c**), (**e**–**f**) Louisiana Sea Grant Program; (**d**) R.T. Bauer; (**g**) National Oceanic and Atmospheric Administration (NOAA); (**h**) William Folsom (NOAA)

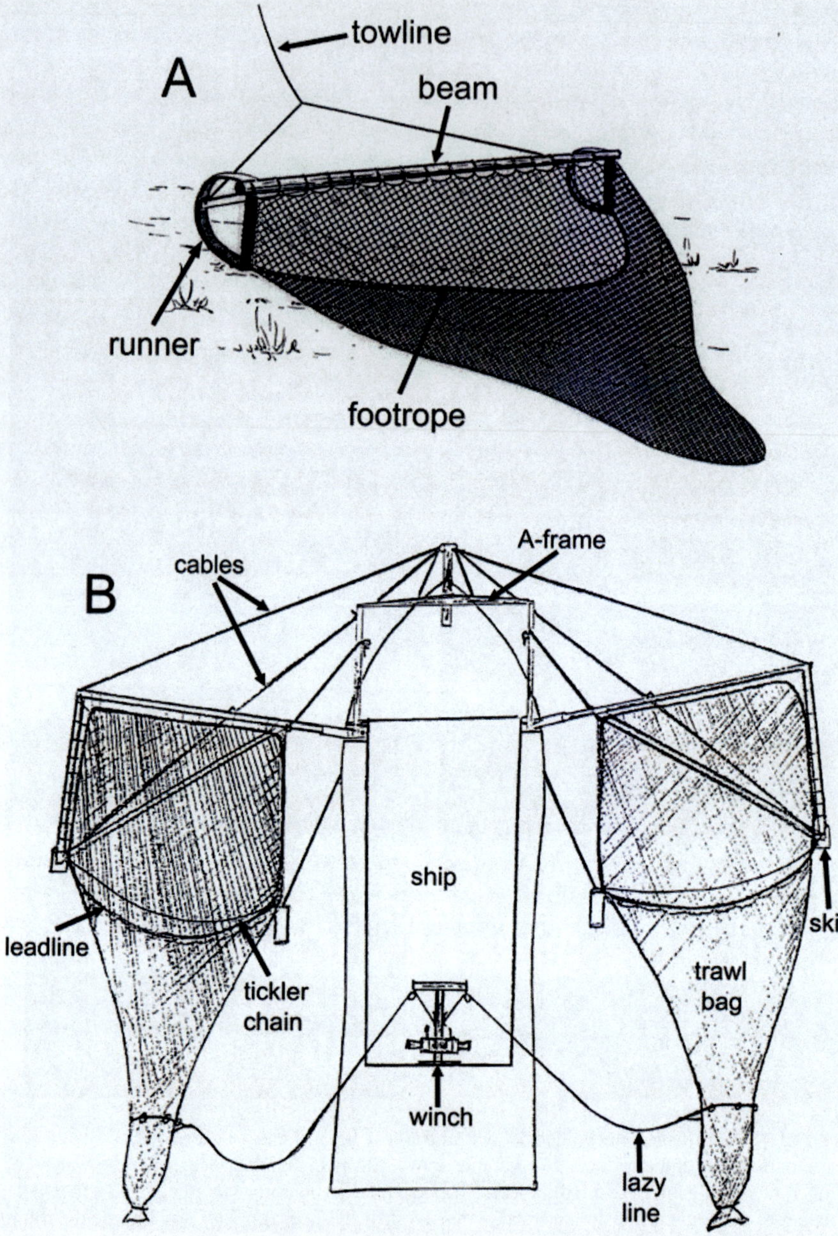

Fig. 11.7 (**a**) Beam trawl. (**b**) Twin skimmer trawls on vessel. From (**a**) Dumont and Sundstrom (1961) courtesy of NOAA and (**b**) Hein and Meier (1995) courtesy of NOAA

heavier skimmer net, the outer frame arm terminates in a skid which glides over the shallow bottom (Fig. 11.7b, Plate 11.1c). In Louisiana bays and estuaries, skimmer nets are as equally productive as otter trawls (Bourgeois et al. 2015). Some vessels use twin skimmer nets while dragging a single otter trawl from the stern (Hein and Meier 1995) (Fig. 11.7b). Butterfly nets are pushed just beneath the surface of the water by boats but also can be attached to wharves, piers, and other structures; shrimps enter the nets during the flow of ebb tide.

11.8.4 Fyke Nets and Traps (Pots)

Sergestid fisheries (mainly *Acetes*, especially *A. japonicus*) are the only major fisheries in which trawling is not used (Omori 1975; Xiao and Greenwood 1993). Sergestid aggregations are mainly fished in very shallow waters (<5 m) in river mouths and estuaries. Bag-like nets of various configurations (e.g., fyke or stow nets) fixed to the sea bottom utilize tidal flow to capture these small shrimps (Fig. 11.8a). Other fishing methods for *Acetes* are push, lift, cast nets and haul seines which one or two fishers can utilize either by wading or from a small craft (Gillett 2008). Stow (fyke) nets are also used extensively in fishing for *Crangon crangon* in European estuaries.

Baited traps (pots) are used on rocky bottoms where trawls of any kind would quickly snag or destroy bottom fauna. Baited traps come in a variety of shapes and designs, but all have funnel openings with the narrow end inside the trap (Fig. 11.8b). Shrimps can enter with ease but escape back up the funnel is difficult. The traps are equipped with a lid or cover that can be opened to take out the catch. Traps are attached by a line to surface buoys to identify location and raise them to the surface. One productive trap fishery is that for the large "spot shrimp" *Pandalus platyceros* from the northwest Pacific coast of North America (Roberts 2005; Wargo et al. 2013). Various attempts have been made to develop trap fisheries for various species of the pandalid *Heterocarpus*, a deeper water genus, in the Indo-West Pacific (Iversen et al. 1993), Hawaii (Polovina 1993) and in South America (Wehrtmann et al. 2012). However, these have not had much success due to factors such as rapid decline of initially high catch rates, frequent loss of traps in deep waters and lack of established markets for them (Polovina 1993). Other species fished by trap for food or bait are mentioned in Holthuis (1980).

11.9 Management of Shrimp Fisheries

As in other fisheries, uncontrolled shrimp fishing eventually leads to overexploitation and collapse of the fishery, with serious impacts not only on shrimp populations and ecosystems but also to human society (economies, employment, and culture). In most countries, government agencies attempt to regulate shrimp fisheries

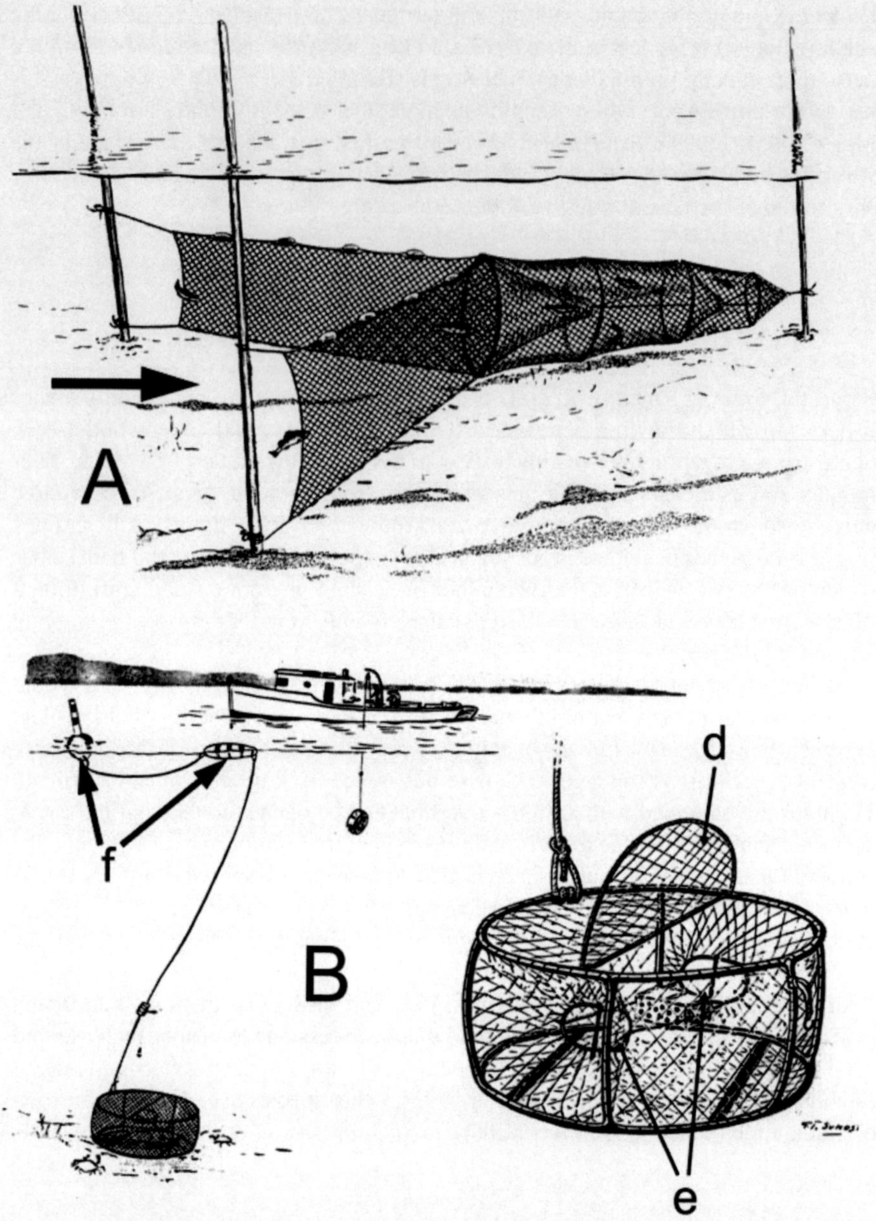

Fig. 11.8 (**a**) Fyke net; arrow in (**a**) indicates entrance and water flow through the trap. (**b**) Ship with shrimp trap (pot) on bottom; enlargement of trap at right. d, door to bait and take out catch; e, entrance to trap; f, floats. Modified from Sundstrom (1957) courtesy of NOAA

with variable degrees of success. Goals of fishery management include prevention of overfishing; conservation of shrimp populations; optimal biological and economical sustainable yields; and protection of other fisheries and their ecosystems. Capture production of shrimp fisheries appears to have reached its maximum in the early 2000s (Fig. 11.1). However, the catch per unit effort (CPUE) is declining in many countries as the number of fishers, number and size of fishing vessels and their efficiency have increased while shrimp populations in general have not (Gulland and Rothschild 1984; Macfayden et al. 2013). Useful reviews on the management of shrimp fisheries can be found in Gulland and Rothschild (1984), Garcia and Le Reste (1986), Gillett (2008) and Bauer (2020).

11.9.1 Overfishing

A challenge to both fishery management and shrimp fishers themselves is how to avoid overfishing. Shrimp populations may suddenly decline from one year to the next due to natural environmental changes. However, a fishing fleet, accustomed to good catches, will go out and exert full fishing pressure, unintentionally damaging the shrimp stock so badly that it may not recover—like kicking someone when they are already down. Changes in the size-frequency distributions of fished populations may indicate overfishing by a reduction in the average size of fished shrimp stocks. For example, important pandalid species are protandric hermaphrodites, so that the larger more desirable individuals are females. When overfished, size-frequency distributions become skewed towards the smaller male-phase shrimps and the size at which sex change occurs decreases (Hannah and Jones 1991) (Chap. 7). In addition to yields with smaller shrimps, the reproductive output of larvae and subsequently recruitment may be decreased by fishing out the female-phase individuals.

Another signal of overfishing is the serial depletion of species, with fishing first concentrated on the more desirable larger and/or more economically valuable species. When these species are fished out, fishers switch to smaller, less desirable species. This phenomenon was clearly demonstrated in the Gulf of Alaska fisheries by Orensanz et al. (1998). Crustacean fisheries first focused on red king crab (*Paralithodes camtschaticus*). When catches on that valuable species declined, effort and catches then increased and declined successively from the pink shrimp *Pandalus eous* to a series of other crab species. Initially, shrimp fishers were able to obtain good catches nearshore, but as those populations were depleted, fishing vessels had to go farther and farther offshore. In coastal penaeid shrimp fisheries, larger size species are targeted when a trawl fishery is first developed in an area. As these populations decline from overfishing, fishing effort shifts to smaller, commercially less-desirable species (Gillett 2008).

The type of life-history strategy of a shrimp species is a factor in its vulnerability to overfishing and unfavorable environmental conditions. Longer-lived K-selected species are more susceptible because of their much slower growth to a fishable size (years) and older age of reproductive maturity, e.g., pandalid shrimps. Fortunately, penaeid and sergestid species, which are the major component of shrimp fisheries globally, are relatively resilient to overfishing because of (a) their short life span (generation time) and very rapid growth. In addition, recruitment of the next generation is more a function of environmental factors affecting the larval and early benthic stages for a particular year than the abundance of reproducing adults in the previous year. Penaeid and sergestid shrimp fisheries are often referred to as an "annual crop," similar to annual plant species in agriculture (Garcia and Le Reste 1986; Gillett 2008). Penaeid shrimp fisheries can recover rapidly after catastrophic environmental events. For example, production in the Gulf of Mexico shrimp fisheries after major hurricanes in 2005 and 2008, as well as the disastrous 2010 *Deepwater Horizon* oil spill, showed no appreciable decline in yields in the years after these events (Bourgeois et al. 2015; NMFS 2015).

11.9.2 Shrimp Fishery Forecasting and Regulation

It is safe to say that every fishery manager's dream is to be able to predict the next year's abundance of fishable shrimps so that the fishing effort can be regulated to produce maximal catches without damaging the population. However, the massive literature on shrimp fishery management shows that a spawning stock-recruitment relationship for commercially important shrimps is difficult to establish and seldom observed (e.g., Gulland and Rothschild 1984; Pauly et al. 1984; Garcia and Le Reste 1986; Garcia 1996; Bergström 2000). However, considerable research does show that correlated environmental and climatological factors prior to or during a particular fishing year, such as temperature, salinity, dissolved oxygen, rainfall, tidal and wind patterns and strengths, as well as the extent of coastal and estuarine habitat, are important in forecasting fishing stock abundance.

For example, the production of brown shrimps (*Farfantepenaeus aztecus*) in the Louisiana (USA) penaeid shrimp fisheries is positively correlated with spring water temperatures in the estuaries (nursery grounds) as well as the expanse of marsh area with salinities greater than 10 ppt (reviewed in Bourgeois et al. 2015). After a dry and warm spring, production is generally good but not when the spring is wet and cold. Although white shrimp (*Litopenaeus setiferus*) postlarvae and young juveniles tolerate lower salinities than *Ft. aztecus* and penetrate further up into the marshes, lower-than-normal river discharges in estuarine and nearshore habitats are necessary for optimal production in the fishery. Tidal conditions affected by wind strength and direction are another factor in the recruitment and movement of estuary-dependent shrimps. Strong southerly winds produce higher tides during spring in the Northern Gulf of Mexico, which promotes movement of postlarval shrimp into nursery areas. However, late spring weather fronts with cold northerly (offshore) winds produce low water temperatures and dry up shallow nursery habitats, decreasing recruitment. During the latter event, developing juveniles temporarily move out of estuarine areas, exposing them to higher predation rates (Bauer 2018). In the Western white shrimp *L. occidentalis,* the recruitment of postlarvae into coastal estuarine nurseries along the coastal Colombian Pacific is inversely correlated with the strength of Trade Winds. Strong offshore winds push water offshore, impeding the movement of postlarvae into coastal estuaries (Díaz-Ochoa and Quiñones 2008). Recruitment is positively correlated with precipitation and river runoff which produces the more favorable lower salinity in the nursery grounds. Models which include both environmental parameters and the abundance of pre-adult life stages are promising in predicting future shrimp catches. For example, Diop et al. (2007) developed a model for predicting abundance and subsequent catch of adult Northern white shrimps (*L. setiferus*) based on lagged (earlier) counts of juveniles, coupled with environmental parameters such as temperature, river discharge, and cumulative wetland loss.

Fishery managers use other types of models to measure the effects of fishing on shrimp catches and populations. The goals of such models are to maximize catches while leaving a healthy, sustainable population for the future. One type of model is based on the logistic-curve concepts of carrying capacity for a species (population size that a particular environment can support) and its intrinsic rate of natural increase. These production models use CPUE as data in calculating maximum sustainable yields (MSYs). Fishing effort (number of fishers, size of vessels, length of fishing season) can be adjusted to take the best advantage of the shrimp's maximum population growth, which is about half of species carrying capacity for a particular habitat. Adjustment of yield to less than the MSY is a more moderate approach for not surpassing it, which may damage the shrimp population and future harvests. Models based on CPUE assume stable population levels, which is often not realistic given the natural population fluctuations of shrimps. As discussed, recruitment is not well correlated with spawning stock abundance but rather a complex of variable climatological and other environmental factors (Gulland and Rothschild 1984; Garcia and Le Reste 1986; Maunder et al. 2006). However, CPUE is data that is usually readily available for fishery management. Yield/recruit (Y/R) models analyze the effects of fishing on mortality and growth of different age/length groups to regulate fishing effort at or below MSY level. Just as it is judicious to fish a population somewhat below the maximum growth rate, it is likewise essential to harvest individuals or life stages at when their individual growth rates are highest. Smaller younger shrimps grow faster than larger older ones. The dilemma is whether to harvest individuals when they are smaller, more abundant, but usually less valuable; or to wait and let them grow to a larger size when they will be less plentiful due to mortality but higher in value because of their larger size. Such decisions are particularly relevant to estuary-dependent penaeid shrimps which typically have sequential fisheries. There is often one fishery for larger juvenile and subadults in the estuarine part of the life cycle and yet another fishery for adults offshore in the open sea. However, Y/R models do not forecast future fishing catch because the data used is "postmortem" (Garcia and Le Reste 1986), i.e., the data are collected from a population already fished. However, Y/R models are useful decisions about fishing effort in future years.

Y/R models use age-specific growth and mortality data. The age of fishes can be obtained by observations on growth rings on their ear bones (otoliths). Determination of age in shrimps is difficult, almost impossible, because they have few equivalents to fish otoliths. Their exoskeleton is completely molted as they grow so that growth rings usually do not accumulate on cuticular structures and appendages. However, recent studies do show that annual growth rings apparently form in the eyestalks of shrimps and can be counted to determine age in years (Kilada and Acuna 2015). However, this technique is still not widely used, and would be of little use for shrimps of short life spans, like those of most commercial penaeids and sergestids. Direct aging on shrimps grown in the laboratory is certainly possible but does not reflect growth and mortality in nature. Mark-recapture methods are difficult to use with short-lived commercial species as recapture rates are very low, handling and marking of captured shrimps may affect their growth, and massive numbers of shrimps must be marked in fished populations to obtain relatively few recaptures. However, with sufficient resources and manpower, mark-recapture can obtain useful results, as in some penaeid species important in fisheries (e.g., *Litopenaeus setiferus*: Linder and Anderson 1956; Table 8.2 in Dall et al. 1990; Mace and Rozas 2014).

Vital population parameters such as age, growth, and mortality are more commonly estimated by measurements of relative abundance and body size of individuals taken in a series of population samples over time. Growth over time is measured using the von Bertalanffy growth function, while mortality can be assessed by changes in relative cohort abundance and size of the individuals in it (e.g., Mace and Rozas 2014; Castilho et al. 2015). Size-frequency distributions are usually composed of several overlapping peaks representing different pulses of recruitment (cohorts). The means or modes of these peaks can be separated and their progression through time followed by using various computational programs (Pauly et al. 1984). Examples of such programs are MIX (mixture distribution analysis), ELEFAN (electronic length frequency analysis), FiSAT (fish stock assessment tools), and PeakFit. Growth and mortality at different ages can be used in the various Y/R models to determine the body size of capture that yields maximum or optimal sustainable production.

Practical techniques used to manage fisheries have been reviewed by various authors (e.g., Gulland and Rothschild 1984; Garcia 1989; Gillett 2008; Bourgeois et al. 2015). In catch-size regulation, with the goal of maximal growth and avoidance of recruitment overfishing, preseason surveys establish the size and abundance of the target species to decide when fishing seasons are best opened and closed. In Gulf of Mexico penaeid fisheries, for example, repeated sampling of shrimp populations in inshore waters monitors the growth and abundance of juvenile and subadult shrimps leaving nursery marshes into coastal bays on their way to offshore waters (Bourgeois et al. 2015). When these individuals are large enough and sufficiently abundant, the fishery is opened until they become too scarce and/or other shrimp species are still too small.

Regulation of mesh size is one way to prevent overfishing of juvenile shrimps when catches include a mix of juveniles and adults. This is not as important in sequential penaeid fisheries in which juveniles and subadults are taken from smaller vessels using small trawls or butterfly or skimmer nets in inshore waters as the juveniles migrate offshore; the reproductive adults are taken by industrial trawling vessels offshore. However, in fisheries where both small- and large-size shrimps are captured together, the very long antennae, multiple appendages and spiny bodies of shrimps can result in small juvenile shrimps being caught by a relatively large net mesh size. Other methods of regulating fishing effort are limitation of the number of fishing permits issued, the number and size of fishing vessels and trips, and fishing efficiency (e.g., number and types of nets). Extended closure of fishing seasons can protect juvenile shrimps during their maximal growth phase or adults from overfishing when reproductive stocks are low and vulnerable to extinction. Preservation of productive coastal wetlands, such as salt marshes, mangroves, and seagrass meadows, where estuary-dependent species grow and shelter as juveniles, is essential to maintain healthy fisheries (Turner 1977; Turner and Boesch 1987; Primavera 1998; Minello et al. 2003; May-Kú et al. 2014).

11.10 Environmental Impacts of the Shrimp Fishing Industry

Just about any human activity has some negative effect on the natural environment, and shrimp fisheries are no exception. Trawling for hours over vast tracts of continental shelves and shallow slopes for shrimps of commercial interest ("target species") also gathers a considerable amount of other fauna ("nontarget species") which cannot be returned intact and healthy to their habitat. At the same time, the heavy and bulky beam and otter trawls drug over the bottom damage its physical and chemical structure and consequently its inhabitants. Fortunately, not all is lost, as there are judicious means of mitigating, at least partially, this damage as we will see below.

11.10.1 Bycatch and Discards

During shrimp trawling, a considerable biomass and large number of bottom invertebrates and fishes are gathered in the net along with the target shrimp species. Other marine fauna such as sea turtles end up in trawl nets as well. All species collected become tightly packed in the cod end of the net as it moves along over the bottom, along with sand, mud, and other debris. The animals are subject to both large temperature and pressure changes as the net is brought up from the bottom. Most arrive dead or so severely damaged that there is little chance that they can be safely returned to the particular habitat from which they have been displaced (Eayrs 2007).

All species of animals captured in trawls apart from the target shrimps are considered bycatch. The magnitude of bycatch taken by fisheries is staggering; on average in shrimp fisheries, up to 80% of the capture biomass is bycatch (Kelleher 2005). An example of the scale and composition of bycatch in a Gulf of Mexico penaeoid shrimp fisheries was given by Scott-Denton et al. (2012). Bycatch was typically about 2.5 times that of targeted shrimps, with the total catch overall composed of 29% penaeid shrimps, 57% finfishes, 7% other crustaceans, 5% other invertebrates with 1% debris, with a total of 185 species. Most fish bycatch is composed of "trash" fish, i.e., fish not of commercial interest, but also includes the juveniles of species that are of part of other fisheries. The latter is of particular concern for finfish fisheries as new recruits are being taken and killed with subsequent damage to those fisheries.

What happens to the bycatch dumped overboard, termed discards? That depends greatly on where the shrimp fishery is located. In much of the world, the nontarget species are simply tossed over the side—essentially marine garbage. One often sees seagulls, pelicans, and other seabirds in large numbers following fishing vessels with bycatch being discarded. Although a bonanza for the scavenging birds, that is not the case for the discards that sink to the bottom. It is a boon for carrion-feeding bottom species, but large loads of dead biomass arriving on the bottom may soon saturate scavengers, leaving lots of decomposing material that can change the structure and function of the trophic food web in the habitat.

After a trawl is brought up, its contents are unloaded onto sorting tables where target species are separated from the bycatch (Plate 11.1d–f). This is a time-consuming and messy process which reduces the efficiency of fishing. The bycatch often contains spiny and/or venomous fishes and invertebrates that may injure crew and certainly slow down the process of sorting. As one who done lot of trawl sorting during research collecting trips in the Gulf of Mexico, avoiding hardhead sea catfish (*Ariopsis felis*) entangled in the trawl net is not much fun—their spines deliver a painful sting!

Not all bycatch is thrown over the side as discards. This source of marine productivity is not wasted by fishers of some countries, especially those of South and Southeast Asia. On the contrary, much of it is sold to companies that process it into marine meal, oil and food pellets for fish and shrimp aquaculture (Gillett 2008). Some potential discards can be sold or used as bait for other types of fisheries, e.g., for baiting fishing hooks or traps. Nevertheless, governmental fishery management agencies, environmental scientists and some shrimp fishers themselves agree that bycatch and discards from shrimp fisheries need to be reduced. Many countries have bycatch and/or discard quotas or bans that are variably enforced in a number of ways, e.g., once a fishing vessel has produced a certain amount of bycatch and/or discards, it must cease fishing for some period or area (Villasante et al. 2015). Monitoring of bycatch and discards is problematic, either by fisher self-enforcement or observation by fishery management.

Both public and private concern over the plight of sea turtles trapped and drowning in fishing nets has stimulated efforts for decreasing bycatch. A variety of modifications of shrimp trawls have been developed which allow mobile organisms such as sea turtles and larger fishes to escape a net before arriving at its cod end. These are termed "bycatch reduction devices" (BRDs), including one common form, the turtle exclusion device (TED). It consists of a metal (Nordmøre) grid inserted into the net at an angle of 30° to 55° (Fig. 11.6c–d, Plate 11.1g–h). It blocks large swimming animals from passing into the cod end and instead directs them to an escape opening at the bottom or top of the net. Bottom escape hatches are preferred over bottom with considerable debris (e.g., grass, sticks, shells etc.) which damage captured shrimps and make catch sorting more difficult. Over relatively clean bottoms, top-opening TEDs are more efficient in catching shrimps (Mitchell et al. 1995). However, in USA shrimp fisheries overall, bottom-opening TEDs are generally preferred (Scott-Denton et al. 2012). In either case, shrimps and fishes smaller than the width of the grid openings pass back into the cod end. Indeed, the caridoid backward escape response further ensures that shrimps will actively move back into the cod end. Some nets include a fish eye BRD, which consists of a small cone-shaped metal frame located between the TED and cod end; it opens to the top or side of the net (Fig. 11.6d). This BRD exploits differences in shrimp and fish behavior in the net (Eayrs 2007). Strongly swimming fish which have passed through a BRD grid tend to swim forward within the net against the direction of the tow and can escape through the fisheye. Shrimps, in contrast, move in the opposite direction. Although these devices are helpful in avoiding capture of highly mobile animals other than shrimps, they are not effective for slow-moving or sessile animals.

Although BRDs and TEDs are effective in reducing bycatch and discards in shrimp fisheries, the application of these devices varies considerably. They are mandatory in USA penaeid shrimp fisheries in federal waters. Shrimp imports into the USA, Australia, and other countries have to comply with regulations for sea turtle protection (Gillett 2008). Fishers generally dislike TEDs in spite of the benefits of bycatch reduction, such as more rapid and safer sorting time and higher shrimp catch ratios. Many shrimp fishers find TEDs burdensome because of their expense and heavy weight, difficulty of net handling and their perceived or actual decrease in shrimp catch and loss of income. One solution to the problem is for national or regional governments to offer subsidies that cover the cost of purchase and installment of TEDs, as well as economic losses due to their use.

11.10.2 Trawling Disturbance of the Sea Bottom and Its Communities

The effect of dragging a large heavy shrimp trawl across the sea bottom might be likened to driving a military tank over and through a woodland or meadow; trees and other plants are crushed, ripped away and otherwise damaged, and the soil is overturned and churned. Both the otter and beam trawls used in shrimping are hefty pieces of equipment. The ground ropes (leadlines) on the bottom front edge of the trawls are held down by large metal weights and often equipped with heavy rollers ("rockhoppers") that help move the net more easily over the bottom. As the net fishes, its cod end fills with animals, bottom debris and sediment to form a bulky heavy ball that bumps and scrapes along the bottom.

The scale of this problem was particularly brought to the attention of the fishing and scientific community by Watling and Norse (1998). Their review paper details how trawling gear digs into the seafloor, smashing, crushing, and burying sessile marine organisms such as sponges, hydroids, soft and hard corals, and bryozoans, among others. Boulders, cobbles, and other objects give structure habitat for many mobile species. Trawling may cover or displace these objects, exposing epi- and infaunal organisms below. The diversity and abundance of organisms in the habitat can be significantly altered, with a subsequent change in food webs and ecosystem function. Trawling over soft bottoms, where much of shrimp fishing occurs, collapses burrows of infaunal organisms such as polychaete worms, clams, brittle stars, amphipods, and thalassanidean crustaceans. Trawls stir up and mix the upper layers of sediment, disturbing its biochemical properties to which infaunal organisms are adapted (Morys et al. 2021). Shrimp populations feed on the organisms living in or just above the substrate. Changes brought about by trawling are bound to have a negative effect on the shrimps left behind after fishing.

The extent, type, and time of recovery from trawl damage varies with the biological composition and geological structure of the bottom community impacted. Shrimp trawling over expanses of sand- and mud-bottom habitats has less impact on biological communities living there than trawling over hard-bottom habitats which are inhabited by attached sessile organisms. Comparisons made on the sedimentary and community structure of trawled and untrawled areas of similar soft-bottom habitats have shown they may recover somewhat rapidly, from a few to several months (Watling and Norse 1998; Sheridan and Doerr 2005; Simpson and Watling 2006). Recuperation of hard-bottom communities can take much longer because their attached fauna generally has longer life spans and slower growth rates than organisms from soft-bottom communities. For example, Yesson et al. (2017) compared recovery of hard- and soft-bottom communities in the frigid waters off the continental shelf off western Greenland, where otter trawling for *Pandalus borealis* is frequent and extensive. Soft-bottom habitats recuperated about twice as fast (5 versus 10 years) as hard-bottom ones.

The reason for the greater resilience of soft-bottom communities is that shrimp trawling is usually conducted over the shallow continental shelf, which is subject to frequent natural disturbances, such as winter storms, hurricanes, tidal currents, wave action, and foraging and digging by megafauna (e.g., fishes, crabs). These processes tend to keep soft-bottom communities in an early successional state in which recovery back to their pretrawled condition is relatively brief. As a general rule, when the frequency and intensity of trawling in the same area are low compared to its recovery time, then that habitat and its community can return to a pre-trawling state (Watling and Norse 1998; Simpson and Watling 2006). But the problem is that favorable fishing grounds are fished over and over again in rapid succession by multiple vessels during fishing seasons. In this case, communities do not have time to recuperate, no matter how resilient they might be. Watling and Norse (1998) estimated that about ~50% of the world's continental shelves are trawled every year, but with some areas subject to much heavier fishing pressure than others. Management of trawling effort (e.g., number and size of vessels, gear restrictions, fishing quotas, length of fishing seasons) can be used to keep both shrimp fishing viable and bottom communities healthy and productive (Vlachopolou et al. 2013; McConnaughey et al. 2019).

11.10.3 Loss of Coastal Wetlands and Shrimp Fisheries

Several of the most productive shrimp fisheries are based on estuarine-dependent species. Salt marshes, mangroves, and seagrass meadows in or near estuaries are nursery grounds for the postlarvae and juveniles of commercially important shrimps, providing both food and shelter from predators. Many studies have shown the links among measures of estuarine vegetation, abundances of shrimp juveniles and the subsequent fishery catch offshore (e.g., reviewed in Bauer 2020; Pickens et al. 2021). Shrimp catches increase when local wetlands grow in size but decrease when wetlands are lost (Turner 1977; Turner and Boesch 1987). These wetlands, including those important for warm-water penaeid shrimp fisheries, have been and continue to be degraded and lost at ever-increasing rates since the nineteenth century but especially in the last three decades. More than half of all coastal wetlands disappeared in the twentieth century (Li et al. 2018). The primary causes are human activities such as development and urbanization, pollution, aquaculture and agriculture, and construction related to shipping and navigation, all of which are followed by further erosion of wetlands (Newton et al. 2020). Dams, levees and channelization of rivers and streams reduce the deposit of sediment into coastal wetlands that are necessary to maintain them. Natural forces compound coastal loss even further, especially coastal subsidence, hurricanes and sea level rise, the latter two exacerbated by climate change.

Using satellite imagery and radar, Thomas et al. (2017) estimated that vast areas of mangrove forests have been destroyed in tropical areas, especially in Southeast Asia, confirming other studies on mangrove loss (e.g., Primavera 1998; Valiela et al. 2001). About 38% of the world's coastal mangroves were lost prior to 2010, the end date of data collection by the Thomas et al. study. In Southeast Asia, clearing of mangroves for palm oil plantations and rice agriculture are primary culprits. In the USA (Florida), residential and industrial development are the main causes of mangrove removal. Coastal land, with its valuable estuarine salt marshes, has disappeared at an alarming rate in the USA state of Louisiana, which suffered a wetland loss of 25% from the 1930s to 2010 (Couvillion et al. 2011) and which still continues today. Marsh eutrophication, urban and industrial development, and diversion of sediment supply due to human activities are major factors in marsh loss (Deegan et al. 2012). Construction of levees along major rivers and waterways, oil pipeline canals, and transportation waterways block and divert the normal sediment supply to the marsh, vital for its maintenance. Government, industry and the general public need to prioritize remedial actions to prevent further destruction of these vital nursery grounds of estuarine-dependent species, including commercially important shrimps.

11.11 Scale and Growth of Shrimp Aquaculture

Subsistence shrimp cultivation has been carried on in the rice paddies and tidal areas of South and Southeast Asia for centuries. However, industrial shrimp aquaculture did not begin until the 1940s–50s when it was established on a sound scientific basis. It has grown today from modest production through the 1980s with accelerating growth from the 1990s to the present (Fig. 11.9). Since 2000, shrimp culture production has skyrocketed to equal and outdo shrimp capture fisheries in production. While the output of shrimp fisheries reached its current plateau in the early 2000s, production by culture has greatly exceeded the fisheries in the last decade. In

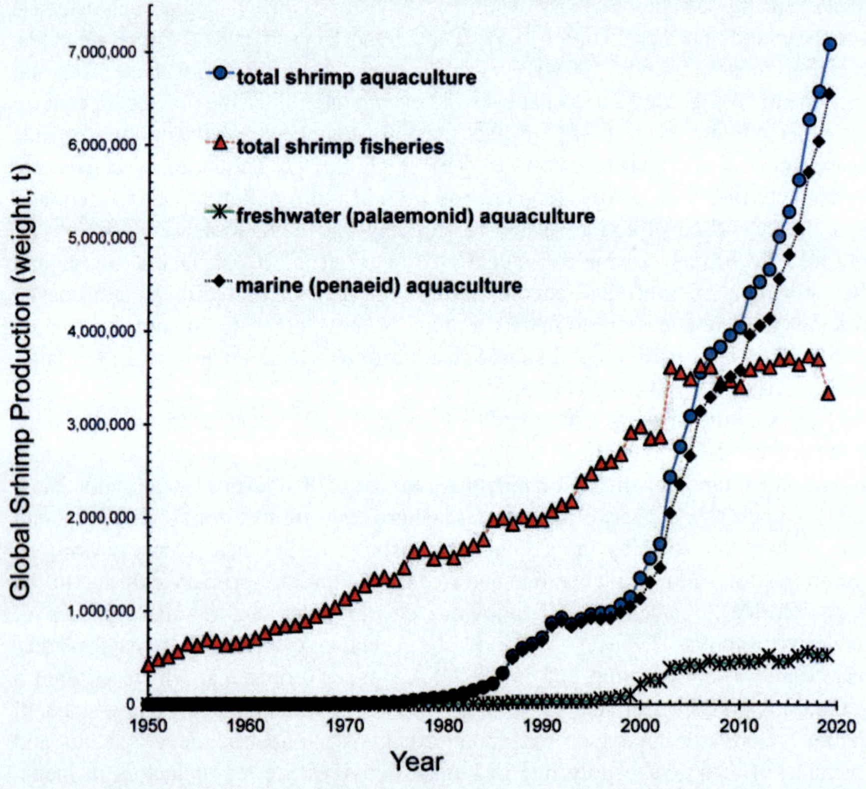

Fig. 11.9 Global shrimp aquaculture production compared to that of capture fisheries from 1950–2019 (FIGIS 2021b). The major taxonomic components of shrimp aquaculture are shown as well: marine/brackish water (penaeid shrimps) and freshwater shrimps (mainly palaemonid carideans, especially *Macrobrachium rosenbergii* and *M. nipponense*)

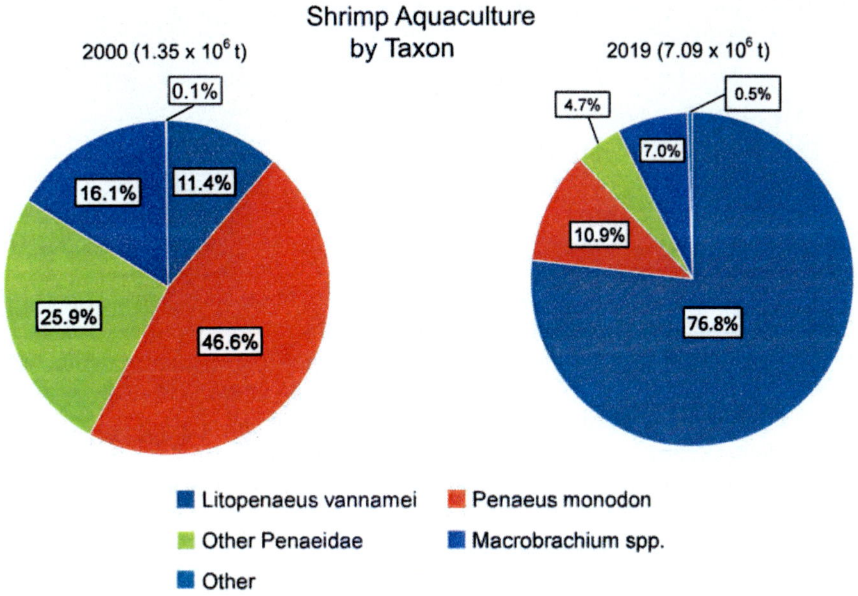

Fig. 11.10 Major taxonomic components of shrimp aquaculture compared between the years 2000 and 2019: Penaeidae (*Litopenaeus vannamei, Penaeus monodon*, other Penaeidae), and Palaemonidae (*Macrobrachium* species, principally *M. rosenbergii* and *M. nipponense*). From FIGIS (2021b)

2019, the shrimp culture crop was more than twice that of the capture fisheries. Some of the factors that have stimulated the growth of shrimp aquaculture are (1) the quality of farm-raised shrimps is more dependable than that of fished shrimps, (2) availability of supply is less seasonal and more reliable, and (3) the species composition and body size of cultured shrimps is more consistent than those supplied to consumers by fisheries (Gillett 2008). However, shrimp fisheries are not in any danger of going out of business because of a flooding of the world's food supply with shrimps—an abundance of high quality, lower priced and appetizing protein is always welcome. Another factor is that the natural environment can provide larger shrimps than aquaculture and a more natural taste that many consumers prefer.

In contrast to the capture fisheries, all shrimp aquaculture is "warm-water;" large-scale culture of cold-water species is simply not viable from an economic standpoint. It just takes too long for cold-water species to grow to a marketable size. The natural distribution of shrimps used in aquaculture is mostly tropical or subtropical r-selected species, although a few are a bit more temperate with somewhat less r-selected life history patterns. Marine and brackish-water penaeids dominate shrimp aquaculture both in production by weight (92.5% in 2019) and first-sale value (90% of the total $45.1 billion). Freshwater shrimp aquaculture, on the other hand, almost completely consists of warm-water palaemonid shrimps of the genus *Macrobrachium* (Figs. 11.9 and 11.10). FAO reported a relatively small amount of sergestid (*Acetes japonicus*) culture from the 1970s until about 2010 but not thereafter (FIGIS 2021b).

11.12 Penaeid Aquaculture

11.12.1 Some History

As related in George Chamberlain's (2010) history of penaeid aquaculture, the foundation of modern shrimp aquaculture was established by the Japanese scientist Dr. Motosaku Fujinaga (Hudinaga), a renowned expert on penaeid reproduction. Techniques for successful mating, spawning, hatching, larval development and growth to adult of the Kuruma prawn, *Marsupenaeus japonicus* were developed in his laboratory in Japan just before World War II. Although his work was interrupted by the war, he made further progress in the 1950s and 60 s, developing the basic techniques of penaeid aquaculture and the establishment of the first commercial-scale shrimp farms. One of his graduate students, Dr. I. Chiu Liao, returned home to Taiwan and concentrated on further developing Fujinaga's techniques applied to the Giant tiger prawn, *Penaeus monodon.* This species soon became and still is one of aquaculture's most productive species. However, the top spot was taken over by the more disease-resistant eastern Pacific Whiteleg shrimp *Litopenaeus vannamei.* Important centers of shrimp aquaculture spread from South Asia to Latin America in the 1980s, first to Ecuador and then to other South American countries such as Colombia, Brazil, Panama, Costa Rica and Honduras. In the USA, penaeid aquaculture has had a spotty record and is not a major player in the world aquaculture scene in terms of production. However, continued research and expertise on culture techniques and penaeid disease by scientists in the USA have been important to the rise of aquaculture in the rest of the world.

The bane of penaeid aquaculture has been viral diseases, which are in large part responsible for the ups and downs of both the aquaculture production by species and by country (Tables 11.4, 11.5). Several cycles of boom and bust have taken place in different countries. Initial efforts with numerous large penaeids were developed and successful but then suffered losses due to disease. Subsequently, other more appropriate species for a particular region or country were identified, and progress was made in techniques for control of disease. As a result, shrimp aquaculture at first complemented but now far exceeds production by capture fisheries.

11.12.2 Principal Species

Several of the same species important in capture fisheries are significant in shrimp aquaculture (Tables 11.2, 11.4a), e.g., *Penaeus monodon, Fenneropenaeus merguiensis, F. chinensis*, *Metapenaeus* species. Almost all are penaeids, with the

Table 11.4 Principal species in global shrimp aquaculture ranked by production weight (metric tonnes) listed by decadal intervals from 1950 to present

Species	Common Name (FAO)	2019	2010	2000	1990	1980	1970	1960	1950
(a) Marine and Brackish water species									
Litopenaeus vannamei (23 cm) (PN) (*e*)	Whiteleg shrimp	1 (5,446,216)	1	3	3	5	8	–	–
Penaeus monodon (34 cm) (PN) (*e*)	Giant tiger prawn	2 (774,484)	2	1	1	2	2	5	–
Penaeus (*s.l.*) spp. (PN)	Penaeus spp. nei	3 (142,317)	3	2	2	1	1	1	1
Metapenaeus spp. (PN) (*e*)	Metapenaeus spp.	4 (72,319)	6	5	5	4	4	11	–
Marsupenaeus japonicus (24 cm) (PN) (*e*)	Kuruma prawn	5 (52,458)	4	7	6	7	6	4	–
Fenneropenaeus chinensis (18 cm) (PN)	Fleshy prawn	6 (38,584)	5	10	14	12	10	–	–
Fenneropenaeus merguiensis (24 cm) (PN) (e)	Banana prawn	7 (18,819)	7	4	4	3	3	–	–
Metapenaeus monoceros (20 cm) (PN) (*e*)	Speckled shrimp	8 (4105)	9	–	–	–	–	–	–
Fenneropenaeus indicus (23 cm) (PN) (*e*)	Indian white prawn	9 (3368)	8	6	8	10	9	–	–
Litopenaeus stylirostris (23 cm) (PN) (*e*)	Blue shrimp	10 (2114)	10	9	7	6	–	–	–
Metapenaeus ensis (16 cm) (PN) (*e*)	Greasyback shrimp	11 (339)	11	11	11	8	5	2	–
Palaemon (*Palaemonetes*) *varians* (5 cm)(PL)	Atlantic ditch shrimp	12 (182)	14	–	–	–	–	–	–

(continued)

Table 11.4 (continued)

Species	Common Name (FAO)	2019	2010	2000	1990	1980	1970	1960	1950
(b) Freshwater species									
Macrobrachium rosenbergii (32 cm) (PL)	Giant river prawn	1 (273,736)	2	1	1	1	1	–	–
Macrobrachium nipponense (7.5 cm) (PL)	Oriental river prawn	2 (225,321)	1	2	–	–	–	–	–
Palaemonidae (PL) (mainly other *Macrobrachium* spp.)	Freshwater prawns, shrimps nei	3 (34,766)	4	3	2	–	–	–	–
Neocaridina (Caridina) denticulata (3 cm) (AT)	Sawtooth caridina	4 (235)	5	4	–	–	–	–	–
Macrobrachium malcolmsonii (20 cm) (PL)	Monkey river prawn	5 (170)	–	–	3	–	–	–	–
Macrobrachium spp. (PL)	River prawns nei	6 (4)	3	–	–	–	–	–	–

Production values are given for 2019 below country rank. Unidentified decapod shrimps (“Natantia nei”) are not included. Maximum size in nature (total length) is given for each species (from Holthuis 1980 and, for *Marsupenaeus japonicus*, Minagawa et al. 2000). Data on production and species names are from FIGIS (2021b); generic names of penaeoids follow Pérez Farfante and Kensley (1997). *e*, estuary-dependent; nei, (FAO “not elsewhere identified”, i.e., not identified to species or other taxon); – no catch reported. AT, Atyidae; PN, Penaeidae; PL, Palaemonidae

Table 11.5 Principal countries in global shrimp aquaculture ranked by production weight (metric tonnes) and listed by decadal intervals from 1950 to present (from FIGIS 2021b)

Country	2019	2010	2000	1990	1980	1970	1960	1950
(a) Marine and Brackish-Water species								
China	1 (2,512,421)	1	1	1	2	8	–	–
Indonesia	2 (937,200)	3	3	3	1	1	–	–
Vietnam	3 (903,129)	4	5	6	6	7	–	–
India	4 (768,608)	6	4	7	7	10	–	–
Ecuador	5 (679,985)	5	7	4	3	9	–	–
Thailand	6 (414,515)	2	2	2	4	3	1	1
Mexico	7 (170,077)	7	9	11	13	11	–	–
Bangladesh	8 (133,748)	8	6	9	5	2	2	2
Philippines	9 (66,560)	11	8	5	10	4	–	–
Myanmar	10 (61,306)	12	17	52	24	–	–	–
Saudi Arabia	11 (60,800)	14	26	37	–	–	–	–
Iran	12 (55,015)	22	20	–	–	–	–	–
Brazil	13 (54,486)	10	10	17	12	–	–	–
Malaysia	14 (53,606)	9	11	16	18	–	–	–
Peru	15 (43,541)	17	31	14	11	–	–	–
Honduras	16 (32,100)	20	15	12	17	–	–	–
Nicaragua	17 (29,490)	15	16	35	–	–	–	–
Venezuela/Bolivia	18 (22,000)	18	14	26	–	–	–	–
Guatemala	19 (19,500)	13	27	21	–	–	–	–
Taiwan	20 (17,438)	16	12	8	8	5	3	3
(b) Freshwater species								
China	1 (391,058)	1	1	–	–	–	–	–

(continued)

Table 11.5 (continued)

Country	2019	2010	2000	1990	1980	1970	1960	1950
Bangladesh	2 (60,835)	2	5	–	–	–	–	–
Thailand	3 (31,345)	4	3	3	3	–	–	–
Vietnam	4 (20,129)	6	6	2	1	–	–	–
Myanmar	5 (9509)	7	–	11	7	–	–	–
India	6 (8702)	3	2	5	–	–	–	–
Taiwan	7 (6780)	5	4	1	–	–	–	–
Indonesia	8 (4600)	8	–	–	–	–	–	–
Americas	9 (280)	10	8	4	2	1	–	–
South Korea	10 (235)	13	11	–	–	–	–	–
Malaysia	11 (206)	9	7	6	4	–	–	–
Cambodia	12 (180)	11	–	–	–	–	–	–
Pakistan	13 (170)	–	–	–	–	–	–	–
Sri Lanka	14 (169)	12	–	–	–	–	–	–
Iran	15 (15)	16	–	–	–	–	–	–
Oceania	16 (15)	15	12	9	6	–	–	–
Brunei	17 (1)	–	–	–	–	–	–	–
Philippines	18 (1)	14	–	–	–	–	–	–

Production weights (metric tonnes) are given for 2019 below country rank. (a) Marine and brackish-water shrimps. (b) Freshwater shrimps. –, no production reported

exception of the marine caridean *Palaemon varians*, a minor component of the industry mainly grown for fish bait as well as human consumption. The majority of productive species in shrimp aquaculture are warm-water, r-selected species which combine the necessary aquaculture traits of fast growth to a large size. Most of the penaeids in Table 11.4a are estuarine-dependent species with a type-2 life history pattern. *Fenneropenaeus chinensis* is an exception as it is a more northerly, cool-temperate species native to the coastal waters of Korea, Japan, and northern China with a type-3 life history pattern. Deep-water k-selected solenocerids such as

Pleoticus muelleri and aristeids are not suitable for shrimp aquaculture because of their slower growth rates and other factors.

The most important aquaculture species at present is the Whiteleg shrimp, *Litopenaeus vannamei,* indigenous to the eastern Pacific from the Gulf of California to northern Peru. Although it sustains major fisheries in that region, it is not as important on a global scale as those top species listed in Table 11.2. Its rise in the ranks in aquaculture (Table 11.4, Fig. 11.10) coincided with severe diseases outbreaks in *P. monodon* in the 1990s and the success of shrimp culture in the Americas and then Asia because of *L. vannamei*'s superior disease resistance, growth rates and adaptability to culture.

The major species supporting the lucrative shrimp fisheries in the USA Gulf of Mexico are *Farfantepenaeus aztecus*, *Ft. duorarum* and *Litopenaeus setiferus*. There was considerable effort in the earlier days in the USA shrimp aquaculture to determine if these native species might be suitable for aquaculture. The advantages of using an indigenous species are that broodstock (mature females for spawning) are readily available and free of exotic viruses. However, none of these species turned out to be favorable for culture. In the end, the nonindigenous *L. vannamei* (Plate 11.2a) is now the species of choice not only in many other parts of the world, but in the USA as well. Important fishery species are not necessarily good candidates for shrimp aquaculture.

11.12.3 Principal Countries

Given that penaeid aquaculture utilizes warm-water species, all the major producers are from subtropical and tropical countries (Table 11.5). Leading the list from the first days of industrial shrimp culture are the South and Southeast Asian countries of China, Indonesia, Vietnam, India, Thailand, as well as Bangladesh, Philippines, Myanmar and Malaysia. Central and South American countries are also on the top list, especially Ecuador, where shrimp aquaculture in the Americas got its start, ranked fifth globally in 2019. Taiwan, the leader in shrimp aquaculture in its initial years, declined considerably due to problems of water supply, pollution and disease with the principal cultured species, *Penaeus monodon*. Most Taiwanese shrimp farmers switched to marine fish culture (Chamberlain 2010). Even in the most successful Asian countries, disease outbreaks have been responsible in great part for the ups and downs of shrimp aquaculture but in recent years, with the proper choice of species and rigorous control of disease (biosecurity), shrimp aquaculture has continued to flourish.

Ecuador became the pioneer and leader to this day of shrimp production in the Americas. The industry was initially begun by clearing (and destruction) of coastal mangrove forests for construction of shrimp ponds with ready access to the sea. The native species, *L. vannamei*, was used from the beginning. Wild-caught PLs (postlarvae, small juveniles) were readily available and used initially for stocking ponds. However, the latter practice is ecologically unsound as it robs the natural

Plate 11.2 Shrimp Aquaculture. (**a**) *Litopenaeus vannamei* from aquaculture pond. (**b**) Penaeid growout ponds at the Waddell Culture Center, South Carolina. (**c**) During harvest, pond water is pumped up into a dewatering tower where shrimps are sieved from water, then delivered through a chute into a large basket for weighing, and then into bins ("totes") for icing and storage. (**d**) Workers picking up shrimps stranded after pond drained during harvest. (**e**) Large indoor raceway (see man walking in raceway for scale). (**f**) Greenhouse raceway in operation. (**g**) *Macrobrachium rosenbergii* large blue-claw male from growout pond, held by Sam Narith, Cambodia. (**h–i**). Examples of color morphs of *Neocaridina davidi;* from left to right: "Blue Dream," "Red Cherry" and "Yellow Neon." a, paddlewheel aerator; b, weighing basket; c, chute; dt, dewatering tower; t, "tote" (shrimp bin). Photographs courtesy of (**a–b, f**) Al Stokes; (**c–d**), David Teichert-Coddington, photography by David Hillegas; (**e**) Leslie Knoesen, Ithuba Shrimp Farm; (**g**) William Daniels; (**h–i**) João Pantaleão

(fished) populations of recruits. As time went on, it became apparent that coastal ponds from cleared mangrove areas were not the best place for ponds because of the poor acidic nature of the substrate. Ponds were constructed further inland, decreasing the pressure on mangrove wetlands, important nursery grounds for both wild shrimp and fish populations. The natural variability in abundance of natural shrimp populations makes the use of wild PLs for stocking ponds undependable. With the support of large companies such as Ralston Purina, hatcheries were developed for rearing of larvae initially hatched from wild females, providing a more secure supply of PLs for ponds. These developments plus eventual control of disease with biosecurity measures have allowed Ecuador to remain a top producer of cultured shrimps.

Although much effort has been expended in the USA to develop shrimp aquaculture, its role on the world stage remains relatively minor. The cost of labor is high compared to Southeast Asian and Latin American countries. Water temperatures along the west coast of the USA, along which flows the cool-water California Current, are too cold for penaeid culture. Although there is extensive coastal land in the southeastern USA where penaeid aquaculture is possible, urban development of this area is a much more lucrative enterprise. Conservation and protection of coastal wetlands is a necessary and important priority of federal and state governments and thus land use is highly regulated. Thus, availability of coastal land for shrimp aquaculture is very limited; however, indoor intensive culture makes penaeid production away from the coast possible (see Techniques, below). The climate of the southeastern USA is subtropical with cool winters, allowing no more than a single crop per year compared to several in tropical countries. Additionally, adequate control of pollution resulting from shrimp aquaculture, highly regulated in the USA, is another expense. As a result of these and other concerns, such as disease outbreaks, USA penaeid culture has not been a major producer compared to other nations, with a peak of ~6000 t in 2003 but declining since then (~2035 t in 2019) (FIGIS 2021b). The most productive state is Texas, followed by other southeastern states (Florida, South Carolina, Alabama) as well as Hawaii, with most consumption in regional restaurants and markets (Treece 2016)

11.12.4 Techniques

Penaeid shrimp aquaculture is practiced at different degrees of economic and labor investment, density of shrimps and production, as described by Boyd and Jescovitch (2020). The oldest form, practiced for centuries in the past and still today in developing countries, is extensive culture. Farmers allow high tides to carry postlarval and juvenile shrimps into coastal ponds in the intertidal zone, then block their escape with weirs and screened gates. The latter prevents fish and other aquatic predators from entering the ponds but still allows water exchange during high tides. The small shrimps grow and feed on the natural food items produced within the ponds. Food is not usually given but fertilization of the ponds with manure or commercial products may stimulate more production of food. Upon reaching a marketable size within 4–6 months (growout period), the ponds or tidal area is harvested by seining the ponds or by opening gates at ebb tides to catch shrimps with nets in the outflow. Shallow (~1 m) ponds may be as small as half a hectare (ha) to 20 ha or more. To give an idea of scale, a hectare has an area of about 2 American football or 1.4 soccer fields. Such ponds can produce one crop per year of a few hundred to about 500 kg (~400–1100 lb)/ha. Extensive culture is not very common today and mainly occurs in countries such as Vietnam, Bangladesh, and Myanmar.

Semi-intensive culture is similar in pond size and depth, but ponds are built above the level of the tides Farms can be very large with more than 50 ponds (Plate 11.2b). To fill ponds, saline water is pumped in from a nearby estuary or sea directly, or first into a reservoir. The pond water is passed through screens to keep out fish and other predators. Daily water exchange is important to flush out excess food and accumulated wastes, as well as to maintain the proper salinity. Ponds have some sort of outlet so that water can be drained back to sea when fresher water is pumped in. Unlike extensive ponds, semi-intensive ponds are stocked, with a low density of PLs (e.g., 5–15/m^2), growth of natural foods is stimulated with fertilizer, and pelleted food is provided a few times per day. After a growout period of 3–4 months, harvest on the order of 1500–2000 kg (~3300–4500 lb)/ha is possible. Such a level of intensity supports a single crop per year.

More intensive forms of shrimp culture require increased labor and financial investment. Ponds are smaller but the reservoirs of seawater are larger to insure adequate flushing and water exchange with the ponds. Reservoir water is first stored for some days to allow sediment to settle out and treated with a disinfectant such as calcium hypochlorite. Ponds are stocked with PLs at a higher density, and ponds are mechanically aerated to insure high levels of oxygenation. Food is supplied to the shrimps several times a day, as they eat slowly and continuously. Feed pellets are composed of plant food such as soybean meal, cereal products; fish meal and oil; and various vitamins and minerals. Considerable research has been done on developing feeds that promote rapid shrimp growth and high production (see "Feeds and Feeding" section in Alday-Sanz 2010). Fertilizers may be added to encourage development of a food chain to produce natural food (e.g., small invertebrates). Production is considerably higher than in extensive and semi-intensive forms of culture, with up to three crops per year in favorable climates every 2–4 months of 3–10 thousand kg (~6600–22,000 lb)/ha per crop. Harvesting is done by draining the ponds through discharge structures or, using fish pumps, through large hoses into sturdy nets or mesh baskets for processing (Plate 11.2c). Shrimps left behind are collected manually (Plate 11.2d). Drained ponds are then dried, cleaned and repaired as needed, and disinfected between crops.

The next level is super-intensive culture, with smaller ponds of 0.2–0.5 ha lined with plastic producing a crop of up to 20,000 kg/ha (~44,000 lb) in 2–3 months with usually no more than 3 crops per year. This level of production is made possible by "biofloc" technology in which waste products are recycled and converted to shrimp food (Krummenauer et al. 2011; Avnimelech 2015). Shrimps are slow and messy eaters. Much of the food supplied is wasted, settles to the pond bottom, and decomposes, with production of ammonia. Further ammonia is added to the water by shrimp and microbial excretion. Ammonia and one of its byproducts, nitrite (NO_2^-) are highly toxic to the shrimps and must be removed. In a balanced system, the ammonia and nitrite will be converted by nitrifying bacteria to nitrate (NO_3^-) which is not toxic. It serves as fertilizer which promotes growth of phytoplankton and heterotrophic bacteria which can utilize uneaten shrimp feed.

Decomposition of unused food reduces pond oxygen that shrimps need. However, in biofloc systems, with intense aeration with pumps and paddlewheels, water bubbles form to which the abundant heterotrophic and nitrifying bacteria adhere, with the help of their secreted mucus, forming small particles or masses 0.1–2.0 mm in size called flocs or bioflocs. Protozoa, phytoplankton, and microinvertebrates may be caught up in the mix. The amount of biofloc produced is a significant addition of nutritious food and protein for the shrimps which readily ingest it. Addition of vertical substrates provides more area for the bioflocs to adhere to and be fed on by the shrimps (Schveitzer et al. 2013). Ammonia and nitrites are cleared or kept at below levels toxic to the shrimps. Biofloc systems also allow super-intensive shrimp culture away from the coast in large raceways within greenhouse and other indoor facilities (Plate 11.2e–f). A seawater supply away from the coast is very expensive but biofloc technology cleans the water so that it can be recirculated and reused.

One of the major hurdles that commercial aquaculture had to overcome was development of reliable and consistent techniques for the mating and insemination of females (broodstock), as well as those for spawning and culture of larvae to produce PLs (small juveniles) to stock shrimp ponds. A basic summary of these techniques can be found in Boyd and Jescovitch (2020), with a detailed treatment in Treece and Fox (1993) subsequently further refined by a voluminous shrimp aquaculture literature (e.g., Juarez et al. 2010). Mating of penaeid shrimps is not as easily arranged as it usually is with caridean shrimps (Chap. 6). In penaeids with a closed thelycum, in which sperm is stored after mating, such as *Penaeus monodon*, females must be induced to molt after which males are attracted and mating occurs. In contrast, in white shrimps (genus *Litopenaeus)*, such as *L. vannamei*, females do not molt before mating and ovarian maturation must occur first (Chap. 6). Females can be mated by artificial insemination. Spermatophores are obtained from the male by electroshocking its genital area, causing contraction of the ejaculatory duct muscles and emission of the spermatophores which are inserted into the female's thelycum (genital area). Artificial insemination is usually practiced in "forced" matings for research and selective breeding. However, this practice often damages and traumatizes the male (no surprise!) and so natural matings are usually the case in shrimp culture facilities.

In both closed and open-thelycum penaeid species, the ovarian maturation that precludes spawning is induced and speeded up by ablation (removal) of one the female eyestalks. These structures contain an important endocrine structure, the X-organ/sinus gland complex. This organ secretes Gonad Inhibiting Hormone (GIH) which suppresses the emission of Gonad Stimulating Hormone (GSH) secreted by the brain and anterior thoracic ganglion. The removal of one eyestalk is sufficient to allow enough secretion of GSH for rapid ovarian development prior to spawning and, in *Litopenaeus* species, brings the female into mating receptivity as well. Broodstock females must be maintained on high-protein natural foods such as polychaete worms and squids, with careful control of environmental factors such as temperature, pH, salinity, and illumination. Mating tanks must be large and circular to allow the active premating chasing by males and coyful flight by females that precede coupling in many penaeids. Years of trial and error combined with careful research have overcome this obstacle (Treece and Fox 1993).

Inseminated mature females later spawn many thousands of fertilized embryos ("eggs") which will hatch with several hours of being spawned. The embryos are carefully disinfected and transferred to hatching tanks, where they will emerge as nauplii several hours later. They are then transferred to the larval-rearing tanks of various sizes and designs. High-quality seawater water is essential for successful larval rearing, e.g., finely filtered to remove microorganisms as well as treated with ultraviolet light to eliminate bacteria and viruses. Nauplii do not feed, but after a series of molts attain the protozoeal stages, which are fed with diatoms that must be cultured apart under rigorous conditions. Formulated feeds which contain essential nutrients are may also supplied in the form of microcapsules. Upon reaching the mysis stages, animal food is supplied in the form of *Artemia* (brine shrimp) nauplii and rotifers. Two to three weeks after hatching, the planktonic mysis molts to the benthic postlarval stage. These the small shrimps are transferred to larger tanks or small ponds and fed enriched *Artemia* and manufactured foods for a few weeks before being delivered to farms for growout. Intensive and super-intensive facilities in tropical areas can deliver up to 3 cycles (crops) of production. Culture of more temperate-adapted species (e.g., *Fenneropenaeus chinensis*), or tropical penaeid species reared in temperate-latitude locales, usually deliver only a single crop because of temperature restraints on growth.

11.13 Freshwater (*Macrobrachium*) Shrimp Aquaculture

Although freshwater shrimp farming only makes up about 10% of the total shrimp aquaculture production (FIGIS 2021b), an industry with a global output in 2019 of over 534,000 t of shrimp with a first-sale value of about 4.5 billion US dollars is certainly significant (Table 11.4b). Freshwater shrimp culture is composed of just a few *Macrobrachium* species, with a small input by one other caridean, an atyid, on the list of species reported by FAO (Table 11.4b). Although the major habitat of *Macrobrachium* species is fresh water, i.e., rivers, streams, and lakes, the important cultivated species are amphidromous, i.e., the larvae must develop in water with some salinity (Chap. 9, Life Histories). The sexual dimorphism, morphology, and social organization of cultured *Macrobrachium* species are quite different than that of penaeoid shrimps. Although the many techniques used for culture of *Macrobrachium* species are similar to those used for penaeids, the biological differences between them pose distinct challenges to aquaculture operators, as will be shown below.

11.13.1 Principal Species

Macrobrachium rosenbergii, the Giant river prawn, has dominated commercial freshwater culture since its beginning in the late 1960s. However, in the last two decades, the Oriental river prawn *M. nipponense* has risen to coequal status (Table 11.4b). *M. rosenbergii* is native to Southeast Asia in rivers and streams that discharge into the sea. Females of this amphidromous species migrate to river estuaries, where larvae are hatched and develop before migrating back upstream to their adult habitat as juveniles (Chap. 9, Life Histories). Unlike most large carideans, it is definitely an r-selected species, with rapid growth to a very large size, particularly the dominant blue-clawed (BC) males (Plate 11.2g); the smaller females have very high fecundity (Table 11.1). The other major species, *M. nipponense*, native to subtropical and temperate latitudes of East Asia, is much smaller in size and less r-selected than its bigger tropical congener (Table 11.1). Cultured stocks *M. nipponense* are derived from a native amphidromous population which, through selective breeding, has been adapted to complete all life stages in fresh water (Kutty and Weimin 2010). In spite of its smaller size, its taste and texture are preferred by the inhabitants of its major producer and consumer, China, where it is a native species. *M. malcolmsonii*, the Monsoon river prawn, indigenous to Indo-West Pacific countries such as Pakistan, India, and Bangladesh, is also cultured although it is still a minor actor on the aquaculture scene. The Amazon river prawn, *M. amazonicum,* which naturally occurs in Atlantic drainages from Venezuela to Paraguay (Holthuis 1980), has been the subject of considerable research by a group led by Wagner Valenti in Brazil (Moraes-Valenti and Valenti 2010). Similar to *M. nipponense*, it is smaller than *M. rosenbergii* and preferred by locals over *M. rosenbergii*, which is also cultured there. Several other *Macrobrachium* species, such as *M. carcinus,* the Painted or Big-Claw prawn (West Atlantic); *M. vollenhovenii*, the African river prawn, as well as a *Macrobrachium* relative, *Cryphiops caementarius* (Changallo river prawn) from Peruvian and Chilean rivers, are in various stages of research and development as aquaculture species (Kutty and Valenti 2010; Morales and Meruane 2013). *M. carcinus,* in particular, is a very large, heavy lobster-like species, with one male recorded at a whopping 850 g and a female of 190 g. However, attainment of such large sizes takes years in these species (6 years in the case of the 190 g female), which is not a favorable culture characteristic.

One atyid, the sawtooth Caridina (*Neocaridina denticulata*), is used dried as a fish flavoring, particularly in China, and also as fish bait (Holthuis 1980). This colorful species, also known as the "red cherry shrimp," is a popular "ornamental" shrimp for freshwater aquarium enthusiasts (see below). Although its culture is relatively low compared to other *Macrobrachium* species, it has fluctuated in fourth or fifth rank by FAO since the mid-1990s (Table 11.4b).

11.13.2 Principal Countries and Some History

Since the mid-90's and particularly since 2000, China has overwhelmingly held sway over other countries in freshwater shrimp culture with over 70% of the total production in 2019, about equally divided between the larger tropical *Macrobrachium rosenbergii* and smaller, more temperate-latitude *M. nipponense* (Tables 11.4b, 11.5b). The tropical countries of Bangladesh, Thailand, Vietnam, Myanmar, India, Taiwan, and Indonesia produce significant quantities of *M. rosenbergii*. Currently, the USA and South American countries are not big producers on a global scale, but freshwater culture of shrimps can be locally important in the USA with great potential in South America (Moraes and Valenti 2010).

M. rosenbergii culture has been going on in Southeast Asia for many centuries. Natural or man-made bodies of water such as ditches, rice paddies, irrigation canals, ponds and reservoirs are closed off and seeded with small shrimps collected elsewhere and allowed to grow until harvest, i.e., primitive extensive culture (Valenti et al. 2010). The modern basis of freshwater shrimp culture began in Malaysia by Shao-Wen Ling who was able to work out and complete the life cycle of *M. rosenbergii* in the laboratory (Ling 1969). A major obstacle was rearing of larvae, which will not develop in fresh water although it is the adult habitat. Very little was known at that time about the life history of amphidromous shrimps such as *M. rosenbergii* (Chap. 9, Life Histories). Ling knew that females migrated down rivers into brackish water estuaries to hatch larvae, and it must have occurred to him that larvae needed water with some salinity. An apocryphal anecdote related by Hanson and Goodwin (1977) is that Ling was eating lunch one day and sprinkled some soy sauce (a very salty condiment) into a larval culture upon which they soon molted to the next larval stage. That led to rearing them in the laboratory (with saline water, no soy sauce!) from embryo to adult. However, the roots of industrial culture must be credited to Takuji Fujimura and coworkers in Hawaii, who developed hatchery techniques for larval culture and mass production of PLs for stocking in ponds (New 2010). *M. rosenbergii* culture blossomed, especially in China, and pond culture spread from Southeast Asia to the Americas, and with major production beginning in the 1990s.

11.13.3 Techniques

Similar to penaeid culture, freshwater shrimp culture is conducted on extensive, semi-intensive, intensive, and super-intensive levels, with corresponding degrees of costs, labor, feeding, water exchange and water quality management. Production can be from 560–2500 kg (~1200–5500 lb)/ha/yr. in temperate latitudes where only a single crop per year is possible, as growing seasons vary from 3–5 months per year. In tropical latitudes where year-round culture is possible, 3 crops of 4 months are possible, typically totaling 1500–4500 kg (~3300–10,000 lb)/ha/yr. depending on the level of intensity. High-intensive culture in tropical China has yielded up to 11,000 kg (~24,000 lb)/ha/year (Valenti and Flickinger 2020). Although *Macrobrachium* shrimps are usually farmed alone (monoculture), polyculture is also practiced with noncarnivorous fishes such as tilapias, carps, and mullets (Zimmermann et al. 2010).

As with penaeid shrimps, farming of *Macrobrachium* can be broken down into four principals phases: reproduction, hatchery, nursery, growout, and harvesting (New 2002). Broodstock can be obtained from the wild or more commonly from pond culture. Care must be taken to introduce females from another source from time to time, as inbreeding with subsequent genetic deterioration and production in stocks has periodically been a problem (New 2010). Mating is usually not a problem in carideans such as *Macrobrachium*, where males readily mate with prespawning, recently molted females. Dominant blue-claw males are preferred as breeders as they are likely to have superior genetic qualities and higher insemination success. Well-fed ovigerous females are kept isolated until spawning, after which larvae are carefully transferred from the freshwater hatching tanks into the larval-rearing tanks with saline (brackish) water, a necessary step in larval development of these amphidromous shrimps.

Hatchery facilities can be outdoors, but most commercial hatcheries are indoors for greater environmental control, especially temperature (see Valenti et al. 2010 for an extensive treatment of hatcheries and their management). Hatcheries must have a source of seawater for larval development, either artificial seawater (a mix of freshwater with a concentrated sea-salt slurry) or a natural supply of good quality seawater (unpolluted, filtered, disinfected). Hatchery systems may have a continuous flow of water with replacement provided that a sufficient supply is available. Alternately, recirculating systems reduce the need for a plentiful water supply. Debris and waste products are removed by mechanical and biofilters, the latter with nitrifying bacteria to detoxify nitrogenous wastes from larval excretion and decomposition of unused food and feces. Larvae require natural salinity, usually 10–15 ppt (psu) depending on the species cultured. Newly hatched *Macrobrachium* larvae are much more developed zoeae than penaeid nauplii and are carnivorous from the start. High-quality *Artemia* nauplii remain the mainstay for larval food in the early stages with supplement of prepared feeds in the later stages. Once larvae make the final molt to the benthic postlarval stage, they are transferred to nursey tanks or raceways for acclimation to fresh water and growth to a larger size for shipping to growout ponds (Coyle et al. 2010).

Macrobrachium growout presents different challenges than those of penaeid culture because of their very different morphology, behavior, and social structure. These palaemonid shrimps have large cheliped weapons, and they are ready to use them. They are very aggressive and cannibalistic, so it behooves other individuals to space themselves out and hide away as much as possible when they are molting and temporarily defenseless. Addition of artificial substrates such as vertically-oriented, plastic-mesh, safety fencing or other structures give shelter and increases the effective density possible in the pond or nursery tank. Larger *Macrobrachium* dominate smaller ones which translates to physiological stress and slow growth in these "laggards" (Karplus and Sagi 2010). These factors necessitate a lower stock density, which results in lower productivity. However, periodic harvesting (culling) of the larger dominant prawns can reduce the problem, as laggard growth rates pick up when dominant individuals are removed.

Growout ponds are generally slightly sloping earthenware or concrete ponds about 1–2 m in depth and somewhat smaller in size (0.2–0.6 ha) compared to the much large penaeid ones. Several thousand juveniles are stocked in the well-aerated ponds; as they grow, they feed both on pelleted food and natural foods that develop in the ponds as bottom vegetation and associated biota grow (e.g., Tidwell et al. 2002; Valenti and Flickinger 2020). Fertilization with manure or other fertilizers encourages the growth of these natural foods. Several feedings of smaller amounts of pelleted or other foods several times a day is preferable to less frequent feeding of larger amounts. Shrimps are slow inefficient feeders and unused food decomposes, uses up oxygen, and generates toxic nitrogenous compounds. After 4–7 months, depending on local climatic conditions, shrimps have grown to a harvestable size of tens of grams, e.g., in the 40–60 gm range (a nice-sized shrimp) depending on the intensity of culture. Harvesting technique depends on the type of culture. In continuous batch culture, the larger individuals are culled at intervals by seining. This stimulates growth in smaller individuals and provides the farmer a periodic source of income. However, harvesting itself imposes handling and temporary stress on individuals left behind. In batch culture, juveniles are stocked, reared, and harvested all together. During harvesting, ponds are drained or pumped out and shrimps collect in depressions (sumps) and are netted out or collected with large mesh baskets. Batch culture provides the farmer with a larger but more seasonal crop which may oversaturate the market. This is an economic problem that can be avoided by wise planning and arrangements between shrimp farmers and their markets.

Culture of *M. rosenbergii* in temperate-latitude areas is limited primarily by temperature, as exemplified by that of the southeastern USA, with farms in states such as Kentucky, Tennessee, Alabama, South Carolina, and Florida (Tidwell et al. 2002). *M. rosenbergii* will not grow well at less than 20 ° C and dies below 16°, with optimal growth between 25–32 °C. Thus, in the southeastern USA, broodstock, hatchery and nursery operations are all indoor operations. Broodstock is maintained indoors during the winter, with mating, spawning, hatchery, and nursery operations taking place in the spring. Juveniles are stocked in early summer with harvest in the early fall, yielding a single crop per year. On the other hand, *Macrobrachium nipponense*, cultured extensively in temperate areas of China, is more adapted to cooler temperatures with optimal growth at no more than 25 °C. Two crops can be produced per year by stocking of ponds in March, with harvest in late June/early July for a first crop, followed by a second stocking after first harvest and the second crop harvested in December. Shrimps below marketable size stay in the ponds to overwinter, although they will not start growing again until the spring (Kutty and Weimin 2010).

11.14 Shrimp Diseases

11.14.1 Penaeids

Cultivated shrimps are exposed to stresses and diseases not encountered in their native habitat. They are overcrowded in densities far beyond those experienced in the natural environment, with frequent contact among individuals. Different life stages are transferred from one tank to another in hatcheries, and from one pond or raceway to another during growout. Water quality is often poor or suboptimal in aquaculture facilities. These conditions promote disease. The extensive trade of and interchange in broodstock and PLs among producers, regionally or from one country to another, has been a primary cause of disease outbreaks that can devastate production. Healthy shrimps may be carriers of many types of disease-causing organisms. In their natural environment, the shrimp's innate immune systems can suppress the increase of the infective bacterium or virus so that disease symptoms are less likely to develop. However, in the unnatural, stressful environment of the aquaculture facility, their immune systems can fail with subsequent disease. The economic and social consequences of disease epidemics in shrimp aquaculture have led to considerable research and a voluminous literature on the nature and control of disease-causing organisms of cultured shrimps (see reviews such as Lightner 1992, 2011a, b; Flegel 2006; Chamberlain 2010; Walker and Winton 2010). Stocks from one geographic area are introduced to another, carrying with them their endemic viruses, bacteria, and parasites. The impact of pathogens has been the most serious threat to the shrimp aquaculture industry throughout its modern commercial history.

Shrimps suffer from a variety of ailments caused by viruses, bacteria, fungi such as *Lagenidium*, *Fusarium*, and microsporidians, as well protozoans such as haplosporidians, gregarines, and ciliates (Overstreet 1978; Johnson 1995). Although penaeids have bacterial diseases, the most serious threat in their aquaculture is from viruses. Dozens of disease-causing viruses have been identified, and the most important ones have been described (Lightner 1992, 2011b; Walker and Winton 2010; Chamberlain 2010). Transmission may be "vertical," i.e., from the female to its offspring, or "horizontal," i.e., from one individual to another by physical contact; by cannibalizing other shrimps which are molting, dying, or dead in the facility; as well as by virions free in the water or introduced by birds or other animals into ponds.

White Spot Syndrome Virus (WSSV) began to show in aquaculture facilities in Asia in the early 1990s and spread to the Americas. It has been one of the most destructive diseases, causing the 1990s collapse of the industry in China and elsewhere in Asia, and then spreading to the Americas. Infected shrimps eat little, are lethargic and develop white spots on the inner surface of the carapace, a result of excessive deposits of calcium. All farmed penaeids are very vulnerable to WSSV; once an outbreak begins in ponds, most of the shrimps die off with a period of several days. A similar plague hit the industry also in the 1990s but came from the Americas—Taura Syndrome Virus (TSV). The virus first emerged in *L. vannamei* farms in Ecuador and then spread worldwide. It first appears in small juveniles in nurseries and growout ponds with rapid onset of high mortalities. One of the horizontal routes of infection is by birds; dying shrimps tend to swim to the surface where they are picked off by birds who can then spread the virus by feeding and defecating in other ponds. Yellow Head Virus (YHV) is one of the most virulent of shrimp diseases, causing rapid and high mortality in ponds. Its disease first appeared in Thailand; one of the notable symptoms is a yellow-colored "head" (cephalothorax).

Penaeids with Infectious Hypodermal and Hematopoietic Necrosis Virus (IHHNV) develop a disease characterized by cuticular deformities of the appendages and general body surface. It is endemic to *P. monodon* which does not display symptoms, but the virus greatly affects other species. It does not cause mass mortalities in *L. vannamei* but may generate cuticular deformities in the general body surface and appendages, reducing growth, production, and appearance affecting market appeal. However, the disease is devastating to *L. stylirostris*, another eastern Pacific species. There are indications that viral spread from shrimp farms to wild populations is responsible for the partial collapse of the *L. stylirostris* fishery in the Gulf of California although other factors may have been involved as well (Morales-Covarrubias et al. 1999).

Yet another serious viral agent is Infectious Myonecrosis Virus (IMNV), first showing up in *L. vannamei* in the Americas and then elsewhere. The disease is recognized by necrosis of tail muscles, turning the abdomen opaque or reddish in color, followed by high mortality. It seems to be triggered by stresses in the shrimp's environment (handling during netting, changes in feeding or inferior environmental parameters). Although the list of penaeid viruses and diseases seems endless, one other group of disease agents, the baculoviruses must be mentioned. These viruses attack the shrimp's digestive system, especially the hepatopancreas. Postlarvae and juveniles in hatcheries are especially prone to baculoviruses which cause high mortality.

A more recent dangerous disease which is NOT caused by a virus is Early Mortality Syndrome (EMS), or, more formally, Acute Hepatopancreatic Necrosis Disease (AHPND) (Zorriehzahra and Banaederakhshan 2015). The infective agent is a strain of a rather common bacterium in coastal waters, *Vibrio parahaemolyticus,* infected by a bacteriophage. It first showed up in China around 2009, rapidly spreading to farms in Southeast Asia. As the name implies, it infects the digestive tract and produces a toxin that destroys the hepatopancreas. It strikes the younger stages, especially about a month after pond stocking. Symptoms include a pale empty gut (lack of feeding, tissue destruction), reduced growth and sluggish, spiral swimming. Species such as *P. monodon*, *L. vannamei* and *F. chinensis* are especially susceptible. Outbreaks have been attributed to poor management of ponds, e.g., feeding and sanitation, as well as stresses on PLs when transported to the ponds for growout. Unlike viruses, which are unphased by antibiotics, this bacterium is susceptible to them. However, use of antibiotics in shrimp ponds is discouraged because of environmental impacts (see below).

On the bright side, considerable research efforts on shrimp pathogens have resulted in development of biosecurity, i.e., techniques for controlling the spread of diseases in shrimp aquaculture (Treece and Fox 1993; Lightner 2005; Chamberlain 2010; Walker and Winton 2010; "Biosecurity and Shrimp Health Management" section in Alday-Sanz 2010; Boyd and Jescovitch 2020). Although the USA is not a major figure in shrimp culture production, research there has led to development of a "Specific Pathogen Free" (SPF) *Litopenaeus vannamei* made available to other countries (Lightner 2005, 2011a). The major crisis and decline in shrimp aquaculture due to shrimp diseases were reversed by technology for producing SPF shrimps with a resurgence and expansion of the industry.

11.14.2 **Macrobrachium** ***Species***

Compared to penaeid shrimps, *Macrobrachium* shrimps in culture are much less plagued by disease, especially viral ones. Only one serious viral disease, "White Tail" or "White Muscle Disease" (WTD) seriously impacts the industry. It was first detected in Caribbean farms in 1995 but since has spread to other areas of the world by the usual route, i.e., import and export of infected broodstock or PLs (Sahul Hameed and Bonami 2012). It afflicts larvae, PLs and juveniles, which become sluggish, develop opaque tails from necrotic striated muscle, and then perish with several days. It is caused by "*Macrobrachium rosenbergii* Nodavirus" (MrNV) and XSV ("Extra Small Virus-Like Particles"). Adults are little affected but are carriers. As with other shrimp viral diseases, there is no cure. The best prevention is use of SPF broodstock using the same methodology as described above for penaeids (see below).

Most of the other common diseases are bacterial or fungal, usually a result of poor hatchery or pond management (Johnson 1995; Pillai et al. 2010). The responsible organisms are ever present in nature, but outbreaks take place in cultured shrimps, both in *Macrobrachium* and penaeid shrimps, due to stresses from overcrowding, excessive handling, poor sanitary practices and inferior water quality. Infection in many cases is due to mechanical breaks in the shrimp cuticle due to handling and small injuries when the shrimp bumps or scrapes against the walls of their enclosures. Minor cuts, abrasions, and injuries from handling or aggressive encounters from conspecifics allow pathogens a pathway through the external defenses of the shrimp. *Macrobrachium* shrimps, which are quite aggressive with dangerous cheliped weaponry, are much more likely to inflict and receive such injuries than penaeids. Deposition of melanin as part of the immune response to such cuticular breaks creates unsightly black blemishes on the exoskeleton ("black spot disease") unattractive to consumers.

Fungi such as *Lagenidium* and *Fusarium* enter thorough breaks in the cuticle where their hyphae spread and destroy shrimp tissues and organs. An immune response by cells in the gills of shrimps to *Fusarium* infection results in deposition of black pigment, giving rise to "black gill disease." Chitinoclastic bacteria are able to derive nourishment from and thus erode crustacean exoskeleton. Various ailments of unknown origin (idiopathic) cause breakdown of tail muscle with an opaque appearance and texture unsuitable for the market. Similarly, suboptimal water quality and overhandling of stock are thought to be the cause of " branchiostegal blister disease," an abnormal swelling of gill filaments and the inner branchiostegite in *M. rosenbergii*. The condition produces a bulge in the carapace and appendage deformities with subsequent body fouling (Pillai et al. 2005). Fouling of the body and gills by filamentous bacteria such as *Leucothrix,* as well as sessile ciliates and suctorians gives shrimps an undesirable appearance to consumers as well as interfering with gas exchange and excretion. Although shrimps do have mechanical antifouling defenses (Chap. 4), water high in organic matter and other nutrients can generate enough microbial and protozoan fouling that such defenses are overcome. Excessive fouling in culture entangles swimming and feeding appendages of larvae with subsequent high mortality.

11.15 Biosecurity

One crucial factor for preventing disease in shrimp culture is good sanitation and high water quality (Pillai et al. 2010). Broodstock, mating, and spawning tanks, larval-rearing facilities and growout ponds must be properly prepared, cleaned, and disinfected before use. Antimicrobial and antiviral compounds such as calcium oxide (burnt lime; quicklime) or hydrated lime (calcium hydroxide) are used to prepare pond bottoms. Equipment should be cleaned with disinfectants containing chlorine, iodine, or other antimicrobials. Facility staff and visitors at a secure facility should pass through disinfectant foot baths and avoid other sources of contamination from other ponds and facilities. Water supplies should be filtered and disinfected with some combination of antimicrobials and ozone or ultraviolet radiation. Natural fresh- or seawater should be as pollution free as possible.

A second cornerstone of good biosecurity is the use of resistant or pathogen-free broodstock for larval production and stocking of nurseries and growout ponds (Lightner 2005, 2011a). Up until about 2000, most broodstock and PLs for seeding ponds were wild caught with no determination of whether or not they carried viruses or other disease-producing organisms. Even today, this is the case in some countries. When pathogens cannot be excluded from the water used or there are other sources of contamination in facilities, specific pathogen resistant (SPR) broodstock are a good option for supplying PLs for pond stocking. Candidate broodstock are tested for pathogens of interest using molecular methods such as PCR (polymerase chain reaction) or immunoassays with monoclonal antibodies. Broodstock are then quarantined under culture conditions to see if a particular disease develops. If not, they can be used to produce disease-free larvae and PLs for stocking. A more secure way is the use of SPF broodstock completely free of a particular pathogen or group of pathogens. As with SPR broodstock, candidate SPF females (F_0 generation) are quarantined and tested for the virus(es) of interest. If negative, they are placed in a second quarantine facility to generate offspring (F_1 generation) which are checked for pathogens. If negative, the F_1 broodstock then generate PLs (F_2). If the latter are negative, then they can be reared as broodstock and sent to mating and hatchery facilities for use in producing PLs for stocking in farms (Lightner 2005).

11.16 Monosex Populations in Shrimp Aquaculture

In species with pronounced sexual dimorphism in size and behavior, stocking of ponds with same-sex individuals would be a significant advance in shrimp aquaculture. Not only could a larger crop of the desired sex be produced, but the shrimps would be more homogeneous in size, a desirable trait for both the producer and the buyer. In *Macrobrachium* species, in particular, some males grow much larger and are more aggressive than others, resulting in stunted growth in other individuals as well as injuries and deaths from agonistic encounters and cannibalism. Shrimp producers either must settle for rearing a crop at lower densities to reduce interactions, or periodically cull and sell the larger males, which requires injurious handling of other individuals in the population as well as expensive labor. However, the large, impressive males do appeal to a specialty market. On the other hand, a larger crop of smaller—but still large-sized—females can be reared at a higher density because problems associated with size intimidation and aggression by males will be avoided (Levy et al. 2016).

Years of research on sex determination of shrimps, especially *Macrobrachium rosenbergii* in particular, have made the possibility of monosex populations for culture a reality. It will be recalled (Chap. 7, Sexual Systems) that the male androgenic gland (AG) secretes an insulin-like male hormone (IAG) which promotes male characteristics in shrimps. Based on this knowledge, a group led by Amir Sagi developed the methods for producing monosex populations (Ventura et al. 2012; Levy et al. 2016, 2019). Unlike humans and other mammals, male shrimps are the homogametic (ZZ) sex, while females are heterogametic (WZ). Expression of the IAG gene in young males can be blocked by silencing the IAG gene (injection of Mr-IAG rsRNA), which, over a series of molts, transforms the males into a fully functional ZZ "neofemales" (Fig. 11.11a). All offspring of ZZ-neofemales mated by ZZ males are males—voila, a male monosex population!

The research group then turned their attention to techniques for routinely generating an all-female population (Levy et al. 2016, 2019). Ablation of a male eyestalk provokes hypertrophy of its androgenic glands. A cell suspension from its AGs (AGCS) is then injected into juvenile WZ females, which then develop as WZ "neomales." These neomales are mated with normal WZ females, and 25% of the offspring will be WW females (Fig. 11.11b). Individual genotypes are verified using

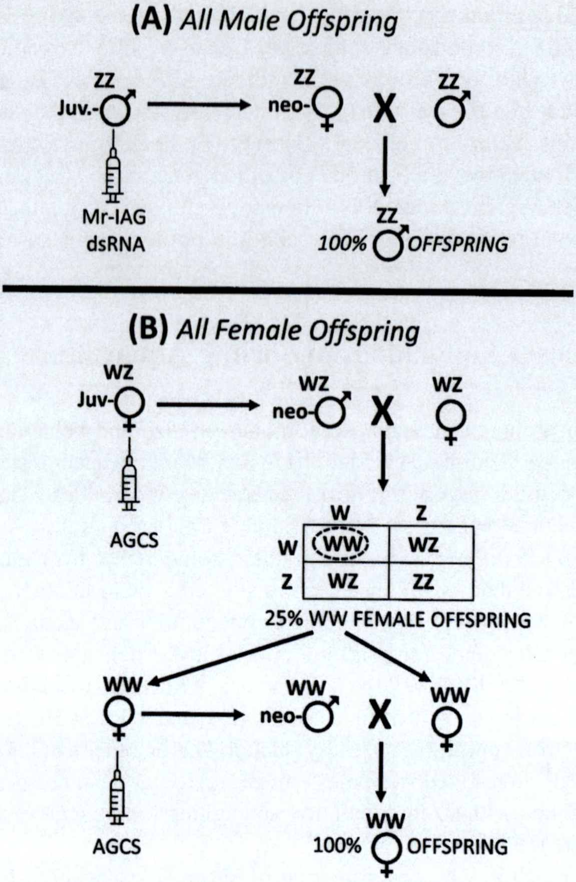

Fig. 11.11 Monosex populations: methodology for producing (**a**) all male and (**b**) all female offspring for shrimp aquaculture, based on methods given in Ventura et al. (2012) and Levy et al. (2016, 2019). (**a**) Juvenile males (ZZ) are injected with an RNA silencer (Mr-IAG dsRNA) which blocks expression of male androgenic hormone so that the individual develops as a phenotypically normal neo-female with a ZZ genotype. ZZ neofemales mated with normal ZZ males produce all male offspring. (**b**) A juvenile WZ female is injected with cell suspensions of hypertrophied androgenic gland (AGCS); they mature as phenotypically normal neomales with WZ genotype. Broods spawned by a normal WZ female mated with a WZ neomale contain 25% WW females. Some of them are injected with AGCS and they mature as WW neomales. Matings between WW neomales and WW females produce 100% female offspring

sex-specific genomic markers. Some of these WW females are injected with AGCS which causes them to mature as WW neomales with typical male traits. When WW-neomales are crossed with WW females, all offspring are female (Fig. 11.11b). These WW females are completely normal, and they spawn successful broods of embryos as do natural, untreated WZ females. The "holy grail" of completely female progeny has been thus attained in *M. rosenbergii*, at least in the laboratory, with all the benefits of female monosex culture.

In theory, the same techniques might be used with other shrimp taxa to yield stocks with sexually-dimorphic size but relatively similar weaponry and aggressiveness, i.e., most penaeoids and the caridean genus *Pandalus*, both highly important in the capture fisheries. Female penaeids are generally larger than males, although the differences are not as striking as in *Macrobrachium* species. However, it might be economically worthwhile to use the techniques developed for *M. rosenbergii* to produce consistently larger individuals for all female pond population. For cold-water carideans such as pandalids, which do have significantly larger females, it may not be economically feasible because of their much slower growth rates and the complications of their protandric hermaphroditism. However, silencing of the IAG gene in the young male-phase individuals does lead to a more rapid sex change to the larger female-phase individuals, which would make their culture much more feasible (Levy et al. 2020).

11.17 Environmental Impacts of Shrimp Aquaculture

As with any agricultural operation, pollution is a major concern for shrimp aquaculture (Páez-Osuna 2001). Growout ponds and other shrimp aquaculture facilities use copious amounts of water in which the activities of the shrimps generate considerable wastes from excretion, defecation, and decomposition of unused foods. Although it behooves shrimp aquaculture to reduce the volume of used water discharged, nutrients, and organic matter will still be released into nearby marine or freshwater habitats. These compounds overstimulate phytoplankton and other plant growth (eutrophication) which stifles and upsets ecosystem function. Bacterial decomposition of excess organic matter robs habitat waters of oxygen needed by other organisms. Efficient treatment and reuse of discharged water, as well as biofloc technology, can reduce these problems.

Although now discouraged by the industry, a wide variety of antibiotics and other antimicrobrials are routinely used by many shrimp farms and hatcheries to reduce bacterial and fungal growth that can cause disease. However, the indiscriminate use of antibiotics greatly encourages the evolution of resistant bacteria, including those responsible for human diseases. Residues of antibiotics, including toxic ones such as chloramphenicol and nitrofurans, can accumulate in shrimps that will be sold, and thus pose a threat to humans. Some countries have banned or seized imports of farmed shrimps contaminated with chloramphenicol, nitrofurans, and other toxic products used in aquaculture such as melamine (Chamberlain 2010; Pillai et al. 2010). A better option is prophylactic (preventive) treatment of facilities with antimicrobials, i.e., disinfectants such as formalin and iodine, as well as good cleaning and sanitation during culture. Limited use of antimicrobials will then be sufficient to control disease with less risk of microbial resistance (Sahul Hameed et al. 2003; Pillai et al. 2010).

Destruction of coastal wetlands (salt marshes, mangroves, seagrass meadows) for construction of ponds in penaeid aquaculture has been mentioned previously. These wetlands serve as nursery grounds for finfishes, shrimps, and many other marine organisms. They also serve as important buffers against the destructive forces of hurricanes, storm surges and winter storms. Approval of coastal sites for shrimp aquaculture should be carefully regulated, as they are in many countries. Inland culture using artificial seawater or inland sources of saltwater, combined with zero-discharge recirculating biofloc technology, are valid alternatives to coastal ponds.

Invasive (exotic) species introduced from their natural range into other geographic areas are a major problem in the modern world. They may outcompete, crowd out and hybridize with native species, sometimes leading to their extinction. Exotic species often introduce their diseases and parasites to native species. This has already taken place with cultured penaeids with a resultant spread of diseases that have devastated the industry in different places and times, as discussed above. Ecological consequences may occur when cultured shrimps are inadvertently released or escape from shrimp farms. Escape or unintentional release has occurred and is to be expected in the future, no matter how effective preventive measures might seem. Human error during pond operations such as water exchange, harvest, and transport of stock are sources of escape from farms and culture facilities (Kongkeo 2021).

A number of cultured shrimp species are now exotics in different parts of the world. For example, *Macrobrachium rosenbergii* is an invasive species which have been introduced from farms in several South American countries. It is now so established that it supports artisanal fisheries in some areas (Valenti and Flickinger 2020; Moraes-Valenti and Valenti 2010). Although this may seem a benefit for some locals, it cannot be beneficial for native *Macrobrachium* or other freshwater species. Several foreign penaeid escapees from farms have been reported or established in marine waters, e.g., the Asiatic *Penaeus monodon* in the Gulf of Mexico, Caribbean, and West Africa (Fuller et al. 2014; Gillett 2008); *Litopenaeus vannamei* from the west coast of the Americas into the Gulf of Mexico, southeastern USA and Brazilian waters (Balboa et al. 1991; Wenner and Knott 1992; Barbieri et al. 2016). *Farfantepenaeus aztecus*, native to the Gulf of Mexico, is well established and part of fisheries in the Mediterranean, although its origin as an escapee from shrimps farms is not certain (Galil et al. 2016).

The opening of the Suez canal in 1869 began an unintentional, large-scale experiment—a test of the hypothesis that exotic penaeids can outcompete and replace native penaeids. The canal allows entry of waters from the Indo-Pacific (IP) via the Red Sea into the eastern Mediterranean. At least 8 IP penaeid prawns have invaded the Mediterranean Sea, with definite reduction of indigenous populations of another penaeid, *Melicertus kerathurus,* especially in the eastern parts of the sea (Otero et al. 2013; Galil 2007). The IP species *Marsupenaeus japonicus, Metapenaeus monoceros* and *M. stebbingi* have been particularly successful in replacing the native penaeid and replacing it as principal fishery species. This "experiment" does show that introduction of exotic penaeids via human activities, including unintentional release from shrimp farms, is something that must be prevented.

11.18 Ornamental Shrimps

Numerous species of marine and freshwater shrimps are maintained, reared, and traded by aquarium enthusiasts, both by private individuals and public aquaria. Their small size, spectacular bright color patterns, and interesting behaviors are the primary features that make these aptly termed "ornamental" shrimps so attractive. However, some of them also provide biocontrol services within aquaria primarily stocked with fishes and other aquatic organisms. In marine aquaria and aquaculture facilities, cleaner shrimps such as *Lysmata* species (Chap. 10) reduce or eliminate ectoparasites from fishes, as well as remove diseased or extraneous material from their bodies and gills. For example, Militz and Hutson (2015) found that *Lysmata amboinensis* was very effective at keeping an ornamental fish species free not only of an ectoparasitic monogenean flatworm, but also of the parasite's oncomiracidium larvae and highly sclerotized egg cases. Ornamental atyid shrimps are noted for their cleaning of unsightly microalgae and biofilms that build up on the walls and bottoms of freshwater aquaria. As it may be recalled (Chap. 3), atyids can use the setal brushes on their chelipeds to brush and scrape food from hard substrates.

The collection and sale of marine ornamental fishes, shrimps, other decapods and invertebrates is big business which has promoted considerable research (Calado 2008, 2020). Although data on the industry is sketchy compared to that of the capture fisheries and shrimp aquaculture, much information can be found on the "Marine aquarium biodiversity and trade flow" website (MABTF 2019). The most popular marine shrimps by far are colorful cleaner shrimps, especially *Lysmata amboinensis*, made famous as Jacques in the popular movie "Finding Nemo." Besides "Jacques," *L. ankeri*, *L. debelius,* and *Stenopus hispidus* are in the top 20 invertebrate species imported into the USA, a major consumer of marine ornamentals (Rhyne et al. 2017). However, numerous other carideans are also popular, such as *Alpheus randalli*, a gobiid fish associate; the hippolytids *Saron marmoratus* and *Thor amboinensis;* other *Lysmata* species, palaemonids such fish cleaners from the genera *Ancylomenes, Periclimenes, Urocaridella*; other palaemonids such as *Hymenocera* species, *Gnathophyllum*, and the rhynchocinetid *R. durbanensis* (Calado 2020). Note that only carideans and stenopodideans are part of the trade in ornamental shrimps but no penaeoids or sergestoids. These latter groups are usually not very colorful nor behaviorally as interesting as aquarium pets.

The quantities of ornamental shrimps that are imported into major consumers such as the USA and Europe are staggering. In 2011 for example, 330,000 *L. amboinensis* were imported into the USA alone, and, together with other *Lysmata* species, composed about 12% of all invertebrates brought into the USA. This figure is probably low, as data reporting from exporting countries is not very regular nor reliable. And many specimens die as a result of collection injuries, lack of proper aeration and other causes during transport. Most ornamentals are tropical species from coral reef habitats in major exporting countries, with the Philippines, Indonesia, Sri Lanka, Haiti, and Vietnam at the top of the list of Indo-Pacific and Caribbean countries. Ornamental marine shrimps do not generally occur in groups or aggregations, with a few exceptions (e.g., *L. seticaudata, L. californica; L. wurdemanni*). Collections are done by individual divers using SCUBA who carefully herd individual shrimps into hand nets and bags. The ecological damage to coral reef systems from the removal of cleaner and other ornamental shrimps has not been carefully assessed but it is probably significant. Regulation of the industry by exporting countries is not very rigorous.

One obvious solution is to culture rather than collect marine ornamentals, and substantial progress has been made in some species (Calado et al. 2008). However, the most collected and traded species have extended planktonic development with several delicate larval stages. For this reason, aquaculture of marine ornamentals is still its infancy. Calado (2020) estimates that only 1% of marine ornamentals sold are produced by culture rather than collected from natural habitats.

Shrimps are popular with freshwater aquarium enthusiasts as well (López Greco et al. 2021). Except for some *Macrobrachium* species, almost all commercial freshwater ornamentals are atyids, especially *Neocaridina, Caridina, Halocaridina, Atya*, *Atyopsis,* and *Atyoida*. Species of the first three genera can be quite colorful, and breeding and hybridization of closely related species has produced some amazing color morphs (Plate 11.2h–i). In *N. davidi,* for example, different varieties bear names such as "Red, Red Rili, Yellow, Yellow Rili, Black Rose, Carbon Rili, Green Jade, Blue Dream, Blue Velvet, Yellow Neon and Orange Sakura." Females of *Neocaridina, Halocaridina* and some of the *Caridina* species have abbreviated development, i.e., spawn a small number of very large embryos with lots of yolk that hatch into nonfeeding lecithotrophic larvae. Culture can be completed in the same aquarium as the adult shrimps. Thus, much of the trade, especially among fellow enthusiasts, is cultured stock. However, even in these species, there is often overcollection from native habitats. *Neocaridina denticulata* (FAO *Caridina*) is not only an ornamental species but also collected, reared, and sold for making food condiments and as fish bait.

The scale of the freshwater ornamental trade is very poorly documented but is sizable. For example, Patoka et al. (2015) reported that about 40,000 specimens of *Caridina multidentata* and 20,000 *Neocaridina denticulata* were imported per MONTH from Indonesia and elsewhere into the Czech Republic for sale to other countries. Besides overcollection and the ecological disturbance it causes to native habitats, the world-wide commerce and private trade of ornamentals has become a source of invasive species. Unfortunately, like aquarium pets, owners who want or need to dispose of their ornamentals often do so irresponsibly. Release into nearby natural habitats may seem like a humane solution, but it only makes the released shrimps exotics. Ornamental atyids with abbreviated development and nonfeeding lecithotrophic larvae (Chap. 9, Life Histories) can rather easily establish themselves in new areas if compatible with the environmental parameters of the release habitat. As a result, several of these atyids are now exotics in freshwater habitats of many countries of the world (e.g., Patoka et al. 2015; Han et al. 2019; Levitt-Barmats et al. 2019; Weiperth et al. 2019).

Atyid invaders may interfere with the ecological equilibrium of the invaded habitat. In laboratory microcosm experiments, grazing by *Neocaridina davidi* altered the structure of the freshwater food web by ingestion of meiofauna, an important link between microbial production and macroinvertebrates and juvenile fishes (Weber and Traunspurger 2016). Exotic ornamentals may carry and introduce their diseases into the invaded habitat. This may be a particular concern with imported atyids originally cultivated in aquaculture ponds, where high densities of shrimps promote growth of parasites and epibionts, e.g., parasitic and/or commensal flatworms, fungi, sessile ciliates and rotifers (Maciaszek et al. 2018). Such "hitchhikers" can be avoided by detailed inspection of imported shrimps, quarantine and treatment as needed, sterilization of discharged aquarium water, and education of consumers to prevent release of exotics into surrounding habitats (Patoka et al. 2015).

Even marine ornamental shrimps with extended planktonic development can become established in foreign habitats if released into coastal waters, e.g., several species of *Lysmata*. For example, *Lysmata vittata*, an ornamental from coral reefs in the Indo-Pacific, has become established in similar habitats in New Zealand, Brazil, Chesapeake Bay and perhaps elsewhere (Alves et al. 2019; Aguilar et al. 2022). Increased public awareness, industry self-discipline and governmental regulation are needed to resolve this problem.

11.19 Climate Change: Impacts on Fisheries and Aquaculture

Variation in regional and global climate over the earth's history has been documented by numerous scientific studies. The earth has warmed and cooled many times but usually on a time scale of thousands to hundreds of thousands of years. Causal factors include regular cyclic variations in solar radiation due to the earth's position in respect to the sun. These Milankovitch cycles are (1) changes in the shape of the earth's orbit around the sun (eccentricity) (100,000 year period); (2) the tilt angle between the earth's rotational and orbital axes (41,000 years); (3) the change in direction of the rotational axis (precession or "wobble"). Their combined action dictates in large part the warming and cooling of the earth. Other important factors are variations in solar output and volcanism, the latter producing aerosol clouds that block sunlight and emit heat-trapping greenhouse gases, primarily carbon dioxide (CO_2) but also methane among others. These natural processes have caused the earth and ocean to warm and cool between frigid glacial periods and steamy warm tropical conditions throughout history with concomitant fluctuations in the composition and abundance of ecosystems and their inhabitants.

In the geological past, climate change occurred over long stretches of time, giving many species the time to evolve and adapt to new conditions. However, in the last several decades, our global environment has undergone unprecedented rapid and, least for some centuries, irreversible alterations. The earth has warmed by 1 °C since the inception of the industrial revolution (~1800) and is estimated to increase by 1–2 °C by the end of the twenty-first century (IPCC 2021). Although this temperature shift might seem slight, the warming it represents has had a tremendous impact on world and regional climates. Global warming is now driven by the emission and accumulation of greenhouse gases, especially carbon dioxide (CO_2) and methane (CH_4), from the burning of fossil fuels (coal, oil, natural gas) to produce energy, as well as expansion of land use for agriculture and urban development. Discharge of CO_2 into the atmosphere has increased 40% from that of the pre-industrial era, with the oceans absorbing over 30% of the gas and 90% of the excess heat (Bahri et al. 2018). The actual and potential changes in oceanic habitats and ecosystems are negative, if not disastrous (Pörtner et al. 2014). Both the capture fisheries and shrimp aquaculture will almost certainly be affected. Climate change alone has been projected to decrease oceanic production of fisheries by 3–12% by 2050 (Barange et al. 2018).

What are some of the numerous consequences of climate change on shrimps? Like all species, shrimps have thermal optima and ranges. When waters warm beyond these, shrimps and their various life stages will become less productive, die, or have to migrate elsewhere. Ocean warming will certainly expand or contract the geographic ranges of many species. The subtropical and tropical penaeoid species must shift into higher latitudes and/or greater depths to escape the heat. Cold-water pandalid shrimps will be squeezed into a more restricted distribution at higher latitudes and greater depths. Regional fisheries and economies will be great affected. Climatic oscillations in the Gulf of Alaska (GOA) have demonstrated how warming waters influence pandalid shrimps populations and their fisheries (Orensanz et al. 1998; Anderson and Piatt 1999). In this region, the oceanic climate cycles between warm and cold regimes because of shifts in the atmospheric Aleutian Low Pressure System. From 1947–1977, cooler waters favored populations of pandalid shrimps and other crustaceans which supported productive fisheries. Larval release by the shrimps occurs in late spring and early summer when zooplankton food supplies are at their peak. In 1977, the Aleutian Low shifted southward, bringing in warmer waters into the GOA, provoking a double blow to shrimp populations. The zooplankton peak took place earlier than previously, reducing the food supply to shrimp larvae and subsequent recruitment. However, this temperature shift also led to increased abundance in a group of warmer water fish predators of shrimps, e.g., pollock and halibut. The larval peak of these predatory fishes coincided with the new zooplankton peak. Consequently, shrimp populations declined due to reduced recruitment and increased fish predation. Recent marine heatwaves (2014–2016) further exacerbated the climatic condition for fisheries in the GOA and perhaps represent the new normal (Suryan et al. 2021). A similar situation has been reported in *Pandalus borealis* and its fishery on the other side of the continent in the Gulf of Maine (Richards and Hunter 2021).

Basic life processes, such as molting, growth, respiration, and immune function are all affected by temperature changes, and excessively high temperatures stress shrimps and make them more susceptible to pathogens and disease (Shields 2019). Ocean warming decreases the life-giving oxygen supply, as warmer water holds less oxygen than cooler waters. Stratification of warmer surface waters over cooler deeper waters retards mixing of oxygenated waters into the depths. Oxygen concentration naturally varies with depth, with hypoxic oxygen minimum zones at ~200 to 1500 m, but these zones will expand with increased ocean temperature, as will nearshore anoxic areas ("dead zones") around the world. Shrimp populations must move out of these areas or suffer the consequences. Marine waters absorb much of the excess atmospheric CO_2. The gas enters the water and reacts with water to form carbonic acid which dissociates into bicarbonate and hydrogen ions, decreasing pH, i.e., increasing the acidity of the water. The exoskeleton of shrimps is calcified to various degrees, and acidification makes the deposition of carbonate into shrimp cuticle more energy expensive (Shields 2019). Another stressor that oceanic warming may impose is change in habitat salinity. Shrimps, like all marine organisms, are adapted to a particular range of salinity, particularly their stenohaline larvae. Regulation of internal osmotic pressure uses metabolic energy, so that salinity change may be another drain on the shrimp's health and reproductive success. All of these stressors will affect shrimp aquaculture, especially in developing countries with less advanced infrastructure and technology.

Another outcome of global warming has been and will continue to be a rise in sea level and weather extremes. Global Mean Sea Level rose 0.2 m from 1900 to 2018 and, depending on the projection used, is reasonably anticipated to climb another 0.3–0.5 m by the end of this century (IPCC 2021). What does this mean for capture fisheries and shrimp aquaculture? Flooding and erosion of coastal wetlands, made worse by more powerful hurricanes and winter storms, reduce these vital nursery grounds for estuarine-dependent penaeids so important in aquaculture. Coastal shrimp ponds and facilities will be forced to relocate inland at great expense and economic disruption. Optimal water temperatures in shallow shrimp ponds will become more difficult to control as more extreme heat events become more common.

All this gloom and doom may seem overstated, yet years of international cooperative studies on greenhouse gas emissions, global warming and its multiple nefarious effects show that some are already here, will continue, and even accelerate (Pörtner et al. 2014; Barange et al. 2018; IPCC 2021). On the bright side, control of the climate-change beast is certainly possible if world governments and their national populations have the will to do it, with environmentalists, naturalists, scientists, good world citizens and, hopefully, readers of this volume leading the way.

11.20 Past, Present, and Future State of Shrimp Fisheries and Aquaculture

Shrimp fisheries have grown considerably since the 1950s as have the world fisheries overall (Figs. 11.1 and 11.3) with an increase in the size and number of fishing vessels. Technological advances in the type and efficiency of fishing gear and storage (refrigeration, freezing) of catch at sea have been another factor. The total shrimp fishery catch in 1950 was 420,000 t and has increased 800% to over 3.3 million t in 2019. However, note in Fig. 11.1 that the total world fisheries production began to level off in the 1990s followed by shrimp fisheries in the early 2000s. This occurred in spite of a doubling of the world fishing fleet between 1950 and 2015 and which is predicted to keep increasing with a corresponding reduction in catch per unit fishing effort (Rousseau et al. 2019).

These events were predicted in a classic paper by John Rhyther (1969) based purely on (1) the known average photosynthetic productivity of major oceanic zones and (2) the efficiency of energy (biomass) food chain transfer in those zones. His projection of sustainable fishery production was that the sea could yield no more than 100 million t of biomass without overexploitation and decline of fished stocks. His forecast appears incredibly good, perhaps a bit high, as the total global fish production has not quite reached that level, leveling off between 90 and 100 million t. Unfortunately, there is a similar asymptotic trend in shrimp fisheries yields as well (Figs. 11.1 and 11.3).

Different views on the future of fisheries and aquaculture have been put forth by prominent fisheries scientists and ecologists Daniel Pauly and Dirk Zeller (Pauly and Zeller 2017). They objected to the rather rosy view of fisheries and aquaculture present by the FAO's 2016 SOFIA report, i.e., that global fishery production was and would continue as stable. They suggested that catches were actually declining but the statistics presented to the contrary by FAO were due to inflated data reported by some south Asian countries for political expediency. Concern was raised that FAO analysis lacked the reconstruction process, in which gaps in reported data are corrected retroactively as new data or information becomes available. FAO recognized these problems in their 2020 SOFIA report, and a backdated reconstruction of the suspect data was done. The result was an adjustment of the FAO estimates of about a 2% decrease for global capture fisheries during 2009–2016 and 5% for global aquaculture production. However, FAO still maintains its view that capture fisheries may increase somewhat in the future—but still below Rhyther's 100 million t limit predicted by basic ecological principles.

As the world's populations and economies grow, so will the demand for an affordable seafood supply. The FAO SOFIA (2016) report opined that food biomass from aquaculture would fill this need. Shrimp aquaculture seems at first glance to bear this out. The rise in shrimp aquaculture since 1950 has been meteoric, from only 1325 t in 1950 to 7.09 million t in 2019, an increase of 535,000% (Fig. 11.9)! However, Pauly and Zeller propose that a continued increase is ecologically untenable. The reason is that fish and shrimp species reared in aquaculture facilities require a high-protein diet which is supplied, in large part, by about 20–30% of capture fish production in the form of fish meal and oil from low-trophic species such as clupeids and anchovies, as well as other less valuable (trash) fishes and fishery bycatch (Pauly and Zeller 2017; Boyd and Jescovitch 2020). This is akin to "robbing Peter (fisheries) to pay Paul (aquaculture)." Pauly and Zeller give the example of using 4 t of wild-caught anchovies to produce 1 t of salmon; there are not 5 t of nutritious seafood available for human consumption but rather just 1 t of expensive seafood.

Given the additional burdens of climate change and environmental deterioration on aquatic habitats, the future of shrimp fisheries and aquaculture might be viewed as somewhat bleak. However, the increasing recognition of and action on these problems by the general public, shrimp fishery and aquaculture scientists and management, cooperation among nations, the good work done by international agencies such as FAO and organizations such as "The Sea Around US" (http://www.seaaroundus.org) make an optimistic future possible. As importantly, the productive life-history adaptations of the major shrimp fishery and aquaculture species will help sustain human populations for a long time to come. Humans just need to give shrimps and other species a chance by maintaining a healthy environment and implementing wise management policies.

References

Aguilar R, 10 coauthors (2022) Unresolved taxonomy confounds invasive species identification: the *Lysmata vittata* Stimpson, 1860 (Decapoda: Caridea: Lysmatidae) species complex and recent introduction of *Lysmata vittata sensu stricto* in the western Atlantic. J Crustac Biol 42:1–18

Alday-Sanz V (ed) (2010) The Shrimp Book. Nottingham University Press, Nottingham

Allen JA (1966) The dynamics and interrelations of mixed population of Caridea found off the north-east coast of England. In: Barnes H (ed) Some contemporary studies in marine science. George Allen & Unwin, London, pp 45–66

Alves DFR, López Greco LS, Barros-Alves SP, Hirose GL (2019) Sexual system, reproductive cycle and embryonic development of the red-striped shrimp *Lysmata vittata,* an invader in the western Atlantic Ocean. PLoS One 14:e0210723. https://doi.org/10.1371/journal.pone.0210723

Amin SMN, Arshad A, Siraj SS, Sidik BJ (2009) Population structure, growth, mortality and yield per recruit of sergestid shrimp, *Acetes japonicus* (Decapod, Sergestidae) from the coastal waters of Malacca, Peninsular Malaysia. Indian J Marine Sci 38:57–68

Anderson PJ, Piatt JF (1999) Community reorganization in the Gulf of Alaska following ocean climate regime shift. Mar Ecol Prog Ser 189:117–123

Ar KA, Nicholas FF, Long SM, Naqiuddin AS, Esa Y (2016) Fecundity of freshwater prawn (*Macrobrachium rosenbergii*) in selected rivers of Sarawak, Malaysia. Biodiversitas 17:498–502

Aranceta-Garza F, Arreguín-Sánchez F, Ponce-Díaz G, Seijo JC (2016) Natural mortality of three commercial penaeid shrimps (*Litopenaeus vannamei, L. stylirostris* and *Farfantepenaeus californiensis*) of the Gulf of California using genomonic time divisions. Sci Mar 80:199–206

Avnimelech Y (2015) Biofloc technology – a practical handbook, 3rd edn. World Aquaculture Society, Baton Rouge, Louisiana

Aye SS (2018) Population structure and growth characteristics of oriental river prawn, *Macrobrachium nipponense* (De Haan, 1849) in Soon Lun Reservoir, Mandalay Region, Myanmar. IOP Conf Ser: Earth Environ Sci 416:012012 (17 pp). https://doi.org/10.1088/1755-1315/416/1/012012

Bahri T, Barange M, Moustahfid H (2018) Climate change and. aquatic systems. In Barange M, Bahri T, Beveridge MCM, Cochrane KL, Funge-Smith SF, Poulain F (eds). Impacts of climate change on fisheries and aquaculture: synthesis of current knowledge, adaptation and mitigation options. FAO Fisheries and Aquaculture Technical Paper No. 627. Rome, FAO pp. 1–18

Balboa WA, King TL, Hammerschmidt PC (1991) Occurrence of Pacific white shrimp in lower Laguna Madre, Texas. Proc Ann Conf Southeastern Assoc Fisheries Wildl Agencies 45:288–292

Barange M, Bahri T, Beveridge MCM, Cochrane KL, Funge-Smith S, Poulain F (eds) (2018) Impacts of climate change on fisheries and aquaculture: synthesis of current knowledge, adaptation and mitigation options. FAO Fisheries and Aquaculture Technical Paper No. 627. Rome, FAO

Barbieri E, Coa F, Rezende KFO (2016) The exotic species *Litopenaeus vannamei* (Boone, 1931) occurrence in Cananeia, Iguape and Ilha Comprida Lagoon Estuary Complex. Bol Inst Pesca 42: 479–485

Bauer RT (2004) Remarkable shrimps: adaptations and natural history of the carideans. University of Oklahoma Press, Norman

Bauer RT (2018) Life-cycle and seasonal migrations. In: Wellborn G, Thiel M (eds) The Natural History of the Crustacea, Volume 5: Life histories. Oxford University Press, New York, pp 203–229

Bauer RT (2020) Shrimp fisheries. In: Lovrich G, Thiel M (eds) The natural history of the Crustacea: Volume 9. Fisheries and aquaculture. Oxford University Press, New York, pp 91–116

Bergström B (2000) The biology of *Pandalus*. Adv Mar Biol 38:1–245

Boschi EE (1989) Biología pesquera del langostino del litoral patagónico de Argentina (*Pleoticus muelleri*). Contribución Instituto Nacional de Investigación y Desarrollo Pesquero 646:5–71

Bourgeois M, Landry L, Lightner J, Marx J, Semon K (2015) Louisiana shrimp: Fishery management plan. Louisiana Department of Wildlife and Fisheries, Baton Rouge, Louisiana

Boyd CE, Jescovitch LN (2020) Penaeid shrimp aquaculture. In: Lovrich G, Thiel M (eds) The natural history of Crustacea. Vol. 9. Fisheries and Aquaculture. Oxford University Press, New York, pp 233–258

Calado R (2008) Marine ornamental shrimp: biology. Aquaculture and Conservation, Wiley-Blackwell, Oxford

Calado R (2020) Marine ornamental decapods—collection, culture, and conservation. In: Lovrich G, Thiel M (eds) The natural history of the Crustacea: Volume 9. Fisheries and Aquaculture. Oxford University Press, New York, pp 313–340

Calado R, Pimentel T, Vitorino A, Dionisio G, Dinis MT (2008) Technical improvements of a rearing system for the culture of decapod crustacean larvae, with emphasis on marine ornamental species. Aquaculture 285:264–269

Castilho AL, Bauer RT, Freire FAM, Fransozo V, Costa RC, Grabowski RC, Fransozo A (2015) Lifespan and reproductive dynamics of the commercially important sea bob shrimp *Xiphopenaeus kroyeri* (Penaeoidea): Synthesis of a 5-year study. J Crustac Biol 35:30–40

Chamberlain GW (2010) History of shrimp farming. In: Alday-Sanz V (ed) The Shrimp Book. Nottingham University Press, Nottingham, pp 1–34

Couvillion BR, Barras JA, Steyer GD, Sleavin W, Fischer M, Beck H, Trahan N, Griffin B, Heckman D (2011) Land area change in coastal Louisiana from 1932 to 2010. U.S. Geological Survey Scientific Investigations Map 3164:1–12

Coyle SD, Alston DE, Sampaio CMS (2010) Nursery systems and management. In: New MB, Valenti WC, Tidwell JH, D'Abramo LR, Kutty MN (eds) Freshwater prawns: biology and farming. Wiley-Blackwell, Oxford, pp 108–126

Crespi V, New MB (2009) *Macrobrachium rosenbergii*. In FAO Cultured aquatic species fact sheets. Text by edited and compiled by Valerio Crespi and Michael New. CD-ROM (multilingual)

Dall W, Hill BC, Rothlisberg PC, Sharples DJ (1990) The biology of the Penaeidae. Adv Mar Biol 27:1–489

Deegan LA, Johnson DS, Warren RS, Peterson BJ, Fleeger JW, Fagherazzi S, Wollheim WM (2012) Coastal eutrophication as a driver of salt marsh loss. Nature 490:388–392

Dıaz-Ochoa JA, Quinones RA (2008) Relationship of precipitation, freshwater input, and sea level height with the abundance of the white shrimp (*Litopenaeus occidentalis* Street, 1871) off Buenaventura, eastern tropical Pacific. Fish Res 92:148–161

Diop H, Keithly WR, Kazmierczak RF Jr, Shaw RF (2007) Predicting the abundance of white shrimp (*Litopenaeus setiferus*) from environmental parameters and previous life stages. Fish Res 86:31–41

Dumont WH, Sundstrom GT (1961) Commercial fishing gear of the United States. United States Fisheries and Wildlife Service Circular No. 109. US Government Printing Office, Washington, DC

Eayrs S (2007) A guide to bycatch reduction in tropical shrimp-trawl fisheries. Revised edition. FAO, Rome

Enin UI, Lowenberg U, Kunzel T (1996) Population dynamics of the estuarine prawn (*Nematopalaemon hastatus* Aurivillius 1898) off the southeast coast of Nigeria. Fish Res 26: 17–35

FAO (2016) SOFIA: The State of World Fisheries and Aquaculture 2016. Contributing to food security and nutrition for all. FAO, Rome, 200 pp

FAO (2020) SOFIA: The State of World Fisheries and Aquaculture 2020. Sustainability in action. FAO, Rome. https://doi.org/10.4060/ca9229en

FAO (Food and Agriculture Organization of the United Nations). 2020. Annual yearbook of Fishery and Aquaculture Statistics 2018. FAO, Rome, Italy

FAO Cultured Aquatic Species Information Programme Penaeus vannamei (Boone, 1931) (2021). http://www.fao.org/fishery/culturedspecies/Penaeus_vannamei/en

FIGIS (Fisheries Global Information System) (2021a) Global capture production 1950–2019. FI Institutional Websites. FAO Fisheries and Aquaculture Department, Rome, Italy

FIGIS (Fisheries Global Information System) (2021b) Global aquaculture production 1950–2019. FI Institutional Websites. FAO Fisheries and Aquaculture Department, Rome, Italy

Flegel TW (2006) Detection of major penaeid shrimp viruses in Asia, a historical perspective with emphasis on Thailand. Aquaculture 258:1–33

Fuller PL, Knott DM, Kingsley-Smith PR, Morris JA, Buckel CA, Hunter ME, Hartman LD (2014) Invasion of Asian tiger shrimp, *Penaeus monodon* Fabricius, 1798, in the western north Atlantic and Gulf of Mexico. Aquat Invasions 9:59–70

Galil B (2007) Loss or gain? Invasive aliens and biodiversity in the Mediterranean Sea. Mar Pollut Bull 55:314–322

Galil BS, Innocenti G, Douek J, Paz G, Rinkevich B (2016) Foul play? On the rapid spread of the brown shrimp *Penaeus aztecus* Ives, 1891 (Crustacea, Decapoda, Penaeidae) in the Mediterranean, with new records from the Gulf of Lion and the southern Levant. Mar Biodivers 47:979–985

Garcia S (1989) The management of coastal penaeid shrimp fisheries. In: Caddy JF (ed) Marine invertebrate fisheries: their assessment and management. Wiley, New York, pp 281–306

Garcia SM (1996) Stock-recruitment relationships and the precautionary approach to management of tropical shrimp fisheries. Mar Freshw Res 47:43–58

Garcia S, Le Reste L (1986) Ciclos vitales, dinámica, explotación y ordenación de las poblaciones de camarones peneidos costeros. FAO Documento Técnico De Pesca 203:1–180

Garibaldi L (2012) The FAO global capture production database: a six-decade effort to catch the trend. Mar Policy 36:760–768

Gillett R (2008) Global study of shrimp fisheries. FAO Fisheries Technical Paper No. 475. FAO Fisheries and Aquaculture Department, Rome, Italy

Gulland JA, Rothschild BJ (1984) Report of the workshop on the scientific bases for the management of penaeid shrimp. In: Gulland JA, Rothschild BJ (eds) Penaeid shrimp—their biology and management. Fishing News Books, Farnham, Surrey, pp 9–29

Han C-C, Hsu K-C, Fang LS, Cheng I-M, Lin H-D (2019) Geographical and temporal origins of *Neocaridina* species (Decapoda: Caridea: Atyidae) in Taiwan. BMC Genet 20:86. https://doi.org/10.1186/s12863-019-0788-y

Hannah RW, Jones SA (1991) Fishery-induced changes in the population structure of pink shrimp *Pandalus jordani*. Fish Bull 89:41–51

Hanson JA, Goodwin HL (1977) Shrimp and Prawn Farming in the Western Hemisphere. Dowden, Hutchison, and Ross, Stroudsburg, Pennsylvania

Hein S, Meier P (1995) Skimmers: their development and use in coastal Louisiana. Mar Fish Rev 57:17–24

Holthuis LB (1980) FAO species catalogue. Volume 1—Shrimps and prawns of the world. An annotated catalogue of species of interest to fisheries. FAO Fisheries Synopsis 125. FAO Fisheries and Aquaculture Department, Rome, Italy

Hufnagl M, Temming A, Pohlmann T (2014) The missing link: tidal-influenced activity a likely candidate to close the migration triangle in brown shrimp *Crangon crangon* (Crustacea, Decapoda). Fish Oceanogr 23:242–257

IPCC (The Intergovernmental Panel on Climate Change) (2021) Climate change 2021: the physical science basis. Contribution of Working Group I to the Sixth Assessment Report of the Intergovernmental Panel on Climate Change. In Masson-Delmotte V, Zhai P, Pirani A, Connors SL, Péan C, Berger S, Caud N, Chen Y, Goldfarb L, Gomis MI, Huang M, Leitzell K, Lonnoy E, Matthews JBR, Maycock TK, Waterfield T, Yelekçi O, Yu R, Zhou B (eds) Cambridge University Press, Cambridge

Iversen ES, Allen DM, Higman JB (1993) Shrimp capture and culture fisheries of the United States. Wiley, New York

Johnson SK (1995) Handbook of shrimp diseases. Department of Wildlife and Fisheries Sciences. Texas A&M University 90-601, 27 pp

Juarez LM, Moss SM, Figueras E (2010) Maturation and rearing of the Pacific white shrimp, *Penaeus vannamei*. In: Alday-Sanz V (ed) The Shrimp Book. Nottingham University Press, Nottingham, pp 305–352

Karplus I, Sagi A (2010) The biology and management of size variation. In: New MB, Valenti WC, Tidwell JH, D'Abramo LR, Kutty MN (eds) Freshwater prawns: biology and farming. Wiley-Blackwell, Oxford, pp 316–345

Kelleher K (2005) Discards in the world's marine fisheries. An update. FAO Fisheries Technical Paper No. 470: 1–131. FAO, Rome, Italy

Kilada R, Acuña E (2015) Direct age determination by growth band counts of three commercially important crustacean species in Chile. Fish Res 170:134–143

Kongkeo H (2021) Cultured aquatic species information programme. *Penaeus monodon*. FAO Fisheries Division [online], Rome

Krummenauer D, Peixoto S, Cavalli RO, Poersch LH, Wasielesky W Jr (2011) Superintensive culture of white shrimp, *Litopenaeus vannamei,* in a biofloc technology system in southern Brazil at different stocking densities. J World Aquacult Soc 42:726–733

Kutty MN, Valenti WC (2010) Culture of other freshwater prawn species. In: New MB, Valenti WC, Tidwell JH, D'Abramo LR, Kutty MN (eds) Freshwater Prawns: biology and farming. Wiley-Blackwell, Oxford, pp 502–523

Kutty MN, Weimin M (2010) Culture of the Oriental River Prawn *Macrobrachium nipponense*. In: New MB, Valenti WC, Tidwell JH, D'Abramo LR, Kutty MN (eds) Freshwater prawns: biology and farming. Wiley-Blackwell, Oxford, pp 475–484

Levitt-Barmats Y, Yanai YZ, Cohen TM, Shenkar N (2019) Life history traits and ecological characteristics of the ornamental shrimp *Neocaridina denticulata* (De Haan, 1844), recently introduced into the freshwater systems of Israel. Aquat Invasions 14:684–702

Levy T, Rosen O, Eilam B, Azulay D, Aflalo ED, Manor R, Shechter A, Sagi A (2016) A single injection of hypertrophied androgenic gland cells produces all-female aquaculture. Mar Biotechnol 18:554–563

Levy T, Rosen O, Manor R, Dotan S, Azulay D, Abramov A, Sklarz MY, Chalifa-Caspi V, Baruch K, Shechter A, Sagi A (2019) Production of WW males lacking the masculine Z chromosome and mining the *Macrobrachium rosenbergii* genome for sex chromosomes. Sci Rep 9:12408. https://doi.org/10.1038/s41598-019-47509-6

Levy T, Tamone SL, Manor R, Bower ED, Sagi A (2020) The protandric life history of the Northern spot shrimp *Pandalus platyceros*: molecular insights and implications for fishery management. Sci Rep 10:1287. https://doi.org/10.1038/s41598-020-58262-6

Li X, Bellerby R, Craft C, Widney SE (2018) Coastal wetland loss, consequences, and challenges for restoration. Anthropocene Coasts 1:1–15. https://doi.org/10.1139/anc-2017-0001

Lightner DV (1992) Shrimp viruses: diagnosis, distribution and management. In: Wyban J (ed) Proceedings of the Special Session on Shrimp Farming, Orlando, Florida, 22–22, May 1992. World Aquaculture Society, Baton Rouge, Louisiana, pp 238–253

Lightner DV (2005) Biosecurity in shrimp farming: pathogen exclusion through use of SPF stock and routine surveillance. J World Aquacult Soc 36:229–248

Lightner DV (2011a) Status of shrimp diseases and advances in shrimp health management. In: Bondad-Reantaso MG, Jones JB, Corsin F, Aoki T (eds) Diseases in Asian Aquaculture VII. Fish Health Section. Asian Fisheries Society, Selangor, Malaysia, pp 121–134

Lightner DV (2011b) Virus diseases of farmed shrimp in the Western Hemisphere (the Americas): a review. J Invertebr Pathol 106:110–130

Linder MJ, Anderson WW (1956) Growth, migrations, spawning and size distribution of shrimp, *Penaeus setiferus*. Fish Bull 56:555–645

Ling SW (1969) The general biology and development of *Macrobrachium rosenbergii* (De Man). FAO Fishery Rep 57:589–606

Lloyd AJ, Yonge CM (1947) The biology of *Crangon vulgaris* L. in the Bristol Channel and Severn Estuary. J Mar Biol Soc UK 26:626–661

López Greco LS, Tomas AL, Farinelli Pantaleão JA & Bauer RT (2021) *Neocaridina davidi*: Una especie ornamental de alta plasticidad biológica con múltiples aplicaciones. Especialización en Pesca y Producción Acuícola, Facultad de Ciencias Veterinarias, Universidad de Buenos Aires, Argentina. Digital book, 47 pp., pdf. ISBN: 978-987-88-1834

MABTF (Marine aquarium biodiversity and trade flow) (2019) Roger Williams University, New England Aquarium, Boston, Massachusetts, USA. http://www.aquariumtradedata.org/

Macchi GJ, Iorio MI, Aubone A (1998) Estimaci6n de la fecundidad del langostino argentino *Pleoticus muelleri* Bate, 1888 de Patagonia (sur de Argentina). Bol Inst Esp Oceanogr 14:19–29

Mace MM III, Rozas LP (2014) Estimating natural mortality rates of juvenile white shrimp *Litopenaeus setiferus*. Estuar Coasts 38:1580–1592

Macfadyen G, Banks R, Davies R (2013) Tropical shrimp trawling: developing a management blueprint and adapting and implementing it in specific countries and fisheries. Mar Policy 40: 25–33

Maciaszek R, Kamaszewski MW, Strużyński, Łapa P (2018) Epibionts of ornamental freshwater shrimps bred in Taiwan. Anim Sci 57:133–142

MacLean MH, Brown JH (1991) Larval growth comparison of *Macrobrachium rosenbergii* (de Man) and *M. nipponense* (de Haan). Aquaculture 95:251–255

Mashiko K (1990) Diversified egg and clutch sizes among local populations of the fresh-water prawn *Macrobrachium nipponense* (de Haan). J Crustac Biol 10:306–314

Maunder MN, Sibert JR, Fonteneau A, Hampton J, Kleiber P, Harley SJ (2006) Interpreting catch per unit effort data to assess the status of individual stocks and communities. ICES J Mar Sci 63: 1373e1385

May-Kú MC, Criales MM, Montero-Muñoz JL, Ardisson PL (2014) Differential use of *Thalassia testudinum* habitats by sympatric penaeids in a nursery ground of the southern Gulf of Mexico. J Crustac Biol 34:144–156

McConnaughey RA, Hiddink JG, Jennings S, Pitcher CR, Kaiser MJ, Suuronen P, Sciberras M, Rijnsdorp AD, Collie JS, Mazor T, Amoroso RO, Parma AM, Hilborn R (2019) Choosing best practices for managing impacts of trawl fishing on seabed habitats and biota. Fish Fish 21:319–337

Militz TA, Hutson KS (2015) Beyond symbiosis: cleaner shrimp clean up in culture. PLoS One 10(2):e0117723. https://doi.org/10.1371/journal.pone.0117723

Minagawa M, Yasumoto S, Ariyoshi T, Umemoto T, Ueda T (2000) Interannual, seasonal, local and body size variations in reproduction of the prawn *Penaeus* (*Marsupenaeus*) *japonicus* (Crustacea: Decapoda: Penaeidae) in the Ariake Sea and Tachibana Bay, Japan. Mar Biol 136:223–231

Minello TJ, Able KW, Weinstein MP, Hayes CG (2003) Salt marshes as nurseries for nekton: testing hypotheses on density, growth and survival through meta-analysis. Mar Ecol Prog Ser 246:39–59

Mitchell JF, Watson JW, Foster DG, Caylor RE (1995) The turtle excluder device (TED): a guide to better performance. NOAA Technical Memorandum NMFS-SEFSC, p 35p

Moraes-Valenti P, Valenti WC (2010) Culture of the Amazon River Prawn *Macrobrachium amazonicum*. In: New MB, Valenti WC, Tidwell JH, D'Abramo LR, Kutty MN (eds) Freshwater Prawns: biology and farming. Wiley-Blackwell, Oxford, pp 485–501

Morales MC, Meruane J (2013) The northern river shrimp *Cryphiops caementarius* (Decapoda, Palaemonidae), research chronology between 1958 and 2008, II: Aquaculture Research and Development in Northern Chile. Crustaceana 86:1452–1467

Morales-Covarrubias MS, Nunan LM, Lightner DV, Mota-Urbina JC, Garza-Aguirre MC, Chávez-Sánchez MC (1999) Prevalence of IHHNV in wild broodstock of Penaeus stylirostris from the upper Gulf of California, Mexico. J Aquatic Anim Health 11:296–301

Morys C, Brüchert V, Bradshaw C (2021) Impacts of bottom trawling on benthic biogeochemistry in muddy sediments: removal of surface sediment using an experimental field study. Mar Environ Res 69:105384. https://doi.org/10.1016/j.marenvres.2021.105384

Motoh H (1981) Studies on the fisheries biology of the giant tiger prawn, *Penaeus monodon* in the Philippines. Aquaculture Department Southeast Asian Fisheries Development Center Technical Report No. 7, 128 pp

New MB (2002) Farming freshwater prawns. 2002. A manual for the culture of the giant river prawn (*Macrobrachium rosenbergii*). FAO Fisheries Technical Paper 428, 212 pp

New MB (2010) History and global status of freshwater prawn farming. In: New MB, Valenti WC, Tidwell JH, D'Abramo LR, Kutty MN (eds) Freshwater Prawns: biology and farming. Wiley-Blackwell, Oxford, pp 1–11

New MB, Valenti WC, Tidwell JH, D'Abramo LR, Kutty MN (eds) (2010) Freshwater Prawns: biology and farming. Wiley-Blackwell, Oxford

Newton A and 22 coauthors (2020) Anthropogenic, direct pressures on coastal wetlands. Front Ecol Evol https://doi.org/10.3389/fevo.2020.00144

Nilssen EM, Aschan MM (2009) Catch, survey and life-history data for shrimp (*Pandalus borealis*) off Jan Mayen. Deep-Sea Res II Topical Stud Oceanogr 56:2023–2036

NMFS (National Marine Fisheries Service) (2020) Fisheries of the United States, 2018. U.-S. Department of Commerce, NOAA Current Fishery Statistics No. 2018. https://www.fisheries.noaa.gov/national/commercial-fishing/fisheries-united-states-2018

NMFS (National Marine Fisheries Service) (2015) Fisheries of the United States, 2014. United States Department of Commerce, NOAA Current Fishery Statistics 2014, 135 pp. NMFS, Silver Spring, Maryland, USA. https://www.fisheries.noaa.gov/national/sustainable-fisheries/fisheries-united-states

Ogawa Y, Ono H, Hashimoto H, Kakuda S, Gushima G (1991) On the growth and life span of the population of oriental river prawn *Macrobrachium nipponense* (De Haan) in the Ashida river [Hiroshima, Japan]. J Faculty Appl Biol Sci Hiroshima Univ Jpn 30:43–53

Omori M (1975) The systematics, biogeography, and fishery of epipelagic shrimps of the genus Acetes (Crustacea, Decapoda, Sergestidae). Bull Ocean Res Inst Univ Tokyo 7:1–91

Orensanz JM, Armstrong J, Armstrong D, Hilborn R (1998) Crustacean resources are vulnerable to serial depletion—the multifaceted decline of crab and shrimp fisheries in the Greater Gulf of Alaska. Rev Fish Biol Fish 8:117–176

Otero M, Cebrian E, Francour P, Galil B, Savini D (2013) Monitoring marine Invasive Species in Mediterranean Marine Protected Areas (MPAs): a strategy and practical guide for managers. International Union for Conservation of Nature, Malaga, p 136

Overstreet RM (1978) Marine maladies? Worms, germs, and other symbionts from the northern Gulf of Mexico. Mississippi–Alabama Sea Grant Consortium, Publication MASGP-78-021

Páez-Osuna F (2001) The environmental impact of shrimp aquaculture: a global perspective. Environ Pollut 112:229–231

Patoka J, Bláha M, Devetter M, Rylkov K, Čadková Z, Kalous L (2015) Aquarium hitchhikers: attached commensals imported with freshwater shrimps via the pet trade. Biol Invasions 15: 125–141

Pauly D, Zeller D (2017) Comments on FAOs State of World Fisheries and Aquaculture (SOFIA 2016). Mar Policy 177:176–181

Pauly D, Ingles J, Neal R (1984) Application to shrimp stocks of objective methods for the estimation of growth, mortality and recruitment-related parameters from length-frequency data (ELEFAN I and II). In: Gulland JA, Rothschild BJ (eds) Penaeid shrimps—their biology and management. Fishing News Books, Farnham, Surrey, pp 220–234

Penn JW (1984) The behavior and catchability of some commercially exploited penaeids and their relationship to stock and recruitment. In: Gulland JA, Rothschild BJ (eds) Penaeid Shrimps – their biology and management. Fishing News Books, Farnham, pp 173–186

Pereira RT, Almeida AC, Teixeira GM, Bueno AAP, Fransozo A (2017) Reproductive strategy of the shrimp *Nematopalaemon schmitti* (Decapoda, Caridea, Palaemonoidea) on the southeastern coast of Brazil. Nauplius 25. https://doi.org/10.1590/2358-2936e2017003

Pérez Farfante I, Kensley B (1997) Penaeoid and sergestoid shrimps and prawns of the world. Keys and diagnoses for the families and genera. Mémoires du Muséum national d'Histoire naturelle 175:1–233

Pianka E (1970) On r- and K-selection. Am Nat 104:592–597

Pickens BA, Carroll R, Taylor JC (2021) Predicting the distribution of penaeid shrimp reveals linkages between estuarine and offshore marine habitats. Estuar Coasts 44:2265–2278. https://doi.org/10.1007/s12237-021-00924-3

Pillai D, Nair CM, Salin KR, Marques A, Widada JS, Bonami JR (2005) Gross signs and histopathology of branchiostegal blister disease (balloon disease): an idiopathic disease of farmed *Macrobrachium rosenbergii* (De Man). J Fish Dis 28:473–478

Pillai D, Johnson SK, Bueno SLS (2010) Health management. In: New MB, Valenti WC, Tidwell JH, D'Abramo LR, Kutty MN (eds) Freshwater Prawns: biology and farming. Wiley-Blackwell, Oxford, pp 256–277

Polovina JJ (1993) The lobster and shrimp fisheries in Hawaii. Mar Fish Rev 55:28–33

Pörtner H-O, Karl DM, Boyd PW, Cheung WWL, Lluch-Cota SE, Nojiri Y, Schmidt DN, Zavialov PO (2014) Ocean systems. In: Field CB, Barros VR, Dokken DJ, Mach KJ, Mastrandrea MD, Bilir TE, Chatterjee M, Ebi KL, Estrada YO, Genova RC, Girma B, Kissel ES, Levy AN, MacCracken S, Mastrandrea PR, White LL (eds) Climate change 2014: impacts, adaptation, and vulnerability. Part A: Global and sectoral aspects. Contribution of Working Group II to the Fifth Assessment Report of the Intergovernmental Panel on Climate Change. Cambridge University Press, Cambridge, pp 411–484

Primavera JH (1998) Mangroves as nurseries: shrimp populations in mangrove and non-mangrove habitat. Estuar Coast Shelf Sci 46:457–464

Rhyne AL, Tlusty MF, Szczebak JT, Holmberg RJ (2017) Expanding our understanding of the trade in marine aquarium animals. Peer J 5:e2949. https://doi.org/10.7717/peerj.2949

Richards RA, Hunter M (2021) Northern shrimp *Pandalus borealis* population collapse linked to climate-driven shifts in predator distribution. PLoS One 16(7):e0253914. https://doi.org/10.1371/journal.pone.0253914

Roberts S (2005) Wild-caught coldwater shrimp. United States, Monterey, California, Seafood Watch Seafood Report. Monterey Bay Aquarium, California, USA

Rousseau Y, Watson RA, Blanchard JL, Fulton EA (2019) Evolution of global marine fishing fleets and the response of fished resources. PNAS 116:11238–12243

Ryther JH (1969) Photosynthesis and fish production in the sea. Science 166:72–76

Sahul Hameed AS, Bonami J-R (2012) White tail disease of freshwater prawn, *Macrobrachium rosenbergii*. Indian J Virol 23:134–140

Sahul Hameed AS, Rahaman KH, Alagan A, Yoganandhan K (2003) Antibiotic resistance in bacteria isolated from hatchery-reared larvae and postlarvae of *Macrobrachium rosenbergii*. Aquaculture 217:39–48

Scelzo MA (2016) Biología reproductive del langostino y del camarón de las aguas marinas Argentinas. El Mar Argentino y sus Recursos Pesqueros 6:71–88

Schveitzer R, Arantes R, Baloi MF, Costódio PFS, Arana LV, Seiffert WQ, Andreatta ER (2013) Use of artificial substrates in the culture of *Litopenaeus vannamei* (Biofloc System) at different stocking densities: effects on microbial activity, water quality and production rates. Aquac Eng 54:93–103

Scott-Denton E, Cryer PF, Duffy MR, Gocke JP, Harrelson MR, Kinsella DL, Nance JM, Pulver JR, Smith RC, Williams JA (2012) Characterization of the US Gulf of Mexico and South Atlantic penaeid and rock shrimp fisheries based on observer data. Mar Fish Rev 74:1–27

Sheridan P, Doerr J (2005) Short-term effects of the cessation of shrimp trawling on Texas benthic habitats. Am Fish Soc Symp 41:571–578

Shields J (2019) Climate change enhances disease processes in crustaceans: case studies in lobsters, crabs, and shrimps. J Crustac Biol 39:673–683

Shumway SE, Perkins HC, Schick DF, Stickney AP (1985) Synopsis of biological data on the pink shrimp, *Pandalus borealis* Kröyer, 1838. NOAA Technical Report NMFS 30, FAO Fisheries Synopsis No 144: 1–57

Simpson AW, Watling L (2006) An investigation of the cumulative impacts of shrimp trawling on mud-bottom fishing grounds in the Gulf of Maine: effects on habitat and macrofaunal community structure. ICES J Mar Sci 63:1616–1630

Sobrino I, Silva C, Sbrana M, Kapiris K (2005) A review of the biology and fisheries of the deep water rose shrimp, *Parapenaeus longirostris*, in European Atlantic and Mediterranean waters (Decapoda, Dendrobranchiata, Penaeidae). Crustaceana 78:1153–1184

Sundstrom GT (1957) Commercial fishing vessels and gear. United Fisheries and Wildlife Service Circular 48:1–45

Suryan RM, Arimitsu ML, Coletti HA, et al. (46 other coauthors) (2021) Ecosystem response persists after a prolonged marine heatwave. Sci Rep 11:6235. https://doi.org/10.1038/s41598-021-83818-5

Temming A, Damm U (2002) Life cycle of *Crangon crangon* in the North Sea: a simulation of the timing of recruitment as a function of the seasonal temperature signal. Fish Oceanogr 11(1): 45–58

Thomas N, Lucas R, Bunting P, Hardy A, Rosenqvist A, Simard M (2017) Distribution and drivers of global mangrove forest change, 1996–2010. PLoS One 12(6):e0179302. https://doi.org/10.1371/journal.pone.0179302

Tidwell J, Coyle S, Durborow RM, Dasgupta S, Wurts WA, Wynne F, Bright LA, Van Arnum A (2002) KSU Prawn Production Manual. Kentucky State University Aquaculture Program, 45 pp

Treece G (2016) The rise and decline in us shrimp farming, from a Texas perspective. Treece and Associates Report for the Aquaculture Association website

Treece GD, Fox JM (1993) Design, operation and training manual for an intensive culture shrimp hatchery (with emphasis on *Penaeus monodon* and *Penaeus vannamei*). Texas and A & M Sea Grant Publica 93–505, 187 pp

Turner RE (1977) Intertidal vegetation and commercial yields of penaeid shrimp. Trans Am Fish Soc 106:411–416

Turner RE, Boesch DF (1987) Aquatic animal production and wetland relationships: insights gleaned following wetland loss or gain. In: Hook B (ed) Ecology and management of Wetlands. Croon Helms Ltd., Beckenham, pp 25–39

Valenti WC, Flickinger DL (2020) Freshwater caridean culture. In: Lovrich G, Thiel M (eds) The natural history of Crustacea, Fisheries and Aquaculture, vol 9. Oxford University Press, New York, pp 217–232

Valenti WC, Daniels WH, New MB, Correia ES (2010) Hatchery systems and management. In: New MB, Valenti WC, Tidwell JH, D'Abramo LR, Kutty MN (eds) Freshwater Prawns: biology and farming. Wiley-Blackwell, Oxford, pp 55–85

Valiela I, Bowen JL, York JK (2001) Mangrove forests: one of the world's threatened major tropical environments. Bioscience 51:807–815

Vance DJ, Rothlisberg PC (2020) The biology and ecology of the banana prawns *Penaeus merguiensis* de Man and *P. indicus*. H. Milne Edwards. Adv Mar Biol 86:1–139

Ventura T, Manor R, Aflalo ED, Weil S, Rosen O, Sagi A (2012) Timing sexual differentiation: full functional sex reversal achieved through silencing of a single insulin-like gene in the prawn, *Macrobrachium rosenbergii*. Biol Reprod 86(90):1–6. https://doi.org/10.1095/biolreprod.111.097261

Villasante S, Guimeráns CP, Rodrigues J, Antelo M, Rodríguez R, Da Rocha JM, Coll MS, Pita C, Pierce G, Hastie L, Sumaila R (2015) Small-scale fisheries and the zero discard target. European Parliament Policy Department B: Structural and Cohesion Policies, Fishery Study, 66 pp

Vlachopolou EI, Wilson AM, Miliou A (2013) Disconnects in EU and Greek fishery policies and practices in the eastern Aegean Sea and impacts on *Posidonia oceanica* meadows. Ocean Coast Manag 76:105–113

Walker PJ, Winton JR (2010) Emerging viral diseases of fish and shrimp. Vet Res 41:51. https://doi.org/10.1051/vetres/2010022

Wargo L, Ayres D, Cheng YW (2013) Washington coastal spot shrimp fishery. Washington Department of Fish and Wildlife Fish Program Report Number FPT 13-01, 79 pp

Watling L, Norse EA (1998) Disturbance of the seabed by mobile fishing gear: acomparison to forest clearcutting. Conserv Biol 12:1180–1197

Weber S, Traunspurger W (2016) Influence of the ornamental red cherry shrimp *Neocaridina davidi* (Bouvier, 1904) on freshwater meiofaunal assemblages. Limnologica 59:155–161

Wehrtmann IS, Arana PM, Barriga E, Gracia A, Pezzuto PR (2012) Deep-water shrimp fisheries in Latin America: a review. Lat Am J Aquat Res 40:497–535

Weiperth A, Gábris V, Danyik T, Farkas A, Kuříková P, Kouba A, Patoka J (2019) Occurrence of non-native red cherry shrimp in European temperate waterbodies: a case study from Hungary. Knowl Manag Aquatic Ecosyst 420:9. https://doi.org/10.1051/kmae/2019002

Wenner EL, Knott DM (1992) Occurrence of Pacific white shrimp, *Penaeus vannamei* in coastal waters of South Carolina. In: DeVoe MR (ed) Proceedings of the conference and workshop, introductions and transfers of marine species: achieving a balance between economic development and resource protection. South Carolina Sea Grant Consortium, Charleston, pp 173–181

Xiao Y, Greenwood JG (1993) The biology of *Acetes* (Crustacea, Sergestidae). Oceanogr Mar Biol Annu Rev 31:259–444

Yesson C, Fisher J, Gorham T, Turner CJ, Hammeken Arboe N, Blicher ME, Kemp KM (2017) The impact of trawling on the epibenthic megafauna of the west Greenland shelf. ICES J Mar Sci 74: 866–876

Zimmermann S, Nair CM, New MB (2010) Grow-out systems–polyculture and integrated culture. In: New MB, Valenti WC, Tidwell JH, D'Abramo LR, Kutty MN (eds) Freshwater Prawns: biology and farming. Wiley-Blackwell, Oxford, pp 195–217

Zorriehzahra MJ, Banaederakhshan R (2015) Early mortality syndrome (EMS) as new emerging threat in shrimp industry. Adv Anim Veter Sci 3(Special Issue 2):64–72

Chapter 12
Evolutionary History and Relationships of Shrimps

12.1 Shrimp Chronicles

Shrimps, like all organisms, have a long history. A description and understanding of their evolution through time establishes a framework to explain both their origin and the selective (environmental) pressures that have resulted in their various adaptations. The type of evidence documenting the geological and biological history of a group of organisms is their fossil record as well as information gleaned from the characteristics of living (extant, Recent) species. The pattern of diversification of ancestors into descendants can be depicted in a phylogeny, like a human genealogy. This branching tree (dendrogram; cladogram; phylogram) is extremely helpful for making and proposing and testing hypotheses about the processes which produced it. The proposal and testing of phylogenetic hypotheses about crustaceans and decapod shrimps has been and continues to be a favorite and important activity of carcinologists and other biologists as will be seen below.

12.2 The Fossil Record

Fossils, the remains or traces of ancient organisms, can be an excellent and actual record of the history of organisms. A fossil may consist of hard parts, such as bone or shell which have been preserved intact or replaced by minerals. Crustacean exoskeletons and soft parts, i.e., tissues and organs, usually decay away through microbial decomposition but sometimes are replaced by minerals to form fossils that show their structure. Fossilization is usually a fortuitous event in that the dead organism is not eaten or torn apart by scavengers or does not decompose due to unique environmental conditions. Sudden burial by floods, mudslides, or volcanic activity often results in deposits with good fossils. Animals with calcified skeletons or shells such as fishes, reptiles, dinosaurs, and mammals may yield excellent fossils.

Similarly, marine organisms such as the extinct trilobites and some other crustacean-like arthropods with highly calcified exoskeletons, cephalopods with massive shells (ammonites, nautiloids, belemnites), bivalved molluscs, corals, echinoderms of various kinds, all have left quite good fossil records. The sequence of geologic strata in which a fossil is found can give it a relative date. More precise absolute dating of the fossil or strata in which it is found can be provided with radiometric methods, i.e., those based on decay of radioactive isotopes.

Unfortunately, these techniques do not do much good with shrimps, which have a very poor fossil record. Shrimps generally have flimsy, poorly calcified exoskeletons (cuticles); consequently, the best shrimp fossils come taxa in which some body parts are highly calcified. Shrimp palaeontologists usually work with very poor material, distorted and broken by the fossilization process: incomplete specimens, parts, and pieces. Reconstructions are often made and illustrated by combining the fossil evidence at hand plus inferences drawn from the morphology of related living species. Such reconstructions are thus hypotheses based on the evidence available, and some should be viewed with a skeptical and open mind. De Grave et al. (2009) give a list of fossil genera for the Decapoda, including shrimps, known prior to that publication.

The oldest reliable decapod shrimp fossils come from the Triassic, 200–250 million years ago (mya), and perhaps not surprisingly, they are penaeoid dendrobranchiates, a relatively primitive group (Burkenroad 1963). Two Triassic genera are *Aeger* (Aegeridae) and *Antrimpos* (Penaeidae), whose external morphology is often well preserved (Fig. 12.1a–b). Burkenroad proposed that the dendrobranchhiate ancestor of the Sergestoidea and Penaeoidea most probably resembled *Aeger*. Most caridean fossil taxa are from the Jurassic (145–200 mya) (Davis et al. 2018), e.g., *Harthofia polzi* (Fig. 12.1c) but at least one has been found from Triassic strata, *Parapalaemonetes thuringiacus* (Brandt and Schulz 2013). Stenopodideans appeared to be absent from the fossil record until the discovery of the Cretaceous fossil *Jilinicaris*, thought to be a spongicolid (Schram et al. 2000), followed by the Cretaceous *Phoenice* (Garassino 2001). However, the time scale for a decapod shrimp was potentially set far back into the Paleozoic, at the border between the Devonian and Carboniferous periods (~360 mya)) by the description of a stenopodidean *Devonostenopus pennsylvaniensis* (Jones et al. 2014). However, the purported stenopodidean fossils are poorly preserved and their identification as stenopodideans are hypotheses that need further verification. Before these more recent finds, the poor fossil record suggested a Triassic appearance of the dendrobranchiates, followed by a Jurassic caridean and a late Jurassic to Cretaceous appearance of the stenopodideans. However, newer evidence is more in line with that suggested by Burkenroad (1963) that the dendrobranchiates, carideans, and stenopodideans diverged in the Paleozoic, along with the benthic "reptantian" taxa such as the well-documented primitive lobster (Devonian) *Palaeopalaemon newberryi* (Fig. 12.1d) (Jones et al. 2018).

Is there much hope that better shrimp fossils might be found and described? Schram (2009) opined that the likelihood was not very promising. However, hope springs eternal, and we should be cheered by the diverse and spectacular finds of

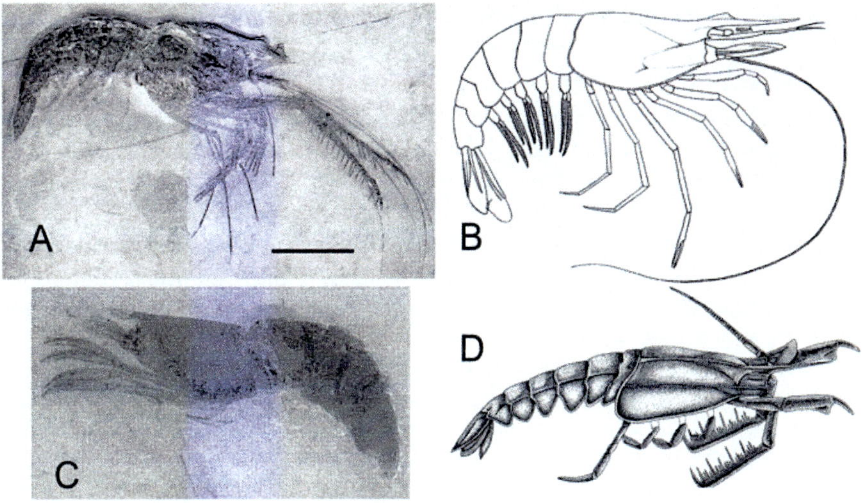

Fig. 12.1 Examples of shrimp and decapod fossils. (**a**) Triassic penaeoid *Aeger tipularius* (Aegeridae). (**b**) Reconstruction of Triassic *Antrimpos germanicus* (Penaeidae). (**c**) Jurassic caridean *Harthofia polzi*. (**d**) Reconstruction of *Palaeopalaemon newberryi* from Devonian shales, the first recognized fossil decapod crustacean. Scale bar in (**a**) represents 34 mm in (**a**), 11 mm in (**b**), 4 mm in (**c**) and 14 mm in (**d**). Modified from (**a**) Huang et al. (2013), (**b**) Brandt and Schulz (2013) Creative Commons CC BY 4.0, (**c**) Schweigert (2011), and (**d**) Jones et al. (2018), courtesy of the Journal of Crustacean Biology

strata termed "Lagerstätten," such as the Burgess Shale of Canada, that contain a diverse and unusually well-preserved invertebrate fossil fauna. Lagerstätten were formed when an ancient fauna was suddenly buried in anoxic or other environmental conditions which prevented the usual decay of dead organisms. Beautifully documented and illustrated in Stephen J. Gould's 1989 book "Wonderful Life," the true nature of its Cambrian fauna, discovered at the beginning of the twentieth century, was not truly revealed until the 1970s and beyond. Besides a vast array of soft-bodied invertebrate taxa, some extinct, some still with us, the Burgess Shale contains many lightly calcified crustacean-like arthropods never seen before. Since then, other similar invertebrate Lagerstätten have been worked and yielded similarly spectacular fossil finds, such as the Chengjiang biota of the Maotianshan Shales in southern China. Although not a recent find, many fossil shrimps, both dendrobranchiate and caridean, have been described from a "shrimp Lagerstätte," the Jurassic Solnhofen limestone deposits (e.g., Schweigert 2011). Thus, although the fossil record does not now give us a good account of decapod shrimp evolution, the seemingly endless stream of new or better-preserved fossils being reported by palaeontologists is a hint of what may be coming to fill in the many gaps. Even so, the reliably identified and dated fossils now available are critical for calibrating the chronology of decapod evolution (see below, e.g., Porter et al. 2005; Wolfe et al. 2019).

12.3 Classification and Systematics

Much of the work on shrimps deals with the related fields of taxonomy and systematics. Taxonomy is the naming and classification of animals, plants, and other organisms into hierarchical categories (taxa), e.g.., phylum, class, order, family, genus, and species. This classification serves as a system of information storage and retrieval. From the name of the taxon, one can extract all kinds of information about the organisms in it, especially their unique characteristics. For example, if an animal is in the Class Mammalia, we know immediately from its diagnostic traits that it has hair or fur at some time during its life, has three middle-ear bones and will suckle milk from its mother's mammary glands when young. Similarly, a creature with jointed appendages, a chitinous exoskeleton, and two pairs of antennae must belong to the Phylum Arthropoda, Subphylum Crustacea. However, depending on the methods used to form taxa, purely taxonomic classifications give little knowledge about the evolutionary history (phylogeny) of the group. Systematists study and describe the phylogeny of a group by trying to determine the pattern of ancestor-descendant relationships using characteristics that can be observed and measured. Phylogenies are summarized and visualized using branching tree diagrams (dendrograms, phylograms) to display a hypothesized genealogy of group. Phylogenies displayed in this way show the pattern and direction of evolution, but do not reveal much about the processes (natural selection) which gave rise to the pattern (Losos 2011; Losos et al. 2012). That type of information must be obtained by observations and experiments on the biology of the organism, e.g., on its morphology, physiology, development, behavior, and ecology.

Different approaches have been used in the taxonomy and systematics in crustaceans and other organisms. Phenetic classifications simply group organisms by the degree of overall similarity in their observable characteristics. In a very simple example, all animals with wings might be put in one taxon, while all swimming animals in another. Obviously, insects and birds do not have a close genealogical relationship nor do fishes and sea lions. Phenetic and numerical classifications might be helpful for some purposes, but not for studying phylogeny and evolution. Similarity in characters may be due to a common ancestry (homologous) but may very well be the result of the action of similar environmental pressures, i.e., convergent or parallel evolution. Some well-known examples of the latter include the independent evolution of succulent stems (water conservation) of spines and spiny leaves (defense against herbivores) in North American cacti and North African euphorbias. In zoology, the major phylogenetic approaches are evolutionary ("orthodox," "traditional," "Darwinian") and cladistic ("phylogenetic") systematics. Both are similar in that taxa are grouped according to their phylogeny (genealogy). However, in evolutionary systematics, taxa are grouped not only on ancestor-descendant relationships but also on the degree of evolutionary change (divergence) among the groups. The problem with evolutionary systematics is that it can be somewhat subjective and often results in the formation of paraphyletic taxa, i.e., taxa which do not include all the descendants of the most recent common ancestor.

In recent decades, the preferred method of constructing phylogenies and classification, phylogenetic systematics, uses the cladistic method proposed and developed by Willi Hennig, first published in German in 1950 and later translated to English (Hennig 1966). It differs from evolutionary systematics in that the formation of taxa is strictly based on ancestor-descendant relationships without consideration of the degree of divergence (dissimilarity) in morphology or other phenotypic traits. The method assumes that an ancestor splits into monophyletic descendant taxa dichotomously, as the probability of splitting into three or more descendant taxa simultaneously is quite small. Taxa in the various branches (clades) of the phylogeny contain only the descendants of the most recent ancestor. Care must be taken not to form polyphyletic taxa in which descendants from different common ancestors are included in the same clade. The two descendants of a common recent ancestor are termed "sister taxa" and form a monophyletic clade. Some terminology: characteristics or traits of an ancestor are termed "primitive" (plesiomorphic); when these change through evolution, they are labeled "derived" (advanced, apomorphic). A clade is formed only by shared derived characters (synapomorphies). All descendants of a group may share some primitive characters but these plesiomorphic traits do not help to sort out or resolve the genealogical relationships among or within clades, as will be seen below. To construct a proper cladistic phylogeny, there must be enough derived traits in the taxa being examined to sort them out. It is also assumed that exclusion of taxa (e.g., unknown or extinct) will not lead to incorrect results. Homologous traits, i.e., those inherited from a common ancestor, and their polarities (direction of evolution from primitive to derived) must be determined correctly (Arnold 1981).

Many types of characters can be used in creating a phylogeny: morphological, developmental, amino-acid composition of proteins, gene sequences from both nuclear and mitochondrial DNA and RNA. However, most crustacean phylogenies are based either on morphological structures or molecular data (RNA and DNA gene sequences). Up until the 1990s, phylogenies of shrimps and other crustaceans were almost exclusively based on morphology. But since that time, there has been an explosive publication of molecular phylogenies, which are often considered superior or more objective than morphological ones. However, both types have their pros and cons. Morphological phylogenies allow one to make good hypotheses on the function of morphological traits and suggest how they might be altered by environmental pressures (natural selection). On the other hand, morphology is often subject to parallel and convergent evolution, which leads to erroneous depiction of phylogeny. Many of the morphological traits used in phylogenies may seem and may be of minor or no functional significance. However, they may carry a phylogenetic signal useful in establishing genealogical relationships.

Molecular data is based directly on the genotype which ultimately produces the phenotype, although the path from gene sequence to phenotype and natural selection is long and usually little understood. There are only four "letters" in the genetic code, i.e., **A**denine, **G**uanine, **C**ytosine, and **T**hymine in DNA, which is a very small character set. However, an organism has many genes, so there is a lot of possible data although not all genes are equally appropriate. Some genes or gene sequences are more useful for examining relationships among higher taxa (phyla, classes, orders, families) because they evolve (mutate) at a slower rate than others which are more variable and better used at lower taxonomic levels. In decapod molecular studies, the mitochondrial gene 16S rRNA, nuclear 28S rDNA, 18S rDNA and protein-coding histone genes are often used. However, more and more genes and gene sequences of myriad taxa are being employed and can be accessed on online gene databases such as GenBank. For example, Wolfe et al. (2019) used 410 different gene sequences in their work on decapod relationships. Mitogenomes, the complete gene sequences of mitochondrial genes, are also being used or incorporated in phylogenetic studies (e.g., Cheng et al. 2018; Wolfe et al. 2019).

In both morphological and molecular phylogenies, the outcome and reliability of the final phylogenetic tree produced is highly dependent on the number and kind of species and characters sampled, as well as the methods of quantitative analysis used (Timm and Bracken-Grissom 2015). The number of possible phylogenetic trees that can be generated from the same data set can be huge, even with relatively few taxa (e.g., with 4 taxa, 15 rooted trees; 8 taxa, 135 trees; 10 taxa, $\sim 34 \times 10^6$ trees and so on!). Different methods of tree building, based on varying assumptions and algorithms, have been devised and are available as computer programs for data analysis, e.g., "neighbor-joining" and "unweighted pair group with arithmetic mean" (UPGMA), maximum parsimony, maximum likelihood and Bayesian inference methods (Hall 2001; Felsenstein 2004). Trees made by these methods basically show the number of changes in gene sequences or morphological character states between ancestors and descendants. Quantitative estimates of the amount of change can be assigned to each branch as well as the degree of confidence or probability of each branch in the tree. Phylogenetic trees are most often rooted, i.e., show the direction of evolution from the earliest ancestors (stem taxa) to the tips of the tree (crown taxa), which are easier to visualize and understand. A cladogram is rooted by using an outgroup, a more distantly related taxon which acts as a reference point of comparison for the ingroup (the group of taxa which are under study). For example, an often-used outgroup to root trees in decapod crustacean studies is the Euphausiacea, the sister group of the Decapoda or, for a group of caridean families, a penaeoid or sergestoid shrimp, which belong to the Dendrobranchiata, the sister group of Pleocyemata, to which the carideans belong.

Phylogenetic trees, both morphological and molecular, show the pattern of evolution in a group. By displaying the various phenotypic (observable) traits of taxa on the tree at different branches, one can get ideas and make hypotheses on how that character evolved and the selection pressures that might be responsible. Some studies work backward from the tips of the tree (most recent taxa) to the ancestor to hypothesize what the characteristics of ancestors might have been like (ancestral state reconstruction). However, phylogenies are just one tool in understanding the evolution and history of a group and should be used in conjunction with the fossil record (if a good one is available) as well as studies on the functional morphology, development, behavior, and ecology of a particular group (Losos 2011). Molecular studies are not necessarily and always superior to morphological ones (Galis et al. 1999; Losos et al. 2012). They have an advantage of tremendous amounts of objective quantitative data, but those data, based on only four character states (the four letters of the genetic code) may be subject to a great amount of unexplained homoplasies, e.g., convergence (Hao et al. 2019) and spontaneous reversals back to the original state.

It is not enough to just gather genetic characters to study a group; in my view, one must first study and experience at first hand the group of organisms being considered for a phylogeny. Without it, interpretations and hypotheses from the phylogeny may be superficial and a rather barren exercise. On the other hand, an expert in the biology of a given group cannot be expected to be a specialist in molecular techniques, although there certainly are some. Today more graduate students are trained both in the group of their interest and molecular methods. Another way to combine expertise in both the biology of a taxon and molecular techniques is a collaboration with colleagues expert in one field or the other. That is more and more the case today, as phylogenetic studies often have several authors sharing their work and credit. Some studies combine both molecular and morphological (or other) characters in constructing the phylogeny and interpreting its findings. Finally, separate morphological and molecular phylogenies can be considered independent tests of phylogenetic hypotheses based on morphology or other traits. If results are concordant, excellent! If not, further study can reveal the reason for the seeming conflict and advance knowledge about the phylogeny of the taxon studied.

12.4 Morphological Characters and Their Evolutionary Polarities

The types of characters used in constructing molecular phylogenies have been briefly discussed. As an organismal biologist and in keeping with the adaption and natural history theme of this book, I will concentrate in this section on morphological characters used in shrimp systematics. Any character used must be variable, i.e., have different forms (character states). Two taxa may have the same trait from inheritance from a common ancestor (homology) or because of convergent or parallel evolution. The latter two phenomena are comparable in that similar environmental conditions (selection pressures) in two relatively unrelated taxa have resulted in analogous adaptations. Similar adaptations in distantly or unrelated taxa (e.g., wings of birds, bats, and insects) are termed convergent, while the same form of a character evolved independently in more closely related groups after separation from a fairly recent common ancestor is considered parallelism. One example of parallelism is the independent reduction of epipod-setobranchs in related genera of thorid shrimps, discussed further below. An inherited adaptation may return to a more primitive state if environmental conditions change (a reversal). However, the latter is probably rare, especially in complex morphological structures, as once evolved, such traits do not de-evolve to their original state in exactly the same way (Dollo's "Law" of Irreversibility, e.g., Galis et al. 1999). Convergence, parallelism, and reversals (homoplasmies) can be determined by detailed studies on structure, function, and ontogeny.

An important decision in using morphological characters for phylogeny is to determine whether a particular character state is primitive (plesiomorphic) or derived (apomorphic). There are various methods or conventions for doing so–none are perfect! Most of these methods are based on evolutionary logic (how selective pressures act to produce change) and parsimony (Occam's razor) in which the hypothesis including the fewest number of steps is most likely the correct one. In the commonality or generality principle, a character state which occurs most frequently among the taxa within a group is likely the primitive state. Another important method is outgroup comparison, in which a more distantly related group, one that does not share a recent common ancestor with the ingroup (the related taxa under study) is chosen as a point of reference. The outgroup allows the phylogeny to be rooted, i.e., gives the direction of evolution in traits within the ingroup from ancestors ("stem taxa") to the most derived ("crown") taxa. To examine relationships among the Decapoda, a choice of outgroup might be or include the euphausiaceans or stomatopods (Burkenroad 1963; Wolfe et al. 2019).

Serial homology is the repetition of similar body parts along the longitudinal axis of a bilateral animal such as a polychaete worm, millipede, or crustacean, all of which are metameric (segmented) animals. In a polychaete worm, such as *Nereis,* the body is composed of a long series of developmentally derived, morphological and functionally similar somites, most of which bear on either side a pair of setiferous appendages, the parapodia. In many other polychaete species, such as the burrowing *Chaetopterus*, groups of somites have fused to form functionally similar units, the tagma, with reduction, loss, or specialization of form and function. This is considered a reduction in serial homology and is phylogenetically advanced from an ancestor more like *Nereis*. Shrimps, like other crustaceans, are metameric animals with a rather high degree of serial homology. Retention in the serial homology of a caridoid-like ancestor is primitive and reductions or changes such as the incorporation of somites into tagma are derived.

Another technique for determining character polarities uses the principle of paradromy (Maslin 1952; Arnold 1981). If there is a cline or gradient in a morphological character in various related taxa, and its primitive and advanced extremes can be identified and are related in parallel to another morphological cline in the same set of taxa, the direction of evolution in the second cline will be the same. As a postdoctoral fellow at the Smithsonian National Museum of Natural History, I curated collections of related thorid taxa, among them *Heptacarpus* and *Eualus*. I noticed that the number and form of epipod-setobranch complexes (Chap. 2, Structure and Function: Fig. 2.8B-D; Chap. 4, Shrimp Hygiene: Fig. 4.9B-C) varied more or less in parallel with body shape. In both *Heptacarpus* and *Eualus*, there are species with short, stout bodies, and short rostra; at the other extreme, some species have elongate bodies and long rostra, and yet others are intermediate in body shapes (see the three types of carapace shape and rostra in Chap. 2, Fig. 2.3B). In

Heptacarpus, at least, these body shapes are also associated with behavior and ecology. The stout species are most cryptic, hiding below and clinging tightly to stones, algae, and sessile invertebrates. They are somewhat sluggish in general behavior. Species with intermediate shapes are active omnivorous shrimps in tidepool, sublittoral rocky and soft-bottom seagrass and algal meadows. The long and slender species lay along blades of seagrass and algae, similar to *Hippolyte* and *Tozeuma* (Hippolytidae). I proposed the direction of evolution in these and other character states using a morphocline in the pereopodal epipod-setobranch complex, a gill-cleaning mechanism. The number of thoracic-limb epipods is a serially homologous feature, and one on each limb except the last is the primitive condition in decapod crustaceans (Burkenroad 1981). A reduction of epipod number is thus derived. In the Caridea, the maximum number of epipod-setobranch complexes is 5 but among *Heptacarpus* and *Eualus* species it varies from 4 to none. A reduction of epipod-setobranchs is, using the principle of serial homology, more advanced than the primitive condition of 4 in these genera. Studies on functional morphology show that epipods are lost from posterior to anterior and that their gill-cleaning function is taken over by gill-brushing with the second (grooming) chelipeds (Bauer 1979). The direction of evolution can be hypothesized in other morphoclines correlated with trends in epipod-setobranch complexes, e.g., carapace and abdominal somite shape; rostral, eye, and walking leg length; presence-absence of a branchiostegal setal fringe (pre-gill filter) as well as other traits (Bauer 1984). Thus, the stout-bodied long-legged species with short rostra and a branchiostegal fringe are judged primitive while the elongate species with stubby legs and no branchiostegal fringe are considered advanced in *Heptacarpus* and *Eualus*. Deductions made from these general principles do not alone constitute proof. Independent evidence from development and functional morphology should be sought to test hypotheses about the evolutionary polarity of a morphological character.

12.5 Origin of Shrimps from a Decapod Ancestor

A combination of evidence from the fossil record, morphology and, more recently, molecular techniques allow an overview of the descent of the major shrimp taxa from their ancestors. Likewise, the relationships of the shrimp taxa among other major decapod groups have become less controversial, although disagreements about the details continue. However, I say this with caution. In my career, from graduate school to the present, once accepted (consensus) views have been upended by new evidence and interpretations of previous conclusions. Certainly, views about phylogeny within major taxa, including shrimp taxa, remain as unsettled and changeable as ever for the present. On the bright side, similar to human genealogies, the more the phylogeny is studied and new evidence is gathered, the more is learned about the life (biology) of the organisms involved–which is the ultimate objective.

Hypotheses on the characteristics of ancestral species or taxa serve as reference points upon which the evolutionary polarities of the character states can be proposed.

A characteristic of the ancestor is inferred from those of the presumed descendant taxa using the methods given above. Which character states are most primitive, i.e., could have given rise to all the derived conditions in descendants? A hypothetical ancestor has didactic value and allows consideration of the selective pressures that may have led to the evolution of its progeny.

Burkenroad (1963, 1981) proposed a decapod ancestor after consideration of the distribution of features in living and the few fossil shrimps, as well as those of other decapod taxa. Like its eucarid sister taxon, the Euphausiacea, the ancestor would have been a long-tailed, shrimp-like crustacean with well-developed caridoid facies, the primitive body form of the Class Malacostraca. It would have all the unique features of the Decapoda, e.g., enclosure of the gills by a lateral extension of the carapace, the branchiostegite, to form a branchial chamber. To move water through the branchial chamber, an exite or branch of the second maxilla was expanded to form a gill bailer (scaphognathite). The first three thoracic appendages were modified and more closely associated with the mouth as maxillipeds. The maxilliped 1 exopod would not be elaborated into a lobe, as in many carideans. In living decapods, there is considerable variability in the number of pairs of chelipeds, from none to three. This is understandable given the different usage of chelipeds as agonistic, feeding, and grooming appendages, leading to a high probability of convergence. Thus, Burkenroad concluded that a decapod ancestor, a probable bottom scavenger, lacked claws (achelate), with all pereopods similar in form, composed of seven articles and ending in simple dactyls. This is very much like the primitive shrimp infraorder Procarididea which was not discovered until after Burkenroad's 1963 paper. The shrimp-like ancestor would have had a large flexible abdomen in which the first somite was just as well developed as the following two, with the pleuron of the first somite not overlapping that of the second. The junctures between each of the six abdominal somites were linked by pleonic hinges (5 total). The pleopods were large and biramous, with the two branches joined by appendices internae to facilitate coordinated paddling. Like euphausiaceans, the male second pleopod would have had an appendix masculina, a sexual or copulatory process (Chap. 6, Reproduction). However, according to Burkenroad, males would not have had a copulatory petasma on the first pleopod. He considered this a structure derived independently in the euphausiaceans from the appendix interna and in dendrobranchiates from the entire endopod.

As noted in Chap. 2, a decapod ancestor would have had a very full branchial formula consisting of an epipod plus podobranch on the 1st to 7th thoracic somites, one arthrobranch on the 1st thoracic somite, a full set of body gills (2 arthrobranchs and 1 pleurobranch) on each of the 2nd to 7th somites, and only a single gill, a pleurobranch, on the last (8th) thoracic somite. Burkenroad changed his views from 1963 to 1981 about the primitive gill type, from dendrobranchiate to trichobranchiate, following some newer studies (see Felgenhauer and Abele 1983). Martin et al. (2007) showed that, within the dendrobranchiates, there is considerable variation with some gills approaching a phyllobranch condition. Similarly, within the reptant groups, both trichobranch and phyllobranch gills occur, and in some species, there are gills intermediate between the two. Based on these observations, I would go with a euphausiid-like filamentous or a trichobranchiate gill as the primitive condition within the Decapoda. Given the distribution and development among decapods of a gastric mill, the food-grinding and sorting chitinous ossicles in the decapod foregut (stomach) (Felgenhauer and Abele 1983, 1989), these were likely well developed in a decapod ancestor. Reduction in the number and kind ossicles in considered a derived feature.

Reproduction in a decapod ancestor would have been similar to euphausiaceans (krill), the decapod sister group, which has been retained in the dendrobranchiate (penaeoidean and sergestoidean) decapods. In these groups, the eggs are fertilized externally at spawning and set free in the water, with hatching as a typical crustacean nauplius, the primitive crustacean larva (Chap. 1, The Shrimps: An Overview). Males of the ancestral decapod probably deposited a spermatophore on the external surface of the female, not inside a sperm-storage structure (spermatheca or seminal receptacle) as in many extant decapods (Bauer 1986). External sperm deposition is found in carideans (Chap. 6), stenopodidean, and many dendrobranchiate shrimps.

12.6 Relationships Among Major Shrimp Groups and Other Decapods

Older ideas about the phylogeny of shrimp and reptant groups were often based on overall similarity, especially the use of shared primitive characters (symplesiomorphies) rather than shared derived ones (synapomorphies). For example, early workers such as Boas (1880) and Borradaile (1907) grouped the penaeoids, sergestoids, carideans, and stenopodideans into a monophyletic group, the suborder Natantia sensu *lato* (*s.l.*) (Fig. 12.2a). This was based on their overall similarity, e.g., their shrimp-like characteristics, a complex of plesiomorphic caridoid traits (Chap. 1) mostly related to forward and backward swimming. All the other heavy-bodied, benthic decapods (lobsters, crabs, etc.) were considered another monophyletic group, the suborder Reptantia. An acceptance of the Natantia *s.l.* as a natural group was maintained by numerous carcinologists for many years up until the 1960s and even beyond (e.g., Holthuis 1993).

However, in 1963, Martin Burkenroad wrote a seminal paper that showed that the Natantia is not a natural group, i.e., is polyphyletic, containing taxa with different ancestors. He proposed that basic life history traits clearly showed that the sergestoids and penaeoids formed a monophyletic clade, the Dendrobranchiata. The other shrimp groups (Caridea and Stenopodidea) were considered more closely related to the Reptantia than to the Dendrobranchiata (Fig. 12.2b) (the Procarididea were not discovered until 1972 by Fenner Chace and Raymond Manning). Like many other Crustacea, the dendrobranchiates spawn fertilized eggs freely into the water where embryonic development to hatching as nauplius larvae takes place (Chap. 1). The nauplius is the primitive larval type of Crustacea, found in many disparate taxa such as remipedes, branchiopods, ostracods, cirripedians (barnacles)

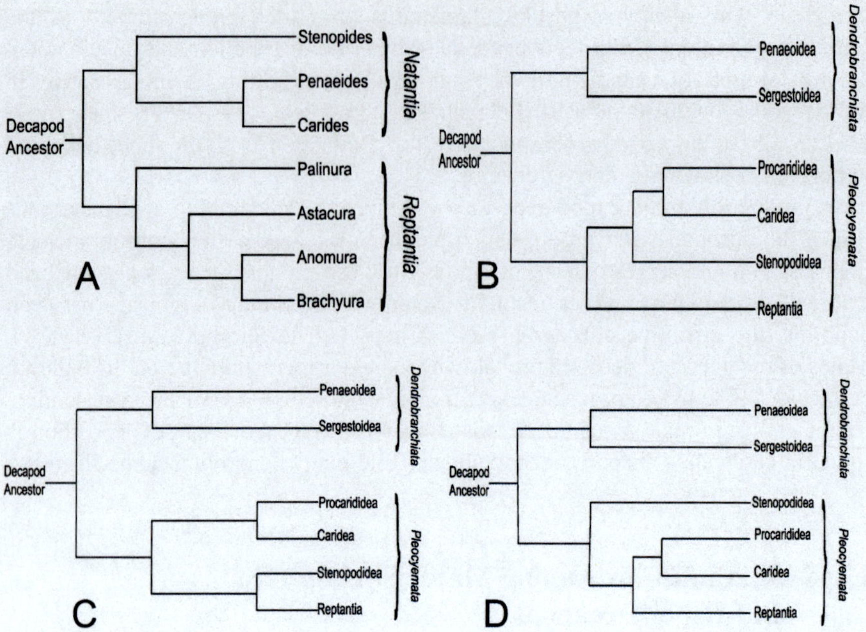

Fig. 12.2 Various views of the relationships of shrimp groups to other major decapod taxa based on various authors. (**a**) Morphology-based view of Borradaile (1907) and Boas (1880). (**b**) View of Burkenroad (1963, 1981) based on morphology, supported by molecular data in Wolfe et al. (2019). (**c**) Molecular phylogeny of Abele (1991), supported by Bracken et al. (2009a). (**d**) A somewhat different viewpoint by Christoffersen (1988a) using an extensive morphological data set

as well as the sister group of the Decapoda, the Euphausiacea. The dendrobranchiate's derived characters include the type of gills (dendrobranchs); ontogeny of the gills (arthrobranchs develop before the pleurobranchs); the first abdominal pleuron overlaps the second; there are 3 pairs of subequal chelipeds; and appendices internae are lost except on male pleopods 1–2. The Caridea, Stenopodidea, and Reptantia (together, the Pleocyemata), on the other hand, have extended development with incubation (brooding) of embryos beneath the female abdomen and hatching as at an advanced larval stage, the zoea or beyond (Chap. 9, Life Histories). The naupliar and pre-zoeal stages are passed within the embryo. Some carcinologists regard the Reptantia as a natural (monophyletic) group while others do not (see discussions in Martin and Davis 2001; Wolfe et al. 2019). For the purposes of this section, I will follow Wolfe et al. and treat it as monophyletic.

It may seem strange that carideans, procarideans, and stenopodideans, with their shrimp-like body form, are considered by Burkenroad (1963) and most current phylogenies to be closely related to decapods with very different morphologies such as lobsters and crabs. However, the shrimp-like body form is primitive in the Decapoda, but it is derived or apomorphic traits that indicate common ancestry. More recent morphological and molecular decapod phylogenies accept the Dendrobranchiata and Pleocyemata, but some find that the Stenopodidea is closer to the reptantians than the Caridea *s.l.* (e.g., Abele and Felgenhauer 1983, 1986; Abele 1991; Bracken et al. 2009a) (Fig. 12.2c). Characters linking stenopodideans to reptants include the lack of a caridean lobe on the first maxilliped, as well as a uniramous first pleopod and reduced first abdominal somite. For Burkenroad (1963, 1981), the key synapomorphy for linking the Caridea sensu stricto (*s. s.*) + Procarididea (Caridea *s.l.,* Abele 1991), which he termed "Eukyphida," with the Stenopodidea ("Euzygida") is their pattern of gill development, i.e., the pleurobranchs appear before the arthrobranchs. In reptantians, the anterior pleurobranchs are absent or the posterior arthrobranchs are delayed in development. In the Stenopodidea and Caridea *s.l.*, one or more pleonic hinges are lost, allowing for easier flexion of the abdomen during the backward escaped response. In the more benthic Reptantia, the reduced abdomen has a full set of 5 pleonic hinges (and a clumsier, less effective tail flip). Wolfe et al. (2019) weighed in with molecular data, concluding that the Stenopodidea and Caridea *s.l.* are one pleocyemate clade while a monophyletic Reptantia is its sister group. Another, more disparate view on this was offered by Christoffersen (1988a), using different synapomorphies, who concluded that the Caridea *s.s.* and Procarididea are more closely related to the Reptantia than to the Stenopodidea (Fig. 12.2d).

There is little doubt that the currently recognized shrimp infraorders Caridea *s.s.* and Procarididea are more closely related to each other than to the Stenopodidea. Important synapomorphies of both groups are the overlap of the second abdominal pleuron with the first and third, loss of a pleonic hinge only between abdominal somites 3–4, and epipod-setobranch complexes for gill cleaning in several families. The Procarididea differs from the Caridea *s.s.* in their lack of pereopod chelae and a third maxilliped of 7 articles (plesiomorphic) compared to a reduced number (4–5) in the Caridea *s.s.* due to the fusion of articles (e.g., Bauer 1977). Most molecular phylogenies show them as sister groups (e.g., Abele 1991; Bracken et al. 2009b, 2010; Wolfe et al. 2019). The stenopodidean synapomorphies include pleonic hinges only between somites 4–5 and 5–6, a reduced, uniramous first pleopod, and a massive third cheliped weapon (Burkenroad 1981; Goy 2010).

12.7 Chronological Estimates of Taxa Origins

As noted above, the fossil evidence alone tells us relatively little about the chronology of major decapod taxa, i.e., the approximate timing of their divergence from a decapod ancestor before the Devonian Period in the early Paleozoic (Burkenroad 1963; Schram 2009). Fortunately, the fossil record does allow chronological calibration of phylogenies using molecular (DNA, RNA) evidence. The basic idea is that the rate of genetic changes (mutations) can be used as a molecular clock. The clock is based on the genetic distance between the same gene sequences of two taxa. The first molecular clocks assumed a constant rate of gene change, but this is not necessarily so. In those clocks, it could only be assumed that one time period is x % as long as another depending on the branching pattern and branch lengths of the phylogeny.

However, a molecular phylogeny calibrated with reliably dated fossils ("phylochronology") can be viewed with more confidence.

To construct a phylochronology of decapods, Porter et al. (2005) and Wolfe et al. (2019) used newly developed statistical methods and clock models. Reassuringly, their molecular phylogeny agreed with morphological studies showing that the major decapod clades (Dendrobranchiata and Pleocyemata) are monophyletic. Their phylochronology estimated that extant decapod taxa originated in the late Ordovician Period (485–444 mya) and their various lineages diverged in the Triassic (252–220 mya) and Jurassic (202–146 mya). The dendrobranchiates diverged in the late Devonian (418–365 mya) into its two main clades, Sergestoidea and Penaeoidea, with both clades diversifying in the Pennsylvanian (323–298 mya). The phylochronology estimated that the Stenopodidea and Caridea *s.l.* split in the Silurian (444–418 mya). The basal (stem) Caridea *s.l.* clade divided into the Procarididea and Caridea *s.s.* in the Pennsylvanian with the crown Caridea appearing in the late Triassic. Divergence of the Caridea as well as that of the other two most speciose and diverse decapod taxa (Anomura, Brachyura) into superfamilies occurred in the late Triassic. Further diversification into families took place in the Jurassic and continued in the Cretaceous, together with and perhaps due to the rise of modern reef-building corals. These coral reefs provided many varied habitats which promote cladogenesis and speciation. The reader may be reminded that these types of studies and their interpretations depend on the number and kind of taxa, gene sequences, statistical methods, and models utilized. This is certainly not the end of the story. Still, it is encouraging that fossil, morphological and molecular evidence continue to converge on a similar view of decapod evolution.

12.8 Relationships Within the Major Shrimp Taxa

12.8.1 Dendrobranchiata

There has been little disagreement that the two superfamilies of the Dendrobranchiata, the Sergestoidea and the Penaeoidea, are monophyletic sister clades. Both morphological and molecular phylogenies are firmly in agreement on this issue. Burkenroad (1963, 1983) speculated that the common ancestor of the two superfamilies was somewhat like the penaeoid solenocerids, as the latter appear intermediate in form between penaeoid aristeids and the fossil genus *Aeger* (Fig. 12.3). It might be recalled (Chap. 3, The Families of Shrimps) that the sergestoids are mainly a pelagic group while the penaeoids have a greater diversity of lifestyles ranging from benthic to pelagic. One major diagnostic feature distinguishing the two groups is the number of respiratory structures, i.e., gills and epipods. In penaeoids, some thoracic somites bear at least 3 gills and no less than 11 total on each side of the body, while in sergestoids there are never more than 2 gills per somite and no more than 7–8 per side (Pérez Farfante and Kensley 1997). Sergestoids are further characterized by a reduction and/or loss of other external

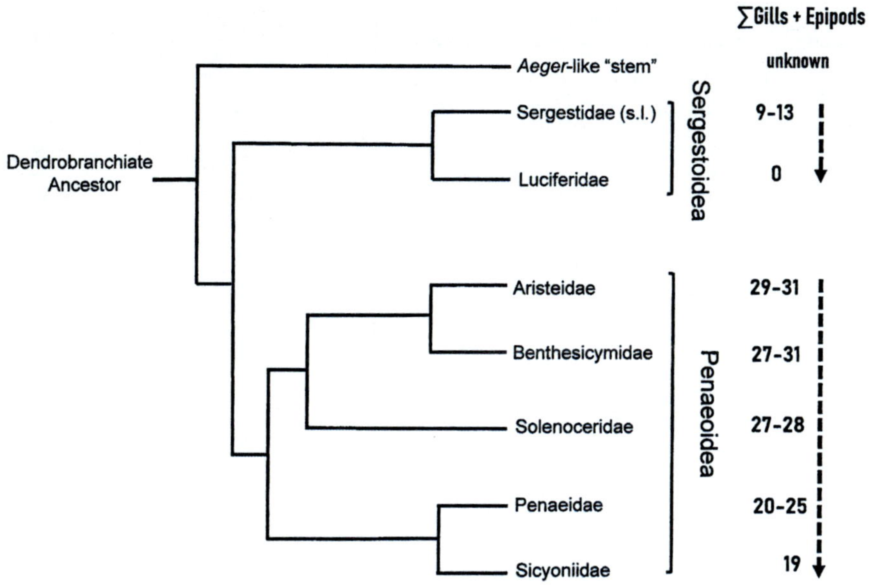

Fig. 12.3 Phylogeny of dendrobranchiate families based on Burkenroad (1983). I have updated the taxonomic terminology used by Burkenroad and have added the number of branchial elements (minimum, maximum sum of body gills and epipods) in each family (branchial formulae for genera from Pérez Farfante and Kensley 1997). Note the trend in a reduction in branchial structures within both the Sergestoidea and Penaeoidea. *Aeger* is a penaeoid fossil genus considered by Burkenroad to be close to the dendrobranchiate ancestor

characters, e.g., posterior pereopods and chelae. A more complete list of distinguishing characters and synapomorphies for dendrobranchiate superfamilies is given in Tavares et al. (2009) and Tavares and Martin (2010).

12.8.2 Sergestoidea

The reduced morphology of this group, particularly extreme in the Luciferidae, is an adaptation to pelagic life. Sergestoid traits such as a membranous, little-calcified exoskeleton and a reduced, "stripped-down" morphology save energy and reduce nutritional needs, as well as provide buoyancy (lower body density) in the aqueous pelagic habitat. The morphology and taxonomy of this group have received considerable attention for many years (e.g., Burkenroad 1983; Judkins and Kensley 2008; Tavares and Martin 2010 and references therein; see Chap. 3). However, there has been relatively little progress on their phylogeny until the recent spate of publications by Alexander Vereshchaka et al. (2014, 2016a, b) and Vereshchaka (2017). They performed cladistic analyses on a very large array of traits from general external morphology (carapace, appendages), copulatory apparatus, and photophore

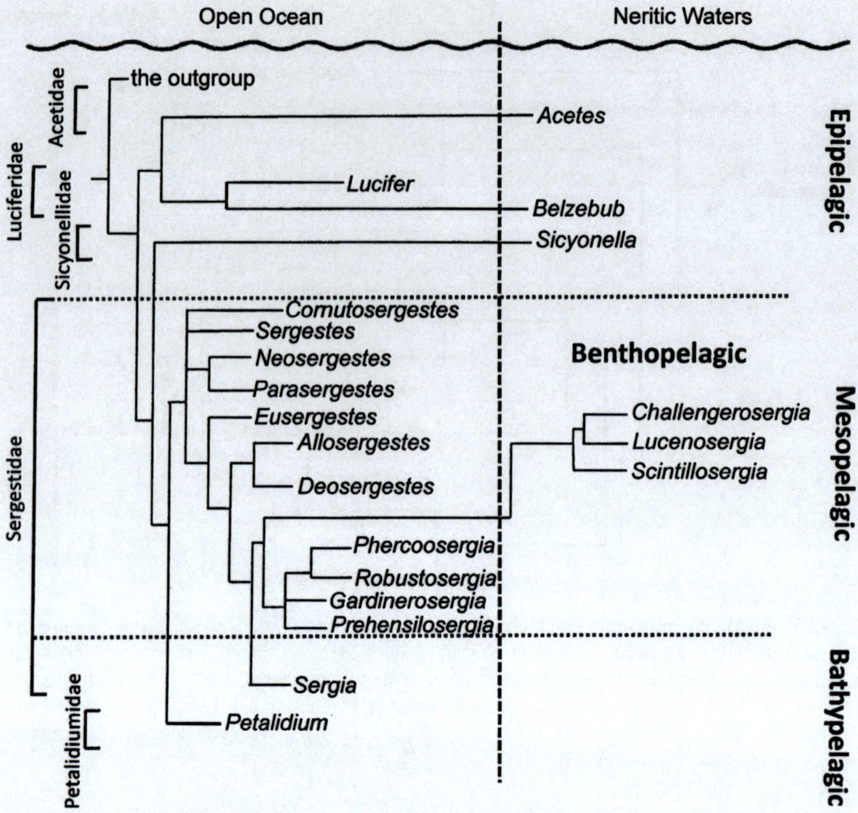

Fig. 12.4 Phylogeny of sergestoid genera related to their distribution in marine habitat zones: neritic vs oceanic and, by depth, the epipelagic, mesopelagic and bathypelagic. The benthopelagic genera are associated with continental slopes and submerged seamounts at mesopelagic depths. Modified from Vereshchaka (2017) Creative Commons BY 4.0

type and distribution. The phylogeny generated was then interpreted with ecological, behavioral and other information to considerably increase understanding about the evolution of this group.

Two major clades were identified, one containing the Luciferidae and Acetidae, and other with the Sicyonellidae, Sergestidae, and Petalidiumidae (Fig. 12.4). *Belzebub* (Luciferidae), Acetidae and Sicyonellidae are neritic, with one species (*Acetes paraguayensis*) occurring in fresh water. These taxa are benthopelagic, found near the bottom during the day but ascending into the water column at night. Additionally, as the depth range over the continental shelf corresponds to the epipelagic of the open ocean, these species can be considered epipelagic. One genus of this clade (*Lucifer s.s.*) as well as *Cornutosergestes* (Sergestidae *s.s.*) are oceanic in the epipelagic zone. Thirteen of the Sergestidae *s.s.* genera are mesopelagic, with three of those forming a benthopelagic clade occurring near the bottom around seamounts and continental shelves. The latter three are characterized by dermal photophores with thick lenses. This equipment favors communication among individuals in the dim "nepheloid" layer of turbid water near the bottom in these deep habitats. Two genera, *Sergia s.s.* and *Petalidium* are bathypelagic.

What was the primitive lifestyle of ancestral sergestoids—benthic or pelagic? Both Burkenroad (1983) and Vereshchaka et al. (2014) suggest that the neritic benthopelagic genus *Sicyonella* is the most primitive within the extant genera of sergestoids. It has the best-developed morphology of the group, e.g., well-formed albeit tiny chelae and pereopods 4–5 as well as elaborate petasmata (Vereshchaka et al. 2016b). The evolutionary hypothesis of Vereshchaka (Fig. 12.4) proposes that descendants of a benthic ancestor first invaded the epipelagic realm and, from there, its descendants expanded down into the mesopelagic and then the bathypelagic. Horizontal radiation within these three major depth zones occurred by sympatric speciation, i.e., reproductive isolation mediated by petasma divergence among populations (Fig. 12.5). The morphological phylogenies generated by the Vereshchaka group, and their interpretation need to be tested independently with molecular phylogenies. Evidence on the geologic timing of a pelagic invasion by a benthonic sergestoid ancestor is slight because divergence (molecular clock) studies have not yet been done. Sergestoids, with their flimsy, poorly calcified cuticles, do not fossilize well. Without a good, well-dated fossil record, the necessary calibration of a molecular clock is not yet available. However, geologic evidence indicates that oxygenation of the meso- and bathypelagic waters occurred after the Mesozoic era, making possible its occupation by pelagic crustaceans (Vereshchaka 2017).

The close relationship between the Luciferidae and the Acetidae (Vereshchaka et al. 2016b) is based on their diminished morphology, even more so than in other sergestoids. However, the reduction of morphological features (e.g., gills, chelae, pereopods) in *Acetes* is less extreme than in luciferids. Burkenroad (1983) suggested that the highly abridged features of luciferids might be rudimentary rather than reduced, i.e., paedomorphic, a retention of larval features in the reproductive adult. If so, *Acetes* may represent an intermediate stage in the evolution of paedomorphic luciferids. The alternate hypothesis is that the morphological features of *Acetes* and luciferids are a culmination of adaptive reductions related to a pelagic existence found in sergestoids. Comparisons of embryonic and larval development between Luciferidae, *Acetes*, and other sergestoids may help to resolve this issue.

12.8.3 Penaeoidea

The most widely recognized and cited work on penaeoid phylogeny is that of Burkenroad (1983). It is remarkable that his general scheme of relationships based on morphology (Fig. 12.3) is still supported, for the most part, by both molecular and other morphological phylogenies published since that time. His phylogeny has been criticized, and rightly so, for being intuitive without citation of a specific data set (character matrix) (Tavares and Martin 2010). However, Burkenroad's vast experience with penaeid morphology (e.g., his 1934 and 1936 monographs) and brilliant

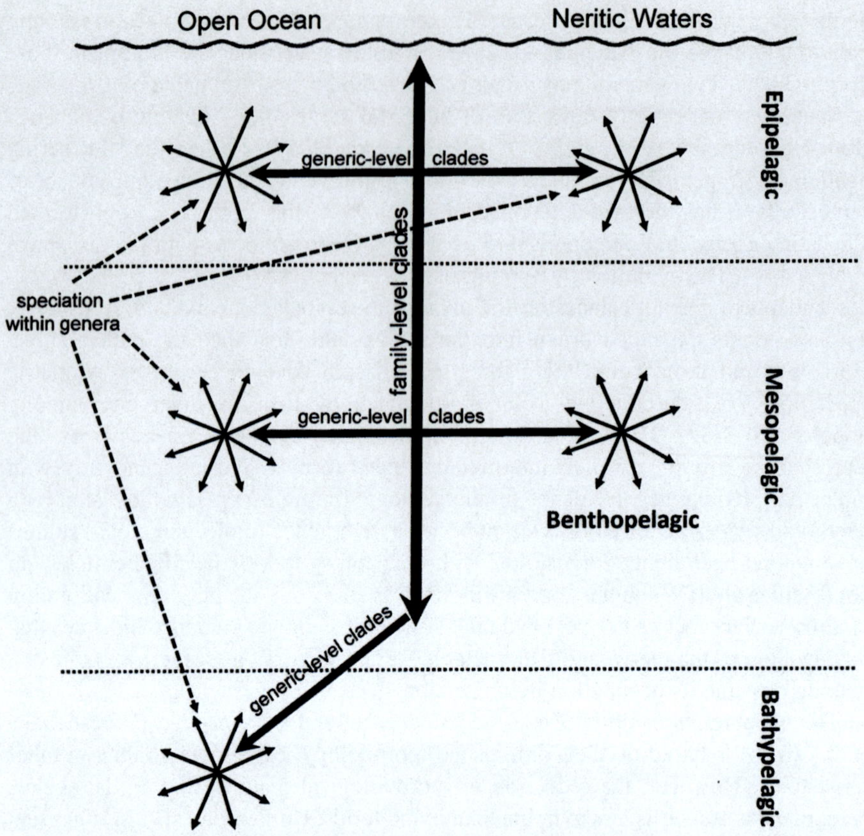

Fig. 12.5 Pattern of sergestoid radiation into habitat zones and speciation within them. Morphological and phylogenetic information indicates that a benthopelagic *Sicyonella*-like ancestor first occupied the epipelagic, with descendant species invading the mesopelagic and subsequently the bathypelagic zone. Within each neritic and oceanic depth zone, speciation occurred by reproductive isolation mediated by variation in genitalia. Modified from Vereshchaka (2017) Creative Commons BY 4.0

mind gave later investigators a sound basis on which to build. Note in Fig. 12.3 that the hypothesized advanced taxa have a lower number of respiratory elements (gills plus epipods). This reduction in serial homology of gill elements supports the view that they are more derived taxa.

In Burkenroad's view, the Aristeidae and Solenoceridae form one clade, considered more primitive, and the Penaeidae and Sicyoniidae another. Note that Burkenroad placed the Benthesicymidae as part of or perhaps a sister group of the Aristeidae. He also used a slightly different spelling for various penaeoid taxa than that in use today (without the "a" in names with the root "penae-), e.g., Peneidae for Penaeidae. He noted that Peneios, a Thessalian river god, and his descendant Aristaios are the basis of so many penaeoid taxonomic names. Burkenroad hypothesized that the sicyoniids are a paedomorphic offshoot of the Penaeidae, i.e., they become sexually mature at an earlier developmental stage. He also grouped genera of the Penaeidae into tribes, a taxonomic rank below family but above genus. These tribes and their composition are essential to understand not only evolution within the Penaeidae but also their relation to other penaeoid families. His tribe Peneini consists of genera such as: *Penaeus s.l., Heteropenaeus* and *Funchalia*; the Parapeneini, with *Parapenaeus, Penaeopsis, Metapenaeopsis*; and the Trachypeneini with *Mangalura* (now *Metapenaeus*), *Macropetasma, Trachypenaeopsis, Atypopenaeus, Protrachypene, Xiphopenaeus, and Trachypenaeus* (*s.l.*). Burkenroad indicated that the Parapeneini and Trachpeneini are more closely related, with the Peneini the most basal of the three tribes.

A preliminary morphological phylogeny was done by Tavares et al. (2009) and summarized with further comments by Tavares and Martin (2010) (Fig. 12.6a). It is exemplary in that it was done with an extensive character set plus modern cladistic and phylogenetic methods of analysis. Like many phylogenetic studies, the total number of genera and species was limited. I mentioned earlier that the methodology often influences the outcome of phylogenetic analyses. The Tavares et al. papers clearly show this, with three variations using different outgroups resulting in somewhat different results. In the analysis that the authors thought most realistic, the phylogeny agreed with some of the major points of the Burkenroad scheme, e.g., the Penaeoidea and Sergestoidea are natural groups, as is the Solenoceridae. On the other hand, it raised doubts about the monophyly of Penaeidae, Benthesicymidae, and Aristeidae, and suggested a different composition of major clades within the Penaeidae. Unlike Burkenroad, they found Solenoceridae closer to Penaeidae and Sicyoniidae than to the Aristeidae. They supported the close relationship between Sicyoniidae and Penaeidae.

A series of molecular phylogenies and divergence-time studies generated by T.-Y. Chan and various colleagues have been done on the Penaeoidea with somewhat different methodologies and objectives. Chan et al. (2008) examined relationships within the family Penaeidae using the mitochondrial 16S rRNA gene. Chan (2012) and Ma et al. (2009) used two nuclear genes (Fig. 12.6b) while Robalino et al. (2016) used both molecular (three nuclear genes) and phenomic (morphological, behavioral, ecological) data. Cheng et al. (2018) studied the superfamily Penaeoidea with mitogenomic data (entire mitochondrial genomes) (Fig. 12.6c). All these studies, which were based on good sample sizes, can be considered as a series in which the most recent built upon the previous studies. The latter three studies used molecular clock methods, calibrated with available fossil evidence to estimate divergence times among major clades.

These studies agree overall with much of Burkenroad's (1983) assessment of penaeoid evolutionary history. The Aristeidae, Solenoceridae, Benthesicymidae, and Sicyoniidae *s.s.* are considered monophyletic, but unlike Burkenroad's view, the Penaeidae is variously described as poly- and/or paraphyletic. The three tribes of Penaeidae proposed by Burkenroad are supported, although the relation of one to the other varies with the study. Both Ma et al. and Cheng et al. recommended raising the taxonomic rank of Burkenroad's tribes to family level. One difference from

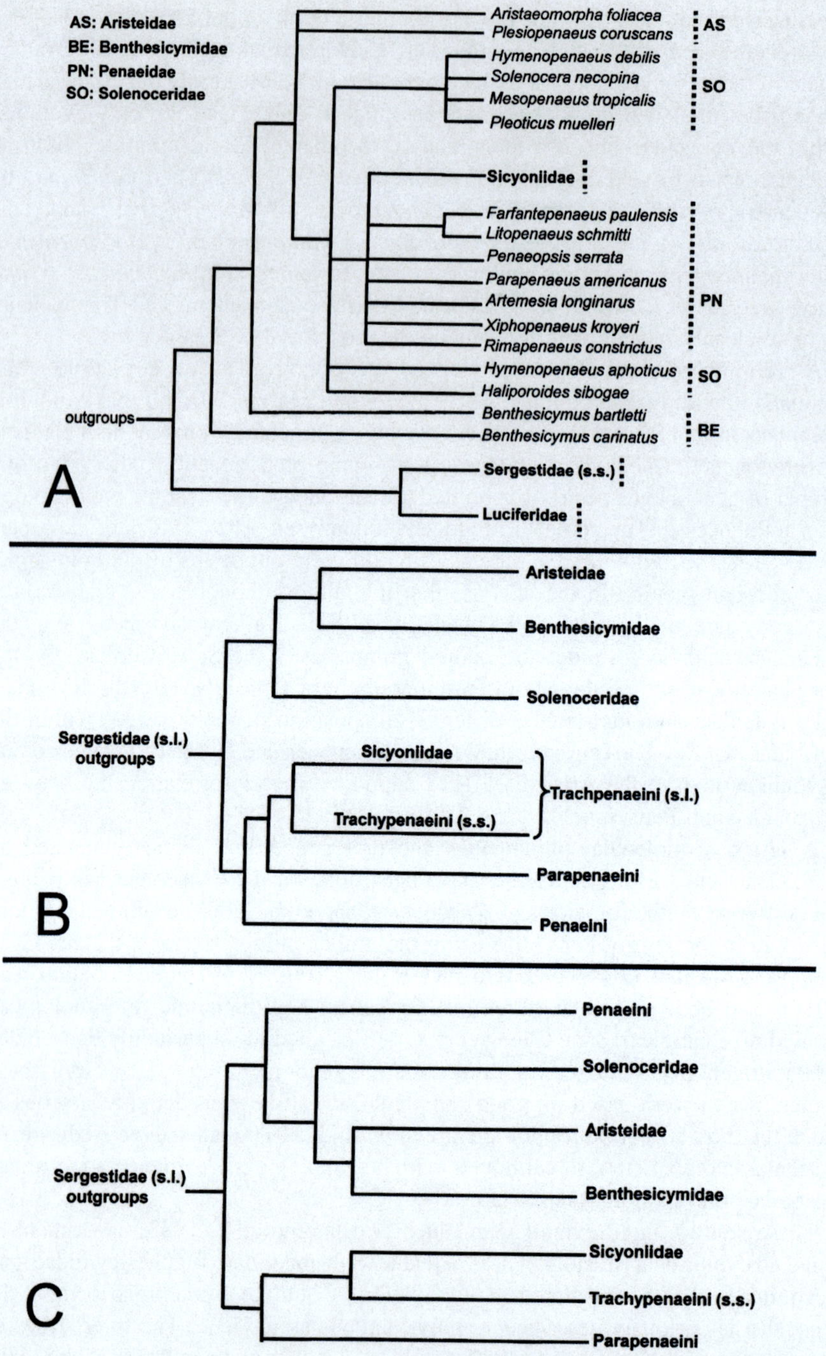

Fig. 12.6 Phylogenetic hypotheses about the Penaeoidea: (**a**) together with the Sergestoidea, based on morphological characters, with nephropidean, stenopodidean, and caridean outgroups; (**b**) based on nuclear genes; (**c**) based on mitochondrial genes. Modified from (**a**) Tavares et al. (2009), (**b**) Chan (2012), courtesy Journal of Crustacean Biology and (**c**) Cheng et al. (2018)

Burkenroad's views is that the Sicyoniidae are considered part of tribe Trachypenaeini rather than the more advanced Penaeini. In contrast, Burkenroad had suggested that the Penaeini was primitive with respect to the other two penaeid tribes. The results of Ma et al. (2009) found all three tribes (with Sicyoniidae) a monophyletic group, with the Penaeini also the most basal of the three (Fig. 12.6b). However, the Cheng et al. study (Fig. 12.6c) yielded novel results, i.e., that the Parapenaeini + Trachypenaeini (including *Sicyonia* species) clade is basal to a more advanced Penaeini + Solenoceridae + Aristeidae + Benthesicymidae clade, more in line with Robalino et al.'s (2016) fossil-calibrated phylogeny.

Together with the phylogenies, an understanding of shrimp evolution also requires information from earth history, functional morphology, and ecology of extant species. According to the divergence-time studies, the penaeoid line began diversification into major clades in the late Devonian, with the pre-Penaeini lineage and the aristeid line (Aristeidae, Benthesicymidae, Solenoceridae) parting ways in the Carboniferous. Diversification of families in the aristeid lineage began in the Jurassic but not until the late Jurassic in the Penaeini. In today's world, the aristeidean families inhabit the deeper waters of the outer continental shelf, continental slope, abyssal plains, and meso-and bathypelagic waters while the penaeoids and sicyoniids dominate shallower waters of the continental shelves. Cheng et al. proposed that ancient shallow-water penaeoids colonized the deep sea. The deep-sea aristeids, benthesicymids, and solenocerids have a larger number of respiratory structures (gills, epipods) than do extant nearshore penaeoids (penaeids and sicyoniids) (Fig. 12.3). At mesopelagic depths, there is an oxygen minimum zone. Deeper than that, oxygen concentration increases again, but the high pressures of the deep sea make oxygen diffusion across gills difficult. The ancestral deep-sea colonizer would have had to evolve a larger number of respiratory structures to obtain enough oxygen in deeper waters.

I would argue that the ancient shallow-water penaeoid ancestor that invaded the deep-sea must have had numerous gills like the decapod ancestor. According to basic evolutionary principles, the reduction of a serially homologous trait, such as gills and epipods, is a derived (apomorphic) feature. Therefore, there may have been a reduction of superfluous ancestral gills and epipods in the evolution of more recent shallow-water penaeids because of the higher oxygen availability in their nearshore habitats. To invade the deep-sea, the ancestor of the aristeids and solenocerids had to retain a large number of gills and epipods. If the ancestral penaeoid had had the relatively low number of gills of its present-day shallow-water descendants, the aristeidean families would have had to "re-evolve" the gills lost from the ancestor. Recall the earlier discussion on "Dollo's Law": evolution of lost structures in exactly or nearly the same form is very unlikely. However, the penaeids, aristeids, and solenocerids all have very similar, "classical" primitive dendrobranchiate gills (Martin et al. 2007). Additionally, they have generally retained a more primitive reproductive morphology, i.e., open petasmata in males and open thelyca on which external spermatophores are deposited in females (especially the Aristeidae, Benthesicymidae, and Solenoceridae). On the other hand, many Penaeidae and all

Sicyoniidae have more derived genitalia, i.e., closed petasmata, thelyca, and internalized sperm storage (Chaps. 3, 6).

12.8.4 Caridea and Procarididea

The Caridea *s.s.* is the most diverse of all shrimp groups with nearly 40 families and a number of genera and species far greater than the dendrobranchiates and stenopodideans taken together. It is no wonder that their evolutionary relationships are far from being resolved. Ideally, the classification of taxa into higher categories, e.g., superfamilies and families, is an expression of their phylogenetic relationships. Both Lipke Holthuis (1955, updated in 1993) and Chace (1992) conceded that their classifications were probably not completely natural ones. They proposed that important morphological characters (e.g., mouthparts, appendages, gills, sexual structures) and larval development were not well known enough in sufficient species to identify monophyletic higher taxa. Chace felt that research time was better spent in first getting more data on the group. The Chace–Holthuis classification scheme remained relatively unchanged up to the beginning of this century, mainly in deference to the experience and knowledge of these two carcinological titans (e.g., Martin and Davis 2001). As we shall see, the application of molecular methods to caridean phylogeny by research groups such as Bracken et al. (2009b) and Li et al. (2011) has led to reappraisal and proposals for change of the Chace–Holthuis classification but with retention of many of its features.

12.8.5 Traditional Views of Caridean Relationships

One of the first modern proposals about relationships among caridean superfamilies and families was done by Thompson (1967). He also opined that perhaps not enough was known about sufficient taxa to make such proposals. However, Thompson thought this view ultraconservative and proceeded to give it a try. He gave a table of characters for half of the caridean families in the Chace–Holthuis scheme, families Thompson considered the most important ("major") ones. The table includes a typical set of morphological characters considered essential for evolutionary analysis, such as gill formulae, mandible and appendage traits, and arrangement of telson spines. Based on standard evolutionary principles about decapod crustaceans, he hypothesized that a caridean ancestor would have a full branchial formula, mandibles with a fused incisor and molar process and a palp; exopods on all pereopods; chelae ending in interdigitating pegs or spines, and a telson with terminal and lateral spines.

I have modified his dendrogram on caridean phylogeny into a more standard form in Fig. 12.7 for discussion. He envisioned that a pelagic oplophoroid ancestor, one with the most generalized and primitive morphological traits, gave rise to most of the

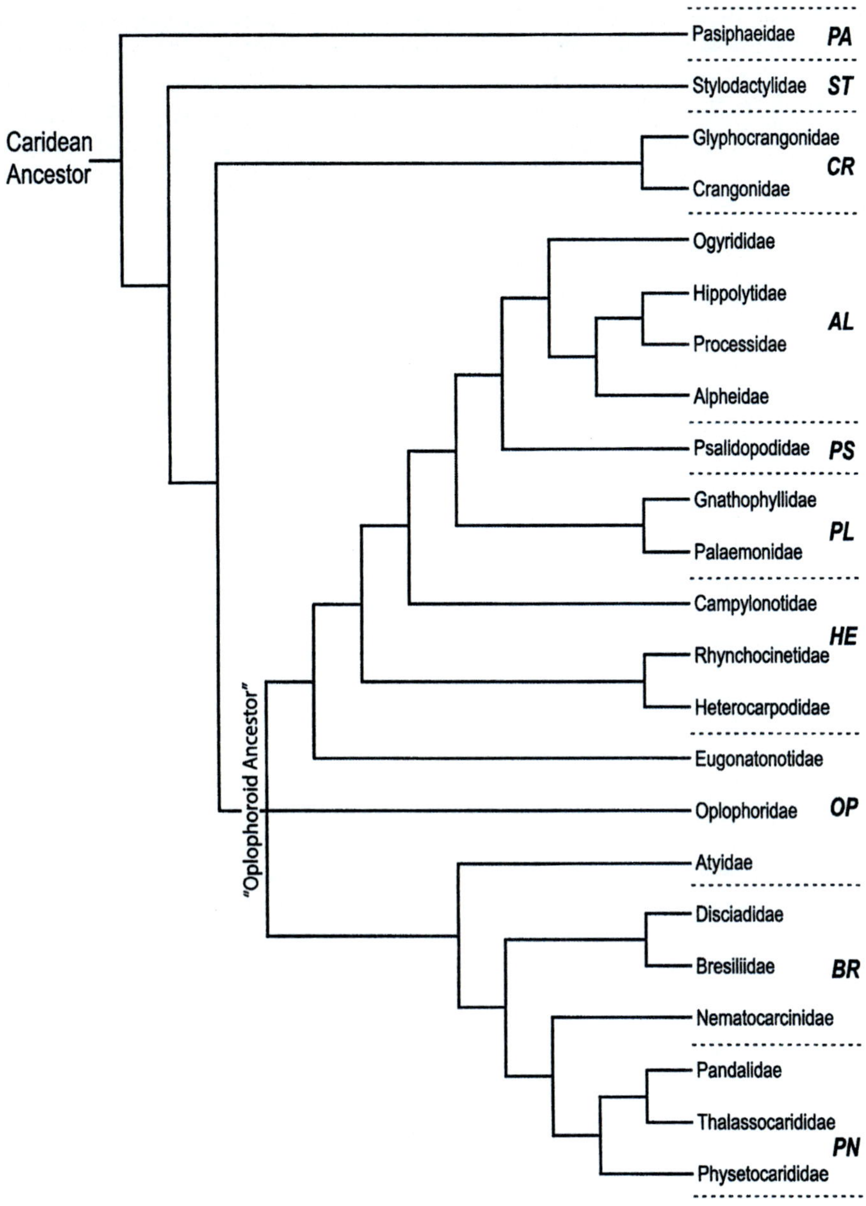

Fig. 12.7 Phylogeny of the Caridea, from data in Thompson (1967), using dichotomous branching where possible. To preserve Thompson's composition of superfamilies, the order of families, from top to bottom, is the same except for placing the Pasiphaeidae first, which does not change its branching sequence. Superfamilies are separated by dotted lines; abbreviations for superfamilies are the same as Fig. 12.8, with the addition of "HE," the Heterocarpodoidea

caridean families except the Pasiphaeidae, Stylodactylidae, Crangonidae, and Glyphocrangonidae. He conjectured that those latter families diverged from the caridean ancestor before the oplophoroid ancestor did. Within the descendants of the oplophoroid ancestor, he envisaged two major lineages, one a palaemonid-alpheoid group and the other an Atyidae-Bresiliidae-Nematocarcinidae-Pandaloid clade. In Thompson's phylogeny, the polycarpidean families (those with subdivided carpus on pereopod 2) were placed into four widely separated superfamilies: Crangonoidea; Alpheoidea, Heterocarpodoidea (a superfamily not recognized since) and Pandaloidea. This indicates the independent evolution of the polycarpidean trait four times in the Caridea rather than a single evolutionary event as envisaged by later caridean phylogenies. In the Thompson scheme, the subchelae of the Crangonidae and Glyphocrangonidae are shown as primitive compared to the chelae of most other carideans. On functional grounds, the evolution of subchelae from primitive achelate pereopods appears more likely, e.g., like those of the Procarididea (which were not first reported and described until 5 years after Thompson's paper).

Thompson's phylogeny and grouping of superfamilies are certainly not cladistic and conventional by today's standards. Some of his families are para- and/or polyphyletic. His phylogeny is somewhat subjective and intuitive, like Burkenroad's (1983) of the Dendrobranchiata. However, it has not withstood the test of time as well as Burkenroad's work. Nonetheless, some of his major proposals have been supported by later work, as will be seen below. Thompson made a clear list of phylogenetically important characters, and his paper promoted later thought and discussion.

Fenner Chace began his 1992 work on caridean classification as a personal aid to help him understand caridean evolutionary relationships among the major taxa of Caridea, which included the Procarididea at that time. He concluded that not enough was yet known about the morphology of enough caridean species to accomplish that. Nevertheless, his useful dichotomous key to superfamilies and families does implicitly express his views about phylogeny. In Bauer (2004), I constructed a phylogram based on his key as a basis for discussion on the Chace–Holthuis traditional views (Fig. 12.8). I have numbered all the branches, named taxa or unnamed ancestors, to aid in the following discussion.

The Procarididae first come off the caridean ancestral line first because they share several important traits with Caridea *s.s.* as noted earlier. Furthermore, they differ little from a hypothetical decapod ancestor in which chelae had not evolved on the first two pairs of pereopods and in which none of the articles of maxilliped 3 were fused. Other primitive procaridid characters include more developed and fully functional natatory exopods than any caridean *s.s.*, as well as a very generalized mandible with fused incisor and molar processes as well as a large palp. Functionally, procaridids are nektobenthic, equally comfortable with agile sustained swimming or perching on a substrate. Such a shrimp could give rise to both pelagic and benthic descendants. However, the procaridids, now restricted to anchialine subterranean environments, are a bit too specialized in other aspects to be the caridean ancestor. Their branchial formula is not very full and there is loss of appendices

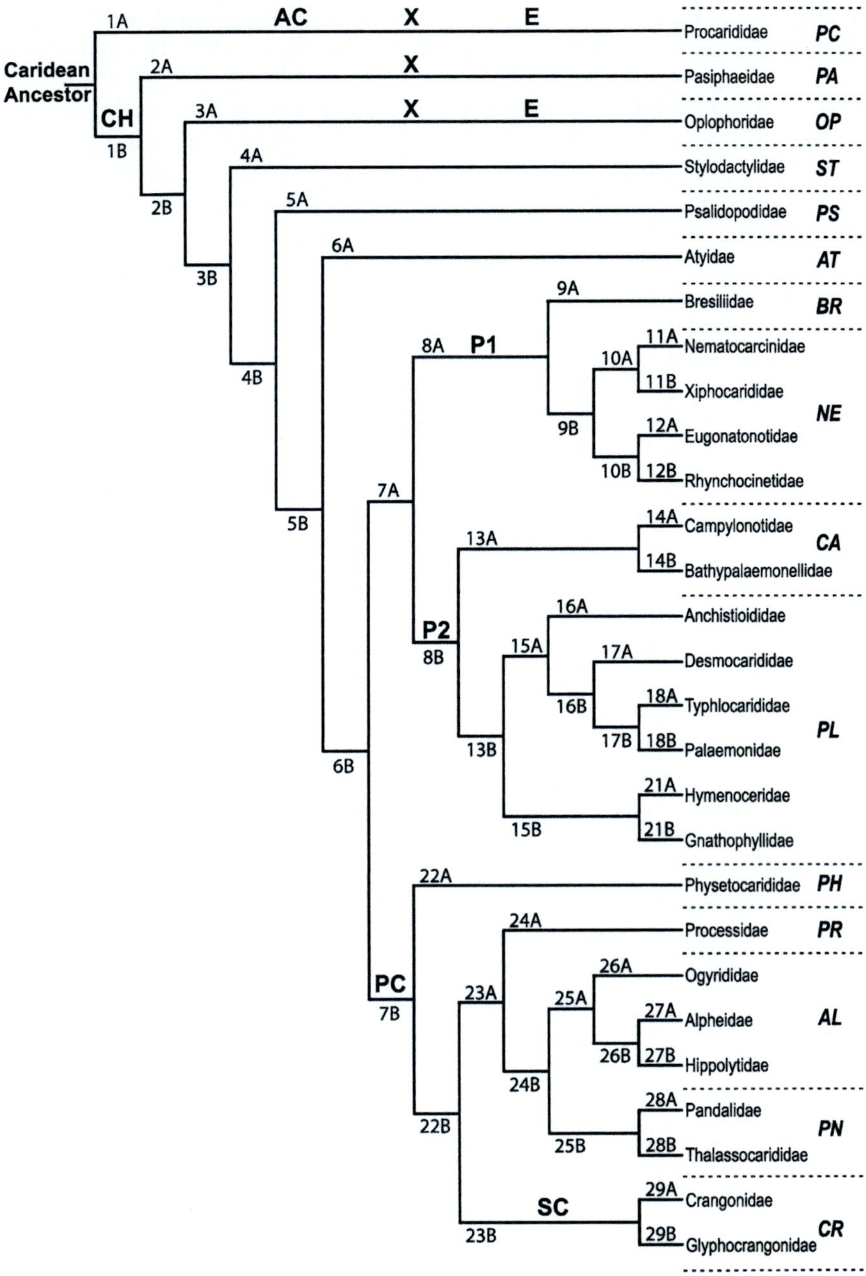

Fig. 12.8 A dendrogram of phylogenetic relationships among caridean families generated from the classification given in Chace (1992). Key for numbered branches can be found in Chace's key or in Appendix A in Bauer (2004). Superfamily abbreviations: *AL* Alpheoidea, *AT* Atyoidea, *BR* Bresilioidea, *CA* Campylonotoidea, *CR* Crangonoidea, *NE* Nematocarcinoidea, *OP* Oplophoroidea, *PA* Pasiphaeoidea, *PC* Procaridoidea, *PH* Physetocaridoidea, *PL* Palaemonoidea, *PN* Pandaloidea, *PR* Processoidea, *PS* Psalidopodoidea, *ST* Stylodactyloidea. Bold-faced abbreviations for phylogenetically important traits: *AC* achelate, *CH* chelate, *E* epipods with dorsal appendix, *P1* first

internae on some or all of the pleopods. The development of chelae (ancestor 1B in Fig. 12.8) was a big morphological innovation within the Caridea *s.s.*, together with fusion of articles of the third maxillipeds. Today, most phylogenies indicate that the Procarididae are certainly much nearer the Caridea *s.s.* than to other decapod groups. Recent phylogenies show them as a sister clade of the Caridea *s.s.* but sufficiently distinct to have equal infraorder status (Bracken et al. 2009b; Li et al. 2011).

As in the Thompson scheme, the pelagic Pasiphaeidae come off the ancestral line sooner than the Oplophoridae. Pasiphaeids have a mixture of primitive traits (well-developed exopods) and derived ones, such as a reduced branchial formula and long, slender pectinate fingers on the chelae, the latter an adaptation for catching prey. The next basal group is the Oplophoridae, a pelagic group with well-developed exopods and a primitive mandible. Like the procaridids, they have dorsal extensions from the epipods protruding up into the branchial chamber, although smaller than the massive procaridid ones. These dorsal appendices are absent in all descendant carideans with epipods. The next three families that separate out, the Stylodactylidae, Psalidopodidae, and Atyidae, each have some specialized (derived) characters that distinguish them from other carideans. Two major sister clades emerge next (Fig. 12.8); one (7B) is the "polycarpidean" ancestor in which the major derived trait is a subdivision of the carpus of pereopod 2. The polycarpidean families include the alpheoid, pandaloid, and crangonoid families as well as the Physetocarididae and Processidae. The "non-polycarpidean" clade (7A) retains the undivided pereopod 2 carpal article and gives rise to two descendant groups. In the first lineage (8A), the first chelipeds are stouter than the second and some exopods are retained, leading to the bresiliod and nematocarcinoid families. In lineage 8B, the first chelipeds are more slender than the second; this is the line leading to the Campylonotoidea and the great group Palaemonoidea. Ancestor 8A splits into two groups, each with different derived mandibular characters. In one, the Bresiloidea (9A), the pereopodal epipods are lost but in the other, the nematocarcinoid ancestor (9B), these epipods are retained. The latter gives rise to two descendants, each of which itself branches into two, terminating in four families: the Nematocarcinidae, Xiphocarididae, Eugonatonotidae, and Rhynchocinetidae (Fig. 12.8).

Ancestor 8B engenders the campylonotoid ancestor (13A) which splits into two families, the Campylonotidae and Bathypalaemonellidae, with the latter family (14B) losing its pereopodal epipods and developing highly asymmetrical second chelipeds. The ancestral palaemonoid (13B) loses arthrobranch gills, the mandible is more specialized, and the dorsal (outer) antennular flagellum bears an accessory branch. This ancestor initiates the superfamily Palaemonoidea, which includes several families, foremost of which is the very large and diverse Palaemonidae.

Fig. 12.8 (continued) cheliped larger, more robust than second cheliped, *P2* Second cheliped larger, more robust than first cheliped, *PC* polycarpidean condition (pereopod 2 with multiarticulate carpus), *SC* subchelate, *X* well-developed natatory pereopodal exopods. Modified from Bauer (2004)

Let us return to the "polycarpidean" ancestor (7B), defined by a major caridean innovation, the subdivision of the pereopod 2 carpus. This is an important adaptation giving extra flexibility to the second pereopods, useful in grooming (Chap. 4) as well as in probing and exploring the surroundings. The polycarpidean ancestor first branches off into the highly specialized pelagic group, the Physetocarididae (22A). It continues, now as 22B, giving off one branch (23B) in which the chela of the first pereopod becomes modified into the subchelate conditions in the Crangonidae and Glyphocrangonidae (29A-B). The undivided pereopod 2 carpus of the Crangonidae represents a reversal back to the pre-polycarpidean condition; the reduction or loss of the second pereopods in some crangonid genera supports this point of view.

After giving rise to the two subchelate families, the polycarpidean ancestor proceeds as ancestor 23A, branching off into the Processidae and continuing as 24 B. This latter group splits into two lineages, one (25B) leading to the Pandalidae and Thalassocarididae (the latter recently renamed with a different generic arrangement as the Chlorotocellidae: Komai et al. 2019). In these two families, the chelae of the first pereopods are reduced or lost. The other lineage (25A), the alpheoid ancestor, gives off a branch leading to the Ogyrididae (26A), with a principal derived feature of elongate eyes, and to another branch (26B) in which the first pereopods are markedly more robust than the second pereopods. This is the ancestor to the sister *group* Alpheidae and hippolytid (*s.l.*) families, both large and diverse taxa (Chap. 3).

12.8.6 Christoffersen's Cladistic Phylogeny

For several years, Brazilian zoologist Martin Christoffersen studied carideans and their taxonomy, gathering an impressive list of morphological characters which could be used to unravel genealogical relationships of the Caridea *s.s.* He did not include the Procarididae within the Caridea. He agreed with previous assessments (Felgenhauer and Abele 1983; Abele 1991) that the procaridids were close to the Caridea *s.s.* but sufficiently distinct to merit separate infraorder status. As in a dichotomous key, he presented the characters as two-state, either-or choices. For each character, he proposed explicitly or implicitly their evolutionary polarities, primitive (plesiomorphic) or derived (apomorphic), based on the evolutionary principles given earlier in this chapter. His results and analyses were presented in a series of papers (Christoffersen 1986, 1987, 1988b, 1989, 1990), the last of which synthesized the previous papers into an overall phylogeny of caridean families and superfamilies. Unlike the Chace–Holthuis scheme, the Christoffersen one is strictly cladistic, with only similarities which are shared derived characters as the basis for determining a monophyletic clade. Primitive characters carried along from basal ancestors show similarity but not descent from a recent common ancestor. Superfamilies and families are identified as monophyletic groups by one or more shared apomorphies (synapomorphies).

As with any cladistic phylogeny, Christoffersen presents the divergence and splitting of groups from a common ancestor as dichotomously branching clades. In

Table 12.1 Christoffersen's (1990) sequenced phylogenetic classification of caridean superfamilies and families (see also Fig. 12.9). The order of superfamilies ("-oidea") and families ("idae") in the table shows their order of branching. *, families not currently recognized by WoRMS (2022); a, those newly erected by Christoffersen and still accepted; b, those not recognized by Chace but resurrected by Christoffersen and still accepted

Atyoidea	**Nematocarcinoidea**	**Alpheoidea**
Oplophoridae	Nematocarcinidae	Merhippolytidae*
Atyidae	**Pandaloidea**	Nauticarididae*
Pasiphaeidae	Pandalidae	Alopidae*
Agostocarididae[b]	Plesionikidae*	Bythocarididae[a]
Alvinocarididae[a]	Heterocarpidae*	Thoridae[b]
Bresiliidae	Heterocarpoididae*	Hippolytidae
Psalidopodidae	Dorodoteidae*	Pterocarididae*
Disciadidae[b]	Thalassocarididae*	Ogyrididae
Stylodactyloidea	Physetocarididae	Alpheidae
Stylodactylidae	**Crangonoidea**	
Campylonotidae	Barbouriidae[a]	
Eugonatonotoidea	Lysmatidae[b]	
Eugonatonotidae	Merguiidae[a]	
Palaemonoidea	Processidae	
Rhynchocinetidae	Glyphocrangonidae	
Palaemonidae	Crangonidae	

theory, each clade could be given a name and rank in a taxonomic hierarchy. However, this results in a copious and cumbersome number of taxonomic categories which would make a primary aim of classification (easy retrieval of information) difficult. To present his phylogeny, Christoffersen used a technique known as "phyletic sequencing" in which the named taxonomic categories (in this case, superfamilies, and families) are shown in their order of branching (Table 12.1). In the table, Christoffersen's basal superfamily is listed first, followed by the superfamilies in evolutionary order. Within each superfamily, the first family listed is the most basal (primitive), followed by the second family, then the third and so on, with the last family being the most derived. In Fig. 12.9, Christoffersen's phylogeny of superfamilies is displayed as dichotomously branching clades with their diagnostic characters (synapomorphies).

Christoffersen viewed the caridean ancestor and its basal descendants as a typical caridoid shrimp, pelagic with well-developed thoracic exopods for swimming. More derived families invaded benthic habitats, with reduction and loss of natatory exopods. Those with epipods lost the ancestral dorsal extensions into the branchial chamber. The primitive mandible had incisor and molar processes, plus a palp of three articles, but became modified by reduction, change in form, or loss of its various components in more advanced families. Although Christoffersen did not discuss epipod-setobranch systems (Chap. 4), the most basal ancestor must have had

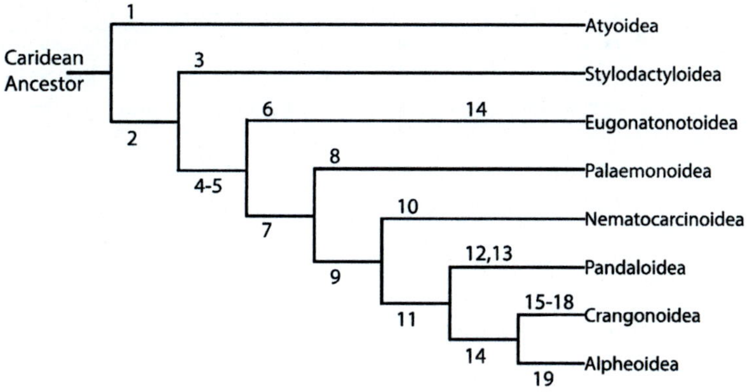

Fig. 12.9 Phylogram of caridean superfamilies according to Christoffersen (1990). Component families are shown in Table 12.1. Numbers represent shared derived characters (synapomorphies) for branching points of superfamily clades: 1, exopod of maxilliped 1 with distal lash shorter than caridean lobe; 2. exopods of pereopods 1–5 not all longer than ischium; 3, mandibular palp with less than 3 articles; 4, incisor process perpendicular to molar process; 5, epipods without distal appendix; 6, abdominal somite 3 with bifid dorsal carina; 7, perpendicular incisor process well developed; 8, basal segment of antennule with distolateral tooth; 9, carpus of pereopod 2 equal to merus + ischium; 10, scaphognathite with ventral (posterior) lobe triangular; 11, pereopod 2 with carpus multiarticulated; 12, pereopod 1 with chela microscopically small or absent; 13, left pereopod 2 longer than right; 14, pereopod 1 stronger than pereopod 2; 15, right pereopod 2 slightly longer than left; 16, perpendicular incisor absent; 17, telson with no more than 2 pairs of dorsal spines; 18, ocellus absent from base of cornea; 19, multiarticulated carpus of pereopod 2 with no more than 16 segments. From Christoffersen (1990)

them, like the sister group of the Caridea *s.s.,* the Procarididea. Some descendants retained them, while others lost them, with replacement by slender cleaning chelipeds used in brushing of gills. An innovation in several families using gill brushing was the subdivision of the pereopod 2 carpus into several to many articulated divisions. This "polycarpidean" condition gives the cheliped 2 extra flexibility for grooming activities (Chap. 4).

It is useful to compare the Chace–Holthuis (C-H) classification, an implied phylogeny (Fig. 12.8) with that of Christoffersen's (Fig. 12.9, Table 12.1). In the C-H arrangement, the pelagic families Pasiphaeidae and then the Oplophoridae *s.l.* first split off from the caridean ancestor, followed by the Stylodactylidae, Psalidopodidae, and finally the Atyidae plus the ancestor to a large clade of multiple families (Fig. 12.8). Each of the previous single families was given superfamily status. Christoffersen, on the other hand, created a superfamily Atyoidea within which the oplophoroids are most primitive, followed by the pasiphaeids and all the bresilioid families recognized currently by WoRMS (Chap. 3), which also agrees with C-H in a monofamilial Atyoidea. Christoffersen then lists his Stylodactyloidea which includes the Campylonotidae. The latter is given superfamily status by C-H, in which the genus *Bathypalaemonella* was incorporated as it was by Christoffersen, but without the family status given that genus today. The Eugonatonotoidea is a monofamilial superfamily in Christoffersen's view, while C-H has the family correctly, in my opinion and that of more current classifications (De Grave and Fransen 2011; WoRMS Editorial Board 2022), within the Nematocarcinoidea (Fig. 12.8). Christoffersen's Palaemonoidea agrees more or less with that of C-H, except for the somewhat strange inclusion of the Rhynchocinetidae, more properly placed in the Nematocarcinoidea by C-H and recent classifications.

Both C-H (Fig. 12.8) and Christoffersen (Figs. 12.9 and 12.10; Table 12.1) approximately agree that the polycarpidean families are indeed more closely related with each other than with other caridean groups. Thereafter, the two classifications diverge in several ways. The C-H scheme has the processids, alpheoids, and pandaloids as one clade and the crangonoids as another. Christoffersen shows the pandaloids as basal to the alpheoids and crangonoids and includes the processids in the latter. The composition of the Pandaloidea by Christoffersen is quite different than that of C-H and current classifications. The alpheoids and crangonoids are sister groups in both C-H and Christoffersen but their family make-up is very different. One reason is that in both clades Christoffersen erected or recognized several monogeneric families, some retained by more recent classifications and others not. In his Alpheoidea, Christoffersen included the currently recognized Thoridae, a revised Hippolytidae, the Bythocarididae, the Ogyrididae, and of course the Alpheidae (compare the C-H superfamily composition of Fig. 12.8 with that of Table 12.1). Some former hippolytid *s.l.* taxa, e.g., those now in the Barbouriidae, Lysmatidae, and Merguiidae, as well as the Processidae, were moved into the Crangonoidea, joining the traditional grouping of Crangonidae and Glyphocrangonidae.

Neither De Grave and Fransen (2011) nor WoRMS Editorial Board (2022) agree with the move of three former hippolytid families into the Crangonoidea. However, the Christoffersen arrangement is intriguing as an evolutionary hypothesis, and one day may be regarded more favorably, as does the Li et al. (2011) molecular analysis (see below). Separately within both Christoffersen's Crangonoidea and Alpheoidea (Figs. 12.10 and 12.11), the component families branch off from a pandaloid ancestor with a typically caridoid body, adapted for agile swimming and epibenthic perching, and evolve into a more infaunal burrowing macruran morphology. In his Crangonoidea, compare a barbouriid or a lysmatid with a crangonid or glyphocrangonid (Fig. 12.11a). The very same trends in habitus and habits occur in the Alpheoidea, i.e., contrast *Merhippolyte* or a thorid with an ogyridid or alpheid (Fig. 12.11b).

Christoffersen's bold cladistic approach to caridean phylogeny was somewhat of a shock to many established carcinologists and proponents of the conservative Chace–Holthuis classification. His ideas and suggestions were met with considerable skepticism when first proposed. He had trouble getting his work published in major carcinological journals because of opposition by established traditionalist reviewers. Although much of his classification and ideas are not accepted today, some are. He erected four new families and resurrected others now considered valid (Table 12.1). As in many disputes about competing phylogenies, the choice of characters and their interpretation, as well as methods of analysis, can make a big difference in the results of phylogenetic research. Time and fresh studies on

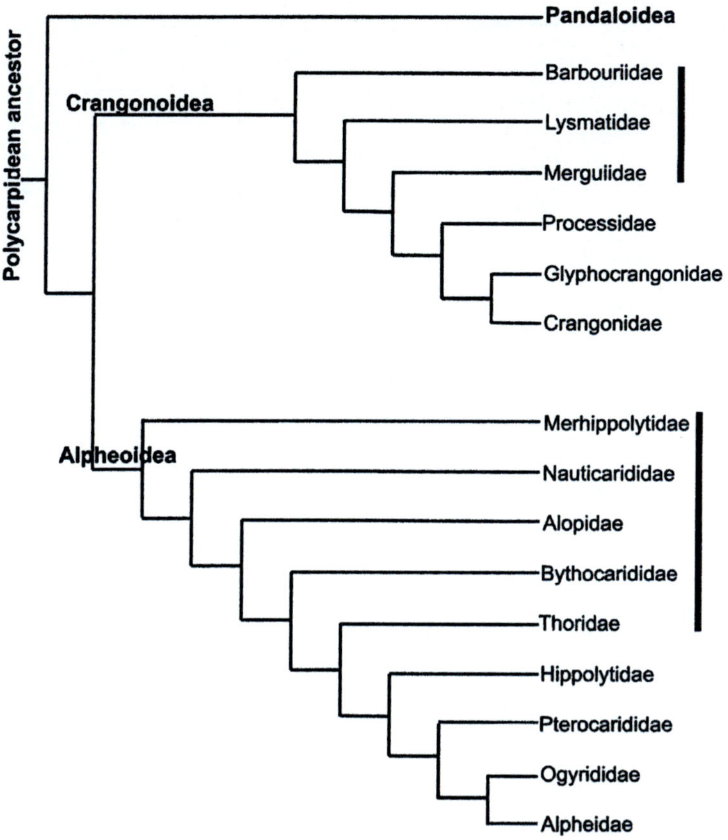

Fig. 12.10 Cladogenesis in the polycarpidean superfamilies of Caridea *s.s.* based on views of Christoffersen (1987, 1990). The relationship of the pandaloids, alpheoids, and crangonoids is shown, as well as the composition of families within the latter two superfamilies. Note that several of these families are not recognized in current classifications (e.g., Martin and Davis, 2001; WoRMS Editorial Board 2022) but others erected or resurrected by Christoffersen are (Table 12.1). The vertical bars to the right denote Christoffersen families that were formerly part of a more diverse and polyphyletic Hippolytidae (e.g., Chace 1992)

morphological traits, together with novel analyses using molecular methods will shed further light on his proposals.

12.8.7 *Molecular Tests of Caridean Phylogeny*

Agreement between morphological- and molecular-based phylogenetic analyses would be ideal, as the latter are separate tests of the former. Alas, this is not nearly the case in the highly diverse Caridea compared to the greater concordance of

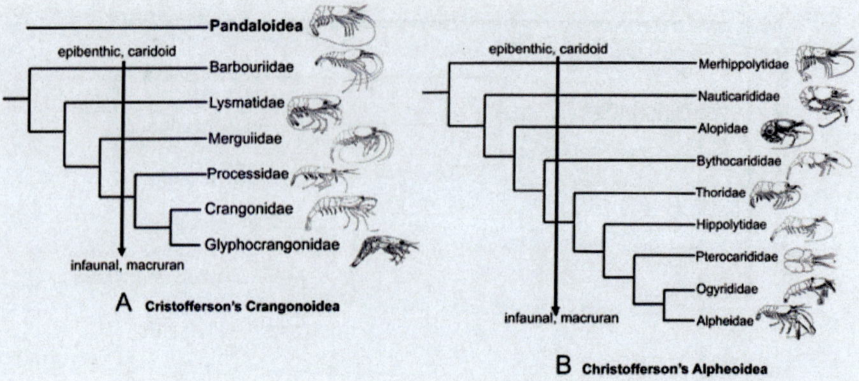

Fig. 12.11 Evolutionary trends in ecology and morphology of the polycarpidean families based Christoffersen's (1987, 1990) views of phylogeny. (**a**) Crangonoidea and (**b**) Alpheoidea. In each superfamily, basal families are epibenthic and more typically caridoid than advanced families which are more infaunal and macruran in body form. The usually epibenthic and natant pandaloids are the polycarpidean sister clade of the crangonoids + alpheoids in this view

dendrobranchiate phylogenies. To date (the final draft of this book before publication), two somewhat comprehensive examinations of caridean relationships using molecular techniques have been performed. They differ in many aspects with each other and previous morphological phylogenies. The Bracken et al. (2009b) study included 14 superfamilies, 30 families, 75 genera, and 104 species with sequences from the 16S mitochondrial and 18S nuclear genes (Fig. 12.12). Li et al. (2011) had a less diverse array of taxa at their disposal (11 superfamilies, 19 families, 28 genera, and 35 species) (Fig. 12.13). However, they used 5 nuclear genes and proposed that these genes are superior in resolving relations among taxa at higher taxonomic levels than those used by the Bracken et al. group. Additionally, they reported that their results have a much higher statistical support than the other molecular study. As noted previously, the results in any phylogenetic study are affected by the coverage of taxa, choice of characters, and statistical methods, in addition to differences in opinion about evolutionary processes.

Comparisons between and among the various molecular and morphological studies are not completely at odds, which is encouraging. All studies confirm the close link between the Crangonidae and Glyphocrangonidae. Similarly, both molecular studies support the Chace–Holthuis (C-H) and Christoffersen view of the polycarpidean families as having close ties, although not necessarily in the same branching pattern. Like Christoffersen, the molecular studies found the Hippolytidae a polyphyletic assemblage, and current classifications have followed Christoffersen's suggestions of raising some the C-H hippolytid genera to familial status, specifically the Barbouriidae, Merguiidae, Thoridae, and Lysmatidae. The results of Li et al. supported a Christoffersen view of the Crangonoidea, with a former Hippolytidae (*Lysmata*) among Christoffersen's crangonoid families (Figs. 12.10 and 12.13). Both Li et al. and Bracken et al.'s molecular studies

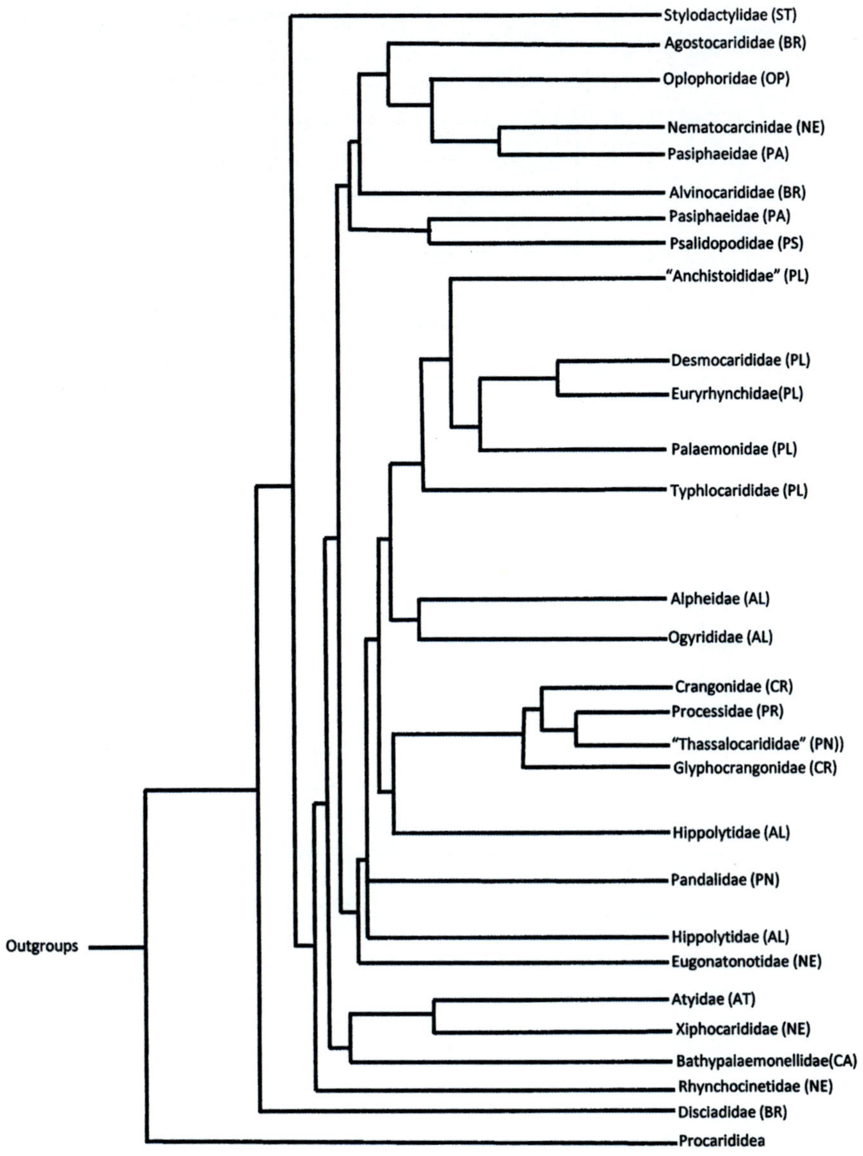

Fig. 12.12 Molecular phylogeny of the Caridea by Bracken et al. (2009b). Note that some families currently in the same superfamily are not in the same clades below. Superfamily composition is similar to that used by Chace (1992) as revised in current classifications (De Grave and Fransen 2011; WoRMS Editorial Board 2022). Superfamily abbreviations in parentheses are the same as in Fig. 12.8. Redrawn and modified from Bracken et al. (2009b)

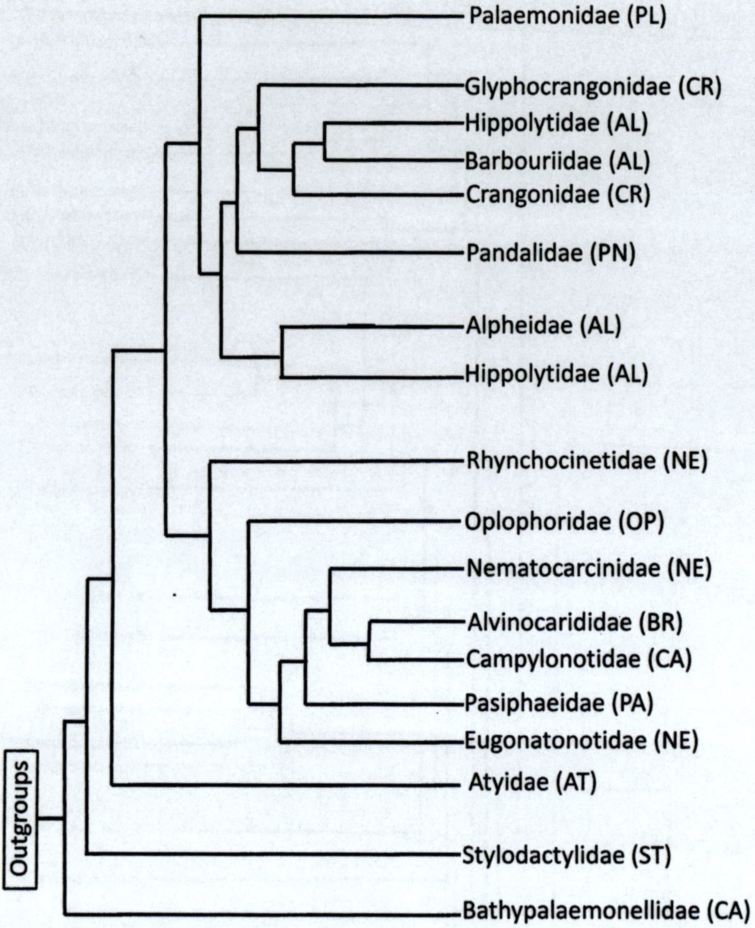

Fig. 12.13 Molecular phylogeny of the Caridea based on Li et al. (2011). Note that some families currently in the same superfamily are not in the same clades. Superfamily composition is similar to that used by Chace (1992) as revised in more currently used classifications (De Grave and Fransen, 2011; WoRMS Editorial Board 2022). Superfamily abbreviations in parentheses are the same given in Fig. 12.8

concluded that various superfamilies and families were poly- and/or paraphyletic and need of revision. In nearly all phylogenies, the Oplophoridae, Atyidae, and Stylodactylidae are basal lineages, reinforcing the view of a pelagic oplophoroid-type ancestor of many caridean families (Thompson 1967; Christoffersen 1990). Very strange is the appearance of families such as the morphologically derived Disciadidae and Bathypalaemonellidae as basal lineages in molecular studies. The differences between and among molecular and morphological hypotheses can be seen by comparisons among Figs. 12.7, 12.8, 12.9, 12.10, 12.11, 12.12, and 12.13. We are still in the early stages of resolving the evolutionary history of carideans, with lots of work to do in studying the distribution and significance of morphological and other non-molecular traits, as well as the phylogenetic value of various types of molecular data and methods. As Li et al. so justifiably stated, given the limited datasets available, it would be "imprudent" to now make changes in the currently accepted classification of decapod shrimps (e.g., De Grave and Fransen 2011; WoRMS Editorial Board 2022) given our present lack of understanding of their relationships. I wholeheartedly agree, but future changes are to be expected.

12.8.8 Stenopodidea

This group, with its three pairs of chelipeds, the last pair a set of menacing weapons, and only 12 genera and 95 species, is not as nearly diverse as the Dendrobranchiata and Caridea (Chap. 3). Its phylogeny has received correspondingly less attention. For many years, all species were grouped in a single family, the Stenopodidae. In his book on Crustacea (1986), Frederick Schram divided the group into two families, the mainly warm- and shallow-water Stenopodidae *s.s.*, with four genera, with *Stenopus* the best known. The other family, the primarily deep-water Spongicolidae, are mostly obligate commensals of hexactinellid (glass) sponges (see Chap. 10, Life Together). One major spongicolid exception is *Microprosthema*, which like the Stenopodidae, is an inhabitant of shallow tropical reef habitats.

The first thorough attempt of any stenopodidean phylogeny was done by Saito and Takeda (2003), who did a cladistic analysis of the Spongicolidae based principally on morphological traits. They also included reproductive, larval, biogeographic, and bathymetric information. They used two species from the other stenopodidean family, *Stenopus hispidus* and *Odontozona minoica*, as outgroups to establish polarity of characters. Their phylogeny showed the shallow-water, non-commensal *Microprosthema* species as basal and having given rise to the deep-sea spongicolid commensals (Fig. 12.14). Interestingly, the polarities in several characters reveal evolutionary trends which are adaptive in the change of habitat from free-living in shallow tropical reefs to life within a sponge host in the dark, cold, food-poor deep sea. There has been selection for a general reduction in morphological features no longer needed in the protected confines of the sponge host, e.g., loss of antipredator carapace spines; lower calcification of exoskeleton; loss of grooming brushes; fewer respiratory gills and epipods; smaller, less pigmented eyes; and switch from extended larval development to abbreviated or direct development (see text details and Fig. 10.5 in Chap. 10). This makes for a very satisfying evolutionary scenario—but is this hypothesis supported by an independent test with molecular data?

A more complex view of stenopodidean evolution is suggested by the molecular phylogeny of the stenopodideans, including not only the Spongicolidae, but also the Stenopodidae and the anchialine cave-dweller Macromaxillocarididae, by Chen et al. (2016). The database was comprised of sequences from two nuclear (NaK and histone H3) and two mitochondrial (16S and 12 rRNA) genes from 31 species

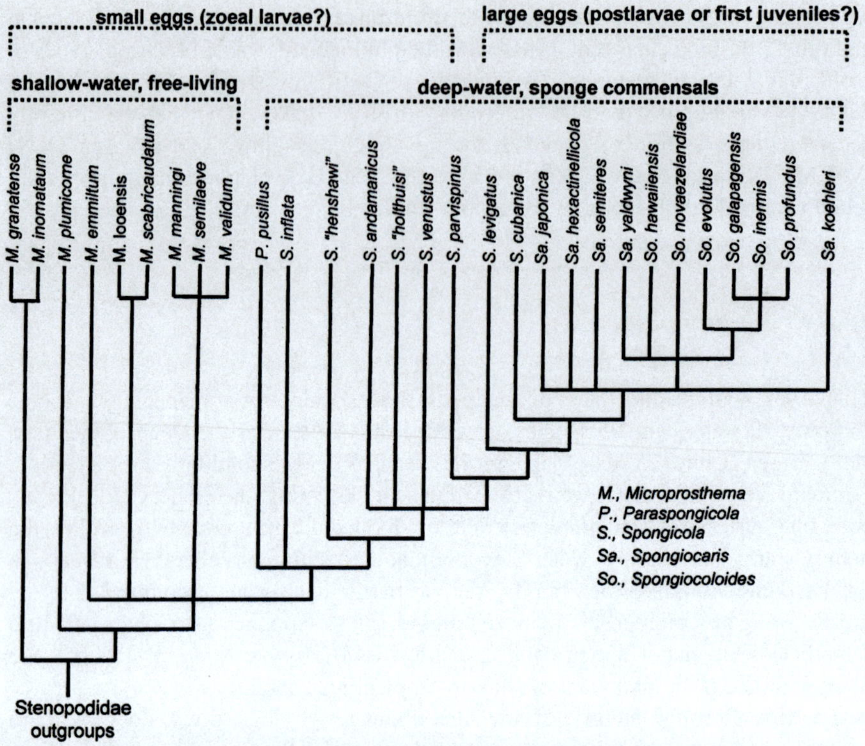

Fig. 12.14 Phylogeny of spongicolid species by Saito and Takeda (2003). Their phylogram has been adapted and modified in accordance with more recent changes in taxonomy and nomenclature documented in WoRMS Editorial Board (2022), e.g., two *Spongicola* species (quotation marks) are now considered synonymous with *S. andamanicus*; switch of a few species from one genus to another; some minor changes in spelling. See Chap. 10 (Fig. 10.5) for other evolutionary trends in spongicolid morphology associated with living free in shallow tropical waters to commensalism with hexactinellid sponges in the deep sea. Saito and Takeda's fig. 3 and text give synapomorphies for each branching

representing all genera. Their results suggest that neither the Stenopodidae nor the Spongicolidae are monophyletic, i.e., some species of one family are more closely related with those of the other. For example, the spongicolid *Globospongia spinulatus* appears more closely related to the stenopodids *Richardina Odontozona spongicola* than with other spongicolids (Fig. 12.15). Several genera are poly- and/or paraphyletic, as seen in the Saito and Takeda phylogram (Fig. 12.14). The Chen group suggested that, instead of single invasion of the deep-sea by *Microprosthema* and a switch from free living to glass sponge or crinoid associations, perhaps there were multiple invasions and change in lifestyle by different shallow-water ancestors. However, this does not negate the clear differences between shallow-, warm-water ancestors and spongicolids living within deep-sea sponges or other sessile

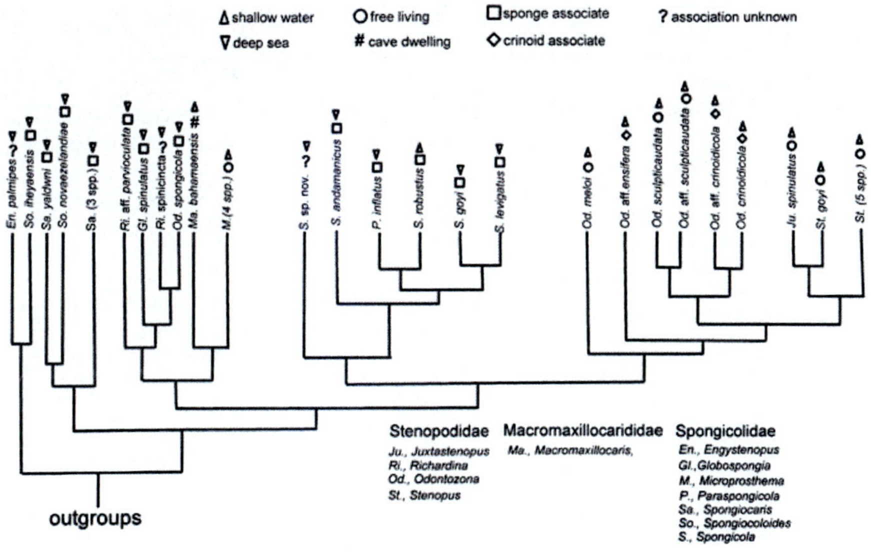

Fig. 12.15 Stenopodidean phylogeny based on Chen et al. (2016). The diagram shows the maximum-likelihood phylogram from the combined mitochondrial 12S, 16S, nuclear H3 and NaK gene sequences of various stenopodidean species. Procarididean and caridean species served as outgroups. I have collapsed subclades of the same genus or species into single clades for simplicity wherever possible. Lifestyles and habitats are shown for each species or species group, as well as the current generic classification of the Stenopodidea (WoRMS Editorial Board 2022). For further details, see fig. 3 and text in Chen et al. (2016)

invertebrates. The same selective pressures acting on shallow-water invaders of the deep-sea symbiotic habitat would have been in operation, with similar results.

Another conclusion of the study was that the unusual stenopodidean cave-dweller *Macromaxillocaris* is nested among the spongicolids with close ties to *Microprosthema*. They suggested that the two genera may share a common ancestor, with one descendant, *Macromaxillocaris,* adapting to anchialine cave habitats. Chen et al. caution that the results are preliminary, and further work is likely to reliably unravel stenopodidean evolution. This is true with all shrimps groups: creative ideas, further study and analyses will bring us closer to the actual evolutionary histories of shrimps.

References

Abele LG (1991) Comparison of morphological and molecular phylogeny of the Decapoda. Mem Queensl Mus 31:101–1108

Abele LG, Felgenhauer BE (1983) Phylogenetic and phenetic relationships among the lower Decapoda. J Crust Biol 6:385–400

Abele LG, Felgenhauer BE (1986) Phylogenetic and phenetic relationships among the lower Decapoda. J Crust Biol 6:385–400

Arnold EN (1981) Estimating phylogenies at low taxonomic levels. Zeitschrift für Zoologische Systematik und Evolutionsforschung 19:1–72

Bauer RT (1977) Antifouling adaptations of marine shrimp (Decapoda: Caridea): functional morphology and adaptive significance of antennular preening by the third maxillipeds. Mar Biol 40:260–276

Bauer RT (1979) Antifouling adaptations of marine shrimp (Decapoda: Caridea): gill cleaning mechanisms and grooming of brooded embryos. Zool J Linnean Soc 65:281–303

Bauer RT (1984) Morphological trends in the genus *Heptacarpus* (Decapoda, Caridea) and their phylogenetic significance. J Crustac Biol 4(201):225

Bauer RT (1986) Phylogenetic trends in sperm transfer and storage complexity in decapod crustaceans. J Crustac Biol 6:313–325

Bauer RT (2004) Remarkable shrimps: adaptations and natural history of the carideans. University of Oklahoma Press, Norman

Boas JEV (1880) Studier over Decapodernes Slaegtskabsforhold. Kongelige Danske Videnskabernes Selskabs Skriffer, Naturvidenskabelig og Mathematisk Afdelig 5(6):25–210

Borradaile LA (1907) On the classification of the decapod crustaceans. Ann Magazine Nat Hist Ser 7(19):457–486

Bracken HD, Toon A, Felder DL, Martin JW, Finley M, Maegan Rasmussen J, Palero and K. Crandall. (2009a) The decapod tree of life: compiling the data and moving toward a consensus of decapod evolution. Arthropod Syst Phylogeny 67:99–116

Bracken HD, De Grave S, Felder DL, D.L. (2009b) Phylogeny of the infraorder Caridea based on mitochondrial and nuclear genes (Crustacea: Decapoda). In: Martin JW, Crandall KA, Felder DL (eds) Decapod crustacean phylogenetics. CRC Press, Taylor & Francis Group, Boca Raton, London, New York, pp 281–305

Bracken HD, De Grave S, Toon A, Felder DL, Crandall KA (2010) Phylogenetic position, systematic status, and divergence time of the Procarididea (Crustacea: Decapoda). Zool Scr 39:198–212

Brandt S, Schulz M (2013) Zwei neue natante Dekapoden aus dem Oberen Muschelkalk (Mittel-Trias, Ladin) des Germanischen Beckens – *Antrimpos germanicus* n. sp. und *Parapalaemonetes thuringiacus* n. gen. n. sp. Vernate 32:67–95

Burkenroad MD (1934) The Penaeidea of Louisiana with a discussion of their world relationships. Bull Am Mus Nat Hist 68:61–143

Burkenroad MD (1936) The Aristaeinae, Solenocerinae and pelagic Penaeiniae of the Bingham oceanographic collection. Bull Bingham Oceanogr Collect 5:1–151

Burkenroad MD (1963) The evolution of the Eucarida (Crustacea: Eumalacostraca) in relation to the fossil record. Tulane Stud Geol 2:3–16

Burkenroad MD (1981) The higher taxonomy and evolution of Decapoda (Crustacea). Transact San Diego Soc Nat Hist 19:251–268

Burkenroad MD (1983) Natural classification of Dendrobranchiata, with a key to recent genera. In: Schram FR (ed) Crustacean phylogeny. Crustacean Issues 1:279–290

Chace FA Jr (1992) On the classification of the Caridea (Decapoda). Crustaceana 63:70–80

Chace FA Jr, Manning RB (1972) Two new caridean shrimps, one representing a new family, from marine pools on Ascension Island (Crustacea: Decapoda: Natantia). Smithson Contrib Zool 131: 18

Chan TY (2012) A new genus of deep-sea solenocerid shrimp (Decapoda: Penaeoidea) from Papua New Guinea. J Crustac Biol 32:489–495

Chan TY, Tong J, Tam YK, Chu KH (2008) Phylogenetic relationships among the genera of the Penaeidae (Crustacea: Decapoda) revealed by mitochondrial 16S rRNA gene sequences. Zootaxa 1694:38–50

Chen C-L, Goy JW, Bracken-Grissom HD, Felder DL, Tsang LM, Chan T-Y (2016) Phylogeny of Stenopodidea (Crustacea: Decapoda) shrimps inferred from nuclear and mitochondrial genes

reveals non-monophyly of the families Spongicolidae and Stenopididae and most of their composite genera. Invertebr Syst 30:479–490
Cheng JT-Y, Chan N, Zhang SS, Sha Z-L (2018) Mitochondrial phylogenomics reveals insights into taxonomy and evolution of Penaeoidea (Crustacea: Decapoda). Zool Scr 47:582–594
Christoffersen ML (1986) Phylogenetic relationships between Oplophoridae, Atyidae, Pasiphaeidae, Alvinocarididae Fam. N., Bresiliidae, Psalidopodidae and Disciadidae (Crustacea Caridea Atyoidea). Boletim Zoologia Universidade São Paulo 10:273–281
Christoffersen ML (1987) Phylogenetic relationships of hippolytid genera, with an assignment of new families for the Crangonoidea and Alpheoidea (Crustacea, Decapoda, Caridea). Cladistics 3:348–362
Christoffersen ML (1988a) Phylogenetic systematics of the Eucarida (Crustacea Malacostraca). Revista Brasileira de Zoologia 5:325–351
Christoffersen ML (1988b) Genealogy and phylogenetic classification of the world Crangonidae (Crustacea, Caridea), with a new species and new records for the south western Atlantic. Revista Nordestina de Biologia 6:43–59
Christoffersen ML (1989) Phylogeny and classification of the Pandaloidea (Crustacea, Caridea). Cladistics 5:259–274
Christoffersen ML (1990) A new superfamily classification of the Caridea (Crustacea: Pleocyemata) based on phylogenetic pattern. Zeitschrift für zoologische Systematik und Evolutionsforschung 28:94–106
Davis KE, De Grave S, Delmer C, Wills MA (2018) Freshwater transitions and symbioses shaped the evolution and extant diversity of caridean shrimps. Commun Biol 1:16. https://doi.org/10.1038/s42003-018-0018-6
De Grave S, Fransen CHJM (2011) Carideorum catalogus: the recent species of the dendrobranchiate, stenopodidean, procarididean and caridean shrimps (crustacea: Decapoda). Zoologische Mededelingen, Leiden 89:195–589
De Grave S, Pentcheff ND, Ahyong ST, Chan T-Y, Crandall KA, Dworschak PC, Felder DL, Feldmann RM, Fransen CHJM, Goulding LYD, Lemaitre R, Low MEY, Martin JW, Ng PKL, Schweitzer CE, Tan SH, Wetzer R (2009) A classification of living and fossil genera of decapod crustaceans. Raffles Bull Zool Suppl 21:1–109
Felgenhauer BE, Abele LG (1983) Phylogenetic relationships among shrimp-like decapods. In: Schram FR (ed) Crustacean phylogeny. A.A. Balkema, Rotterdam, pp 291–311
Felgenhauer BE, Abele LG (1989) Evolution of the foregut in the lower Decapoda. In: Felgenhauer BE, Watling L, Thistle AB (eds) Functional morphology of feeding and grooming in crustacea. A.A. Balkema, Rotterdam, pp 205–218
Felsenstein J (2004) Inferring phylogenies. Sinauer Associates, Inc., Sunderland
Galis F, Arntzen JW, Lande R (1999) Dollo's law and the irreversibility of digit loss in *Bachia*. Evolution 64:2466–2476
Garassino A (2001) New decapod crustaceans from the Cenomanian (Upper Cretaceous) of Lebanon.) Atti della Società Italiana di Scienze Naturali e Del Museo Civico di Storia Naturale di Milano, 141/2000 (II): 237–250
Gould SJ (1989) Wonderful life: the burgess shale and the nature of history. W.W. Norton & Co
Goy JW (2010) Infraorder Stenopodidea Claus, 1872. pp. 215–265. In: Schram FR, von Vaupel Klein JC, Forest J, Charmantier-Daures M (eds) Treatise on zoology – anatomy, taxonomy, biology. The Crustacea, Decapoda. Vol. 9. Part A: Eucarida: Euphausiacea, Amphionidacea, and Decapoda (partim). Koninklijke Brill NV, Leiden
Hall BG (2001) Phylogenetic trees made easy. A how-to manual for molecular biologists. Sinauer Associates, Inc. Sunderland
Hao Y, Qu Y, Song G, Lei F (2019) Genomic insights into the adaptive convergent evolution. Curr Genomics 20:81–89
Hennig W (1950) Grundzuge einer Theorie der phylogenetischen Systematik. Deutscher Zentralverlag, Berlin
Hennig W (1966) Phylogenetic systematics. University of Illinois Press, Urbana

Holthuis LB (1955) The recent genera of the caridean and stenopodidean shrimps (class Crustacea, order Decapoda, supersection Natantia) with keys for their determination. Zoologische Verhandelingen, Leiden 26:1–157

Holthuis LB (1993) The recent genera of the Caridean and Stenopodidean shrimps (Crustacea, Decapoda): with an appendix on the order Amphionidacea. Nationaal Natuurhistorisch Museum, Leiden

Huang JRM, Feldmann CE, Schweitzer HSH, Zhou C, Benton MJ, Zhang Q, Wen W, Xie T (2013) A new shrimp (Decapoda, Dendrobranchiata, Penaeoidea) from the Middle Triassic of Yunnan, Southwest China. J Paleontol 87:603–611

Jones WT, Feldmann RM, Schweitzer CE, Schram FR, Behr R-A, Hand KL (2014) The first Paleozoic stenopodidean from the Huntley Mountain Formation (Devonian–Carboniferous), north-Central Pennsylvania. J Paleontol 88:1251–1256

Jones WT, Feldmann RM, Hannibal JT, Schweitzer CE, Garland MC, Maguire EP, Tashman JN (2018) Morphology and paleoecology of the oldest lobster-like decapod, *Palaeopalaemon newberryi* Whitfield, 1880 (Decapoda: Malacostraca). J Crustac Biol 38:302–314

Judkins D, Kensley B (2008) New genera in the family Sergestidae (Crustacea: Decapoda: Penaeidea). Proc Biol Soc Wash 121:72–84

Komai T, Chan T-Y, De Grave S (2019) Establishment of a new shrimp family Chlorotocellidae for four genera previously assigned to Pandalidae (Decapoda, Caridea, Pandaloidea). Zoosyst Evol 95:391–402

Li CP, De Grave S, Chan T-Y, Lei HC, Chu KH (2011) Molecular systematics of caridean shrimps based on five nuclear genes: implications for superfamily classification. Zool Anz 250:270–279

Losos JB (2011) Seeing the forest for the trees: the limitations of phylogenies in comparative biology. Am Nat 177:709–727

Losos JB, Hillis DM, Greene HW (2012) Who speaks with a forked tongue? Science 338:1428–1429

Ma KY, Chan T-Y, Chu KH (2009) Phylogeny of penaeoid shrimps (Decapoda: Penaeoidea) inferred from nuclear protein-coding genes. Mol Phylogenet Evol 53:45–55

Martin JW, Davis GE (2001) An updated classification of the recent Crustacea in Natural History Museum of Los Angeles County Contributions in Science No. 39, 124 pp

Martin JW, Liub EM, Striley D (2007) Morphological observations on the gills of dendrobranchiate shrimps. Zool Anz 246:115–125

Maslin TP (1952) Morphological criteria of phyletic relationships. Syst Zool 1:49–70

Pérez Farfante I, Kensley B (1997) Penaeoid and sergestoid shrimps and prawns of the world. Keys and diagnoses for the families and genera. Mémoires du Muséum National d'Histoire Naturelle 175:1–233

Porter ML, Pérez-Losada M, Crandall KA (2005) Model-based multi-locus estimation of decapod phylogeny and divergence times. Mol Phylogenet Evol 37:355–369

Robalino J, Wilkins B, Bracken-Grissom HD, Chan T-Y, O'Leary MA (2016) The origin of large-bodied shrimp that dominate modern global aquaculture. PLoS One 11:e0158840. https://doi.org/10.1371/journal.pone.0158840

Saito T, Takeda M (2003) Phylogeny of the Spongicolidae (Crustacea: Stenopodidae): evolutionary trend from shallow-water free-living to deep-water sponge-associated habitat. J Mar Biol Assoc UK 83:119–131

Schram F, Yanbin S, Vonk R, Taylor RS (2000) The first fossil stenopodidean. Crustaceana 72: 235–242

Schram FR (1986) Crustacea. Oxford University Press, New York

Schram FR (2009) On the origin of Decapoda. In: Martin JW, Crandall KA, Felder DL (eds) Decapod crustacean phylogenetics. CRC Press, Taylor & Francis Group, Boca Raton, pp 3–13

Schweigert G (2011) The decapod crustaceans of the upper Jurassic Solnhofen Limestones: a historical review and some recent discoveries. Neues Jb Geol Paläontol Abh:131–140

Tavares C, Martin JW (2010) Suborder Dendrobranchiata Bate, 1888. In: Schram FR, von Vaupel Klein JC (eds) Treatise on zoology – anatomy, taxonomy, biology -the crustacea volume 9 part a. Koninklijke Brill NV, Leiden, pp 99–164

Tavares C, Serejo C, Martin J (2009) A preliminary phylogenetic analysis of the Dendrobranchiata based on morphological characters. In: Martin JW, Crandall KA, Felder DL (eds) Decapod crustacean phylogenetics. CRC Press, Taylor & Francis Group, Boca Raton, pp 261–279

Thompson JR (1967) Comments on phylogeny of section Caridea (Decapoda Natantia) and the phylogenetic importance of the Oplophoroidea. Proc Symp Crustacea Mar Biol Assoc India 1: 314–326

Timm L, Bracken-Grissom HD (2015) The forest for the trees: evaluating molecular phylogenies with an emphasis on higher-level Decapoda. J Crustac Biol 35:577–592

Vereshchaka AL (2017) The shrimp superfamily Sergestoidea: a global phylogeny with definition of new families and an assessment of the pathways into principal biotopes. R Soc Open Sci 4: 170221. https://doi.org/10.1098/rsos.170221

Vereshchaka AL, Olesen J, Lunina AA (2014) Global diversity and phylogeny of pelagic shrimps of the former genera *Sergestes* and *Sergia* (Crustacea, Dendrobranchiata, Sergestidae), with definition of eight new genera. PLoS One 9:e112057. https://doi.org/10.1371/journal.pone.0112057

Vereshchaka AL, Olesen J, Lunina AA (2016a) A phylogeny-based revision of the family Luciferidae. De Haan, 1849 (Crustacea: Decapoda). Zool J Linnean Soc 178:15–32

Vereshchaka AL, Lunina AA, Olesen J (2016b) Phylogeny and classification of the shrimp genera *Acetes, Peisos, and Sicyonella* (Sergestidae: Crustacea: Decapoda). Zool J Linnean Soc 177: 353–377

Wolfe JM, Breinholt JW, Crandall KA, Lemmon AR, Lemmon EM, Timm LE, Siddall ME, Bracken-Grissom HD (2019) A phylogenomic framework, evolutionary timeline and genomic resources for comparative studies of decapod crustaceans. Proc R Soc B 286:20190079. https://doi.org/10.1098/rspb.2019.0079

WoRMS Editorial Board (2022) World Register of Marine Species. Available from http://www.marinespecies.org at VLIZ. https://doi.org/10.14284/170

Subject Index

D

E

F

G

T

Taxonomic Index

B